POLYMER PROCESS ENGINEERING

POLYMER PROCESS ENGINEERING

Eric A. Grulke

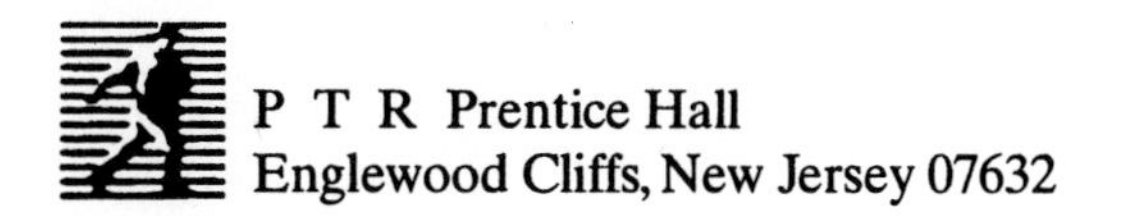

P T R Prentice Hall
Englewood Cliffs, New Jersey 07632

Library of Congress Cataloging-in-Publication Data

Grulke, Eric A.
Polymer process engineering / by Eric A. Grulke.
p. cm.
Includes index.
ISBN 0-13-015397-4
1. Polymers. I. Title.
TP1087.G78 1994
668.9—dc20 93-22178
CIP

Editorial/production supervision and interior design: *Northeastern Graphic Services, Inc.*
Cover design: *Lundgren Graphics*
Buyer: *Mary Elizabeth McCartney*
Acquisitions editor: *Betty Sun*

Prentice-Hall, Inc.
A Paramount Communications Company
Englewood Cliffs, New Jersey 07632

The publisher offers discounts on this book when ordered in bulk quantities. For more information, contact:

Corporate Sales Department
PTR Prentice Hall
113 Sylvan Avenue
Englewood Cliffs, NJ 07632

Phone: 201-592-2863
Fax: 201-592-2249

Printed in the United States of America
10 9 8 7 6 5 4 3 2 1

ISBN 0-13-015397-4

Prentice-Hall International (UK) Limited, *London*
Prentice-Hall of Australia Pty., Limited, *Sydney*
Prentice-Hall Canada Inc., *Toronto*
Prentice-Hall Hispanoamericana, S.A., *Mexico*
Prentice-Hall of India Private Limited, *New Delhi*
Prentice-Hall of Japan, Inc., *Tokyo*
Simon & Schuster Asia Pte. Ltd., *Singapore*
Editora Prentice-Hall do Brasil, Ltda., *Rio de Janeiro*

Contents

Preface

The goal of this text is to provide an introduction to polymer processes for engineers. Polymer processes include the conversion of monomers to polymers, as well as the processing of polymeric liquids and solids into products. The conversion of monomer to polymer, its recovery from the reaction medium, and the processing of the polymer into a finished product are illustrated by the following example.

Some polymer processes might be used to make a polymer and to fabricate a product. One example of this type of application would be a coated wire rack used in dishwashers. Small particles are needed to form a uniform, glossy coating over the wire blanks, so an emulsion polymer with particles less than 1 micron in size would be used. The polymer emulsion is made by feeding monomer, water, initiators, and suspending agents into a series of continuously stirred tank reactors. Residual monomer would be stripped from the latex slurry and recycled. Dry polymer powder is produced by spray drying and screening operations. The dried powder is purchased by the part fabricator. The polymer is combined with additives such as color, stabilizers, lubricants, and plasticizers. The modified polymer is fluidized in an air bed, and heated wire blanks are dipped through the mixed powder. Particles hitting the wire surfaces will sinter and melt. A final baking step might be used to make a high-gloss finish.

The performance and quality of the finished racks depend on all steps in this sequence. Therefore, engineers working at any step in this sequence need to understand the relationships between processes and product performance.

Engineering of polymer processes should be based on the underlying polymer chemistry and physics; the engineering science of polymerization; phase equilibria; flow and mechanical properties; and the integration of transport phenomena, thermodynamics, and kinetics into the design process. Such data and models are available for commodity thermoplastics, thermosets, and elastomers. Fundamental property data and models may not generally be known for specialty and advanced materials. This text is intended to be general and broad. More details on many of the topics treated here are available in graduate texts, research monographs, and referreed journals.

The text is divided into four parts: an introduction to polymer systems with qualitative descriptions of polymer chemistry and physics; a section covering the engineering and physical sciences applying to polymerizations, phase equilibria, property analysis, flow properties, and mechanical properties; a section describing some polymer processes and illustrating design calculations; and appendices containing polymer property data. This material is directed toward senior undergraduate or first-year graduate students. However, it has been taught to students in their junior year of chemical engineering.

The first chapter, a primer of polymer science and engineering, is intended to provide a general overview of polymers for engineers. Chapter 2 has an inventory of different types of polymers, showing their structures and containing brief comments about properties and uses. In later chapters, specific polymers will be discussed without lengthy reference to their structures. Chapter 3 describes the physical state of polymer systems, beginning with polymer solutions and continuing through composites and liquid crystals.

The next section of the text covers engineering science of polymer systems. Chapter 4 demonstrates how to develop a set of rate expressions that simulate a given polymerization and solve these for batch processes. Polymer-solvent phase equilibria (Chapter 5) are critical to polymer precipitation, crystallization, and recovery from the reaction mixture. The measurement of molecular weight is sufficiently important to deserve a separate chapter (Chapter 6). Methods for thermal and chemical analyses are discussed in Chapter 7. It includes a section on qualitative methods for polymer evaluation that can be very useful to the practicing polymer engineer. The flow and mechanical properties are discussed in Chapter 8.

Polymerization processes and polymer fabrication are described in Chapters 9 and 10. The discussion emphasizes aspects of polymerization processes that are different from gas and liquid phase reactions of simple organic and inorganic compounds. These include constraints imposed by the polymerization system, mixing effects, and solvent and monomer removal.

The appendices are an integral part of the text. These data should help the student compare different polymer systems. Instructors can use the data to develop additional homework and example problems. Appendix B has a general properties list for a number of commercial polymers and monomers. Appendix C includes special properties that are important for fiber and elastomer applications. Appendix D contains information needed to do kinetic analyses of polymerization systems. Appendix E contains physical and processing properties such as viscosity and thermal properties. Appendix F has other physical property data.

Trademarks

Acrilan—Monsanto Chemical Co.
Acrylite—The Gilman Co., Inc. (acrylic sheet)
Acrylite GP, FF, PMMA & Plus—Cyro Industries
Araldite—Ciba-Geigy Corp.
Bakelite—Union Carbide Corp.
Caprolan—Allied Chemical Corp.
Celcon—Hoechst Celanese Corp., Advanced Materials Group
Creslan—American Cyanamid Co. (Fibers Div.)
Dacron—E.I. DuPont De Nemours & Co., Inc.
Delrin—DuPont
Dynel—Union Carbide Corp.
Epon—Shell Oil Co.
Kapton—Allied Signal, Inc. (plastic); DuPont (polyimide film)
Kevlar—New England Ropes, Inc.
Kraton—GLS Plastics, Shell Chemical Co.
Kynar—ATOCHEM North America, Inc. (plastics); OK Industries, Inc. (wire wrapping wire)
Lexan—General Electric Co.
Lucidol—ATOCHEM
Lucite—DuPont and PPG Architectural Finishes, Inc. (paint)

Lycra—DuPont
Marlex—FNT Industries, Inc.
Melmac—American Cyanamid Corp.
Merlon—Mobay Chemical Co.
Mylar—DuPont
Nomex—DuPont
Noryl—General Electric
Nylon II—Parkway Fabricators, Inc. (rubber diving suits)
Orlon—DuPont
Perkadox ACS & JPP—Akzo Chemicals Inc.
Perlon—Perlon Industries, Inc.
Perspex—ICI Acrylics Canada Inc.
Plexiglas—Rohm & Haas Company
Rilsan—ATOCHEM
Ryton—Phillips Fibers Corporation
Saran—The Dow Chemical Company
Saran Wrap—Dow Brands (Household Products Division)
Styron—The Dow Chemical Company
Styrofoam—Dow Chemical U.S.A.
Super Glue—Pro Seal Products; Loctite Corporation (Automotive Division)
Teflon—DuPont
Thiokol—Thiokol Corp. [Chemical Division]
Vazo—Dupont Co., Dupont Chemicals
Vectra—Ocli-Optical Coating Laboratories, Inc. (sunglasses); Veeco Manufacturing Co. (beauty aids)
Vulkollan—Miles Inc.
Xylan—The Lannett Co., Inc.

1

A Primer of Polymer Science and Engineering

This book is an introduction to polymers and polymer processes for engineers. The material is divided into three sections: an overview of polymer science, engineering analysis of polymer systems, and descriptions of polymer processes and processing. This chapter presents some of the fundamental concepts and language of polymer science, including nomenclature, molecular weight, chemical bonding and entanglements, and thermal transitions. It is intended to act as a primer for the rest of the text.

1.1 GENERAL TYPES OF POLYMERS

This section describes some of the scientific classifications of polymers. These are related to the reaction steps for converting the starting materials, usually liquids at reaction conditions, into very viscous liquids or solids. A *polymer* is a large macromolecule made up of many small, repetitive units. "Poly" is the Greek word for many and "mer" is the Greek word for unit, so polymer means "many units." The word was first used by Berzelius, the Swedish chemist, in 1833. Although styrene was polymerized in 1839, and poly(ethylene glycol) and poly(ethylene succinate) were made in the 1860s, the long chain nature of these materials was not understood until much later. Most of the first polymer products were derivatives of cellulose. Nitrated cellulose, which was called nitrocellulose, was used as gun cotton.

Macromolecule is a synonym for polymer and applies to synthetic and biological materials. Most commercial polymers are polymerized from simple molecules called *monomers*. Polymers with solid-like properties usually have thousands of repeating units in each *chain*. These individual chains associate with each other to make up the polymer product. The physical state of the polymer may vary from *amorphous*, no repeating structure, to *crystalline*, regular repeating structure throughout much of the material, to *cross-linked*, chemical links between chain segments so that the "chain" is endless and has infinite molecular weight.

Polymers and Polymerizations. Polymers can have strikingly different properties depending on their chemical structure and chain morphology. Low molecular weight chains, *oligomers,* usually have liquid-like properties. The *degree of polymerization* describes the number of *repeat units* in the average chain. The repeating unit may be a monomer, or it may be a combination of several reacted units.

With the notable exception of the polysilicones, most of the specialty and commodity polymers produced have carbon atoms in their *chain backbone*. The types of polymers we use have been greatly influenced by the available sources of carbon compounds and their costs. Table 1.1 shows elemental analyses of the major hydrocarbon resources.

The raw materials used for today's polymers—natural gas, crude oil, and coal—contain very little oxygen. Their carbon to hydrogen ratio increases from natural gas to crude oil to coal. Renewable materials such as wood, algae, and sea kelp may be the carbon source for future polymers. These all contain significant amounts of oxygen in their chemical structure. Commodity polymers based on these oxygenated materials would be much different from the commercial polymers used today.

About 5% of the crude oil used in the United States is converted to polymers. The rest is made into liquid and gaseous fuels. Reuse or recycling of the polymeric materials could reduce the amount of the hydrocarbon resources used for materials. If commodity polymers were burned to recover energy at the end of their useful life, the energy contained by these materials could be reused in part. This could contribute a significant amount to the energy resources of the country.

There are several methods for polymerization. The two major methods are *chain (addition)* and *step (condensation)*. Commodity polystyrene is manufactured using a chain polymerization process. Styrene monomer is made using benzene and ethylene, both of which are derived from petroleum. The long polymer chains are described as having n repeating units (n is equivalent to the degree of polymerization). Chain polymerizations have a few reacting sites in the polymer phase at any given time. Long

TABLE 1.1 Hydrocarbon Resources

Type		Weight Ratio, C : H : O
Gas	Natural gas	3 : 1 : 0
Liquid	Crude oil	6 : 1 : 0
Solid	Coal	14 : 1 : 0
	Cellulose	6 : 1 : 5.3
	Hemicellulose	6 : 1 : 8
	Lignin	6.8 : 1 : 3

polymer chains add monomers one unit at a time and grow very quickly. After the reactive site terminates, stable polymer remains.

By contrast, step polymerizations are based on the reactions between two different monomers. Nylon 6,6 can be made from hexamethylene diamine and adipic acid. Both of these starting materials have two end groups. The amine end groups react with the acid end groups in a stepwise fashion. Many of the stepwise systems can be thought of as condensation reactions, in which the reactants "condense" and lose a low molecular weight group. Many reactions take place quickly, but high molecular weight chains are not present as significant fractions of all polymeric material until very high conversions are reached.

The previous examples are called *homopolymers* because there is one type of repeating unit in the chain. Several monomers or repeating units can be combined into one polymer. These materials are called *copolymers* (two monomers) or *terpolymers* (three monomers). When the starting units are present in the reaction medium at the same time, *alternating* or *random* copolymers are formed.

A-B-A-B-A-B-A-B-A-B- A-A-B-A-B-B-B-A-A-B-A-B-A-
alternating random

When the polymer segments are formed sequentially from each type of repeating unit, *block* or *graft* copolymers are made. Figure 1.1 shows configurations of random, block, and graft copolymers. Polymers can be classified as *linear*, *branched*, or *network*. Linear polymers have all the repeating units aligned sequentially in the chain. Branched polymers have segments "branching" off of the main chain as in the case of graft polymers. Network polymers are cross-linked. They can be made by cross-linking the chains after polymerization or by using multifunctional monomers with

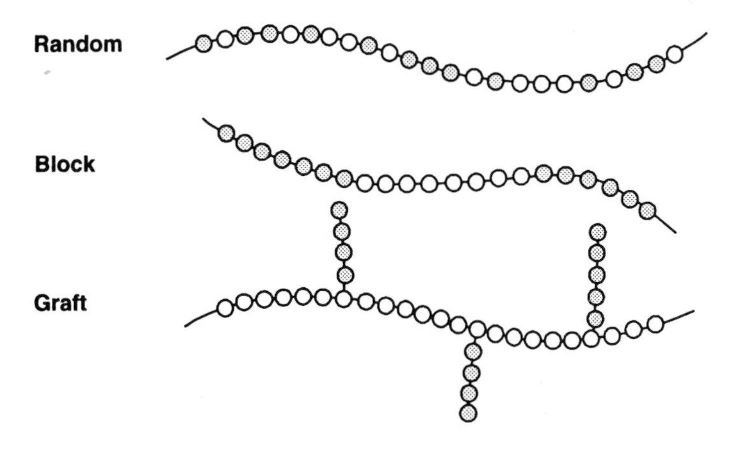

Figure 1.1 The arrangement of the monomers in a copolymer made with both monomers present in the reactor usually is **random. Block** or **graft copolymers** usually are synthesized using sequential monomer additions.

three or more functional groups per repeating unit, which form cross-links as they polymerize.

Classifications of Polymers. There are three main classes of solid polymers: *thermoplastics, thermosets*, and *elastomers.* These classifications relate to thermal performance and mechanical properties. All three types of materials can be used as solids. However, thermoplastics can be solidified by cooling and remelted by heating. Examples include polystyrene and polyethylene. The cooling/heating cycle can be repeated many times with little loss in properties. Thermosets retain their shape during cooling/heating cycles. They are cross-linked, so that postpolymerization heating softens the solid but does not permit the material to flow. Examples include phenolic resins, urea-formaldehyde resins, epoxies, cross-linked polyesters, and many polyurethanes. Elastomers have elastic properties: They deform readily with applied force and can recover their original shape after the force is removed. The term *rubber* may be applied to elastomers from natural sources, while elastomer may refer to material from synthetic sources. In this text, elastomer is used to refer to all materials with elastic properties.

Commercial products can be combinations of these classifications. Important products of the rubber industry—natural and synthetic rubber—actually are thermoset elastomers, because the final product has been cross-linked in order to retain its shape under use conditions. There are also thermoplastic elastomers, which are block copolymers of an elastomer with a thermoplastic and are considered part of the plastics industry.

Comparison of Polymers to Other Materials. Before the 1930s most household items, durable goods, and industrial products were made of wood, metals, glass, paper, leather, and vulcanized natural rubber. Since that time, plastics have replaced many of those materials because of their ease of forming and low cost. Polymer parts can be nearly isotropic compared to wood or some fabricated metals with respect to material properties.

Plastics are able to compete with other structural materials on a strength-per-unit weight basis (Table 1.2). The *elastic modulus* (E) and the *strength* (S) of materials have been used to characterize them for structural applications. Table 1.2 shows that polymer materials do not have the high modulus of metals or the moderate modulus of wood. Except for PEEK, an advanced thermoplastic, the polymers also do not have the strength of metals or wood.

However, most parts depend on the strength of their material as well as their three-dimensional shape for their performance properties. The stiffness of a given structure depends on the fundamental strength of the material, as well as the volume of the structure. In many cases, the weight of the structure is important. Here, the ratio of the modulus to the material density, or the strength of the material per unit density (ρ), will affect the part's performance. In these cases, the E/ρ and S/ρ ratios can be used to compare material choices. As shown in Table 1.2, polymers compete very well with metals when strength-per-unit weight is considered.

TABLE 1.2 Typical Properties of Materials

Material	Elastic Modulus (E) 10^6 psi	Strength (S) 10^3 psi	Specific gravity (ρ)	E/ρ	S/ρ
Mild steel	29	60	7.8	3.7	7.7
Aluminum	10	18	2.7	3.7	6.7
Brass	13	50	8.4	1.6	6.0
Zinc	13	13	6.8	1.9	1.9
Douglas fir*	1.4	10	0.5	2.8	20.0
Low density polyethylene LDPE	0.03	2	0.92	0.03	2.2
Polypropylene PP	0.2	5	0.91	0.22	5.5
Nylon†	0.42	8	1.14	0.37	7.0
Polyetheretherketone PEEK	0.56	12	1.31	0.43	9.2

*Air-dried, with the grain.
†As-molded, 0.2% moisture.
SOURCE: J. F. Carley, "A Plastics Primer'" in *Modern Plastics Encyclopedia* (R. Juran, Ed.), McGraw-Hill, NY, 1989.

Compared to the traditional structural materials, polymers exhibit a significant amount of creep under load. For many structural applications, this deficiency can be corrected by adding fillers, particularly fibrous materials that are either bonded or wetted to the polymer. The fibers reduce elongation, making the materials more brittle but stronger.

Polymers do not undergo rapid oxidative attack in the environment, as compared to the ferrous metals. Colors and light stabilizers can be used to provide uniform color throughout a part. Most polymers have reasonable solvent resistance.

Thermoplastics. Table 1.3 shows some plastics grouped by their performance properties. There are four tiers of thermoplastics: commodity, intermediate, engineering, and advanced. Over 75% of the total volume of thermoplastics are the commodity resins: low density polyethylene (LDPE), polypropylene (PP), polystyrene (PS), and poly(vinyl chloride) (PVC). These are made in standard formulations by a number of resin companies all over the world. Because of their low cost, they are considered first for most applications.

The intermediate group has slightly higher heat deflection temperatures compared to the commodity plastics. In this test, a standard-sized bar of the material is supported at each end and loaded in the center. The temperature of the system is raised until the center of the bar deflects a specified distance. This temperature is the heat deflection temperature. The engineering plastics can be used in boiling-water applications, a key criteria in the durable-goods markets. Their strengths and moduli are about an order of magnitude higher than the commodity resins. Polycarbonate has a low heat deflection temperature but is the toughest thermoplastic and it is transparent. It can therefore be used as lens material for eyeglasses. Filled nylon 6,6 has a very high heat deflection temperature, making it useful for a number of "under-the-hood" automotive applications.

TABLE 1.3 Plastics Grouped by Performance Properties

Group	Polymer	Properties			
		Cost $/in^3	Flexural modulus 10^3 psi	Strength 10^3 psi	Deflection temperature (264 psi), °F
THERMOPLASTICS					
Commodity					
	low density polyethylene (LDPE)	.014	40	1.6	< 100
	polypropylene (PP)	.0153	200	7	130
	polystyrene (PS)	.022	430	12	180
	poly(vinyl chloride) (PVC), rigid, pipe grade	.019	400	13	155
Intermediate					
	poly(methyl methacrylate) (PMMA)	.04	380	14	180
	acrylonitrile-butadiene-styrene terpolymer (ABS)	.052	350	11	230
	cellulose acetate butyrate (CAB)	.055	150	4	230
	thermoplastic olefin elastomer (TEO)	.04	10	5	n.a.
Engineering					
	polyoxymethylene (acetal)	.079	400	13	230
	nylon 6,6 w/30% glass fiber	.094	1000	30	350
	polycarbonate (PC)	.094	340	13.5	270
	poly(phenylene sulfide) (PPS)	.12	550	14	275
Advanced					
	liquid crystal polymer	.40	1800	19	650
	polytetrafluoroethylene (PTFE)	.45	80	1.7	100
	polyetheretherketone (PEEK)	1.10	560	16	320
	polyethersulfone (PES)	.25	360	18	395
THERMOSETS					
	alkyd polyester	.037	540	12	80
	epoxy, general purpose	.06	350	16	350
	phenolic, general purpose	.027	1200	10	370
	urea-formaldehyde, black	.034	1400	10	270

SOURCE: J. F. Carley, "A Plastics Primer" in *Modern Plastics Encyclopedia* (R. Juran, Ed.), McGraw-Hill, NY, 1989.

The advanced thermoplastics have extreme properties for polymers. Their moduli are within an order of magnitude of the moduli for metals (Table 1.2), and their E/ρ ratios are the same order of magnitude. Polytetrafluoroethylene (PTFE) is included in this list because it is very inert and has a low coefficient of friction, even though its mechanical properties and thermal properties are modest. Most thermoplastics can be polymerized and formed into beads or pellets for easy transportation from producers to fabricators. The fabricators remelt the pellets and form the melt into the final product shape. Excess material from the forming operations often can be recycled.

Thermosets. Compared to the thermoplastic group, the thermosets have relatively high moduli, high flexural strengths, and high heat deflection temperatures at moderate costs. Because they are cross-linked, thermoset products retain their shapes and strengths through many heating cycles. However, cross-linking tends to make the solids brittle and they have low impact strengths. Because they retain mold shapes precisely, thermosets usually are reacted into their final form by the fabricator. The fabricator may buy prepolymers (low molecular weight oligomers that are partially reacted) to speed cycle time and improve product uniformity. Many thermosets polymerize by stepwise or condensation mechanisms. Excess material from part fabrication is difficult to recycle because it is cross-linked.

1.2 POLYMER PRODUCTS AND INDUSTRIES

There are several steps required to convert monomers to useful parts and products. The first step, polymerization, has been discussed in the previous section. Two additional steps, formulation and fabrication, will be discussed in this section.

Formulation. The polymer leaving the reactor may not have all the properties needed for good performance in the final product. Additional materials may be added to the polymer in order to improve its properties or lower its cost. Some examples are

1. *Additives* such as heat and light stabilizers, lubricants, colorants, flame retardants, plasticizers, or foaming agents,
2. *Fillers* and *reinforcements* such as particulate minerals, hollow glass spheres, and fibers of glass, carbon, or polymers, and
3. Other polymers to form a polymer *blend*, polymer *alloy*, or layered material.

Additives can make improvements in the processing or performance properties of the polymer. Some might improve surface finish to make a shiny, smooth part. Stabilizers reduce degradation during processing as well as in the application. Particulate fillers can improve the high temperature performance of the material, as well as reduce its cost. Foaming agents are added to generate bubbles during fabrication, resulting in permanent voids in the final product. *Foams* have lower weight and, in some cases, can have good mechanical properties. There are *open-cell* and *closed-cell* foams, as well as *rigid* and *flexible* products.

Polymers with fiber reinforcements are grouped in a special class, *composites.* Fibers give improved strength along the direction of the filaments and allow the material to be used in structural applications. Fibers can be incorporated by simple mixing, or by the more complex methods of lay-up, winding, and weaving.

Packaging materials can consist of several layers of different polymers. Each layer will contribute different properties to improve the overall performance of the laminate. Examples include multilayer bottles and laminated films.

Formulation Operations. Most thermoplastics are made as particulate solids or melts. It is usual to convert these forms to pellets about the size of steel shot for easy transportation and *compounding*. Pelletizing is done using extruders, in which the melt is forced through pelletizing dies. The molten strands are chopped into pellets that are cooled, often in a water stream. Dried pellets can be stored, shipped, or compounded. Commodity polymers often are purchased by compounders, who blend them with a variety of additives for specific product types. The compounded pellets are sent to processors, who convert them into products and parts.

Thermoset resins also may be compounded. These materials often are molded with reinforcing material such as reinforcing mats, cloth, or chopped fiber glass rovings. Compounders buy liquid resins and add fillers, pigments, processing aids, cross-linking catalysts and reinforcements. Their product may be a prepreg, a mixture of reinforcing material with matrix material that may be partially polymerized. The *prepreg* reduces time in the final molding operation and may be prepared in the form of sheets or tapes.

Fabrication. The final form of the polymer product may require special processing. The elastomer, plastic, fiber, coating, adhesive, and adhesive industries all use polymers but have much different product requirements. The general product requirements of these industries are given in Table 1.4. However, a number of polymers can be made into several types of products. Polyethylene can be made into large parts, into films, and into fibers.

1.2.1 Plastics

Plastics are rigid and will maintain their shape under load at use temperatures, but flow viscously during fabrication. The two categories of plastics are thermoplastics and thermosets. One of the earliest examples of their use was the substitution of compression-molded nitrated cellulose and camphor balls for ivory as billiard balls. The less expensive balls not only popularized the game of billiards, but saved a lot of elephants. Bakelite plastics, made from phenol and formaldehyde, were some of the

TABLE 1.4 Polymer Industry Requirements

Industry	General Product Requirements
Elastomers	Large deformation and recovery
Plastics	Stable deformation under static load
Fibers	High strength per cross-sectional area, may require orientation
Coatings	Film-forming
Adhesives	Strong surface forces
Foams	Light weight components, low thermal conductivity
Composites	Structural materials

first commodity polymers made (1909). A typical use of this plastic is as an automotive distributor cap cover. Polyethylene, polystyrene, and poly(vinyl chloride) are some of the main commercial polymers used today. Table 1.5 lists commodity thermoplastics and some of their applications. Low density polyethylene (LDPE) has very good impact properties making it suitable for a variety of coating applications and bottles. High density polyethylene (HDPE) is a little stiffer and is used when flexural strength is needed (pipe and sheet). Polypropylene is highly crystalline material with low density. Its high strength-to-weight ratio and good surface gloss are advantages in many applications. Poly(vinyl chloride) can be compounded to give rigid (pipe applications) or flexible (wire insulation) properties. Polystyrene has the highest flexural modulus and is excellent in foam applications.

Thermoplastics are processed in two general ways: *profile extrusions* and *thermoforming* operations. Profile extrusions are used for products that have the same cross-sectional area throughout their length. Hoses, pipes, sheet, film, wire coating, and rods are some products made by this technique. *Extruders* (Fig. 1.2) melt the pellets and force the melt through dies to form the sheet or profile. Extruders can be coupled to *film-blowing* or *wire-coating* equipment (Fig. 1.3 and 1.4). Floor covering and plasticized film can be made using *calender* (Fig. 1.5), in which the polymer is formed into thick sheets on large rolls. Thermoforming operations include *molding*, *casting*, and other fabricating operations. *Injection-molding* operations use extruders to melt the pellets. The screw can act as a ram during part of the cycle to force the melt into molds. The parts are released from the mold and the cycle is repeated. Polymer sheets can be formed into three-dimensional shapes in other molding operations. Refrigerator door liners, automotive parts, and packages often are made by thermoforming. *Blow molding* (Fig. 1.6) is used to make bottles and hollow parts. Rotational casting, or *rotomolding*, is used to make large items.

Thermoset resins also can be formed into parts using molding operations. The resins enter the molds as viscous materials and are polymerized to take the mold

TABLE 1.5 Commodity Plastics and Their Uses

Type	Major Uses
Low density polyethylene (LDPE)	Packaging firm, wire and cable insulation, toys, flexible bottles, housewares, coatings
High density polyethylene (HDPE)	Bottles, drums, pipe, conduit, sheet, film, wire and cable insulation
Polypropylene (PP)	Automobile and appliance parts, rope, cordage, webbing, carpeting, film
Poly(vinyl chloride) (PVC)	Construction, rigid pipe, flooring, wire and cable insulation, film, sheet
Polystyrene (PS)	Packaging (foam and film), foam insulation, appliances, housewares, toys

Figure 1.2 A twin screw plastics extruder system. The barrel section has been removed to reveal the twin screw elements. Published with the permission of the APV Corporation.

shape. *Compression molding* (Fig. 1.7) is the simplest process for making thermoset products. Phenolic resins can be molded starting with preformed slugs or molding compound. These resins contain polar groups, which are easy to heat using microwave energy. These materials are heated to temperatures just below the molding temperature and then placed in the heated mold. As the mold parts come together, the compound is heated and pressed into the desired shape. The polymer is cured in the mold, the pressure is released, and the finished part is removed.

Transfer molding (Fig. 1.8) can be done with thermosets by methods similar to injection molding. The compound must have lower viscosity and less cure in order to tolerate longer processing times without set-up. Table 1.6 lists some thermoset materials and their uses. As shown in Table 1.3, epoxies have high flexural strength, which makes them good for adhesives and coatings. Phenol-formaldehyde materials are compatible with wood and are used as binders and adhesives. The polyesters are less expensive than epoxies and their moderate heat deflection temperature makes them adequate for applications like boat hulls.

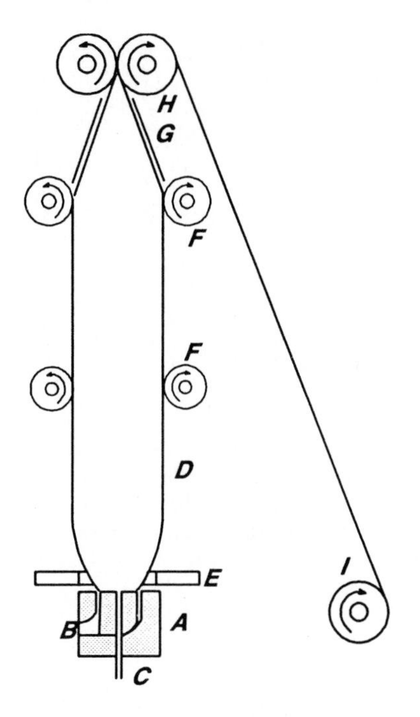

Figure 1.3 Schematic view of a film-blowing operation. The melt enters the blown film die, **A**, through inlet **B**. Air is injected into the bubble using **C**, an air hole and valve system. Air pressure expands the plastic tube (bubble), **D**. An air ring, **E**, cools the melt. Guide rolls, **F**, maintain the bubble shape. A collapsing frame, **G**, and pull rolls, **H**, form a flat sheet and seal in the air. The collapsed bubble is taken up on a wind up roll, **I** (Richardson, 1974).

1.2.2 Elastomers

Elastomers can recover elastically from large deformations at their use temperature. Most elastomers are cross-linked to form materials of essentially infinite molecular weights. Elastomers have been used by man for a long time. Rubber *latex* was harvested from trees by Indians in Central and South America. It may have been used

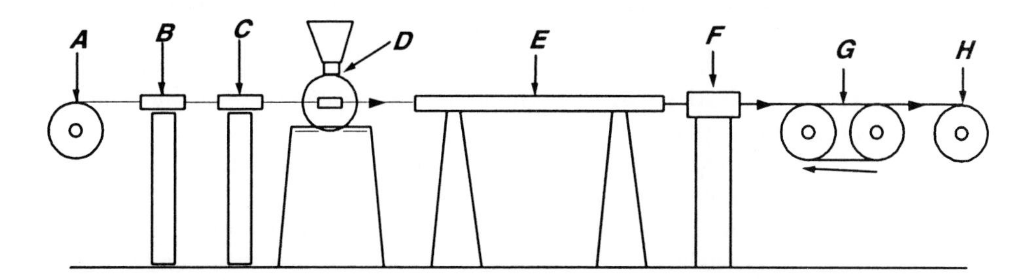

Figure 1.4 Schematic view of a wire coating line. Wire is unwound from the payoff roll, **A**; straightened, **B**; heated, **C**; and fed to the extruder die, **D**. Polymer melt coats the wire and is solidified in the water trough, **E**. The coating thickness is tested **F** and coated wire is wound by a capstan, **G**, and wind-up roll, **H**, system (Richardson, 1974).

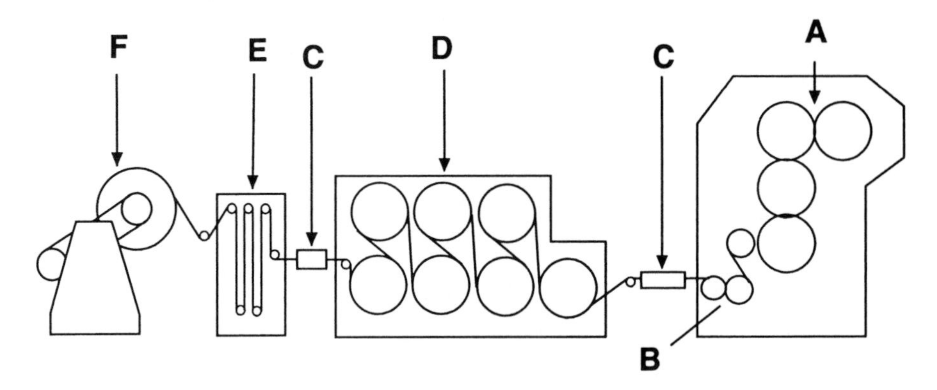

Figure 1.5 Schematic view of an inverted "L" calender plant for the production of plastic sheeting. Molten polymer is dropped into the calender system at **A**. Continuous sheet is formed by passing the melt between the rolls. One of the surfaces is given texture by an embossing roll, **B**. Thickness gauges, **C**, provide data for control of the process. The sheet is cooled, **D**, tensioned using wind-up accumulators, **E**, and accumulated on the wind-up roll, **F** (Holmes-Walker, 1975).

as a foot coating, and it served as the material used for the ball in a Mayan team game resembling basketball. *Vulcanization*, the cross-linking process in which sulfur is used to link double bonds, was discovered in 1839 and led to nontacky materials. The careful control of vulcanization allowed the rubber industry to make tires that were both elastic and long-lasting.

The growth of the rubber industry paralleled the growth of the automobile industry. In fact, this industry played a major role in World War II. The Japanese quickly seized Southeast Asia at the start of the war, not only for its value as a military position, but because it was a major source of natural rubber needed for tires and other war material. The United States responded with a crash program to develop synthetic rubber. The successful development of styrene-butadiene rubber (SBR) by the Rubber Reserve had a major impact on the war effort and the polymer industry in the United States. The development of a substitute for a well-known natural material gave a boost to the search for new and different polymers. Table 1.7 and Table 2 in Appendix C describe some commercial elastomers. Notice that a number of these materials are copolymers.

The processing of elastomers to make solid products is complex. Automotive tires start with complex blends of fillers (carbon black is a major component) and polymers, including natural and synthetic elastomers. Without cross-linking, elastomers creep extensively. The tire compound is formed into sheets so that various compound types—including side-wall, tread, and liner materials—can be used to lay up the tire. The crude tire laminate is placed in a tire mold, where heat is applied for about 20 minutes in order to make the elastomer flow into the mold cavity and to cross-link it.

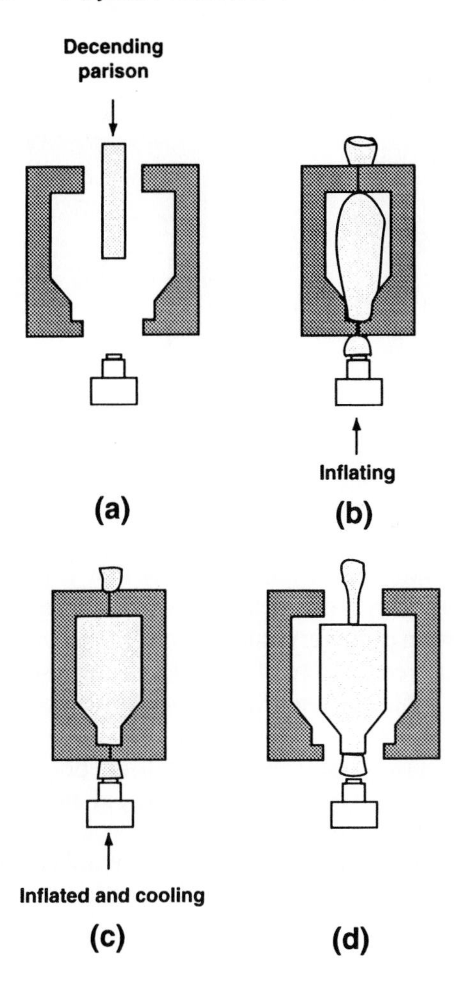

Figure 1-6 Schematic view of the blow-molding process. **(a)** A molten blob of polymer (parison) is extruded into the open bottle mold. **(b)** The mold closes and air is injected into the parison, causing it to inflate. **(c)** The melt expands until it hits the mold walls, where it cools and solidifies. **(d)** The mold is opened and the bottle is ejected for trimming.

1.2.3 Fibers

The first important commercial fibers were made around 1900. Two fibers derived from cellulose—rayon, which is essentially regenerated cellulose, and cellulose acetate—are still important polymers in today's markets. Cellulose acetate is used for yarn, photographic film, and other plastics. Rayon is used in a number of fiber blends. These polymers are based on naturally occurring polysaccharides. The cellulose in the cell walls of plants is made of long chains of glucose sugars, which are linked through a β glucoside link to the C_4 hydroxyl of the next glucose unit. The molecular weight of natural cellulose is high. For example, cotton averages 3000 repeat units per chain. All the natural fibers from biomass—cotton, flax, hemp, and jute—are cellulose.

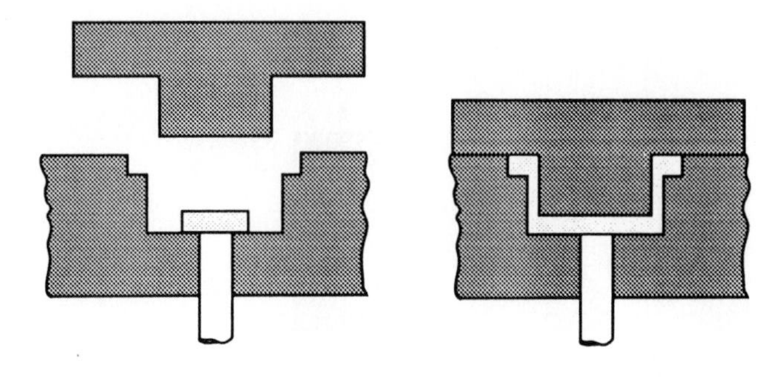

Figure 1.7 Schematic view of the compression-molding operation. A heated polymer form is placed into the mold cavity (left). Closing the mold forces molten polymer throughout the cavity (right). The high temperature of the mold causes cross-linking reactions and cures the part.

Commercial fibers are spun into small filaments that can be stretched and twisted into yarns. The drawing process is critical to attaining the high modulus needed in fiber products. The orientation causes crystallization of the polymer chains, leading to very high strength-per-unit-area. Yarn twisting and weaving also contribute to the properties of the materials. Table 1.8 and Table 1 in Appendix C describe some synthetic fibers.

Silk and wool are two naturally occurring fibers. Silk is a polypeptide containing glycine, L-alanine, L-serine, and L-tyrosine as the principal amino acids. It has an oriented crystalline structure, and the polypeptide chains occur in sheets, an example of the three-dimensional structure of polymers (Fig. 1.9). Wool is more complicated chemically than silk and contains cystine. Cystine can react to form disulfide bonds

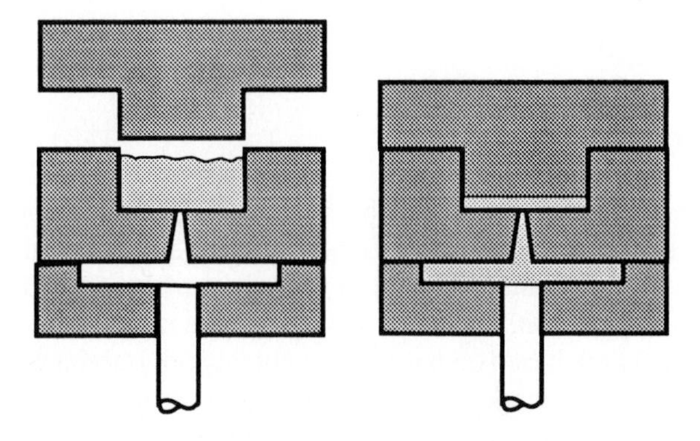

Figure 1.8 Schematic view of a transfer-molding operation. Low viscosity is placed in the upper cavity (left). The polymer is transferred into the cavity by pressure (right). The part is ejected after curing.

TABLE 1.6 Thermoset Polymers and Their Uses

Type	Typical Uses
Phenol-formaldehyde (PF)	Electrical and electronic equipment, automobile parts, utensil handles, plywood adhesives, particleboard binder
Urea-formaldehyde (UF)	Similar to PF polymers; also treatment of textiles, coatings
Unsaturated polyester (UP)	Construction, automobile parts, boat hulls, marine accessories, corrosion-resistant ducting, pipe, tanks, etc., business equipment
Epoxy (EP)	Protective coatings, adhesives, electrical and electronics applications, industrial flooring, highway paving materials, composites
Melamine-formaldehyde (MF)	Similar to UF polymers; also decorative panels, counter and table tops, dinnerware

between chains, making the chains stiff relative to chain rotation and movement. Reduction of these disulfide linkages with reagents such as ammonium thioglycolate makes wool more pliable. Hair permanents usually start with the reduction of the disulfide linkages to make the hair more pliable, continue with the desired curling of the hair, and finish with formation of new double bonds through the use of an oxidizing agent.

TABLE 1.7 Descriptions of Some Elastomers

Type	Description
Styrene-butadine	Copolymer of the two monomers in various proportions depending on properties desired; called SBR for styrene-butadiene rubber
Polybutadiene	*cis*-1,4 polymer
Ethylene-propylene	Often abbreviated EPD for ethylene-propylene-diene monomer; made up principally of ethylene and propylene units with small amounts of a diene to provide unsaturation
Polychloroprene	Poly(1-chloro 1-butenylene), good resistance to nonpolar organic solvents, also called neoprene rubber
Polyisoprene	*cis*-1,4 polymer, sometimes called synthetic natural rubber, good resistance to nonpolar organic solvents
Nitrile	Copolymer of acrylonitrile and butadiene, mainly the latter
Butyl	Copolymer of isobutylene and isoprene, with only small amounts of latter
Silicone	Contains inorganic backbone of alternating oxygen and methylated silicon atoms; also called polysiloxane
Urethane	Elastomers prepared by linking polyethers through urethane groups

TABLE 1.8 Descriptions of Some Synthetic Fibers

Type	Descriptions
CELLULOSIC	
Acetate rayon	Cellulose acetate
Viscose rayon	Regenerated cellulose
NONCELLULOSIC	
Polyester	Principally poly(ethylene terephthalate)
Nylon	Includes nylon 6,6, nylon 6, and variety of other aliphatic and aromatic polyamides
Olefin	Includes polypropylene and copolymers of vinyl chloride, with lesser amounts of acrylonitrile, vinyl acetate, or vinylidene chloride ($CH_2{=}CCl_2$) (copolymers consisting of more than 85% vinyl chloride are called *vinyon* fibers)
Acrylic	Contain at least 80% acrylonitrile; included are *modacrylic* fibers comprising acrylonitrile and about 20% vinyl chloride or vinylidence chloride

1.2.4 Coatings

Coating materials readily form films on solid surfaces. Alkyd resins, developed by Dow, are styrene-butadiene latexes containing modified natural oils and are excellent film-formers. Paints (mostly based on emulsion latexes from vinyl acetate and acrylic resins) are also a major part of the coatings market.

1.2.5 Adhesives

Adhesives are able to form strong bonds with surfaces and often form films that are oriented at interfaces. SUPER GLUE is an example of a polymeric adhesive and is based on cyano-acrylate chemistry.

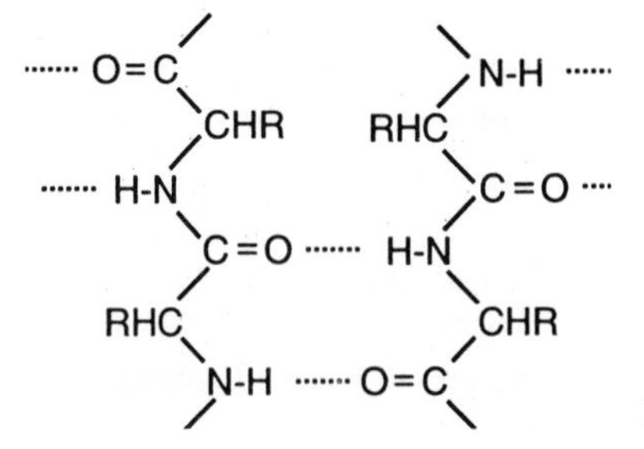

Figure 1.9 Structure of silk fibroin.

1.2.6 Foams

Foam materials now are a significant fraction of all polymer products. Foam specific gravities can range from near that of the polymer to as low as 0.05 g/cm^3. Many industrial and retail uses have been found for polystyrene (PS) foam. Isopentane is a common foaming agent for this product, because it is soluble in the polymer and is retained for long periods of time. PS foams can be made in a two-stage process. In the first stage, polymer beads containing isopentane are expanded slightly to make small balls. The balls are aged for 24 hours to permit the isopentane to redistribute evenly through them. Parts are made in low pressure molds by heating the beads with steam. The steam resoftens the polystyrene and increases the vapor pressure of the isopentane. The balls expand to fill the mold and fuse together.

Foam systems can be extruded. Inert gases such as nitrogen and carbon dioxide can be used as the blowing agent. These gases can be generated in situ by decomposition of special chemicals. Polyurethane foams are made in situ by using reactive monomers that generate a volatile leaving group. A linear polyester and an isocyanate will react to form polymer, with carbon dioxide as the volatile. The gas inflates the foam, which is cross-linked in place as the reaction continues. In reaction injection molding (RIM) (Fig. 1.10), the mixture is injected into a heated mold under modest pressure. The inner core of the material is foamed, but the outer surface has few bubbles and is near the density of the unfoamed material. This technique is used to make furniture pieces and automotive parts.

1.2.7 Composites and Fiber-reinforced Materials

Composite materials are composed of a reinforcing material—often a fiber—surrounded by a matrix material, which bonds the whole mass together. They are used in structural parts and give high strength for low weight. Composites such as fiber glass–reinforced epoxies are used in the automobile industry.

Many thermoplastic resins can be purchased in compounds containing reinforcing fibers. Fine glass fibers (diameters of 0.0004 in.) are chopped to 1/8 in. lengths and incorporated with the plastics. These materials can be formed in conventional processes: injection molding, compression molding, and transfer molding. Carbon fibers and aramid fibers (Kevlar) also can be used. Metal powders and fibers are used to improve electrical and conductive properties, providing radio frequency shielding.

Sheet-molding compound (SMC) refers to a class of reinforced thermoset materials. The sheets of fibers and matrix are aged, and cured using compression-molding techniques. The flow of SMC in the mold is complex and limits the shapes that can be formed. SMC's using thermoplastic resins overcome some of the forming difficulties of traditional SMC's.

High performance composite materials can be made using *filament-winding* techniques. A resin-soaked filament of about 200 fibers is wound around a *mandrel*,

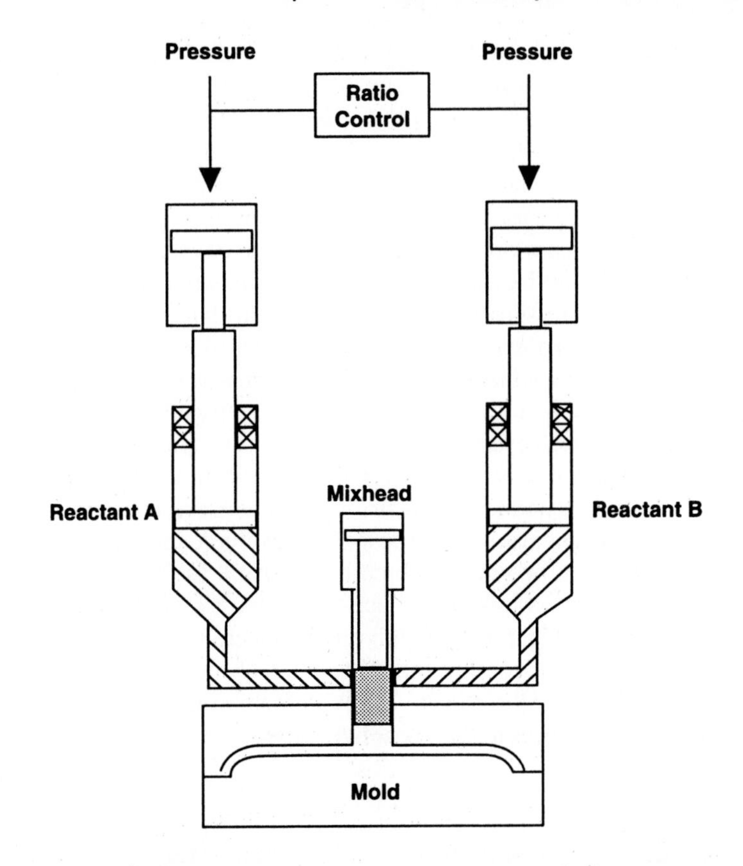

Figure 1.10 Schematic view of a reaction injection molding (RIM) machine. When the mixhead valve opens, two liquid reactants, **A** and **B**, flow at high pressure (100–200 bar) into the mixhead chamber. There they mix by impinging flow and begin to polymerize as they flow into the mold (C.W. Macosko, 1989).

which is the male form of the object. Computer-controlled winders can make complex, three-dimensional shapes. The material can be cured in autoclaves.

1.3 POLYMER NOMENCLATURE

Polymer nomenclature is widely varied between industrial and academic literature. A number of trademarks, such as Teflon, have become synonymous with certain classes of polymers. Often, it may be more convenient to use the common name rather than the technically correct name.

Brandrup (1989) has summarized IUPAC rules for naming polymers. The preference of these rules is to use common names when they are precise but to use formal names when common names require many modifiers to describe the polymer.

For example, poly(vinyl chloride) is preferred to polychlorethamer. However, poly(4,4′-isopropylidenediphenylene carbonate) is preferred scientifically to trade names such as polycarbonate from Bisphenol A: Few people use the preferred name in the case of polycarbonate. The rules given below are for regular, single-strand organic polymers. For copolymers and multiple-strand polymers, it is wise to consult the IUPAC rules.

1.3.1 Rules

1. The polymer is named according to its source when it is derived from a single original or hypothetical monomer, or when it is derived from two or more components that are built into the polymer in a random order.

   Examples: poly(vinyl alcohol)
   poly(styrene-co-butadiene)
   polyformaldehyde, not polyoxymethylene
   poly(ethylene oxide), not poly(ethylene glycol)

2. The polymer is named according to its structure whenever its *constitutional repeating unit* (CRU) is composed of several monomeric components following each other in a regular fashion. The CRU is independent of the way in which the polymer was prepared.

   Examples: poly(hexamethylene adipamide)
   poly(ethylene terephthalate)

3. Copolymers are treated using Rule 1. Specific morphologies for monomers A, B, and C can be described by the conventions given in Table 1.9.
4. The prefix, poly, is separated from the name by parentheses whenever the name is long or contains two words.

Initially, Rule 1 was adequate for naming polymers. Now, Rule 2, the structure-based name, is the preferred method for selecting names. For many common polymers, the structure-based name will not replace the source-based name. However, new polymers should be named according to Rule 2. Tables 1.10 and 1.11 give the chemical

TABLE 1.9 Rules for Nomenclature

Type	Connective	Example
Unspecified	-co-	Poly(A-co-B)
Statistical	-stat-	Poly(A-stat-B)
Random	-ran-	Poly(A-ran-B)
Alternating	-alt-	Poly(A-alt-B)
Periodic	-per-	Poly(A-per-B-per-C)
Block	-block-	Poly A-block-poly B
Graft	-graft-	Poly A-graft-poly B

source-based names of commercial polymers along with their common trade names and structure-based names. It is important to recognize that there are alternative naming schemes in the literature. For example, alternative names for polyethylene are polyethene and ethene, homopolymer.

1.4 CONSEQUENCES OF LONG CHAIN LENGTH

Although natural polymers (fibers, asphalt, and amber) have been used by man for centuries, the characteristics of polymers that are responsible for their unique properties have been understood only recently. A major stumbling block for modern chemistry was discovering that polymers had high molecular weights. In the mid-1800s, macromolecules were not distinguished from colloidal materials. Colloids themselves were difficult to explain because they did not follow Raoult's Law. Polymers were thought to be physical aggregates of small molecules. The empirical formula of natural rubber, C_5H_8, was known in 1826, and isoprene was known as a component of the destructive distillation of natural rubber in 1860. However, isoprene was not identified as the repeating unit until the early 1900s. Traditional chemistry identification techniques, such as end-group analysis, did not work well with high polymers, and polymers were thought to be ring structures that associated like colloids.

Staudinger and others established the long chain nature of polymers in the 1920s. Carothers' work at DuPont on nylon condensations helped verify that long chains were involved. The dilemma associated with the properties of macromolecules is illustrated by considering the alkane series. Table 1.12 shows the properties of alkanes as a function of chain length. Polyethylene has useful properties at high molecular

TABLE 1.10 Source and Structure Names. Polymer Chains Linked with Carbon-carbon Bonds

Source-based Name	Structure-based Name	Trade Name/Abbreviation
Polyethylene	Polymethylene	PE, LDPE, HDPE, LLDPE
Polypropylene	Poly(propylene)	PP
Polyisobutylene	Poly(1,1-dimethylethylene)	PIB
Polystyrene	Poly(1-phenylethylene)	Styron, Styrofoam
Poly(vinyl chloride)	Poly(1-chloroethylene)	PVC
Poly(vinylidene chloride)	Poly(1,1-dichloroethylene)	Saran
Polytetrafluoroethylene	Poly(difluoromethylene)	Teflon
Poly(vinyl acetate)	Poly(1-acetoxyethylene)	PVAC
Poly(vinyl alcohol)	Poly(1-hydroxyethylene)	PVAL
Poly(methyl methacrylate)	Poly(1-methoxycarbonyl-1-methylethylene)	Lucite, Plexiglass, PMMA
Polyacrylonitrile	Poly(1-cyanoethylene)	Orlon, Acrilan fibers
Polybutadiene	Poly(1-butenylene)	BR rubber
Polyisoprene	Poly(1-methyl-1-butenylene)	NR rubber
Polychloroprene	Poly(1-chloro-butenylene)	Neoprene

TABLE 1.11 Source and Structure Names. Polymer Chains Linked with Bonds Other than C—C

Source-based Name	Structure-based Name	Trade Name/Abbreviation
Polyformaldehyde	Poly(oxymethylene)	POM
Poly(ethylene oxide)	Poly(oxyethylene)	PEO
Poly(ethylene glycol adipate)	Poly(oxyethylene oxyadipoyl)	Polyester 2,6
Poly(ethylene terephthalate)	Poly(oxyethylene oxy-terephthaloyl)	Dacron, PET
Poly(hexamethylene adipamide)	Poly(iminoadipoyl imino-hexamethylene)	Nylon 6,6
Poly(ε-caprolactam)	Poly(imino[1-oxohexamethylene])	Nylon 6
Polyglycine	Poly(imino[1-oxoethylene])	Nylon 2

weights. Crystallinity does not completely explain its properties, since wax is also crystalline.

Polymers with similar bonds along their chain backbone often have much different properties. Polyethylene is a crystalline solid thermoplastic that is flexible. Polystyrene is a brittle thermoplastic. The carbon-carbon bonds along the backbone of each polymer have similar strengths. However, the properties of the two polymers are quite different. One consequence of long chain length is that secondary valence forces on the macromolecule become a significant factor in the ability of a polymer solid to resist imposed forces. Van der Waals (dispersion) forces act between neighboring molecules of wax and polyethylene. The long chain length of polyethylene compared to wax means that many more intermolecular secondary valence bonds contribute to the ability of one molecule to resist movement. Similarly, the benzene ring on polystyrene leads to different secondary bonding than hydrogen alone, so polystyrene should have different properties from polyethylene.

A second consequence of long chain length is *physical entanglement* of polymer chains. At chain lengths of 600 atoms or more, the chance of one loop of a polymer wrapping around another polymer chain (entanglements) is large. Wax is highly crystalline; it would be expected to have few crystal defects and essentially no entanglements, due to the short chain length. Entanglements reduce the ability of one molecule to slip past another and can act as cross-links between chains.

TABLE 1.12 Properties of Alkanes

Carbon Atoms	State/Properties	Use
1–4	Gas	Gaseous fuel
5–11	Low-viscosity liquid	Gasoline
9–16	Medium-viscosity liquid	Kerosene
16–25	High-viscosity liquid	Oil and grease
25–50	Simple solid (crystalline)	Paraffin wax
1000–3000	Plastic solid (crystalline and amorphous)	Polyethylene

Linear polyethylene has crystalline and amorphous regions. Entanglements can occur in the amorphous regions and would help anchor neighboring crystalline regions together. Linear polystyrene has little crystalline content, and its chain is somewhat "stiff" (extended conformation) because of steric hindrances to rotation around the carbon-carbon backbone. The probability of entanglements for similar chain lengths of these two polymers should be different. Entanglements affect solid and fluid properties.

1.5 CHEMICAL BONDING IN POLYMERS

Most of the commercial polymers have covalent bonds along the backbone of the polymer. As discussed in the previous section, secondary valence forces contribute significantly to cohesive bonding in polymers since chain lengths are long.

1.5.1 Types of Primary Bonds

Covalent Bonds. Covalent bonds occur when pairs of valence electrons are shared between two atoms, resulting in stable electronic shells. Bonds between two dissimilar atoms may be polar covalent. Bond angle and ease of rotation around the bond affect the conformation of the polymer. Since most commercial polymers contain carbon-carbon bonds along their backbone, their backbone bond angles are between 105° and 113° (close to the 109° 28′ of the tetrahedral angle for sp^3 orbitals). The silicone polymers, containing O-Si-O bonds, are an obvious exception to this rule. These bonds can have angles from 104° to 180° and are quite flexible when small groups are attached to the silicon. For example, poly(dimethylsiloxane) has the lowest T_g (glass transition temperature) of any elastomer, −110°C. Glass transition temperature defines when the polymer changes form solid-like to fluid-like behavior (see Section 1.7).

Ionic Bonds. An ionic bond occurs when one atom donates an electron to another. This process results in the outermost shell of one of the atoms being filled with electrons. Each atom is electrostatically charged, which provides an attractive force between the two atoms. These types of bonds do not occur in most commercial polymers. Divalent ions have been used to provide cross-links between carboxyl groups in natural resins.

Ionomer is a generic term for a class of thermoplastics with carboxyl group cross-links between chains. In general, these are copolymers of α-olefins with carboxylic acid monomers. The polymers are partially neutralized with a metal cation, which results in cross-links between chains. The cross-links are labile at processing temperatures, but "melt out" at high temperatures. Ionomers are usually transparent, tough, and flexible.

Coordinate Bonds. Coordinate bonds are found in some inorganic and semiorganic polymers. In addition to silicones, there are metal chelate polymers in which polydentate ligands are linked with metal ions. Some new polymers based on Si-P, P-N, and B-N bonding along the backbone may become commercially important.

1.5.2 Secondary Bonding Forces

The forces between different molecules or different segments of the same molecule contribute to their chemical properties. The unusual properties of polymers are often the result of a large number of small attractive forces acting at different points along the chain, in addition to physical entanglements and other physical interactions. Factors affecting these secondary bonding forces include

1. The nature of the secondary inter- and intramolecular bonding forces,
2. The molecular arrangement, or configuration, and cross-linking,
3. The molecular weight and molecular weight distribution, and
4. The conformation and micromorphology of the polymer.

While secondary bonding forces are much weaker than primary bond forces on a per bond basis, they can dominate the total sum of forces on a macromolecule due to the large number of repeating units in a given polymer chain. Consider a macromolecule under tension (Fig. 1.11).

The macromolecule is imagined to be confined in a tube and under tension over the cross-sectional area of the tube. Primary bonds carry some of the load, but each

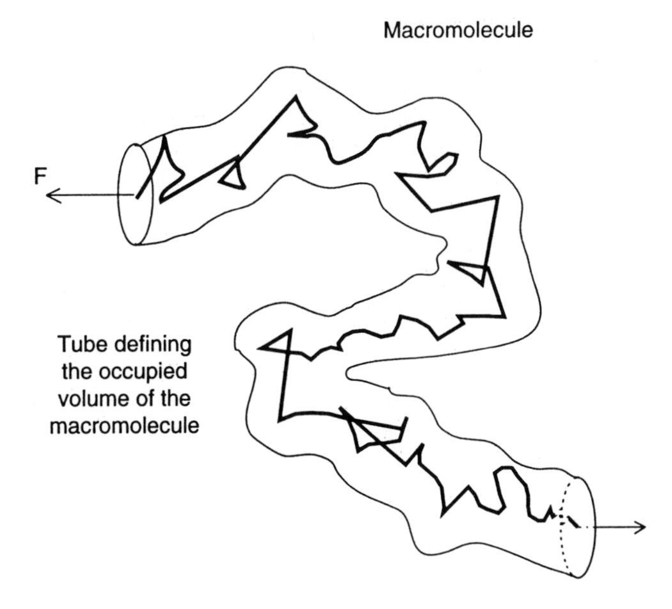

Figure 1.11 Macromolecule under tension. Resistance to the applied force occurs by stretching of the covalent bonds along the polymer backbone, and by stretching of the secondary valence bonds over the surface area of the molecule.

is under a similar tensile force and there is one per tube cross-sectional area. Secondary forces also can help carry the extensional load by resisting the motion of the macromolecule through the tube area. The area across which these secondary forces act is the cylinder wall. The longer the macromolecule, the greater is its ability to resist deformation, as more secondary bonds are available to restrain the motion. Secondary valence forces affect many polymer properties, including T_m (crystalline melting point), T_g, crystallinity, tensile strength, and density.

There are four categories of secondary (van der Waals) bonding forces:

1. *Polar Dipole*: Equal and opposite electric charges in different segments of one molecule will attract or repel each other. The strength of polar dipole forces is temperature-dependent.
2. *Polar Induction*: Polar segments can induce charges in neighboring segments or atoms. The strength of polar induction forces is usually less than polar dipole forces and is also temperature-dependent.
3. *Nonpolar Dispersion*: These are time-varying dipole forces that integrate to zero over time and are the result of perturbation of nearby atoms' electric structure. Nonpolar dispersion forces are temperature-dependent.
4. *Hydrogen Bonds*: They occur between two functional groups. The bond occurs between proton donors and electron acceptors. Hydroxyl, carboxyl, amine, or amide groups are acidic and act as proton donors. Carbonyl, ether, hydroxyl, nitrogen, and halogens may be electron acceptors.

Table 1.13 shows the approximate strengths of different types of bonding forces. The secondary bonding forces vary in strength and range of action. In general, hydrogen bonds are the strongest and decrease only with the inverse of the square of the distance between proton donor and electron acceptor.

Hydrogen bonds can be classified with respect to their relative strength. Differences in bonding strengths can account for some of the nonideal solution behavior of polymers in either polymer-solvent or polymer-polymer mixtures. Table 1.14 lists five classes of hydrogen bonding in order of descending strength.

TABLE 1.13 Intermolecular Bonding Forces

Type of Bond	Bond Energy, kcal/mol	Range of Action, A°
PRIMARY BONDS		
Ionic	140–250	2–3
Covalent	15–170	1–2
Metallic	27–83	
SECONDARY BONDS		
Dispersion	0.1–5.0	3–5 (r^{-6})
Dipole-dipole	0.5–5.0	1–2 (r^{-3})
Dipole-induced dipole	.05–0.5	1–2 (r^{-6})
Hydrogen bond	1.0–12	2–3 (r^{-2})

EXAMPLE 1.1 COMPARISON OF COHESIVE ENERGY DENSITY OF TWO POLYMERS

Cohesive energy density, the amount of energy per mole of monomer or repeat unit, can be used to illustrate the effects of hydrogen bonding on polymer properties. Polyethylene should have only dispersion forces acting between polymer chains. Poly(vinyl chloride) has Class IV hydrogen bonds acting between chlorine and hydrogens on the polymer backbone. The cohesive energy density of polyethylene is 61 kcal/cm^3. By contrast, the cohesive energy density of poly(vinyl chloride) is 91 kcal/cm^3. Figure 1.12 shows typical secondary bonding, which might occur between adjacent chains of these homopolymers. In addition to dispersion forces, PVC might have one hydrogen bond (Cl- - -H) per monomer unit. The Class IV hydrogen bonds are strong compared to dispersion forces and contribute to the 50% increase in cohesive energy between PE to PVC.

The covalent bonds of the polymer backbone can affect some polymer use properties. Thermal stability and oxidation are dependent on the strength of the carbon-carbon bonds along the chain. Differences in substitution at one particular carbon atom—such as branching points, cross-links, chain ends, or other defects—give it different reactive properties.

TABLE 1.14 Classification of Groups According to Hydrogen Bonding Potential

Class	Description	Examples
I	Groups capable of forming 3-D networks of strong hydrogen bonds	Water, glycols, glycerol, amino alcohols, amides, hydroxylamines, hydroxyacids polyphenols
II	Groups containing both active hydrogen atoms and donor atoms (O, N, and F)	Alcohols, acids, phenols, amines, oximes, nitro and nitrile compounds with α-hydrogen atoms, ammonia, hydrazine, hydrogen fluoride, and hydrogen cyanide
III	Groups containing donor atoms but no active hydrogen	Ethers, ketones, aldehydes, esters, tertiary amines (including pyridine types), nitro and nitrile compounds without α-hydrogen atoms
IV	Groups containing active hydrogen but no donor atoms that have two or three Cl atoms on the same C atom as the H atom, or one Cl on the same C atom and one or more Cl on adjacent carbon atoms	$CHCl_3$, CH_2Cl_2, CH_3CHCl_2, CH_2Cl-CH_2Cl, $CH_2Cl-CHCl_2$, $CH_2Cl-CHCl-CH_2Cl$
V	All other molecules have neither active hydrogen atoms or donor atoms	Hydrocarbons, CS_2, sulfides, mercaptans, and halohydrocarbons not in Class IV

SOURCE: R. H. Ewell, J. M. Harrison, and L. Berg, Ind. Eng. Chem. *36*, 871–875 (1944).

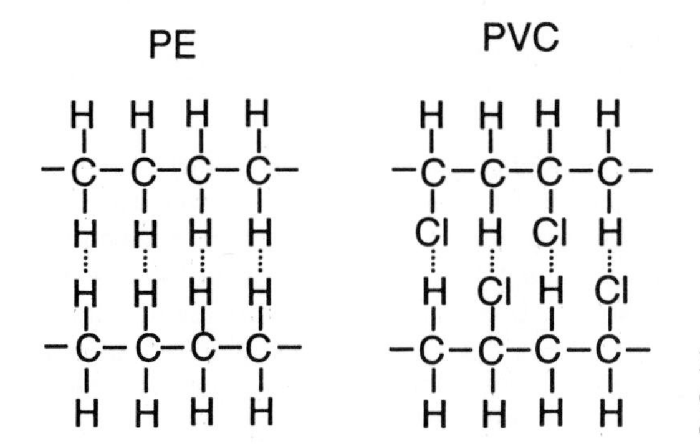

Figure 1.12 Secondary valence forces: dispersion forces in polyethylene and hydrogen bonding in poly(vinyl chloride).

1.6 MOLECULAR WEIGHT AND MOLECULAR WEIGHT DISTRIBUTION

Prior to Staudinger's work on macromolecular compounds, it was known that starch and cellulose were polymers of glucose, but the differences between these and crystalline di-, tri-, and polysaccharides were unexplained. Osmotic measurements had shown that cellulose solutions consisted of particles the size of colloids. Therefore, the properties of cellulose were thought to be due to secondary valence forces causing aggregation of about ten glucoside units. Staudinger felt that the colloidal particles in cellulose solutions were actual molecules of those materials. He proved the macromolecular concept using a series of reactions called polymer analogous conversions. Cellulose was nitrated to give a material with different solubility properties but with the same molecular weight. The fact that the original cellulose had the same molecular weight as the nitrated cellulose implied that there was a polymer molecule whose main valence linkages were unaffected by the reaction. It is highly unlikely that the particle size of a secondary valence aggregate in water solution would be the same as the esterified particle in organic solution.

In contrast with low molecular weight compounds, polymers do not have a unique molecular weight. In fact, two homopolymers differing in molecular weight may not even be miscible in each other. Polystyrene of low molecular weight is a viscous liquid, becomes a brittle polymer at 20,000 molecular weight, and is a hard plastic at 250,000 molecular weight. As will be shown in Chapter 4 on polymerization, the molecular weight of material produced in batch polymerizations changes with time. Both the average molecular weight and the breadth of the distribution affect polymer properties. For this reason, much attention has been paid to measuring and understanding molecular weight distributions.

Molecular weight distributions are often available as frequency plotted versus molecular weight. The first moment of each curve, normalized by its area, defines an average molecular weight.

Number Average Molecular Weight:

$$\overline{M}_n = \frac{\Sigma n_i M_i}{\Sigma n_i} \tag{1.1}$$

Weight Average Molecular Weight:

$$\overline{M}_w = \frac{\Sigma w_i M_i}{\Sigma w_i} \tag{1.2}$$

The number and weight fractions are simply connected.

$$w_i = n_i M_i \tag{1.3}$$

The ratio of the weight average to the number average molecular weight, $\overline{M}_w / \overline{M}_n$, is the polydispersity. This ratio describes the breadth of the distribution. The polydispersity for most commercial polymers is two or greater.

Figure 1.13 shows molecular weight distributions for two polymer samples of low density polyethylene. The distributions are shown as weight frequency versus molecular weight. The samples have similar $\overline{M}_n$'s but much different $\overline{M}_w$'s, so their polydispersities are different. The narrow distribution sample has a polydispersity of 5.7, while the broad distribution sample has a polydispersity of 14.6. These samples should exhibit different processing (flow) properties and performance (tensile and appearance) properties.

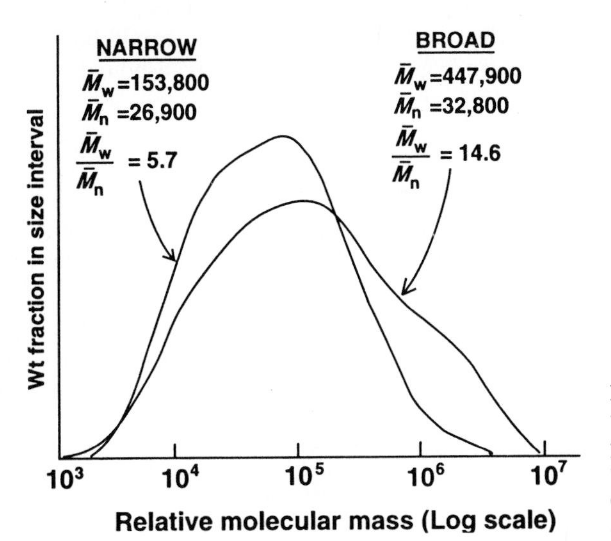

Figure 1.13 Weight frequency versus relative molecular mass for two low-density polyethylenes determined by gel permeation chromatography: one with a narrow distribution, the other broad (R.B. Staub and R.J. Turbett, 1973).

1.7 THERMAL TRANSITIONS

The mechanical properties of polymers change similarly with temperature for many polymer systems. There are five regions of viscoelastic behavior, as shown in Figure 1.14. The main curve shows the logarithm of the elastic modulus over a wide range of temperature for an amorphous polymer. At low temperatures, the polymer behaves as a glassy solid, and the elastic modulus is fairly constant as temperature changes (Curve **A**). Curve **B** marks the glass transition temperature, T_g. The modulus decreases rapidly with increasing temperature, and the polymer begins to exhibit liquid-like properties. Below T_g, only movements of short chain segments are thought to occur. At T_g and above, chain movements are thought to involve 20 to 50 atoms along the chain. Crystalline polymers may show a smaller change in modulus (due to their lower amorphous content), and their modulus may be fairly constant over wide ranges of temperature (Curve **F**). At their crystalline melting point, they will also exhibit a rapid decrease in modulus with temperature (Curve $\mathbf{F}^1$). In the rubbery plateau region (Curve **C**), the modulus is again fairly constant, and material deforms like an elastomer to imposed forces. Lightly cross-linked elastomers will show rubbery plateau regions over wide ranges of temperature (Curve **G**). Curve **D** shows a second rapid decrease in modulus with increasing temperature, the rubbery flow region. Polymers in this region behave as viscoelastic fluids. Viscous response is liquid-like, and elastic response is solid-like. If the polymer does not degrade as the temperature is raised further, there may be a fluid flow region in which the polymer acts like a Newtonian fluid with no elastic recovery.

Engineering Use Temperatures. Many polymer systems exhibit the behavior shown in Figure 1.14. Elastomers are used well above their glass transition in their rubbery plateau region. There must be high segment mobility in order to have elastic response to deforming forces. Amorphous structural polymers, which rely on rigidity, should be used well below T_g. Tough, leather-like polymers, such as plasticized PVC, can be used near T_g. Highly crystalline and oriented polymers should be used well below T_m. Semicrystalline polymers, such as polyethylene, can be used in between T_g and T_m.

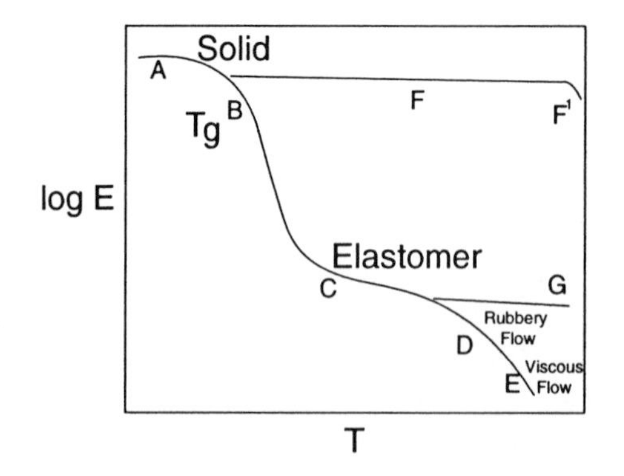

Figure 1.14 Regions of polymer behavior. At low temperatures, the modulus is high and the material behaves as a glassy solid, **A** At the glass translation temperature. **B**, the modulus falls rapidly with temperature. In the rubbery plateau region, **C**, the modulus is fairly constant and then falls again in the rubbery flow region, **D**, and the fluid flow region, **E**. Crystalline materials retain high moduli over large temperature ranges, **F**. Cross-linked elastomers have rubbery moduli, **G**.

Figure 1.15 shows the dependence of the shear modulus (G) for three materials: an elastomer, an amorphous thermoplastic, and a crystalline polymer. The elastomer, natural rubber, is used at room temperature or above (the C region of Fig. 1.14). If the temperature drops below −50°C, natural rubber will become a glassy solid. PVC is an amorphous polymer that can be used as a solid up to about 80°C. Nylon 6, the crystalline material, has a glass transition temperature of about 40°C, but has a high modulus all the way up to its melt temperature, 210°C. The heat deflection temperature of nylon 6 is about 185°C.

1.8 PHYSICAL PROPERTIES

Polymeric materials are selected for their performance and processing properties. There are three property areas that are very important to polymers: elasticity, viscoelasticity, and yield/fracture.

1.8.1 Elasticity

As shown in Figure 1.14, there are five regions of mechanical behavior. Every amorphous polymer above its glass transition temperature displays the following properties: high extensibility under low mechanical stress over short times, and recovery of original shape after mechanical deformation. These two properties are due to deformation-induced changes in chain entropy.

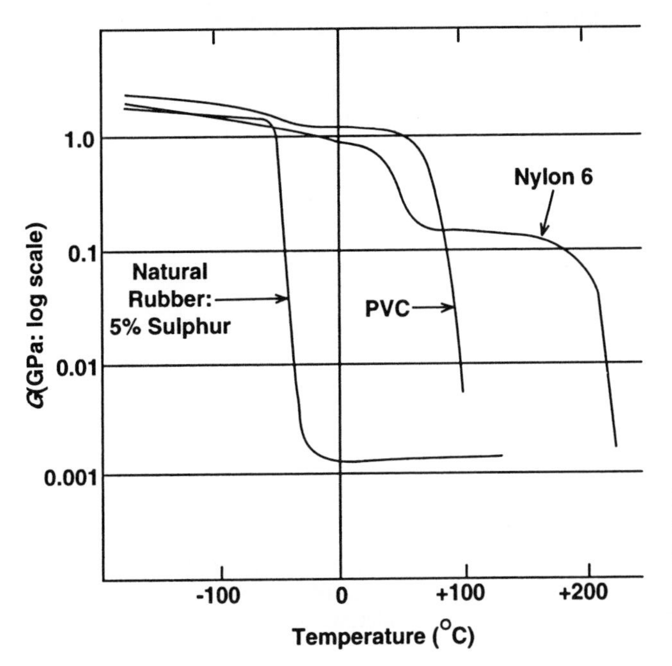

Figure 1.15 Dependence of the shear modulus on temperature for three representative engineering polymers: natural rubber (cross-linked), PVC (essentially amorphous and not cross-linked), and nylon 6 (crystalline). McCrum, et al., 1988.

1.8.2 Viscoelasticity

Consider the simple experiment of a weight suspended by a polymer film or fiber. Under a constant load over long periods of time, the strain of the sample will increase, that is, the film will stretch. This effect, creep, is due to molecular rearrangements in the film induced by the stress. When the stress is released, the film may slowly recover the original conformation of the molecules and the strain goes to zero. Creep is a result of the general property of polymeric solids, viscoelasticity. Polymer solids are elastic and can recover their original dimensions after rapid mechanical deformation. However, they also are viscous since they flow under long-term loads.

Polymers are viscoelastic at all temperatures and over all mechanical deformation times. Over short deformation times they appear to be elastic, and over long times they appear to be viscous. In the glassy state, the entire chain may be "immobile," but long chain segments can move. In the liquid state, the entire chain can move through other chains by reptation.

1.8.3 Yield and Fracture of Solids

Figure 1.16 shows stress(σ)-strain(ε) curves for two plastics, polystyrene and polyethylene, in a tensile test. At low strains (<.1%), most plastics behave elastically and the stress-strain curve is linear with a slope equal to Young's modulus (E). As strains under load become higher, one of two failure modes occur.

Polystyrene undergoes brittle failure. The yield point ($d\sigma/d\varepsilon = 0$) is reached at moderate strains and the sample fails by cracking. Careful examination of the failed polymer reveals crazes, local yielding over a small region near the tip of the crack. In polystyrene, there are fibrils of 200 Å that span the cracks and bear the imposed stress. Molecular entanglements between chains stabilize the fibrils. Low molecular weight polystyrene materials are very brittle, because the chains are too short to stabilize the fibrils.

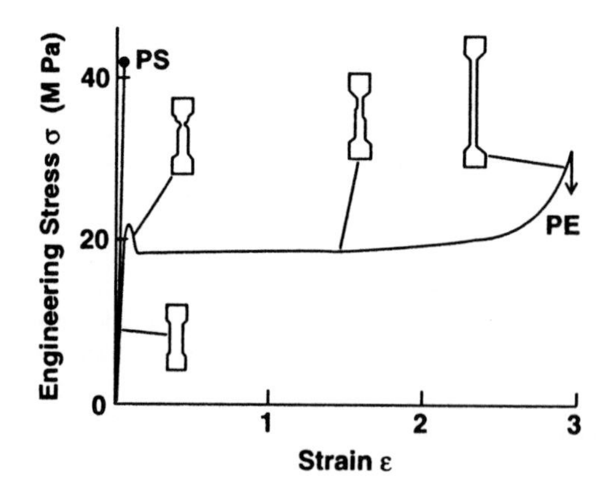

Figure 1.16 Stress-strain curves for polystyrene (PS) and polyethylene (PE) (McCrum et al., 1988).

Polystyrene under compression is ductile as are other thermoplastics and lightly cross-linked thermosets. Temperature, strain rate, loading type, sample geometry, and environmental factors (solvents, water) will affect the brittle failure mechanism.

Polyethylene, a crystalline material, is ductile and can be drawn as high as 25 times its original length before the sample fails by breaking. It is elastic below the yield point. After the yield point, many ductile polymers show a drop in load due to a combination of strain softening and localized necking (the sample cross section becomes smaller).

Above the yield point, elasticity is difficult to measure. A sample may not be elastic over short periods of time, but may recover much of its original shape by allowing long recovery times or by increasing the temperature so that chain segments can respond quicker to the internal strains imposed by the new shape.

NOMENCLATURE

Symbol	*Definition*
M_n	number average molecular weight
M_w	weight average molecular weight
n_i	number frequency
T	temperature
T_g	glass transition temperature
T_m	melting temperature
w_i	weight frequency

PROBLEMS

Students may wish to consult tables in Chapters 2, 3, and the Appendix to work these problems.

1.1 Name these polymers

(a) S-S-S...-S-B-B-B-B...B-B-S-S-S...S-S
where S = styrene and B = butadiene

(b) S-S-S-S-S-...S-S-S-S | B-B-B...-B-B
where S = styrene and B = butadiene

(c) $[\text{-NH}(CH_2)_6\text{NH-CO-}(CH_2)_8\text{-CO-}]_n$

(d) $[\text{-CH}(CH_3)\text{- O-}]_n$

(e) $[\text{-}CH_2\text{-NH-CO-}]_n$

(f) $[\text{-}CH_2\text{-C}(CH_3)_2\text{-}]_n$

(g) $[\text{-CH}_2\text{-NH-R-}\overset{\overset{\text{R}''}{|}}{\text{N}}\text{-R}'\text{-NH-}]_n$

1.2 Nitrile rubber and neoprene have good resistance to gasoline and other organic solvents compared to other elastomers. Rationalize their solvency performance based on secondary bonding considerations. What solvents might swell these elastomers?

1.3 Complete the following table:

Polymer	*Polymerization Method*	T_g	T_m	*Crystal Form*
Poly(methyl methacrylate)				
Poly(vinylidene fluoride)				
Nylon 6,6				

1.4 Find tensile moduli, densities, and processing temperatures for the polymers of Problem 1.3.

1.5 Categorize the following copolymers with respect to their secondary valence forces:
Polyester 2,6
Urea-formaldehyde resin
Cellulose
Copolymer made by first polymerizing butadiene, then polymerizing styrene
Copolymer made by polymerizing butadiene and styrene together

1.6 Explain the differences in T_g and T_m for the following polymer pairs:
cis-Polyisoprene/Polychloroprene, Poly(vinyl ether)/Poly(vinyl ethyl ether).

1.7 Find the chemical names and industrial suppliers for the following polymers: Noryl, Lexan, Bakelite, Kralon, Kevlar, and Delrin. Consult a listing of trade names such as the *Modern Plastics Encyclopedia* to determine these. List typical end uses for each material.

1.8 Write one sentence definitions of the terms below. The meaning of these terms should be known by you before continuing with this text.

(a) Thermoplastic
(b) Elastomer
(c) Thermoset
(d) Single screw extruder
(e) Injection molding
(f) Cross-linking
(g) Copolymerization
(h) Engineering plastics
(i) Graft copolymer
(j) Block copolymer
(k) Addition polymerization
(l) Condensation polymerization
(m) Monomer
(n) Macromolecule
(o) Oligomer
(p) Linear polymer
(q) Blow molding
(r) Constitutional repeating unit
(s) Spinning
(t) Thermoforming
(u) Fiber
(v) Adhesive

1.9 Determine the degree of polymerization for polymers having $M_w = 50{,}000$ and named
(a) Nylon 6,6
(b) Polypropylene
(c) Teflon
(d) Polycarbonate
(e) Polystyrene

1.10 Propose some monomers that can be used to make the polymers in Problem 1.1.

1.11 Sketch the structures of
(a) Poly(propylene-co-[methyl methacrylate])
(b) Polystyrene-block-(butadiene-alt-vinyl acetate)
(c) Poly(vinyl acetate-per-butadiene-per-acrylonitrile)
(d) Polystyrene-graft-polybutadiene

1.12 Engineering use temperatures are often related to the glass transition temperature of the polymer. Match the polymer category with the appropriate temperature range and explain your choice.

___ (a) Elastomers	1. Below T_m and T_g
___ (b) Oriented polymers	2. Above T_g
___ (c) Tough, flexible (leathery)	3. Near T_g

1.13 Match each polymer with its appropriate classification and explain your choice.

(a) Silicone rubber	
(b) Melamine molding resin	1. Thermoplastic
(c) Nylon 6,6	2. Thermoset
(d) Neoprene	3. Elastomer
(e) Acrylic fibers	

1.14 Define T_m and T_g.

1.15 Polymers with only carbon-carbon bonds in the backbone chain are usually polymerized from vinyl monomers using free radical catalysts. The phenol-formaldehyde resins are an exception to this rule (see Table 2.9). What differences in structure might you expect between polyacrylonitrile and a phenolformaldehyde resin.

1.16 It is thought that many adhesives cause bonding via secondary valence forces. Phenolic adhesives are often used in the manufacture of plywood. A critical property of laminate interfaces is their water uptake. Using the general structures of cellulose and phenolic resins, explain why.

1.17 Find the cohesive energy densities for poly(vinylidene chloride) and poly(vinyl chloride). Justify the difference based on secondary valence forces.

REFERENCES

J. BRANDRUP, "Nomenclature Rules," in *Polymer Handbook* (J. Brandrup and E.H. Immergut, Eds.), Wiley, N Y, 1989.

J.F. CARLEY, "A Plastics Primer" in *Modern Plastics Encyclopedia*, (R. Juran, Ed.), McGraw-Hill, N Y, 1989.

R.H. EWELL, J.M. HARRISON, and L. BERG, *Ind. Eng. Chem. 36*, 871–875 (1944).

W.A. HOLMES-WALKER, *Polymer Conversion*, Halsted Press, London, 1975.

C.W. MACOSKO, *Fundamentals of Reaction Injection Molding*, Hanser Publishers, NY, 1989.

N.G. MCCRUM, C.P. BUCKLEY, and C.B. BUCKNELL, *Principles of Polymer Engineering*, Oxford Univ. Press, NY, 1988.

P.N. RICHARDSON, *Introduction to Extrusion*, Society of Plastics Engineers, Inc., Greenwich, CT, 1974.

R.B. STAUB and R.J. TURBETT, *Modern Plastics Encyclopedia*, McGraw-Hill, NY, 1973.

2

Molecular Architecture

Because polymers can have complex structures, more information is needed to define their properties than is needed for simple organic and inorganic materials. Polymer properties depend on the chemistry and physics of individual macromolecules, *molecular architecture,* and associations between the macromolecule with other constituents in real systems, the *physical state* of the polymer. This chapter covers molecular architecture while Chapter 3 covers some physical states of polymers. Most of the polymers discussed in Chapter 2 are synthetic. The natural polymers will be discussed in Chapter 3.

Macromolecular properties depend on their detailed three-dimensional structure. Three pieces of information are usually needed to adequately define the structure of a macromolecule. They are

1. *Constitution*: the type of atoms in the chain, the type of side groups, the type of end groups, the sequence of monomers, the type and size of branch units, the molecular weight, and the molecular weight distribution,
2. *Configuration*: the arrangement of neighboring atoms or groups about a specific atom, and
3. *Conformation*: the arrangement of the chain in space, associated with the rotation of groups around bonds of the macromolecule.

Constitution and configuration may be used interchangeably in some references. They are established by the polymer's formation process. Conformation is a function of the polymer's environment.

EXAMPLE 2.1. MOLECULAR ARCHITECTURE OF LOW DENSITY POLYETHYLENE (LDPE)

The above definitions will be illustrated by describing the molecular architecture of a commercial polymer product: low density polyethylene. LDPE is a homopolymer historically made by polymerizing ethylene at high pressure in a free radical, addition polymerization process.

Constitution. The chain structure of polyethylene is about as simple as polymers can get. The monomer is ethylene and the repeating unit is $-CH_2-$, resulting in the preferred IUPAC name of poly(methylene).

$$H_2C = CH_2 \qquad\qquad \left[CH_2 - CH_2 \right]_n$$

ETHYLENE POLYETHYLENE [POLYMETHYLENE]

While most of the addition reactions add monomer to the free radical site in a linear fashion, side reactions are possible. The nature of the free radical process used to make LDPE permits random, short chain rearrangements at the reacting end. The short chain branches range in size from two to eight carbon atoms long and average four carbon atoms. The reactor product is analogous to a copolymer of ethylene and a mixture of alpha-olefins, primarily 1-hexene.

Another side reaction is caused by free radical attack on hydrogen atoms along the carbon chain. This leads to the formation of long chain branches, as the new reaction site adds monomer.

Peroxide initiators or oxygen may be used in commercial systems. Fragments of these initiators should be incorporated into the molecule at the chain ends.

Configuration. Because LDPE is a homopolymer and its monomer is symmetrical (the methylene group is not a chiral center), the configuration of the polymer backbone is unremarkable. The location and distribution of the short and long side chains can affect its properties. Because of the random nature of these branching reactions, the configuration is best described by the average number of branches per chain. Commercial LDPE's have 15 to 25 short chain branches per 1000 carbon atoms. The number of long chain branches containing more than five carbon atoms ranges from 0.5 to 4.1 per 1000 carbon atoms along the chain. The orientation of the branches is random.

Conformation. Segments of LDPE molecules in dilute solution with a poor solvent could assume a zig-zag chain conformation, with the carbon atoms lying in the same plane. The linear conformation represents a minimum energy state of the molecule for large numbers of chain segments. Branch sites would change the minimum energy conformation near them, distorting the molecule from the planar zig-zag form. LDPE would have some crystallinity because of its stereoregular structure.

Effects of Molecular Architecture on Properties. The molecular architecture of LDPE affects its processing and performance properties relative to other polyethylenes. Polyethylene molecules containing few defects (side chains, unsaturation sites, etc.) tend to be highly crystalline solids. The side chains do not fit into the crystal lattice of linear polyethylene. Linear polyethylene has 70% to 90% crystallinity and a density of 0.965 g/cm^3, with LDPE having 45% to 55% crystallinity and densities between 0.916 and 0.930 g/cm^3. Linear polyethylene has a sharp melting point near 135°C, with LDPE having a melting point range between 105°C and 115°C.

The commercial advantages of LDPE are its improved flexibility and ease of processing. These seem to be a result of its lower crystallinity. In some products, crystallinity is reduced further by copolymerizing ethylene with vinyl acetate or ethyl acrylate to give better impact strength. Careful control of reaction conditions and the production of copolymers can give a range of marketable properties for LDPE.

This illustration of the role of constitution, configuration, and conformation on the bulk properties is typical of a number of polymer systems. Polymers with different chemical building blocks will have very different processing and performance properties. Polyethylene is quite different in strength, solvent resistance, and flex from polyacrylonitrile for example. However, within one polymer family, the bulk properties can be varied by subtle changes in the physical structure of the macromolecule. Copolymers containing modest weight percents of a second monomer often provide a better balance of properties for a given application than the homopolymer alone. Compounding also can be used to modify homopolymer properties.

Polymers are used as bulk materials and rarely as individual molecules. Therefore, the structure of the macromolecule may not be sufficient for understanding the performance of various polymers as liquids and solids. Macromolecules associate with themselves and other materials in the liquid and solid states. These associations can lead to higher order structures that affect the performance properties of the material. In Chapter 3 the physical states of polymers are described as a convenient way to discuss intermolecular associations.

2.1 SYNTHETIC POLYMER CONSTITUTIONS AND CONFIGURATIONS

It is convenient to group polymers by their chemical composition in order to compare properties. Homopolymers will be described in Sections 2.1-1 through 2.1-5. Copolymers are discussed in Section 2.1-6.

Most of the polymers discussed will be based on carbon atoms in the chain or as side groups. Carbon, nitrogen, and oxygen are the most important chain atoms for commercial polymers. However, a number of other elements can form macromolecular chains. These elements are in the groups IVb, Vb, and VIb of the periodic table, in addition to boron (see Table 2.1).

Within group IVb, carbon forms chains of infinite length. In addition, carbon occurs in the natural polymeric forms of graphite and diamond. Silanes form chains

TABLE 2.1 Elements that Form Macromolecules*

IIIb $2s, 1p$	IVb $2s, 2p$	Vb $2s, 3p$	VIb $2s, 4p$	VIIb $2s, 5p$
5 B, ~5	6 C, ∞	7 N, ∞?	8 O, ∞?	9 F, 2
13 Al, 1	14 Si, 45	15 P, >4	16 S, 30,000	17 Cl, 2
31 Ga, 1	32 Ge, 6	33 As, 5	34 Se, ?	35 Br, 2
49 In, 1	50 Sn, 5	51 Sb, 3	52 Te, ?	53 I, 2
81 Tl, 1	82 Pb, 2	83 Bi, ?	84 Po, ?	85 At, 2

*The figures at the right are the highest chain-link numbers so far observed for isochains after isolation.
SOURCE: H.B. Elias, *Macromolecules*, Vol. 1, Plenum Press, NY, 2nd Ed., 1984.

up to 45 units long, while germanium chains are only six atoms long. Boron occurs as a polymer in its solid state. The elements in the first row of Table 2.1 can form no more than four σ bonds per atoms (in the sp^3 hybridization). Elements to the right of carbon, nitrogen, and oxygen, can act as electron donors if other atoms in the chain act as electron acceptors. Therefore, they form heterochains readily with carbon, which has a higher bonding energy to itself. Elements in the second row of Table 2.1 can participate in π bonding. The hybridizations of silicon, phosphorous, and sulfur occur, and these materials can be polymers in their solid states. Heterochains of inorganic elements also can form. Elias (1984) has a good discussion of these materials.

There are a variety of ways to categorize organic polymers. One common method is to group polymers by the types of atoms along their backbones, such as carbon-carbon chains, carbon-nitrogen chains, and carbon-oxygen chains. There are subcategories of polymer families within each major category. Each subcategory is illustrated in a table containing structures of monomers or reactants.

2.1.1 Carbon-Carbon Chains

Most of the carbon-carbon chain polymers are based on ethylene as a building block. Many of these materials can be made via addition polymerizations.

Polyacrylics. (See Table 2.2.) This group is based on polymers with an polyethylene chain backbone and one side group per ethylene unit based on acrylic acid or similar structures. Acrylic acid, methacrylic acid, and esters of these acids are included in this group. Poly(acrylic acid) (PAA) is water soluble and can be very linear

TABLE 2.2 Structures of Polyacrylics

Common Name	Monomer	Polymer Structure
Poly(acrylic acid)	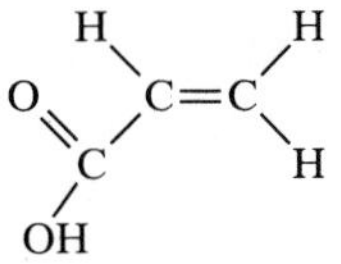	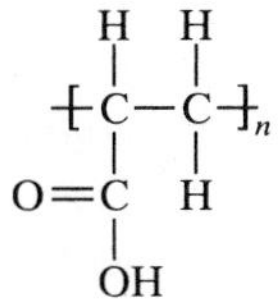
Polyacrolein	$CH_2 = CH - CH = O$	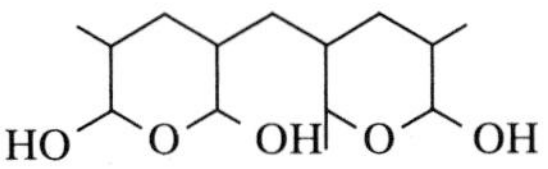 Free radical 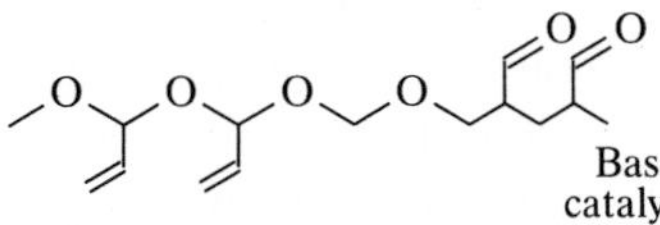 Base catalysis
Polyacrylamide	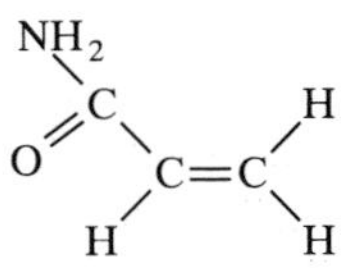	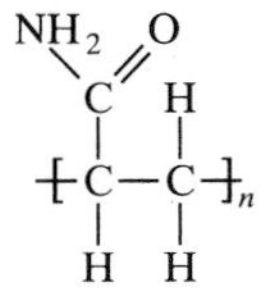
Polyacrylonitrile Orlon, Acrilan	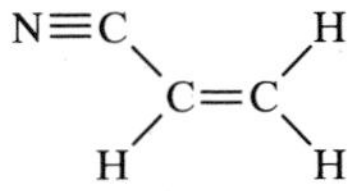	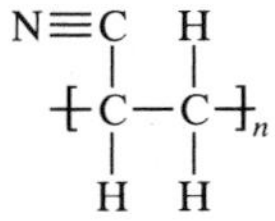
Poly(methyl methacrylate) Lucite, Plexiglas	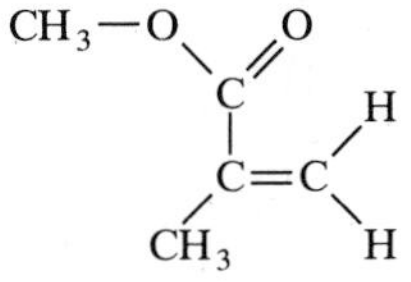	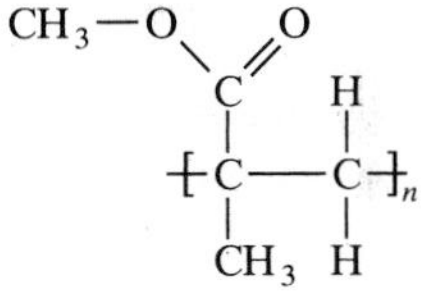
Poly(2-hydroxyethyl-methacrylate)(HEMA)	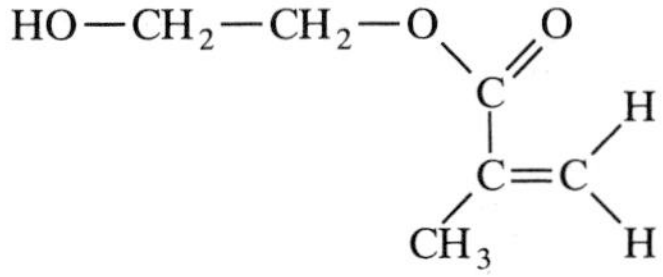	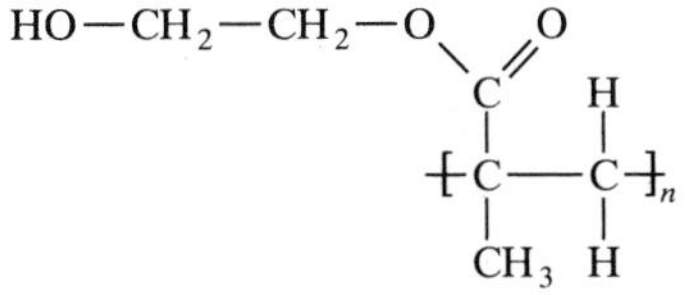

or compact depending on solution pH. Polyacrolein is polymerized from acrolein, but can have several structures depend on polymerization conditions. Polyacrylamide is water soluble and is used as a surfactant, a precipitating agent, and a flocculent.

Polyacrylonitrile (PAN) is soluble only in polar organic liquids and tends to be crystalline. It can be spun into fibers with high strength, which have good weathering properties. Acrylonitrile can be copolymerized with vinyl chloride or vinylidene chloride to make modacrylic fibers. These contain at least 85% acrylonitrile and have the trademarks, Orlon, Acrilan, Creslan, and Dynel. Acrylonitrile combines with butadiene and styrene in the ABS (acrylonitrile-butadiene-styrene) class of copolymers.

Poly(methyl methacrylate) (PMMA) is the best known material in this group. Plexiglas, Lucite, Perspec, and Acrylite are well-known trade names of commercial products. It is clear and tough, making it useful for hard contact lenses as well as lenses over lighting panels in automotive and architectural applications. A related material, poly(2-hydroxyethyl methacrylate), is used to make soft contact lenses.

Polydienes. (See Table 2.3.) Dienes contain two unsaturated carbon-carbon bonds per molecule and form polymers containing one unsaturate double bond per repeat unit. These materials are elastomers and are easy to cross-link after the initial polymerization. Natural rubber is *cis*-1,4-polyisoprene and can be obtained from the rubber tree and the guayule shrub.

TABLE 2.3 Structures of Polydienes

Common Name	Monomer	Polymer
Polybutadiene Poly(1,4-butadiene) BR Rubber	$H_2C{=}CH{-}CH{=}CH_2$	$-[CH_2{-}CH{=}CH{-}CH_2]_n-$
Polyisoprene NR Rubber	$H_2C{=}C(CH_3){-}CH{=}CH_2$	$-[CH_2{-}C(CH_3){=}CH{-}CH_2]_n-$
Polychloroprene Neoprene	$H_2C{=}C(Cl){-}CH{=}CH_2$	$-[CH_2{-}C(Cl){=}CH{-}CH_2]_n-$
Polynorbornene	(norbornene ring)	$-[C_5H_8{-}CH{=}CH]_n-$ (cyclopentane ring–C(H)=C(H)–)
Poly(pentenamer)	(cyclopentene ring)	$-[CH{=}CH{-}(CH_2)_3]_n-$

A number of 1,3-dienes can be used to make polymers having one unsaturated double bond per repeat unit. The most common monomers are butadiene, isoprene, and chloroprene. Four major repeating units (configurations) can be formed during these polymerizations (Fig. 2.1).

The polymerization method and conditions can be varied to control the stereochemistry of the product. Table 2.4 shows conditions of commercial polymerizations that lead to varied polymer structures. Both the initiator and the solvent affect the product distribution. *Cis*-1,4-polybutadiene (BR, butadiene rubber) is preferred for tire manufacturing. Syndiotactic 1,2-polybutadienes can be made into tear-resistant films with good gas permeability. Natural rubber (polyisoprene, NR) is found in the *cis*-1,4 and *trans*-1,4 configurations. Synthetic polyisoprenes (IR) can be produced in a range of configurations.

Polychloroprene (neoprene, CR) can be polymerized in bulk or in emulsion (particles < 1 micron) polymerization processes. The early work on the bulk process showed that oxygen was detrimental to polymerization rate. Adding sodium hypodisulfite to scavenge the oxygen lead to high accelerations in the polymerization rate. This led to the discovery of redox catalysts. Many polydienes are highly viscous liquids at room temperature. However, vulcanization, the cross-linking of polymer C=C bonds with sulfur, gives solids that are highly elastic.

Ring-opening polymerizations of cycloolefins lead to poly(1-alkenylenes). Poly(pentenamer) is an example. Cyclopentene polymerizes by a metathesis mechanism, whereby exchange or disproportionation reactions occur between cycloolefins.

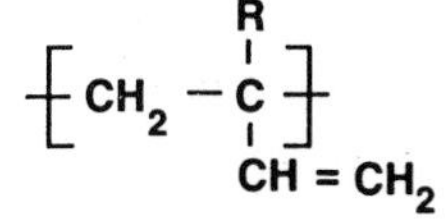

1,2

$-[CH_2-CH]-$ with pendant $C(R)=CH_2$

3,4

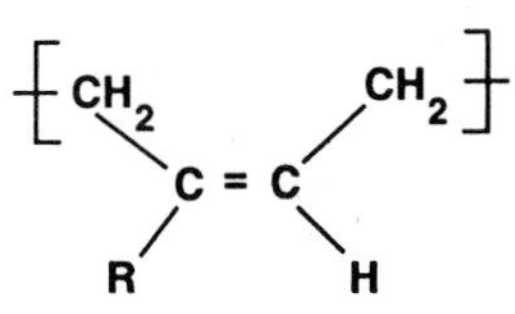

cis - 1,4

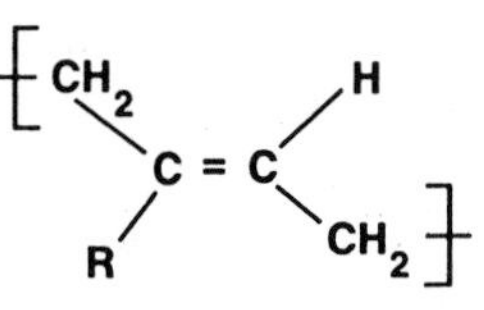

trans - 1,4

Figure 2.1 Four major configurations for diene polymers.

TABLE 2.4 Constitution and Configuration of Commercial Polydienes

	Polymerization		Percentage Structures			
Monomer	Initiator	Medium	1,4-*cis*	1,4-*trans*	1,2	3,4
Butadiene	Sodium	—	10	25	65	—
	Lithium ethyl	THF	0	9	91	—
	Lithium ethyl	THF/benzene	13	13	74	—
	Lithium ethyl	Benzene/triethylamine	23	40	37	—
	Lithium	Hexane	38	53	9	—
	Lithium ethyl	Toluene	44	47	9	—
	Titanium compounds	—	95	3	2	—
	Cobalt compounds	—	98	1	1	—
	Nickel compounds	—	97	2	1	—
Butadiene/ styrene	Free radical	Emulsion, 70°C	20	63	17	—
	Free radical	Emulsion, 5°C	12	72	16	—
	Anionic	Solution	40	54	6	—
Isoprene	Lithium alkyls	Solution	93	0	0	7
Chloroprene	Free radical	—	11	86	2	1

SOURCE: H.B. Elias, *Macromolecules*, Vols. 1(a) and 2(b), Plenum Press, NY, 2nd Ed., 1984.

The *trans* product is an all-purpose rubber with properties similar to those of natural rubber and *cis*-polybutadiene. Polynorbornene was developed for use as a powder rubber, with particle sizes in the range of 100–1500 microns. The particle form is easy to store and process.

Polyhalogen Hydrocarbons. (See Table 2.5.) Polymers containing fluorine and chlorine are less dependent on the cost of ethylene for their price, since they contain halogens that are recovered from brines or minerals. The energies of carbon-halogen bonds are: C-H = 377, C-F = 461, C-Cl = 293, C-Br = 251, and C-I = 188 kJ/mol. Fluorocarbons will be the most stable of these materials because of the high C-F bond energies. Polytetrafluoroethylene (PTFE) has high thermal stability, is resistant to oxygen attack, has low flammability, and is very resistant to attack by solvents and reactive chemicals. The melt viscosity of the homopolymer is very high, so solid parts are made by compression and sintering or by plasticizing the polymer and devolatilizing the parts. Because of the processing problems, a range of copolymers have been developed. Comonomers include hexafluoropropylene, perfluoropropylvinylether, and ethylene.

Polytrifluorochloroethylene competes with PTFE. It melts at 220°C rather than 327°C and can be processed in the range of 250°C to 300°C. Extruder parts need to be corrosion resistant because of the polymer degradation that can occur during processing. Poly(vinylidene fluoride) (PVDF) has excellent weathering properties and is used as a coating for architectural panels. Poly(vinyl fluoride) (PVF) has film applications.

Poly(vinyl chloride) (PVC) is one of the high volume, commodity thermoplastics. PVC has low crystallinity and forms stable, dry, flexible solutions with a variety

TABLE 2.5 Structures of Polyhalogen Hydrocarbons

Common Name	Monomer	Polymer
Poly(vinyl fluoride)	$\mathrm{(F)(H)C{=}C(H)(H)}$	$\mathrm{+C(F)(H)-C(H)(H)+_{n}}$
Poly(vinylidene fluoride) Kynar	$\mathrm{(F)(F)C{=}C(H)(H)}$	$\mathrm{+C(F)(F)-C(H)(H)+_{n}}$
Poly(fluorotrichloroethylene)	$\mathrm{(Cl)(Cl)C{=}C(Cl)(F)}$	$\mathrm{+C(Cl)(Cl)-C(Cl)(F)+_{n}}$
Poly(chlorotrifluoroethylene)	$\mathrm{(F)(F)C{=}C(F)(Cl)}$	$\mathrm{+C(F)(F)-C(F)(Cl)+_{n}}$
Polytetrafluoroethylene Teflon	$\mathrm{(F)(F)C{=}C(F)(F)}$	$\mathrm{+C(F)(F)-C(F)(F)+_{n}}$
Poly(vinyl chloride)	$\mathrm{(Cl)(H)C{=}C(H)(H)}$	$\mathrm{+C(Cl)(H)-C(H)(H)+_{n}}$
Chlorinated PVC	none	$\mathrm{+C(H)(Cl)-C(Cl)(H)+_{n}}$ random
Poly(vinylidene chloride) Saran	$\mathrm{(Cl)(Cl)C{=}C(H)(H)}$	$\mathrm{+C(Cl)(Cl)-C(H)(H)+_{n}}$

of liquids. These systems are used for coatings, tubing, hoses, and film products. PVC can be compounded with stabilizers and lubricants to give it good performance in a variety of products. PVC can be chlorinated to give a product with a high T_g, which can be used for hot water piping systems.

Poly(vinylidene chloride) (PVDC) is highly crystalline and is the major component in packaging films (Saran). It normally is copolymerized with other monomers such as vinyl chloride or methyl acrylate.

Polyolefins. (See Table 2.6.) The polyolefin family is based on monomers with aliphatic and aromatic substituents on ethylene. Polyethylene (PE) is a commodity polymer that can be made with a wide range of properties. Table 2.7 shows several polymerization systems for PE and the physical properties of each product. The processes to make the high density product (HDPE) are based on stereoregular catalytic methods. The gas-phase process often makes copolymers as a method for controlling molecular weight. The polyethylenes are used for film and bottles as well

TABLE 2.6 Structures of Polyolefins

Common Name	Monomer	Polymer
Polyethylene	$H_2C=CH_2$	$-\!\!\left[CH_2-CH_2\right]_n\!\!-$
Chlorinated PE	PE + 30-45 wt% Cl_2	$-\!\!\left[CHCl-CH_2\right]_n\!\!-$
Polypropylene	$CH_3CH=CH_2$	$-\!\!\left[CH(CH_3)-CH_2\right]_n\!\!-$
Poly(1-butene)	$CH_3-H_2C-CH=CH_2$	$-\!\!\left[CH(CH_2-CH_3)-CH_2\right]_n\!\!-$
Polyisobutylene	$(CH_3)_2C=CH_2$	$-\!\!\left[C(CH_3)_2-CH_2\right]_n\!\!-$
Polystyrene	$C_6H_5-CH=CH_2$	$-\!\!\left[CH(C_6H_5)-CH_2\right]_n\!\!-$
Poly(2-vinyl pyridine)	$C_5H_4N-CH=CH_2$ (2-pyridyl)	$-\!\!\left[CH(C_5H_4N)-CH_2\right]_n\!\!-$

TABLE 2.7 Typical Industrial Polymerization Procedures for Ethylene

		High pressure		Med. pressure		Low pressure	
		ICI	BASF	Standard Oil	Phillips	Ziegler	UCC
Pressure	Bar	1500	500	70	40	4	14
Medium	—	Bulk	Emulsion in CH_3OH	Solution in xylene	Solution in xylene	Solution in lubricating oil	Gas phase
Temperature	°C	180		<200	130	70	<100
Initiator	—	O_2	Peroxide	Part. red. MoO_3 on Al_2O_3 or aluminum silicates	Part. red. chromium oxide on Al_2O_3	$TiCl_4$/ R_2AlCl AlR_3	$TiCl_4$/ $Mg(OC_2H_5)_2$/ AlR_3
Yield	%	20		100	100	100	50–100
Density	g/cm_3	0.92		0.96	0.96	0.94	0.92–0.94
Melting temp.	°C	108		133	133	130	
Methyl per C	—	0.03		<0.00015	<0.00015	0.006	
Polymer type		LDPE	Wax	HDPE	HDPE	HDPE	LLDPE (copolymer)

SOURCE: H.B. Elias, *Macromolecules*, Vols. 1(a) and 2(b), Plenum Press, NY, 2nd Ed., 1984, p. 889.

as tubing and cable coating. PE can be chlorinated (CPE) and sulfochlorinated (SC-PE) in postpolymerization reactions. Chlorinated polyethylene has about the same chlorine level as PVAC but can be processed with fewer stabilizers. CPE is used for hot water pipe, while the sulfochlorinated materials are used for coating with good weatherability. Polypropylene (PP) also is a commodity polymer. It crystallizes in a helical form, giving it a higher melting temperature and higher tensile strength than polyethylene. PP tends to be brittle at low temperatures. Some of its deficiencies can be reduced by polymerizing propylene to high conversion and adding ethylene to complete the process. The small amount of copolymer does not reduce the polymer's strength from that of homopolymer and improves processing and low temperature performance.

Poly(1-butene) has high tensile strength and good corrosion resistance. It is used for plastic pipes and films. Poly(4-methyl pentene-1) has a low glass transition temperature (40°C) but a high softening temperature (179°C). It is clear and is used for graduated laboratory apparatus for aqueous systems. Polyisobutylene (PIB) has a low glass transition temperature (−70°C) and only crystallizes under stress. It can be used as an elastomer at room temperature.

Polystyrene (PS) is a clear, glassy commodity polymer. It is a thermoplastic molding resin and can be used for foam insulation. A large fraction of styrene goes into copolymer systems, particularly those with acrylonitrile and butadiene (the ABS system). These systems are described in Section 2.4 as an example of co- and terpolymers. Poly(vinyl pyridines) are used in specialty chemicals (Figure 2.2).

Other Vinyl Polymers. (See Table 2.8.) These materials can be made by polymerizing vinyl monomers ($CH_2 = CHR$) or by reactions on vinyl polymers.

Poly(vinyl acetate) (PVAc). Vinyl acetate is polymerized with free radical initiation in several polymerization systems. The bulk process leads to branched polymer. Emulsion and suspension products are made also. The homopolymer is used for adhesives and glues but is susceptible to hydrolysis. Copolymers with vinyl stearate or vinyl pivalate are more resistant to hydrolysis because the bulky side groups provide some steric protection.

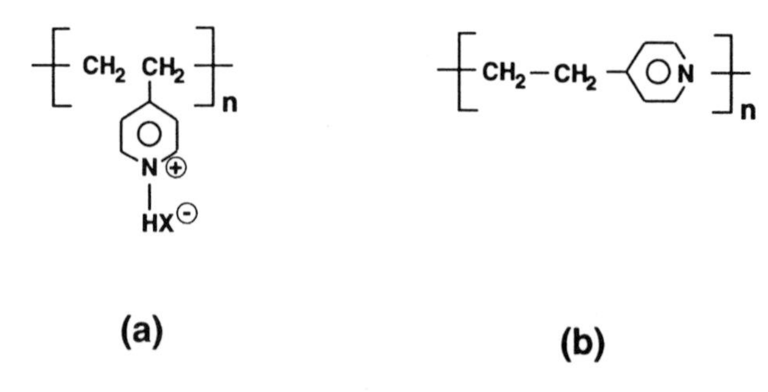

Figure 2.2 Repeating unit structures for poly(4-vinyl pyridine).

TABLE 2.8 Structures of Other Vinyl Polymers

Name	Monomer/Reactants	Repeating Unit
Poly(vinyl acetate)	$CH_2 = CH - O - OC - CH_3$	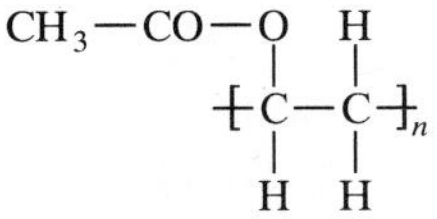 CH_3-CO-O H; $-[C-C]_n-$; H H
Poly(vinyl alcohol) Poly(acetoaldehyde)	Poly(vinyl acetate) + methanol	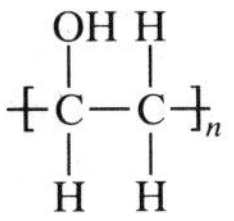 OH H; $-[C-C]_n-$; H H
Poly(vinyl formal)	Poly(vinyl acetate) + formaldehyde	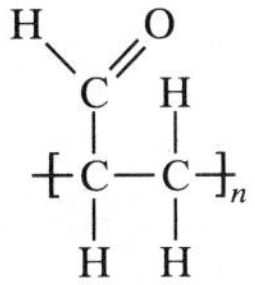 H–C=O, H; $-[C-C]_n-$; H H
Poly(vinyl butyral)	Poly(vinyl alcohol) + $CH_3CH_2CH_2COH$	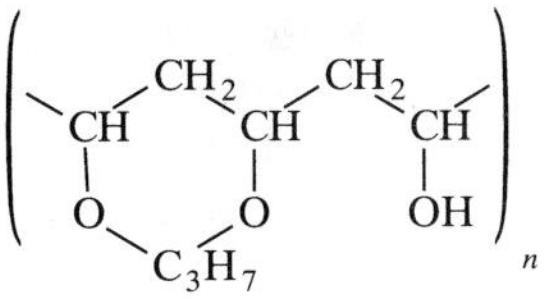 $(-CH-CH_2-CH-CH_2-CH-)_n$; O, O (C_3H_7), OH
Poly(vinyl methyl ether)	$CH_3 = CH - O - CH_3$	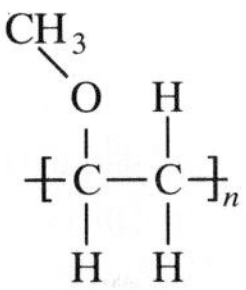 CH_3–O H; $-[C-C]_n-$; H H
Poly(2-vinyl pyrrolidone)	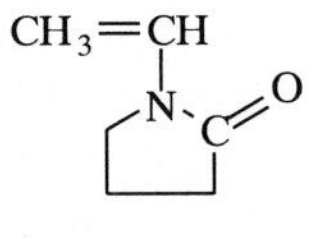 $CH_3{=}CH$–N(ring)–C=O	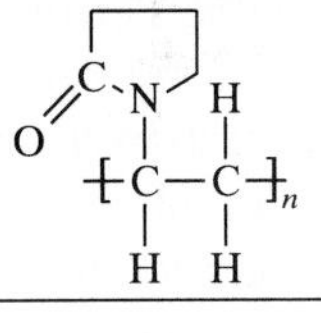 O=C–N(ring) H; $-[C-C]_n-$; H H

Poly(vinyl acetals). These polymers are produced from poly(vinyl acetate) or poly(vinyl alcohol) (PVAL) by postpolymerization reactions. None of the reactions can go to complete conversion so there are always mixtures of acetate, acetal, and hydroxyl groups. Poly(vinyl butyral) is used as a film binder for safety glass, "sandwiched" between two glass sheets.

Poly(vinyl ethers). Vinyl ethers can be polymerized at low temperatures with cationic initiation. The polymers are soft and are used for adhesives and plasticizers.

Aromatic Hydrocarbon Chains. (See Table 2.9.) Poly(*p*-xylylenes) can be polymerized on surfaces and have low gas and water permeabilities. They can be used to provide protective coatings for delicate biological specimens.

Phenolic Resins. Formaldehyde and phenol can react with acid or base catalysis to form large macromolecules. Acid catalysis makes novolacs, condensation polymers that require addition of a multifunctional amine in order to cross-link. Resoles (or Bakelite) are made with base catalysis and can cross-link with heat. These resins are used as bases for lacquers and paints. Pure novolacs are soluble in alcohol solvents only. By substituting various functional groups on the polymer, a wide range of solvency properties can be achieved. Other uses are for fibers for nonflammable clothes, ablative shields for space capsules, and ion exchange resins.

2.1.2 Carbon-Nitrogen Chains

Polyamides. (See Table 2.10.) Polymers with amide groups (-NH-CO-) in the main chain can be divided into two categories. In the Perlon series, the repeating unit and the monomer are the same. In the nylon series, the repeating unit is formed from two monomer units.

$$\left(\text{NH-CO-R}\right) \qquad \left(\text{NH-R-NH-CO-R}^1\text{-CO}\right)$$

PERLON NYLON

The R and R^1 groups may be aliphatic, aromatic, or heterocyclic groups. Commercial products of both types are called nylons by the trade.

The numbers of carbon atoms in the chain are used to identify the various types. The Perlon series can be characterized by one number; for example, polymer made by the ring-opening reaction of ε-caprolactam is called nylon 6 because of the six carbon atoms between amide groups. The first number of the nylon series is the number of carbon atoms in the diamine and the second number is the number of carbon atoms in the diacid. Polyamides often are spun into fibers. Nylon 6,6 was used to make the first completely synthetic fiber. The major commercial polyamides are nylon 6,6, nylon 6,10, and nylon 6,12. The polymerization reactions follow an equilibrium mechanism. The nylon series has the advantage that the equilibrium is far to the right for high polymer products. The concentration of monomers and oligomers are low. By contrast, the ring-opening polymerizations of lactams often have significant concentrations of monomer in the reaction media. Also, the Perlon series often degrade directly to lactams at high temperatures.

Poly(α-amino) acids have been synthesized and have useful properties. However, the monomer costs are very high, and most products are not cost-competitive.

Polyazoles. (See Table 2.11.) Polyazoles have five-member heterocyclic rings in the main chain that contain at least one tertiary nitrogen atom. Polybenzimidazoles

TABLE 2.9 Structures of Polymers with Aromatic Hydrocarbon Chains

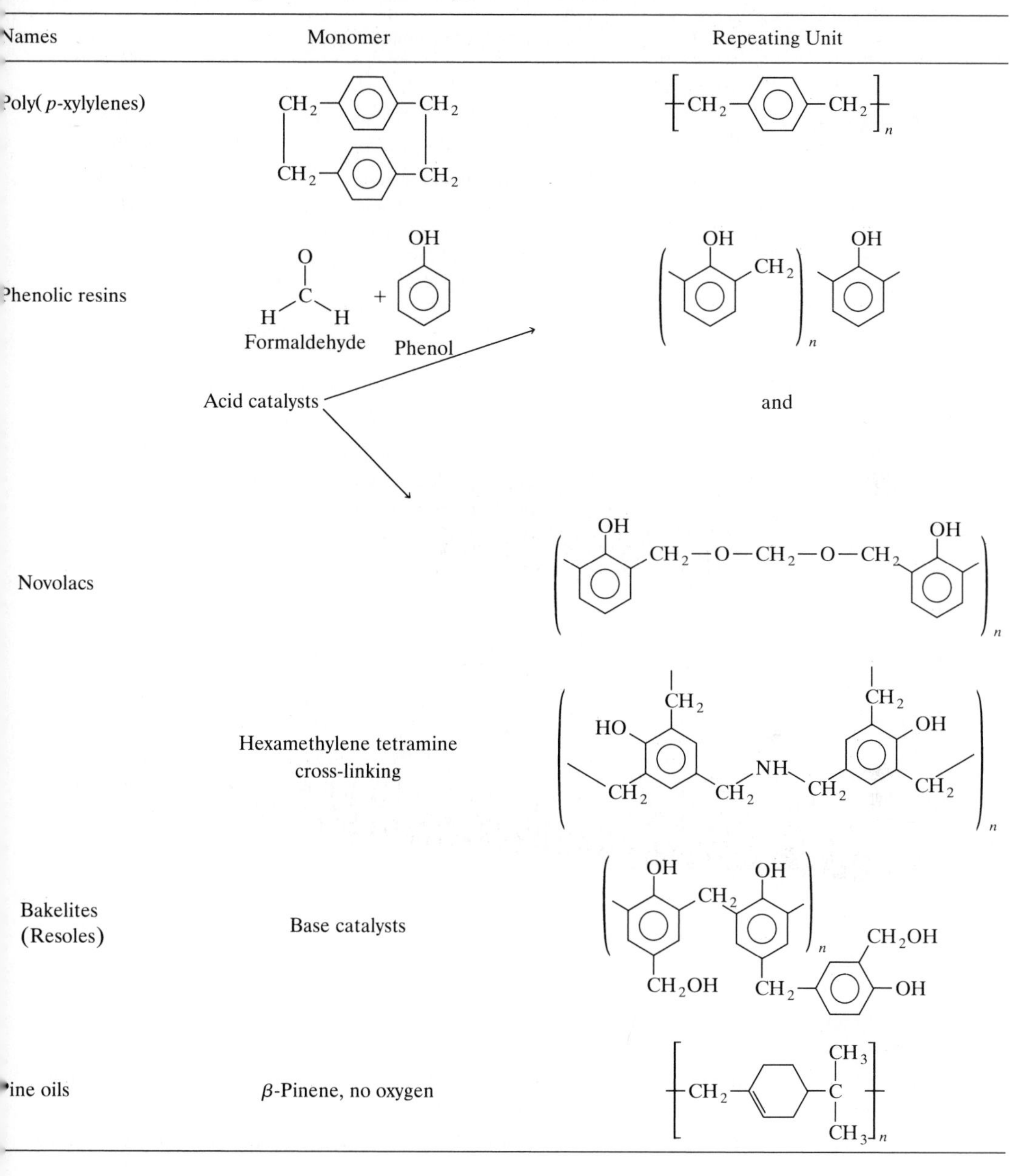

Names	Monomer	Repeating Unit
Poly(*p*-xylylenes)		
Phenolic resins	Formaldehyde + Phenol; Acid catalysts	and
Novolacs		
	Hexamethylene tetramine cross-linking	
Bakelites (Resoles)	Base catalysts	
Pine oils	β-Pinene, no oxygen	

TABLE 2.10 Structures of Nylons and Other Polyamides

Structure	Generic and/or Common Name	Type*
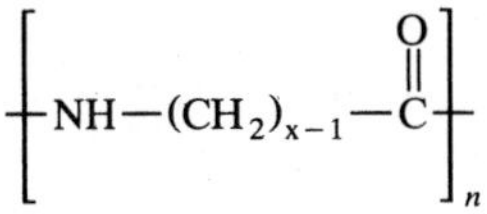 $\left[NH-(CH_2)_{x-1}-\overset{O}{\overset{\parallel}{C}} \right]_n$	NYLON X	
	Nylon 6 (polycaprolactam)	F, P
	Nylon 7 (poly[7-heptanoamide])	F, P
	Nylon 8 (polycapryllactam)	F, P
	Nylon 9 (poly[9-nonanoamide])	F
	Nylon 11 (poly[11-undecanoamide])	P
	Nylon 12 (polylauryllactam)	P
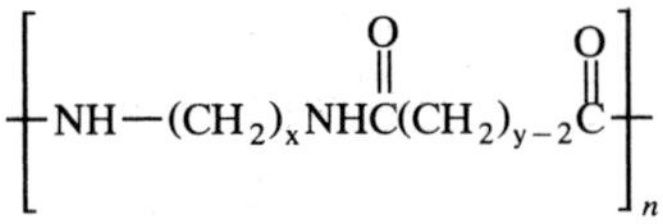 $\left[NH-(CH_2)_xNH\overset{O}{\overset{\parallel}{C}}(CH_2)_{y-2}\overset{O}{\overset{\parallel}{C}} \right]_n$	Nylon X, Y	
	Nylon 4,6(poly[tetramethylene adipamide])	F, P
	Nylon 6,6 (poly[hexamethylene adipamide])	F, P
	Nylon 6,9 (poly[hexamethylene azelamide])	P
	Nylon 6,10 (poly[hexamethylene sebacamide])	P
	Nylon 6,12 (poly[hexamethylene dodecanedioamide])	P
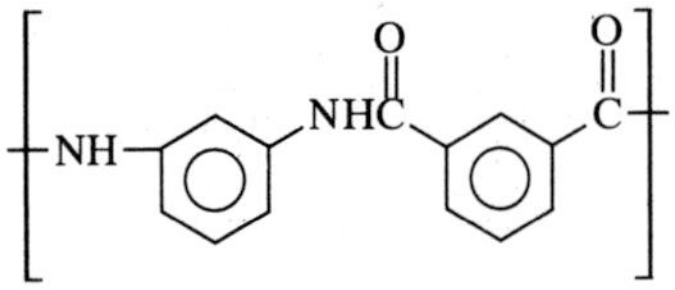 [structure: –NH–(m-phenylene)–NHC(=O)–(m-phenylene)–C(=O)–]	Poly(*m*-phenylene isophthalamide) Nomex	F
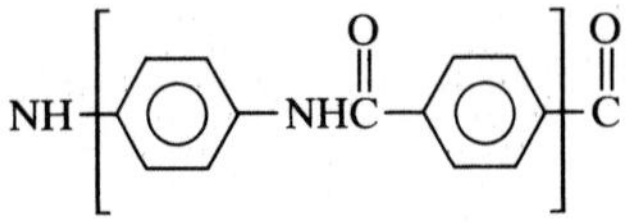 [structure: NH–(p-phenylene)–NHC(=O)–(p-phenylene)–C(=O)]	Poly(*p*-phenylene terephthalamide) Kevlar	F

*F = fiber, P = plastic

are specialty polymers made by polycondensation with phenol as the leaving group. Partial ring closure reactions occur. These polymers are used for thermally stable protective clothing and as starting materials for the production of graphite fibers. Hollow fibers and films can be used for desalination of sea water by reverse osmosis.

Ladder-type polymers can be made by replacing the dicarboxylic acids with tetracarboxylic acids. The thermal stability of these materials is higher, but they have to be processed in very aggressive solvents, such as polyphosphoric acid.

Polyimides. (See Table 2.12.) Nylon 1 is the simplest polyimide (-CO-NR-CO-) and can be made by the polymerization of isocyanic acid. Commercial poly-

TABLE 2.11 Structures of Polyazoles

Ring System	Polymer Type	Functional Groups Required	Intermediate Stage
	Polybenzimidazole	Phenyl ester + diamine	Polyimine
	Polybenzoxazole	Phenyl ester + *o*-aminophenol	Polyamide
	Polybenzothiazole	Phenyl ester + *o*-aminothiophenol	Polyamide
	Polyimidazopyrrolone	Anhydride + diamine	Polyamide
	Poly(1,2,4-triazole)	Hydrazide + amine	Polyhydrazide
	Poly(parabanic acid)	Isocyanate + HCN	Polyurea

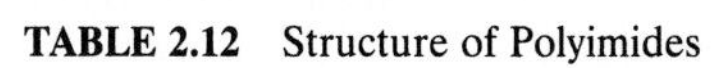
TABLE 2.12 Structure of Polyimides

Names	Monomer	Repeating Unit
Poly(isocyanic acid) Nylon 1	H — N = C = O	$\left[-\overset{H}{N}-\overset{O}{\overset{\Vert}{C}}- \right]_n$
Kapton	Pyrometallic dianhydride + H_2N–C₆H₄–O–C₆H₄–NH_2 4,4′-Diaminophenyloxide	

imides contain imide groups bound to cyclic compounds. They have good thermal properties and high temperature strength, making them useful for films and fibers.

Polyimines. Polyimines have the repeating unit, -NH-CHR-. They can be produced by the polymerization of nitriles followed by hydrogenation. Poly(ethylene imines) are used as adhesives and flocculating aids. These materials are branched to help improve their performance. Poly(carbodiimides) are related to polyimines. They are made by elimination of CO_2 from isocyanates to give open-celled rigid foams that can be formed by compression molding. The reaction is

$$\text{OCN-R-NCO} \rightarrow (\text{-N-C} = \text{N-R-})_n + n\ CO_2$$

Polyurethanes. The reaction of diisocyanates with dialcohols gives polymers with a urethane linkage.

$$\text{O=C=N-R-N=C=O} + \text{HO-R}^1\text{-OH} \rightarrow [\text{-R-NH-CO-O-R}^1\text{-}]$$

DIISOCYANATE DIALCOHOL URETHANE

Urethanes have a number of applications because of the range of isocyanate and alcohol groups that can be used. Some of the different product types are described in Table 2.13.

Polyureas. A commercial polymer is produced from the reaction between urea and nonamethylene diamine, having a repeating structure of $[-(CH_2)_9-NH-CO-NH-]_n$. The reaction can be run in phenol or in a molten monomer mixture. T_m is 240°C and the polymer has good resistance to alkali. The reaction follows an equilibrium mechanism. A number of polyureas have been synthesized, but most suffer from side reactions to undesirable products.

2.1.3 Carbon-Oxygen Chains

Polyacetals. The repeating unit of polyacetals is -CHR-O- (Table 2.14), and they can be produced by polymerization of aldehydes or cyclic trimers. Polyoxymethylene can be produced from formaldehyde or trioxane. Either polymerization can be done using ionic initiation. The chain ends are stabilized using ethylene oxide. The polyacetals are engineering thermoplastics used in engine, appliance, and plumbing parts; electronic components; zippers; and buckles. They absorb very little water and have good abrasion and wear resistance. Higher aldehydes, such as acetaldehydes, can be polymerized but usually are difficult to process.

Polyethers. The polyethers have the repeating unit - R - O - (Table 2.14), and they can be divided into those with aliphatic chains and those with aromatic chains.

TABLE 2.13 Polyurethane Systems

Product Type	Typical Monomers	Processing
Fibers and films	Hexamethylene diisocyanate + 1,4-butane diol	Injection molding, spinning
Paints, lacquers	Triisocyanates with polyester, pentaerythritol or cellulose acetate chains	—
Adhesives	Triisocyanates + polyester diol	Hydrogen bonding to glass, removing the water film on substrate surface by polyurea formation, reaction with surface OH groups
Foams	Toluene diisocyanate (2,4 and 2,6) with polyester or polyetherdiols with water	Fast reaction using prepolymers with CO_2 leaving groups providing gas
Elastomers	Rigid segment (aromatic isocyanates) + Flexible segment (aliphatic polyesters or polyethers; poly(propylene glycol) or polytetrahydrofuran)	Two-step reaction: linear diisocyanates are polymerized and then cross-linked
Image reproduction	Aromatic carboxylicazide esters in the presence of poly(vinyl alcohol)	Light degrades the esters, making cross-linked products. Uncrosslinked poly(vinyl alcohol) is dissolved away.

Aliphatic Polyethers. Poly(ethylene oxide) is soluble in a range of organic solvents, as well as water. It can be used to make water-soluble packaging films, for textile sizing and for a variety of surfactants. Poly(propylene oxide) can be made as a stereoregular product or as an atactic product. The homopolymer is used for polyurethane intermediates, lubricants, and surfactants. Propylene oxide can be copolymerized with ethylene oxide to make water-soluble detergents. It can be polymerized with diene monomers to give oil-resistant elastomers that can be vulcanized (cross-linked with sulfur).

Polytetrahydrofuran is used for thermoplastic elastomers and artificial leather. Polymers of about 2000 molecular weight can be used as the soft segment for elastic polyurethane fibers or polyether ester elastomers. Homopolymers of epichlorohydrin are resistant to oils and ozone and can be used as elastomers. These materials can be cross-linked by amine reactions with the chlorine groups.

Epoxy materials are used for coatings, laminated circuit boards, adhesives, composites, bridge road surfaces, and highway rumble strips. Aromatic epoxides have greater stability than those with aliphatic chains because there is less rotation around the chain atoms. Unsaturated rings in the chain are stable but do not react as well during curing. The aromatic materials are suitable for engineering thermoplastic

TABLE 2.14 Structures of Polyacetals and Polyethers

Names	Monomer	Repeating Unit
Polyacetal (homopolymer)	$H{-}C({=}O){-}H$	$+CH_2{-}O+_n$ or $+CH(OH)+_n$
Polyacetal (copolymer), Polyoxymethylene	O O O (ring)	$+CH_2{-}O+_n$
Poly(ethylene oxide), Poly(oxyethylene)	$CH_2{-}CH_2$ bridged by O	$+CH_2{-}CH_2{-}O+_n$
Poly(propylene oxide)	$CH_2{-}CH_2{-}CH_2$, first two bridged by O	$+CH_2{-}CH(CH_2){-}O+_n$
Polytetrahydrofuran	O (ring)	$+(CH_2)_4{-}O+_n$
Polyepichlorohydrin	$CH_2{-}C(Cl){-}CH_2$, first two bridged by O	$+C(H)(CH_2Cl){-}CH_2{-}O+_n$

applications. Their higher thermal stability and mechanical strength can justify the higher monomer costs and longer cure times.

There are a variety of epoxide resins, but the highest volume material is based on bisphenol A. The commercial products contain plasticizers, diluents, and pigments. Other products include epoxidized phenol/formaldehyde, cresol/formaldehyde, and heterocyclic structures. Cross-linking reactions lead to higher glass transition temperatures. However, not all the groups react and complete networks are not formed.

Poly(phenylene oxides) are known as poly(oxyphenylenes) and poly(phenyl ethers). Poly(phenylene oxide) based on 2,6-dimethylphenol is an engineering thermoplastic that is miscible with polystyrene (Noryl).

Phenoxy resins contain secondary hydroxyl groups and are excellent primers in automotive applications. The secondary hydroxyl groups lead to cross-linked polymers unless epichlorohydrin is in excess. The reaction is carried out in two stages to produce high molecular weight material.

Poly(etherether ketone) (PEEK) is a semicrystalline polymer with a T_g of 144°C and a T_m of 334°C. This is an example of a commercial plastic prepared by nucleophilic aromatic substitution occurring when the leaving group is activated by electron-withdrawing groups. Its excellent high temperature properties make it an advanced engineering thermoplastic. Polyetherimide is polymerized by a nucleophilic aromatic substitution similar to PEEK and also has good high temperature properties.

Phenolic Resins. Phenolic resin can be classified as carbon chain or carbon-oxygen chain polymers, depending on their structure.

Polyesters. The polyester class is subdivided into aliphatic, cross-linked, and aromatic polyesters.

Aliphatic Polyesters. (See Table 2.15.) These polymers have the ester group in the main chain, -COO -, and an aliphatic segment in the repeating unit. Polyesters are typical of step polymerizations since there are a number of reaction pathways for synthesizing the polymers: ring-opening polymerization of lactones, self-condensation of α- and Ω-hydroxy acids, transesterifications, polycondensations of diols with either dicarboxylic acids or diacyl chlorides, copolymerization of acid anhydrides with cyclic

TABLE 2.15 Structures of Aliphatic Polyesters

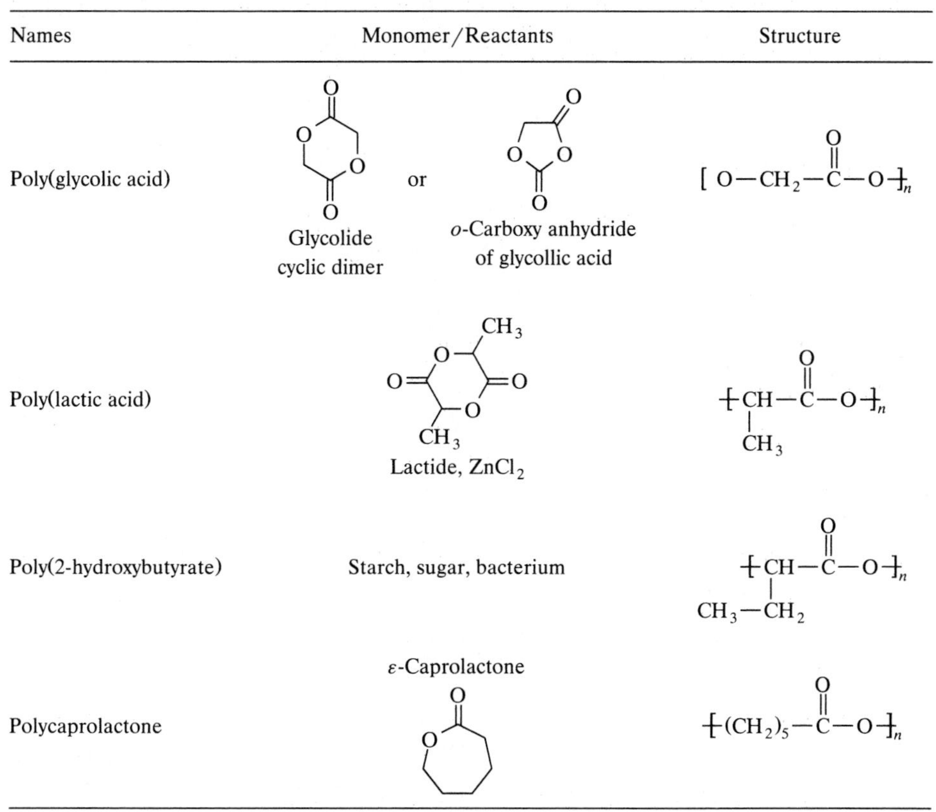

Names	Monomer/Reactants	Structure
Poly(glycolic acid)	Glycolide cyclic dimer or *o*-Carboxy anhydride of glycollic acid	$[O-CH_2-C(=O)-O]_n$
Poly(lactic acid)	Lactide, $ZnCl_2$	$[CH(CH_3)-C(=O)-O]_n$
Poly(2-hydroxybutyrate)	Starch, sugar, bacterium	$[CH(CH_2-CH_3)-C(=O)-O]_n$
Polycaprolactone	ε-Caprolactone	$[(CH_2)_5-C(=O)-O]_n$

ethers, and polymerization of o-carboxy anhydrides of α- and β-hydroxycarboxylic acids. (See Table 2.16.)

Poly(glycolic acid) is the simplest of the aliphatic polyesters and can be made in two different ways. Either the cyclic dimer or the anhydride of glycollic acid can be anionically polymerized (the latter loses carbon dioxide during polymerization). The polymer is susceptible to hydrolysis, so this material can be used for surgical thread that disintegrates in the body. Poly(lactide) (or poly[lactic acid]) and poly(2-hydroxybutyrate) have similar properties. The latter is polymerized by a bacterium and is deposited in granules in the cytoplasm. The polymers have degrees of polymerization of 23,000, and commercial processes are available. Poly(ϵ-caprolactone) is used to improve the impact strength of polyolefins.

Cross-linked Polyesters. (See Table 2.17.) Unsaturated polyesters provide one method for developing cross-linked polyester polymer networks. The other is the use of multifunctional monomers that cross-link during esterification. A typical network is that made by polymerizing phthalic anhydride with glycerol to produce a highly cross-linked structure. These reactions are normally done in two stages. The first stage starts the reaction to make a viscous liquid. This liquid is further reacted in a mold to provide a structure to the part. Oil-modified alkyds used to be the major coating material used prior to the development of latex paints. Polyols and oil can be transesterified, then mixed with dibasic acid or anhydride to give the complex polymer. Unsaturated oils, such as linoleic and linolenic acids, can be used so that the paint cross-links during drying by reaction with oxygen.

TABLE 2.16 Reaction Mechanisms for Polyester Formation

Direct esterification

$$RCO_2H + R^1OH \rightleftharpoons RCO_2R^1 + H_2O$$

Transesterification

$$RCO_2R^{11} + R^1OH \rightleftharpoons RCO_2R^1 + R^{11}OH$$

Diol with diacyl chloride (nearly irreversible)

$$RCOCl + R^1OH \rightleftharpoons RCO_2R^1 + HCl$$

Anhydrides

$$(RCO)_2O + R^1OH \rightleftharpoons RCO_2R^1 + RCO_2H$$

Acidolysis

$$RCO_2H + R^1CO_2R^{11} \rightleftharpoons RCO_2R^{11} + R^1CO_2H$$

Carboxylic acids with epoxides

$$RCO_2H + \overset{}{CH_2}\overset{O}{\frown}CH_2 \rightarrow RCO_2CH_2CH_2OH$$

Nucleophilic displacement

$$RCO_2^- + R^1Br \rightarrow RCO_2R^1 + Br^-$$

Unsaturated polyesters are used for fiber glass reinforcing (sheet molding compound). One of the difunctional monomers contains a double bond that can be cross-linked. A linear unsaturated polyester is prepared with moderate molecular weight. This material is mixed with styrene or other vinyl monomers. The reaction is initiated by free radical initiators in a heated mold.

Aromatic Polyesters. (See Table 2.17.)

Polycarbonates (PC). Polycarbonates are polyesters based on carbonic acid. Since carbonic acid is not stable, its derivatives such as phosgene and urea are used in the polymerization. General Electric and Farbenfabriken Bayer both developed polymers based on phosgene and Bisphenol A. Commercial products are called Lexan (General Electric) and Merlon (Mobay). The polymer is an engineering thermoplastic, is transparent, and has very high impact strength. It is used for bulletproof windows, safety shields, gears, bushings, and automotive parts. Several blends of polycarbonate with other engineering thermoplastics have been developed, including poly(ethylene terephthalate) and ABS.

Poly(ethylene terephthalate) (PET). Poly(ethylene terephthalate) is the most widely used linear polyester and can be formed by the condensation of ethylene glycol and terephthalic acid. Dimethyl and acid chloride derivatives may also be used. The polymer is melt-spun into fibers that have excellent wear and washing properties. Injection molding is possible, but the processing window for this material is narrow.

Poly(butylene terephthalate) (PBT). This polyester can be processed at lower temperatures than PET, but has a lower glass transition temperatures and poorer properties. Block copolymers of PBT and polytetrahydrofuran, which forms flexible units, are thermoplastic elastomers.

Poly(4-hydroxybenzoate). p-Hydroxybenzoate can be condensed at temperatures above 200°C. The homopolymer has a very high melting temperature (550°C) and high thermal stability. The homopolymer can be worked only by hammering, sintering, or spraying. Copolymers with other acids have been made (terephthalic, isophthalic acids, and Bisphenol A). These materials are called aromatic polyesters or arylates and have the properties of engineering plastics.

2.1.4 Carbon-Sulfur Chains.

There are some commercial polymers based on carbon-sulfur chains (see Table 2.18).

Polysulfides. These polymers are the sulfur analogs of polyethers and can be prepared in similar reactions. A variety of products have been made, but the most successful commercial product is Ryton, a poly(phenylene sulfide). This polymer is highly crystalline with a T_m of 288°C. It qualifies as an engineering thermoplastics and can be made highly conducting by adding dopants.

Poly(alkylene polysulfides) are important elastomers (Thiokol rubbers). They can be prepared from dihalides and sodium polysulfides. Cross-linked materials can

TABLE 2.17 Structure of Cross-linked and Aromatic Polyesters

Name	Monomer/Reactants	Structure
Cross-linked polyesters Saturated alkyd resin	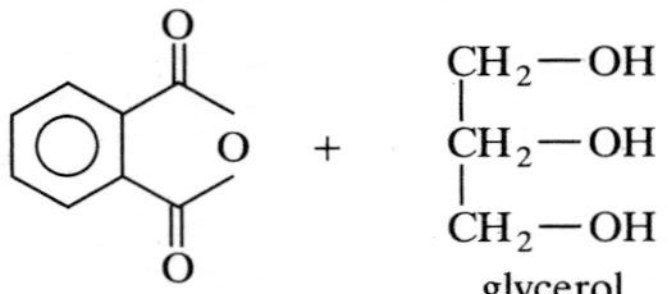 phthalic anhydride + $CH_2{-}OH$ / $CH_2{-}OH$ / $CH_2{-}OH$ glycerol	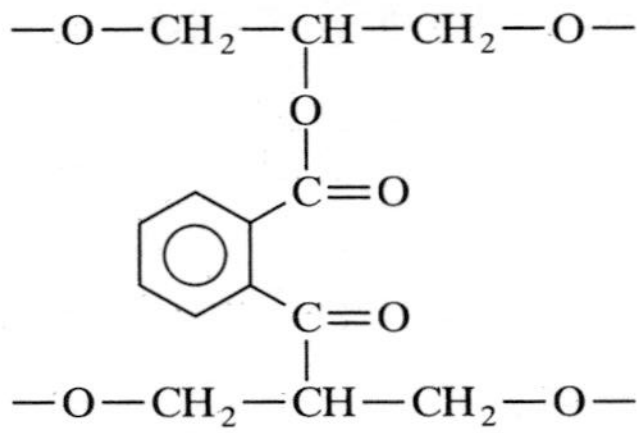 $-O-CH_2-CH(-O-C(=O)-C_6H_4-C(=O)-)-CH_2-O-$ … $-O-CH_2-CH-CH_2-O-$
Unsaturated polyester	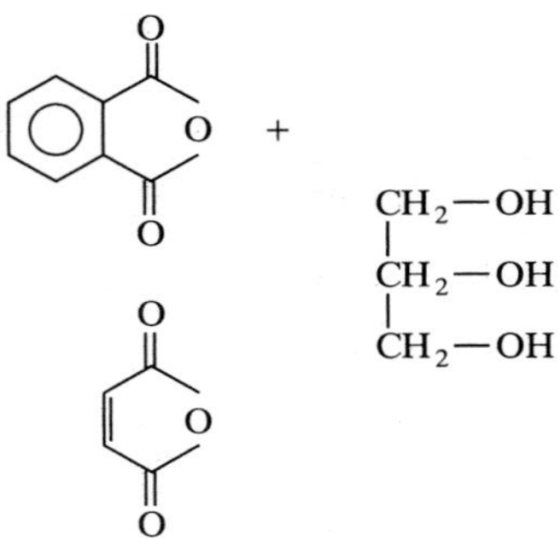 phthalic anhydride + maleic anhydride + $CH_2{-}OH$ / $CH_2{-}OH$ / $CH_2{-}OH$	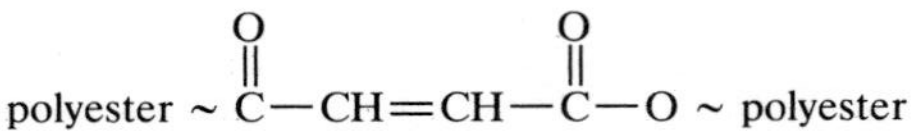 polyester ~ $C(=O)-CH{=}CH-C(=O)-O$ ~ polyester

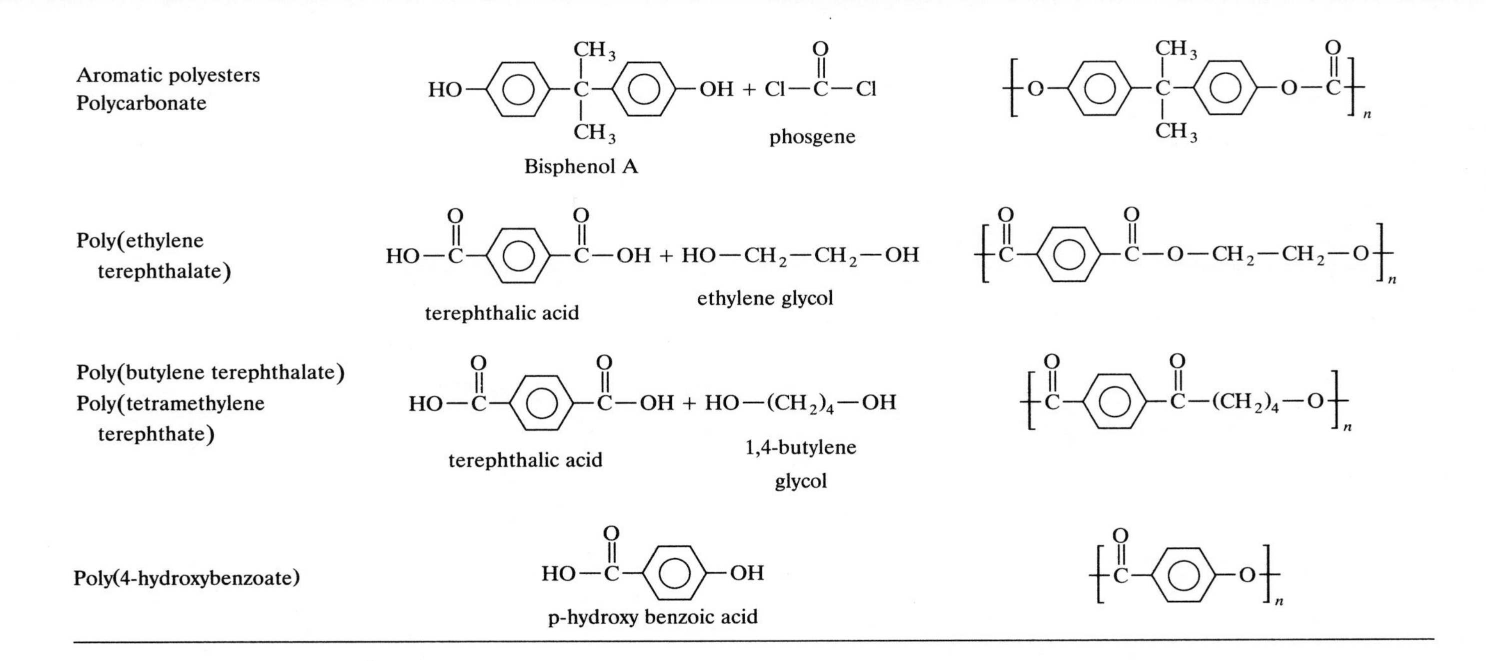
Aromatic polyesters
Polycarbonate
Bisphenol A
phosgene
Poly(ethylene terephthalate)
terephthalic acid
ethylene glycol
Poly(butylene terephthalate)
Poly(tetramethylene terephthate)
terephthalic acid
1,4-butylene glycol
Poly(4-hydroxybenzoate)
p-hydroxy benzoic acid

TABLE 2.18 Structures of Some Carbon-Sulfur Polymers

Name	Reactants	Structure
Poly(phenylene sulfide) Ryton	$Cl-C_6H_4-Cl + Na_2S$ *p*-Dichlorobenzene sodium sulfide	$-[C_6H_4-S]_n-$
	$Br-C_6H_4-S^-Na^+ + NaOH$ *p*-Bromothiophenol	
Poly(alkylene sulfides) (Thiokol rubbers)	$Cl-(CH_2)_2-O-(CH_2)_2-Cl + Na_2S_x$ Bis(2-chloroethyl)formal Sodium polysulfide	$-[(CH_2)_2-O-(CH_2)_2-S_x]_n-$
Polysulfone	Bisphenol A + $Cl-C_6H_4-SO_2-C_6H_4-Cl$	$-[O-C_6H_4-C(CH_3)_2-C_6H_4-O-C_6H_4-SO_2-C_6H_4]_n-$

be made by using small amounts of polyhalides and reacting the polymer with sodium hydrosulfide sodium sulfite. The poly(alkylene sulfides) have unpleasant odors due to the presence of low molecular weight mercaptans and disulfides. They do have good oil resistance and weathering resistance.

Polysulfones. Polysulfides can be oxidized to polysulfones with hydrogen peroxide. The resulting polymers have higher melting temperatures than the corresponding polysulfides. Aromatic polysulfones have engineering thermoplastic properties and are prepared by nucleophilic substitution reactions, similar to PEEK and polyetherimide polymerizations. Polysulfone from Bisphenol A is amorphous with a T_g of 220°C. It has good creep behavior and hydrolytic resistance. It can be used for reverse osmosis membranes in desalination applications.

2.1.5 Inorganic Polymers

Polymers not containing carbon in the backbone are classified as inorganic polymers (Table 2.19). Several inorganic elements form bonds with other species that have higher energies than carbon-carbon bonds (320 kJ/mol). These include B-C (370 kJ/mol), Si-O (370 kJ/mol), B-N (440 kJ/mol), and B-O (500 kJ/mol). These bonds might be expected to have higher thermal stability than carbon-carbon bonds. Since most of these bonds are polarized, they are susceptible to attack by oxygen and water. Their strength can best be utilized in inert environments, such as outer space. The

TABLE 2.19 Structures of Some Inorganic Polymers

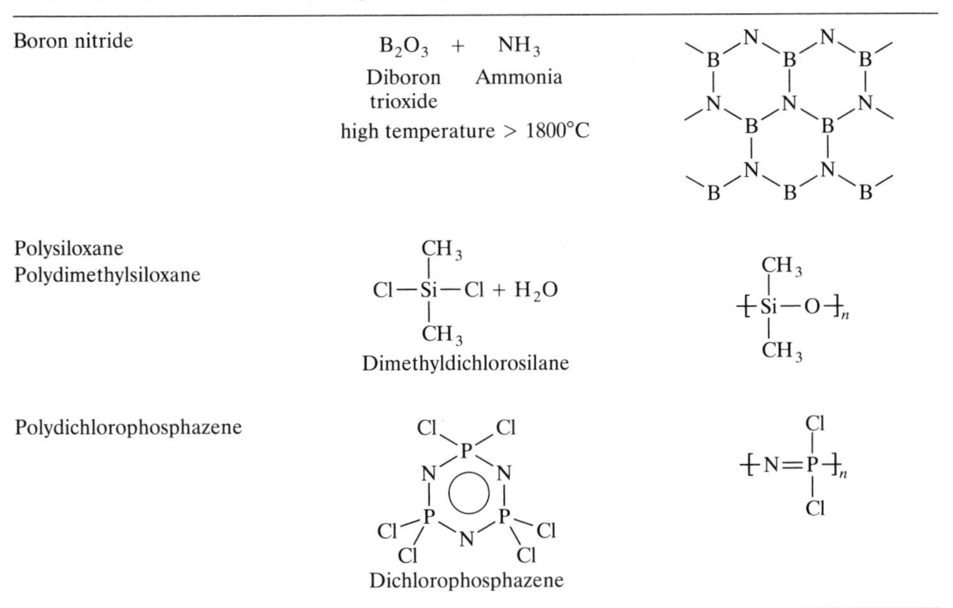

Boron nitride	B_2O_3 + NH_3 Diboron trioxide, Ammonia high temperature > 1800°C	(B–N hexagonal network)
Polysiloxane Polydimethylsiloxane	$Cl-Si(CH_3)_2-Cl + H_2O$ Dimethyldichlorosilane	$-[Si(CH_3)_2-O]_n-$
Polydichlorophosphazene	$(NPCl_2)_3$ ring Dichlorophosphazene	$-[N=PCl_2]_n-$

attack of environmental agents on these bonds can be reduced when the main chain is cross-linked (ladder, parquet, or lattice polymers). Boron nitride forms a parquet polymer that can be used at temperatures up to 2000°C.

Figure 2.3 shows some structures of silicates. Glasses have properties that are directly related to their chain structure. Sodium silicate glass with 49% SiO_2 has a glass transition temperature of 420°C. Natural silicates (glass, mica, asbestos, montmorillonite) have been used for centuries. E-glass, which is alkaline free, has been used as a reinforcing fiber for many structural composites.

Polysiloxanes are known by the popular name of silicones. Polymers can be prepared by hydrolyzing dichlorosilanes to give linear polymers. Cross-linked polymers can be made by using some trichlorosilanes. Polysiloxanes have very low glass transition temperatures (−127°C), due in part to the great flexibility around the Si-O bond. A variety of oils, elastomers, and resins can be made, leading to a broad range of applications. They are physiologically inert and are used in a number of medical applications. Inorganic glasses can be modified with polysiloxanes to give a class of materials called ceramers.

Phosphonitrilic polymers have been developed fairly recently. The basic reaction is that of a ring-opening polymerization. Polydichlorophosphazene has no applications because it is susceptible to hydrolysis, but products in which the chlorines are

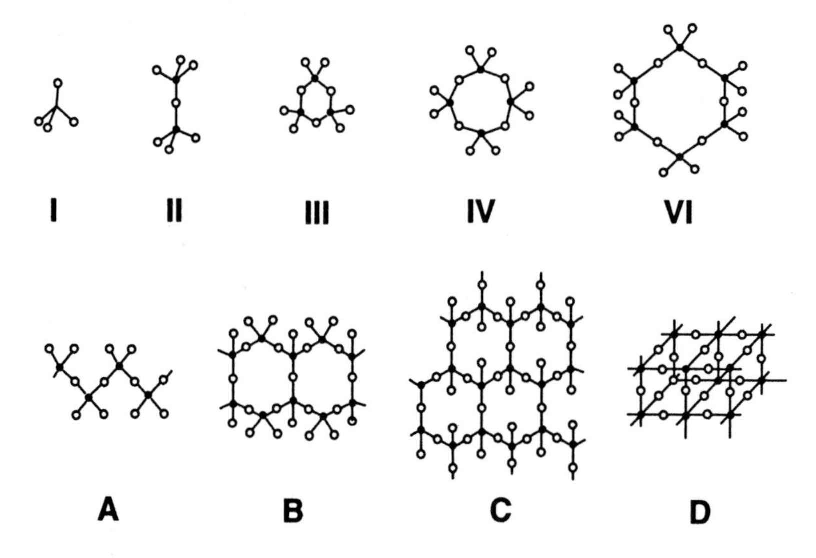

Figure 2.3 Structures of silicates, with examples of naturally occuring silicates: ●, silicon atoms; ○, oxygen atoms. Some of the oxygen atoms carry a negative charge (not shown). Left to right in the top row (low molecular weight structures): **(I)** tetrahedral (olivine, granate topaz); **(II)** double tetrahedral (mellilith group); **(III)** six-membered ring (three silicon atoms) (wollastonite); **(IV)** eight-membered ring (neptunite); **(V)** twelve-membered ring (beryl). The lower row consists of macromolecular structures; **(A)** linear chains (augite); **(B)** ladder or double-strand polymers (hornblende); **(C)** parquet polymers (glimmer, talc); **(D)** spatial or network polymers (quartz) (Elias, 1984).

replaced by alkoxy, aryloxy, or amino groups are useful. They have low glass transition temperatures and can be vulcanized to give useful elastomers.

2.1.6 Copolymers

Copolymers can be made in a number of constitutions. Random copolymers can be made from a mixture of vinyl monomers polymerized by free radicals. For a perfect, alternating copolymer to be made by such a process, the free radical of each monomer would need to have a zero probability of reacting with its own monomer. Copolymer nomenclature is given in Table 2.20.

The connective, -co-, is used to describe a polymer for which the sequence arrangement is unknown. Statistical copolymers (-stat-) have sequential distributions that conform to those expected from the polymerization rate constants. Most copolymers prepared by polymerizing from monomer mixtures are statistical. A random copolymer has a sequence such that the probability of occurrence of either monomer unit does not depend on the previous one (the copolymerization rate constants are equal, $k_{11} = k_{12} = k_{21} = k_{22}$, as discussed in Section 4.2). An alternating copolymer has a perfect sequence of one monomer unit followed by the other. This can occur when a reactive center of monomer A will not add monomer B. An example of an alternating copolymer is poly(styrene-alt-[maleic anhydride]). An alternative copolymer is the simplest case of a periodic copolymer. Several copolymer structures are shown in Figure 2.4.

TABLE 2.20 Copolymer Nomenclature

Type	Connective	Example
SHORT SEQUENCES		
See Table 1.9		
LONG SEQUENCES		
Block	-block-	Poly A-block-poly B
Graft	-graft-	Poly A-graft-poly B
Star	-star-	Star-poly A
Blend	-blend-	Poly A-blend-poly B
Starblock	-star-. . . -block-	Star-poly A-block-B
NETWORKS		
Cross-linked	-cross-	Poly(cross-A)
Interpenetrating	-inter-	Poly(cross-A)-inter-poly(cross-B)
Co-terminous	-cross-	Poly A-cross-poly B

SOURCE: W. Ring, *Source-Based Nomenclature for Copolymers*, IUPAC, 1983; L.H. Sperling, *Source-Based Nomenclature for Polymer Blends*, Interpenetrating Polymer *Networks* and *Related Polymers*, Nomenclature Committee of the Polymer Chemistry Division of the American Chemical Society, 1984.

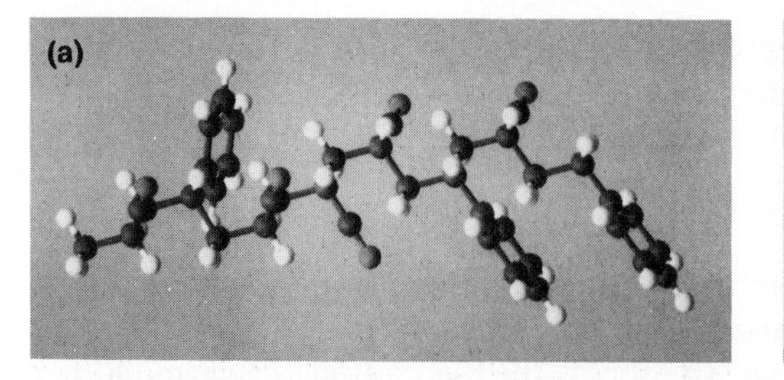

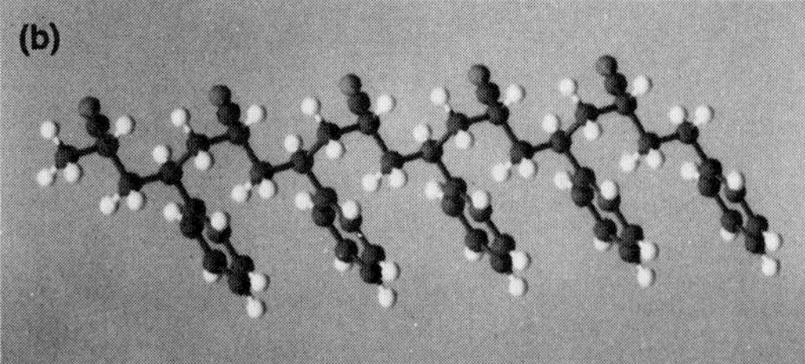

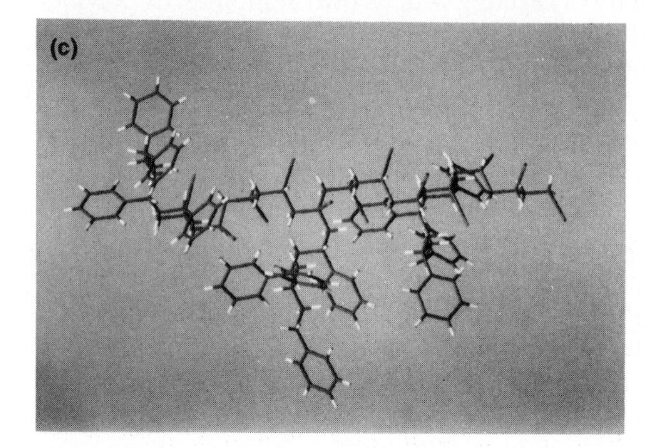

Figure 2.4 Copolymer configurations. POLYGRAF simulation models for **(a)** poly (acrylonitrile-ran-styrene); **(b)** poly(acrylonitrile-alt-styrene); **(c)** poly(acrylonitrile-g-styrene).

2.2 GENERAL TYPES OF POLYMER CONSTITUTIONS AND CONFIGURATIONS

A number of polymer configurations have already been discussed in the previous sections on polymer constitutions. This section summarizes different types of configurations for linear homopolymers, nonlinear homopolymers, and copolymers.

Linear Homopolymers. Homopolymers are those materials having one type of repeating unit. Both chain and step polymerizations yield homopolymers by this definition. Chain polymerizations make homopolymers in which the repeating unit is the same as the monomer. Step polymerizations make homopolymers in which the repeating unit is a combination of the two monomers used. Nylon 6,6 would be called a homopolymer even though it is made from a dichloride and a diamine. Polymers that have long ordered regions are said to be stereoregular, and often can be crystalline. Polymer crystallinity usually improves elastic modulus and other physical properties.

A number of important commercial polymers have a carbon-carbon backbone, and there are a number of categories that refer to this class of materials. These polymers are usually called vinyl polymers, since they can be polymerized (or named) for a substituted ethylene monomer. Vinyl monomers can be polymerized in chain reactions, and there are two ways in which order can occur along the polymer chain:

head-to-tail and head-to-head organization. Figure 2.5 shows these two configurations for a vinyl polymer.

The properties of two such polymers made from the same monomer may be quite different. For example, polyisobutylene made in the head-to-head conformation has a T_m of 187°C (1983). Head-to-tail polyisobutylene can be crystallized only under stress and has a T_m of 5°C. A number of polymers with a carbon-carbon backbone polymerize in the head-to-tail configuration, since substitute groups tend to sterically hinder the addition in the head-to-head configuration. Addition of monomer in a random fashion, alternating between head-to-tail and head-to-head, leads to atactic polymer with essentially no crystalline regions.

Tacticity is a description of the stereoregular arrangement of carbon-carbon backbone polymers. For polymers with head-to-tail configurations, there are two stereoregular forms: isotactic and syndiotactic. These stereoregular forms (Figure 2.6) tend to form crystals in the solid state. Except for head-to-tail poly(vinyl alcohol), atactic forms of the α-olefins are amorphous. Poly(vinyl alcohol) is unique because the hydroxyl group will easily fit into the polyethylene crystal structure conformation without distortion.

The isotactic form has the substituted group on the same side of the polymer chain. In the syndiotactic form, the substituted group alternates from one side of the chain to the other. Figure 2.6 also shows an atactic polymer segment which has added in the head-to-tail configuration. This material will not be highly crystalline because the substituted group does not appear in a regular order along the chain. Therefore, there should be few long segments of stereoregular polymer in the chains.

Elastomers based on carbon-carbon backbones are usually polymerized from dienes and have -C=C- along their backbone. Since there is no rotation around this bond, there are two different configurations: *cis* and *trans*. These are shown in Figure 2.7. Rotation around the single carbon-carbon bonds will not lead to similar organization.

Nonlinear Homopolymers. A number of polymers are not linear and have a variety of branching architectures. Table 2.21 lists some categories with sketches and examples. Random branching may occur in short segments or long segments, as discussed in Example 2.1. The two different types give much different melt rheologies.

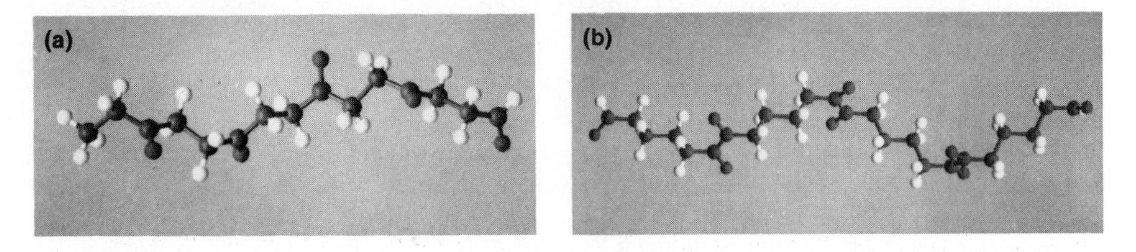

Figure 2.5 Head-to-tail and tail-to-tail configurations of a vinyl polymer. POLYGRAF simulation models for **(a)** head-to-tail polypropionaldehyde and **(b)** head-to-head polyprionaldehyde. The head-to-tail configuration has the carbonyl groups occurring every three carbon atoms along the chain. The head-to-head configuration has carbonyl groups adjacent to each other.

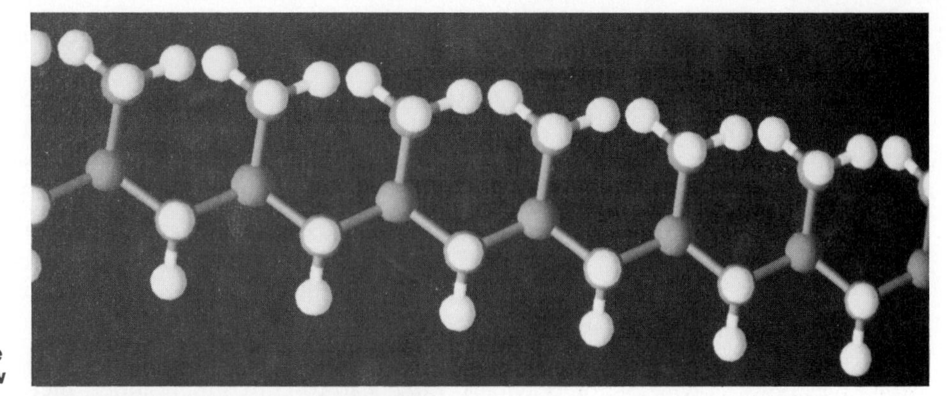

(a)
side
view

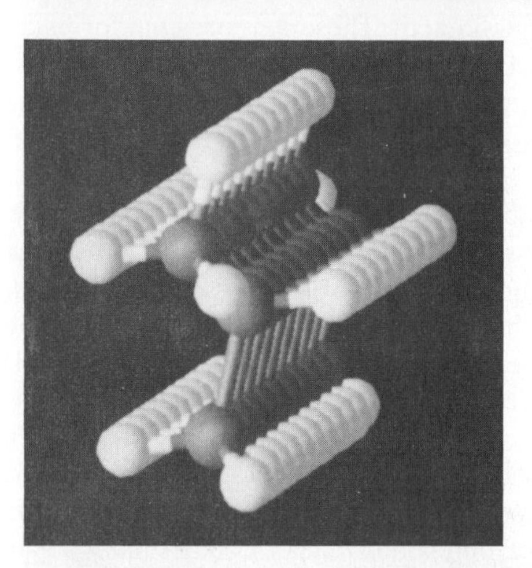

(a)
end
view

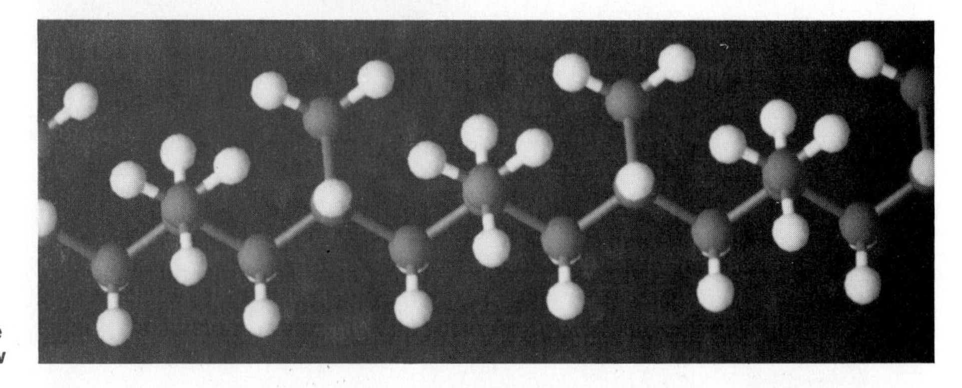

(b)
side
view

Figure 2.6 Stereoregularity in a vinyl polymer: three different molecular configurations **(a)** isotactic, **(b)** syndiotactic, and **(c)** atactic. The molecular shape may be changed from the planar zig-zag by rotation around C-C bonds. This will change the **conformation** of the molecular; it does not change the **configuration**, which is established at the instant of polymerization.

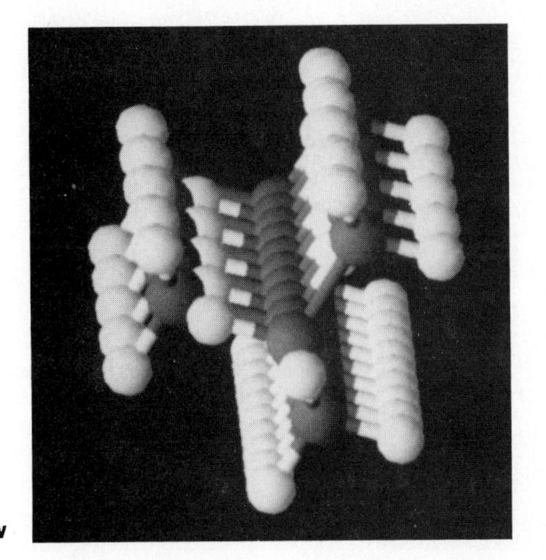

(b)
end
view

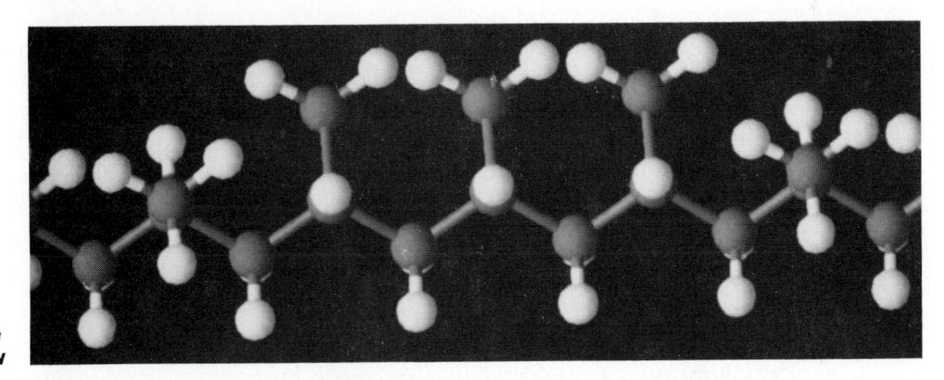

(c)
side
view

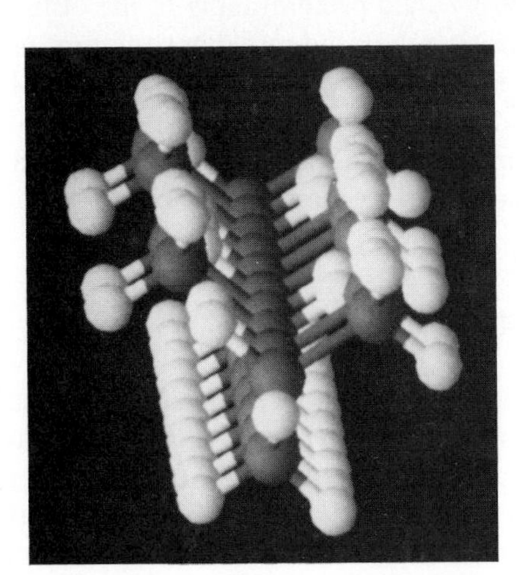

(c)
end
view

Figure 2.6 *(continued)*

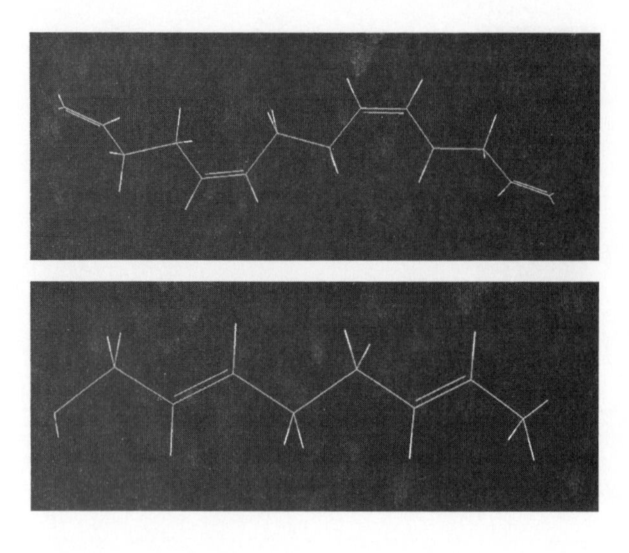

Figure 2.7 Diene configurations for 1, 4-polybutadiene (POLYGRAF vector representations): **(a)** *cis*-1,4-polybutadiene has the chain continuing on the same side of the double bond; **(b)** *trans*-1,4-polybutadiene has the chain continuing on the opposite side of the double bond.

Polymerization of some monomers results in a comb-like structure, with short chain segments attached at regular intervals to the backbone of the polymer. Imidazoles give a cross-linked, or ladder, structure in the chain itself. Star polymers usually are built with monomer units having three or more "arms." The sketch shows a trimer of this star polymer. One application of these materials is as nucleation agents.

Network polymers are represented by the general classification of thermosets. Interpenetrating polymer networks can be formed by two different network polymers made in situ or one after the other. Semi-interpenetrating networks refer to linear polymers trapped in networks.

Copolymers. Copolymers can be made by reaction or blending. In most vinyl copolymerizations, one monomer is added preferentially to the other, and the copolymer has a different bulk composition than the monomer mixture used in its synthesis. Block copolymers usually are made by polymerizing one polymer to completion, adding a second monomer, and continuing the polymerization with the second monomer. Graft copolymers are made by initiating a radical site along the backbone of one polymer and adding a second monomer.

2.3 CONFORMATIONS OF SINGLE MACROMOLECULES

A polymer molecule can assume a variety of shapes in space. The shape of a macromolecule in real systems depends on its environment, but can have a big effect on the polymer's performance and use properties. When polymer solutions are very dilute, most macromolecules interact with solvent molecules rather than each other. Dilute solvent-polymer systems are discussed in this chapter as a

TABLE 2.21 Types of Polymer Chain Structures

Type	Sketch	Example
Linear		Poly(vinyl chloride) Polystyrene
Branched short chain		Low-density polyethylene
Long chain branching		Low-density polyethylene
Comb		Polyacrolein
Ladder	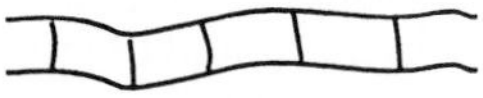	Poly(imidazole pyrrolones)
Semiladder		
Star (trimer)	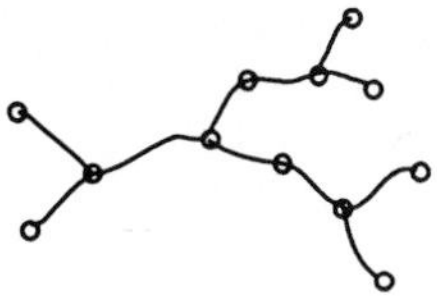	
Network	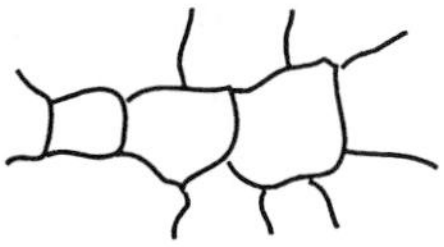	Phenol-formaldehyde resins
Interpenetrating polymer network (IPN)		Two cross-linked polymers not bonded to each other
Semi-interpenetrating polymer network		Cross-linked epoxy with vinyl polymer

prelude to the discussion of polymer conformations of solids and liquids in Chapter 3. Except for unusually large molecules like DNA, the conformation of individual polymer chains is very difficult to observe. However, studies of polymers in dilute solutions gives some indications on how individual chains are affected by their environment.

Noncrystalline, Linear Polymers. In an infinitely dilute polymer solution, the polymer molecule is awash in a sea of solvent molecules. Because the polymer molecules occupy a low volume fraction of the solution, the probability that chain segments of different molecules interact is low. Therefore, intermolecular bonding forces are very low compared to the interaction of polymer segments with the solvent and with other chain segments. Intramolecular forces are low also, since if they were not, the polymer would tend to self-associate and precipitate from solution. If the solvent is a good one, then there may not be much difference between the enthalpies of all secondary bonds in the solution state compared to the solid state. There will be big differences between the conformation of the macromolecule in the solid state and the solution. The free energy difference between a polymer molecule in the solid state and in infinite dilution in a solvent will be most affected by the increase in the entropy of the system.

A key factor in polymer chain conformation is rotation about the backbone chain of the molecule. Figure 2.8 shows some rotation possibilities for a four-carbon chain segment. When the first bond is fixed, the second bond can rotate through 360° while retaining the appropriate tetrahedral bond angle. When the second bond is fixed, the third bond can rotate. Determining the *average* conformation of a macromolecule requires finding the conditional probabilities that each bond has a specific conformation.

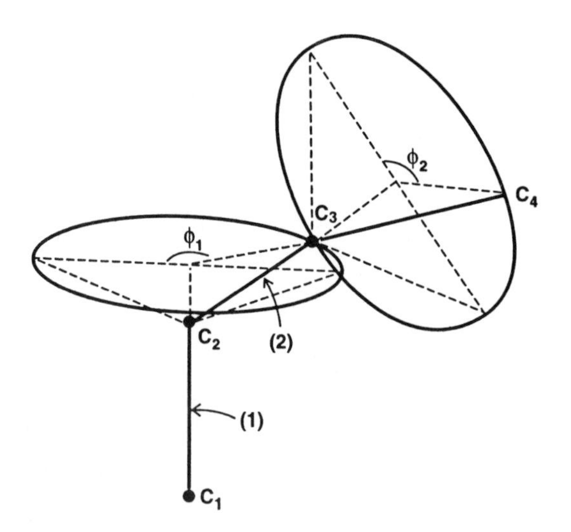

Figure 2.8 Rotational freedom in an idealized chain of four carbon atoms: C_1, C_2, C_3, and C_4.

The most probable form for a macromolecule in solution is not a sphere, or a straight rod, but a random coil. The coiling of macromolecules in solution was first recognized by Kuhn and Mark. A number of important polymer properties can be deduced by applying random flight statistics to coil conformation. The average coil density, ρ_{ave}, in solution can be related to the molecular weight of the chain

$$\rho_{ave} = \frac{6M_w}{\pi d^3 N_a} \tag{2.1}$$

where M_w is the molecular weight, d is the diameter of the coil, and N_a is Avogadro's number. The density of the coil can also be related to the average end-to-end distance of the coil by

$$\rho_{ave} = \frac{kM_w}{(R_o^2)^{3/2}} \tag{2.2}$$

where k is a proportionality constant and R_o is the average end-to-end distance for the coil. Since the average end-to-end distance is proportional to the square root of the molecular weight, the density of the average coil is inversely proportional to the square root of the molecular weight. This result has been experimentally confirmed and was a major step forward to understanding polymer solutions.

It is not possible to assign specific shapes to the coils, since their shapes are dynamic and are continuously changing with time as secondary valence bonds are broken and reformed. The most probable form is a bean-like structure, since there would be the highest number of possible formations of such a shape. By contrast, only a few conformations would lead to a spherical or rod-like molecule. Rod-like molecules occur in solution but are usually due to either severe hindrance to chain rotation, helical coil structure, planar chain molecules, or molecules for which enthalpic forces predominate and that have groups that repel each other. The axial ratios of the most statistically probable form are 1.36:0.78:0.50 (oblate ellipsoid).

Theta Solvents. The polymer-polymer and polymer-solvent interactions are always in dynamic equilibrium with each other. For some polymer-solvent pairs at specific temperatures, the effects balance out and the polymer conformation approaches an ideal statistical coil. Such a solvent is called a theta solvent, and the temperature at which this effect occurs is called the theta temperature. At temperatures slightly removed from the theta temperature, the polymer coils either precipitate or no longer behave like ideal statistical coils. At theta conditions, the density of the polymer coil is proportional to the square root of the molecular weight.

In good solvents, small amounts of polymer added to solution result in large viscosity increases. This is because the polymer tends to associate with the solvent and the coils become elongated. Electrostatic effects between polymer segments can also result in rod-like molecules and give large viscosities for dilute polymer solutions (less than 1wt%). One example is hair-conditioning products that are water-based gels. These contain dilute amounts of poly(acrylic acid), whose structure is shown in Figure 2.9. In water solution, the acid groups tend to either repulse each other or associate with solvent so strongly that the polymer chains are almost linear.

Globular proteins tend to assume conformations similar to random coils. Proteins are polypeptides made from the 24 common amino acids. The α-amino acids have the general formula HO-CO-CRH-NH_2, where R can be one of 24 functional groups. The larger proteins often have cross-links, formed by the S-S bonds of cysteine, cystine, or methionine, which help them maintain their shape over ranges of ionic strength, pH's, and ion types. Some of the functional groups are hydrophilic, while others are hydrophobic. Shape studies of enzymes have shown that, in dilute solution, the natural conformations have hydrophilic groups

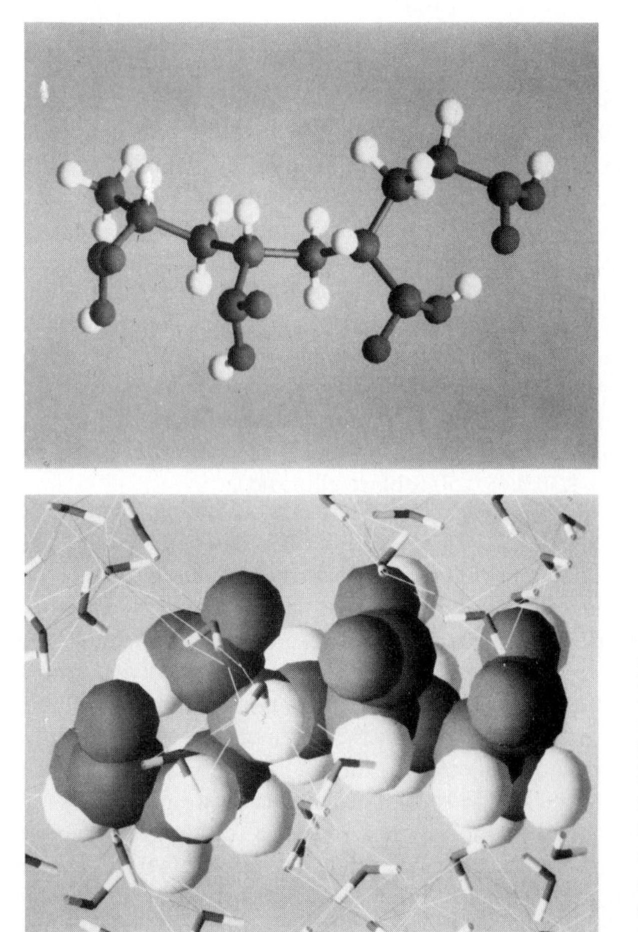

Figure 2.9 Poly(acrylic acid). **(a)** Carboxyl groups tend to repulse each other and hinder rotation. The chain backbone is not a planar zig-zag conformation. **(b)** Hydrogen bonding between water (cylindrical representation) and poly(acrylic acid) (space-filling representation). Hydrogen bond vectors are seen between adjacent water molecules and between some water and polar groups on the polymer chain segments.

facing the exterior of the molecule and hydrophobic groups associating with themselves in the interior. Placing a protein in a solvent that associates better with the exterior groups does not necessarily result in significant expansion of molecular size. Apparently, nonpolar-nonpolar interactions result in considerable binding forces. These forces can be enhanced by poor solvents.

Crystalline, Linear Polymers. Polymers that are stereoregular can have crystalline segments even when they are in solution. Figure 2.10 shows four different kinds of macromolecule conformations in solution. The random coil is most likely to be assumed by polymers that are not crystalline. The extended chain or rod conformation would be assumed by polymers containing repulsive side groups along the chain. The helix and folded chain conformations would be assumed by polymers that tend to crystallize. Table 2.22 shows some chain conformation and solid crystal systems for some stereoregular polymers.

The crystalline conformations will be low energy shapes for molecules that crystallize. Polymers that tend to crystallize include

- isotactic and syndiotactic asymmetric substitution addition polymers (polypropylene)
- unsubstituted linear addition polymers (polyethylene, polytetrafluoroethylene, poly [oxymethylene])
- addition polymers with di-substituted repeating units (poly[vinylidene chloride])
- straight chain and symmetrical ring-containing condensation type polymers (poly[hexamethylene adipamide] and poly[phenylene terephthalamide])
- some nonstereoregular asymmetric addition polymers (poly[vinyl alcohol] and poly[monochlorotrifluoroethylene])

Polyethylene molecules tend to align in a planar zig-zag conformation (Fig. 2.11). The molecule is in a minimum energy conformation in this form, as there are the minimum steric and bonding hindrances when the hydrogen atoms alternate as shown.

Figure 2.12 shows two vinyl polymers. R groups that repel each other on isotactic segments of vinyl polymers can cause the chain to twist (Fig. 2.12a). Bulky R groups can lead to helix formation (Fig. 2.12b). Other polymers also form helices as minimum energy conformations (Fig. 2.13).

Several polymers have more than one low-energy conformation, so that changes in temperature, pressure, or solvent can induce transitions from one form to another. At room temperature in solution, most polymers that can crystallize will have amorphous segments in their chains. Figure 2.14 shows single crystal lamellas for polyethylene.

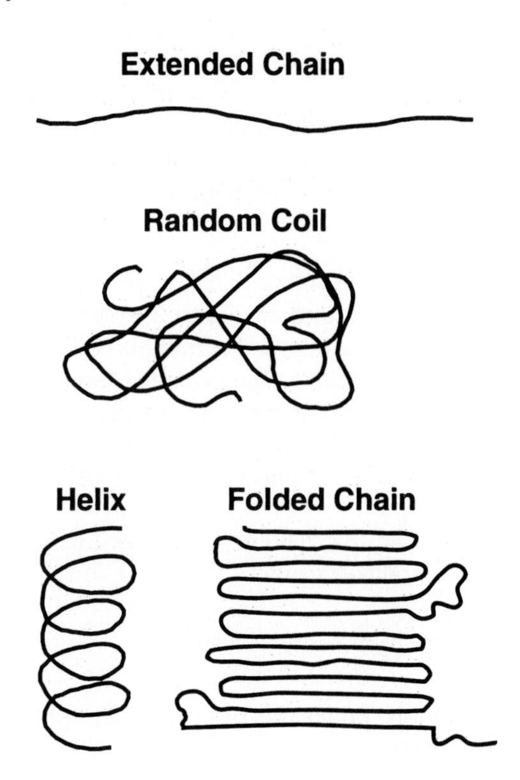

Figure 2.10 Extended chain, random coil, helix and folded chain conformations.

TABLE 2.22 Crystal Structures of Selected Polymers at 25°C

Polymer	Crystal System	Chain Conformation
Polyethylene	Orthorhombic	Folded chain
Polytetrafluoroethylene I	Triclinic	13,6 helix
Polytetrafluoroethylene II	Trigonal	15,7 helix
Polyoxymethylene	Trigonal	9,5 helix
Poly(ethylene oxide)	Monoclinic	7,2 helix
α-*trans*-1,4-polyisoprene	Monoclinic	*trans*
β-*trans*-1,4-polyisoprene	Orthorhombic	*trans*
α-Poly(hexamethylene adipamide)	Triclinic	Extended (trans)
Poly(*p*-phenylene terephthalamide)	Monoclinic	Extended (trans)
Cellulose I	Monoclinic	Fringed micelle
Isotactic polypropylene	Monoclinic	3,1 helix
Isotactic polybutene-1 II	Tetragonal	11,3 helix
Isotactic polystyrene	Trigonal	3,1 helix

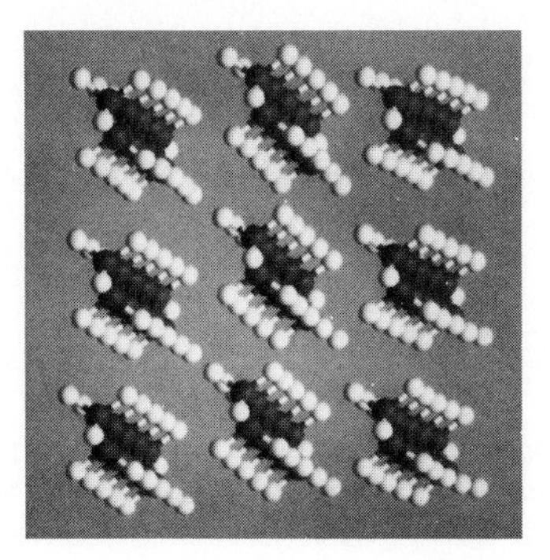

Figure 2.11 Planar zig-zag conformation of polyethylene chain segments. A 3 X 3 array of polyethylene chain segments in the planar zig-zag conformation. The chains are fairly close-packed in this arrangement. Solvent molecules such as hexane do not penetrate to the interior of the array in a POLYGRAF simulation.

2.4 RELATIONSHIPS BETWEEN POLYMER MORPHOLOGY AND USE PROPERTIES

It is important to try to relate the polymer chemistry and physics to the end-use properties of the polymers. The ABS (acrylonitrile-butadiene-styrene) family of polymers will be used as an example case. The homopolymers have different properties: polyacrylonitrile is a crystalline polymer; polystyrene is glassy; polybutadiene is an elastomer; and the co- and terpolymers have some unique properties. This system will also help show the effects that polymerization has on polymer properties.

Polyacrylonitrile (PAN). The structure of the homopolymer from acrylonitrile is shown in Table 2.2. The secondary valence forces for this molecule include London dispersion forces between neighboring hydrogen atoms along the carbon

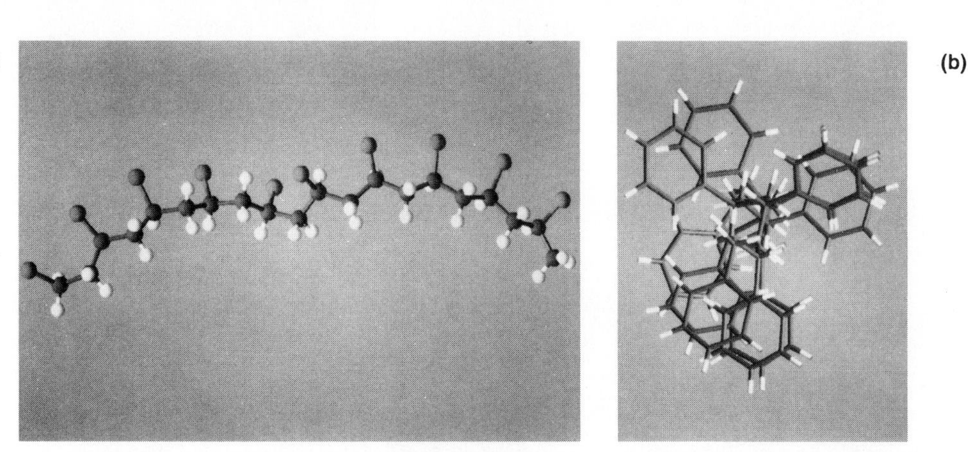

Figure 2.12 Vinyl polymer conformations. Vinyl subsitution groups are often too large to fit into the planar zig-zag conformation as a minimum energy state. **(a)** An isotactic chain segment of poly(vinyl chloride) shows curvature caused by the repulsion of the chlorine atoms. **(b)** An axial view of a syndiotactic polystyrene segment shows that chain rotation results in a different helix.

(a)

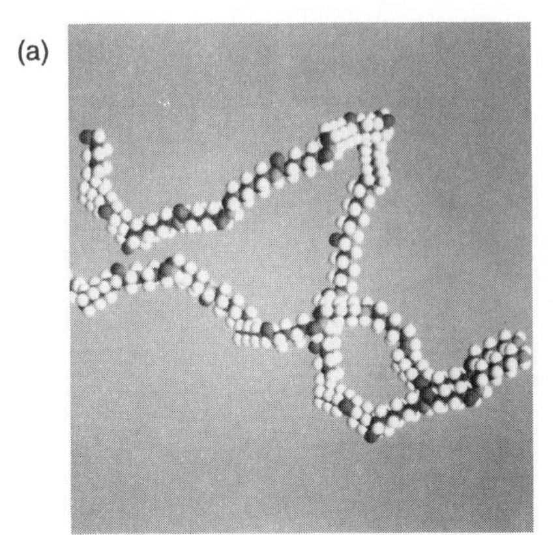

Figure 2.13 Helical conformations. **(a)** Two different helical conformations of nylon 6,6. Both chains start at parallel positions on the left. The tighter helix has a lower energy associated with its structure. **(b)** A side view of poly(vinylidene fluoride) shows that the pairs of fluorine atoms tend to repel each other. **(c)** An axial view of poly(vinylidene fluoride) shows that the chain segment forms a helix.

(b) (c)

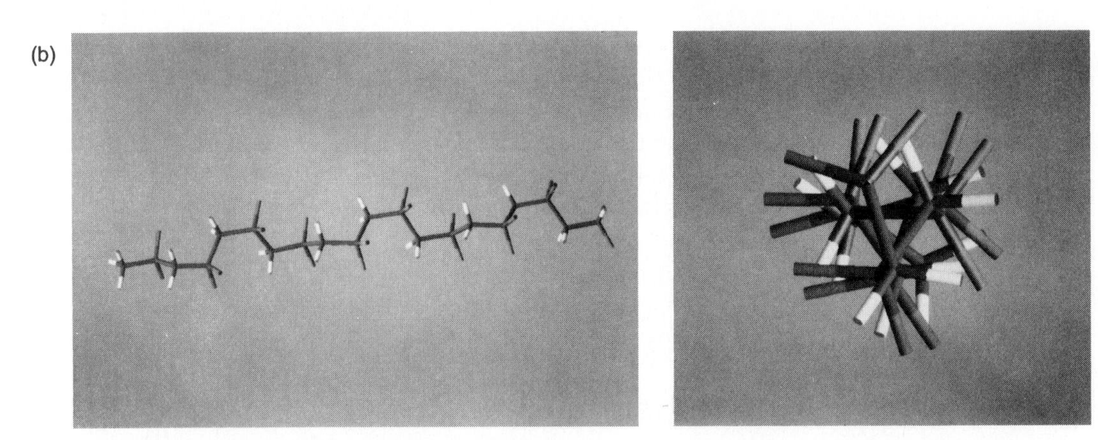

Figure 2.14 Scanning force micrograph (Topometrix TMX 2000) of a polyethylene lamella. Published with the permission of F.F. Lin and D.J. Meier, Michigan Molecular Institute.

chain backbone and hydrogen bonds between hydrogens and the cyanide group. The hydrogen bonds should be strong, as they fall in Class II of Table 1.14. Polyacrylonitrile should have a high cohesive energy density relative to polyethylene, and it does (237 cal/cm3 vs. 62 cal/cm3) (Small, 1953; Billmeyer, 1984).

Stereoregular polyacrylonitrile would be expected to be crystalline and have a high modulus, making it a hard and tough polymer. However, a stereoregular polymer usually requires a coordination complex catalyst to force monomer units to add to the chain in the same head-to-tail manner every time. Most commercial polyacrylonitrile is made using free radical initiators that do not cause all monomers to add to the growing chain in the same way. Rotation can occur around the carbon-carbon bond at the active site, and any orientation of adding monomer would have to depend on the repulsive and attractive forces of the side groups. Apparently, monomer does not add in a completely random fashion since commercial polyacrylonitrile does contain significant amounts of crystallinity.

The amount and nature of crystallinity in a polyacrylonitrile sample will also depend on the history of the material from polymerization to final fabrication step. Crystallization is a rate process and depends on temperature, melt viscosity, and external forces that can orient the polymer, such as uniform shear fields. The molecular weight distribution of the polymer chains also is important since chains of widely different molecular weight may phase separate.

Polyacrylonitrile is used for fibers and is a good substitute for wool in carpet and upholstery applications. Commercial acrylic fibers are usually copolymers of acrylonitrile and other monomers for improved processing and dyeing. Adding monomers with basic or acidic groups allows the fiber to be dyed using conventional dye technology. The copolymers can be processed at lower temperatures and have lower viscosities than the homopolymer.

Since polyacrylonitrile contains a polar group, it tends to be soluble in polar solvents such as dimethylformamide and dimethyl sulfoxide. It can also be dissolved in some concentrated inorganic and quaternary ammonium salt solutions. The T_g for completely amorphous polymer is 110°C (87°C has been measured for crystalline samples), and the T_m is 240°C. Both syndiotactic and isotactic polyacrylonitrile can be made.

Polystyrene (PS). Polystyrene (Table 2.6) and polyacrylonitrile are fairly easy to compare since they both are vinyl polymers. The benzene ring will not participate directly in hydrogen bonding but could have π-bonds with neighboring atoms. This should result in a cohesive energy density greater than that of polyethylene, but less than that of polyacrylonitrile (74 cal/cm3).

Isotactic polystyrene can be made but is not important commercially. The isotactic form has a 3,1 helix conformation when crystallized from solution. The commercial atactic polymer is brittle, has a low modulus, and is optically clear. The polymer chains probably take the random coil conformation. The T_g of the amorphous material is about 100°C, and the isotactic material has a T_m of 240°C. Polystyrene is soluble in a number of organic solvents, particularly those with aromatic rings, such as toluene and ethylbenzene.

Polybutadiene (PBR). Polybutadiene (Table 2.3) can be made in several forms from the butadiene monomer (Fig. 2.1). Elastomers are usually used well above their glass transition temperature and retain their form through cross-links. The *cis* material is most useful since its melting temperature is 2°C, while the T_m for the *trans* material is 145°C. The 1,2 polymer has melting points of 126°C for the isotactic material and 156°C for the syndiotactic material.

The catalyst chosen for the polymerization makes a big difference in the amount of *cis*, *trans*, and 1,2 adduct in the product (Table 2.4). Polybutadiene is soluble in a variety of organic solvents. However, the commercial polymer usually contains cross-links, so that polymer samples are only swollen in the presence of solvent.

Co- and Terpolymers of Acrylonitrile, Butadiene, and Styrene. Figure 2.15 shows a three component phase diagram for ABS co- and terpolymers. The copolymers are located along the lines joining the homopolymers. The phase diagram is interesting because it includes a crystalline, a glassy, and a rubbery polymer. Polyacrylonitrile has high order, a well-defined morphology, and a high modulus. PAN samples usually have a low elongation to break, but this depends on their initial orientation. Polybutadiene is a typical elastomer with a low modulus of elasticity but high elongation to break. It is typically used in a cross-linked form. Glassy polymers

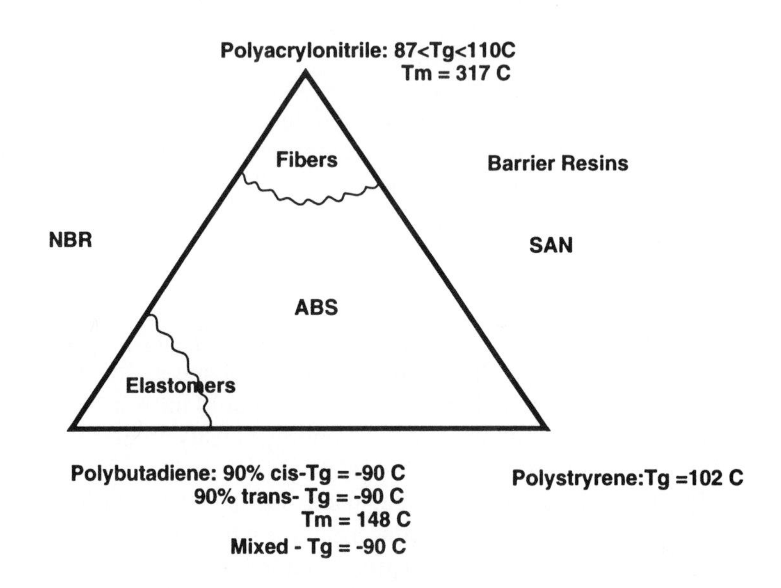

Figure 2.15 Phase diagram for polyacrylonitrile, polybutadiene, and polystyrene.

like polystyrene have a random coil structure, are brittle, have low moduli of elasticity, and are easy to form.

Product fabricators try to choose the correct balance of crystalline, amorphous, and elastomeric properties for a given application. However, the properties of these polymer blends are not necessarily linearly related to the weight fraction composition of the individual polymer components.

Nitrile butyl rubber (NBR) is an elastomer with high resistance to organic solvents. NBR is made with ratios of acrylonitrile to butadiene monomers of one to five. The solvent resistance is due to the polarity of the acrylonitrile portion of the polymer chain. These elastomers are usually cross-linked and have good elongation properties. Elastomers occur at compositions of acrylonitrile between 15 and 55wt%.

Styrene butadiene rubber (SBR) was originally used to make latex paints. It has now been replaced by vinyl and acrylic latexes.

Acrylic fibers (Orlon, Acrilan, Creslan, and Dynel) are copolymers of about 85wt% acrylonitrile with some butadiene and styrene. The fibers are prepared by wet or dry spinning techniques.

Styrene-acrylonitrile (SAN) is a clear plastic with a modulus of elasticity greater than that of polystyrene but less than that of polyacrylonitrile. It is a clear material with slightly better mechanical properties than polystyrene.

ABS terpolymers have unique combinations of toughness and elasticity and can be processed relatively easily. The terpolymers can be made by either mechanical blending of the individual polymers or by copolymerization. ABS plastic can be made by copolymerizing styrene and acrylonitrile around polybutadiene latex particles.

NOMENCLATURE

Symbol	*Definition*
d	diameter of polymer coil
M_w	molecular weight
N_a	Avogadro's number
ρ_{ave}	average density of a polymer coil
R_o	average end-to-end distance of a coil

PROBLEMS

2.1 Define constitution, configuration, and conformation.

2.2 Polymer configuration is established by the covalent bonds in the polymer backbone. Draw "three-dimensional" bonding pictures of polystyrene in an isotactic, an atactic, and a syndiotactic form. Which forms would you expect to be most probable for a free-radical process?

2.3 Secondary valence forces are responsible for many of the unique properties of polymers. For poly(vinylidene chloride), poly(methyl methacrylate), poly(ε-caprolactam), and poly(dimethyl siloxane), indicate which secondary bonding forces (dipole, induction, dispersion, or hydrogen bonding) would be important in establishing polymer conformation. Rank these forces in order of importance for each polymer. A sketch of each material may help complete the answer.

2.4 Draw the primary structure of the following polymers: Polyethylene, polybutadiene, poly(hexamethylene adipamide), cellulose, melamine. Indicate whether higher order structures exist and sketch them. Which of these would be characterized as network polymers, as homopolymers, as copolymers?

2.5 Give an example of commercial random copolymers, alternating sequence copolymers, and block copolymers. Draw the structures of the repeating units. How could each sample be made?

2.6 Silk fiber (Fig. 1.9) has a constant high modulus to the break point. Of the polymers poly(hexamethylene adipamide) and poly(vinylidene chloride), which would have fiber properties most similar to silk? Justify your answer in terms of configuration, conformation, and higher order structures. Start by describing silk in these terms.

2.7 Categorize the polymers listed in Table 1.10 with respect to their generic classifications (main headings in Sections 2.1-1 to 2.1-4).

2.8 Categorize the polymers listed in Table 1.11 with respect to their generic classifications (main headings in Section 2.1-1 to 2.1-4).

2.9 Write a short description of polymerization techniques used to prepare one of the commercial polydienes. How do the polymerization conditions affect the product's constitution, configuration, and conformation?

REFERENCES

F.W. BILLMEYER, JR., *Textbook of Polymer Science*, 3rd ed., Wiley, NY, 1984.

H.B. ELIAS, *Macromolecules*, Vols. 1(a) and 2(b), Plenum Press, NY, 2nd Ed. 1984.

W. RING, *Source-Based Nomenclature for Copolymers*, IUPAC, 1983.

L.H. SPERLING, *Source-Based Nomenclature for Polymer Blends, Interpenetrating Polymer Networks and Related Polymers,* Nomenclature Committee of the Polymer Chemistry Division of the American Chemical Society, 1984.

P.A. SMALL, *J. Appl. Chem. 3*, 71 (1953).

3

Physical States of Polymers

The previous chapter on molecular architecture described the constitution, configuration, and conformation of individual macromolecules. Polymer systems, both bulk liquids and bulk solids, are composed of many associating molecules. This chapter describes different ways in which macromolecules interact to give the performance and processing properties we observe in large samples.

The physical states of polymers are described here by examining the properties of extremely dilute polymer solutions (few polymer molecules in large amounts of solvent) and considering how the properties change as the polymer becomes more concentrated. The physical state of the polymer changes from infinitely dilute solution, to concentrated solutions, to the gel state (which exhibits some properties of elastomers), to the solid state (which may include ordered and disordered regions). Sections 3.1 to 3.5 describe polymer solutions, the gel state, supercritical solutions and gels, the rubber-elastic state, and the solid state. Natural polymers are discussed in Section 3.6. The final sections of the chapter cover polymer blends, polymer additives, liquid crystalline polymers, and structured polymers and composites.

3.1 POLYMER SOLUTIONS

A number of important processing problems require knowledge of polymer solutions. They include

- phase equilibria and phase separation
- cleaning polymer processing equipment
- choosing a volatile carrier for polymeric adhesives
- precipitation of polymer from solution after polymerization
- developing polymer resistance to solvent swelling
- devolatilization of polymers

In order to design the processes above, the polymer engineer would like to predict which solvents can be used to dissolve a polymer, the effects of temperature and pressure on solubility, volume change on mixing, and multicomponent phase equilibria.

Brownian motions in liquids are rapid with respect to most experimental measurements, so it is difficult to obtain direct evidence of their three-dimensional structure. However, there is indirect evidence provided by viscometry and thermodynamic behaviors that helps provide conceptions of the solution states of polymers. The application of thermodynamics and statistical mechanics to polymer solutions has been one of the triumphs of polymer chemistry. Simple models for polymer-solvent phase equilibrium are presented in Chapter 5.

Amorphous Polymer Dissolution. The free energy of polymer dissolution, ΔG, is determined by the changes in system enthalpy and entropy during the process.

$$\Delta G = \Delta H - T\Delta S \qquad 3.1$$

When ΔG of the solution process is negative (the right-hand side of Eq. 3.1 is negative), the polymer will dissolve in the solvent. System entropy changes when the coil conformation changes through rotation, stretching, or other mechanisms. ΔS always increases (has a positive sign) for dissolution processes. System enthalpy changes when secondary valence forces are broken and formed. If ΔH is less than or equal to zero, then dissolution will occur spontaneously because the free energy of the mixing process will always be negative. Polymers with positive heats of solution, or negative ΔH, will be miscible in all proportions.

Many polymer-solvent pairs exhibit limited miscibility, that is, there are some conditions of composition, temperature, and pressure at which two phases exist. This can occur for systems in which the enthalpy change on mixing is positive. As ΔS on mixing becomes smaller, the magnitude of $T\Delta S$ approaches the magnitude of ΔH, ΔG goes to zero, and the dissolution process stops. At this point the polymer has reached an equilibrium solubility in the solvent and the solution is saturated. Adding polymer to the system at this point will result in two phases: a solvent-rich phase and a polymer-rich phase.

The polymer dissolution process is illustrated in Figure 3.1. As the polymer dissolves in the solvent at the bulk polymer surface, the secondary valence forces between polymer chain segments in the solid are replaced by secondary valence forces between the polymer segments and the solvent molecules. When the entire polymer chain has been solvated, it can diffuse away from the polymer solid. This process is distinctly different from the solution of low molecular weight inorganic salts, in which the molecules can diffuse from the solid surface after relatively few atoms have associated with solvent. For example, the dissolution of $CaCl_2$ into water solution requires the breaking of a few ionic bonds per molecule of 110 g/mol. The dissolution of a typical polystyrene molecule of 250,000 molecular weight requires that many of the 2500 chain repeating units associate with the solvent.

At the polymer solid surface, there is always a large concentration of macromolecules because of the slow diffusion rate and slow dissolution rate. However,

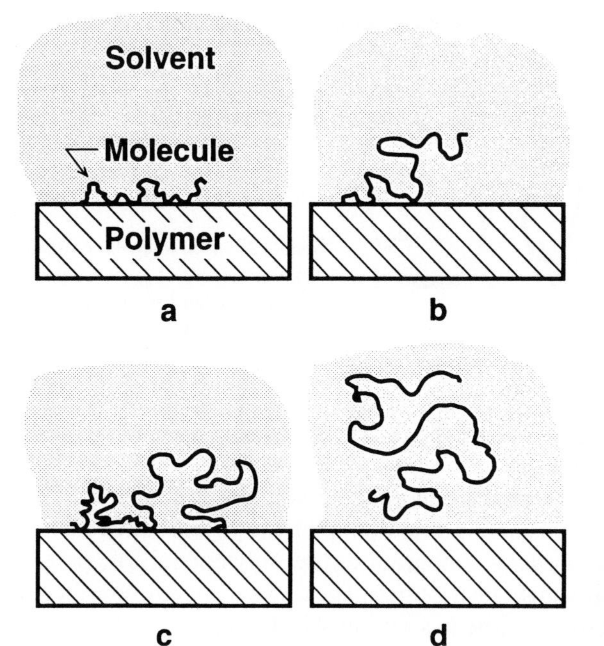

Figure 3.1 The polymer dissolution process. **(a)** Small loops of the polymer are dissolved. **(b)** One long chain segment is solvated. **(c)** Most of the molecule is disolved. **(d)** One entire chain has left the solid surface and is diffused into the bulk solution.

polymer segment solvation does not occur solely at the interface. Solvent can penetrate the polymer solid and solvate some chain segments. If there is dynamic interaction between dissolved chain segments and undissolved chain segments in the bulk solid, there may be little net dissolution. In poor solvents, the polymer phase may only swell and very little polymer will dissolve into the solvent phase.

In dilute solutions, a good solvent will disrupt most of the intramolecular associations of the solid macromolecule, while a poor solvent may disrupt only a fraction of these associations. For most solvent-polymer pairs, intramolecular cross-linking will exist (Fig. 3.2). Intramolecular cross-links are secondary valence bonds that occur between different chain segments of one macromolecule. They result in a decrease in the entropy of the system, and the random coils contract. In more concentrated solutions, intermolecular cross-links occur between chain segments of different macromolecules. The type of cross-linking favored depends on the shape of the random coil in solution and the polymer concentration. Intermolecular cross-linking is favored when the decrease in entropy caused by intermolecular bonding is matched or exceeded by the decrease in entropy caused by coil contraction due to intramolecular bonding. At high polymer coil concentrations, intermolecular cross-linking occurs because further intramolecular cross-linking (which would also increase coil density) would increase the internal stress of the coil segments. Large amounts of intermolecular cross-linking in solution can lead to aggregation of polymer coils. Dilute solution viscosity experiments

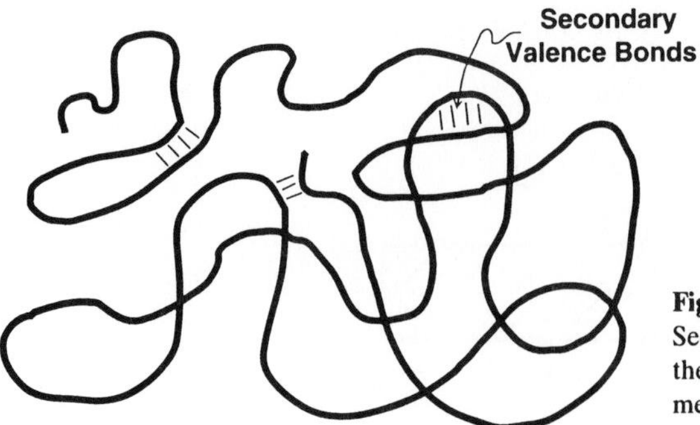

Figure 3.2 Intramolecular cross-linking. Secondary valence bonds, designated by the thin lines, can form between chain segments in solution.

suggest that some commercial polymers are almost always aggregated in solution (poly(vinyl chloride) for example).

A conceptual sketch of a very dilute polymer solution is shown in Figure 3.3. The amorphous polymer chains assume random coil conformations, which are dynamic and are affected by system temperature, the polymer-solvent pair, system pressure, and the molecular weight distribution. The dashed line around each macromolecule suggests its effective volume in solution. Most of the space of the effective coil volume is occupied by solvent molecules. This system is considered dilute since the coils are not interacting with each other. Under these conditions, the coils should behave ideally with respect to viscous properties, which are related to how the solution responds to shear forces, and thermodynamic properties, which are related to the conformation assumed by the coils.

The balance between polymer-polymer and polymer-solvent forces changes for different size molecules. Equation 2.2 suggests that the density of an ideal coil is proportional to the inverse of the square root of M_w. The volume of the ideal coil is proportional to the square root of M_w. Consider two polymer chains of the same constitution with different molecular weights, 10,000 and 90,000, at the same wt% concentration in dilute solution. The ratio of the coil volumes for these molecules would be the square root of 90,000/10,000, or 3. Intermolecular forces become very important when polymer coils begin to interact physically. Physical interactions of the coils will occur at lower weight fraction of polymer in solution for the high molecular weight material. The increased probability of intermolecular forces for long chains means that they tend to aggregate or precipitate at lower solution concentrations than short chains. The higher solubility for smaller molecules forms the basis for fractional precipitation processes, an important method of separating and purifying polymers of differing molecular weight.

Crystalline Polymer Dissolution. Polymers containing highly oriented or crystalline regions may retain some crystallinity in the "solution" process. These systems can be considered to have two polymer phases: a crystalline phase and an amorphous phase. The amorphous material will dissolve based on the principles

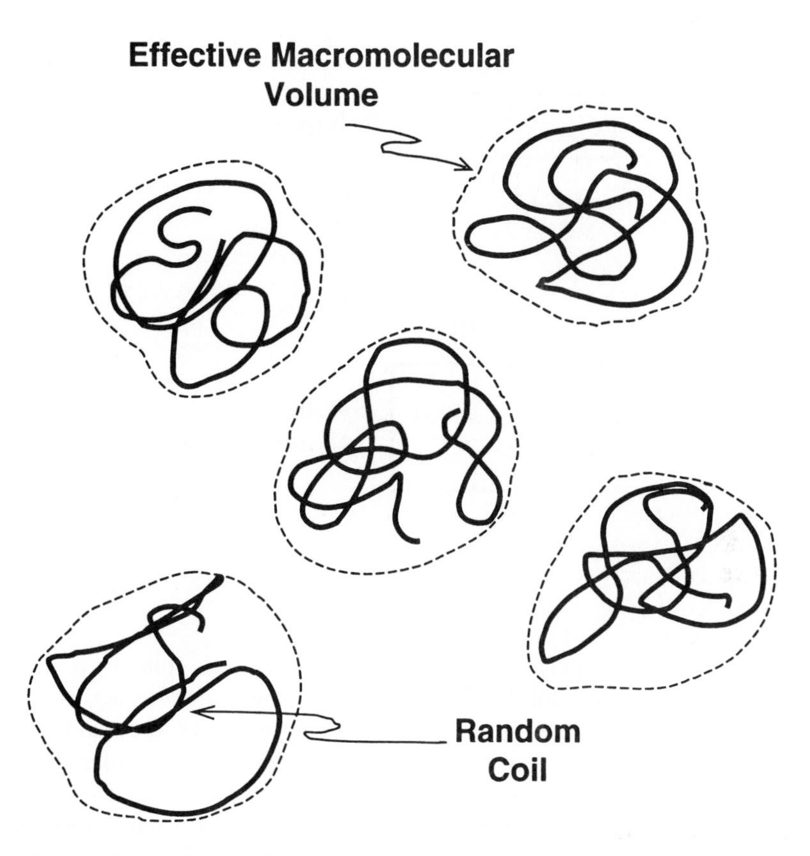

Figure 3.3 Sketch of a very dilute polymer solution.

previously described. The crystalline material must "melt," or lose some of its crystallinity in order to dissolve. The crystalline material has high intermolecular cross-links that must dissociate for the chain segments to associate with solvent.

Swelling of Cross-linked Polymers. Network polymers do not dissolve when exposed to solvents, but may swell (Fig. 3.4). The solvent penetrates through the chain segments and expands the network by causing the chains to elongate relative to their previous conformations. The polymer does not dissolve due to the cross-linking points between chains. The amount of swelling depends on how well the solvent associates with chain segments at the system temperature and the amount of cross-linking between chains.

3.2 THE GEL STATE

The gel state can be distinguished from concentrated polymer solutions by the fact that the coils no longer move as distinct units and interchange places with each other.

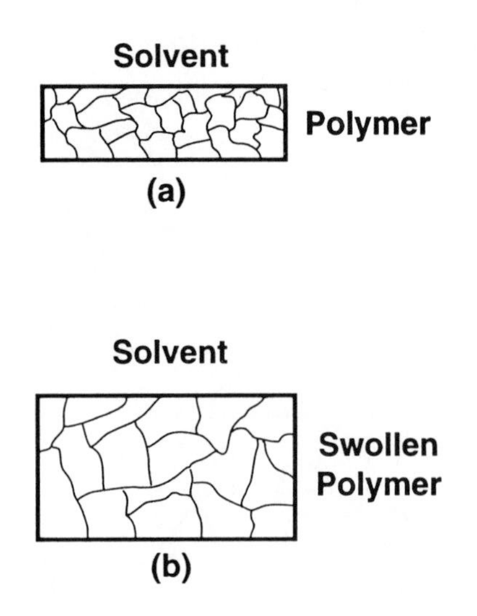

Figure 3.4 Solvent swelling of a network polymer: **(a)** start of the swelling process, **(b)** the swollen cross-linked polymer expands.

The viscosity of a gel is very high, and it behaves like a solid. By analogy, the transition from concentrated solution to a gel is similar to the transition of a liquid to a solid. However, large segments of a polymer chain can move about in the included solvent. The linkages between coils can be covalent bonds or secondary valence forces (Fig. 3.5). When gels are formed by cross-linking polymer chains in solution, the gel point of the system can be observed as the time when the mixture suddenly loses fluidity and bubbles no longer rise quickly through the liquid. Cross-linked gels may show solid-like properties with only a few weight percent of polymer.

Polymers that form secondary valence gels are cellulose nitrate, plasticized PVC (for example, PVC with several percent dioctyl phthalate), gelatins, pectins, and starch. The secondary valence gels are thermoreversible; on warming they

Figure 3.5 The gel state. Inter- and intramolecular secondary valence bonds can form to stabilize the conformations. Only the intermolecular bonds are shown.

become solutions. Poor solvents are used to form secondary valence gels so as not to saturate all secondary bonding sites with solvent. The solution equilibrium of the system should be temperature dependent. The mixture should behave like a viscous solution on the mill or mixing equipment and should be solid at lower temperatures, but not too solid so as to be brittle. The plasticizers should be colorless, odorless, nontoxic and have a low vapor pressure, low diffusivity, and high solubility.

Mechanical Properties of the Gel State. Gels have solid-like mechanical properties. They do not flow, but react elastically to deforming forces. In normal solids such as metals, applied forces strain the bonds connecting individual atoms, changing their potential energy and the enthalpy of the system. Gel deformation occurs with little change in the distance between atoms or in valence angles. This implies that there is no change in potential energy of the atoms. Instead, chain segments align in a parallel fashion (Fig. 3.6). The coils are forced to a less probable state and there is a decrease in entropy. Rubbers and elastomers have entropic elasticity also.

The amount of micro-Brownian motion is influenced by the degree of cross-linking more than the chain structure. In general, the character of the solvent-free polymer is maintained in the gel state. Brittle or glassy polymers form brittle gels on cross-linking. Strong mechanical gels, such as the secondary valence gels, have applications as rubbery or leathery materials.

Cross-linking thermoreversible gels causes them not to redissolve upon heating. The position of the cross-links on the macromolecules will affect their response to deforming forces, and therefore its elasticity. Inter- and intramolecular cross-links are not equivalent in terms of their effect on the modulus of elasticity.

3.3 SUPERCRITICAL SOLUTIONS AND GELS

As solvent is removed past the critical point in solutions or gels, two events can occur: (1) the individual coils become denser, or (2) the coils interpenetrate. Polymer melts can be considered supercritical gels: Molten polymer flows but individual chain segments are very close together and there is little free volume in the liquid. Current concepts of polymer rheology suggest that considerable intermeshing of polymer chains occur. The intermeshing can lead to *entanglement* of polymer chains: Entan-

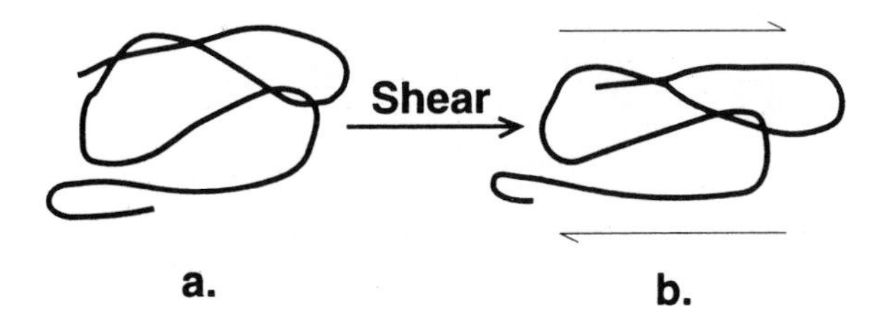

Figure 3.6 Deformation of the gel state: **(a)** the random coil of a polymer gel at rest; **(b)** shearing deforms only a few primary bonds, and long chain segments move in parallel with each other.

glements represent locations at which polymer chains cannot be physically separated. Entanglements may be caused by one polymer chain looping around another or by strong, high density secondary valence forces caused by lack of solvent. Figure 3.7 shows a sketch of several polymer chains with physical entanglements. Over long experimental times, the loops might become disentangled while continuous stress was placed on the macromolecules. However, over short or moderate experimental times (minutes or less), the physical entanglements act as chemical cross-links, keeping the chains bound together.

Entanglements between polymer chains should be statistical and depend on their molecular weights. The melt viscosity of homopolymers versus molecular weight often shows a sharp break in the curve as molecular weight is increased. Figure 3.8 shows the log of melt viscosity versus degree of polymerization plotted for nine polymers. The curves have been separated and plotted on a relative basis to demonstrate the effect of entanglements. Below the critical molecular weight, the melt viscosity increases linearly with increasing molecular weight and the slope of the curves are one. Above the critical molecular weight, the melt viscosity scales with molecular weight to the 3.4 power. This power dependence has been modeled using entanglement theory.

Free Energy of Mixing of Polymer Blends: Incompatibility. It usually is difficult to achieve uniform blends of two different polymers (or sometimes two different samples of the same polymer with different molecular weights) on the molecular scale. This problem is easy to understand when the free energy of mixing for the two polymers is considered. Entropy provides a significant amount of the free energy of mixing change for solvent-polymer pairs. However, the entropy difference between the two-phase state of a polymer pair and a uniform mix at a molecular level of the pair is orders of magnitude smaller than it is for a solvent-polymer pair. The

Figure 3.7 Entanglements in supercritical gels. Entanglements are indicated by dashed circles, showing chain segments looped over neighboring segments.

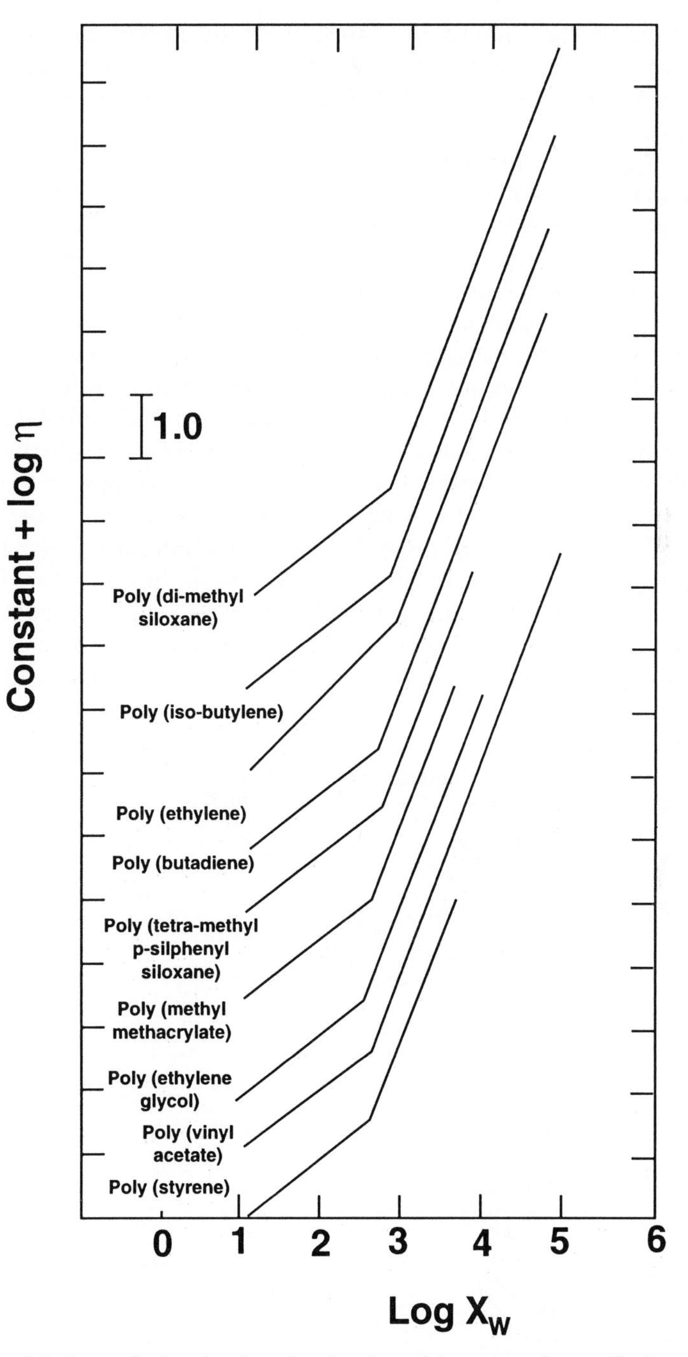

Figure 3.8 Log melt viscosity plotted against the weight-average degree of polymerization for nine different polymers. The lines have slope 1.0 below the critical molecular weight and 3.4 above the critical molecular weight (Berry and Fox, 1968).

entropy increase on mixing is a function of the number of particles. Since the degree of polymerization of commercial polymers is 2000 to 5000, the number of molecules being mixed is lower by three to four orders of magnitude compared to solvent-polymer mixtures.

Equation 3.1 shows that complete compatibility occurs only if the entropic term is greater than the enthalpic term. ΔH is usually positive and ΔS is very small for polymer mixtures because the number of molecules being mixed is small. Therefore, molecular solutions of polymers occur only when ΔH is negative, that is the attractive forces between the polymers are high.

The usual result of mixing two polymers is discontinuous clumps of one polymer dispersed in a continuous phase of the other. The morphology of the dispersed phase depends on the intensity of mixing, its volume fraction, and the interactions of the two polymers. Incompatible polymers may have poor mechanical integrity if there is not good bonding at the polymer-polymer interface to help distribute stresses through both phases. If the thermodynamics do not favor mixing, then the interpenetration of coils from two different polymers at the interface rarely will occur.

Figure 3.9a and b shows sketches of the interface regions of polymer blends. Figure 3.9a shows a compatible blend, with an interfacial region (or interphase). Because the polymers are compatible, there is significant mixing of chains in this region. The intermixing of chains can help transfer mechanical stresses between the phases. Figure 3.9b shows the interface for an incompatible blend. This thinner region

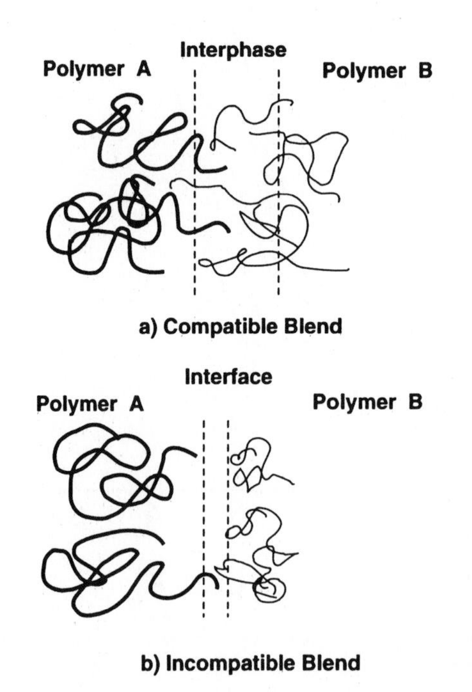

Figure 3.9 Sketch of polymer blends. **(a)** The compatible blend has a large interphase volume in which the two different polymer chains mix. **(b)** The incompatible blend has a thin interference with few chains extended into the opposite phase.

shows fewer interactions between the chains of the two different polymers, and the blend should have lower tensile strength than the compatible blend because the force is carried primarily by the continuous phase.

3.4 THE RUBBER-ELASTIC STATE

There is not a precise temperature at which a solid polymer softens and becomes a liquid. Both micro-Brownian (chain segment) and macro-Brownian (entire coil) motions can occur in the solid. In the glassy state, both types of motion are "frozen in." Near the melting temperature, each macromolecule is linked to its neighbors by a few secondary valence forces. Rapid deformations will not cause movements of the coils relative to each other but will only cause deformations of small sections of the coils. The associated change in coil shape causes a decrease in entropy. This behavior—large strains under small deforming forces and elastic recovery—was first observed for rubbers. At room temperature, rubbers are well above their glass transition and melting temperatures.

Rubber elasticity is not a special behavior. All macromolecules exhibit this property above their softening temperature (T_g). The elasticity of polymers above their softening temperature is not perfect. Deformations for long times allow molecular repositioning to occur. This process is called relaxation. Rubber-elastic melts are viscous liquids with long relaxation times. Thermoplastics are those polymers whose softening temperature is above room temperature. Elastomers have softening points below room temperature and usually are lightly cross-linked in the final product form. The low cross-linking allows deformation, but not viscous flow. Polymers made from monomers with conjugated double bonds have one double bond per repeat unit that can be used as a cross-linking site.

Even polymers that are normally considered solids exhibit viscoelastic properties in certain regions of temperature, pressure, and concentration. Cold flow is the plastic deformation of polymers at low temperatures. Injection molding and other thermoforming processes orient thermoplastics, and reheating often allows relaxation. Internal stresses in the polymer results in lower tensile strength. The higher the molecular weight, the broader the rubber-elastic region is. Polymers under 10,000 molecular weight are usually viscous liquids.

The concept for deformation of an elastomer is similar to that of a gel (Fig. 3.6). The polymer chains deform and elongate under stress. At low stresses, there is no elongation, which implies that a certain minimum amount of energy must be expended before the coils will move. The weakest cross-links deform first. Since the coils move slightly past each other as deformation continues, orientation can occur in the stretched state. This increase in orientation increases the tensile strength of the sample. It also has the unusual property of increasing the strength of the elastomer at higher temperatures. If a rubber band stretched under constant load is heated, its length *decreases*. Since the material stores energy as entropy rather than enthalpy, elevations in temperature increase the free energy of the process and the strength of the sample.

The effects of deformation on orientation depend on the cross-linking density of the material. Materials with high cross-linking density may have less orientation due to the shorter distance each chain can move. Materials that are lightly cross-linked will show higher strains for a given applied force.

3.5 THE SOLID STATE

Crystalline polymers are solids below their melting points, while amorphous polymers are solid below their glass transition temperatures. Low density polyethylene is a solid below its melting point of 130°C, even though its glass transition temperature is −25°C. Polystyrene is a solid below its glass transition temperature of 102°C. Crystallization of organic and inorganic molecules is usually easy to accomplish, results in solids of high order, and is often used as a method of purification. Crystals of low molecular weight materials have high strength, but can be brittle and have low strains at their yield points. In contrast, polymeric solids can have good tensile strength with larger strains.

The glassy state of amorphous polymers marks the transition point between solid and liquid properties. Crystalline polymers may exhibit both a glass transition temperature and a crystalline melting temperature.

3.5.1 The Glassy State

If an atactic polymer sample is cooled below the melting point (T_m) of the isotactic or syndiotactic material, no abrupt change in physical properties normally associated with crystallization is observed. At a lower temperature, the melt solidifies to a glass. This temperature is called the glass transition temperature, T_g. The ratio of T_g to T_m on absolute scales is about 0.6 for many polymers.

The motions of the macromolecule gradually "freeze" without giving rise to any change in structure. When the solid state forms, not all the free volume associated with the liquid state is lost, since the chain segments might not be able to respond under the experimental time frame to fill the voids. Therefore, the form of the solid state depends on the cooling rate. The glassy state often contains several volume percent of free volume, which is important to properties such as the permeation of gases through glassy solids and polymeric membranes.

The glassy state is not detected for some polymers that decompose before they melt (proteins and polysaccharides are examples). T_g can be determined by detecting property changes, such as changes in the specific volume, as a function of temperature. The temperature at which the property exhibits a change in temperature dependence is T_g. Refractive index, specific heat, shear modulus, thermal conductivity, and dielectric constants change at T_g. Solvents, plasticizers, molecular weight distribution, and most other additives can affect the glass transition temperature of the polymer sample.

3.5.2 The Crystalline State

The crystal structures of single molecules have been discussed in Chapter 2. This section covers the structures of crystals in bulk states. In order to crystallize, a polymer chain must have a regular composition and be sterically regular. The exceptions to this rule are atactic polymers that can fit into crystal lattice conformations of their related polymers. Poly(vinyl alcohol) is an example, as it fits into the crystal structure of polyethylene. The chains of some crystallizable polymers may be so hindered in mobility near the melt temperature that further cooling at normal rates causes glassy solidification. Polymer samples often exhibit a melting point range. This can be caused by crystallites of different size and by crystallites of different purity.

In polymeric solids, there seem to be no pure crystals, and there are always some noncrystalline domains. The chain ends contribute two defects in the crystal structure per linear molecule. Therefore, most crystalline polymers have no more than 90% crystallinity, and even this amount of orientation may require careful processing of the sample.

Crystallization from Solution

Lamellar Structures. When stereoregular polymers in dilute solutions (< 0.1 weight per volume) are supercooled, thin lamella of 5 to 20 nm in thickness can form. These lamella can have many shapes and sizes depending on the polymer, the solvent, the polymer concentration, and the temperature. Lamellae shapes include diamond, pyramidal, rectangular, hexagonal, and square. Lamellar morphology depends on the crystal lattice and the mechanisms for crystal growth. One example was shown in Chapter 2. Branching, twinning, and interlocking growth also can occur. In these dilute solutions, crystal formation can be observed using optical microscopy. Crystallization can be affected by shear and pressure.

The thickness of a lamella is very sensitive to crystallization temperature. Conditions just below the supercritical temperature lead to the thickest lamellae. The chain folding model is consistent with single lamella formation. Polymer chain segments align with each other perpendicular to the thickness of the lamella. There are three morphologies: tight adjacent, loose adjacent, and nonadjacent (Figure 3.10a). At the surface, one of three events can occur. The chain can fold tightly back into the crystal structure (Fig. 3.10b), the chain can fold loosely back into the chain structure, or the chain end can associate with the solvent, with the neighboring segment being provided by a different chain.

Multilamella Structures. Few polymers are used as single crystals, but many are used as bulk materials. Crystallization from solutions at concentrations greater than 0.1% weight per volume leads to multilamella structures. Stacks or sheaves of

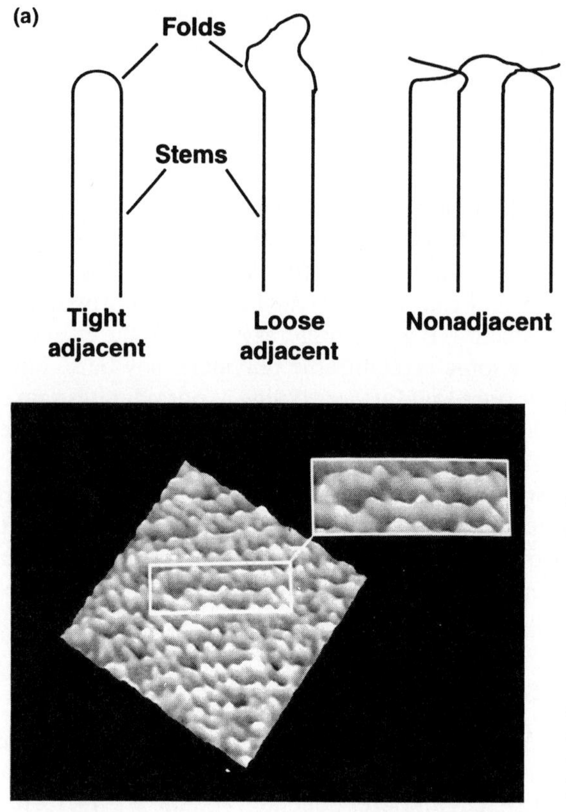

Figure 3-10 Chain folding. **(a)** Three possible structures for the lamella boundry of a chain folding polymer crystal: tight adjacent folds immediately return the chain into the crystal lattice, loose adjacent folds leave a number of chain segments outside the adjacent lattice positions (Woodward, 1989). **(b)** Scanning force micrograph (Topometrix TMX 2000) of a chain fold in polyethylene. Published with the permission of F. F. Lin and D.J. Meier, Michigan Molecular Institute.

lamellae are common. Figure 3.11a shows a sketch of a typical multilamella growth and shows the orientation of the folded chain to one of the arms. Some of the layers may be curved perpendicular to the crystalline chain axis. New lamella can be initiated from the surfaces or ends of existing lamella. The crystal morphology is a combination of nucleations on chains protruding from lamellae surfaces, screw dislocation growth, and epitaxial growth. This structure is thought to be the precursor of spherulitic growth, which is observed in melt crystallization. Multilamella growth has been observed in a number of polymer systems (Woodward, 1989), including balata rubber and poly(vinylidene chloride) (Fig. 3.11b).

Crystallization from the Melt. Crystallization of polymers from the melt results in multilamellae structures. The most common structure is the spherulite. The spherulite is a microscopic "sphere" of crystalline polymer with distinct boundaries, similar to the grain of a metal. Melt crystallization under high temperature and pressure can lead to extended chain structures.

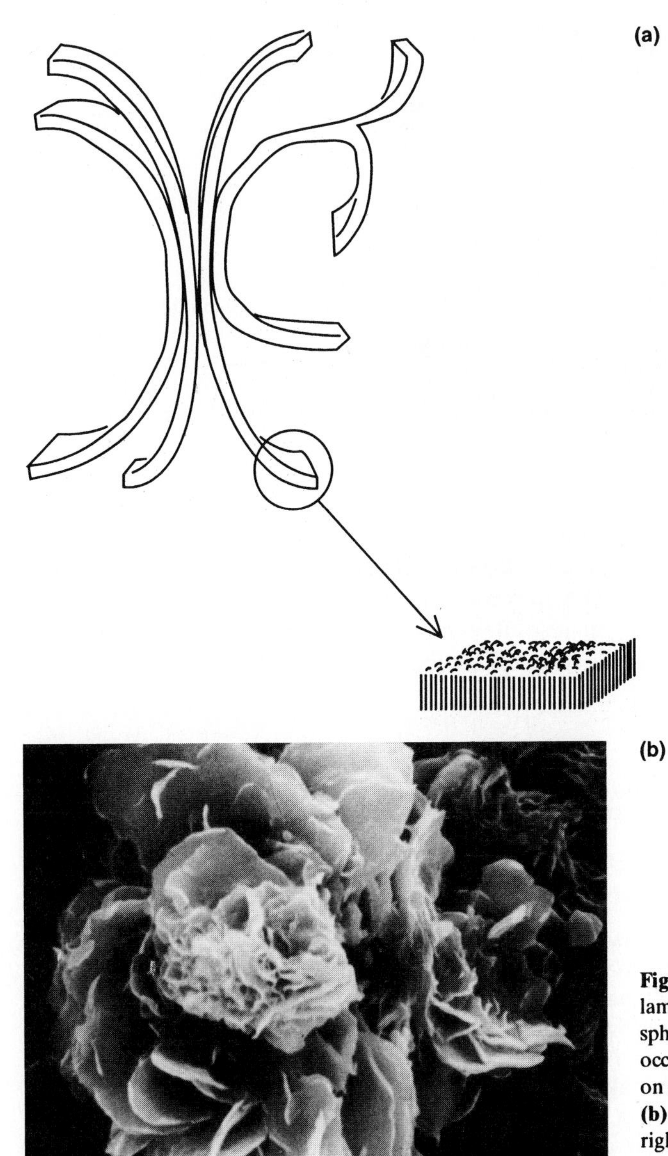

Figure 3.11 Polymer chain orientation in lamellar growth. **(a)** Ribbons of a pre-spherulitic structure. Crystal growth can occur by branching, spurring, or layering on existing surfaces (Woodward, 1989). **(b)** Spherulitic of saran copolymer. Copyright 1992, Dow Chemical. Used with permission.

Figure 3.12 shows spherulitic growth in a polystyrene sample. Crystal formation is initiated at several sites in the melt. The spherulites grow in the radial direction, with branching and spawning occurring. When the surface of a spherulite touches another, its growth stops. Amorphous material will be trapped within the spherulite and at its boundaries. These trapped molecules can crystallize in a secondary process. However, the amorphous regions left after primary growth often contain defects of

Figure 3.12 Spherulite structure of polystyrene. Scanning force micrograph (Topometrix TMX 2000). Published with permission of F. F. Lin and D.J. Meier, Michigan Molecular Institute.

branching in the polymer chain, chain ends, cross-links, or atactic segments. These defect materials may not crystallize. The size of the spherulites depends on crystallization conditions. The rate of crystallization depends on the rate of nucleation and the rate of crystal growth. The maximum rate of crystallization often occurs at supercooling of about 0.2 T_m in absolute temperature. The temperature of maximum crystallization rate, T_1, can be estimated by the rule

$$T_1 = 0.8 * T_m \qquad 3.2$$

EXAMPLE 3.1 EFFECTS OF DEFECTS ON MELTING TEMPERATURE

Branching is a well-known defect in low density polyethylene (LDPE). It makes a big difference in the melting characteristics of LDPE relative to linear polyethylene. The melting temperature of LDPE is 115°C, while that of linear polyethylene is 138°C. The relative volume of these polymers can be used to compare their melting properties. Figure 3.13 shows the change in relative volume versus temperature for each material. Linear polyethylene, Curve **A**, has a sharp melting point (the temperature at which the derivative of the curve is discontinuous) with a rapid drop in relative volume as temperature is decreased from the melting point. LDPE, Curve **B**, has a lower melting temperature. The large contraction of the linear material near the melting point is due to the higher density of the crystalline material. Most of the change in density occurs over a narrow temperature range for the linear polyethylene compared to the branched sample. The total change in volume on melting is higher for linear polyethylene due to its higher percent crystallinity. Because of its higher melting point and narrower melting range, linear polyethylene has better mechanical properties than LDPE above 100°C and is classified as an engineering thermoplastic.

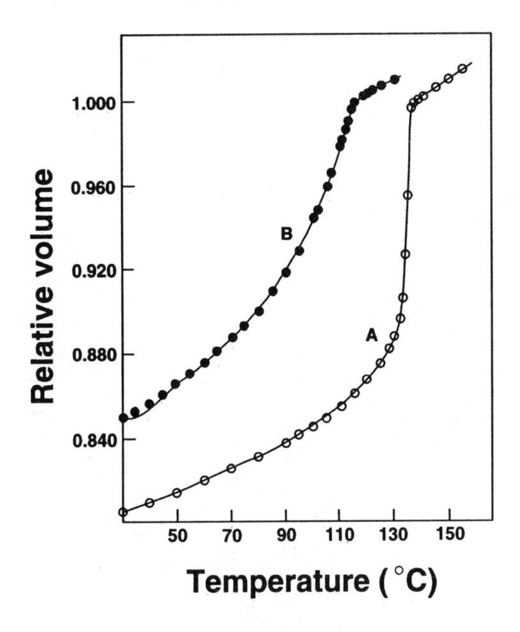

Figure 3.13 Relative volume versus temperature. Curve A is for linear polyethylene. Curve B is for branched polyethylene. The sample relative volume is 1.00 at the melting temperature. Permission granted by the publisher of *Calendaring of Plastics* by R.A. Elden and A.D. Swan. Copyright 1971 by Elsevier Science Publishing Co., Inc.

An old model of crystallinity is the fringed micelle model (Fig. 3.14). This picture does not correlate with the coil character of polymer molecules and does not apply to polymers that crystallize from the melt. Fibrils of cellulosic fibers and other natural polymers fit this model. Protein fibers such as collagen tend to align parallel to the fiber axis. Since the fibers are enzymatically catalyzed in place, the monomers were probably added at a surface of the protein. Fibrils or larger units may consist of bundles of oriented polymer molecules.

The thermodynamics of polymer crystallization can be analyzed by applying Equation 3.1. The free energy for the melting process is given by

$$\Delta G_m = \Delta H_m - T\Delta S_m \qquad 3.3$$

where the subscript m refers to the change in the state property for the melting process. At the melting temperature, the amorphous and crystalline phases are in equilibrium and the free energy of the melting process is zero. This means that the enthalpy of mixing and the entropy of mixing contributions to the free energy are equivalent in magnitude. Since $T = T_m$, Equation 3.3 can be solved to relate the melting temperature to the ratio of the enthalpy and entropy of melting

$$T_m = \frac{\Delta H_m}{\Delta S_m} \qquad 3.4$$

The enthalpy change on melting can be interpreted as the energy needed to overcome crystalline bonding forces. The entropy change on melting should be greater for short

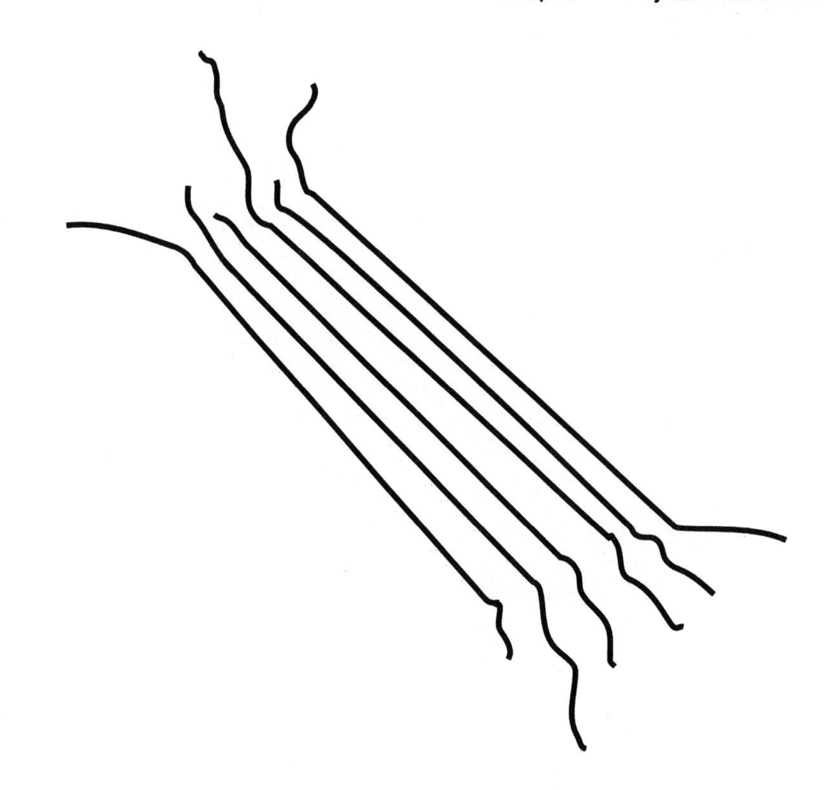

Figure 3.14 Fringed micelle model for crystalline cellulose.

chains and less for long chains. The application of Equation 3.4 to polymer systems suggests that melting temperatures should increase with increasing molecular weight, which is the experimental observation. Melting temperature does not increase linearly with molecular weight, but approaches a limiting value for samples of infinite molecular weight. Polymers with high enthalpy changes on melting (with strong crystalline structure) have high melting temperatures.

Equation 3.4 is not applicable theoretically to the glass transition temperature of polymers. T_g is not a distinct thermodynamic event, as the melting point is. However, the glass transition temperature also increases with chain length and approaches a limiting value for infinite chain length.

The rate of crystallization or melting depends on a number of factors, including the temperature, pressure, and other external forces such as shear and extension. Crystallization rates can be increased in such gradients. Films and fibers are often stretched during cooling to orient the crystal segments. The crystal structure of unstretched fibers is perpendicular to the axis. Stretching orients the crystals parallel to the axis. Some of these effects will be discussed in the sections on polymer processing.

EXAMPLE 3.2 COMPARISON OF POLYMER MELTING TEMPERATURES

The structures of three different polymers that contain the amide bond are shown below.

$[\text{-NH-CO-O-R}_1\text{-O-CO-NH-R}_1\text{-}]_n$	$[\text{-NH-CO-R}_1\text{-}]_n$	$[\text{-NH-CO-NH-R}_1\text{-}]_n$
POLYURETHANE	POLYAMIDE	POLYUREA

These polymers are listed in order of increasing melting temperature. The enthalpies and entropies of melting of these polymers can be evaluated using the repeating unit structures. Both the polyurethane and the polyamide contain one amide bond per segment of R_1 methylene (-CH2-) groups. The oxygen atom in the polymer backbone should not hydrogen bond to any great extent since its valence is completely satisfied along the polymer chain axis. Therefore, the enthalpy of mixing of both of these molecules should be about the same. However, the polyurethane can rotate about the oxygen atom as well as the carboxyl groups, while the polyamide molecule can undergo hindered rotation about the carboxyl groups. The entropy of mixing of the polyurethane should be higher than that of the polyamide per chain segment, so Equation 3.4 suggests that polyurethane would have the lower melting temperature.

The rotation about the chain backbone for the polyamide and the polyurea should be similar, so that their entropies of mixing should be similar. However, the presence of another nitrogen atom in the repeating segment with an unbonded pair can lead to additional hydrogen bonding. Therefore, the heat of mixing of the polyurea should be higher than that of the polyamide, and its melting temperature should also be higher.

3.6 NATURAL POLYMERS

There are a number of naturally occurring polymers that have important properties and applications. There are three major classifications of the natural polymers: proteins, polysaccharides, and polynucleotides. Polynucleotides (DNA for example) will not be discussed. A number of natural polymers are important commercially as fibers or as food components. There are several biological polymers that are important commercially, but do not fit into a single chemical classification.

Globular proteins exist as single macromolecules, but many of the natural polymers useful to man are found in organized structures in the cells of plants and animals. These organized structures establish the physical properties of the organism and can be thought of as naturally occurring composites. For these reasons, the natural polymers are discussed in the chapter covering assemblies of macromolecules, rather than in Chapter 2 with the synthetic polymers and the properties of single macromolecules.

3.6.1 Proteins

Protein architecture is divided into primary, secondary, tertiary, and quaternary structures. Primary structure is equivalent to constitution and configuration. Secondary structure is generated by hydrogen bonding and other secondary valence forces between chain segments (conformation). Tertiary structure is the interaction of residue side chains. Intermolecular interactions leading to aggregates of molecules is defined as quaternary structure.

There are two major groups of proteins: globular and fibrous. The globular proteins (enzymes, hormones, hemoglobin, casein, and albumin for example) often take the shape of coils in solution. However, their shapes (helical, pleated sheets) are not random, but are controlled by their amino acid sequences. Factors affecting protein structure include crystalline chain segments, nonpolar interactions, polar interactions, hydrogen bonding, intramolecular cross-linking, and intermolecular cross-linking.

Most of the polar groups of globular proteins are directed toward the outer "surface" of the molecule, where hydration is readily accomplished. Hydrophobic groups usually are oriented toward the interior. These orientations make globular proteins soluble in water and assist in body functions that depend on the mobility of the molecules.

Fibrous proteins have some commercial applications. Polypeptide chains are held together by hydrogen bonds, leading to molecules with high aspect ratios (L/D). They are insoluble and can act as structural members in biological systems. Keratin, collagen, fibroin, and myosin are examples of fibrous proteins. Silk is a fibrous protein secreted by some insects. There are two protein components of silk: silk fibroin, which is in the form of crystalline fibers, and sericin, a water-soluble matrix material. Fibroin is high in nonpolar amino acids (alanine and glycine), and sericin has high levels of polar amino acids (serine and spartic acid). Fibroin usually is found in antiparallel beta sheets. The repeating sequence is Gly-Ser-Gly-Ala-Gly-Ala and packs as shown in Figure 3.15. The sheets pack together so that glycine groups face glycine groups and alanine faces alanine or serine. The hydrogen bonds between the chains give silk high strength and low elasticity.

Collagen is a protein found in connective tissue and bones, and it constitutes about 6 wt% of human body weight. The basic unit has a molecular weight of 300,000. About one-third of the residues are glycine and another third are proline or hydroxyproline. Typical repeating units are Gly-X-Pro or Gly-X-Hyp. It lacks sulfur-containing amino acids. Collagen is soluble in low pH (3–4) buffers and can be reprecipi-

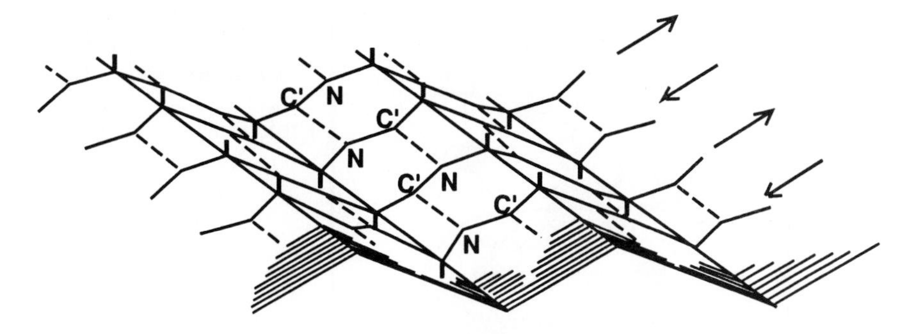

Figure 3.15 Pleated sheet structure of silk fibroin (Schulz and Schirmer, 1979).

tated using caustic. Boiling collagen solutions for long times converts the polymer to soluble proteins (gelatins).

Elastin is present in tissues where elastic function is necessary (aorta, blood vessels, lungs, and ligaments). There appear to be two components: a fibrous unit and an amorphous material. Nonpolar residues of glycine, alanine, and valine make up over 50% of the protein. In its native form, elastin is cross-linked and insoluble.

Keratins are found in hair, wool, and feathers. There are two different types: α-keratin and β-keratin. Proline and cysteine, polar amino acids, are present at high levels. They tend to be cross-linked by S-S bonds. Figure 3.16 shows three levels of organization for keratin fibers. α-Helix polypeptides associate into a three-strand rope. Assemblies of eleven ropes associate in a protofibril that is embedded by matrix material in a macrofibril. The matrix material is high in sulfur and can cross-link to modify the shape of the assembly. Macrofibrils are surrounded by a cortex to form a wool fiber. The α-form can be converted into the β-form by stretching in steam or alkali solution. The β-form microfibril contains a pair of pleated sheets twisted into a left-hand superhelix.

3.6.2 Polysaccharides

A number of commercial polymer products are made from polysaccharides and their derivatives. There are three classifications of polysaccharides: structural polymers, storage polymers, and gel-forming polymers. Table 3.1 shows some of the polymers in each classification. The extended chains of the structural polysaccharides form long fibrils, which act as supporting members in the organism. The storage polymers are branched and have compact structures. They can be converted to monosaccharides to provide nutrition. The gel-forming polymers, such as agar and carrageenan, form stable gels in water solution.

Glucose is the primary monomer unit for the structural and storage polysaccharides. The β-form, β-D-glucose, is the repeating unit for cellulose, and the α-form,

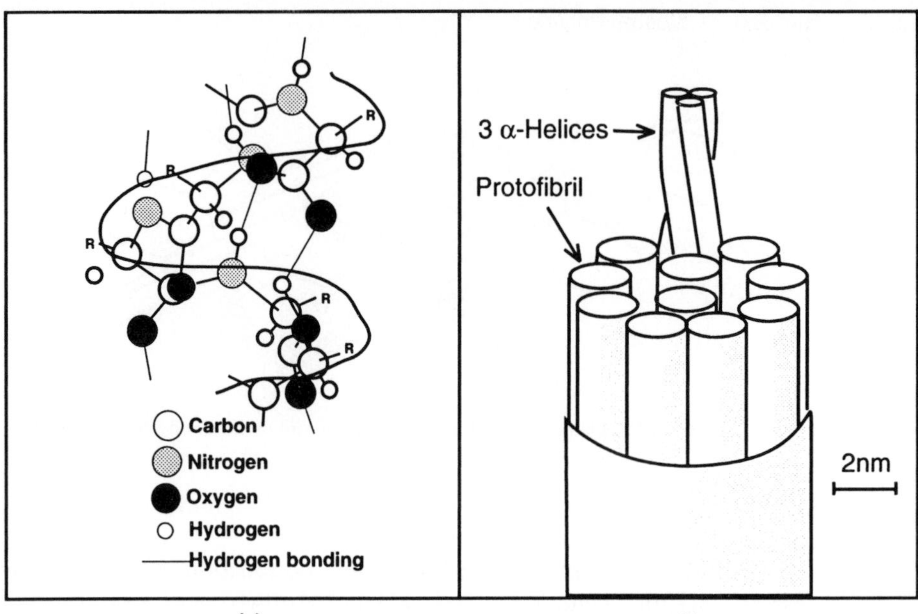

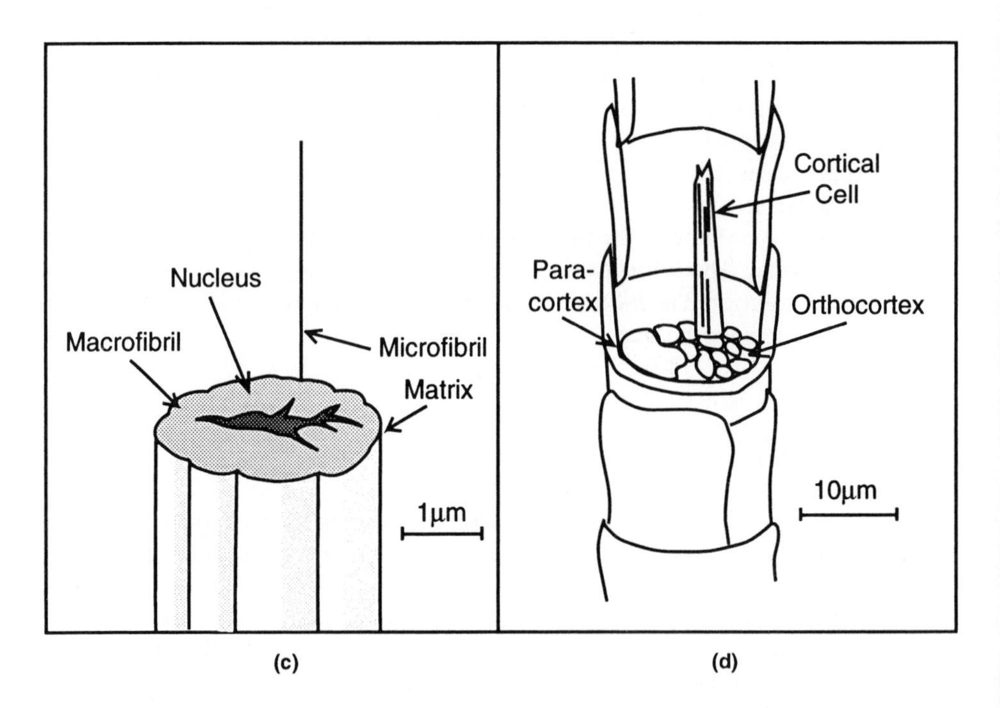

Figure 3.16 Sketch of keratin fiber morphology: **(a)** chemical structure of the α-helix; **(b)** association of helics into protofibrils; **(c)** association of microfibrils into macrofibrils; **(d)** a keratin fiber (wool). L. Rebenfeld, "Fibers," in Encyclopedia of Polymer Science and Engineering, Vol. 6, 685(1986).

TABLE 3.1 Types of Polysaccharides

Category	Polymers	Monomers and Linkage
Structural	Cellulose	β-(1,4)-D-glucose
	Xylan	β-(1,4)-xylobiose
	Chitin	β-(1,4)-N-acetyl-D-glucosamine
Storage	Amylopectin	α-(1,4)-D-glucose with α-(1,6) branching
	Amylose	α-(1,4)-D-glucose
	Glycogen	α-(1,4)-D-glucose
Gel-forming	Agar	$[A\text{-}(1,3)\text{-}B\text{-}(1,4)]_n$ sequence
	Mucopolysaccharides	$[A\text{-}(1,3)\text{-}B\text{-}(1,4)]_n$ sequence

α-D-glucose, is the repeating unit for amylose. Figures 3.17 and 3.18 show the monomer and repeating sequences for these two polymers.

Structural Polysaccharides

Cellulose. Cellulose is the most abundant natural polymer and is synthesized almost exclusively by plants. About 10^{11} tons are grown every year, which accounts for about 50% of the carbon in plants. It forms the skeletal material of plant cell walls.

The primary structural unit is D-glucose. Cellulose is a linear array of D-glucose units linked by β-(1,4)-glycoside bonds (Fig. 3.18). Molecular weights of natural cellulose are high, ranging from at least 50,000 to over one million. Exact values are difficult to obtain since there is some degradation in the processes to recover the polymer from natural sources. The polymer chains form microfibrils that are several hundred angstroms long.

There are at least five polymorphic forms of crystalline cellulose. The native polymer form is known as cellulose I. It can be transformed to cellulose II by treatment with concentrated caustic solution. The treatment, called mercerization after Mercer who developed the process in 1845, expands the unit cell of the crystal structure and increases the fiber strength, while improving its dyeing properties. Cellulose fibers can be recovered from wood and converted by a number of reactions to useful products. Table 3.2 lists several of the products and their uses. Variations in the native cellulose used to make these materials will change their performance.

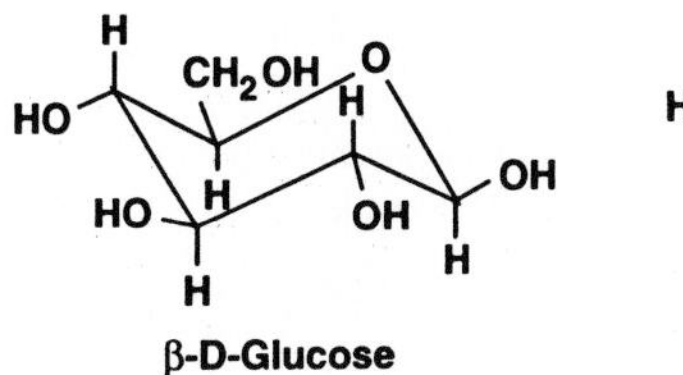

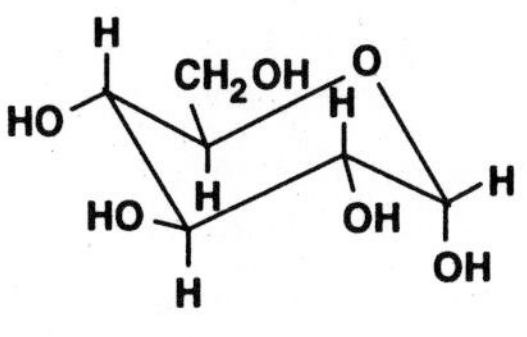

Figure 3.17 Monomers for cellulose (β-D-glucose) and amylose (α-D-glucose).

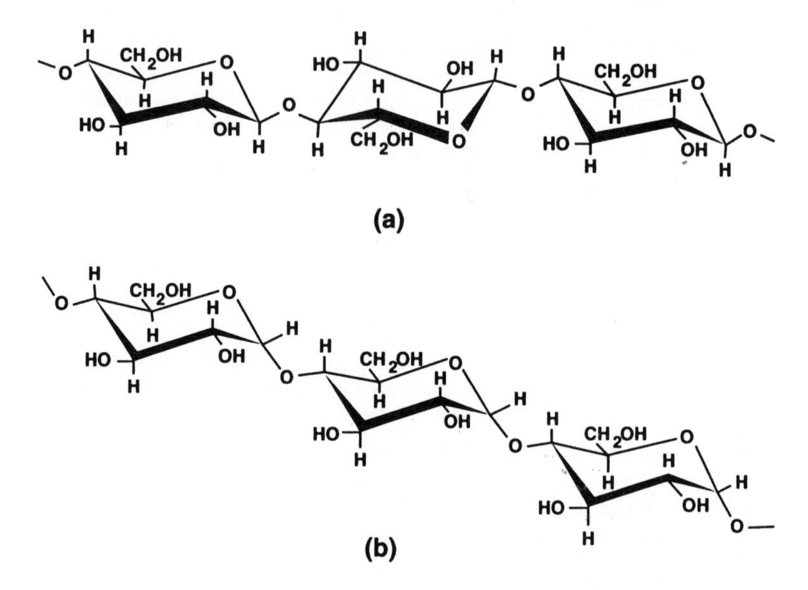

Figure 3.18 Cellulose **(a)** and amylose **(b)** polymers.

Xylan. Xylan polymers are known as hemicellulose. The xylans may be partially acetylated and have side chains of sugar rings. The side chains in β-(1,4)-xylan are L-arabinose and 4-o-methyl-D-glucuronic acid. Xylans do not crystallize without chemical modification and provide a matrix material highly associated with the cellulose fibers in wood.

Chitin. Chitin performs the function of cellulose in lower forms of plant life, such as fungi, and in the arthropods. It is the second most abundant natural polymer and may be produced in the range of 10^9 tons per year. Chitin is the homopolymer of β-(1,4)-N-acetyl-D-glucosamine, as show in Figure 3.19. The repeating unit is similar

TABLE 3.2 Cellulosic Derivatives

Material	Derivative Method	Uses
Cellulose nitrate	Nitration with nitric acid/sulfuric acid	Gun cotton, photographic film, lacquers
Cellulose acetate	Acetylation of cellulose, saponification to remove triacetates	Fibers
Cellulose methyl or ethyl ethers	Reaction of alkali cellulose with methyl, ethyl chloride	Textile finishing
Hydroxypropylcellulose	Reaction of alkali cellulose with propylene oxide	Water-soluble packaging film, suspending agent
Carboxymethylcellulose	Reaction of alkali cellulose with chloroacetic acid	Detergent, food thickener

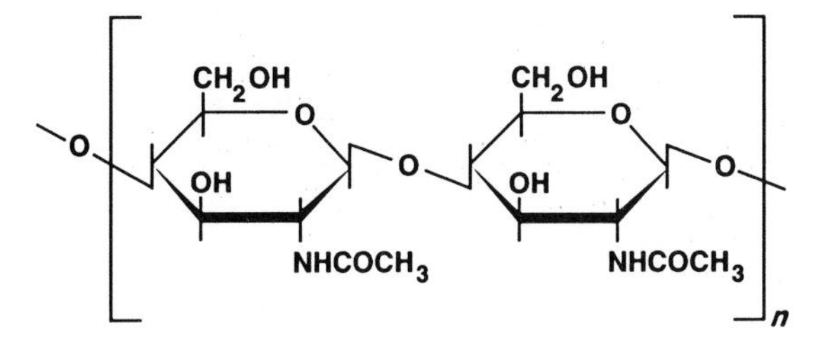

Figure 3.19 Repeating unit of chitin.

to that of cellulose, except that a hydroxyl group has been replaced by an acetamido group. Chitins are highly linked to proteins and inorganic salts in nature and are difficult to separate in pure form. There are a variety of possible morphologies, including cylindrical chitin fibers embedded in a protein matrix and a layered structure with chitin layers separated by protein layers.

Storage Polysaccharides. Table 3.3 lists some of the poly(α-glucoses).

Starch. Starch occurs in plants as microgranules at various sites in the cell. It appears in tubers (potatoes) and stems, as well as in seeds (corn, wheat, barley, rye, and rice). Since it can be converted readily to D-glucose, it provides the major carbohydrate reserve for plants. Amylose and amylopectin polymers are both found in starch.

Amylose is the major component and consists of poly-α-(1,4)-D-glucose (Fig. 3.18). Amylopectin contains the same homopolymer backbone chain, but has various amounts of α-(1,6) branching units (Fig. 3.20). The polysaccharides in starch do not

TABLE 3.3 Classification of Poly(α-glucoses)

Type	Example	Repeating Structure	Conformation
Poly(α-glucoses)	Amylose	α-(1,4)-D-glucose	Helix, linear, $\overline{X}_n = 6000$
	Amylopectin	α-(1,4)-D-glucose	Branched in C^6 position, branches occur every 18–27 glucose residues
	Glycogen	α-(1,4)-D-glucose	Branch points every 8–16 glucose residues
Decomposition products of poly(α-glucoses)	Dextrins	α-(1,4)-anhydroglucose	Cyclic oligomers with 6–12 glucose residues, decomposition of amylose, amylopectin and glycogen to cycloamyloses, molecular weights of 40,000.
	Pullulan	β-(1,5)-D-maltotriose	
	Dextran	α-(1,6)-D-glucose with α-(1,4) branch points	$\overline{M}_w = 200,000$

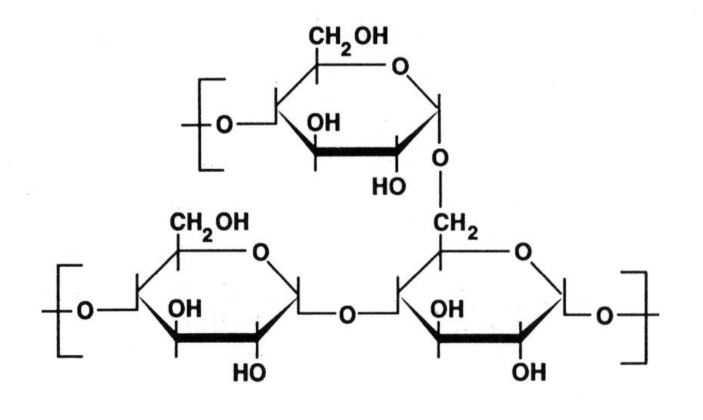

Figure 3.20 Repeating unit of amylopectin.

form fibrous structures. X-ray diffraction studies show crystalline patterns: Cereal starches give one pattern and tuber starches another. Amylose is linear with molecular weights up to one million. Amylopectin is branched with long side chains.

Figure 3.21 shows how amylose and amylopectin chains might be associated in a starch microgranule. There are lamellae with thicknesses between 100 and 200 Å. The chain axes would be oriented perpendicular to the lamellae. There is probably some chain folding in the natural material. The branched structure prevents large regions of crystalline polymer from forming. The lack of long-range crystallinity probably helps in the enzymatic cleavage of the chains to recover glucose. Other polysaccharides are listed in Table 3.4.

Gel-Forming Polysaccharides (A-B polysaccharides). Polygalactose copolymers with the general sequence $[A\text{-}(1,3)\text{-}B\text{-}(1,4)]_n$ form gels (Table 3.5). Mucopolysaccharides are in animal connective tissues, associated with collagen or elastin. They are thought to associate with these protein fibers through covalent bonding. The open, water-carrying structure of the mucopolysaccharides is thought to give flexibility to the protein fibers. As an example, keratin sulfate occurs in the cornea of the eye. It is thought to help with changes in the water content of the cornea, permitting changes in its refractive index. Agar and carrageenan are recovered commercially. Agar forms gels at 1wt% concentration in aqueous solution and is used as culture media. Carrageenans provide creamy texture in foods and are found in yogurt, ice cream, and other dairy products.

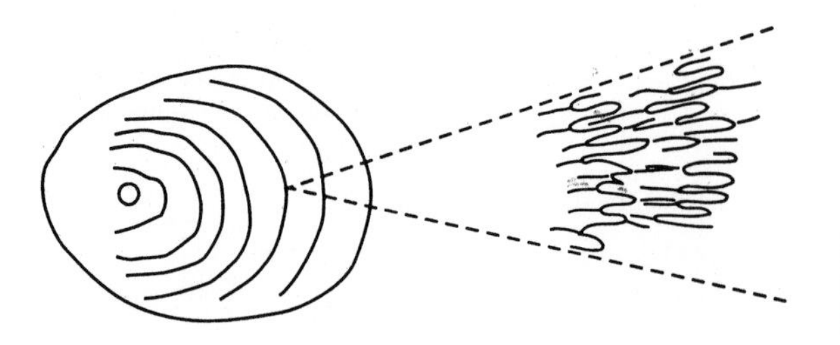

Figure 3.21 Sketch of starch granule layers and polysaccharide (Marchessault and Sarko, 1969).

TABLE 3.4 Other Polysaccharides

Type	Main Repeating Unit	Source	Uses
Guar	β-(1,4)-D-mannopyranosyl	Guava seeds	Thickener
Alginate	Copolymer of β-(1,4)-D-mannuronic acid and α-(1,4)-L-guluronic acid	Brown algae	Suspension and emulsification agent
Xylans	β-(1,4)-D-xylopyranose branched	Plants	Cell walls of plants
Xanthan	Random terpolymer of D-glucuronic acid, D-mannose and D-glucose	*Xanthomonas campestris* digestion of dextrose	Drag-reducing additive

3.6.3 Other Natural Polymers

There are a number of important natural polymers that do not fit into a single chemical classification. The chemical compositions of these materials can vary widely with their source and recovery techniques.

Asphaltenes. Figure 3.22 shows a hypothetical structure for asphaltenes. These resins are recovered as distillation bottoms from the first separation step for petroleum liquids. Their molecular weights range from several thousand to several hundred thousands, and they consist of polynuclear aromatic and cycloaliphatic ring combinations. Because of their high viscosities at temperatures up to 150°F and low water uptake, they are used for roofing, paving, and sealing applications.

Coal. Coals are the fossilized remains of vegetation. Their compositions vary widely, extending between those of peat ($C_{75}H_{140}O_{56}N_2S$) and anthracite coal ($C_{240}H_{90}O_4NS$). Coals with higher carbon to hydrogen ratios have aged longer. A hypothetical structure for coal is shown in Figure 3.23. Coal can be converted into intermediate compounds by pyrolysis, hydrogenation, extraction, or partial combustion (Fischer–Tropsch process). Because of the complex nature of coal, the difficulty

TABLE 3.5 Poly(galactoses) That Form Gels

Type	Main Repeating Unit	Source	Use
Gum arabic	(1,3)-D-galactopyranose	Acacia trees	Thickener for foods, pharmaceuticals, cosmetics
Agar-agar	Copolymer of β-D-galactopyranosyl and 3,6-anhydro-α-L-galactopyranosyl with (1,3) bonds	Sulfuric acid extract of red algae	Bacteria culture and jam thickener
Tragacanth	D-galacturonic acid	Legume exudate	Food thickener
Carrageenan	Copolymer of galactopyranose sulfates	Atlantic red algae	Thickener in foods, gelling agent
Pectins	α-(1,4)-D-galacturonic acid, carboxyl groups are esterified	Fruits, sugar beets	Thickening agent, gelling agent

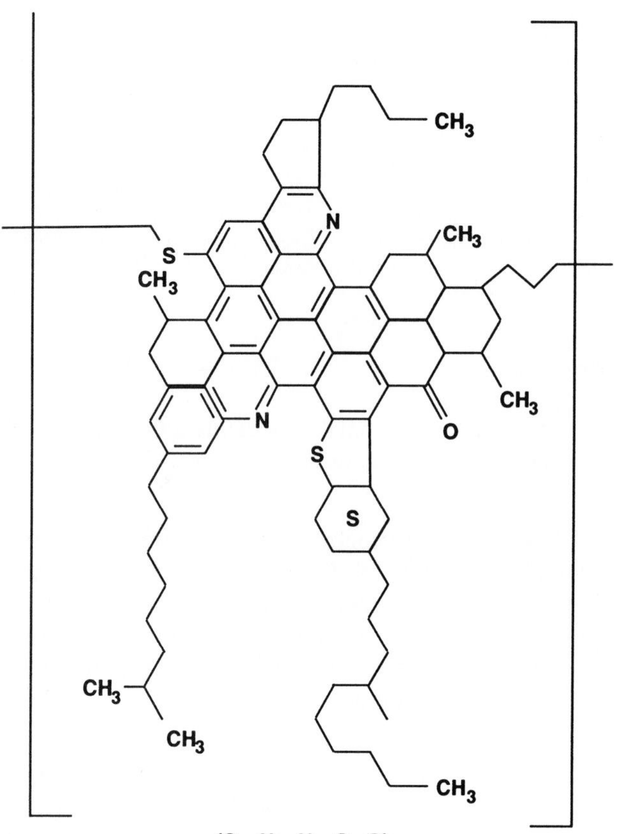

$(C_{79}H_{92}N_2S_2O)_3$

Figure 3.22 Hypothetical structure of a petroleum asphaltene (Speight and Moschopedis, 1981).

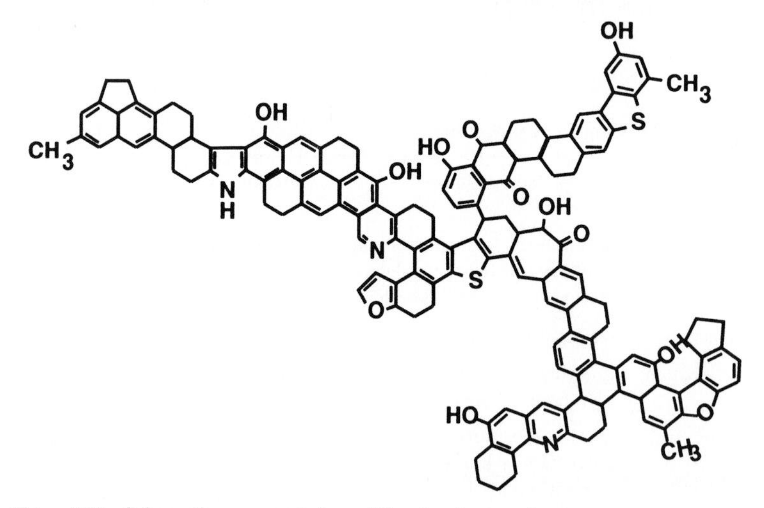

Figure 3.23 Schematic representation of the structure coal.

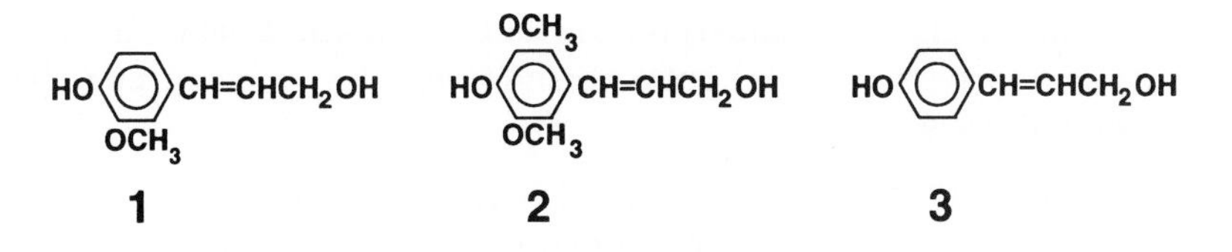

Figure 3.24 Three key monomers for lignin synthesis: 1) *trans*-coniferyl alcohol, 2) *trans*-sinapyl alcohol, 3) *trans-p*-coumaryl alcohol.

in converting it to intermediates, and the availability of other hydrocarbon sources, it is used primarily for fuel and not as a resource for materials.

Lignin. Lignin is the third major component of wood and serves as a "glue" to hold the cellulose and hemicellulose in their microstructures. It is about 25% to 30% of the total mass of wood, but there are few applications for this biopolymer. It is amorphous with a variable structure and composition. Lignin is extracted from wood chips during pulping operations and is burned to provide energy for the process.

Figure 3.24 shows three key monomers of lignin synthesis: *trans*-coniferyl, *trans*-sinapyl, and *trans*-p-coumaryl alcohols. These combine into a complex, cross-linked polymer (Fig. 3.25). Lignin has a number of phenol groups and has been used as a substitute for phenolic compounds, principally in adhesives for wood products. Some modifications have been tried for its conversion to engineering thermoplastics. One is its reaction with propylene oxide to form hydroxypropyl derivatives useful for polyurethanes.

Rubber. Rubber is the most important natural polymer. It is obtained by tapping the sap from *Hevea brasiliensis*. Several other plants produce it, including milkweed, guayule, and dandelion. The natural material is a terpene synthesized by the enzymatic polymerization of isopentenylpyrophosphate. The repeating unit of the chain is *cis*-(1,4)-isoprene.

Natural rubber is harvested from the tree as a latex containing about 35% rubber and 5% other compounds (fatty acids, sugars, proteins, sterols, esters). Rubber is harvested from the guayule shrub by pulping and refining. The natural product has a molecular weight of one million and is amorphous. It can be formed into products by coating, followed by vulcanization. The material used for bulk applications (tire

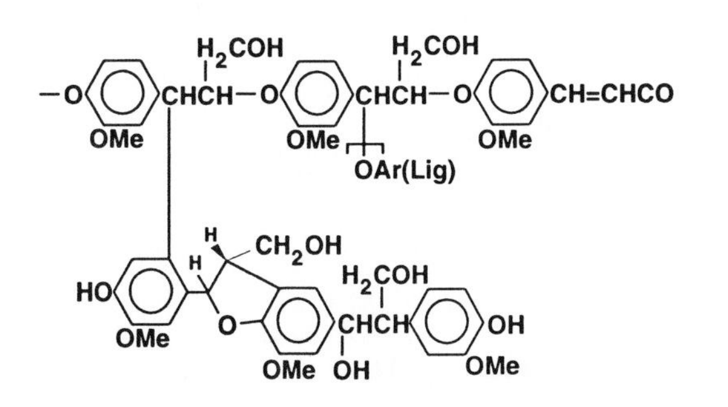

Figure 3.25 Typical structure for normal conifer wood lignin, based on modifications of the spruce lignin formulae by Freudenberg, 1964; 1965.

manufacturing for example) is coagulated with acetic acid. There are several other natural rubbers, including gutta-percha (*trans*-[1,4]-polyisoprene), balata, and chicle (used for chewing gum).

Tall Oil Derivatives. Tall oil is recovered from wood pulping processes. It contains two major components that can be used for polymer synthesis. Rosin is a mixture of 20-carbon fused-ring monocarboxylic acids (Fig. 3.26). These can be made into varnishes. Esterification of the acids with glycerol generates ester gums used in adhesives and lacquers. Rosin is a potential polymer feedstock because it can be converted to polyamideimides.

The tall oil fatty acids have 18 carbons and contain mostly oleic and linoleic acids. These can be recovered from the rosin by distillation. Linoleic acids can be used in the production of some nylons, vinyl plasticizers, and epoxy curing agents.

3.7 POLYMER BLENDS

Two or more polymers can be combined into a polymer blend by mixing processes or by certain copolymerizations. Copolymers can be considered as polymer blends if the two repeating units are segregated in the chain and if the repeating units phase separate from each other under use conditions. These systems act as dispersions of one polymer in the other, with chemical linking between dissimilar chains across the interface. If the copolymer has only one phase, it will act as a homopolymer. The easiest methods for preparing such materials is with graft or block copolymerizations.

Polymer blends can be made by physically mixing two or more polymers. Melt and solution blending methods can give two phase dispersions. Since most polymers are insoluble in each other, physical blends lack chemical cross-links binding the phases together. There are a number of techniques for improving interfacial adhesion, including wetting agents, block copolymers that will dissolve chain ends in each phase, and reactive alloying that will link chains in each phase by chemical reactions at specific end groups.

The styrene-butadiene system provides a classic example of polymer blend systems. Diblock copolymers and triblock copolymers can be prepared as shown in

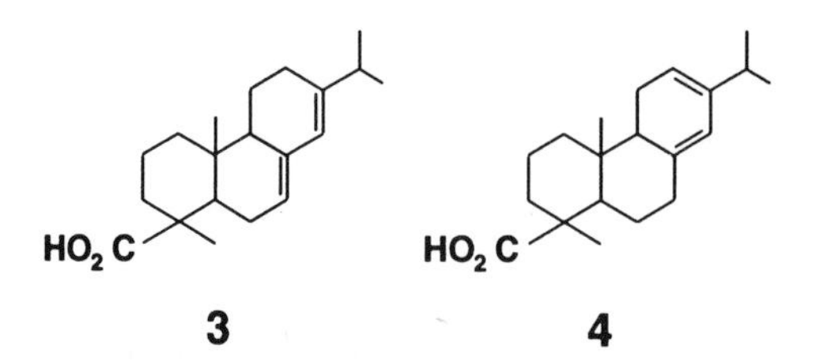

Figure 3.26 Tall oil acids.

Butadiene Block Styrene Block

a) Diblock Copolymer

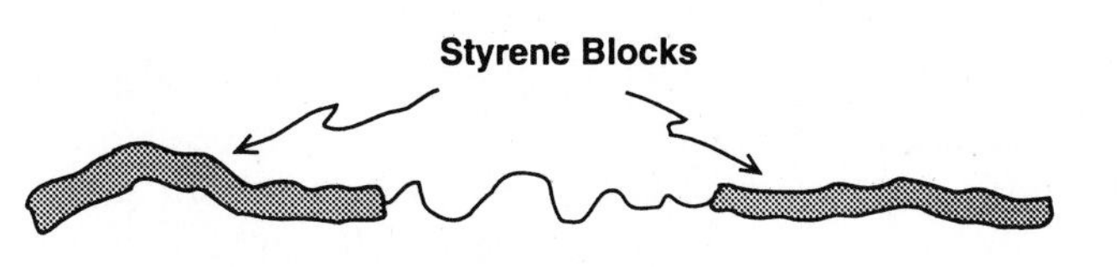

Butadiene Block

b) Triblock Copolymer

Figure 3.27 Sketches of diblock and triblock copolymers.

Figure 3.27. In the solid phase, the butadiene blocks will aggregate with each other, as will the styrene blocks, and the copolymer will exhibit separate phases. A thermoplastic rubber can be made from the triblock material, as shown in Figure 3.28. If the chains of the polybutadiene (PB) are long compared to the polystyrene (PS), PB will form the continuous phase with polystyrene chain segments aggregating in discon-

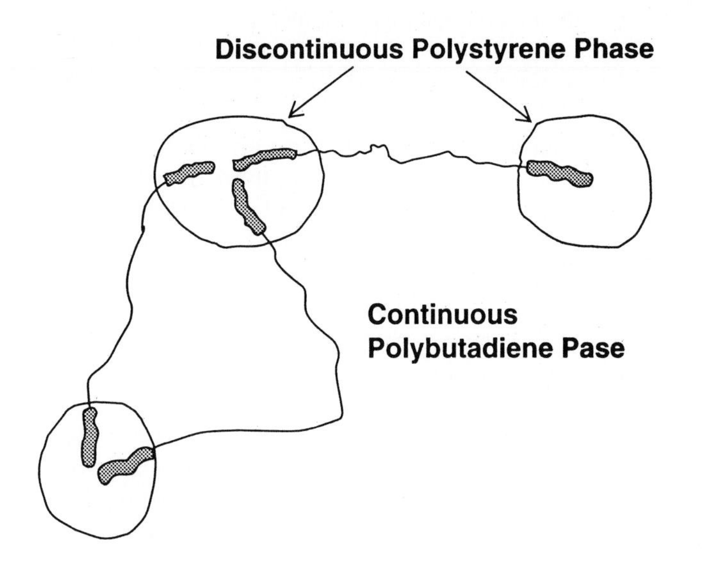

Thermoplastic Rubber

Figure 3.28 Thermoplastic rubber of triblock copolymers.

tinuous phases. The high PB content makes the material an elastomer, while the PS content adds stiffness and wear resistance.

These materials can be processed in conventional equipment for thermoplastics, unlike most thermosetting elastomers. The PB segments are above their T_g, so they can flow at a range of temperatures. When the material is heated above the T_g of polystyrene, these domains liquify and the thermoplastic rubber will flow. When the material cools after processing, the polystyrene domains reform and the material stiffens.

A variety of binary block copolymers can be prepared. The two types shown in Figure 3.27 are AB—a long sequence of A monomer attached to a long sequence of B monomer—and ABA—a long sequence of B monomer attached at both ends to long sequences of A monomers. The properties of these materials can be varied by changing the ratios of the chain lengths. Other types of binary block copolymers are star blocks, in which one chain type has a number of arms that are capped by segments of the other monomer, and segmented blocks, in which the AB sequence is short but is repeated many times. ABC block terpolymers also have been studied.

Ionic polymerizations are convenient for the synthesis of these materials. For example, polybutadiene may be made to specific chain lengths by anionic polymerization, followed by addition of styrene monomer and continued polymerization. In the case of these two monomers, the chain segments act as random coils in their individual domains.

Phase morphology is quite complex. At low volume fractions of the A phase, B polymer forms the continuous phase and A polymer aggregates into spheres. As the volume fraction of the A phase increases, B begins to form continuous cylinders. Lamellae of A and B form near 50vol%. Above this volume fraction, A becomes the continuous phase and the B phases go through a similar sequence until they form spherical phases. Figure 3.29 shows lamellae for Fina 315, a styrene-butadiene diblock system.

Diblock polymers can be used as blending aids for melt-mixing of the two homopolymer phases. The blend morphologies are similar to those discussed earlier.

Random copolymers are much different from block and graft copolymers with respect to their mechanical and thermal properties. This difference is illustrated in Figure 3.30 using the styrene-butadiene system. G is the sheer modulus and measures the elasticity of the sample. λ measures the damping of the sample, and high values are related to T_g. Butadiene homopolymer is shown in the far left, and styrene homopolymer is shown on the far right. Polybutadiene is an elastomer with a low T_g (-70°C). Its modulus is low at all temperatures above T_g. Polystyrene has a high modulus and a high T_g. A random copolymer of these two monomers has thermal properties between the two homopolymers. There is one T_g, as demonstrated by a single peak on the λ curve and the one change in shear modulus. The copolymer forms a single phase, and its properties can be varied by changing the ratio of butadiene to styrene.

The block and graft copolymers form two phases. Each phase has two T_g's near those of the homopolymers. This is demonstrated by two peaks in the damping curve.

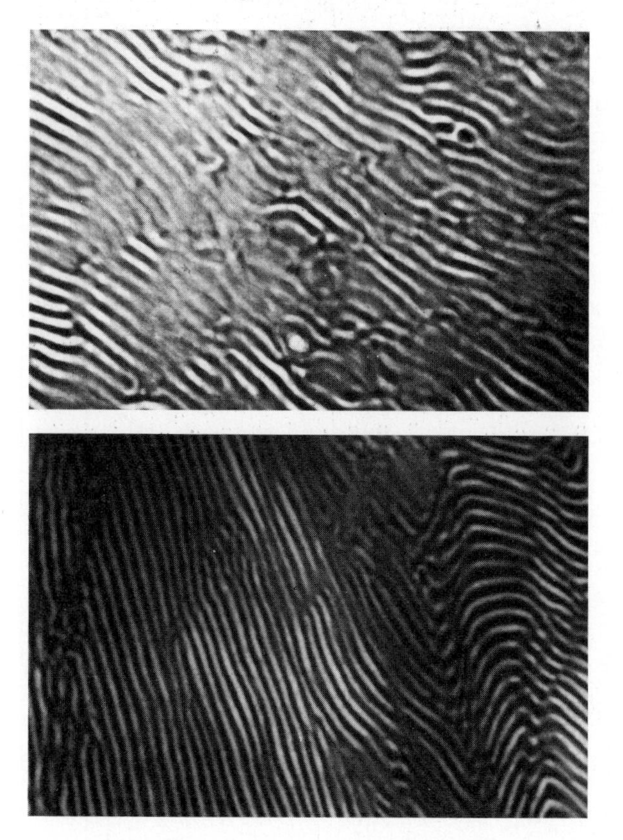

Figure 3.29 Styrene/butadiene diblock morphologies. Transmission Electron Micrographs (TEM'S) of Fina 315 tapered styrene-butadiene diblock annealed for one hour at **(a)** 130°C or **(b)** 200°C. Published with the permission of K. Nichols.

The shear modulus shows two "step" reductions, each associated with the T_g of a polymer phase. The modulus of the block or graft copolymers has an intermediate plateau over normal use temperatures (25°C). This means that the materials will not be as stiff as polystyrene but will not flow like uncross-linked polybutadiene. The result is a thermoplastic elastomer that will act like a cross-linked elastomer at use temperatures, but can be remelted when the temperature is raised above the T_g of the polystyrene phase. The cross-links of these copolymers are the primary bonds of the chains continuing from one phase to another. If the two homopolymers did not phase-separate so that physical cross-links formed, the system would not function as a cross-linked elastomer.

3.8 POLYMER ADDITIVES

Each class of materials—thermoplastics, elastomers, and thermosets—are filled or compounded in order to improve their properties, lower their cost, or both. The significance of this section is that most polymer products contain *mixtures* of polymers

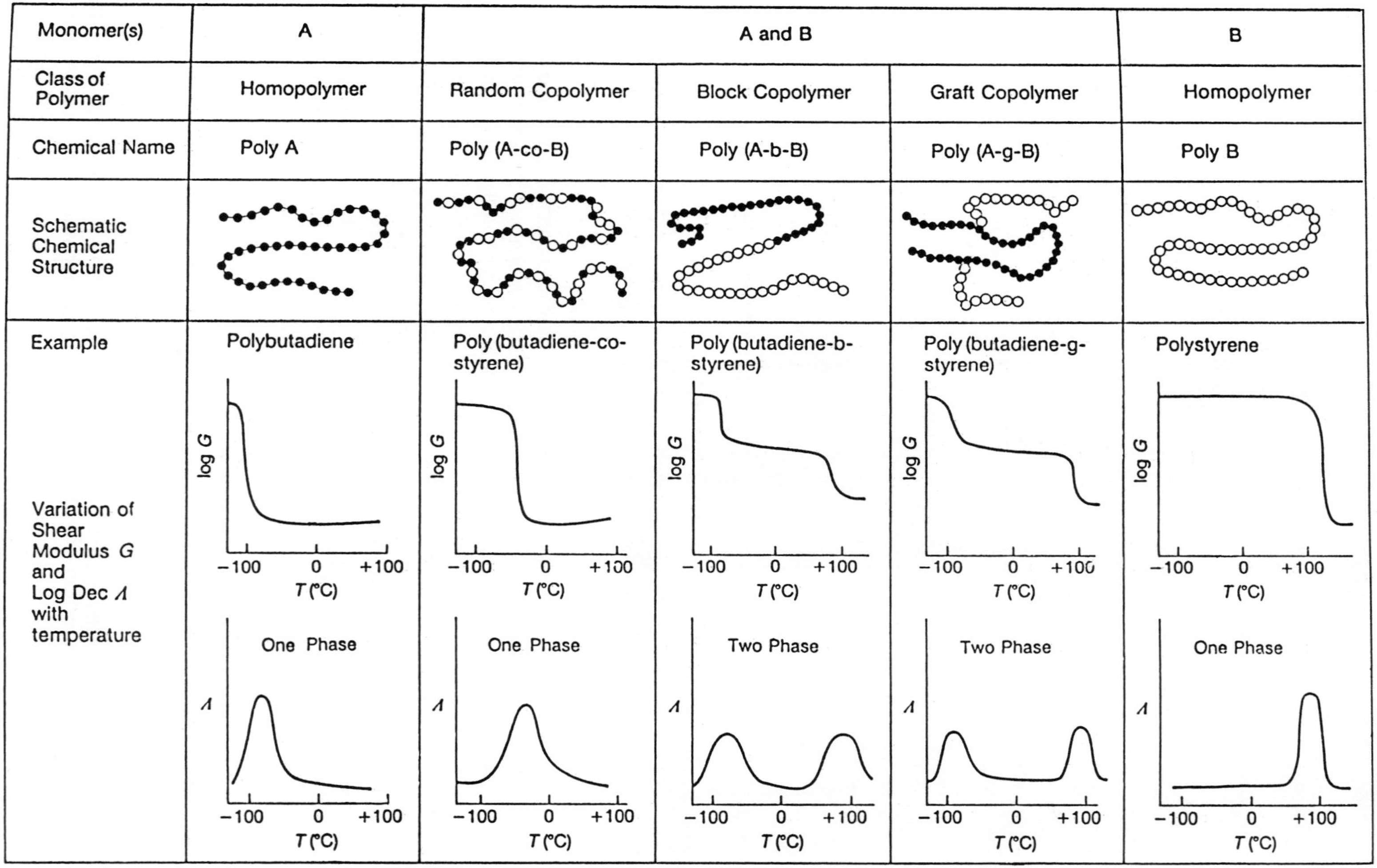

Monomer(s)	A	A and B			B
Class of Polymer	Homopolymer	Random Copolymer	Block Copolymer	Graft Copolymer	Homopolymer
Chemical Name	Poly A	Poly (A-co-B)	Poly (A-b-B)	Poly (A-g-B)	Poly B
Schematic Chemical Structure					
Example	Polybutadiene	Poly (butadiene-co-styrene)	Poly (butadiene-b-styrene)	Poly (butadiene-g-styrene)	Polystyrene
Variation of Shear Modulus G and Log Dec Λ with temperature	One Phase	One Phase	Two Phase	Two Phase	One Phase

Figure 3.30 Comparison of the mechanical and thermal properties of copolymers—The Styrene-Butadiene System (McCrum et al., 1988).

with additives. The additives change the processing and performance properties of the material. Thermoplastic and elastomer additives will be discussed in this section and the thermoset additives will be discussed in the composites section.

3.8.1 Thermoplastic Additives

Few thermoplastics products contain pure homopolymer. Most contain one or more of the following materials.

Reinforcing Agents. There are two types of reinforcing agents: oriented (fibers) and unoriented (powders or particles). Oriented reinforcing agents are classified as composites. Particulate reinforcing agents can be used to improve the impact strength of thermoplastics. As an example, polybutadiene particles in the range of one micron in diameter are added to polystyrene to improve its impact strength. This blended copolymer is tougher and has higher impact resistance than the pure polystyrene (Fig. 3.31). The elastomer, an impact modifier, reduces the tensile strength of the material but gives the part much ductile deformation, converting a brittle thermoplastic into a tough one. The reduction in tensile strength can be compensated for by increasing the part cross section. Other commercial systems are high impact polystyrene (PBS) and ABS.

Inert Fillers. Inert fillers are low-cost additives and are used to extend thermoplastics and thermosets. Calcium carbonate, talc, and mica are commonly used. These fillers reduce cost and improve heat resistance. Their lower coefficients of thermal expansion reduce shrinkage and warpage of plastic parts after molding. Hollow glass beads have been used to reduce the density of certain compounds. In general,

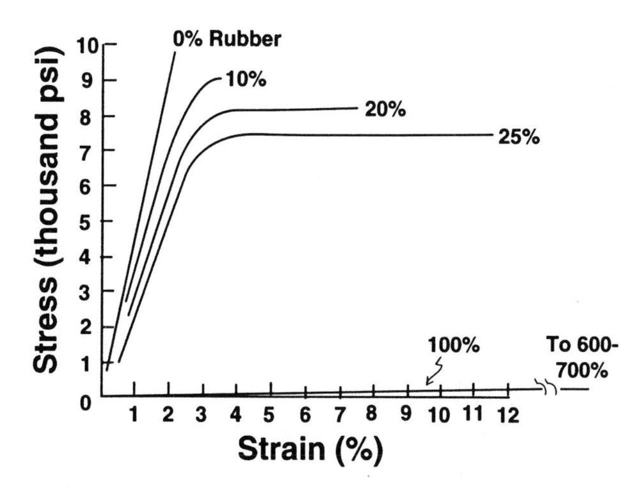

Figure 3.31 The influence of discrete, micron-sized rubber particles on the high-speed (133/in/in min) stress-strain curves of polystyrene (Garvin, 1976).

fillers with higher hardness and modulus than the thermoplastic will increase the hardness and modulus of the compound, but elongation and toughness will be reduced.

Inert fillers do not chemically bind with the polymer. As a result, the fracture resistance of filled compounds is not improved. Figure 3.32a shows a fracture surface for a polystyrene-polyethylene blend. The polyethylene beads do not adhere to the polystyrene continuous phase, so most of the mechanical load is still carried by the continuous polymer phase. In this sense, inert fillers can act as defects and reduce the failure properties of the compound. Figure 3.32b shows a fracture surface of the blend with 0.5wt% diblock copolymer. Copolymer fibrils between the phases help transmit the mechanical load.

Stabilizers. Many polymers are susceptible to degradation from environmental conditions, such as heat, oxygen, and ultraviolet radiation. Stabilizers are added to reduce the degradation rate of polymers, but are selected based on the

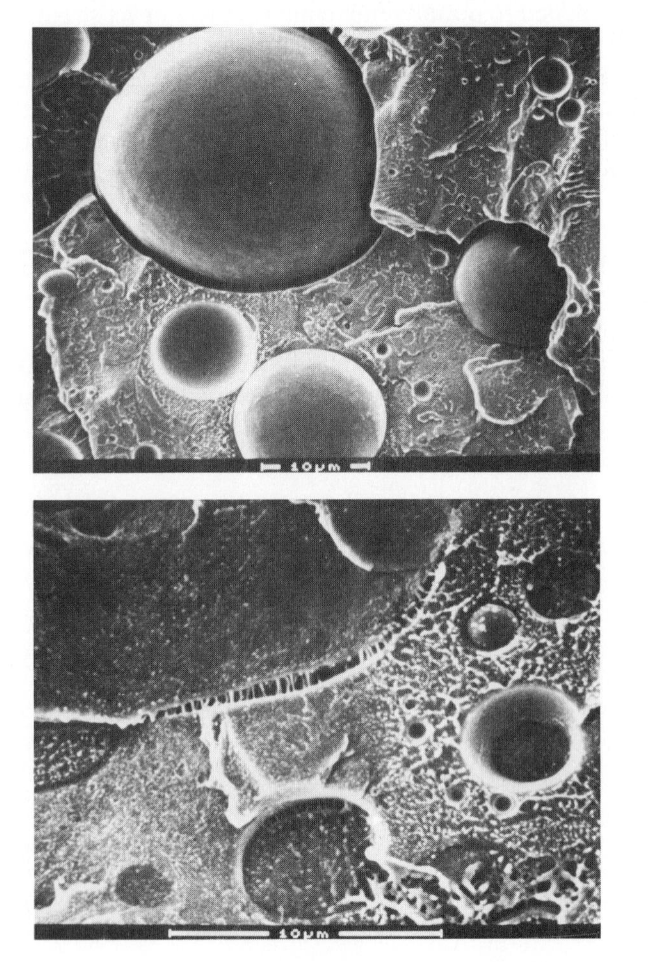

Figure 3.32 Scanning electron micrograph of a fracture surface of a polystyrene/polyethylene blend: **(a)** neat; **(b)** with 0.5wt% diblock copolymer. Diblock fibrils at the interface help transfer energy between the phases. Published with the permission of T. Imai and D.J. Meier, Michigan Molecular Institute.

polymer and its degradation mechanism. For example, PVC degrades at high temperatures by the loss of HCl from one site on the chain, followed by the "unzipping" of HCl from long sequences of the polymer. Thermal stability of PVC can be improved by adding metal oxides, which react with the HCl to form stable compounds.

Ultraviolet attack of polymer chains in thermoplastics is related to free radical attack by activated oxygen or peroxide. Most UV stabilizers are chemicals that can complex with free radicals and react to form inert products. Quinones are one class of materials that are effective for UV stabilization.

Pigments and Dyes. One of the best marketing features of polymers in the consumer and durable-goods industries is their flexibility for being colored and dyed. Pigments are finely dispersed solids that are mixed in the plastic to form uniform color throughout the part. Titanium dioxide, calcium carbonate, and carbon black are fillers that also color the compound.

Dyes are colored organic chemicals that function either by dissolving in bulk clear plastics or by reaction with groups on the polymer. The highest use of dyes is in the fiber industry. There are two major mechanisms for dye action: the dye forms strong secondary bonds with polar groups on the polymer or reacts to form covalent bonds with functional groups on the polymer. Fibers are dyed after spinning, so the dye must be soluble in the polymer in order to penetrate some depth from its surface. Crystalline regions are not dyed to any great extent as this would disrupt their orientation. Since only amorphous regions accept dye, many synthetic polymer fibers contain small amounts of comonomer, which lead to reproducible levels of amorphous content. This approach works well for polar polymers such as polyacrylonitrile, which has strong cyano groups that tend to be inaccessible to dyes in the homopolymer. Polypropylene, which has no polar groups, is not dyed, but is pigmented. Fine pigments are added to the melt prior to fiber spinning. Copolymers of polypropylene containing polar groups can be dyed.

Plasticizers. Plasticizers are used to reduce the brittleness of a compound. They should be miscible with the polymer and form supercritical solutions that have glass transition temperatures near the use temperature. These materials usually are low molecular weight organic compounds.

Lubricants and Processing Aids. There are two types of lubricants: external and internal. External lubricants are not very soluble in the compound and phase-separate under processing conditions. The phase separation causes them to migrate to the polymer surface, where they form thin films between the compound surface and the processing equipment. The surface film has a lower viscosity than the polymer melt and improves flow in channels while reducing sticking in molds. External lubricants remain on the part after processing and are usually not used with parts that need to be painted or plated. Stearic acid salts are common external lubricants.

Internal lubricants reduce the viscosity of the compound, rather than change the surface viscosity. They are soluble in the polymer. Polymers may be added to improve the processing of some compounds. The mechanism by which they function is similar to that of external lubricants since most polymer additives will phase-separate. As an example, low molecular weight polyethylene can be added to PVC to improve surface finish and increase extrusion output at the same extruder power input.

Blowing Agents. Blowing agents for thermoplastic foams are usually liquids that will volatilize under processing conditions. Chlorofluorocarbons (CFC's) and low molecular weight alkanes have been popular choices, but CFC's are being replaced due to concern about their effects on the ozone layer. Polyurethane thermosets are blown with chemical blowing agents that decompose at processing conditions to give gaseous compounds.

Flame Retardants. Many polymers are flammable and will continue to burn once they have been ignited. Flame retardants are added to reduce flammability and work by several mechanisms. Some contain halogens that quench the free radical processes occurring in flame propagation. Hydrated aluminum contains 35% water, which is released on heating and helps absorb energy. Other materials form chars, which inhibit flame spreading and prevent oxygen from getting to the unburned polymer.

Miscellaneous. Biocides are added to reduce microorganism attack in humid environments. Antistatic agents are used to reduce surface charging on records and bottles.

3.8.2 Elastomer Additives

Tire compounds are typical of the kind of modifications made in elastomers by additives. Table 3.6 shows materials added to synthetic elastomer to make compounds for tire tread. Less than 60% of the material is polymer. Rubber compounding is extremely complex, as will be shown.

Reinforcing Fillers. Carbon black has been used as a reinforcing filler to give good tensile strength and abrasion resistance. Because it is a reinforcing filler, the morphology of carbon black is critical to developing the mixture's properties. Rubber forms strong secondary bonds and covalent bonds with functional groups on the surface of the carbon black. These can be acidic groups (carboxylic) or basic groups (hydroxyl). The polymer chains also adsorb on the porous surface of the carbon particles. It can be produced by burning methane in an oxygen-deficient atmosphere. The process makes spherical particles in the range of 0.01 to 1 micron in diameter. Better reinforcement seems to be obtained when the carbon black aggregates into chain-like structures. Inert fillers may be added; the most common choice is calcium carbonate.

TABLE 3.6 Typical Recipe for Tire-tread Elastomer

Ingredient	Parts by Weight
GR-S 1000 (75/25 butadiene/styrene emulsion copolymer)	100
HAF black	50
Zinc oxide (promoter)	5
Stearic acid (promoter)	3
Sulfur	2
Santocure (accelerator)	0.75
Circosol 2XH (extending oil)	10

SOURCE: G. S. Garvin, *Polymerization Processes* (C. E. Schildknecht, Ed.), Wiley-Interscience, NY, 1976.

Extending Oils. Hydrocarbon oils perform two functions in rubber compounds. They plasticize the polymer to make it softer and easier to process. They act as inert fillers to reduce the cost of the compound. Lower cost rubber compounds can be made by using elastomers with higher molecular weight than is needed, reducing the molecular weight during blending and mixing operations using extending oils to reduce the melt viscosity, and improving the modulus with carbon black.

Curing (Vulcanization). The standard method for cross-linking the polymer chains is vulcanization, in which sulfur is added to react with carbon-carbon double bonds. Sulfur curing is slow and probably acts by forming sulfide or disulfide linkages between the chains. Accelerators, promoters, and activators can be used to improve the efficiency of this process. In some systems, free radical initiators are used to form radicals on the chain backbone, which can lead to cross-linking.

Antioxidants and Stabilizers. Elastomers contain double bonds that are particularly susceptible to attack by ozone and oxygen with ultraviolet radiation. Oxidative attack can be seen as thin "slits" and checks in tire surfaces after exposure to the environment. The stabilizers scavenge for free radicals in order to slow the process.

Pigments. Carbon black is the major pigment for tires, but other pigments are used to provide colors other than black.

3.9 LIQUID CRYSTALS

Normal liquids are *isotropic* and have the same properties in all directions. Crystalline solids are *anisotropic* and have disparate properties in different directions relative to the crystal axes. Liquid crystals occur when molecules align while in the liquid state and the liquid exhibits anisotropic behavior. The ordered regions of the liquid are called *mesophases*. Liquid crystals have been known in low molecular weight materials for about a century, but the applications in polymers have been recent.

Polymers that are crystalline in the liquid state have very high moduli and can be made into ultrahigh modulus fibers. In the 1970s a new class of materials was developed that led to fiber strengths an order of magnitude greater than those available before.

Flory suggested that rod-shaped molecules should be highly oriented at low concentrations. A two-dimensional model of the phenomenon is given by toothpicks floating on a water surface. Moving a pencil through the liquid causes the toothpicks to align with the flow direction. At the end of the motion, the toothpicks retain their alignment. Without adhesion between the toothpicks, the time needed to return to random orientation is very long because of the steric hindrances to motion.

The two major classifications of liquid crystals are *lyotropic* and *thermotropic*. Lyotropic liquid crystals form in solvents, while thermotropic liquid crystals form in the melts. There are three crystal habits for these materials: *nematic*, *smectic*, and *cholesteric* (Fig. 3.33). The smectic state has parallel chains that are arranged in layers. The molecular axis is perpendicular to the layers. In the nematic state, the molecules are parallel to each other, but the chain ends are not aligned and do not form layers. The cholesteric state has layers of chains in which the molecular axis is parallel to the layer plane. Some A-B block copolymers form liquid crystals when their phases separate. When the blocks aggregate into lamellar, spherical, or cylindrical domains, the domains can become ordered into micellar, cubic, or hexagonally packed crystalline mesophases.

Molten polymers with smectic or nematic crystal habits are liquids because either the layers or the randomly aligned chains can be displaced with moderate shear. However, because the chains are oriented, these materials are optically anisotropic and show a range of colors under visible light illumination.

Copolyesters were among the first polymers known to be liquid crystals. The copolymer of terephthalic acid, *p*-hydroxybenzoic acid, and ethylene glycol forms liquid crystals as the amount of *p*-hydroxybenzoic acid is increased. Figure 3.34 shows the melt viscosity of these copolyesters with varying *p*-hydroxybenzoic acid (HBA)

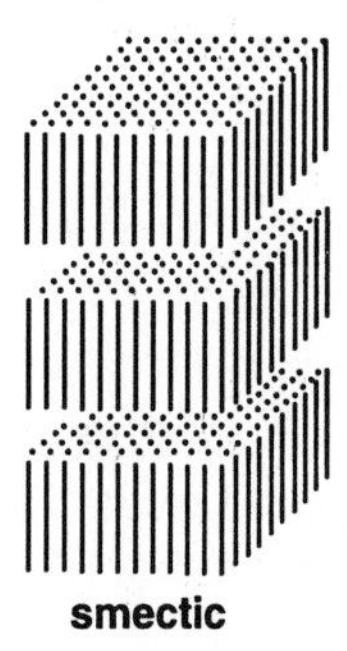

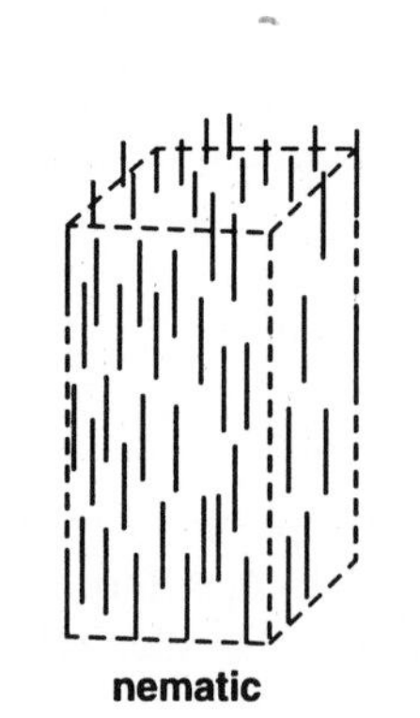

Figure 3.33 Liquid crystal habits (Elias, 1984).

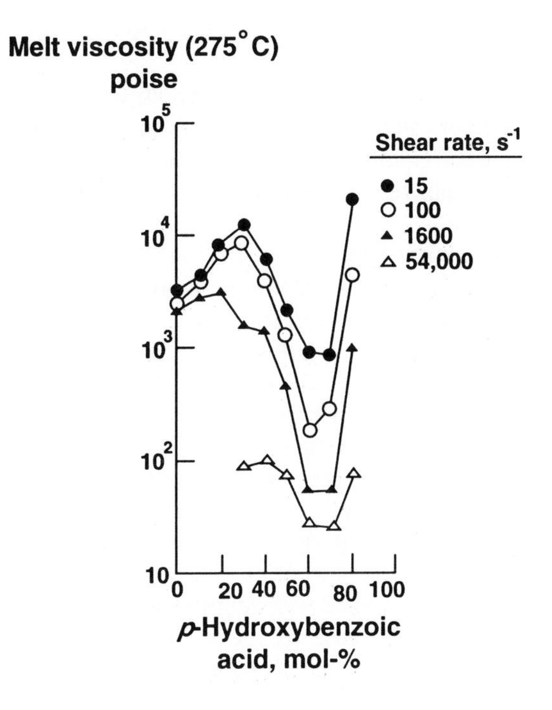

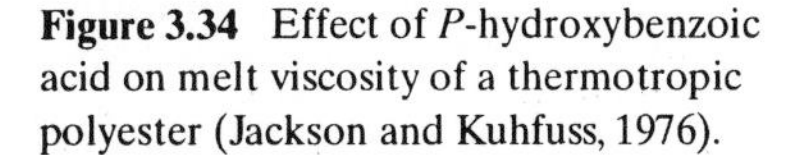
Figure 3.34 Effect of *P*-hydroxybenzoic acid on melt viscosity of a thermotropic polyester (Jackson and Kuhfuss, 1976).

content. As HBA is added up to 30 mol%, the melt viscosity increases due to the steric hindrance to rotation caused by its addition. Above 30 mol%, the melt viscosity decreases and reaches a minimum at about 70 mol%. As the melt viscosity decreases, the melt changes from clear to opaque, suggesting the onset of a liquid crystal phase. The viscosity reduction occurs because the rigid molecules align in the flow direction and reduce frictional drag. In this thermotropic material, the ordered molecules retain their alignments on cooling.

Several aromatic polyamides, such as poly(1,4-benzamide), form rod-shaped molecules. Spinning nematic phase polymer solutions gives fibers with very high orientation. The polymer chains align parallel with the flow direction, reducing the fluid's viscosity in die flow. Poly(1,4-benzamide) (PBA) has a tensile strength four times that of nylon. As is typical of highly oriented fibers, these materials have low elongations at break compared to other polymeric materials with less orientation.

Kevlar is a lyotrophic liquid crystal of an aromatic polyamide. The polymer is spun from a solution of sulfuric acid. Removal of the solvent (sulfuric acid) results in a fiber with high alignment. Because the chains in a liquid crystal fiber are aligned with the fiber direction and are fully extended due to their rod-like conformation, tensile loads are borne directly by the bonds of the chain backbone. When the fiber breaks, backbone bonds are broken. The stress is transferred along the fiber direction through secondary bonding between chains.

3.10 STRUCTURED POLYMERS/COMPOSITES

The addition of fibers to thermosets and thermoplastics gives materials with very high tensile strengths and moduli. The fiber can be chemically linked to the matrix material so that both share the tensile load. Reinforced materials are those that contain reinforcing particulate fillers. Composites contain fibers that make the material anisotropic.

Commercial structural composites are similar in concept to some of the natural fiber systems. The microfibrils of α-keratin in wool are assemblies of crystalline polymers bonded together with a matrix material. The structure of wood includes crystalline cellulose regions bonded with hemicelluloses and lignin. The properties of these natural composites are different from those of the individual components and depend on the orientation of the fibers in the matrix, the ratio of matrix material to fiber, and the chemical bonding between the two components. In natural fibers these factors vary little, since the components are formed in situ through a variety of enzyme reactions. By contrast, synthetic composites can be made with a range of orientations, fiber contents, and interphase bondings.

Most thermoplastic polymers can be mixed with reinforcing particulates and fibers. The most common groups of materials to use are the engineering thermoplastics and the advanced engineering thermoplastics. Typical property changes for reinforced and composite materials can be found in Table 1 in Appendix B.

Both the fiber length and the fiber type affect the properties of the material. Table 3.7 shows such effects on the mechanical properties of nylon 6,6 composites containing 30wt% glass. Glass fiber reinforcements are popular because of their cost, high strength-to-weight ratio, and chemical stability. Mats, rovings, and cloths may also be used. Carbon fibers are stronger and make the material stiffer and more brittle.

TABLE 3.7 Effect of Fibers on Mechanical Properties of Reinforced Nylon 6,6

	Filler Type			
	30% glass fiber		30% carbon fiber	30%
	Long	Chopped	Long	Glass beads
Tensile strength, MPa	193	172	282	76
Elongation, %	2.5	2.5	—	4
Tensile modulus, GPa	9.6	9.0	24.1	4.3
Flexural strength, MPa	276	248	372	124
Flexural modulus, GPa	9.0	9.0	21.3	4.3
Izod impact, J/m	214	107	139	32
Specific gravity	1.40	1.40	1.28	1.37
Linear mold shrinkage, 3.2-mm section, %	0.3	0.2	—	1

SOURCE: S. Gerbig, *The Performance Polymer People*, Wilson-Fiberfil Int., Evansville, IN.

TABLE 3.8 Properties of Glass Fiber–reinforced Thermoset (RTS) Polyesters

Polyester Types	Characteristics	Applications
General purpose	Rigid moldings	Trays, boats, tanks, boxes, luggage, seating
Flexible resins and semirigid resins	Tough, good impact resistance, high flexural strength, low flexural modulus	Vibration damping: machine covers and guards, safety helmets, electronic part encapsulation, gel coats, patching compounds, auto bodies, boats
Light-stable and weather-resistant	Resistant to weather and ultraviolet degradation	Structural panels, sky-lighting, glazing
Chemical-resistant	Highest chemical resistance of polyesters, excellent in acid, fair in alkali	Corrosion-resistant pipe, tanks, ducts, fume stacks
Flame-resistant	Self-extinguishing, rigid	Building panels (interior), electrical components, fuel tanks
High heat distortion	Service up to 260°C, rigid	Aircraft parts
Hot strength	Fast rate of cure (hot), moldings easily removed from die	Containers, trays, housings
Low exotherm	Void-free thick laminates, low heat generated during cure	Encapsulating electronic components, electrical parts, switch gear
Extended pot life	Void-free and uniform, long flow time in mold before gel	Large complex moldings
Air dry	Cures tack-free at room temperature	Pools, boat, tanks
Thixotropic	Resists flow or drainage when applied to vertical surfaces	Boats, pools, tank linings

Thermoplastic composites can be fabricated by injection molding, extrusion, pultrusion, and resin transfer molding.

Thermoset resins—such as polyester, phenolic silicone, and epoxy—have been combined with glass, graphite, carbon, and aramid fibers to make composites. The highest volume thermoset composites contain chopped glass fibers in polyester matrices. Advanced composite refers to the use of high moduli fibers—such as boron, graphite, and aramid—in the material. The orientation of the reinforcement makes a big difference in the material properties. Fiber glass–reinforced polyesters have a wide variety of uses. A partial list is given in Table 3.8.

NOMENCLATURE

T	temperature
T_g	glass transition temperature
T_m	melting temperature
ΔG	free energy
ΔH	enthalpy
ΔS	entropy

PROBLEMS

3.1 Definitions
(a) Secondary bonding forces
(b) The gel state
(c) T_m
(d) T_g
(e) Liquid crystal
(f) Triblock copolymer
(g) Composite
(h) Lamellar
(i) Spherulite
(j) Elastic modulus
(k) Fillers

3.2 If a sample of PET (Mylar) were allowed to crystallize from the melt and then cooled to room temperature, the T_g of the polymer could still be observed. Explain why.

3.3 Differentiate between the gel state and the rubber-elastic state.

3.4 Proteins are polypeptides and may have specific amino acid sequence configurations.
(a) Describe the expected secondary structure of a protein that does not have repeating peptide sequences in its backbone chain.
(b) Which molecular forces are responsible for the conformation, and what are their relative strengths?
(c) $10\ \overline{M}$ urea causes many proteins to denature and precipitate by disrupting the associations between chain segments. Should $10\ \overline{M}$ urea be considered a Theta solvent for proteins? Why or why not?

3.5 Cellulose in wood is one of the few materials that has a fringed micelle structure. Cellulose acetate, nitro-cellulose, and rayon are made by chemical reactions with functional groups on the cellulose molecules. How will the physical structure of cellulose affect these chemical treatments?

3.6 Wood can be considered a composite material. Find a picture of the structure of wood and identify the matrix and fiber components.

3.7 The glass transition temperature (T_g) and the melting temperature (T_m) are frequently used to define polymer properties. Determine which of the following statements are correct and explain your answer.

(a) Polymer chains with steric hindrance to rotation around main chain bonds will generally have higher values for T_m.

(b) T_m of a polymer is the temperature at which an abrupt change occurs in properties such as specific volume.

(c) The glass transition temperature is always a sharply defined point.

(d) Below the glass transition temperature, large chain segment motions occur frequently.

3.8 Polymer gels and cross-linked rubbers have some similar properties. Determine which statements apply to each physical state and explain your answers.

Polymer Gels	*Cross-linked Rubbers*
________	________
________	________
________	________
________	________

(a) Secondary valence forces are the main attractive force between polymer coils.

(b) On warming, these can return to the solution state.

(c) The forces holding coils together are covalent bonds.

(d) Crystallization often occurs in the deformed state.

3.9 Describe the polymer sample morphology and the structure of the polymer chains for polyethylene crystallized from the melt.

3.10 Sticky plant secretions that harden when exposed to the environment are called resins. The fossilized form of tree resins is called amber. Tree resin composed of terpenoids can be generated by the tree from acetate through isoprene as an intermediate. Tree resins consist of monoterpenes (10 carbon atoms), sequiterpenes (15 carbon atoms), diterpenes (20 carbon atoms), and triterpenes (30 carbon atoms). Sequiterpenes do not seem to polymerize and will volatilize with time. Photochemical studies of labdane diterpenes showed that UV radiation initiates free radical polymerization of the monomers. Speculate on how this system of terpenes can be transformed into a stable polymer capable of lasting millions of years (Langenheim, 1990).

3.11 The choice of a plastic material for a specific consumer or industrial application requires careful comparison of price, properties, availability, and ease of processing. You are working on component design for a major automotive manufacturer, and your boss has asked you to choose a specific plastic for the one-piece roof of a 4-wd, off-road vehicle. The critical product criteria are rigidity, price, and weight.

For your engineering analysis

(a) Construct a table showing specific gravity, impact strength, flexural strength, heat deflection temperature, and the effects of sunlight on the engineering thermoplastics: high-impact ABS, unfilled poly-carbonate, and unmodified polypropylene. Include units, cite your references and page numbers (see *Modern Plastics Encyclopedia*).

For your financial analysis

(b) Construct a table showing price per pound, total pounds polymer required to make the roof, and the total price of the roof. To make a 5′ × 7′ roof of equivalent strength, assume that 3400 cubic inches of poly-propylene, 2500 cubic inches of ABS, or 1300 cubic inches of polycarbonate are required. See *Chemical Marketing Reporter* or *Plastics World*. Cite your sources.

(c) Choose a material to make the roof and justify your answer. Do not limit yourself to your boss's criteria.

REFERENCES

G.C. BERRY and T.G. FOX, *Fortschr. Hochpolym.-Forsch.* (*Adv. Polymer Sci.*) *5*, 261 (1968).

H.G. ELIAS, *Macromolecules*, Vol. 1, Plenum Press, NY, 1984.

K. FRIEDRICH, R. WALTER, H. VOSS, and J. KARGER-KOCSIS, *Composites 17*, 205 (1986).

G.S. GARVIN, *Polymerization Processes*, (C.E. Schildknecht, Ed.), Wiley-Interscience, NY, 1976.

S. GERBIG, *The Performance Polymer People*, Wilson-Fiberfil Int., Evansville, IN

W.J. JACKSON, JR. and H.F. KUHFUSS, *J. Polym. Sci.* (*Polym. Chem. Ed.*) *14*, 2043 (1976).

J.H. LANGENHELM, *Amer. Scientist, 78*, 16–24, 1990.

L. MANDELKERN, M. HELLMAN, D.W. BROWN, D.E. ROBERTS, and F.A. QUINN, JR.: *J. Am. Chem. Soc.*, *75*, 4093 (1953).

R.H. MARCHESSAULT and A. SARKO, *J. Polym. Sci. C28*, 317 (1969).

N.G. MCCRUM, C.P. BUCKLEY, and C.B. BUCKNELL. *Principles of Polymer Engineering*. Oxford University Press, NY, 1988.

K.L. NICHOLS, Ph.D. Dissertation, Michigan State University, 1988.

G.E. SCHULZ and H.A. SCHIRMER, *Principles of Protein Structure*, Springer-Verlag, NY, 1979.

J.G. SPEIGHT and S.E. MOSCHOPEDIS, "Chemistry of Asphaltenes" *Advances in Chemistry*, Series (J.W. Bunger and W.C. Li, Eds.), 195, ACS, Washington, DC, 1981, pp. 1–15.

A.E. WOODWARD, *Atlas of Polymer Morphology*, Oxford University Press, NY, 1989.

4

Theory of Polymerization Reactions

Molecular weight and molecular weight distribution are important to the engineering properties of polymers. These characteristics often are established during polymerization. Crystallinity, side reactions, residual materials (monomer, solvent, surfactants, swelling agents, chain transfer agents), and cross-linking can affect product properties. Tacticity, discussed in Chapters 1 and 2, is affected by the polymerization environment. The presence or lack of crystallinity can be controlled to some extent by suitable choices of reaction temperature, initiator/catalyst, and solvents. For example, in the production of linear, high molecular weight polymers, there are a number of side reactions that cause differences in the final product. These side reactions include branching (long and short chains), chain transfer (which usually decreases the average chain length), inhibition, and retardation—all of which can be controlled using the polymerization environment. Network polymers are cross-linked, and the amount of cross-linking and its location affects their properties.

There are many different types of polymerization reactions used in making commercial polymers. Each type of polymerization involves different reaction rate expressions. Furthermore, the reaction environment—that is, the temperature, pressure, solvents, catalysts/initiators—often has subtle but important effects on the reaction rate, molecular weight distribution, and product morphology. The purpose of this section is to develop quantitative rate expressions for several important polymerization types and to illustrate some of the choices the engineer has for designing a given system.

Polymerization Classifications. Carothers, a researcher for du Pont, suggested classifying polymers and polymerization reactions according to the stoichiometry of the reaction. His two categories were condensation and addition

TABLE 4.1 Differences between Chain and Step Polymerizations

Chain Polymerization	Step Polymerization
1. Only species with active centers add monomer units	1. Any two potentially reactive end groups can react
2. Monomer concentration decreases steadily	2. Monomer depletion occurs very rapidly
3. High molecular weight polymer forms at once	3. Polymer molecular weight increases slowly with time
4. The concentration of reacting chains is usually low compared to the non-reacting monomer and polymer	4. Any size species can react with another and many chains are reacting at one time

polymerization. Condensation polymerization involves the splitting of a small molecule (such as water) from the reactants as they polymerized. Nylon 6,6 is polymerized by condensation, and one molecule is lost per two reacted end groups. Addition polymerizations do not result in groups leaving the growing polymer chain (polystyrene is an example of this type).

P.J. Flory (1952) revised these classifications to step and chain polymerization. This classification emphasizes the mechanism by which the two processes proceed. Nylon 6,6 polymerization proceeds stepwise by intermolecular condensation of reactive groups. Polystyrene grows by a chain reaction promoted at an active site on the molecule. The differences between these two mechanisms are given in Table 4.1

Figure 4.1 shows the molecular size distribution in a two-dimensional space-filling matrix of monomer and oligomer (short chain) molecules. Long macromolecules are made immediately in a chain polymerization, but only a few of the monomer

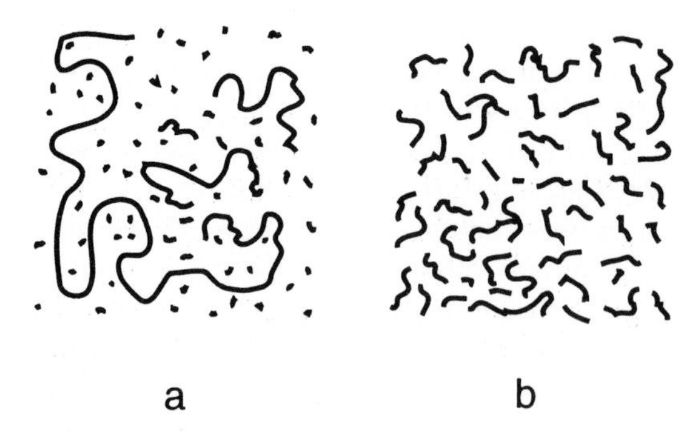

Figure 4.1 Molecular size distributions for chain and step polymerizations. **(a)** Long chains form quickly in chain polymerizations, but the monomer concentration (designated by dots) is high througout the reaction. **(b)** Monomer quickly reacts to dimmer and trimmer in step polymerizations.

molecules have reacted. In a step polymerization, most monomers have reacted within a short time, but the average chain length is small.

This chapter reviews free radical chain polymerization, chain copolymerization, stepwise polymerization, ring-opening polymerization, stereospecific and coordination complex polymerization, and ionic polymerization. Depolymerization and degradation depend on the type of polymer, its environment, and impurities or defects that initiate the process. Models and theories for these processes are often specific to the process and product. One objective of the chapter is to demonstrate how to develop models of polymerization kinetics based on reaction systems and expected mechanisms.

4.1 FREE RADICAL CHAIN POLYMERIZATION

Free radical chain polymerization occurs with monomer units having double bonds (for example, vinyl, divinyl, and 1,3-diene molecules). Examples of these monomer types are shown in Table 4.2. Table 4.3 lists some commercial polymers made from vinyl monomers by free radical polymerization. Free radical polymerizations can be carried out in bulk, emulsion, suspension, and solution processes. Emulsion products have fine particle sizes and can be used for paints, coatings, and plastisols. Suspension processes generate free-flowing powders that are easy to process by extrusion or calendaring techniques. Bulk polymerizations can be used when the polymer does not form cross-linked gels or wall deposits, which would contaminate continuous polym-

TABLE 4.2 Examples of Vinyl Monomers

No Homopolymerization without Initiators	Typical Homopolymerization	Rapid Homopolymerization
α-methyl styrene	ethylene	hydroxy methyl vinyl ketone
$CH_2{=}C(CH_3){-}C_6H_5$	propylene	$CH_2{=}CH{-}CO{-}CH_2{-}OH$
allyl alcohol	isobutylene	vinylidene cyanide
$CH_2{=}CH{-}CH_2{-}OH$	butadiene	$CH_2{=}C(CN)_2$
stilbene	styrene	acrylic acid
$H_5C_6{-}CH{=}CH{-}C_6H_5$	vinyl chloride	acrylic anhydride
dichloroethylene	tetrafluoroethylene	methylene malonate
maleic acid	vinyl esters	$CH_2{=}C(CO{-}O{-}R)_2$
$HOOC{-}CH{=}CH{-}COOH$	$CH_2{=}CH{-}O{-}CO{-}R$	nitroethylene
	vinyl ethers	$CH_2{=}CH{-}NO_2$
	$CH_2{=}CH{-}O{-}R$	
	acrylic acid esters	
	$CH_2{=}CH{-}CO{-}O{-}R$	
	acrylonitrile	
	vinyl pyrrolidone	

TABLE 4.3 Commercially Important Vinyl Polymers Prepared by Free Radical Polymerization

Polymer or Copolymer	Method of Manufacture*	Major Uses
Polyethylene, low density	HP, B, Sol	Packaging film, wire and cable insulation, toys, flexible bottles, housewares, coatings
Poly(vinyl chloride)	B, S, Sol, E	Construction, rigid pipe, flooring, wire and cable insulation, film and sheet
Polystyrene†	B, S, Sol, E	Packaging (foam and film), foam insulation, appliances, housewares, toys
Polychloroprene (neoprene rubber)	E	Tires, wire coatings, belting, hoses, shoe heels, coated fabrics
Poly(methyl methacrylate)	B, S	Automotive parts, molding, compositions, decorative panels, skylights, glazing
Polyacrylonitrile	S	Textile fibers, food packaging
Poly(vinyl acetate)	B, S, Sol, E	Water-based paints, adhesives, conversion to poly(vinyl alcohol)
Poly(acrylic acid) and poly(methacrylic acid)	Sol, E, HP	Adhesives, thickening agents, ionomers
Polyacrylamide	E	Thickening agent, flocculent
Polytetrafluoroethylene	HP, S, Sol	Electrical insulation, gaskets, bearings, bushings, valves, nonstick cooking utensils
Polytrichlorofluoroethylene	S, Sol	Gaskets, tubing, wire insulation
Poly(vinylidene fluoride)	S, Sol, E	Protective coatings, gaskets, pipe
Poly(vinyl fluoride)	S, Sol	Protective coatings
Allyl resins‡	B, Sol	Lenses, electronics parts

*HP = high pressure; B = bulk; E = emulsion; S = suspension; Sol = solution.
†Poly(*p*-methylstyrene) also available.
‡Cross-linked diallyl and triallyl esters and ethers.

erization equipment. Solution polymerizations are becoming less popular as environmental regulations increase, since the solvent must be removed from the polymer prior to use. Some high pressure polymerizations (ethylene for example) are done at supercritical conditions and may have two phases in the reactor, a gas phase and a polymer solution in monomer.

4.1.1 Reaction Mechanisms

In free radical polymerizations, a polymer molecule of high molecular weight (1000 repeat units) can be produced from an active center in a very short time (one second or less). There are three major steps in the process: the initiation of the chain, the propagation of the chain as monomers add to the reactive site, and the termination of the reactive site to give completed macromolecules. Other mechanisms that can affect this process include the inhibition or retardation of free radical initiation, chain transfer of the radical center to other molecules in the reaction medium, and branching reactions.

The reaction begins with an initiator decomposing to form molecules with free radicals. Chain polymerizations can be done using catalysts but are discussed in a later section. These free radicals can add monomer so that the free radical, or reactive center, is at one end of a linear molecule. The active center then continues to add monomer and the macromolecule lengthens until a termination reaction occurs. The free radical center on the polymer chain can be destroyed by reacting with a similar center on a nearby chain. Termination can also occur via side reactions, including nonproductive reactions with monomer; reactions with impurities, solvents, initiators, and polymer chain segments.

Initiation. Initiators and catalysts function differently. Initiators become incorporated into the polymer chain, usually at one end, and are consumed during the reaction. Table 4.4 shows some examples of free radical initiators. Their dissociation, or decomposition, is usually modeled as a first order process, and the dissociation rate constant varies with temperature. Coordination complex catalysts, such as the Friedel–Crafts catalyst used to copolymerize isobutylene and isoprene, function by inserting monomer at the end of the chain. This usually occurs in a stereospecific manner, leading to polymers with stereoregular structure (crystallinity). The catalysts can be said to perform chain polymerizations, but are not incorporated into a specific, fixed location on the chain.

Residual catalyst or initiator in the polymer has the potential to reduce the thermal stability of the product. The Ziegler–Natta catalysts are usually deactivated with water and may not contribute significantly to degradation at high temperatures. Because initiators become a chain end, they may reduce the thermal stability of the polymer if their residues decompose at lower temperatures than the rest of the chain. Thermal degradation of initiator residues is often considered while choosing an initiator system. Other factors are its solubility in the reaction system and its decomposition characteristics.

Initiators tend to produce polymers with less tacticity than coordination complex catalysts, unless there are strong steric effects influencing the orientation of the monomer at the reactive center or the monomer is symmetrical. A typical initiator dissociates to form two free radicals, 2 R^*.

$$I \xrightarrow{k_d} 2R^* \tag{4.1a}$$

The radicals initiate the polymerization by reacting with monomer.

$$R^* + M \xrightarrow{k_i} R - M^* \tag{4.1b}$$

Example: Benzoyl peroxide (I) initiating styrene (M).

$$R\text{-}CO\text{-}O\text{-}O\text{-}CO\text{-}R \rightarrow 2\,R\text{-}CO\text{-}O^*;\ R = \text{benzene ring},\ k = k_d$$

$$R\text{-}CO\text{-}O^* + H_2C{=}CH\text{-}C_6H_5 \rightarrow R\text{-}CO\text{-}O\text{-}CH_2\text{-}CH(C_6H_5)^*;\ k = k_i$$

TABLE 4.4 Examples of Free Radical Initiators

Name	Formula	Suitable Polymerization Temperature C
Potassium persulfate	$KO{-}S(=O)_2{-}O{-}O{-}S(=O)_2{-}OK$	40–80
Dibenzoyl peroxide (Lucidol)	$C_6H_5{-}C(=O){-}O{-}O{-}C(=O){-}C_6H_5$	40–90
Cumene hydroperoxide	$C_6H_5{-}C(CH_3)_2{-}O{-}O{-}H$	50–100
Cyclohexanone peroxide	$HO{-}C_6H_{10}{-}O{-}O{-}C_6H_{10}{-}O{-}O{-}H$	20–80
Di-t-butyl peroxide	$CH_3{-}C(CH_3)_2{-}O{-}O{-}C(CH_3)_2{-}CH_3$	80–150
Azo-bis-isobutyronitrile (Vazo, AIBN)	$(CH_3)_2C(CN){-}N{=}N{-}C(CN)(CH_3)_2$	20–100
Cyclohexylsulfonyl-acetyl peroxide (Percadox ACS)	$C_6H_5{-}S(=O)_2{-}O{-}O{-}C(=O){-}CH_3$	0–40
Diisopropyl percarbonate (Percadox JPP)	$(CH_3)_2CH{-}O{-}C(=O){-}O{-}O{-}C(=O){-}O{-}CH(CH_3)_2$	40–80

Rotation can occur around the carbon-carbon bond at the end of the growing chain. This rotation event is random and leads to atactic polymer in the absence of steric hindrance to force stereoregular addition.

Propagation. Rapid growth of the chain occurs by monomer reacting with the active center on the chain, generating a new active center. The concentration of active centers in the polymerization medium is usually kept low (less than 0.1wt% initiator). This practice results in low numbers of polymerizing chains and a high probability that the radical center will react with monomer rather than other molecules. On the average, the chains can grow to long macromolecules before terminating. The mechanism is

$$M_n{*} + M \xrightarrow{k_p} M_{n+1}{*} \qquad 4.2$$

Example: Growing polystyrene chain reacting with styrene. Head-to-tail configuration.

$$M_n{-}CH_2{-}CHR^* + H_2C{=}CHR \rightarrow M_n{-}CH_2{-}CHR{-}CH_2{-}CHR^*$$

Termination. Chain termination occurs when two active centers come in close proximity and react with each other by combination or disproportionation. Both reactions yield completed macromolecules that no longer propagate chains. The termination mechanism is a function of the monomer, the solvent, the temperature, and the viscosity of the medium. Combination occurs when the two chain ends approach along their lines of centers and a carbon-carbon bond is reformed between them.

$$M_n{*} + M_m{*} \xrightarrow{k_{tc}} P_{n+m} \qquad 4.3a$$

Example: Two polystyrene chains terminating by coupling.

$$M_{n-1}{-}CH_2{-}CHR{*} + {*}RHC{-}CH_2{-}M_{m-1} \rightarrow$$

$$M_{n-1}{-}CH_2{-}CHR{-}\ RHC{-}CH_2{-}M_{m-1}$$

The activation energy for combination termination is low because one bond is formed from two radicals. Disproportionation occurs when the reactive center on one chain abstracts a hydrogen from the carbon atom neighboring the reactive center of a second molecule. The abstracted chain reforms a carbon-carbon double bond at its end.

$$M_n{*} + M_m{*} \xrightarrow{k_{td}} P_n + P_m \qquad 4.3b$$

Example: Two polystyrene chains terminating by disproportionation.

$$M_{n-1}-CH_2-CHR^* + {}^*RHC-CH_2-M_{m-1} \rightarrow M_{n-1}-CH_2= CHR + RH_2C-CH_2-M_{m-1}$$

Free radical chains also stop propagation and form polymer after undergoing chain transfer reactions. In this mechanism, the reactive center is not eliminated, but is transferred to another molecule in the medium, such as an initiator, a monomer, a solvent, or another polymer molecule. Chain transfer reactions are used intentionally to limit the molecular weight of a polymer chain. The rate expressions are similar to those of the propagation step and will be discussed in detail in the next section.

4.1.2 Kinetic Rate Expressions

The mechanisms described by Equations 4.1 through 4.3 can be transformed into kinetic rate expressions. It is usually assumed that the rate constants are independent of chain length. At high conversions, however, the diffusion rate of chain ends can slow dramatically, resulting in very low termination rates. At this point, the rate constants for propagation and termination depend on local mass transfer rates in the concentrated polymer solution (the gel effect). Since chain length affects diffusion of the polymer, rate constants at the gel point may be chain length dependent.

It is further assumed that the change in concentration of free radical species during the reaction is very small compared to the changes in the concentrations of other reacting species (the quasi-steady-state assumption). Both of these assumptions are probably valid for most free radical polymerizations in the time period between several seconds after initiator addition to the onset of the gel effect at very high conversion. There are only a few commercial reactions that occur in extremely short times (these are often ionic polymerizations). The quasi-steady-state assumption may not be appropriate at the onset of free radical initiations or for some ionic polymerization.

Quasi-Steady-State Assumption. It has been shown experimentally that the change in free radical concentration with time, dM*/dt, is small compared to the reaction rate of the propagating chain. Figure 4.2 shows typical concentrations of radicals and monomer as functions of time.

The rate-limiting step of free radical polymerizations often is the propagation step (Eq. 4.2), so the rate of polymerization is given by

$$R_p = -\frac{dM}{dt} = k_p \, M \, M* \qquad 4.4$$

A simple relationship for the free radical concentration, M^*, is needed to integrate Equation 4.4. The quasi-steady-state assumption can be used to generate another

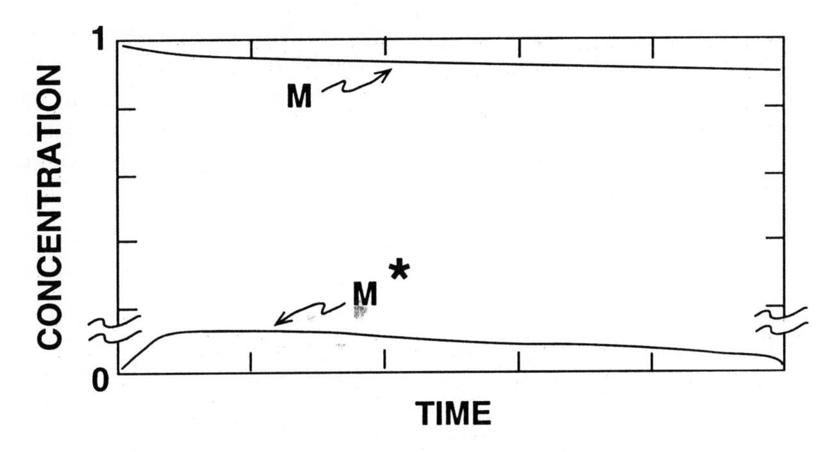

Figure 4.2 Monomer (M) and monomer radical (M^*) concentration as a function of time during chain polymerization.

equation. It states that the rate of radical generation is approximately balanced by the rate of radical destruction throughout the polymerization. This implies that the derivative of M^* with respect to time is zero, and that the concentration of free radicals changes slowly during the polymerization. The equation for M^* can be used to link the three major steps for chain polymerizations given in Equations 4.1 through 4.3.

The material balance equations follow a common convention in kinetic modeling (Carberry, 1976) and take the rate of an elementary reaction as proportional to the concentrations (mol/l) of reacting species. For the reaction of a moles of Species A with b moles of Species B to form c moles of Species C

$$a\,A + b\,B \xrightarrow{k_p} c\ C \tag{4.5}$$

with the rate equations

$$-\frac{1}{a}\frac{d\,A}{dt} = -\frac{1}{b}\frac{d\,B}{dt} = +\frac{1}{c}\frac{d\,C}{dt} = k_p\,A^a\,B^b \tag{4.6}$$

Equations 4.5 and 4.6 are used to complete the material balance on M^* and solve for M^* as a function of initiator concentration and rate constants. The initiator decomposition mechanism (Eq. 4.1a) leads to

$$R_i = -\frac{dI}{dt} = k_d\,f\,I = +\frac{1}{2}\frac{dR*}{dt} \tag{4.7}$$

where f is the efficiency of the initiator decomposition. The efficiency ranges between zero and one and is $0.5 \leq f < 0.8$ for many systems. The f factor is intended to account for the decomposition products that recombine to give initiator again. An alternate mechanism is initiator dissociation followed by the reaction of a free radical in a nonproductive manner, that is, so as not to start a chain. This mechanism could be modeled by including another rate equation. This type of mechanism could be important in solution polymerizations at low monomer concentrations.

For termination by disproportionation (Eq. 4.3b).

$$R_t = -\frac{1}{2}\frac{d\ M*}{dt} = k_{td}\ M*^2 \tag{4.8}$$

The steady-state assumption is

$$\frac{d\ M*}{dt} = 0 = \frac{d\ M*\ (init)}{dt} - \frac{d\ M*\ (term)}{dt} \tag{4.9a}$$

or

$$0 = 2\ k_d f I - 2\ k_{td} M*^2 \tag{4.9b}$$

and

$$M* = [k_d f I / k_{td}]^{1/2} \tag{4.9c}$$

Equation 4.9c can be substituted into Equation 4.4 to give an integrable equation for the change in monomer concentration with time.

$$R_p = -\frac{dM}{dt} = k_p\ M\left[k_d\ f\ I\ /\ k_{td}\right]^{1/2} \tag{4.10}$$

For this simple model, the rate of polymerization depends on the square root of the initiator concentration, and the rate is first order in monomer concentration. The polymerization of many common synthetic polymers—such as polystyrene, poly(vinyl chloride), and polyethylene—can be described by equations similar to 4.7 through 4.10 in bulk, solution, or suspension reaction systems. Emulsion polymerizations of free radical systems follow different kinetic models because the spatial dimensions of the monomer phase (100–8000 Å) are such that the free radical concentration is not considered to be continuously distributed through the particles.

For the case of thermal initiation, the polymerization rate is second order in monomer concentration. In practice, the polymerization rates deviate from first, zeroth, and second order in monomer concentration and half order in initiator concentration. The dependence of the rate of polymerization on monomer concentration may vary during the reaction rate with conversion (opposite the predictions of the rate expressions). Sometimes this variation is the result of an autoacceleration called the gel or Trommsdorff effect. At this point in the reaction, the diffusion rate of the radical chain ends of large molecules is decreasing and the termination rate falls. Chain end movement is limited as the solution becomes increasingly viscous during polymerization.

Several vinyl monomers are only sparingly soluble in their homopolymers. This effect complicates the physical description of these reactions and can lead to complicated rate expressions. In practice, semiempirical rate expressions may be used to model the processes. In the case of poly(vinyl chloride), a model that includes polymerization in the monomer and polymer phases has been successful.

Initiation, Initiators, and Decomposition Rate Constants. The initiation rate is composed of two steps (Eq. 4.1a and 4.1b). Usually the addition of a primary radical to a monomer (Eq. 4.1b) is much faster than the decomposition of the initiator. Therefore, initiator decomposition (Eq. 4.1a) is the rate-limiting step for the rate of chain initiation. Typical values for the dissociation rate constants and the activation energy for decomposition (E_d) are given in Table 2 of Appendix D. The temperature dependence of the initiation rate constant, k_d, typically follows an Arrhenius-type relation. The initiation rate constant, k_d, and the activation energy for initiation, E_d, may be determined independently from a polymerization system. These parameters are usually considered unique to a particular initiator molecule. Not every initiator radical produced will generate a chain. The initiator radical may recombine with another initiator radical or it may react with oxygen and other inhibitors. The effectiveness factor, f in Equation 4.7, depends on the polymerization environment and may depend on monomer concentration. This parameter is best determined during an actual polymerization. Odian (1981) describes methods for determining the effectiveness factor.

Propagation and Termination Rate Constants. Assuming that the initiator decomposition rate constant is known, the material balance on a batch polymerization will give the collection of coefficients, fk_p^2/k_{td} (Eq. 4.10). The study of propagation and termination rate constants usually requires unsteady-state experimen- tation, since these rates change during the course of the reaction. The rates change because the concentrations of reactants change. The rate constants that describe these reactions do not vary over normal ranges of concentrations. Propagation and termination rates are easier to study if free radical initiation can be started and stopped at will. This should allow the investigator to initiate a reaction, stop the initiation process, and watch the reactions slow as free radicals are terminated.

The rotating sector method is one way to perform such an experiment. Light is used to initiate free radicals of the monomer. If the light beam is broken on a periodic basis (by being blocked by a rotating, slotted disk), then decay of free radicals can be observed. Some method for measuring the concentration of free radicals with time must be used. The average lifetime of a growing radical is given by its steady-state value divided by its steady-state rate of disappearance. Flash photolysis can also be used.

Table 1 in Appendix D shows propagation and termination rate constants for several vinyl monomers. The group of terms, $f\,k_p^2/k_t$, is often reported as the experimentally measured rate constant for various temperatures and conversions. The values for this group may be accompanied by exponents for the initiator and monomer concentration dependence of the rate expression. The group of terms mentioned above and a rate constant for initiator decomposition are all that are needed to define a simple free radical polymerization system.

An analysis of activation energies for the initiation, polymerization, and termination processes helps define the way the overall reaction rate will change with temperature. The propagation, termination, and initiation rate constants can be represented by Arrhenius functions,

$$k_p = A_p \exp(-E_p/RT) \qquad 4.11$$

where k_p is the propagation rate constant, A_p is the collision frequency factor, and E_p is the activation energy for polymerization. Similar expressions can be written for the termination and initiation rate constants. As discussed above, the rate of polymerization is proportional to the group, k_p^2/k_t (or $k_p/(k_t^{1/2})$). The relationship between this group and the temperature of polymerization is

$$k_p / (k_t^{1/2}) = [A_p / (A_t^{1/2})] \exp[-(E_p - E_t/2)/RT] \qquad 4.12$$

where the variables subscripted with t refer to the rate constant, collision frequency, and activation energy for the termination process.

The value of the group, $E_p - E_t/2$, will define the change in the rate of polymerization with temperature. For photoinitiated reactions, the value of this group is about 5 to 6 kcal/mole. This means that the polymerization rate will change 30% to 35% for every 10°C difference near room temperature. For chemically initiated reactions, the appropriate group is $k_p (k_d/k_t)^{1/2}$ (which includes the initiator decomposition), with a value of 30 kcal/mol for many initiators. The activation energy for thermally initiated

EXAMPLE 4.1 MODEL FOR A METHYL METHACRYLATE POLYMERIZATION

Big Bucks Chemical vs. The IRS

Big Bucks Chemical Company has claimed as an income tax loss a tank car full of methyl methacrylate. The monomer did not have inhibitor in it and BBCC claimed that the carload polymerized while it was "lost" on a siding for two weeks in Orange, Texas. The ever-watchful IRS agent in charge of the BBCC audit has challenged this deduction. Assuming that the temperature in Orange was never below 30°C during the incident, determine how such polymer might have been made in two weeks and whether the deduction should be allowed. Assume that the polymer terminates by coupling and that in six months 5% of the monomer would react to form radicals.

The rates of free radical termination and chain propagation are given by Equations 4.4 and 4.8. The monomer can generate free radicals by a bimolecular mechanism, resulting in two free radicals.

(continued)

Initiation:

$$2M \xrightarrow{k_d} 2M*$$

The kinetic rate expression will be

$$\frac{1}{2}\frac{dM*}{dt} = k_d M^2$$

Termination:

$$-\frac{1}{2}\frac{dM*}{dt} = k_{tc}\left(M*\right)^2$$

Propagation:

$$-\frac{dM}{dt} = k_p \; M \; M*$$

This monomer material balance is incomplete because it should include the monomer lost to radical formation. For long chain lengths, the loss of monomer to free radical generation can be ignored. The steady-state assumption results in

$$\frac{dM*}{dt} = 2 \; k_d \; M^2 - 2 \; k_{tc} \; M*^2$$

The concentration of free radicals is

$$M* = \left(k_d / k_{tc}\right)^{1/2} \; M$$

This expression can be substituted in Equation 4.4 to give the rate expression for polymerization.

$$-\frac{dM}{dt} = k_p \left(k_d / k_{tc}\right)^{1/2} \; M^2$$

The thermal initiation rate constant can be evaluated from the information given in the problem. The rate expression for the bimolecular decomposition is

$$-\frac{1}{2}\frac{dM}{dt} = k_d \; M^2$$

The change in monomer concentration is integrated from the initial concentration (8.46 gmol/l, found by dividing the monomer density by its molecular weight) to 95% of that. The time period for this change is six months, so that the decomposition rate constant is

$$k_d = 1.80 \times 10^{-10} \text{l}/mol\text{–}s$$

(continued)

The rate constants at 60°C are $k_p = 0.705 \times 10^3$ and $k_t = 2.55 \times 10^7$, with activation energies of 4700 and 1200 kcal/mol. Equations 4.11 or 4.12 can be used to scale the rate constants with temperature.

$$\frac{k_{p,T_2}}{k_{p,T_1}} = \frac{\exp(-E_p/RT_2)}{\exp(-E_p/RT_1)} = \exp\left[-\frac{E_p}{R}\left(\frac{1}{T_2} - \frac{1}{T_1}\right)\right]$$

$$k_{p,T_2} = .705 * 10^3 \exp\left[-\frac{4700}{1.987}\left(\frac{1}{303} - \frac{1}{333}\right)\right]$$

$$= 349$$

$$k_{t,T_2} = 2.55 * 10^7 \exp\left[-\frac{1200}{1.987}\left(\frac{1}{303} - \frac{1}{333}\right)\right]$$

$$= 2.09 * 10^7$$

$$\frac{k_p}{k_t^{1/2}} = .0763 \frac{l}{mol - s}$$

Integrating the polymerization rate expression gives the amount of monomer remaining in the tank car at the end of two weeks.

$$\int -\frac{dM}{M^2} = \int k_p(k_d / k_t)^{1/2} dt$$

$$\frac{1}{M_2} - \frac{1}{M_1} = k_p\left(\frac{kd}{kt}\right)^{1/2} t$$

$$M_2 = 4.7 \frac{mol}{l}$$

About 45% of the original monomer has been converted to polymer. The IRS clearly should allow the deduction.

reactions is similar, but the collision frequency is very low since the generation of a free radical depends on a bimolecular collision (bimolecular mechanism).

4.1.3 Chain Length and Chain Transfer Reactions

Polymer molecular weight establishes many end-use properties. This section treats the relationship of molecular weight to polymerization conditions. Reactor operating conditions can have great effects on polymer molecular weight and molecular weight distribution. Some of these effects can be deduced from the kinetic rate expressions.

The kinetic chain length, λ_k, is defined as the average number of monomer molecules consumed per initiating radical. Over a specific time interval, the kinetic chain length is the ratio of the rate of polymerization to the rate of initiation. In other words, the average length of the chains is given by the ratio of how fast they add monomer to how fast new chains are started. Since the rate of initiation is equal to the rate of termination, the kinetic chain length is also the ratio of the polymerization rate to the termination rate.

$$\lambda_k = \frac{\frac{dM}{dt}}{\frac{dM*(term)}{dt}} \qquad 4.13$$

The polymerization rate is Equation 4.4, and the initiation and termination rates of Equations 4.7 and 4.8 can be substituted into Equation 4.13.

$$\lambda_k = \frac{k_p(fk_d I / kt)^{1/2} M}{2fk_d I} \qquad 4.14a$$

or,

$$\lambda_k = k_p M/[2(f k_d I k_t)^{½}] \qquad 4.14b$$

Using the expression for R_t,

$$\lambda_k = \frac{k_p\ M\ M*}{2k_t\ M^2} = \frac{(k_p\ M)^2}{2k_t\ R_p} \qquad 4.15$$

These equations show that the kinetic chain length is directly proportional to monomer concentration and inversely proportional to initiator concentration and polymerization rate. Higher free radical concentrations and faster polymerization rates should result in shorter polymer molecules. The number average molecular weight of a polymer is the molecular weight of a repeat unit multiplied by either λ_k or 2 λ_k, depending on whether termination is by disproportionation or by combination.

Equations 4.13 through 4.15 are only approximately correct, since other factors also influence chain length (thermal initiation, the gel effect, and chain transfer). Often the molecular weight of a polymer is much less than that predicted. Growing polymer chains can be terminated by the transfer of hydrogen, another atom, or a group of atoms to the growing polymer chain. The substituent may come from any compound in the system—monomer, initiator, polymer, or solvent. In synthetic rubber manufacture and other commercial processes, molecular weight is controlled by addition of a chain transfer agent. The rate mechanism can be written as

$$M_n * + XA \xrightarrow{k_{tr}} M_n - X + A* \qquad 4.16$$

where XA is the transfer agent and X is the group that transfers. The radical generated by the chain transfer reaction, A^*, can often reinitiate polymerization. If reinitiation cannot occur, then XA is called an inhibitor. The rate of chain transfer is

$$R_{tr} = -\frac{d\ XA}{dt} = k_{tr}\ M*\ XA \qquad 4.17$$

The reinitiation mechanism and rate equations are

$$A* + M \xrightarrow{k_a} M* \qquad 4.18$$

$$R_a = -\frac{d\ A*}{dt} = k_a\ A*\ M \qquad 4.19$$

The effect of chain transfer on the propagation rate depends on the rate of reinitiation by A^*. Table 4.5 shows the effects of chain transfer rate constants on the polymerization rates and the average chain length.

When significant chain transfer occurs, the number average degree of polymerization is no longer given by λ_k or $2\lambda_k$. The chain transfer reactions are new contributors to the termination rate, so that the degree of polymerization, $\overline{X}_n$, is given by the ratio of the polymerization rate to the sum of all termination rates.

TABLE 4.5 Effect of Chain Transfer on R_p and $\overline{X}_n$

Case	Transfer, Propagation, and Reinitiation Rate Constants	Type of Effect	Effect on R_p	Effect on $\overline{X}_n$
1	$k_p >> k_{tr}, k_a \simeq k_p$	Normal chain transfer	None	Decrease
2	$k_p << k_{tr}, k_a \simeq k_p$	Telomerization	None	Large decrease
3	$k_p >> k_{tr}, k_a < k_p$	Retardation	Decrease	Decrease
4	$k_p << k_{tr}, k_a < k_p$	Degradative chain transfer	Large decrease	Large decrease

$$\overline{X_n} = \frac{R_p}{R_t + k_{tr,m}M\ M* + k_{tr,s}M*\ S + k_{tr,i}M*\ I} \tag{4.20}$$

Defining the constants:

$$C_m = k_{tr,m}/k_p \tag{4.21a}$$

$$C_s = k_{tr,s}/k_p \tag{4.21b}$$

$$C_i = k_{tr,i}/k_p \tag{4.21c}$$

and inverting Equation 4.20

$$\frac{1}{\overline{X_n}} = \frac{k_t\ R_p}{\left(k_p\ M\right)^2} + C_m + \frac{C_s S}{M} + \frac{C_i\ R_p^2\ k_t}{f\ k_d\ k_p^2\ M^3} \tag{4.22}$$

$$\frac{1}{\overline{X_n}} = \frac{k_t\ R_p}{\left(k_p M\right)^2} + C_m + C_s\frac{S}{M} + C_i\frac{I}{M} \tag{4.23}$$

The above equations are quadratic in R_p, so that a plot of $1/\overline{X}_n$ versus R_p gives the effect of chain transfer. At low rates of polymerization, the relationship is nonlinear.

The value of the transfer constant for specific chain transfer agent varies greatly for reactions with different monomers. However, the order of reactivity of these agents usually remains the same. Carbon tetrachloride and carbon tetrabromide are very effective transfer reagents because of their weak carbon-halogen bonds.

Chain transfer to polymer can also occur. This leads to the generation of branched polymers. Chain transfer to polymer does not affect measurements of C_i, C_m, and C_s but does become important at high conversion. It can play a significant role in the physical properties of the polymer, since it can lead to branched chains. In order to evaluate the transfer constant, the number of branches produced in a polymerization relative to the number of monomer molecules polymerized must be known. Tables 4, 5, 6, and 7 in Appendix D list some values of chain transfer constants.

Chain transfer to polymer broadens the molecular weight distribution, but the average molecular weight stays about the same. The reactive center abstracts a hydrogen from the polymer chain. This terminates the previously growing chain and generates a radical site on the polymer chain, which then can continue to propagate. The branching site results in a much larger polymer molecule.

Polyethylene can have long and short chain branching. Example 4.2 describes branching reactions for this material.

EXAMPLE 4.2 POLYMERIZATION OF LOW DENSITY POLYETHYLENE

Polyethylene made by the high pressure free radical process has short chain branches, which form by molecular rearrangements of the polymer backbone. The branches are two to eight carbon atoms long and do not fit into the crystal structure of polyethylene. As a result, branched polymer has a lower melting point, crystallinity, and density than unbranched polymer. A comparison of commercial polymer properties is shown below.

Polymer	*Melting Pt.*	*Crystallinity*	*Density, g/cm³*
LDPE (branched)	105–115°C	45–55%	0.916–0.930
Unbranched PE	135°C	70–90%	0.960–0.970

A polyethylene free radical polymerization model could consider the following steps:

Initiation.
Either organic peroxides or oxygen can be used to initiate chains. The organic peroxide mechanism is the same as shown by Equations 4.1a, 4.1b, and 4.7 can be used for the initiation rate.

Propagation.
The propagation mechanism is similar to Equation 4.2, and Equation 4.4 describes the rate. Ethylene pressure can be used instead of ethylene concentration if desired.

Termination.
Termination can occur by disproportionation or combination (Eq. 4.6).
Five side reactions are important in LDPE polymerization and affect the molecular weight. They include chain transfer to ethylene, chain transfer with solvents, intramolecular chain transfer, intermolecular chain transfer, and β-scission of polymer radicals.

Chain Transfer Reactions.
The chain transfer reactions with monomer (ethylene) and solvent follow the mechanism of Equation 4.16 and the rate expression of Equation 4.17. Polyethylene radicals are very reactive and can extract hydrogen or halogen atoms from many organic compounds, even when present in small amounts.

Intramolecular Chain Transfer.
Short chain branching is the result of this "backbiting" mechanism. The free radical end abstracts a hydrogen from a carbon atom a short distance back on the chain (usually the fifth carbon back), creating a radical with a long chain on one side and a butyl group on the other. This radical can add monomer to continue the chain.

$$R-(CH_2)_4-CH_2\cdot \rightarrow R-\dot{C}H-C_4H_9$$

$$R-\dot{C}H-C_4H_9 + H_2C{=}CH_2 \rightarrow R-CH(C_4H_9)-CH_2-CH_2\cdot$$

(continued)

Rate of backbiting:

$$R_b = \frac{dM_n^*}{dt} = k_b M_n^*$$

Intermolecular Chain Transfer.
Polyethylene radicals can abstract a hydrogen from a neighboring chain. This forms a free radical with two long chains on either side. The new radical can add monomer to form a long chain branch. The mechanism is

$$R' - CH_2 - R'' + M_n^* \xrightarrow{k_{tr,p}} P_n - H + R' - \overset{*}{C}H - R''$$

The rate expression is

$$R_{tr,p} = k_{tr,p}\,(M_n^*)\,P$$

β-Scission of Polymer Radicals.
The polymer radicals formed by inter- or intramolecular chain transfer mechanisms may undergo cleavage to form a terminated chain and a new radical, which forms a new chain. The mechanism is

$$R - CH_2 - \overset{*}{C}RCH_2 - R_1 \rightarrow R^* + H_2C{=}CR_2(CH_2 - R_1)$$
$$\rightarrow R_1^* + (R - CH_2)R_2C{=}CH_2$$

The rate of β-scission is

$$R_\beta = k_\beta\,(M_n^*)$$

LDPE is made at very high pressures, often 2000 to 3000 atm, and at temperatures between 100°C to 200°C. Several effects become important at these conditions. The high pressure reduces the decomposition rate for organic peroxides. The reduced free volume of the polymer melt at high pressures probably increases the recombination rate of free radical fragments. At these pressures, initiator decomposition rates may be lowered by a factor of two. The polymerization mixture can separate into polymer-rich and monomer-rich phases. Ethylene fugacity may be used instead of ethylene pressure to model the kinetics.

Rate Constants.
Temperature and pressure affect the rate constants. They have been modeled as

$$k_i = A_i \exp\left(-\frac{E_i}{RT} - \frac{P\Delta V^*}{RT}\right)$$

where P is the pressure and ΔV^* is the activation volume in cm^3/mol. Doak (1989) gives activation volumes for some of the chain transfer reactions.

4.1.4 Inhibition and Retardation

Inhibitors and retarders react with initiating and propagating radicals and convert them to nonradical species or radicals with low reactivity. Inhibitors stop every radical they encounter, and polymerization is completely halted until they are consumed. Retarders are less efficient and only stop a portion of the radicals. The difference is merely one of degree. Monomer impurities may act as inhibitors or retarders. Inhibitors are intentionally added to monomers to prevent polymerization during purification, shipment, and storage. Inhibitors are removed before polymerization. Many free radical reactions exhibit an induction period, during which inhibitors and retarders are reacted away. Figure 4.3 shows the effects of inhibitor concentration of the conversion of monomer as a function of time.

In a normal polymerization, monomer conversion has a high initial rate and then continuously slows down until high conversions are reached. A retarder will slow the rate until it is consumed. Then, the conversion versus time curve may show the high rate expected for short times followed by a gradual decrease. By contrast, an inhibitor is usually much more efficient in stopping radicals. The initial reaction rate will be very low, almost negligible. Once significant portions of the inhibitor have been consumed, the reaction rate will increase. Because of the loss of initiator during this period, the rest of the polymerization proceeds quite slowly. Quinones, hydroquinones, and dihydroxybenzenes with oxygen (such as butylcatechol) can act as inhibitors. Other materials, such as O_2, S, C, and FeCl, can inhibit polymerizations. Molecular oxygen reacts with radicals to form unreactive peroxy radicals. Oxygen is known also as an initiator, probably acting via thermal decomposition of the peroxy radical.

$$R{-}M* + O_2 \longrightarrow R{-}M{-}O{-}O* \qquad 4.24$$

Inhibition reactions often follow the mechanism

$$R{-}M* + Z \xrightarrow{k_z} R{-}M + Z* \qquad 4.25$$

where Z is the inhibitor and $Z*$ is the unreactive Z radical.

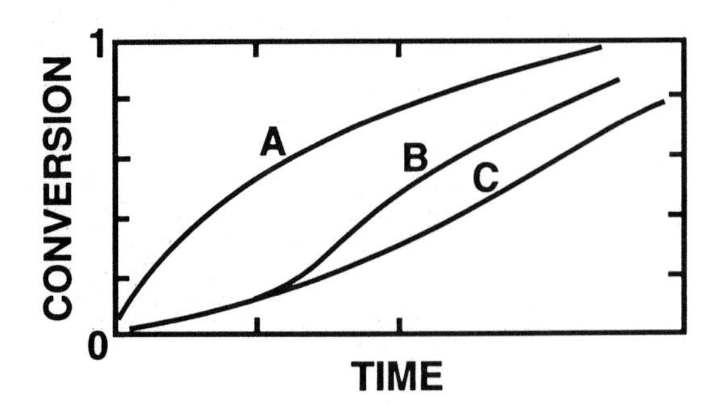

Figure 4.3 Fractional monomer conversion as a function of time for **A**, a normal chain polymerization; **B**, a polymerization in the presence of a retarder; and **C**, a polymerization in the presence of an inhibitor.

The inhibition constant, C_z, is the ratio of k_z to k_p. An interesting case of inhibition is the autoinhibition of allylic monomers. This is a consequence of degradative chain transfer to monomer. Table 3 in Appendix D lists some inhibitor constants.

EXAMPLE 4.3 THE EFFECTS OF IMPURITIES ON $\overline{X}_n$ FOR POLYETHYLENE

Light initiated reaction at 83°C with azoisopropane

Although there are a number of important chain transfer reactions for polyethylene, it can be difficult to find sets of self-consistent rate constants. This problem shows the effects of an impurity on the number average degree of polymerization for polyethylene made at a low temperature.

Laita (1959) and Laita and Macháček (1959) have a consistent set of rate constants for ethylene polymerized in benzene. The initiator, azoisopropane, does not have a high thermal dissociation rate at 83°C, but can be activated with light. The low pressure suppresses a number of the possible reactions. For example, termination by disproportionation occurs at a much higher rate than termination by coupling. Chain transfer to monomer (ethylene) and solvent (benzene) also seem to be suppressed. The dominant mechanism for controlling chain length seems to be chain transfer to initiator.

The kinetic rate constants found by Laita are

$$k_p = 4.7 * 10^2 \frac{l}{mol-s}$$

$$k_{tc} = 1.05 * 10^{-9} \frac{l}{mol-s}$$

$$k_{td} = 1.85 * 10^2 \frac{l}{mol-s}$$

$$k_d = 1.14 * 10^{-6} \frac{mol}{l-s}$$

The polymerization rate is given by

$$-\frac{dM}{dt} = k_p \, M* M$$

with

$$M* = \left(\frac{k_d \, I}{k_t}\right)^{1/2}$$

(continued)

When the initiator concentration is held constant ($I = 26.3*10^{-3}$ mol/l), a plot of the polymerization rate should be linear with monomer concentration. Figure 4.4 shows that the rate equation describes the data well. The degree of polymerization is given by

$$\overline{X_n} = \frac{R_p}{\Sigma R_t} = \frac{k_p\, M*\, M}{k_{td}(M*)^2 + k_{tr,e}(M*)M + k_{tr,s}\, M*S + k_\beta(M*) + k_{tr,i}\, M*\, I}$$

because disproportionation is the predominate mechanism. If chains terminated by disproportionation alone, then the degree of polymerization would be

$$\overline{X_n} = \frac{k_p\, M*M}{k_{td}(M*)^2} = \frac{k_p\, M}{k_{td}(M*)} = \frac{470\cdot 6.08}{185 * (1.27\times 10^{-5})}$$

$$\overline{X_n} = 1.22\times 10^6$$

This is extremely high and is not observed in practice. Therefore, the chain transfer mechanisms must be controlling the molecular weight. β-scission becomes important at higher temperatures but is not important at 83°C (Ehrlich and Mortimer, 1970). The chain transfer coefficients to ethylene and benzene are some of the lowest for the polyethylene system. The chain transfer to initiator may be significant. If azoisopropane is the primary chain transfer agent, then $\overline{X_n}$ should be approximated by

$$\overline{X_n} = \frac{k_p\, M*\, M}{k_{td}(M*)^2 + k_{tr,i}\, M*\, I + k\, M*C}$$

where C is a constant for constant monomer and solvent concentration. The inverse of $\overline{X_n}$

is

$$\frac{1}{\overline{X_n}} = \frac{k_{td}M*}{k_p M} + \frac{k_{tr,i}\, I}{k_p M} + \frac{k\, C}{k_p\, M}$$

In terms of initiator concentration, the inverse of $\overline{X_n}$ is

$$\frac{1}{\overline{X_n}} = \frac{(k_{td}\, k_d)^{1/2}}{k_p\, M} I^{1/2} + \frac{k_{tr,i}}{k_p\, M} I + C'$$

A plot of $1/\overline{X_n}$ versus I should vary in order between half at low initiator concentrations to one at high initiator concentrations. Figure 4.5 shows a plot of this function compared with data.

The model fits the data well; however, the value of the chain transfer coefficient, $k_{tr,i}$, has been adjusted slightly from the reported value.

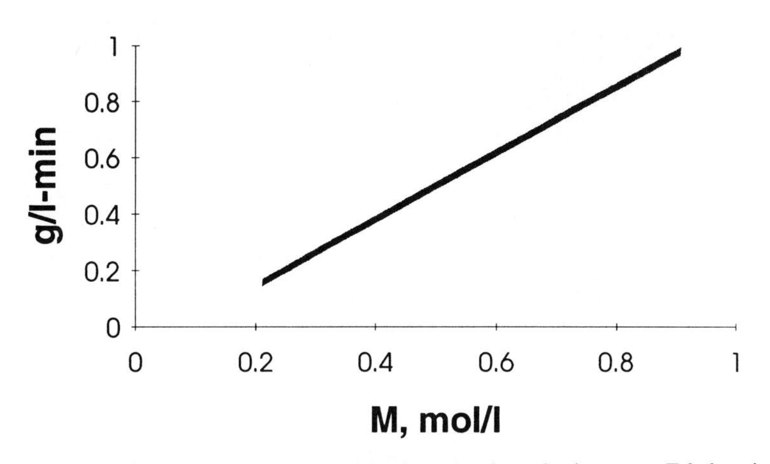

Figure 4.4 Effect of monomer concentration on polymerization rate—Ethylene in benzene (Laita, 1959). Data are squares and model is the curve.

4.1.5 Molecular Weight Distribution

Polymer produced by commercial processes has a distribution of molecular weights. The specific molecular weight distribution has a big effect on the final properties of the polymer. The polymerization kinetics and design and operation of the reactor help control this distribution.

The equations describing kinetic chain length (Eq. 4.13–4.15) show that the average molecular weight of the polymer will change during the polymerization as the concentrations of monomer and initiator change. Molecular weight distributions

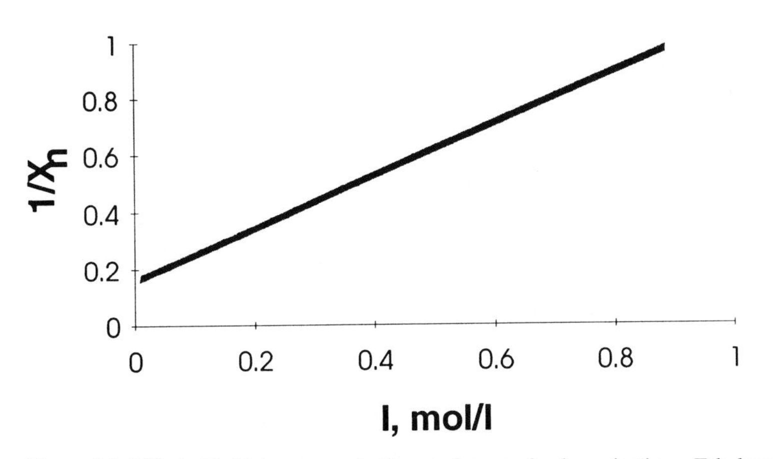

Figure 4.5 Effect of initiator concentration on degree of polymerization—Ethylene in benzene (Laita, 1959). Data are squares and model is the curve.

are easily calculated for low conversion cases where all of the parameters are nearly constant (M, I, k_p, k_t).

The molecular weight distribution for chain polymerizations in the absence of chain transfer can be derived by considering the probability that a given chain will have a specific molecular weight. Assume that termination is by disproportionation so that one polymer molecule is made from each chain. Let p = probability that a radical will continue to propagate. For a polymer with x repeat units, there will be $x-1$ groups that have propagated and 1 that has not. The probability of finding $x-1$ propagated groups in a molecule is p^{x-1}, and the probability of finding one unpropagated end group is $1-p$. Therefore, the probability of finding a polymer with x repeat units, N_x, is

$$N_x = p^{x-1}(1-p) \qquad 4.26$$

The sum of all N_x over all sizes x will be one, so N_x is equivalent to the mole fraction of molecules which are x-mers. The total number of molecules which are x-mers is:

$$N_{tx} = N_t N_x = N_0(1-p)N_x = N_0(1-p)^2 p^{x-1} \qquad 4.27$$

where N_t is the total number of molecules and N_0 is the number of monomer molecules initially present.

Equations 4.26 and 4.27 can be converted to a weight-fraction basis

$$W_x = x\,N_{tx}/N_0 \qquad 4.28$$

$$= x(1-p)^2 p^{x-1} \qquad 4.29$$

where x is the size of the molecule.

The number average and weight average degree of polymerizations can be derived from the number and weight fractions. The number average degree of polymerization, $\overline{X}_n$, is

$$\overline{X}_n = \frac{\Sigma\, x\, N_{tx}}{\Sigma\, N_{tx}} \qquad 4.30$$

$$= \Sigma\, x\, p^{x-1}(1-p) = (1-p)^{-1} \qquad 4.31$$

Since the number average molecular weight, $\overline{M}_n$, equals the number average degree of polymerization times the monomer molecular weight, Equation 4.31 can be used to determine $\overline{M}_n$

$$\overline{M}_n = M_o/(1-p) \qquad 4.32$$

The weight average degree of polymerization is obtained by summing the product of the size of the polymer, x, and its weight fraction, W_x overall sizes.

$$\overline{X}_w = \Sigma\, x\, W_x = \Sigma\, x^2(1 - p)^2\, p^{x-1} \tag{4.33}$$

$$\overline{X}_w = (1 + p)/(1 - p) \tag{4.34}$$

The average molecular weight is the monomer molecular weight times the weight average degree of polymerization.

$$\overline{M}_w = M_o\, (1 + p)/(1 - p) \tag{4.35}$$

The ratio of the weight average to number average molecular weights is characteristic of the distribution of molecular weights in a given sample and is called the polydispersity.

$$\frac{\overline{M_w}}{\overline{M_n}} = \frac{\overline{X_w}}{\overline{X_n}} = 1 + p \tag{4.36}$$

The polydispersity of polymers usually increases with extent of reaction and approaches two in the limit of large extents of reaction as indicated by Equation 4.36. The probability that a given monomer unit will react is proportional to the ratio of the polymerization rate to monomer rate (that is the ratio of the reactions that lead to polymer of the expected molecular weight to the reactions to polymerization, termination, and transfer). As the probability increases, the expected molecular weight increases.

$$p = R_p \,/\, (R_p + R_t + R_{tr}) \tag{4.37}$$

Equations 4.26 through 4.36 are based on the assumption that the probability of reaction, p, is a constant. Equation 4.20 is another method for computing the number average degree of polymerization. As previously discussed, $\overline{X}_n$ varies as the polymerization proceeds, as will the probability of reaction, p. The above equations should be considered correct for differential times of a polymerization. The actual distribution determined for a reaction starting at 0% conversion and stopping at 90% conversion comes from integrating Equations 4.30 and 4.33 over the extent of reaction.

When termination by coupling occurs, the expected size distribution is narrow compared to the disproportionation case.

$$N_x = (1 - p)^2\, (x - 1)\, p^{x-2} \tag{4.38}$$

$$W_x = 1/2\, x\, (1 - p)^3\, (x - 1)\, p^{x-2} \tag{4.39}$$

$$\overline{X}_n = 2/(1 - p) \tag{4.40}$$

$$\overline{X}_w = (2 + p)/(1 - p) \tag{4.41}$$

$$\frac{\overline{X_w}}{\overline{X_n}} = \frac{2 + p}{2} \tag{4.42}$$

For disproportionation, the limiting polydispersity at high conversions ($p = 1$) is 2; for termination by coupling, the polydispersity limit is 1.5. Polydispersities of 2 to 3 are common in bulk polymerization systems. Suspension polymerizations often have polydispersities of 5 to 7. For cases of chain transfer to polymer, such as occurs with low density polyethylene, the polydispersity may be 13.

4.1.6 Effects of Reactor Conditions of Polymer Properties

Example 4.2 explored some of the effects of reactor conditions on polymer properties. This section expands that discussion, using polystyrene as the model system. Polymer molecular weight is a key factor in setting end-use properties. The molecular weight is established in the reactor. Removing the heat of reaction is one of the key design problems in most polymerizations. For polystyrene, the heat of polymerization is −20.5 kcal/mol. It is not always easy to get this energy out, since the thermal conductivity of the melt is low and its viscosity is high. Rate models can help in the design of the reactor for heat removal.

EXAMPLE 4.4 EFFECTS OF SOLVENT AND TEMPERATURE ON $\bar{X}_n$, STYRENE IN TOLUENE

This example considers the polymerization of styrene in toluene with AIBN as the initiator. AIBN is a low temperature initiator, as shown in Tables 4.4 and Table 2 of Appendix D. AIBN dissociates to give two free radicals and one molecule of nitrogen. Usually, the concentration of AIBN is low enough so that bubble formation is not a problem. In aromatic solvents, AIBN is not 100% efficient and $f = 0.6$.

Polystyrene is known as an atactic polymer, yet stereoregular segments probably form. Based on steric effects, the activation energy for the formation of syndiotactic polymer is 0.5 kcal/mol less than that for the isotactic form. Lower temperatures and poor solvents should increase the amount of syndiotactic polymer formed. Chain transfer to monomer (styrene) can occur (Table 4 of Appendix D).

The rate equations previously derived hold over wide ranges of conversion, but they may not model the polymerization well in the gel phase. When polymer coils are in high concentration, they can form gels in which neighboring chains become cross-linked via entanglements. Under these conditions, the free radical chain ends can continue to add monomer but have much lower termination rates (Trommsdorff effect). The termination mechanism of combination and disproportionation require that free radical chain ends diffuse to each other.

(continued)

The termination step for the bulk polymerization of styrene is diffusion controlled at temperatures below 150°C. As the solution viscosity increases, entanglements suppress diffusion of chain ends and termination reactions are slowed. The effective termination rate constant becomes lower. In the absence of chain transfer reactions, the molecular weight would increase greatly as long as monomer continued to diffuse to reactive chain ends. Styrene dimer, which forms over a range of reaction temperatures, acts as a chain transfer agent.

The toluene solvent may slow the onset of the Trommsforff effect but probably will not prevent it. The following list shows some measured values for the radical lifetimes, as well as the relative change in the ratio of the polymerization to the termination rate constant.

These data are for the bulk polymerization of styrene at 50°C.

Conversion%	Radical Lifetime, s	k_p/k_t
0	2.29	1.0
34.7	1.80	1.4
36.3	9.1	8.0
39.5	13.9	13.2
43.8	18.8	21.3

The polymerization model will include chain transfer to dimer and solvent. The average degree of polymerization is

$$\overline{X_n} = \frac{k_p\, M}{\dfrac{k_{tc}(M*)}{2} + k_{tr,s}\, S + k_{tr,m}\, M}$$

and

$$M* = \left(\frac{k_d\, f\, I}{k_{tc}}\right)^{1/2}$$

The initiator concentration is obtained by integrating Equation 4.7.

$$-\frac{dI}{I} = k_d f I$$

$$\int_{I_o}^{I} \frac{dI}{I} = -k_d \int_o^t dt$$

$$\ln \frac{I}{I_o} = -k_d\, t$$

$$I = I_o \exp(-k_d\, t)$$

(continued)

The polymerization rate equation (Eq. 4.10) can be integrated to solve for the monomer concentration as a function of time.

$$-\frac{dM}{dt} = k_p M\left(\frac{k_d f I}{k_{tc}}\right)^{1/2}$$

$$\int_{M_o}^{M} -\frac{dM}{M} = \left(k_p^2 \frac{k_d f}{k_{tc}}\right)^{1/2} \int_o^t I^{1/2}\, dt$$

$$-\ln\frac{M}{M_o} = \left(k_p^2 \frac{k_d f}{k_{tc}}\right)^{1/2} I_o^{1/2}\left(-\frac{2}{k_d}\right) \exp\left(\frac{-k_d t}{2}\right)\Bigg|_o^t$$

$$\ln\frac{M_o}{M} = \left(k_p^2 \frac{2f}{k_d k_{tc}}\right)^{1/2} I_o^{1/2}\left[1-\exp\left(\frac{-k_d t}{2}\right)\right]$$

The rate constants used for these calculations were based on data available in literature reviews on styrene polymers. At 60°C, the values for the coefficients are

$$k_p = 180\ l/mol-s$$

$$k_{tc} = 3.28*10^6\ l/mol-s$$

$$k_{tr,s} = 0.011\ l/mol-s$$

$$k_{tr,m} = 0.010\ l/mol-s$$

$$k_d = 1.12*10^{-5}\ s^{-1}$$

There are a number of free radical initiators that can be used with polystyrene. The practical working temperature range is related to the initiator half-life. AIBN usually is used between temperatures of 60°C and 90°C. At lower temperatures, the polymerization rate may be too slow to be economical, at higher temperatures, the demands on the heat removal system become too great.

Commercial processes are operated at the highest practical rate, which usually means the highest practical temperature. If the temperature is too high and the initial initiator concentration is too low, the reaction mixture can become depleted in initiator before reaching the desired monomer conversion. Figure 4.6 compares the change in monomer concentration with time for two assumptions: a constant initiator concentration and a variable initiator concentration. The constant initiator concentration curve is a straight line on a semilog plot (see Eq. 4.10). The variable initiator model has lower polymerization rates, particularly at longer times when the initiator concentration is lower. High conversions of monomer to polymer are desirable, and the model predictions at long times are very important.

(continued)

Temperature has an important effect on the polymerization rate. Figure 4.7 shows monomer versus time curves for three different temperatures. As expected, high temperatures lead to fast rates. The polymerization rate at 70°C (the slope of the curve) is not linear with time and is greatly reduced at long times. The cause of the rate reduction is low initiator concentration.

The polymers made at these three temperatures will have different properties and molecular weights. Figure 4.8 shows the number average degree of polymerization for each temperature. As expected, the low temperature case (50°C) has the longest chain lengths over the conversion range calculated. The chain length decreases continuously as a result of the lower monomer concentrations at higher conversions.

The high temperature case (70°C) shows an increase in chain length at conversions greater than 40%. This is caused by the depletion of initiator. There is little initiator remaining, so that the ratio of the polymerization rate to the initiation rate increases.

The monomer/solvent ratio affects molecular weight in two ways (Fig. 4.9). Increasing this ratio increases the monomer concentration and the kinetic chain length. However, increasing the monomer concentration leads to greater chain termination by transfer to monomer. The chain transfer coefficients and the transfer agent concentrations are similar for the base case at 60°C, so both mechanisms contribute equally to termination. Also, changing the monomer/solvent ratio does not change the termination rate due to chain transfer. The coupling termination mechanism is not dependent on monomer concentration. With different transfer coefficients, there might be changes in the polymer weight caused by changing the relative importance of the chain transfer mechanisms.

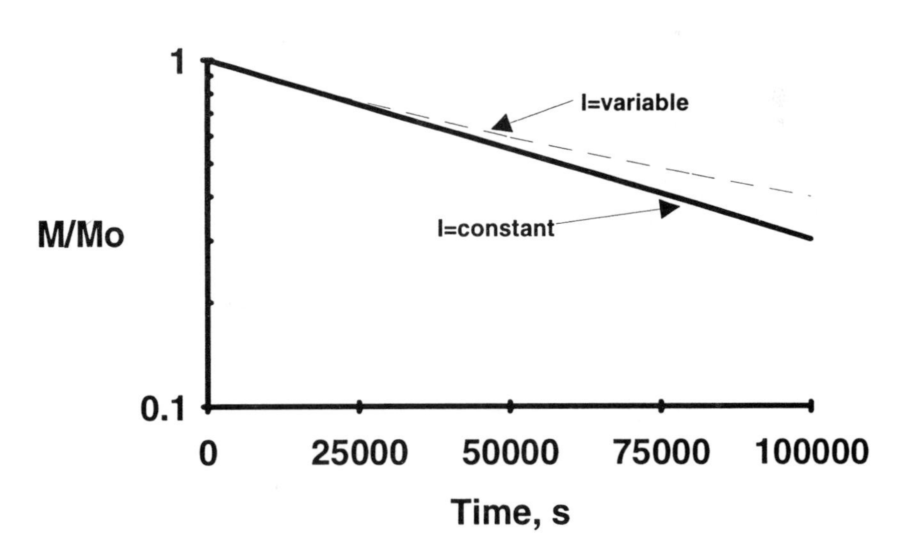

Figure 4.6 M/Mo versus time for variable and constant initiator concentrations. Styrene (4.3$\overline{M}$) in toluene (4.7$\overline{M}$) with AIBN (0.008$\overline{M}$) at 60°C. Constant initiator assumption is the solid curve; variable initiator assumption is the dashed curve.

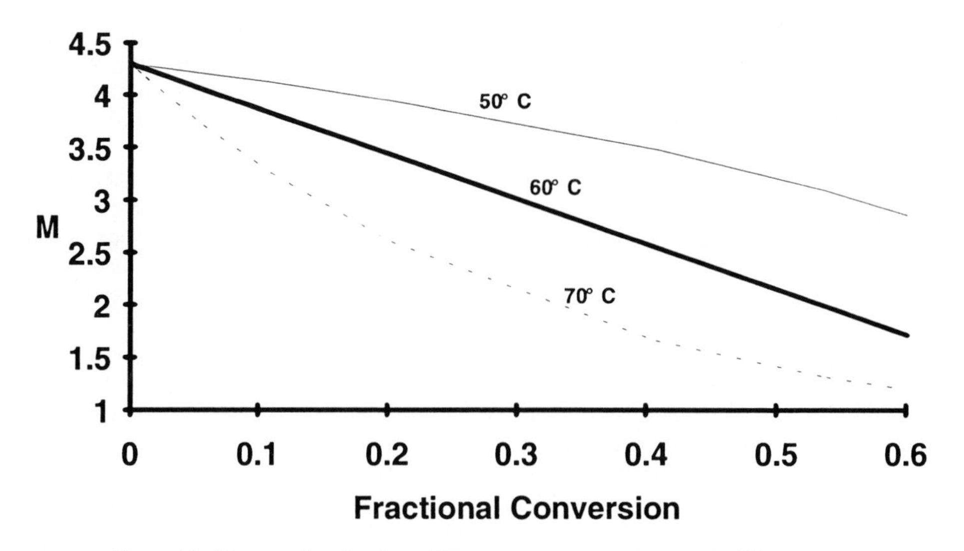

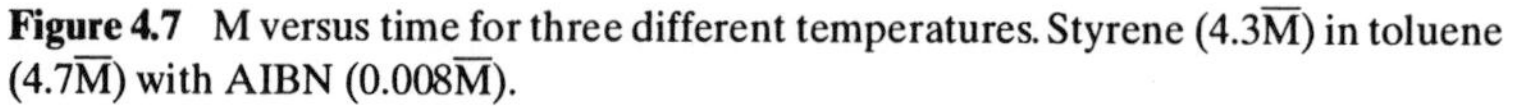
Figure 4.7 M versus time for three different temperatures. Styrene (4.3$\overline{M}$) in toluene (4.7$\overline{M}$) with AIBN (0.008$\overline{M}$).

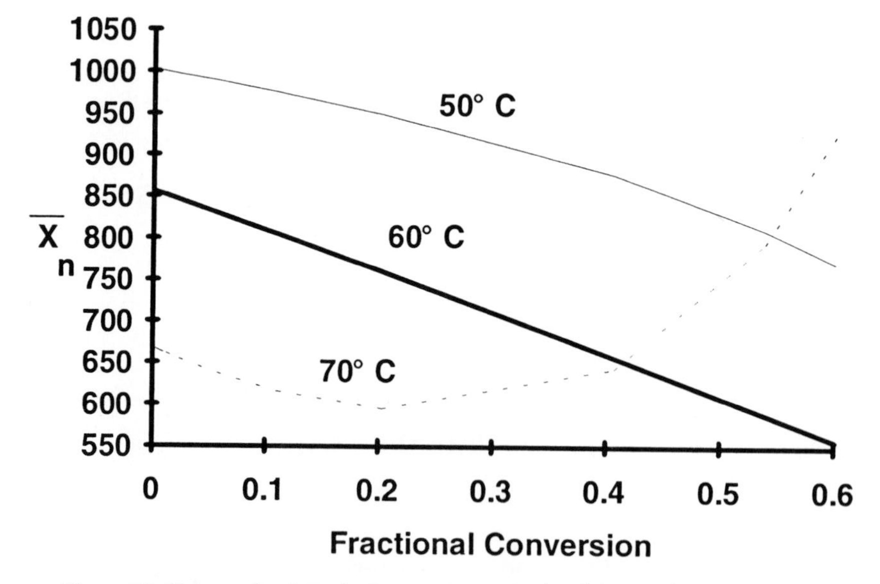

Figure 4.8 Degree of polymerization versus conversion for several temperatures. Styrene (4.3$\overline{M}$) in toluene (4.7$\overline{M}$) with AIBN (0.008$\overline{M}$).

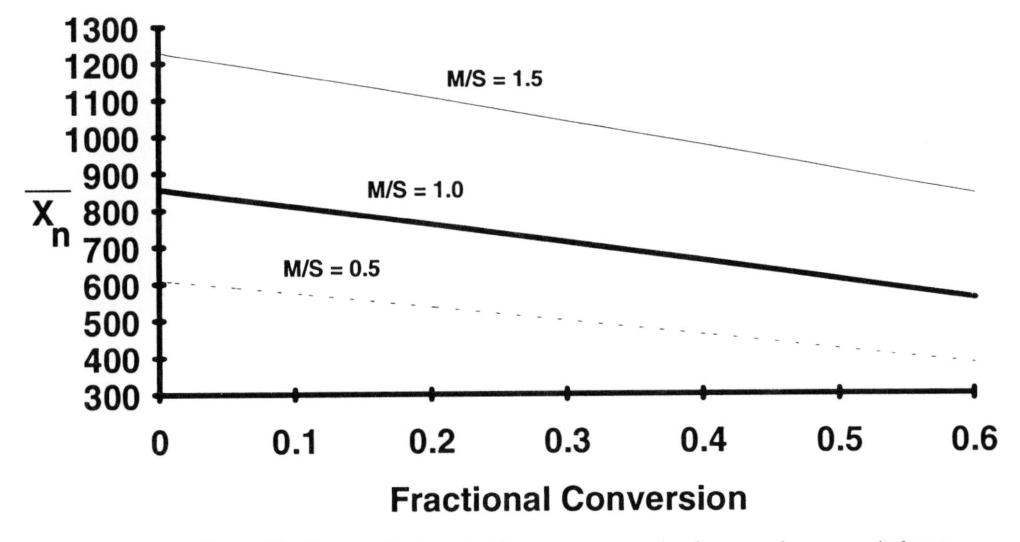

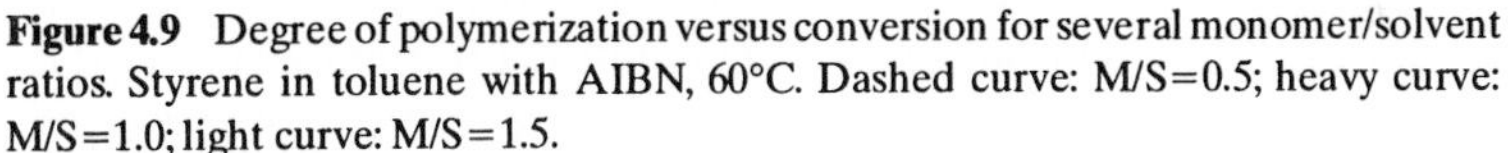
Figure 4.9 Degree of polymerization versus conversion for several monomer/solvent ratios. Styrene in toluene with AIBN, 60°C. Dashed curve: M/S=0.5; heavy curve: M/S=1.0; light curve: M/S=1.5.

4.2 CHAIN COPOLYMERIZATION

Chain polymerization can be used to produce alternating, block, or random copolymers. Alternating copolymers can be made by free radical polymerizations only when the kinetic rate constants have special values. Block and random copolymers have useful commercial properties. Examples include SAN (styrene-acrylonitrile copolymer), ABS (acrylonitrile-butadiene-styrene terpolymer), and Saran (vinyl chloride–vinylidene chloride copolymer). As discussed in Chapter 3, styrene is fairly brittle but is inexpensive. The addition of acrylonitrile improves its tensile properties, while the addition of butadiene improves its elastomer properties. The addition of vinylidene chloride to vinyl chloride results in excellent barrier properties. Table 4.6 shows some important copolymers.

Copolymerizations can be carried out in solvents or in solutions containing the monomers and the polymer. The presence of solvents or additional monomers affects the reactivity of the monomers and the radical chain ends of those monomers. Solving for the overall reaction rate in copolymerizations requires that rate equations must be written for each set of possible reactions and the rate constants must be determined. The copolymer produced will not have the same composition as the monomer mixture from which it is produced (except for special values of the rate constants). Some monomers that will not homopolymerize can be incorporated into copolymers.

4.2.1 Copolymer Composition Equation The major assumption in copolymerization is that the end of the propagating chain has chemical reactivity based on

TABLE 4.6 Commercially Important Vinyl Copolymers Prepared by Free Radical Polymerization

Polymer or Copolymer	Method of Manufacture	Major Uses
Styrene-butadiene rubber (SBR)	Emulsion	Tires, belting, flooring, molded goods, shoe soles, electrical insulation
Butadiene-acrylonitrile copolymer (nitrile rubber)	Emulsion	Fuel tanks, gasoline, hoses, adhesives, impregnated paper, leather and textiles
Acrylonitrile-butadiene-styrene copolymer (ABS)	Emulsion	Engineering plastics, household appliances, business machines, telephones, electronics, automotive parts, luggage
Poly(vinylidene chloride)*	Emulsion, suspension	Food packaging

*Includes copolymers, principally with vinyl chloride.

the last monomer unit and is independent of the rest of the chain. Another assumption is that the rates of initiation and termination are small compared to the consumption of monomers by propagation. The reaction equations for propagation of monomers M_1 and M_2 are

$$M_1* + M_1 \xrightarrow{k_{11}} M_1* \quad \text{homopropagation} \tag{4.43}$$

$$M_1* + M_2 \xrightarrow{k_{12}} M_2* \quad \text{cross-propagation} \tag{4.44}$$

$$M_2* + M_1 \xrightarrow{k_{21}} M_1* \quad \text{cross-propagation} \tag{4.45}$$

$$M_2* + M_2 \xrightarrow{k_{22}} M_2* \quad \text{homopropagation} \tag{4.46}$$

M_1* and M_2* are long chains of m units ending in monomer units M_1 and M_2. The rates of disappearance of the two monomers are

$$\frac{-dM_1}{dt} = k_{11}(M_1*)\, M_1 + k_{21}(M_2*)\, M_1 \tag{4.47}$$

$$\frac{-dM_2}{dt} = k_{12}(M_1*)\, M_2 + k_{22}(M_2*)\, M_2 \tag{4.48}$$

The ratio of rates at which the monomers enter the copolymer is

$$\frac{dM_1}{dM_2} = \frac{k_{11}(M_1*)\ M_1 + k_{21}(M_2*)\ M_1}{k_{12}(M_1*)\ M_2 + k_{22}(M_2*)\ M_2} \qquad 4.49$$

The steady-state assumption is used.

$$\frac{dM_1*}{dt} = \frac{dM_2*}{dt} = 0 \qquad 4.50$$

While the concentrations of the two chain end species, M_1* and M_2*, are rarely equal, their values change only slowly with time. Equation 4.50 becomes

$$\frac{-dM_1*}{dt} = k_{12}(M_1*)\ M_2 - k_{21}(M_2*)\ M_1 = \frac{-dM_2*}{dt} \qquad 4.51$$

and solving for M_1*

$$M_1* = \frac{k_{21}(M_2*)\ M_1}{k_{12}\ M_2} \qquad 4.52$$

Substituting Equation 4.52 into Equation 4.49 gives an equation that is proportional to M_2* in every term and results in

$$\frac{dM_1}{dM_2} = \frac{(k_{11}k_{21}\ M_1^2)/(k_{12}M_2) + k_{21}M_1}{k_{21}\ M_1 + k_{22}\ M_2} \qquad 4.53$$

Defining r_1 and r_2 as the monomer reactivity ratios

$$r_1 = k_{11}/k_{12} \qquad 4.54$$

$$r_2 = k_{22}/k_{21} \qquad 4.55$$

Equation 4.53 becomes

$$\frac{dM_1}{dM_2} = \frac{M_1(r_1\ M_1 + M_2)}{M_2(M_1 + r_2\ M_2)} \qquad 4.56$$

Equation 4.56 is the copolymerization equation. A similar result can be obtained by a statistical method. The monomer reactivity ratios will be between 0 and 1 if the monomers tend to polymerize. When $r_1 > 1$, M_1* tends to add M_1 preferentially to M_2. When $r_1 < 1$, M_1* tends to add M_2. When $f_1 = 0$, M_1 does not homopolymerize. Maleic anhydride, stilbene, and fumaric esters are examples of monomers that copolymerize but do not homopolymerize. Equation 4.56 can be expressed in terms of mole fractions. If f_i is the mole fraction of monomer i in the reacting mixture and F_1 is the mole fraction of monomer i in the copolymer for a short-time interval then

$$f_1 = 1 - f_2 = \frac{M_1}{M_1 + M_2} \qquad 4.57$$

$$F_1 = 1 - F_2 = \frac{dM_1}{dM_1 + dM_2} \tag{4.58}$$

The mole fraction of monomer 1 in the polymer can be related to the monomer concentrations in the reaction mixture,

$$F_1 = \frac{r_1 f_1^2 + f_1 f_2}{r_2 f_2^2 + 2 f_1 f_2 + r_1 f_1^2} \tag{4.59}$$

Equation 4.59 relates the instantaneous copolymer composition, F_1, to the instantaneous monomer composition, f_1. Figure 4.10 for $r_1 = r_2 = 1$ shows that $F_1 = f_1$ through the entire mole fraction range. When $r_1 r_2 = 1$, then

$$\frac{F_1}{1 - F_1} = r_1 \frac{f_1}{(1 - f_1)} \tag{4.60}$$

Plotting F_1 as a function of f_1 is analogous to an $x-y$ composition plot of vapor-liquid equilibria. When the product of the reactivity ratios equals one, the F_1-f_1 curve never crosses the 45° line. When $r_1 r_2$ is not equal to one, the F_1-f_1 curve can cross the diagonal in Figure 4.10. The intersection of the curve and the diagonal represents an *azeotropic* polymerization, in which the composition in the polymer equals the composition in the monomer phase. Once this condition is reached, the copolymer composition will not change with percent conversion. Alternating copolymers can be made in the special case when $r_1 = r_2 = 0$ or when the monomers will not homopolymerize. A practical example of a system that should form an alternating copolymer

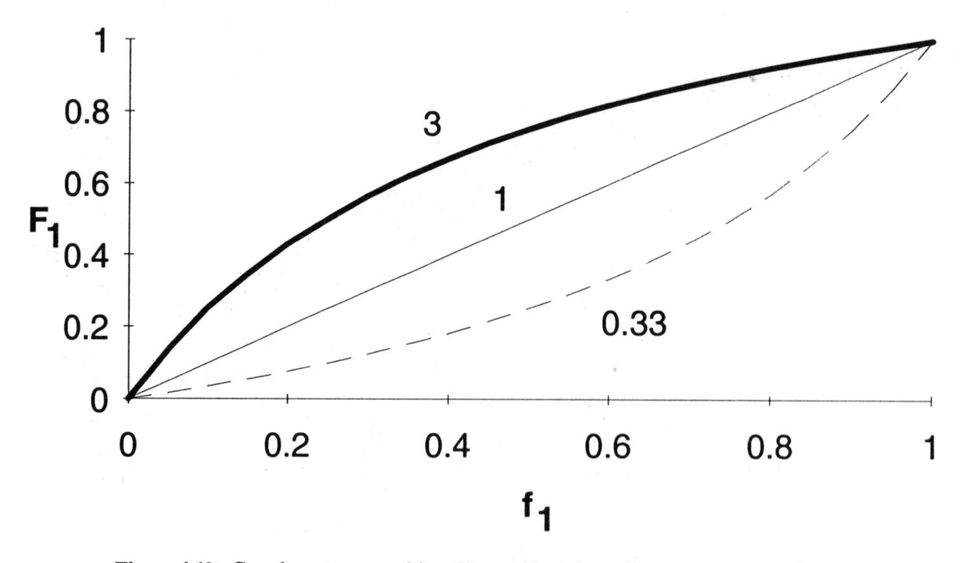

Figure 4.10 Copolymer composition, F_1, as a function of unreacted monomer fraction, f_1 for the product of the reactivity ratios equal to 1 and several choices of r_1.

TABLE 4.7 Q and e Values of Monomers

Monomer	Q	e
Butadiene	2.39	−1.05
Styrene	1.00	−0.80
Vinyl chloride	0.044	0.20
Vinylidene chloride	0.22	0.36
Methyl methacrylate	0.74	0.40
Acrylonitrile	0.60	1.20

is acrylonitrile-butadiene, for which $r_1 r_2 = 0.006$. An ideal random polymer will be formed if the product of the reactivity ratios equals one. A practical example is the system acrylonitrile and methyl vinyl ketone, for which $r_1 r_2 = 1.1$.

4.2.2 Reactivity Ratios and the Alfrey–Price Relationship

Some obvious problems with modeling copolymerizations are the measurement, correlation and prediction of the various rate constants. Alfrey and Price (Alfrey et al., 1952) worked out predictive relationships that allow the estimation of r_1 and r_2 from parameters associated with the individual monomers. Each monomer can be characterized by its general reactivity, Q, and its polarity, e. Q_1 is proportional to the relative stability of the radical of monomer 1. Resonance-stabilized monomers have high values of Q (for example, butadiene). Monomer reactivity ratios are listed in Table 8 of Appendix E, and values of Q and are listed in Table 4.7. The rate constant for the r_{12} reaction (Equation 4.44) is modeled by

$$k_{12} = P_1\, Q_2 \exp(-\, e_1\, e_2) \tag{4.61}$$

where P_1 is a measurement of the general reactivity of the radical, Q_2 measures the general reactivity of the monomer (and both define the resonance effects of the pair), and e_1, e_2 measure the polarities of monomers 1 and 2. The monomer reactivity ratios are given by

$$r_1 = (Q_1/Q_2) \exp[\,-e_1\,(e_1 - e_2)] \tag{4.62}$$

$$r_2 = (Q_2/Q_1) \exp[\,-e_2\,(e_2 - e_1)] \tag{4.63}$$

This empirical approach has been successful. Equations 4.62 and 4.63 are used to compute the cross-propagation rate constants from the homopropagation rate constants.

Reaction conditions will affect the copolymerization rate constants. Changing

EXAMPLE 4.5 COMPARISON OF CALCULATED RATE CONSTANTS WITH EXPERIMENTAL VALUES

The system is styrene copolymerizing with butadiene. Data are shown below.

1 = styrene, 2 = butadiene

Monomer	Q_i	e_i
1	1.00	−0.80
2	2.39	−1.05

Calculations:

$$r_1 = \frac{Q_1}{Q_2} \exp[-e_1(e_1 - e_2)]$$

$$r_1 = \frac{1.00}{2.39} \exp[.8 *(-.80 + 1.05)]$$

$$r_1 = 0.511$$

$$r_2 = \frac{2.39}{1.00} \exp[1.05\,(-1.05 + .80)]$$

$$r_2 = 1.84$$

The actual numbers at 50°C are $r_1 = 0.58$ and $r_2 = 1.35$. In this case, the styrene radical tends to add butadiene and the butadiene radical tends to add butadiene.

the temperature can change the selectivity of the free radical end group, but these effects usually are modest. Increasing the temperature often results in reduced selectivity. Increasing reaction pressure may reduce selectivity, but less is known about this phenomenon.

Equations 4.43 through 4.46 are rate mechanisms for chemically controlled termination steps. While these are reasonable, considerable research is being done now on diffusion-controlled termination in homopolymerizations.

4.2.3 Production of Uniform Composition Copolymers

Polymers of different chemical composition tend to be incompatible and form two-phase systems when mixed. Even copolymers of the same monomers with different compositions exhibit this behavior. Since the properties of such mixtures often are inferior to single-phase copolymers, methods for achieving uniform composition copolymers are important.

When r_1 and r_2 are both less than unity, the $F_1 - f_1$ curve crosses the line representing $F_1 = f_1$. At these compositions and rates, copolymerization occurs without a change in feed composition (azeotropic copolymerization). This method is commonly employed in the production of styrene-acrylonitrile or ABS resins.

For some monomer pairs, the azeotropic composition does not yield the most useful polymers (Saran for example). Here, uniform compositions are produced by stopping polymerization, recovering copolymer after the composition change becomes excessive, or by continuously adding the more reactive monomer to the reactor. Analysis of individual situations is often aided by construction of copolymer composition plots. These plots are prepared by integrating the monomer consumption equation, Equation 4.56. Recall that the equations presented so far have given instantaneous copolymer compositions. Consider a batch polymerization and the material balance on component 1,

$$\text{In} - \text{Out} = \text{Accumulation} \qquad 4.64$$

$$M f_1 - (M - dM)(f_1 - df_1) = F_1 dM \qquad 4.65$$

with M equal to the total moles of both monomers. Equation 4.65 simplifies to

$$M df_1 + f_1 dM = F_1 dM \qquad 4.66$$

Separating variables leads to

$$\int_{f_1(i)}^{f_1} \frac{df_1}{F_1 - f_1} = \int_{M_i}^{M_f} \frac{dM}{M} = \ln\left(M_f / M_i\right) \qquad 4.67$$

where $f_1(i)$ and M_i are the mole fraction of monomer 1 and the total moles of monomer at $t = 0$. This equation can be integrated numerically and has been integrated to a closed form that relates fractional conversion to monomer composition changes.

$$1 - \frac{M}{M_1} = 1 - \left(\frac{f_1}{f_1(i)}\right)^{\alpha}\left(\frac{f_2}{f_2(i)}\right)^{\beta}\left(\frac{f_1(i) - \delta}{f_1 - \delta}\right)^{\lambda}$$

$$\alpha = \frac{r_2}{1 - r_2}\ ; \beta = \frac{r_1}{1 - r_1}\ ; \delta = \frac{1 - r_2}{2 - r_1 - r_2}\ ; \lambda = \frac{1 - r_1 r_2}{(1 - r_1)(1 - r_2)} \qquad 4.68$$

Equation 4.68 can be applied to radical, cationic, and anionic chain copolymerizations.

EXAMPLE 4.6 SARAN COPOLYMER

Saran is a copolymer of vinylidene chloride and vinyl chloride. Suppose that your company wants to make a copolymer with the following criteria: the cumulative weight percent of vinyl chloride monomer should be less than 5% and no molecules should have more than 12% of vinyl chloride. Suggest appropriate conditions for achieving these properties in the copolymer from a batch reaction.

The first step is to compute the reactivity ratios (Table 4.7) using the Alfrey–Price relations and generate a plot of F_1–f_1 to determine the location of any azeotropes.

Using Equations 4.62 and 4.63:

$$r_1 = (.044/.22)\exp[-.20(.20-.36)] = .207$$

$$r_2 = (.22/.044)\exp[-.36(.36-.20)] = 4.72$$

$$r_1 r_2 = .977$$

The reactivity ratios, r_1 and r_2, indicate that both radical chain ends tend to add vinylidene chloride monomer. Equation 4.59 can be used to generate a table of f_1, F_1 data.

f_1	F_1	f_1	F_1
.01	.00214	.60	.240
.02	.0043	.75	.400
.05	.011	.90	.653
.10	.023	.999	.995
.25	.066	.9999	.9999 (azeotrope)
.40	.123		

The table of $f_1 - F_1$ data for this system indicates that the concentration of the monomer phase must be less than 40mol% to prevent copolymer with greater than 12mol% vinyl chloride. This can be solved by a trial and error procedure using Equation 4.68. Consider starting with 90% VCl_2 and 10% VCM and reacting to $f_1 = .40$. The total conversion of both monomers to polymer can be computed to determine the total fraction of VCM in the polymer (which must be less than 5%).

$$\alpha = \frac{r_2}{1-r_2} = \frac{4.72}{1-4.72} = -1.27$$

$$\beta = \frac{r_1}{1-r_1} = \frac{.207}{1-.207} = .261$$

(continued)

$$\delta = \frac{1 - r_2}{2 - r_1 - r_2} = \frac{1 - 4.72}{2 - .207 - 4.72} = 1.27$$

$$\lambda = \frac{1 - r_1 r_2}{(1 - r_1)(1 - r_2)} = \frac{1 - .977}{.793(-3.72)} = -.00058$$

$$1 - \frac{M}{M_i} = 1 - (.40/.10)^{-1.27}\,(.60/.90)^{.261}\left[(.10 - 1.27)/(.40 - 1.27)\right]^{-.00788}$$

$$\frac{M}{M_i} = .1546$$

Suppose that $M_i = 10$ moles with the initial amount of VCM at 1 mole and the amount of VCl_2 at 9 moles. At the end of the polymerization, there are .618 moles of VCM and .927 moles of VCl_2 remaining. The total moles reacted are $1 - .618 + 9 - .927 = 8.073$ moles. The fraction of VCM in the polymer is .382/8.07 = .047, which is less than 5%. The amount of VCM in the polymer can be increased by increasing the amount of VCM in the initial monomer phase.

4.2.4 Rate of Copolymerization

Initiation, termination, and propagation rates determine the copolymerization rate. If the termination is chemically controlled, there are three termination steps.

$$M_1{*} + M_1{*} \xrightarrow{k_{t11}} M_x \qquad 4.69$$

$$M_2{*} + M_2{*} \xrightarrow{k_{t22}} M_y \qquad 4.70$$

$$M_1{*} + M_2{*} \xrightarrow{k_{t12}} M_z \qquad 4.71$$

Equations have been derived for the polymerization rate, which depends on the sum of the four propagation rates and diffusion-controlled termination.

4.3 STEPWISE POLYMERIZATION

In stepwise polymerizations, two monomers react with the elimination of a small molecule such as water. They may also occur by scission or pseudocondensation, which occurs in the polymerization of polyurethanes. Condensation is another term applied to this type of polymerization. Esterifications, amidations, and formation of urethanes are all step polymerizations. Table 4.8 shows some examples of stepwise polymerizations, and Table 4.9 shows monomers used. Many of the monomers are solids at room temperature. Usually, two different functional groups are involved, with only one type being present on one monomer molecule. Monomers must have more than one functional group to form high molecular weight polymers. Difunctional monomers lead to linear polymer molecules. Monomers with more than two functional groups form branched polymers and can form cross-linked networks. The end groups are usually chosen so that reactivity between like end groups will be very small compared to reactivity between dissimilar end groups.

4.3.1 Examples of Stepwise Polymerization Mechanisms

The reactions of hexamethylene diamine with adipic acid (nylon) and the reaction of an amino acid with itself are typical examples of step polymerizations.

Nylon 6,6:

$$n\,H_2N{-}(CH_2)_6{-}NH_2 + n\,HO{-}CO{-}(CH_2)_4{-}CO{-}OH \rightarrow$$

$$H{-}[NH{-}(CH_2)_6{-}NH{-}CO{-}(CH_2)_4{-}CO]_nOH + 2n{-}1\,H_2O$$

TABLE 4.8 Examples of Stepwise Polymerizations

Reaction	Type of Chain	Name of Product
Dicarboxylic acids + glycols Dicarboxylic acid esters + glycols Dicarboxylic acid anhydrides + glycols	Polyesters	Dacron, Mylar, polyester-casting resins
Phthalic anhydride + glycerol	Polyesters	Alkyd resins (raw materials for coatings)
Dicarboxylic acids + diamines	Polyamides	Nylon 6,6
ω-Aminoacids + diamines	Polyamides	Rilsan
Diisocyanates + dicarboxylic acids	Polyamides	Lycra (elastic fiber)
Lactams	Polyamides	Nylon 6 (Caprolan)
Urea + formaldehyde	Polyureas	Aminoplastics (foam for insulation)
Melamine + formaldehyde	Polyamines	Melamine resins (Melmac, etc.)
Dichlorosilane + water	Polysiloxanes	Silicones (oils, elastomers, resins)
Bisphenols + phosgene	Polycarbonates	Lexan
Bischlorocarbonates + diamines	Polyurethanes	Elastomers, foams
Diisocyanates + glycols	Polyurethanes	Vulcollan (elastomers)
Diisocyanates + water	Polyureas	Foams (Moltopren)
Bisphenols + bisepoxides	Polyethers	Epoxy-resins (Epon, Araldite)
Cyclic ethers	Polyethers	Penton, polyethyleneoxide
Formaldehyde	Polyacetals	Polyformaldehyde, polyoxymethylene (Delrin, Celcon)

TABLE 4.9 Monomers Used for Stepwise Reactions

Monomer	Melting point, °C	Boiling point, °C	Application
Adipic acid	152	205	Nylon 6,6
Sebacic acid	134	295	Nylon 6,10
Maleic anhydride	53	202	Polyester casting resin
Phthalic anhydride	131	285	Polyester alkyds
Terephthalic acid			Polyester fibers
Hexamethylene diamine	40.7	200	Nylons
Ethylene glycol	−11	197	Polyester fibers
Glycerol	20	290	Polyester alkyds
Dimethyl dichlorosilane	−76	70	Silicones
Diethyl dichlorosilane		130	Silicones
4,4′-Diaminodicyclo-diphenyl sulfone	149		Polysulfone
4,4′-Diaminodiphenyl ether	187		Polyimides (Kapton)
2,6-Dimethyl phenol	49	212	Poly(phenylene oxide) Noryl

Polyamide:

$$n\, H_2N{-}R{-}CO{-}OH \rightarrow \mathrm{H}{-}[NH{-}R{-}CO{-}]_n OH + (n-1)\, H_2O$$

Step reactions can be represented in a shorthand form.

$$n(A{-}M_1{-}A) + n(B{-}M_2{-}B) \{ \quad A{-}(M_1{-}M_2){-}B + (2n-1)A{-}B$$

where A and B are the end groups and M_1 and M_2 represent the carbon chains of the monomers. The M_1-M_2 group is the structural unit of the polymer.

The functionality of a monomer molecule is the number of bonds the molecule may form during a polymerization. As in radical chain polymerization, the reactivity of a functional group is assumed to be independent of chain length and is dependent on the collision frequency of the group. The lower mobility of the polymer does change the time distribution of the collisions and actually increases the reactivity. However, this change in reactivity occurs quite early during the polymerization, and after the chain is several monomer units long, the reactivity changes very little with additional chain length.

There are a variety of reactions that can lead to a given polymer structure, as shown in Table 2.16 for polyester formation. Each of the seven end-group reactions require different monomers and reaction conditions. Each reaction will have different by-products as contaminants, which should be removed for stable products.

The first five methods—esterification, transesterification, diacyl chloride, anhydride, and acidolysis—have leaving groups that must be removed to force the system to high molecular weight polymer. The reaction of the diol with diacyl chlorides is very attractive because the HCl leaving group is strongly associated, making the reverse reaction rate very low. The various mechanisms lead to different intermediates and rate-limiting steps, making it difficult to develop generic models for step polymers similar to those for chain reactions.

4.3.2 Rate of Stepwise Polymerization: Diacid-Diol Example

Step polymerizations often have several known intermediates in the reaction mechanism sequence. The polyesterification of a diacid and a diol is a typical step polymerization, and reactions between the functional groups proceed as shown in Figure 4.11. The carboxylic acid group is protonated to form Species I. Species I can react with an alcohol group to form Species II, which loses water and a proton to form an ester

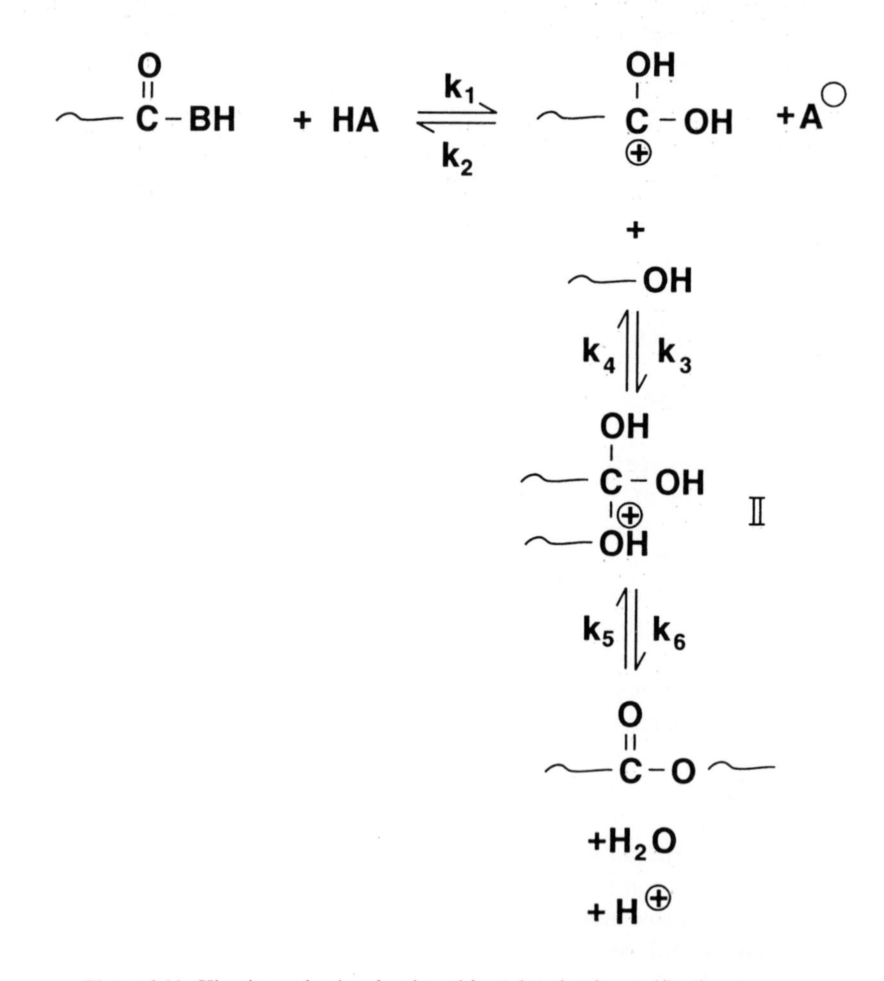

Figure 4.11 Kinetic mechanism for the acid-catalyzed polyesterification.

group in a chain. Any acid group and alcohol group can react with each other on collision.

The reactions shown are equilibrium reactions, with one leaving group and one product per reaction. To shift the equilibrium to the desired product, water must be removed from the reaction medium. The rate of step reactions is usually expressed in terms of the concentration of the reacting functional groups. For this problem, the rate of polymerization can be related to the rate of disappearance of carboxyl groups of the diacid,

$$R_p = \frac{-d\ COOH}{dt} \qquad 4.72$$

The slow step in the esterification is usually the reaction of the carbonium ion with the alcohol to form Species II, so that k_3 is usually much smaller than k_1, k_2, and k_5. When water is removed, k_6 does not occur. Species II usually reacts to form water and product, so that k_4 is not considered. Based on this mechanism, the rate of polymerization is

$$R_p = \frac{-d\ COOH}{dt} = k_3\ C^+(OH)_2\ OH \qquad 4.73$$

The concentration of carbonium ion can be obtained from the equilibrium coefficient for the first reaction, but this is difficult to observe experimentally.

$$K_1 = \frac{k_1}{k_2} = \frac{C^+(OH)_2\ A^-}{COOH\ HA} \qquad 4.74$$

The rate of polymerization can be written

$$R_p = \frac{k_1\, k_3\ COOH\ OH\ H^+}{k_2\ K_{HA}} \qquad 4.75$$

where K_{HA} is the acid dissociation constant.

Without a strong acid catalyst, the diacid will act as its own catalyst. The concentration, HA, becomes COOH and Equation 4.75 becomes

$$R_p = k\ COOH^2\ OH,\ k = \frac{k_1\, k_3}{k_2\ A^-} \qquad 4.76$$

The reaction becomes third order, with a second-order dependence on [COOH]. If the two functional groups are present in stoichiometric proportions, then the reaction rate is

$$R_p = -\frac{d\ M}{dt} = k\ M^3 \qquad 4.77$$

and

$$\int -\frac{d\,M}{M^3} = \int k\ dt \tag{4.78}$$

The integration of Equation 4.78 gives

$$2\,k\,t = \frac{1}{M^2} - \frac{1}{M_o^2} \tag{4.79}$$

If the monomer concentration is converted to extent of reaction by $M = M_o\,(1-p)$, Equation 4.78 becomes

$$2\ M_o^2\ kt = \frac{1}{(1-p)^2} - 1 \tag{4.80}$$

Equation 4.80 indicates that a plot of $(1-p)^2$ versus time should be linear. Figure 4.12 shows that this equation is followed for data between 80% and 93% conversion. In practice, such plots are not linear at low and high conversions. At low conversion, many esterifications are not properly modeled using Equation 4.80 and polyesterifications seem to be no different. At conversions of greater than 50%, there will be a major reduction in the COOH present in the solution. This leads to a big change in the polarity for the polymerization mixture. The change in polarity may affect the efficiency of the intermediate steps. Rate constants are usually measured at small incremental conversions and may vary as the chemistry of the system changes drastically.

In practice, there are other causes for the observed behavior at high conversion. High vacuums are used to remove water and may cause the loss of low molecular weight material. Catalysts are added in commercial process to reduce the rate to a second-order process and speed the reaction.

Other kinetic models have been used for Equation 4.76, including

$$R_p = k\ \mathrm{COOH\ OH} \tag{4.81a}$$

and

$$R_p = k\ \mathrm{COOH}^{3/2}\ \mathrm{OH} \tag{4.81b}$$

These lead to integrated rate expression in which $(1-p)^{-1}$ or $(1-p)^{3/2}$ is proportional to t. Figures 4.13 and 4.14 show these models and the data. Equation 4.81a fits the data best between 50% to 85% conversion, but shows significant deviations at higher conversions. Equation 4.81b does well at conversions below 45%, but shows significant deviations at higher conversions.

Since high molecular polymer is made at high conversion, Equation 4.80 is the best choice for a model. Adding a catalyst improves the situation, as the solution

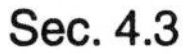

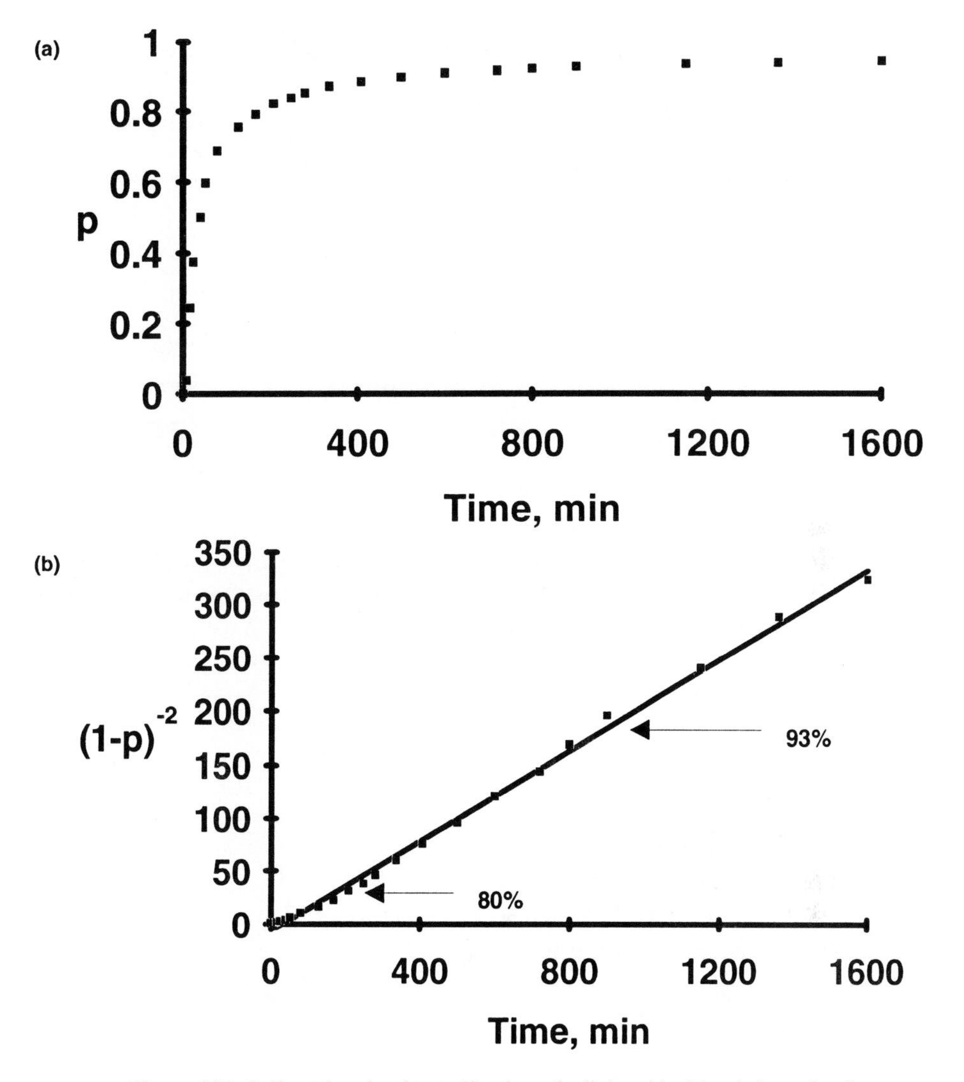

Figure 4.12 Self-catalyzed polyesterification of adipic acid with ethylene glycol at 166°C: **(a)** data for the extent of reaction, p, versus time; **(b)** third order model shown as solid curve. The fit is best between 80% and 93% conversion (Odian, 1981).

polarity is more nearly constant and makes Equation 4.77 second order in carboxyl groups. Figure 4.15 shows the fit of the second-order model to data taken with a catalyst. Commercial polymerization systems can be modeled by simple kinetic models when reaction conditions are carefully controlled. Table D10 in Appendix D shows such a model for the polymerization of a nylon 6,6 salt.

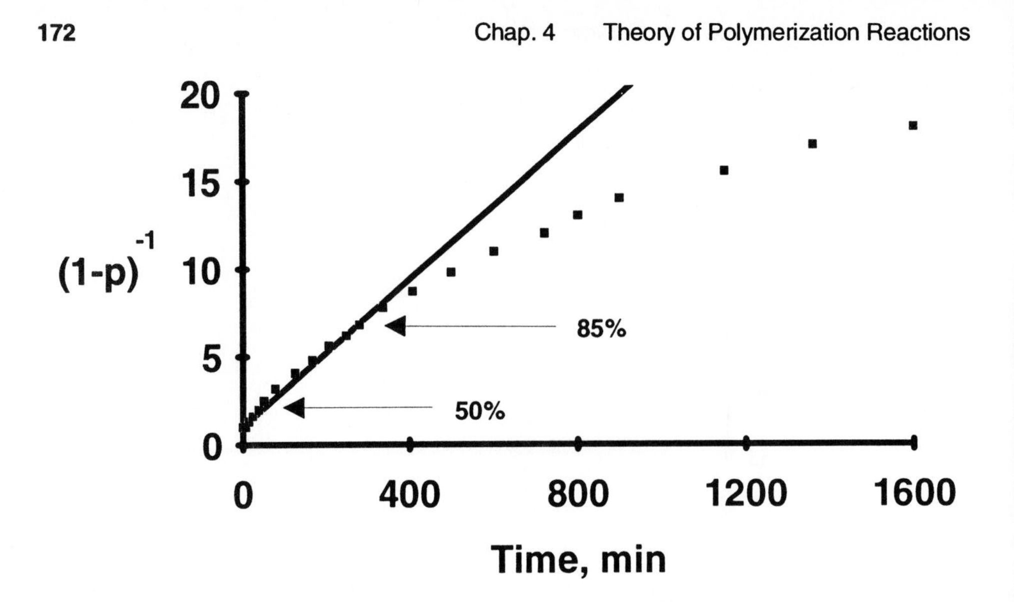

Figure 4.13 Second-order plot of the self-catalyzed polyesterification of adipic acid with ethylene glycol at 166°C. The fit is best between 50% and 85% conversion (Odian, 1981).

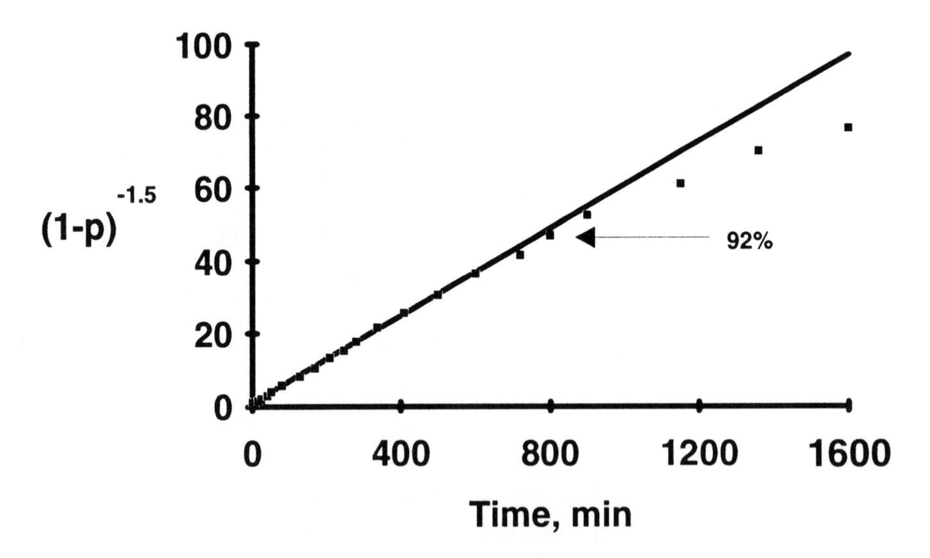

Figure 4.14 Two and one-half order plot of the self-catalyzed polyesterification of adipic acid with ethylene glycol at 166°. The fit is best below 45% conversion (Odian, 1981).

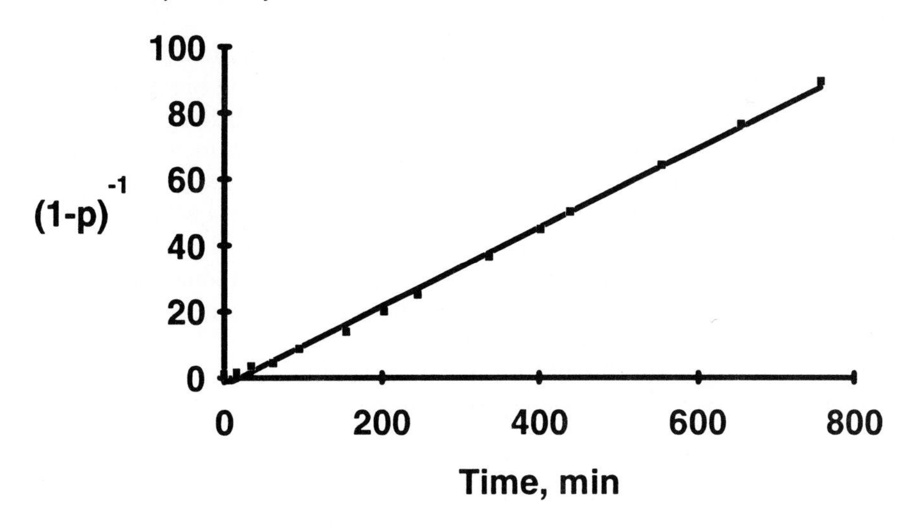

Figure 4.15 Polyesterification of adipic acid with ethylene glycol at 109°C catalyzed by 0.4mol% *p*-toluenesulfonic acid. The addition of an acid catalyst makes the formation of the carbonium intermediate rapid (Odian, 1981).

4.3.3 Functional Groups and Degree of Polymerization

The average molecular weight can be related to the fraction of groups that have reacted, similar to the approach used for chain polymers. Define p as the fraction of *functional* groups initially present that have undergone reaction at any time, t; p can be considered to be the extent of reaction. For equal moles of functional groups of Type A and B

$$p = \frac{A_o - A}{A_o} = \frac{B_o - B}{B_o} \tag{4.82}$$

where A and B are the number of functional groups remaining at time, t, and A_o and B_o are the initial concentrations or numbers of functional groups A and B. The degree of polymerization, D_p or $\overline{X_n}$ is the number of repeating structural units per chain molecule.

$$\overline{X_n} = \frac{No.\ of\ units}{No.\ of\ molecules} = \frac{2\left(A_o/2\right)}{A} = \frac{1}{1-p} \tag{4.83}$$

The number average molecular weight is

$$\overline{M_n} = M_o\,\overline{X_n} = \frac{M_o}{1-p} \tag{4.84}$$

Polymers of commercial interest have high degrees of polymerization. Equation 4.84 suggests that high molecular weights are achieved only when most of the end groups are reacted so that p is nearly one. When p equals 0.5, the average polymer is a dimer. When p equals 0.9, the average polymer is a decamer. Since commercial polymers usually have chain lengths of 1000 or greater, p must be larger than 0.999.

In real systems, impurities, side reactions, polyfunctionality, branching monofunctional units, and unequal stoichiometry all reduce $\overline{X}_n$. Stoichiometry is particularly important. For Equations 4.82 through 4.84 to be valid, the initial concentrations of A and B, A_o and B_o, should be equal. Clearly, if extents of reaction should be greater than 99.9% to accomplish commercial polymer properties, small differences in stoichiometry will make significant changes in the molecular weight of the final product.

Unequal stoichiometry of functional groups can be accounted for by defining r as the ratio of the initial end-group concentrations.

$$r = \frac{A_o}{B_o} \tag{4.85}$$

The number of monomer units available is half of the total moles of functional units. The number of monomer units initially is

$$\frac{A_o}{2} + \frac{B_o}{2} = A_o \frac{(1 + 1/r)}{2} \tag{4.86}$$

The fractions of unreacted A and B groups are

$$A = A_o(1 - p); B = B_o(1 - rp) \tag{4.87}$$

The number of chain ends at some time, t, is the sum of the unreacted A and B functional groups. The number of polymer molecules is half the number of chain ends. The number average degree of polymerization is

$$\overline{X}_n = \frac{No.\,of\,units}{No.\,of\,molecules} = \frac{A_o(1 + 1/r)}{A_o(1 - p) + B_o(1 - rp)}$$

$$= \frac{A_o(1 + 1/r)}{A_o[(1 - p) + (1 - rp)/r]}$$

$$= \frac{r + 1}{1 + r - 2rp} \tag{4.88}$$

At high conversions, p approaches one and the degree of polymerization approaches

$$\overline{X}_n = \frac{1 + r}{1 - r} \tag{4.89}$$

4.3.4 Polyfunctionality and Gelation

Molecules with functionality of greater than two cause branching, cross-linking, and gelation. Cross-linked polymers can form gels at some point during the reaction if there is sufficient cross-linking to cause nearly all molecules to be chemically joined. Such molecules are called gels. At the gel point, the polymer becomes insoluble in the reaction media and the molecule has the dimensions of its container. Polyurethane foams are an example of a commercial material that is cross-linked.

The critical point at which gelation occurs can be calculated by using probabilities. Consider a system of difunction molecules, A-A and B-B, in addition to A_f, a molecule with functionality, f, greater than two. The functionality, f, is the number of A-type functional groups on the molecule. For p_a = fraction of all A groups reacted, p_b = fraction of B groups reacted, and rr = A's in branch units per A's in the mixture, the branching coefficient, a, is

$$a = \frac{p_a\, p_b\, rr}{1 - p_a\, p_b(1 - rr)} \qquad 4.90$$

For some systems, there are more than one type of multifunctional group. For this case, an average functionality is defined

$$f_{ave} = \Sigma f_i rr_i \qquad 4.91$$

The branching coefficient, a, is the probability that a given branch unit leads via a chain to another branch unit. This coefficient will be affected by unequal stoichiometry between A and B groups. Defining $r = p_b/p_a$,

$$a = \frac{r\, p_a^2\, rr}{1 - r\, p_a^2(1 - rr)} \qquad 4.92$$

The critical branching point for gel formation is

$$a_{crit} = \frac{1}{f - 1} \qquad 4.93$$

and the extent of reaction at the gel point is

$$p_{crit} = \frac{1}{\left(r + r(f - 2)\, rr\right)^{1/2}} \qquad 4.94$$

For a trifunctional reagent (f = 3), the critical branching point for gel formation is when $a = 1/2$. The approach to computing the gel point given in Equations 4.91 through 4.94 is based on work by Flory and Stockmayer. Alternate approaches lead to different answers. The experimental values for the critical extent of reaction at the gel point are usually less than those calculated by Equation 4.94. This is thought to be caused by intermolecular cyclization, which is not accounted for in the statistical model. End groups can continue to react and cross-link the polymer further after the gel point is reached, but the gel point is important because it represents the conversion at which the product has solidlike properties.

4.3.5 Molecular Weight Distributions

The number average degree of polymerization can be calculated using an approach similar to that used to derive the molecular weight of chain polymers (see Eq. 4.38–4.42). For p = probability that A has reacted, the probability of finding an x-mer is

$$N_x = p^{(x-1)}(1-p) \qquad 4.95$$

the total number of molecules which are x-mers is

$$N_{tx} = N_t N_x = N_t p^{(x-1)}(1-p) \qquad 4.96$$

the number average degree of polymerization is

$$\overline{X}_n = N_o/N_{tx} = (1-p)^{-1} \qquad 4.97$$

the weight fraction of x-mers is

$$\begin{aligned} W_x &= xN_{tx} / N_0 \\ &= x(1-p)^2 p^{(x-1)} \end{aligned} \qquad 4.98$$

and the weight average degree of polymerization is

$$\begin{aligned} \overline{X_w} &= \frac{\Sigma\, x\, W_x}{\Sigma\, x\, N_{tx}} = \frac{\Sigma\, x^2\, p(x-1)(1-p)}{\Sigma\, x\, p(x-1)(1-p)} \\ &= \frac{1+p}{1-p} \end{aligned} \qquad 4.99$$

The polydispersity is

$$\frac{\overline{X_w}}{\overline{X_n}} = 1 + p \qquad 4.100$$

The polydispersity of step polymerization, Equation 4.100, is identical to the polydispersity of chain polymerizations when termination is by disproportionation only (Eq. 4.36). It only holds when the two monomers are initially present in identical amounts.

EXAMPLE 4.8 CALCULATION OF MOLECULAR WEIGHT DISTRIBUTIONS FOR STEP POLYMERIZATIONS

The molecular weight distribution of stepwise polymers is very sensitive to extent of conversion. This example problem shows several calculations for moderate extents of conversion to demonstrate the effects.

The first calculation shows the number average and weight average distribution for 70% conversion of end groups. As discussed in previous sections, 70% conversion of monomer in a chain polymerization would yield high molecular weight material. By contrast, 70% conversion of end groups in a step polymerization yields oligomers.

The number average and weight average degrees of polymerization are given by Equations 4.97 and 4.100. Figure 4.16 shows the two distributions for 70% conversion. Monomer (chain length = 1) is the highest concentration oligomer on a number average basis. Trimer is the highest concentration oligomer on a weight average basis.

Many step polymers are made in two-stage process. The first stage converts monomer to intermediate molecular weight oligomers. The second stage continues the polymerization after monomer and/or leaving groups have been removed. Equation 4.98 can be used to determine the conversion needed to convert 99wt% of the monomer to longer chains. A conversion of 90% will give a weight fraction of .01 for monomer ($x = 1$).

The weight frequency distribution can be used to optimize the production of specific oligomers. This is done by taking the derivative of Equation 4.98 with respect to x and setting it to zero at x = the desired chain length.

$$\frac{dW_x}{dx} = \frac{d}{dx}\left(x(1-p)^2\, p^{x-1}\right) = 0$$

$$0 = (1-p)^2\left(xp^{x-1}\, 1_n p + p^{x-1}\right)$$

$$0 = x\, 1_n p + 1$$

Figure 4.17 shows the conversion for the maximum frequency for various chain lengths.

4.4 Ring-opening Polymerization

Ring-opening polymerization of cyclic monomers of ethers, acetals, esters, and siloxanes are becoming more important. Some major commercial polymers are shown in

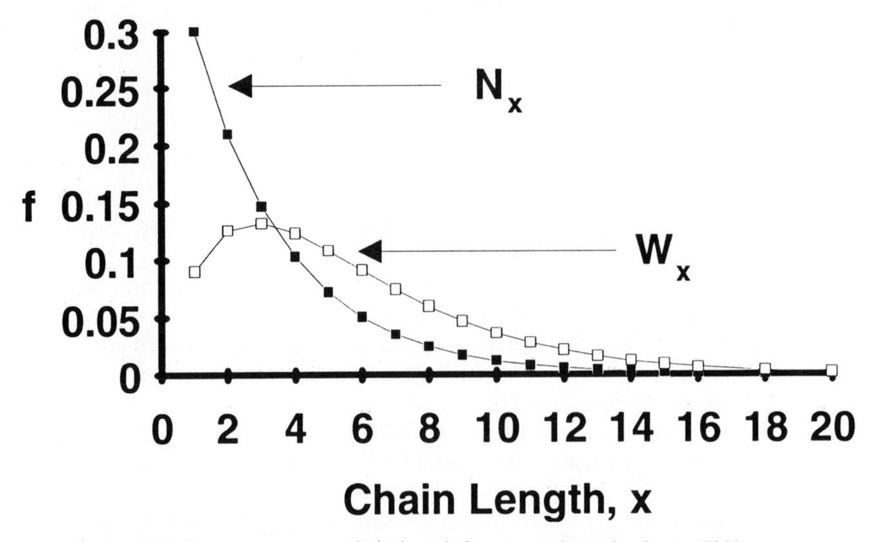

Figure 4.16 Frequency versus chain length for step polymerization at 70% conversion. Solid squares, number frequency; open squares, weight frequency.

Table 4.10. The polymers formed during these polymers may be classified as condensation or step polymers on the basis of their functional groups. Polymerization rates are high for 3 and 4 member rings because of the high strains in these monomers. Rates are also high for rings of 7 to 11 members.

Initiation results in the opening of the ring to form an initiator species, M^*, which is either an ion or a neutral molecule.

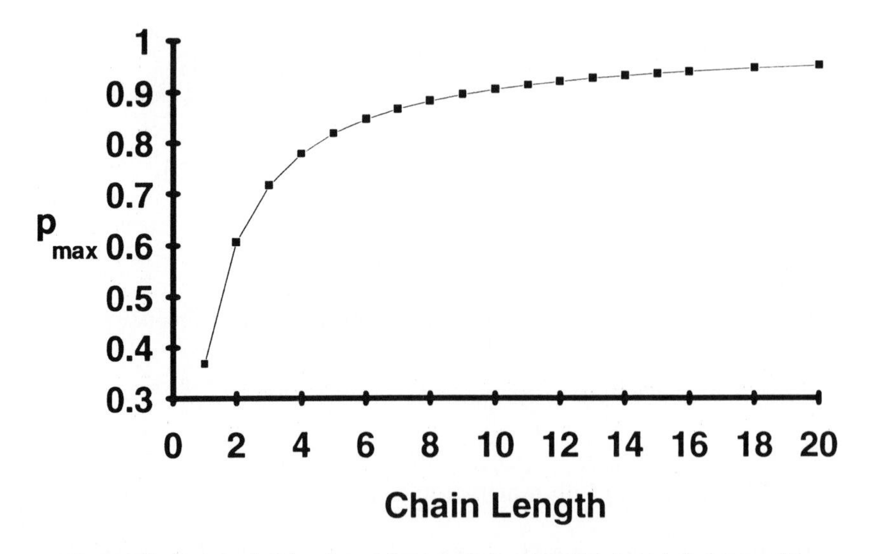

Figure 4.17 Extent of conversion, p, for the maximum production of oligomers with chain length, x.

TABLE 4.10 Commercially Important Polymers Prepared by Ring-opening Polymerization

Polymer Type	Polymer Repeating Group	Monomer Structure	Monomer Type
Polyalkene	$\text{-}[CH{=}CH(CH_2)_x]\text{-}$	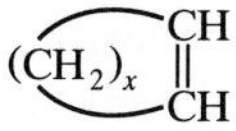 $(CH_2)_x$ ring closed by $CH{=}CH$	Cycloalkene
Polyether	$\text{-}[(CH_2O)]\text{-}$	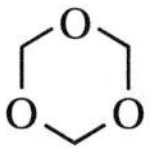 six-membered ring with three O atoms	Trioxane
Polyether	$\text{-}[(CH_2)_xO]\text{-}$	 $(CH_2)_xO$ ring	Cyclic ether
Polyester	$\text{-}[(CH_2)_xC({=}O){-}O]\text{-}$	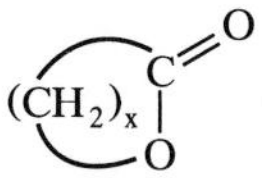 $(CH_2)_x$ ring closed by $C({=}O){-}O$	Lactone
Polyamide	$\text{-}[(CH_2)_xC({=}O)NH]\text{-}$	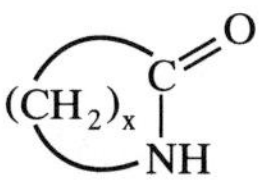 $(CH_2)_x$ ring closed by $C({=}O){-}NH$	Lactam
Polysiloxane	$\text{-}[Si(CH_3)_2O]\text{-}$	 $[Si(CH_3)_2]_x$ ring	Cyclic siloxane
Polyphosphazene	$\text{-}[P(Cl)_2{=}N]\text{-}$	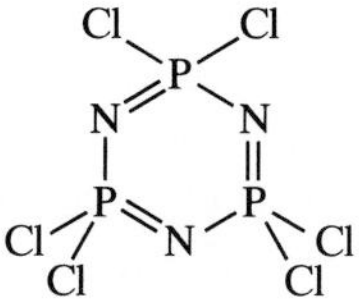 $(NPCl_2)_3$ ring (alternating P and N, two Cl on each P)	Hexachloro-cyclotriphos-phazene
Polyamine	$\text{-}[CH_2CH_2NH]\text{-}$	$CH_2{-}CH_2$ ring closed by NH	Aziridene

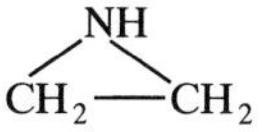

$$R - Z + C \longrightarrow M* \qquad 4.101$$

where *R-Z* represents the ring monomer and C is the catalyst. Water can be a molecular initiator. The initiator species grows by

$$M* + R - Z \longrightarrow M - R - Z* \qquad 4.102$$

Table 9 in Appendix D shows a complete kinetic mechanism for the ring-opening polymerization of nylon 6. It includes the ring-opening of monomer, ε-caprolactam, and dimer. Dimers often form in equilibrium with monomer in these systems. Polymer can grow by addition of monomer or dimer. The reaction can terminate by polycondensations between two polymers or by the reaction of polymer with monofunctional acids added to control molecular weight. Ring-opening polymers made by ionic mechanisms are analyzed in Section 4.6.

4.5 STEREOSPECIFIC AND COORDINATION COMPLEX POLYMERIZATION

Table 4.11 shows a number of stereospecific polymer products. Coordination complex catalysts result in a high degree of regularity in the polymer structure and cause a similar insertion of the monomer at the active group of the catalyst-polymer complex. The very active catalyst allows the polymerization to be run at low temperature, reducing side chain branching and chain transfer reactions. A comparison of properties of isotactic and atactic polymers is shown in Table 4.12. The catalysts may be supported on suspended solids or may be complexes dissolved in the reaction medium.

Reaction Mechanisms: Polypropylene Example. The polymerization rate of propylene with Ziegler–Natta catalysts is given by

$$- d M/dt = k_p C^* M \tag{4.103}$$

where k_p is the propagation constant, C^* is the concentration of active sites, and M is the concentration of monomer. In practice, polymerization rates increase and then decline to stationary levels. The initial increase may be due to the activation of new reactive centers on the catalyst. Because the catalyst is present on a support dispersed through the polymerization media, mixing conditions affect its performance. Polymerization systems are optimized by considering the ratio between the catalyst and the activator, their concentration in the medium, hydrogen concentration, temperature, agitation, and type and amount of Lewis base. The effects differ depending on whether the reaction is carried in liquid or gas phases.

The most common theory on chain propagation is that the monomer inserts into a metal-carbon bond that is polarized on the catalyst. This carbon atom has a weak negative charge. The active site is thought to be an atom of a transition metal with an octahedral shape and a vacant position. The other positions are occupied by alkyl groups from the aluminum alkyl reaction and four chlorine atoms. Figure 4.18 shows the generation of the active center following the reaction of the titanium chloride with the aluminum alkyl and the insertion of the monomer. Monomer coordinates with the transition metal, forming a π complex. The monomer inserts in the weakened $Ti - C$ bond. Next, the vacancy site on the titanium and the growing chain exchange positions

TABLE 4.11 Commercially Available Polymers Synthesized with Complex Coordination Catalysts

Polymer	Principal Stereochemistry	Typical Uses
PLASTICS		
Polyethylene, high density (HDPE)	—	Bottles, drums, pipe, conduit, sheet, film, wire and cable insulation
Polypropylene	Isotactic	Automobile and appliance parts, rope, cordage, webbing, carpeting, film
Poly(1-butene)	Isotactic	Film, pipe
Poly(4-methyl-1-pentene)*	Isotactic	Packaging, medical supplies, lighting
1,4-Polyisoprene	*trans*	Gold ball covers, orthopedic devices
Ethylene-1-butene† copolymer (linear low-density polyethylene, LLDPE)	—	Blending with LDPE, film packaging, bottles
Ethylene-propylene block copolymers (polyallomers)	Isotactic	Food packaging, automotive trim, toys, bottles, film, heat-sterilizable containers
Polydicyclopentadiene‡	—	Reaction injection molding (RIM) structural plastics
ELASTOMERS		
1,4-Polybutadiene	*cis*	Tires, conveyer belts, wire and cable insulation, footware
1,4-Polyisoprene	*cis*	Tires, footware, adhesives, coated fabrics
Poly(1-pentenylene) (polypentenamer)†	*trans*	Tires
Poly(1-octenylene) (polyoctenamer)†	*trans*	Blending with other elastomers
Poly(1,3-cyclo-pentenylene-vinylene) (norbornene polymer)†	*trans*	Molding compounds, engine mounts, car bumper guards
Polypropylene (amorphous)	—	Asphalt blends, sealants, adhesives, cable coatings
Ethylene-propylene copolymer (EPM, EPR)	—	Impact modifier for polypropylene, car bumper guards
Ethylene-propylene-diene copolymer (EPDM)	—	Wire and cable insulation, weather stripping, tire side walls, hose, seals

*Usually copolymerized with small amounts of 1-pentene.
†1-Hexene and 1-octene used in smaller amounts.
‡Synthesized by metathesis polymerization of the corresponding cycloalkene.

(these will not be equivalent in the crystal lattice of the catalyst). This mechanism is termed monometallic, because it requires only the participation of the transition metal. Transfer reactions with hydrogen, monomer, and aluminum alkyl stop chain growth and cause a new chain to be formed. When the catalyst groups are precipitated on a support, polymer grows over the catalyst particle. Figure 4.19 shows a catalyst particle of about 50 μm in diameter and a polypropylene particle of 1600 μm in diameter from the reactor.

TABLE 4.12 Comparison of Isotactic and Atactic Polymer Properties

Polymer	Isotactic (crystalline) Tm · °C	Isotactic (crystalline) Density, g/cm^3	Atactic (amorphous) Tm · °C	Atactic (amorphous) Density, g/cm^3
Polyethylene	135	0.96	105	0.92
Polypropylene	160	0.92	72	0.85
Poly-1-butene	128	0.91	65	0.87
Polystyrene	230	1.08	100	1.06

4.6 IONIC POLYMERIZATION

Ionic polymerizations are used to produce stereoregular polymers with high tacticity. Most monomers containing $-C{=}C-$ bonds will undergo free radical polymerization, but not all can be made to polymerize via ionic mechanisms. The selectivity of ionic polymerizations is due to the requirements for stabilization of the anionic and cationic reactive groups used to propagate the chain. Vinyl monomers containing electron-releasing substituents will undergo cationic polymerization. Examples of electron-releasing substituents include alkoxy, phenyl, and 1,1 dialkyl groups. Anionic polymerizations can be carried out if electron-withdrawing groups—such as nitrile, carboxyl, phenyl, or vinyl—are attached to one of the carbons in the double bond.

While the mechanism of ionic polymerizations is a chain reaction, there are two important differences between ionic polymerizations and free radical polymerizations. The initiation reactions require low activation energies, so that the polymeriza-

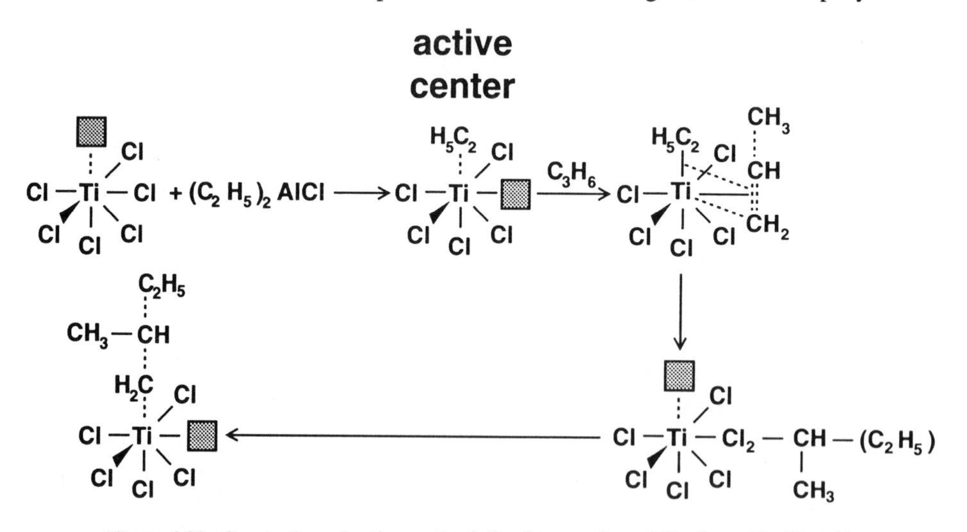

Figure 4.18 Generation of active center following reaction of titanium chloride with aluminum alkyl and the insertion of the monomer.

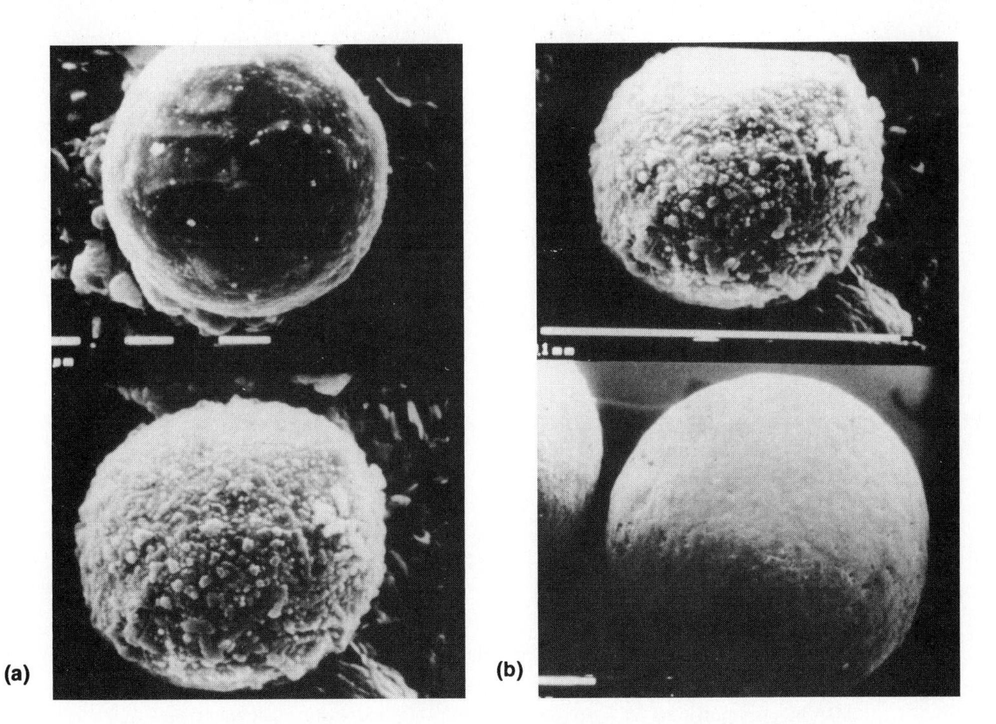

Figure 4.19 Polypropylene growth mechanism with catalyst particles: **(a)** comparison of catalysts before (above) and after (below) the first polymerization step; **(b)** comparison between prepolymer (above) and final polymer (below) morphologies. Published with the permission of HIMONT Incorporated., Wilmington, DE.

tion rate varies only slightly with temperature. The reaction rates can be extremely rapid. In many cases, ionic polymerizations are controlled explosions with monomer being fed on pressure demand to a reactor with internal reflux to provide cooling. The anionic polymerization of styrene in THF occurs at −70°C and the cationic polymerization of isobutylene occurs at −100°C.

Ionic chain ends do not react with each other, so there is no coupling or disproportionation termination reaction. In general, the solvents used do not react with the ionic catalysts. Impurities can be added to control chain length. Additives—including water, alcohols, acids, amines, and oxygen—react with the polymerizing ions to form inert species.

Table 4.13 shows commercially important polymers prepared by ionic polymerization. The dienes probably represent the highest volume of polymers made by ionic reactions (see Table 2.4). Copolymers can be prepared using ionic catalysts. The kinetic models include r_i's to characterize the monomer pair, but these parameters are different from those for free radical copolymerizations.

TABLE 4.13 Commercially Important Polymers Prepared by Ionic Polymerization

Polymer or Copolymer	Major Uses
CATIONIC*	
Polyisobutylene and polybutenes† (low and high molecular weight)	Adhesives, sealants, insulating oils, lubricating oil and grease additives, moisture barriers
Isobutylene-cyclopentadiene copolymer‡ ("butyl rubber")	Inner tubes, engine mounts and springs, chemical tank linings, protective clothing, hoses, gaskets, electrical insulation
Isobutylene-cyclopentadiene copolymer	Ozone-resistant rubber
Hydrocarbon§ and polyterpene resins	Inks, varnishes, paints, adhesives, sealants
Coumarone-indene resins‖	Flooring, coatings, adhesives
Poly(vinyl ether)s	Polymer modifiers, tackifiers, adhesives
ANIONIC¶	
cis-1,4-Polybutadiene	Tires
cis-1,4-Polyisoprene	Replacement for natural rubber
Styrene-butadiene rubber (SBR)**	Tire treads, belting, hose, shoe soles, flooring, coated fabrics
Styrene-butadiene block and star copolymers	Flooring, shoe soles, artificial leather, wire and cable insulation
ABA block copolymers (A = styrene, B = butadiene or isoprene)	Thermoplastic elastomers
Polycyanoacrylate††	Adhesives

*$AlCl_3$ and BF_3 most frequently used coinitiators.
†"Polybutenes" are copolymers based on C_4 alkenes and lesser amounts of propylene and C_5 and higher alkenes from refinery streams.
‡Terpolymers of isobutylene, isoprene, and divinylbenzene are also used in sealant and adhesive formulations.
§Aliphatic and aromatic refinery products.
‖Coumarone (benzofuran) and indene (benzocyclopentadiene) are products of coal tar.
¶*n*-Butyllithium most common initiator.
**Contains higher *cis* content than SBR prepared by free radical polymerization.
††Monomer polymerized by adventitious water.

4.6.1 Cationic Polymerization

Alkenes can be polymerized via cationic methods because the groups on the double bond will release electrons. These reactions can be initiated by protonic acids that generate an ion pair at the chain end or by Lewis acids. Butyl rubber, a mixture of isobutylene and isoprene monomers, is polymerized at −100°C, using Lewis acids as catalysts. A chlorinated solvent is normally used as the reaction medium, and $AlCl_3$ along with water as a cocatalyst are used to initiate the process. The catalyst is thought to react to form an ion pair with water. The carbon with the electron-withdrawing groups becomes a carbonium ion in the pair, and the hydrogen attaches to the other carbon of the double bond. New monomer adds at the ion pair to allow the chain to grow.

EXAMPLE 4.9 CATIONIC POLYMERIZATION OF STYRENE (H_2SO_4 IN ETHYLENE DICHLORIDE AT 25°C)

Polystyrene can be made by cationic polymerization with H_2SO_4. This example develops rate equations and estimates of the degree of polymerization. The acid dissociates immediately in the reaction mixture and initiates a cation on the monomer as soon as it is added.

$$H\text{–}A + M \rightarrow H\text{–}M^{+}A^{-} \qquad 4.104$$

The number of reacting chains is determined by the concentration of the Lewis acid. The chain propagates by adding monomer.

$$H\text{–}M_n^{+}\,A^{-} + M \rightarrow H\text{–}M_{n+1}^{+}\,A^{-} \qquad 4.105$$

A termination step occurs in this process through the combination of propagating carbonium ions. The reaction destroys the ion pair and terminates the chain.

$$H\text{–}M_n\text{–}M^{+}A^{-} \xrightarrow{k_t} H\text{–}M_nMA \qquad 4.106$$

The termination reaction is monomolecular.

$$-\frac{d\left(HM_n^{+}\,A^{-}\right)}{dt} = k_t\left(H\,M_n^{+}\,A^{-}\right) \qquad 4.107$$

The concentration of chain ends is

$$ln\frac{\left(HM_n^{+}\,A^{-}\right)_2}{\left(HM_n^{+}\,A^{-}\right)_1} = -k_t\,t_2 \qquad 4.108$$

where $(HM_n + A^{-})_1 = H_2SO_4$ at the beginning of the reaction. The propagation rate is

$$-\frac{dM}{dt} = k_p\,M{*}\,M = k_p\left(H_2SO_4\right)\exp\left(-k_t\,t\right)M \qquad 4.109$$

and the concentration of monomer as a function of time is

$$M_2 = M_1\exp\left[\frac{k_p}{k_t}\cdot H_2SO_4\cdot\left[\exp(-k_tt) - 1\right]\right] \qquad 4.110$$

(continued)

Rate constants for this process are

$$k_p = 7.6\ 1/mol\text{-}s$$

$$k_{tr,M} = 1.2*10^{-1}\ 1/mol\text{-}s$$

$$k_t = \begin{matrix} 4.9*10^{-2}\ s^{-1} \text{(spontaneous)} \\ 6.7*10^{-3}\ s^{-1} \text{(combination)} \end{matrix}$$

Figure 4.20 shows the concentrations of monomer and cation during the polymerization. Figure 4.21 shows monomer versus time curves for the proposed

4.6.2 Anionic Polymerization

Catalysts used for this type of polymerization include metal amides, alkoxides, alkyls, aryls, hydroxides, and cyamides. Alkali organic compounds—such as phenyllithium, butyllithium, and sodium naphthalene—form an initiator complex with solvents having unshared electron pairs (Lewis acids). An example is phenyllithium reacting with styrene.

$$(C_6H_5)-Li + H_2C{=}CH(C_6H_5) \rightarrow (C_6H_5)-CH_2-C(C_6H_5)\ H^-\ Li^+ \qquad 4.111$$

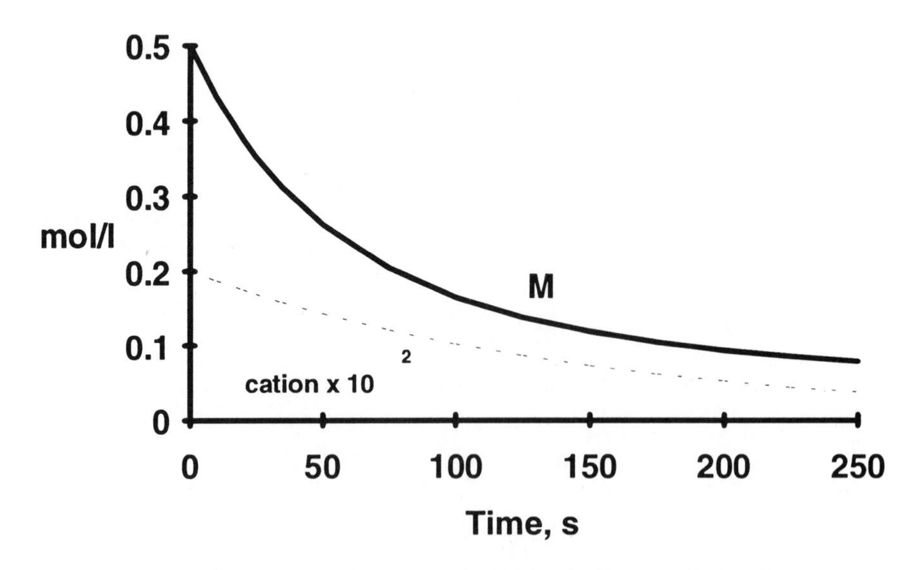

Figure 4.20 Polymer production versus time for M_o =0.5 mol/l. Cationic polymerization of styrene. Dashed curve, cation concentration; solid curve, styrene concentration.

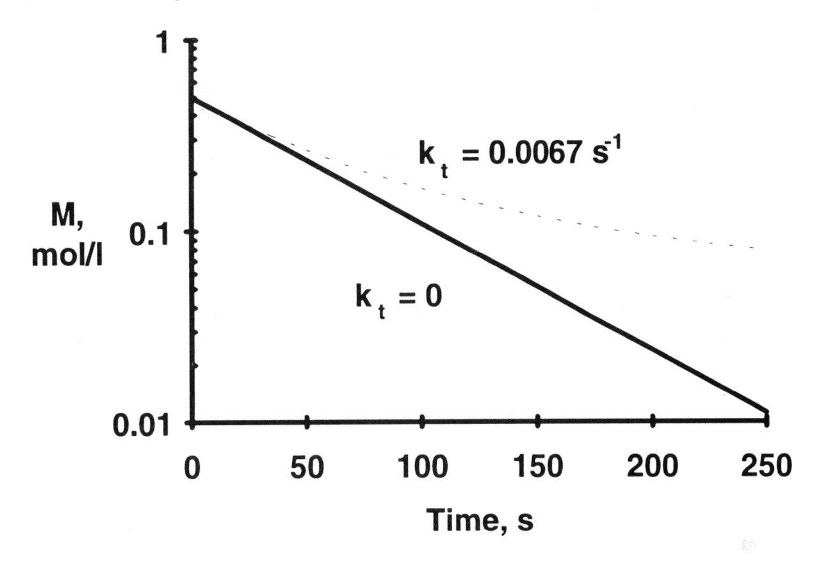

Figure 4.21 Effect of termination rate on monomer depletion.

The carbanion is coupled with the lithium counterion. This ion pair is highly associated in solution and initiates the chain. The cation is often left out of mechanism structures for simplicity. The propagation step is

$$M^{-}_{n}\,L_{i}^{+} + H_2C{=}CH(C_6H_5) \rightarrow M^{-}_{n+1}\,Li^{+} \qquad 4.112$$

Phenyllithium can form anion and radical chain ends with styrene. Both types of ends polymerize and the free radical ends can undergo the expected termination reactions.

The propagation rates of anionic polymerizations are very high and the reactions must be controlled carefully to prevent explosions. The initiating species would be formed by adding the catalyst to solvent. The monomer would be added slowly with agitation. Solvent acts to reduce the monomer concentration and is an energy sink for the heat of polymerization. Solvent condensers could be attached to the reactor to reflux solvent vapors back into the system.

In many systems, the catalyst complex is highly stable. Monomer addition can be stopped for long periods of time and then restarted with the chain ends continuing to add monomer stoichiometrically. Because the polymerization can be restarted, these systems have been called "living polymerizations." Chains can be terminated by adding water, acids, or oxygen to react with the catalyst complex. Cationic systems can also polymerize stoichiometrically in polar solvents such as SO_2(l) or CO_2(l).

Anionic polymerizations provide an excellent method for making diblock copolymers. Since the monomers add stoichiometrically, the chain length of each monomer is known accurately. Styrene-butadiene copolymers are made in this way.

Kinetics of Stoichiometric Polymerizations. If there is only ionic addition (free radicals either do not form or react at low rates compared to the ions), the polymerization kinetics are simple. The formation of initiating centers is completed before monomer is added. The concentration of centers is constant while there are no terminating reactions. The polymerization rate is

$$-\frac{dM}{dt} = k_p A^- M \tag{4.113}$$

where A^- is the anion concentration and M is monomer concentration. The reaction is limited by heat removal. Monomer addition is based on maintaining reactor temperature below safe levels.

Monomer should be uniformly distributed throughout the reaction solution. When this occurs, the chains are all the same length. The degree of polymerization is

$$\overline{X}_n = \frac{M_{added}}{A^-} \tag{4.114}$$

and the number average molecular weight is

$$\overline{M}_n = M_o \cdot \overline{X}_n \tag{4.115}$$

Anionic polymerizations can give monodisperse polymers. For example, the polydispersity ratio of anionically polymerized polystyrene is 1.02 to 1.2, while that of polystyrene from free radical processes is 2 to 5. The compounds in Table 4.14 are

TABLE 4.14 Ionic Solvents, Monomers and Initiators*

Anionic				Cationic
Solvents				
tetrahydrofuran -----------			---------------- CH_3Cl, CH_2Cl_2, $CHCl_3$	
liquid NH_3 ------	----------- hydrocarbons ----------------			----------- liquid SO_2, CO_2
Initiators				
butyl lithium	AlR_3-$Ti(OR)_4$	AlR_3-$TiCl_2$	$Ti(OR)_4$	$H[BF_3OH]$
α-methyl styrene-Na	AlR_2Cl-$CoCl_2$	AlR_3-$TiCl_4$		$[(C_2H_5)_3O]BF_4$
Monomers				
methyl methacrylate -------	----------- propylene -----------			
acrylonitrile -----------	----------- ethylene --------------------			
methacrylic acid esters -----	------------ vinyl ethers --			
---------------- butadiene -----------------	--------------- isobutylene --			
---------------- isoprene ------------------				
------ α-methyl styrene -------			------ α-methyl styrene -----------	
--------- ethylene oxide -----------			---------------- ethylene oxide -----------	
--- styrene ---				

*The dashed lines indicate the polarity range over which the solvents or monomers are used.

arranged according to type from anionic to undissociated catalysts to cationic. This sequence gives a rough estimate of the initiator type, based on the properties of the polymers produced. The dielectric constant of the solvent can affect the catalyst function, particularly for those that dissociate.

NOMENCLATURE

Symbol	*Definition*
a, b, c, d	stoichiometric coefficients (Eq. 4.5)
a	branching coefficient (Eq. 4.90)
a_{crit}	critical branching coefficient at gel formation
A, B, C, D	concentrations of species A, B, C, D; mol/l or kmol/m^3
A_i	collision frequency factor for reaction i; same units as rate constant i
C_i	ratio of chain transfer to propagation rate constant for mechanism i
e_i	monomer polarity (Eq. 4.61)
E_i	activation energy for reaction i; kcal/mol or kJ/mol
f	initiator efficiency (Eq. 4.7)
f_i	mole fraction of i in monomer phase
F_i	mole fraction of i in polymer phase
f_{ave}	average functionality (Eq. 4.91)
I	initiator concentration; mol/l or kmol/m^3
k_d	initiator decomposition rate constant; s^{-1}
k_{td}	disproportionation termination rate constant; l/mol-s or m^3/kmol-s
k_{tc}	combination termination rate constant; l/mol-s or m^3/kmol-s
$k_{tr,i}$	chain transfer rate constant to initiator, solvent, monomer, polymer ($i = i, s, m, p$); l/mol-s or m^3/kmol-s
k_i	rate constant for mechanism i; units as appropriate
k_{ij}	rate constant for the reaction of monomer j with radical center i; l/mol-s or m^3/kmol-s
M	monomer concentration; mol/l or kmol/m^3
M_i	concentration of monomer i in monomer phase; mol/l or kmol/m^3
M^*	monomer radical concentration; mol/l or kmol/m^3
M_i^*	concentration of an i-mer radical; mol/l or kmol/m^3

M_o	initial monomer concentration; mol/l or kmol/m^3
$\overline{M_n}$	number average molecular weight
$\overline{M_w}$	weight average molecular weight
N_x	probability of x-mer
N_o	initial number of structural units
N_t	total number of polymer molecules
N_{tx}	total number of x-mers
p	extent of reaction, fraction of monomer units that have reacted
p_{crit}	extent of reaction at the critical branching point for gelation
p_i	fraction of i groups reacted, step polymerization
P_i	reactivity of radical i
Q_i	reactivity of monomer i
r	ratio of end group concentrations for step polymerizations
rr	ratio of i groups in a branch unit to all i groups in the mixture
r_i	k_{ii}/k_{ij}, ratio of copolymerization rate constants
R_n	rate of mechanism n; mol/l-s or kmol/m^3-s
R^*	free radical concentration; mol/l or kmol/m^3
t	time, s
W_x	weight fraction distribution of N_x
x	chain length
$\overline{X_n}$	number average degree of polymerization
$\overline{X_w}$	weight average degree of polymerization

Greek Symbols

$\alpha, \beta, \delta, \lambda$	coefficients defined by Equation 4.68
λ_k	kinetic chain length

PROBLEMS

4.1 SBR latex paints are based on random copolymers with about 70% styrene. Maintaining a relative constant copolymer composition is important, since the T_g of the paint will be a function of composition (a change in T_g will make the applied paint more or less brittle). Compute the monomer concentration needed to produce a copolymer of 70% styrene. If only a 4% change in copolymer composition can be permitted, compute the monomer composition in a batch reactor at this limit. The data in Table 4.12 can be used for reactivity ratio values.

4.2 In reaction injection molding, two reactants are quickly mixed while they are injected into the mold. An instructor tried to prepare a class demonstration with the polyurethane recipe given below. Unfortunately, the reaction mixture boiled before it started to become tacky. As a result, the polyurethane did not develop into a low-density foam. Given the starting composition for the reaction, what changes would you make to improve the foam?

Material	Amount	Functionality
Polyol	0.6 moles −OH	2.3
Toluene diisocyanate	0.5 moles −N=C=0	2
Methylene chloride	0.3 moles	—

4.3 The kinetics for an acid-catalyzed stepwise polymerization of a polyester are given by

$$-\frac{dCOOH}{dt} = k\ COOH\ OH$$

where COOH refers to the concentration of carboxylic acid groups on the polyprotic acid monomer, and OH refers to the concentration of hydroxyl groups on the glycol.

(a) Assuming that $COOH = OH = C$, and using the equation $C = C_o(1 - p)$, where C_o is the initial concentration of functional groups and p is the extent of reaction (fraction of functional groups that have reacted), show that

$$C_0kt = \frac{1}{1-p} + constant$$

(b) The number average degree of polymerization is given by

$$\overline{X}_n = \frac{C_0}{C}$$

Using the results of part (a), show that x_n is a linear function of time.

(c) The following data were taken for the stepwise reaction of diethylene glycol with adipic acid, catalyzed by 0.4mol of p-toluene-sulfonic acid:

Time t(min)	$\overline{X}_n$
200	20
325	37
400	46

Estimate the time required to form polymer of molecular weight

$$15{,}000\ (\overline{X}_n \simeq 90).$$

4.4 Styrene can be polymerized with potassium amide in liquid ammonia via an anionic polymerization. The following steps are thought to be important.

$$KNH_2 \xrightarrow{k_1} K^+ + NH_2^-$$

$$NH_2^- + M \xrightarrow{k_2} NH_2M^-$$

$$NH_2M_n^- + M \xrightarrow{k_3} NH_2M_{n+1}^-$$

$$NH_2M^- + NH_3 \xrightarrow{k_4} NH_2MH + NH_2^-$$

In a living polymer system, the kinetic chain length equals the rate of polymerization divided by the rate of chain transfer. Derive an equation for the degree of polymerization, $\overline{Xn}$. (HINT: Make sure that the equation contains no anionic species concentration.)

4.5 A diol (2 molar) is being reacted with a tetra-carboxylic acid (1 molar) to form a gel. The critical coefficient has been used to compute the extent of reaction at the gel point ($p_A = 1/3$). The mechanism is

$$-OH + -COOH \underset{k_2}{\overset{k_1}{\rightleftarrows}} -O-CO- + H_2O$$

$k_1 = 0.25$ l/mol-hr, $k_2 = 0.02$ l/mol-hr.

(a) Compute the time required to form a gel if the reaction is carried out under vacuum.

(b) What is the molecular weight of the polymer produced?

(c) Assume that water is present in the reaction media at 0.05mol/l. What will be the extent of conversion for the functional groups?

4.6 Compute the maximum molar excess of adipic acid necessary to produce nylon 6,6 of 100,000 molecular weight.

4.7 Polyacrylamide is prepared by solution polymerization using isobutyryl peroxide as a free radical initiator. Under reaction conditions, the peroxide half-life is 20 hours. The initiator decomposes to two radicals, and termination is by coupling. Estimate the polymerization rate at t = 0. Estimate the number average degree of polymerization for polymer formed at 90% conversion. Justify any simplifying assumptions that you make.

$k_p = 18{,}000$ l/gmol-s
$k_t = 14.5 \times 10^6$ l/gmol-s
$M_o = 1$ gmol/l
$I_o = 10^{-4}$ gmol/l

4.8 Determine the copolymer composition as a function of monomer composition for the following system: vinyl acetate ($r_1 = 0.01$)/styrene ($r_2 = 55$). Are there regions where the polymer might be an alternating, a random, or a homopolymer?

4.9 Adipic acid (A) reacts with glycerol (G) in the presence of base to form a gel. If there is an equal number of function groups of each type, what will be the fraction of G groups reacted at the gel point?

4.10 *Cis*-1,4-Polyisoprene must be highly stereoregular to function well in automobile tires. Which type of polymerization process should be used in its manufacture, and why?

4.11 Nylon 6,6 is made by the stepwise polymerization of hexamethylene diamine with adipic acid. A general reaction mechanism is the reaction of carboxylic acid groups with amino groups. The equilibrium constant for the polyamidation reaction is

$$\frac{-CONH - H_2O}{-COOH - NH_2} = K_1 = B\, exp\left(\frac{\Delta H_a}{RT}\right)$$

where B is a temperature-independent constant.

(a) Given that the enthalpy change in the reaction $-\Delta H_a = 6$ kcal/mol and that $K_1 = 250$ at 280°C, what would the value of K_1 be at 300°C?

(b) What will be the fractional conversions of reactants to products at 280°C and 300°C? Assume that the reaction proceeds to equilibrium, that no components are removed, and that equimolar concentrations of reactant are present at t = 0.

(c) The equilibrium is not as simple as that described by the above equation since the equilibrium coefficient, K_1, varies with water content in the polymer phase. The water content in the polymer depends on the partition coefficient of water between the polymer and gas phase at the temperature and pressure of the reactor. At 280°C and 1 atm, the water content of the polymer is 0.16wt%. Compute the fractional conversion to polyamide under these conditions. (HINT: Using the fractional conversion of part b, compute the weights of polymer and water formed from an equimolar charge. Calculate the amount of amide linkages formed based on a revised number for the water concentration in the polymer phase and assuming no change in K_1).

4.12 If the amounts of monomer added in A-B-A-B condensation polymerizations are not precisely equivalent, an imbalance in the number of end groups and a reduction of number average molecular weight can result. Let T_o = number of end groups initially and T = number of end groups at the end of reaction, then the extent of polymerization

$$p = 1 - \frac{T}{T_o}$$

T is frequently expressed in units of equivalents per 10^6g of polymer. For linear polymers, the number of chains per 10^6 is $T/2$ and the number average molecular weight is

$$\overline{Mn} = \frac{2 \times 10^6}{T}$$

If $COOH\ NH_2 = P$ and $COOH - NH_2 = D$, the values of the end-group concentrations are

$$NH_2 = \frac{-D + (D^2 + 4P)^{1/2}}{2}$$

$$COOH = \frac{D + (D^2 + 4P)^{1/2}}{2}$$

The number average molecular weight is

$$\overline{Mn} = \frac{2 \times 10^6}{(D^2 + 4P)^{1/2} + E}$$

where E is the concentration of monofunctional groups. Compute the drop in molecular weight caused by a 0.001mol% excess of carboxyl end groups in a polyamide polymerization where E is 0.001mol% of acetic acid. You may assume complete conversion of reactants to products. Compare this result to that obtained for the same imbalance but assuming no monofunctional groups are present.

4.13 The equations derived in Section 4.3.4 for the prediction of the gel point are based on a statistical approach. Carothers related the extent of reaction, p, with the number average molecular weight to obtain a value for p_c, the critical extent of reaction. For the case, $f_c > 2$ and $f_A = f_B = 2$,

$$pc = \frac{1 - \rho}{2} + \frac{1}{2r} + \frac{\rho}{f_c}$$

(a) For a mixture of 1,2,3-propane tricarboxylic acid, adipic acid, and diethylene glycol with $p = 0.404$ and $r = COOH/OH = 1.002$, compute the gel point using the Carothers and the statistical approach.

(b) Carothers' approach must yield a value for p_c that is too large since some molecules will have a molecular weight greater than $\overline{M}_n$ and thus will reach the gel point sooner. Yet the statistical approach (which is theoretically correct) yields a value for the gel point at too low an extent of reaction. The observed value often lies between the two predictions. Explain this phenomenon in terms of side reactions.

4.14 Your plant is currently polymerizing vinyl acetate in benzene. The Research Group has recommended switching to cyclohexane for environmental reasons. In the current reaction scheme, monomer is present at a 5 mol/l concentration and benzene is present at 8.6 mol/l. If the monomer concentration is kept the same, the cyclohexane concentration would be 8.2 molar in the new system. If the plant process is now making polymer with a degree of polymerization of 580 at the beginning of the reaction, estimate the change in the degree of polymerization (at $t = 0$) if cyclohexane is the solvent. C_s, benzene $= 1.2 \times 10^{-4}$, C_s, cyclohexane $= 7.0 \times 10^{-4}$. You may assume that the solvent change does not radically affect the other rate constants and that termination is by combination.

4.15 The conversion of commercial chain polymerization is often between 60% and 80% because of the reduction in rate and change in properties at higher conversions. In a test charge, methyl methacrylate is polymerized at 60°C using benzoyl peroxide as an initiator. Assuming that the initial charge contains 100 g methyl methacrylate/l and 0.5g benzoyl peroxide/l compute the number average degree of polymerization at 70% conversion. The following data may be useful:

M_w (methyl methacrylate) $= 102$
M_w (benzoyl peroxide) $= 242$
Initiator decomposition is 100% efficient
$k_p^2/k_t = .010$ ℓ/mol-s
Half-life of benzoyl peroxide @ 60°C = 46 hours
Termination is by combination only
No thermal initiation occurs
Assume that reaction time is much faster than initiator disappearance

4.16 Using the following information, determine the acrylonitrile composition of nitrile rubber when the mole fraction of acrylonitrile is 0.60.

Monomer	e	Q
Butadiene	−1.05	2.39
Acrylonitrile	1.20	0.60

4.17 The addition of multifunctional components to stepwise polymerizations leads to gels. For the following mixture, compute the extent of reaction at the gel point.

Monomer	Moles
Pentaerythritol	.5
Ethylene glycol	1.0
Adipic acid	2.0

Pentaerythritol = $C(CH_2OH)_4$

Suppose the polymerization operator told you that he added 2.2 moles of adipic acid rather than the recipe amount. At what conversion would you expect to have power usage by the agitator increase?

4.18 Methyl methacrylate is polymerized to 60% conversion in a solution process. Compute the reaction time at 60°C. The monomer is initially present at 3 mol/l Transfer to solvent is unimportant. The initiator concentration is 0.1 molar and the half-life is 100 hours. Some rate coefficients are

$$\frac{k_p^2}{k_t} = .010 \frac{l}{mol-s} \text{ at } 60°C$$

Indicate any assumptions.

4.19 In the copolymerization of vinyl acetate (1) with vinyl chloride (2), it is desirable to keep the acetate concentration in the polymer to less than .30 mole fraction. If the reactor initially contains .20 mole fraction of vinyl acetate, at what conversion should the reaction be stopped to achieve this property? At 60°C, $r_1 = .23, r_2 = 1.68$.

4.20 At River Side Chemical, a condensation polymer is made by reacting adipic acid with ethylene glycol to form a linear high molecular weight polymer. A reactor was being charged with ethylene glycol as the midnight shift was coming on. The reactor operator, who was in training, thought that glycerol was called for in the reaction recipe, so he turned off the ethylene glycol and turned on the glycerol.

The operator informed the foreman about the switch 15 minutes after adding the catalyst. The typical reactor charge is 10 kg moles of adipic acid and 10 kg moles of ethylene glycol. The foreman estimates that 7 kg moles of glycol and 2 kg moles of glycerol are in the tank. The reaction rate for this system can be approximated by

$$-\frac{dB}{dt} = -\frac{dA}{dt} = k\,A\,B$$

with k = 2l/mol-h, and A,B are functional groups. What action should be taken and can the operator go on his coffee break before taking it?

4.21 The monomer reactivity ratio, r_1, is the ratio of the rate constant for the radical adding its monomer to the rate constant for the radical adding another monomer.

For each entry in the table below, predict whether the copolymer will alternate or be random, and suggest the overall composition of the copolymer.

Monomer	$r_1 \cdot r_2$
Styrene	0.98
Vinyl chloride	0.31
Methyl methacrylate	0.19
Acrylonitrile	0.0006

Draw an alternating copolymer and a random copolymer.

4.22 Use the data in Appendix D to solve the problems below.

(a) Compute the time required to convert 60% of a charge of methyl methacrylate to polymer using benzoyl peroxide as an initiator in benzene solution at 80°C. Initial charge per liter of solution is 100g methyl methacrylate and 1.0g benzoyl peroxide.

(b) What number average degree of polymerization do you expect at 60% conversion? At 0% conversion?

(c) Suppose this process is limited by reactor capacity. If the conversion is kept the same, but reactor temperature can be increased to 90%, suggest an initiator charge level that will give a similar degree of polymerization in a shorter time.

4.23 The reaction rate of an isocyanate with a primary alcohol may be given by

$$\frac{-dB}{dt} = \frac{-dA}{dt} = k\,A\,B$$

where A and B are the concentrations of isocyanate and hydroxyl groups respectively in moles per liter. For the following mixture, estimate the time required for gel to form. Use $k = 0.40$ l/mol-h.

3,3′ − Dimethyl phenyl methyl methane−4,4′−disocyanate ($C_{17}H_{14}N_2O_2$)	2 moles
2,5′−Diethyl hexanol ($C_8H_{18}O$)	1.4 moles
Trimethylol propane ($C_6H_{14}O_3$)	1.1 moles
Benzene	to make one liter

4.24 Commercial Saran Wrap is a copolymer of 95wt% vinylidene chloride and 5wt% vinyl chloride. If it is produced in a batch suspension polymerization system, estimate the maximum conversion achievable so that

(a) No copolymer molecules contain greater than 12% by weight vinyl chloride.
(b) The cumulative average weight percent vinyl chloride in the copolymer does not exceed 5wt%.

4.25 Aldehyde condensations occur by the following general reaction scheme.

$$R-CO-X + HX \rightarrow R-C(OH)X-H$$

$$R-C(OH)X-H + HX \rightarrow R-CX_2-H + H_2O$$

$$R-C(OH)X-H + R-CO-H \rightarrow R-CHX-O-C(OH)R-H + HX \rightarrow R-CHX-O-CXR-H$$

When formaldehyde and urea are used, $R = H$ and $X = -NH-CO-NH_2$. Can furfural (a derivative of oat hulls) be substituted for one of these reactants, and, if so, which one?

4.26 Estimate the activation energies for
(a) the propagation rate constant of chloroprene,
(b) the termination and propagation rate constants of vinyl chloride, and
(c) the propagation rate constant of ethylene as a function of pressure.

References

T. ALFREY, J.J. BOHRER, and H. MARK, *Copolymerization*, Interscience, NY, 1952.

J.J. CARBERRY, *Chemical and Catalytic Reaction Engineering*, McGraw-Hill NY, 1976.

K.W. DOAK, "Ethylene Polymers," in *Encyclopedia of Polymer Science and Engineering*, Vol. 6, McGraw-Hill, NY, 1989.

P. EHRLICH and G.A. MORTIMER, *Adv. Polymer Sci. 7*, 386–448 (1970).

P.J. FLORY, *Principles of Polymer Chemistry*, Cornell Univ. Press, Ithaca, NY, 1952.

Z. LAITA, *J. Polymer Sci. 38*, 247–258 (1959).

Z. LAITA and Z. MACHÁČEK, *J. Polymer Sci. 38*, 459–469 (1959).

G. ODIAN, *Principles of Polymerization*, 2nd ed., Wiley-Interscience, NY, 1981.

W.H. STOCKMAYER, *J. Chem. Phys. 11*, 45 (1943).

5

Polymer Solution Thermodynamics

The physical properties of polymer solutions are quite different from those of simple liquids and solids. For example, small molar concentrations of a polymer in an organic solvent can result in solutions with much higher viscosities and much lower solvent vapor pressures compared to equivalent molar concentrations of a low molecular weight solute in the solvent. High molecular weight polymers are completely soluble only in a limited number of solvents, and the solubility usually depends on the polymer's molecular weight and the system temperature. Cross-linked elastomers may have increased tensile strength as temperature increases over a modest range. These phenomena have been explained as resulting from the long-chain nature of the material. This chapter describes phase equilibria and shows how to apply a simple model for polymer solution thermodynamics.

5.1 REVIEW OF SOLUTION THERMODYNAMICS

Physicochemical systems are driven toward equilibrium by the tendency toward conditions of minimum energy, minimum free energy, and maximum entropy. Sandler (1989) and Prausnitz (1969) have good discussions of the conditions for equilibrium for ideal and regular solutions. Two phases brought into contact tend to exchange constituents until the composition of each phase attains a constant value. When that state is reached, the phases are in equilibrium. One aim of phase equilibrium thermodynamics is to describe quantitatively the distribution at equilibrium of every component among all the phases present. The chemical potentials, μ_i, of substances between phases are

$$\mu_i = \left(\frac{\partial G}{\partial n_i}\right)_{T,P,n_j} \qquad 5.1$$

where G is the Gibbs free energy and n_i is the number of molecules of component i. At equilibrium, the chemical potentials of each species are equal in all phases.

The relationships among the abstract quantity μ_i, the chemical potential, and physically measurable quantities like temperature, pressure, and composition are defined by the introduction of several quantities: the fugacity, f_i; the activity, a_i; and the activity coefficient γ_i.

$$d\mu_i = RT\, d \ln f_i \qquad 5.2$$

$$\mu_i - \mu_i^\circ = RT \ln f_i/f_i^\circ = RT \ln a_i \qquad 5.3$$

The activity of a substance (Eq. 5.3) is a measure of the difference between the substance's chemical potential at a state of interest and that at its standard state μ_i°. The activity coefficient is defined as the ratio of the activity of component i to its mole fraction.

$$\gamma_i = \frac{a_i}{x_i} \qquad 5.4$$

Criteria for Phase Equilibria: Two phases, a and b, are in equilibrium when the chemical potentials of any component, i, are the same in each phase,

$$\mu_i(a) = \mu_i(b) \qquad 5.5a$$

and

$$f_i(a) = f_i(b) \qquad 5.5b$$

Partial Molar Properties: Properties of mixtures are frequently discussed in terms of partial molar properties and the change in those properties on mixing. Partial molar properties, Z_i, are defined by

$$Z_i = \left(\frac{\partial Z}{\partial n_i}\right)_{T,P,N_j} \qquad 5.6$$

The total value of the property of a mixture is the weighted sum of the partial molar properties. For example, consider the volume of the following system:

$$V = \Sigma\, n_i V_i \qquad 5.7$$

In general

$$Z = \Sigma\, n_i Z_i \tag{5.8}$$

The change in volume, V, or any property, Z, on mixing is given by the difference between the value of the property for the mixture and the value of the property for pure components,

$$\Delta V_m = V_{mixture} - V_{pure\ components} \tag{5.9a}$$

$$\Delta V_m = \Sigma\, n_i(V_i - V_i^\circ) \tag{5.9b}$$

$$\Delta Z_m = \Sigma\, n_i(Z_i - Z_i^\circ) \tag{5.9c}$$

where V_i° and Z_i° are the molar volume and molar property of pure i at the temperature and pressure of the mixture, and V_i and Z_i are the partial molar properties of i in the mixture.

5.1.1 Ideal Solutions

Mixture properties may be classified relative to the concept of an ideal solution. An ideal solution is a simplified, idealized physical and mathematical model; it has the property that the fugacity of every component at constant temperature and pressure is proportional to its mole fraction. An ideal vapor in equilibrium with an ideal liquid obeys Raoult's Law. For this case, Equation 5.5b becomes

$$f_i(\mathrm{v}) = f_i(l) \tag{5.10a}$$

where v denotes the vapor phase and l denotes the liquid phase. The partial pressure of component i is related to the vapor mole fraction, y_i, and liquid mole fraction, x_i, by

$$P_i = y_i P = x_i P_i^{sat} \tag{5.10b}$$

Raoult's Law predicts that the partial pressure of a component in the vapor phase, p_i, is equal to the product of the liquid phase mole fraction of i and the saturated vapor pressure, P_i^{sat}. Nonideal solutions are modeled by an equation similar to Equation 10b, but having an activity coefficient.

Ideal fluids mixtures form from their pure components with no change in volume and without evolution or absorption of heat.

$$\Delta V_{m,\ ideal} = 0 \tag{5.11a}$$

$$\Delta H_{m,\ ideal} = 0 \quad 5.11b$$

The entropy of mixing per mole for ideal fluids is

$$\Delta S_m = -R\,\Sigma\,x_i \ln x_i \quad 5.12$$

In an ideal liquid, the molecules of the components are interchangeable with respect to their size, shape, and energies of interaction. Consider a binary ideal solution formed from n_1 molecules of component 1 and n_2 molecules of component 2. The solution might be formed by mixing the molecules on an imaginary three-dimensional lattice of $n_1 + n_2$ sites. The configurational entropy of mixing is given by the logarithm of the number of possible ways of arranging n_1 molecules of type 1 and n_2 molecules of type 2 on the lattice.

$$S = k \ln \frac{(n_1 + n_2)!}{n_{1!}\, n_{2!}} \quad 5.13$$

This assumes that the molecules are of similar size and can be moved interchangeably from one lattice site to another. Equation 5.13 is simplified by using Sterling's approximation for the logarithm.

$$\ln N! = N \ln N - N$$

This results in the following equation for the entropy of mixing for ideal fluids:

$$\Delta S_m = -R\,(x_i \ln x_1 + x_2 \ln x_2) \quad 5.14$$

5.1.2 Regular Solutions

Regular solutions have been defined as solutions whose entropy and volume changes on mixing are ideal and have nonzero heats of mixing. The nonzero heat of mixing often takes the following form (Hildebrand and Scott, 1950).

$$\Delta H_m = RT\,K\,x_1^2 \quad 5.15$$

where K is an empirical heat of mixing parameter. The entropy of mixing is described by Equation 5.14. The free energy of mixing of regular solutions is therefore represented as

$$\Delta G_m = \Delta H_m - T\,\Delta S_m \quad 5.16a$$

$$\Delta G_m = RT\,[K x_1^2 + \Sigma\, x_i \ln x_i] \qquad 5.16b$$

The activity of a component of a regular binary solution can be computed by using Equations 5.16b, 5.1, and 5.3.

$$\ln a_i = K x_i^2 \qquad 5.17$$

Equations 5.14, 5.15, 5.16b, and 5.17 are useful for computing thermodynamic properties of regular solutions. Nonideal effects of polymer solutions are discussed in the rest of the chapter.

5.2 CONSEQUENCES OF LONG CHAINS FOR POLYMER SOLUTIONS

Amorphous Polymer Solutions. The following discussion is for amorphous polymers in solution and in the solid state. Crystalline polymeric solids behave like crystalline solids in some ways. However, crystalline polymers usually contain amorphous regions.

Ideal behavior is often observed for dilute solutions of low molecular weight components. For macromolecules, mole fraction can be a misleading measure of concentration in liquid solutions. For example, consider a polymer of molecular weight 10^5 in a solvent of molecular weight 100. If the mole fraction of the polymer in the solution is 0.01 (which could be dilute in conventional terms), the mass fraction of polymer in the solution would be 0.91. The transport properties of the polymer solution would be much different than those of the solvent alone. For example, a 1wt% polymer solution is more viscous than a mixture of the same solvent and a low molecular weight solute at 1wt%. Also, the vapor pressure of the solvent above the polymer solution would be different than that of the low molecular weight mixture. The rate of diffusion of the polymer through the solution would be much lower than that for the low molecular weight mixture.

EXAMPLE 5.1 VAPOR PRESSURE OVER POLYMER SOLUTIONS

Differences between regular solutions and polymer solutions can be illustrated by comparing their properties at similar conditions. Suppose that styrene was being reacted to polystyrene with toluene as a solvent to make an adhesive. Figure 5.1a shows the reactor prior to initiation, and Figure 5.1b shows it after the complete reaction of styrene to polystyrene. The vapor and liquid phases of each system are assumed to be in equilibrium. There are two degrees of freedom for each tank. Setting the temperature and liquid phase composition sets the system pressure and the vapor phase composition. In each case,

(continued)

the vapor phase will be ideal because of the low system pressure. Typical properties of the components are given in the following table:

Component	p_i^{sat}, bar	ρ, g/cm³	M_w
Toluene (T)	0.04921	0.866	92.13
Styrene (S)	0.01316	0.903	104.14
Polystyrene (PS)	—	1.02	200,000

Therefore, the vapor phase fugacities can be calculated as

$$f_i(v) = y_i P = P_i$$

The total pressure is given by

$$\rho = P_{toluene} + P_{styrene}$$

The liquid fugacities for the solvent mixture can be modeled using

$$f_i(l) = x_i \gamma_i P_i^{sat}$$

If toluene and styrene form an ideal solution, $\gamma_i = 1$ for each component; the following table gives the liquid and vapor phase compositions:

	Liquid Phase		Vapor Phase	
Component	x_i	ϕ_i	p_i, bar	y_i
Toluene (T)	0.5306	.5105	.02512	0.7959
Styrene (S)	0.4694	.4895	.00644	0.2041
			.03156 bar total pressure	

At the end of the reaction, all the styrene is converted to polymer, which exerts no vapor pressure. If polystyrene were similar in behavior to an inorganic solid, toluene would be the only component in the liquid phase and the reactor pressure would be its pure component vapor pressure at 30°C, 0.049 bar. However, polystyrene is miscible with toluene and changes its fugacity in the liquid phase. The following table gives the liquid and vapor phase compositions:

	Liquid Phase		Vapor Phase	
Component	x_i	ϕ_i	p_i, bar	y_i
Toluene (T)	0.9995	0.5304	0.0444	1
Polystyrene (PS)	0.0005	0.4696	—	

The vapor pressure of toluene is 90% of its pure component value. Regular solution theory would suggest that the liquid phase mixture is ideal because of its high solvent mole fraction.

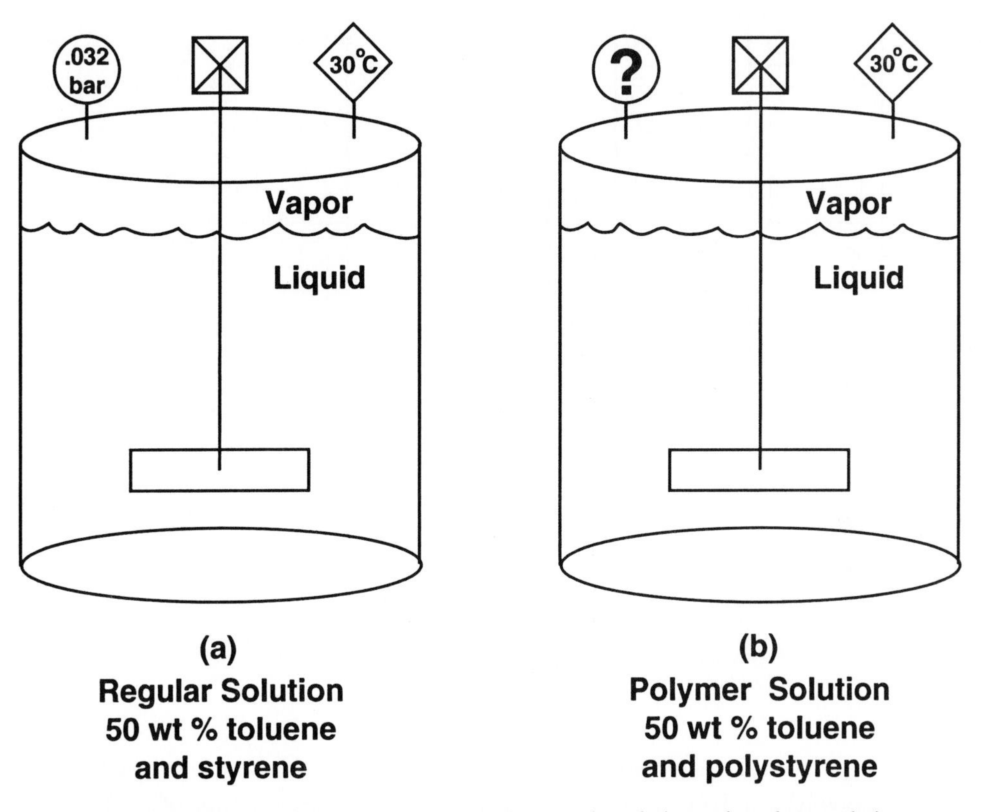

Figure 5.1 Vapor-lined equilibria experiment for a regular solution and a polymer solution.

Deviations from ideal behavior may be caused by nonideal values of the entropy and enthalpy changes on mixing. The extent of the deviations from ideality can be expressed in terms of the excess enthalpy or entropy on mixing, ΔH_m and ΔS_m. A solution that deviates from ideal behavior in both enthalpy and entropy may still seem to exhibit ideal behavior if it is studied when the excess free energy of the process is zero.

Equation 5.13 can be used to calculate ΔS_m for a polymer-solvent mixture. The calculated value of ΔS_m will be small because of the low numbers of polymer molecules in typical solutions. The polymer distribution on lattice sites is constrained since large numbers of molecular segments must move when the position of one segment is changed. Figure 5.2 compares the placement of 18 solute molecules in a lattice with the placement of 18 polymer segments. Each polymer segment must be placed adjacent to its neighbors, which greatly reduces the entropy of mixing.

The enthalpy change on mixing measures variations in the secondary bonding forces between molecules. Nonideal enthalpy changes on mixing are a result of break-

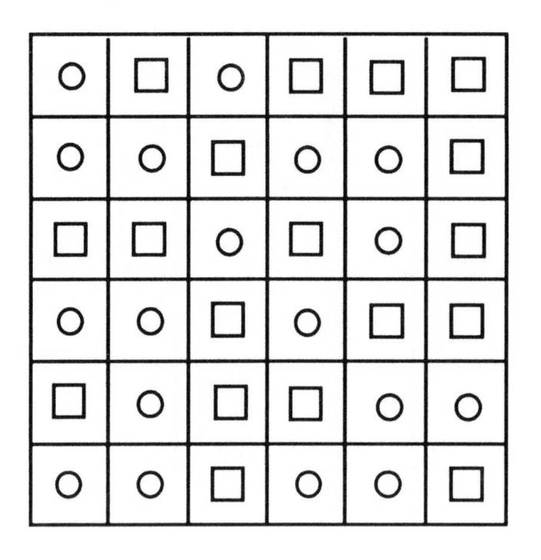

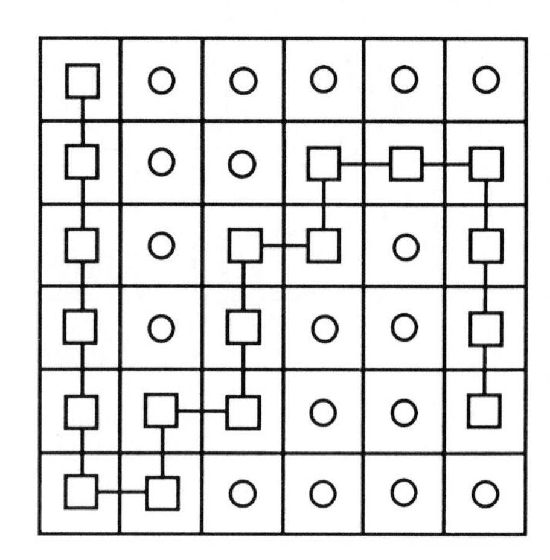

Figure 5.2 Several configurations for mixing equal numbers of two components in fixed lattice sites.

ing solvent-solvent bonds (1-1) and polymer-polymer bonds (2-2) to form solvent-polymer bonds (1-2). The change in energy due to the formation of a 1-2 bond is

$$\Delta W_{1-2} = W_{1-2} - \frac{W_{1-1} + W_{2-2}}{2} \tag{5.18}$$

and ΔH_m will only be zero in the unlikely event that ΔW_{1-2} precisely equals the energy needed to break 1-1 and 2-2 bonds. This equation assumes one secondary valence bond per solvent and polymer molecule; it best describes nonpolar molecules in nonpolar solvents.

ΔH_m and ΔS_m are not necessarily independent: These quantities are coupled to the conformation of the molecule. As the polymer molecule dissolves, conformational entropy is released and the molecule expands. However, due to stearic constraints, the expansion is not entirely free. ΔH_m usually changes because the solvent is different chemically from the polymer. If ΔH_m is negative, then polymer-solvent contacts are favored. This means that the polymer will not be tightly coiled and that ΔS_m is reduced below the value expected for $\Delta H_m = 0$. If ΔH_m is positive, polymer-solvent contacts are discouraged, tight conformations are probable and ΔS_m is again reduced. For any nonelectrolyte solution, the value of ΔS_m calculated for random mixing with $\Delta H_m = 0$ is not only the value giving ideal behavior, but also the maximum value unless the

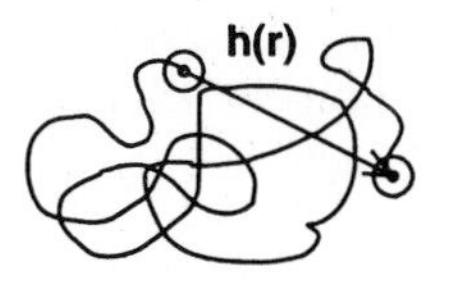

(a) Coil under no load

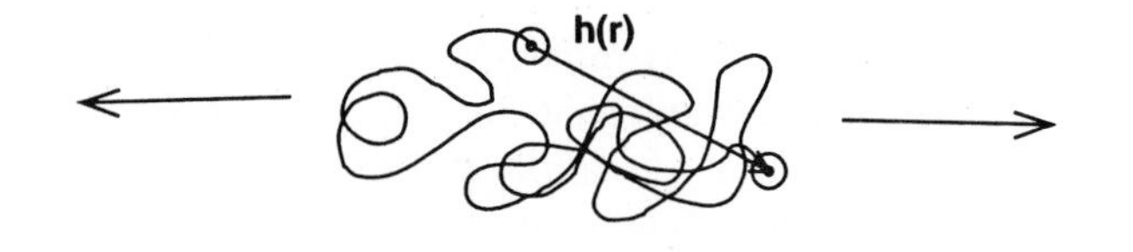

(b) Coil under moderate load

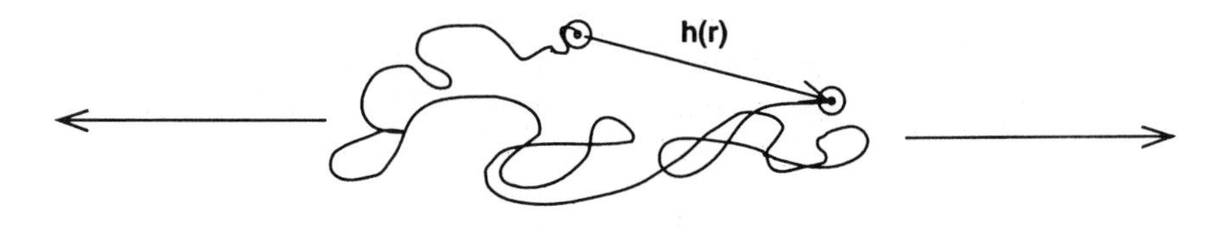

(c) Coil under high load

Figure 5.3 End-to-end distances of chains under different loads.

polymer breaks up ordered liquid structure in the solvent. ΔS_m can be considered to have two components: a combinatorial entropy arising from random mixing and a noncombinatorial entropy from polymer-solvent interactions.

Displacement of Elastomers—Entropic Effects. Cross-linked polymeric elastomers have tensile strengths that increase with temperature and are related to the entropy of the material. Imagine two ends of a chain segment in an elastomer under three load conditions: no load, a moderate load, and a high load. In the no load case (Fig. 5.3a), the amorphous coil can have a number of conformations and still retain the same end-to-end distance between the chain ends. As the distance between the chain ends increases, there are fewer conformations for the chain to assume and retain the same end-to-end distance. In the high load case (Fig. 5.3c), there are very few conformations the chain can take to retain the end-to-end difference.

While the entropy changes between these three conformations may be large, the enthalpy differences between these three cases may be small. None of the chains are extended so that the carbon-carbon chain bonds are under tension. Other contributions to the enthalpies of each case are the differences between secondary bonding of chain segments.

5.3 STATISTICAL POLYMER MODELS: DILUTE POLYMER SOLUTIONS

The conformations of polymer molecules in solution are important to a number of engineering properties. Polymer shape is related to properties such as dipole moment and optical activity. Viscosity, an important engineering transport property, depends on the polymer conformation in solution.

5.3.1 Random Flight Chain Model

The random flight chain, or freely jointed chain, model for polymers in solution was one of the first successful statistical models. This model predicts the smallest average size of the macromolecule in solution. The random flight model constructs a hypothetical chain with linear links, each of length b. The linear links are free to rotate in any direction. The random flight model is an analogy to the random movements of a molecule in Brownian motion. The polymer is thought to follow the vectors describing successive random displacements of each segment in a three-dimensional space.

End-to-End Distance of a Chain. The chemical bonds of real molecules restrict the angles of rotation. In the random flight model, the valence angle and the bond rotation angle can have any values from 0 to 2 π with equal probability. The end-to-end distance, $|h(r)|$, is not directly measurable, but a conformational average, $<h^2>$, can be calculated. If the chain has z links of length b, $h = b_1 + b_2 + \ldots + b_z$, then the dot product of h should give the following length:

$$<h^2> = h \cdot h = (b_1 + b_2 + \ldots + b_z) \cdot (b_1 + b_2 + \ldots + b_z) \qquad 5.19$$

The average value for the dot product over all molecules is

$$<H^2> = \sum_i^z b_1 b_2 + \sum_{i+j}^z \sum_j^z b_j b_j = z b^2 \qquad 5.20$$

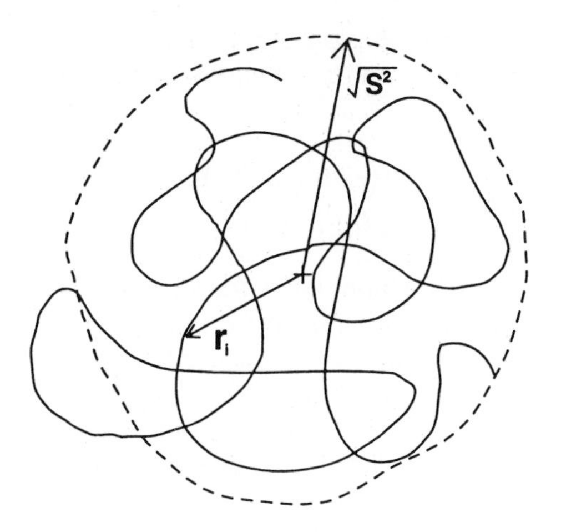

Figure 5.4 Radius of gyration of a polymer chain. r_i is the distance from the center of mass to the ith chain segment. The sphere defined by $r = \sqrt{s^2}$ approximates the volume occupied by the chain.

Equation 5.20 indicates that the average end-to-end distance will be given by the total number of segments times the square of an individual segment length.

Radius of Gyration. The radius of gyration defines the average volume occupied by a molecule (Fig. 5.4). The distance, r_i, is the distance of a segment from the center of mass of the chain. The radius of gyration of the chain, s, will be given by the sum of all the distances, r_i. Taking the square of the distance ensures that all values are positive.

$$s^2 = \frac{1}{z+1} \sum_i^z r_i^2 \qquad 5.21$$

The contour length, L, which is the total length of the molecule if stretched in a straight line, is

$$L = z\,b \qquad 5.22$$

Chain Displacement Distribution. A distribution function that gives the probability that a molecule had assumed a specific conformation can be developed. Averaging this distribution function over all possible shapes should give an average conformation for the molecule. Consider a one-dimensional random walk for a chain of z segments. Each segment has an equal probability of making a + or − step in the x direction. The probability that the chain displacement length has a value in the range $x + dx$, $W(x)$, is given by a Gaussian distribution

$$W(x) = \frac{\exp\left[-x^2/(2\,z\,b^2)\right]}{(2\,\pi\,z\,b^2)^{0.5}} \qquad 5.23$$

where b^2 is the bond length in one dimension. Similar equations can be written for the chain displacement length in the y and z directions, and the distribution can be generalized over three dimensions.

$$W(h) = \frac{\exp[-3\,h^2/(2\,z\,b^2)]}{(2\,\Pi\,z\,b^2\,z/3)^{1.5}} \qquad 5.24$$

The Gaussian density distribution function, $W(h)$, for an arbitrary segment i determines the distribution of all segments of the molecule relative to each other.

Figure 5.5 shows a plot of Equation 5.24 for selected values of z, the number of segments, and b, the segment size. Molecules do not have a high probability of having small or large end-to-end distances. There is a distribution about the average value and a range of molecular sizes that is expected. The polymer chain is dynamic, that is, its conformation will be changing continuously in solution unless it is highly cross-linked. As restrictions are placed on the rotations of chain segments (bond structure, stearic hindrance, etc.), the average end-to-end distance should increase and the chains should occupy a greater volume than predicted by the ideal case.

5.3.2 Real Polymer Molecules

Real polymer molecules deviate from the random flight model in that the direction of a specific bond is usually dependent on the neighboring bonds. The random flight

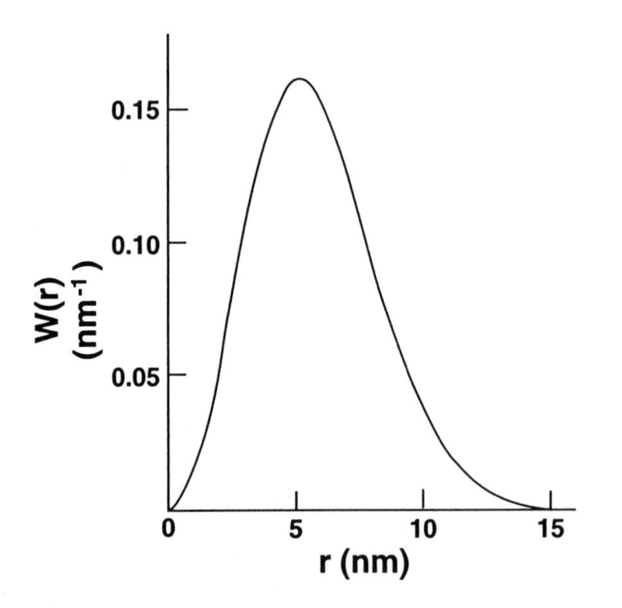

Figure 5.5 Distribution function, $W(r)$, for the distance between chain ends of a statistical coil with $z = 10^4$ and $b = 2.5$ Å (Flory, 1953).

chain model can be modified to consider the effects of short-and long-range interactions between portions of the molecule as well as hindered rotations.

Short-range interactions include the bond structure and the local interactions between atoms and groups of atoms on separate segments. A real chain has restricted motion, since the possible choices for the valence angle and the rotation angle do not range from 0 to 2 π. Valence angles are usually restricted and rotation may be hindered sterically. The result of restricted angles is that a real chain is expanded over a random flight chain of the same contour length, L. The representation of a real chain with short chain interactions is different from a random flight chain. There are fewer and longer segments than the random flight model and the chain cannot self-intersect. The characteristic ratio, C^∞, is a measure of the expansion of a coil in the unperturbed state. C^∞ is defined equal to 1 for a random flight chain. The conformation average end-to-end distance for the random flight chain, $\langle h^2 \rangle_o$, is

$$\langle h^2 \rangle_o = C^\infty z\, b^2 \tag{5.25}$$

Freely rotating chains are those for which the bond lengths and valence angles are fixed but the rotational angles can vary (Flory, 1953). Free rotation means that the rotation angle may take any value with equal probability and equal energy. Freely rotating chains with carbon-carbon backbones have a value of $C^\infty = 2$.

Another level of complexity is added by allowing chain rotation to be hindered by stearic factors. Flory (1953) and Yamakawa (1971) have more detailed discussions of this model. The models begin to account for preferred energy positions caused by repulsive and attractive forces. An independent hindered rotation will depend only on its immediate neighbors. Its value of $C^\infty = 3$.

In real chains hindered rotations are interdependent; they are affected by the conformation of the immediately adjacent chain segments. The interdependent hindered rotation model applied to polyethylene gives a value for C^∞ of 6.9, which is very close to the experimental value of 7.3.

Excluded Volume Effects. The above corrections to the statistics of chain conformation have been based on short-range interactions. There are also some restrictions to chain conformation based on long-range interactions. Long-range interactions occur between segments that are many bond lengths apart and are often called excluded volume effects.

A volume element in a space-filling lattice cannot be occupied twice by two different chain segments. In the random flight chain model, a volume segment can be occupied by any number of segments. The excluded volume effect causes expansion of a chain in space relative to a random flight chain. Figure 5.6 and the following table show a comparison of a random flight chain to an excluded volume chain for a two-dimensional square lattice chain with two bonds.

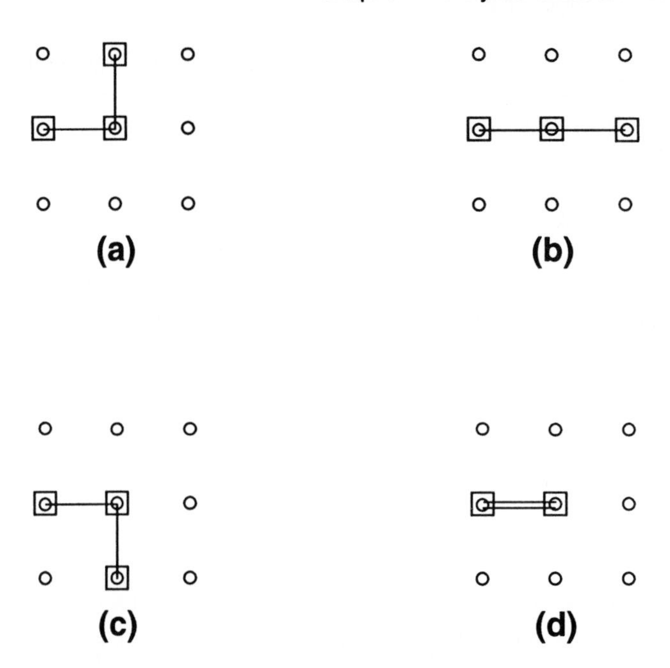

Figure 5.6 Two-bond molecule distributed on a two-dimensional square lattice. An ideal chain can assume all conformations. A real chain cannot assume conformation due to steric hindrances.

h^2	Figure	Random Flight Chain	Excluded Volume Chain
2	5.6a	allowed	allowed
4	5.6b	allowed	allowed
2	5.6c	allowed	allowed
0	5.6d	allowed	not allowed

The average end-to-end distance for the random flight chain is the sum of all the possible values for h divided by the total number of cases.

$$<h^2> = \frac{2 + 4 + 2 + 0}{4} = 2$$

The average end-to-distance for an excluded volume chain can be calculated in a similar way.

$$<h^2> = \frac{2 + 4 + 2}{3} = 2.66$$

The average end-to-end distance of the excluded volume chain is higher than that of the random flight chain. As can be inferred from this simple example, an excluded volume chain will always occupy more volume than a random flight chain since some conformations allowed by the random flight chain will not be allowed by the excluded volume chain.

The solvent in which the polymer is dissolved affects long-range interactions. In a "good" solvent, the energy of interaction of the solvent-polymer system is greater than that of the polymer and the solvent alone. This is due to van der Waals interactions and results in an expansion of the molecule so that the frequency of the polymer-polymer contacts are minimized (the free energy of mixing is reduced). In a "poor" solvent, polymer-polymer contacts are favored and there is a shrinkage of the molecule relative to its size in a good solvent. In a good solvent, the number of conformations decreases as the chain expands. This reduces the entropy change on mixing, and the system equilibrium for molecule size occurs where the enthalpic changes due to solvent association are balanced by the change in the entropic term.

5.4 POLYMER SOLUTION THERMODYNAMICS

5.4.1 Flory–Huggins Theory

Thermodynamic experiments show that polymer solutions exhibit large deviations from Raoult's law. This seems true even for "ideal" systems such as nonpolar molecules in nonpolar solvents where small heats of mixing are expected. Early researchers concluded that polymer solutions mix with an entropy of mixing that is quite different than that predicted by the ideal entropy of mixing expression (Eq. 5.12). The cause of the nonideality in the entropy of mixing is the long chain nature of polymer molecules.

Consider that a solution can be represented by a three-dimensional network of lattice sites that can be occupied by either solvent or polymer. In terms of this model, a polymer molecule of n repeat units occupies n continuous sites. By contrast, an organic solute at a concentration of n units in the same solvent would occupy any n lattice sites with equal probability. Huggins (1958), Flory (1953), and Flory and Krigbaum (1951) have developed this notion into theories for the entropy of mixing of polymer solutions. The entropy of mixing of long chain polymer molecules with small solvent molecules is

$$\Delta S_m = -R(x_1 \ln \phi_1 + x_2 \ln \phi_2) \qquad 5.26$$

where x_1 and x_2 are the monomer and polymer mole fractions, and ϕ_1 and ϕ_2 are the site or volume fractions of solvent and polymer. It is assumed that polymer segments and solvent each occupy one lattice site. The volume fractions are related to the mole fractions by

$$\phi_1 = \frac{x_1}{x_1 + N x_2} = \frac{V_1 x_1}{V_1 x_1 + V_2 x_2} \tag{5.27a}$$

and

$$\phi_2 = \frac{x_2}{x_1 + N x_2} = \frac{V_2 x_2}{V_1 x_1 + V_2 x_2} \tag{5.27b}$$

The quantity, N, is the ratio of molar volumes, V_2/V_1. It is usually different from the degree of polymerization. Equation 5.26 is combined with Equation 5.15 for nonideal heat of mixing to give the following free energy change on mixing.

$$\frac{\Delta G_m}{RT} = \left(x_1 \ln \phi_1 + x_2 \ln \phi_2 + x_1 g \, \phi_2\right) \tag{5.28}$$

where g is a polymer-solvent interaction parameter and the left-hand side of the equation is the reduced molar Gibbs energy of mixing.

The reduced molar Gibbs energy of mixing contains a combinatorial portion (Eq. 5.26) and a residual portion. Equation 5.28 can be differentiated with respect to the moles of solvent, n_1, or the moles of polymer, n_2, to obtain the reduced partial molar Gibbs energy of mixing for each component.

$$\frac{\Delta\mu_1}{RT} = \ln\left(1-\phi_2\right) + \left(1-\frac{1}{N}\right)\phi_2 + \chi \, \phi_2^2 \tag{5.29a}$$

$$\frac{\Delta\mu_2}{RT} = \ln(1-\phi_1) + (1-N)\phi_1 + \varepsilon \, N \phi_1^2 \tag{5.29b}$$

χ and ξ are the interaction parameters for the solvent and polymer, respectively. The three interaction parameters can be related by (Gundert and Wolf, 1989)

$$g = \phi_2 \chi + (1 - \phi_2)\xi \tag{5.30a}$$

$$\chi = g - (1-\phi_2)\frac{\partial g}{\partial \phi_2} \tag{5.30b}$$

$$\varepsilon = g + \phi_2 \frac{\partial g}{\partial \phi_2} \tag{5.30c}$$

In most experiments for measuring phase equilibria, only the chemical potential of the solvent is measured, and χ is the only parameter that can be obtained directly.

The models for the interaction parameters include their dependence on polymer volume fraction (Eq. 5.30a–c). The original formulation of the theory considered χ to be a constant, but measurement of χ for a wide range of systems has shown that χ varies with concentration.

Solvent activity is often the simplest parameter to evaluate for equilibria experiments, and a useful form of the theory is

$$\ln a_1 = \ln \gamma_1 x_1 = \ln (\phi_1) + (1 - \frac{1}{N})\phi_2 + \chi \phi_2^2 \quad 5.31$$

Concentration-Dependence of χ. Equation 5.31 suggests that the Flory–Huggins theory may be tested by plotting $\ln (a_1/\phi_1) - (1 - V_1/V_2) \phi_2$ vs. ϕ_2^2. A straight line should result due to the linear relationship. This does not occur for a number of polymer systems. However, the theory is an improvement over ideal solution theory and can be used to predict phase separation.

EXAMPLE 5.2 CALCULATION OF TOLUENE VAPOR PRESSURE OVER A POLYSTYRENE SOLUTION

Example 5.1 compared the vapor pressures of toluene in equilibrium with the liquid phase for two systems: a regular solution of toluene and styrene and a polymer solution of toluene and polystyrene. This example uses the Flory–Huggins equation to calculate the vapor pressure of toluene over the polymer solution. The composition of the liquid phase is 50wt% toluene, 50wt% polystyrene. The various solution compositions are

	W_i	X_i	ϕ_i
Toluene	.50	.9995	.5304
Polystyrene	.50	.0005	.4696

Equation 5.31 can be used to solve for the solvent activity if the solvent interaction parameter, χ, is known. Interaction parameters are available from a number of sources (Gundert and Wolf 1989). For a toluene-polystyrene mixture at 25°C, χ = 0.277. These calculations ignore the effect of temperature on χ and assume that the ratio of molar volumes, N, is large, so that the term $(1 - 1/N)$, is close to one. The solvent activity is

$$\ln a_1 = \ln \phi_1 + (1 - \frac{1}{N}) \phi_2 + \chi \phi_2^2$$

$$= \ln(0.5304) + (1)(.4696) + .277(.4696)^2$$

$$= -0.1034$$

$$\text{and } a_1 = 0.9017$$

(continued)

The activity of the solvent in the liquid phase equals the activity of the solvent in the gas phase. The reference fugacity for the gas can be taken as the pure component vapor pressure for the system temperature. For toluene at 30°C, $f_1^\circ = 0.04921$ bar. Therefore, the vapor pressure over the polystyrene solution should be 0.0444 bar in a closed tank. Figure 5.7 shows that the toluene vapor pressure over a polystyrene solution is not linear with polymer volume fraction.

5.4.2 The Solvent Interaction Parameter

The solvent interaction parameter originally was considered to be a correction only to the heat of mixing. Now, χ is considered an excess free energy parameter and is affected by excess entropy of mixing and heats of mixing other than endothermic, regular solution type. The term "excess free energy" is used to denote excess with respect to the conformation entropy of mixing as described by the Flory–Huggins lattice model.

Procedures for measuring χ are described in Collins, Bares, and Billmeyer (1973). The techniques differ in degree of experimental difficulty and in accuracy of the result. Osmotic pressure and light-scattering methods are often chosen since they result in reasonably precise measurements from standard experimental equipment. Dilute solution viscosity or equilibrium swelling experiments are used because they are simple to perform. Table 5.1 shows some different methods used for determining χ and the concentration range for their application.

As shown in Figures 5.8 and 5.9, the interaction parameter often varies with concentration. Solvents that are good solvents for the polymer often have interaction

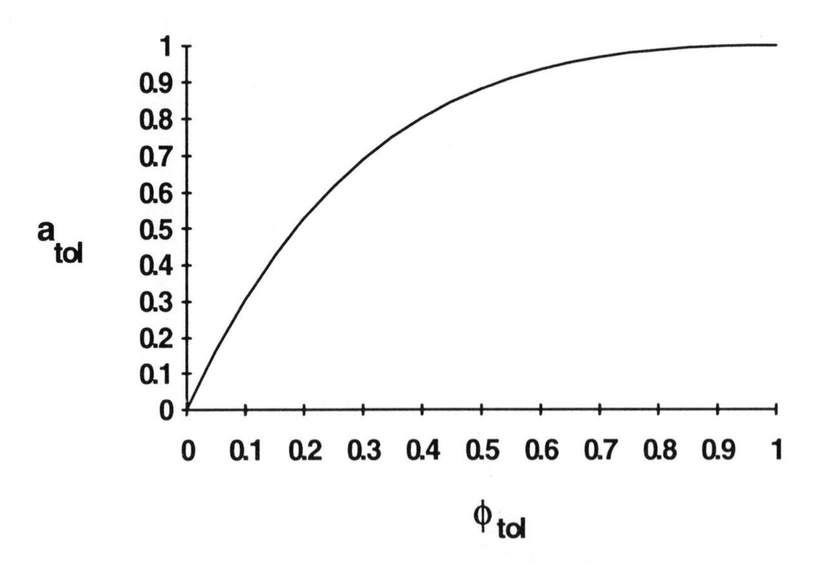

Figure 5.7 Toluene activity over a polystyrene solution at 25°C.

TABLE 5.1 Methods for the Determination of the Polymer-Solvent Interaction Parameter

Typical Range of Concentration	Method
$\phi_2 \rightarrow 0$	Light scattering
$0 < \phi_2 < 0,3$	Osmosis
	Critical miscibility method (only applicable with poor solvents)
$0.3 < \phi_2 < 0.8$	Vapor pressure methods
$\phi_2 \rightarrow 1$	Inverse gas chromatography (gas liquid chromatography)

parameters that are nearly constant over wide volume fraction ranges. The interaction parameter usually increases with polymer volume fraction for poor solvents. For highly exothermic systems, χ decreases with increasing polymer volume fraction. The curves of Figures 5.8 and 5.9 are based on a power series expansion of χ in the polymer volume fraction:

$$\chi = \chi_o + \chi_1 \phi_2 + \chi_2 \phi_2^2 \tag{5.32}$$

The coefficients shown in Table 5.2 are based on literature data. Cyclohexane/polyisobutylene and n-octane/poly(dimethyl siloxane) are examples of good solvents, as χ

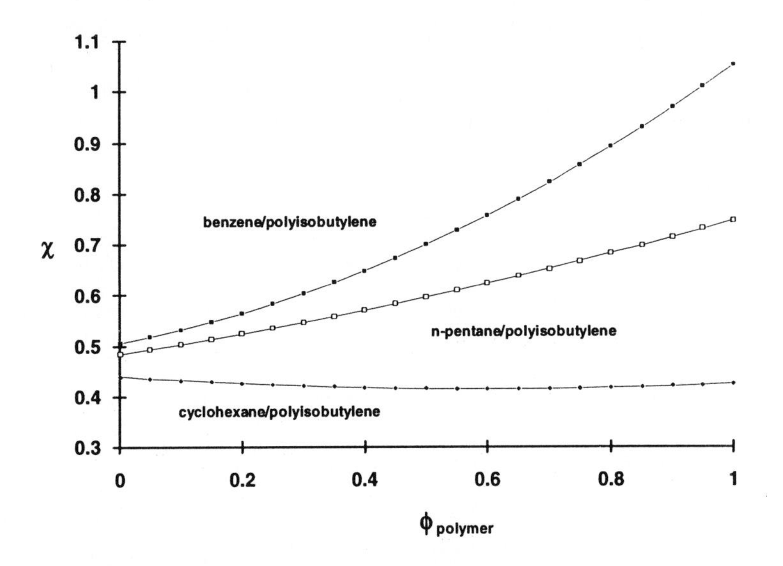

Figure 5.8 Examples for the variations of the Flory–Huggins interatction parameter χ with the volume fraction φ_2 of polymer for the systems benzene/poly isobutylene (PIB), n-pentant/PIB and cyclohexane/PIB (Gundert and Wolf, 1989).

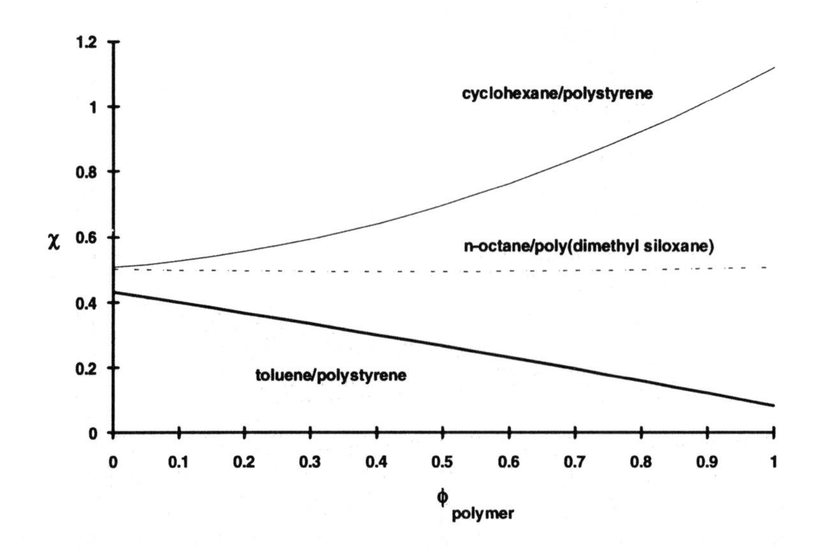

Figure 5.9 Examples for the variations of the Flory–Huggins interatction parameter χ with the volume fraction φ_2 of polymer for the systems cyclohexane/polystyrene (PS), n-octane/poly(dimethyl siloxane)(PDMS) and toluene/PS (Gundert and Wolf, 1989).

is nearly constant over the entire volume fraction range. Toluene/polystyrene is highly exothermal and χ decreases as the polymer volume fraction increases. Ben-zene/poly-isobutylene, n-pentane/polyisobutylene and cyclohexane/polystyrene are poor solvent systems, and χ increases with polymer volume fraction.

An alternative to Equation 5.32 is a relationship based on classical lattice theory of polymer solutions (Koningsveld and Kleintjens, 1971),

TABLE 5.2 Various Power Series Coefficients for χ

System	T_1 °C	χ_0	χ_1	χ_2	References
Benzene/PIB	25	0.506	0.229	0.317	Flory & Daoust (1957); Flory (1943); Eichinger & Flory (1968a)
n-Pentane/PIB	25	0.484	0.184	0.080	Baker et al. (1962); Eichinger & Flory (1968c)
Cyclohexane/PIB	25	0.440	−0.083	0.068	Newman & Prausnitz (1972); Flory (1943); Eichinger & Flory (1968b)
Cyclohexane/PS	34	0.507	0.149	0.464	Krigbaum & Geymer (1959)
n-Octane/PDMS	20	0.501	−0.035	0.042	Chahal et al. (1973)
Toluene/PS	25	0.431	−0.311	−0.036	Noda et al. (1984)

$$\chi = A + \frac{B(1-D)}{(1-D\phi_2)^2}, \; B = B_\infty + \frac{B_t}{T} \tag{5.33}$$

where D is related to the number of nearest neighbors in the lattice model, A can be related to the spinodal criterion, and B is related to the critical or consolute state. The spinodal and critical states will be discussed in the next section. The coefficients for Equation 5.33 can be evaluated from sets of critical miscibility data. The potential advantage of Equation 5.33 over Equation 5.32 is that the former should give an improved representation of separation of the liquid phase. The power series coefficients of Equation 5.32 can be evaluated using vapor-liquid equilibria data, but may not predict liquid phase separation well. Coefficients are available for Equation 5.33 for the system, cyclohexane/polystyrene (Koningsveld and Kleintjens, 1971). The prediction of χ is compared in Figure 5.10 to the power series expansion. The lattice theory predicts an increase in χ with increasing polymer volume fraction, but the values are about one-half of the actual data. The lattice theory coefficients were determined using critical miscibility experiments, while the power series model was fit to data taken by vapor pressure depression experiments. The apparent discrepancy between the two models probably is due to comparing different data sets. Different measurement methods can yield different values for χ. The lattice model can be made to fit the data as shown in Figure 5.10.

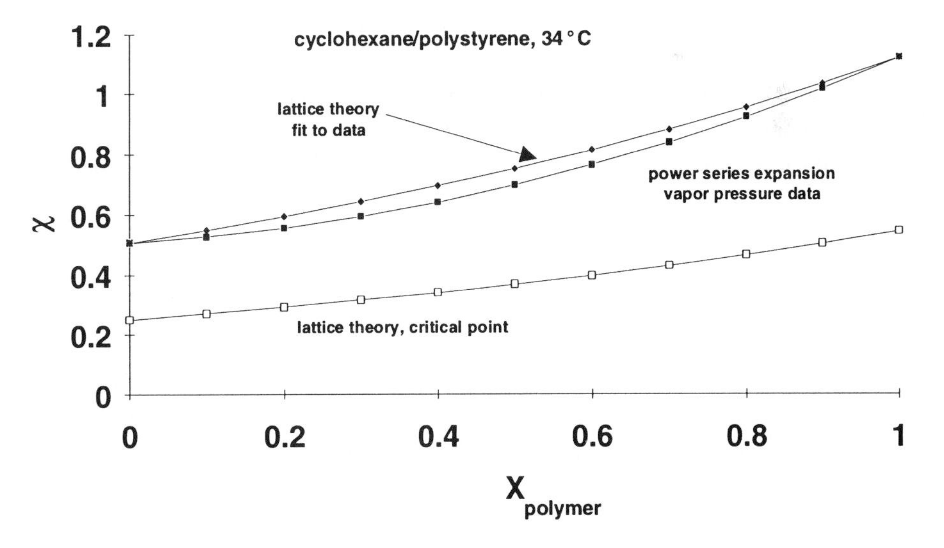

Figure 5.10 Comparisons of Equations 5.32 and 5.33 for cyclohexane/polystyrene at 348C. Heavy line: power series coefficients from Table 5.2. Dashed line: lattice theory coefficients from data of Koningsveld and Kleintjens (1971) thin line: lattice theory fit to power series data

5.4.3 Solubility Parameter and Cohesive Energy Density

Prediction of thermodynamic model parameters from the chemical structure of a system is an important objective for the engineer. The solubility parameter has been used to predict the interaction parameter with some success.

Solubility parameters were developed to describe the enthalpy of mixing of simple liquids (nonpolar, nonassociating solvents), but have been extended to polar solvents and polymers. Hildebrand and Scott (1950) and Scatchard (1949) proposed that

$$\Delta H_m = V\left[\left(\frac{\Delta E_1^v}{V_1}\right)^{1/2} - \left(\frac{\Delta E_2^v}{V_2}\right)^{1/2}\right]^2 \phi_1\phi_2 \qquad 5.34$$

where V = volume of the mixture, ΔE_i^v = energy of vaporization of species i, V_i = molar volume of species i, and ϕ_i = volume fraction of i in the mixture. ΔE_i^v is the energy change upon isothermal vaporization of the saturated liquid to the ideal gas state at infinite volume. The cohesive energy density (CED) is the energy of vaporization per cm^3. The solubility parameter has been defined as the square root of the cohesive energy density; it describes the attractive strength between molecules of the material.

$$\delta_i = \left(\frac{\Delta E_1^v}{V_i}\right)^{1/2} \qquad 5.35$$

The dimensions of δ_i are (cal/cm^3)$^{1/2}$ = 4.187 (J/10^{-6}m $^{1/2}$) = 2.046 × 10^3 (J/m^3)$^{1/2}$ = 2.046 (MPa)$^{1/2}$. The solubility parameter can be considered as the "internal pressure" of the solvent. Equation 5.34 can be rewritten to give the heat of mixing per unit volume for a binary mixture.

$$\frac{\Delta H_m}{V} = (\delta_1 - \delta_2)^2 \phi_1\phi_2 \qquad 5.36$$

In order for ΔG_m to be less than zero, the heat of mixing must be smaller than the entropic term. ΔG_m will always be less than zero for regular solutions if $\delta_1 = \delta_2$ and the components will be miscible in all proportions. In general $(\delta_1 - \delta_2)^2$ must be small for the components to be miscible. Equation 5.36 gives the heat of mixing of regular solutions in which the components mix with (a) no volume change on mixing at constant pressure, (b) no reaction between the components, and (c) no complex formation or special associations. δ_is of Equation 5.35 often are called Hildebrand or cohesion parameters. The solubility parameter of a mixture is modeled as the sum of the products of the component solubility parameters with their volume fractions.

ΔE_i^v is related to the enthalphy of vaporization

$$\Delta E_i^v = \Delta H_i^v - RT \tag{5.37}$$

where R = ideal gas constant. The solubility parameter of solvents can be determined by measuring the enthalpy of vaporization (or using a correlation for ΔH_i^v) and using Equation 5.37.

$$\delta_i = \left(\frac{\Delta H_i^v - RT}{V_i} \right)^{1/2} \tag{5.38}$$

Equation 5.37 is a simplified description of the molar cohesive energy. A complete description includes ΔH_i^∞ = molar increase in enthalpy on isothermally expanding the saturated vapor to zero pressure and $P_i^{sat}\ V_i$ = the pV work of the process. At pressures below 1 atm, the ΔH_i^∞ and $p_s V_i$ terms are usually much less than ΔH_i^v and RT. At the critical point, $\Delta H_i^v = 0$, and Equation 5.37 incorrectly predicts a negative value for the cohesive energy density, while inclusion of the extra terms yields a small positive value.

The Gibbs free energy of mixing can be modified for crystalline polymer phase equilibria by including terms for the heat of fusion and the entropy change associated with the disruption of the crystal lattice. Some crystalline polymers obey the solubility parameter model at temperatures near their melting point, $T \geq 0.9\ T_m$ (van Krevelen, 1990). Solvent swelling experiments with crystalline polymers may fit the model.

Cohesive Energy Parameters for Polar Systems. The solubility parameter describes the enthalphy change on mixing of nonpolar solvents well, but it does not give uniform results when extended to polar systems. Complete miscibility is expected to occur if the solubility parameters are similar and the degree of hydrogen bonding is similar between the components. Table 1.14 lists chemical functional groups by their hydrogen bonding strength. A model for δ is

$$\delta^2 = \delta_d^2 + \delta_p^2 + \delta_h^2 \tag{5.39}$$

where δ_d = dispersive term, δ_p = polar term, and δ_h = hydrogen bonding term. δ_h probably accounts for a variety of association bonds, including permanent dipole–induced dipole. The values of these compounds for solvents were calculated from a large number of solubility data sets. Polymer solubility parameters can also be decomposed to a three-term set. The Hansen parameters give improved agreement with data, but they are still not completely accurate in predicting solution thermodynamics for every system.

Applications of solubility parameters include finding compatible solvents for coating resins, predicting the swelling of cured elastomers by solvents, estimating solvent pressure in devolatilization and reactor equipment, and predicting polymer, polymer-binary, random copolymer, and multicomponent solvent equilibria. Extension to multicomponent systems is described by Olabisi, Robeson, and Shaw (1979).

Solvent Interaction Parameter. The solvent interaction parameter has been written as a sum of an entropic and enthalpic component,

$$\chi = \chi_s + \chi_H \tag{5.40}$$

where χ_s = entropic term and χ_H = enthalpic term of the interaction parameter. χ_s is usually taken to be a constant between 0.3 and 0.4 for nonpolar systems ($\chi_s = 0.34$ is often used). The enthalpic component can be related to the solubility parameters.

$$\chi_H = \frac{V_1}{RT}(\delta_1 - \delta_2)^2 \tag{5.41}$$

$$\chi = 0.34 + \frac{V_1}{RT}(\delta_1 - \delta_2)^2 \tag{5.42}$$

Mixed solvents can be treated as a single solvent by determining the solubility parameter of the solvent mixture and then using this value in Equation 5.42. If both the solvents and the polymers interact, then the description is more complicated.

Calculations of and Correlations for Solubility Parameters. Solubility parameters can be determined by direct measurement, correlations with other physical parameters, or indirect calculations. The solubility parameters of solvents usually can be determined directly. The solubility parameters of polymers can only be determined indirectly and may be affected by variations in their chemical constitutions, that is, the number of cross-links and the distribution of chain branches or substitutive groups along the polymer backbone. Barton (1983) and Grulke (1989) give a number of methods.

EXAMPLE 5.3 SOLVENCY TESTING (SCREENING PROCEDURE)

The usual purpose of these experiments is to find good solvents for commercial product formulations. The solvency properties of a commercial polymer can be defined by determining its solubility parameter range for each hydrogen bonding class of solvents (poor, moderate, and strong). Table 5.3 shows a set of solvents used for this purpose. The procedure is a set of simple phase equilibrium experiments. A gram or two of solid polymer is placed in a test tube and an approximate amount of a selected solvent is added such that the final solution

(continued)

would have about the correct solids content for the expected commercial use (for example, 50% for alkyds, 20% for vinyls, etc.). The exact amount is often unimportant, except for poor solvents.

The mixture may be warmed and stirred to speed up solution, but it should be cooled and observed at room temperature. The soluble mixture should be a single phase, clear and free of gel particles or cloudiness, otherwise the polymer is judged insoluble. The experiments should identify solvents representing the maximum and minimum solubility parameter liquid in which the polymer will dissolve.

TABLE 5.3 Selected Solvents for Use in Polymer Solvency Testing (δ in $MPa^{1/2}$)

Solvent	δ
POORLY HYDROGEN BONDED	
n-Pentane	14.3
n-Hexane	14.9
n-Heptane	15.1
Methylcyclohexane	16.0
Solvesso 150	17.4
Toluene	18.2
Tetrahydronaphthalene	19.4
o-Dichlorobenzene	20.5
1-Bromonaphthalene	21.7
Nitroethane	22.7
Acetonitrile	24.1
Nitromethane	26.0
MODERATELY HYDROGEN BONDED	
Diethyl ether	15.1
Diisobutyl ketone	16.0
n-Butyl acetate	17.4
Methyl propionate	18.2
Dibutyl phthalate	19.0
Dioxane	20.3
Dimethyl phthalate	21.9
2,3-Butylene carbonate	24.8
Propylene carbonate	27.2
Ethylene carbonate	30.1
STRONGLY HYDROGEN BONDED	
2-Ethyl hexanol	19.4
Methyl isobutyl carbinol	20.5
2-Ethylbutanol	21.5
n-Pentanol	22.3
n-Butanol	23.3
n-Propanol	24.3
Ethanol	26.0
Methanol	29.7
Water	42.3

Estimation of δ by Group Contribution Methods. Group contribution methods have been applied to estimating the solubility parameter. The group constants sets of Small (1953), Hoy (1970), and van Krevelen (1990), seem to be most comprehensive. Table 5.4 gives the group molar attraction constants at 25°C.

The group contribution techniques are based on the assumption that the contributions of different functional groups to the thermodynamic property are additive (Eq. 5.8). The energy of vaporization of a solvent or polymer is

$$\Delta E_i^v = \left(\frac{(\sum_j n_j F_j)^2}{V_i} \right) \tag{5.43}$$

where F_j is the molar attraction constant. The solubility parameter is

$$\delta_i = \left(\frac{\Delta E_i^v}{V_i} \right)^{1/2} = \frac{\sum_j n_j F_j}{V_i} \tag{5.44}$$

The molar volume of a solvent or a polymer can be calculated using ρ_i, the density, and M_i, molecular weight. Polymer solubility parameters can be evaluated using the repeating unit structure.

EXAMPLE 5.4 SOLUBILITY PARAMETER OF POLYSTYRENE

Equation 5.44 is applied by building the polystyrene repeating unit using the groups shown in Table 5.4, computing the overall molar attraction factor and dividing by the molar volume of the repeating unit. The structure is a methylene group, a C-H group in the chain, and a phenyl group. The sum of the attraction constants is

$$\Sigma n_j F_j = 1 \times 280 + 1 \times 140 + 1 \times 1517 = 1937 (MPa)^{1/2} \frac{cm^3}{gmol}$$

The molar volume of the polystyrene repeating unit can be estimated by multiplying the inverse of the polymer density by the segment molecular weight. The segment molar volume is 104/1.02 = 102 $cm^3/gmol$. The estimated solubility parameter is

$$\delta_2 = \frac{1937}{102} = 19.0 \ MPa^{1/2}$$

This compares with the experimental value of 18.5

TABLE 5.4 Group Contribution Coefficients for the Solubility Parameter 25°C

.Group	F, $(J/cm^3)^{1/2} \frac{cm^3}{g\ mol}$	Group	F, $(J/cm^3)^{1/2} \frac{cm^3}{g\ mol}$
$-CH_3$	420	—H	140[a]
$-CH_2-$	280	—F	164
		—Cl	471
H —C— |	140	—Br	614
		—I	—
| —C— |	0	—CN	982
		—CH(CN)—	1122
		—OH	754
$-CH(CH_3)-$	560	—O—	256
$-C(CH_3)_2-$	840	—CO—	685
—CH=CH—	444	—COOH	652
		—COO—	512
| —C=CH—	304	—S—	460
$-C(CH_3)=CH-$	724	Cyclopentyl	1384
—O—CO—O—	767	Cyclohexyl	1664
—CO—O—CO—	767	Phenyl	1517
—CO—NH—	1228	*p*-Phenylene	1337
—O—CO—NH—	1483		

[a]estimate

SOURCE: D.W. van Krevelen, *Properties of Polymers, Correlations with Chemical Structures*, Elsevier, NY, 1990.

Other Group Contribution Methods. In the mid-70s, improved models for solution thermodynamic properties were developed. These models were based on group contribution theory, and the enthalpic and entropic contributions of subgroups of molecules, such as -CH_2 groups, were determined by analyzing many thermodynamic data sets. Group contribution models can be used to give a priori predictions of activity coefficients by breaking down the molecule into subgroups and using table values to "build" the appropriate contributions. These models have recently been applied to polymer solutions.

Oishi and Prausnitz (1978) have adapted the UNIFAC model to polymer solutions. UNIFAC is based on statistical thermodynamics and works by reducing a chemical compound into a set of functional groups. By defining a method for approximating the properties of compounds from the known functional group properties, the properties of new compounds can be determined. Such a system might be very useful for polymer solutions, where detailed data are often not available and where solutions are often composed of a range of molecular weight solutes. Oishi and Prausnitz modified the UNIFAC model by adding an extra term to the entropic activity terms that was much like the interaction parameter corrector. Results agree with experiments to within 20%. Chen, Fredenslund, and Rasmussen (1990) have made additional modifications that use a group contribution equation-of-state to model vapor-liquid equilibria of polymer solutions.

The ASOG model has a similar basis and was developed by Derr and Deal (1969). Misovich, Grulke, and Blanks (1985) have modified the ASOG model to apply it to polymer solutions. The ASOG model is about the same as the UNIFAC model in terms of accuracy for low molecular weight solutions, but it has the advantage of resulting in closed-form solutions when applied to nonpolar polymers. To fit data well, a constant must be determined, but this constant can be related directly to simple activity coefficient measurements.

5.4.4 Flory–Krigbaum Theory: Theta Temperature

An improvement on the Flory–Huggins theory is particularly useful in defining the concept of the theta temperature. At the theta temperature for a solvent-polymer pair, the polymer chains are in their smallest conformation (nearly random flight conformation).

One assumption made in deriving the Flory–Huggins equations was that the polymer coils were uniformly distributed throughout the solution volume. If the polymer molecules are considered to be distributed over a three-dimensional lattice, it is easy to show that the probability of finding a second chain of n lattice sites within the volume of the first chain is very low. A dilute polymer solution can be visualized as having volume regions with no polymer and regions with high chain densities (Fig. 3.4).

Flory and Krigbaum assumed that polymer segments were distributed in a Gaussian fashion about their centers of mass in the random coils. The excess chemical potential of the solvent in the polymer volume element is (Eq. 5.29a)

$$(\mu_1 - \mu_1^\circ)_E = RT[\ln \phi_1 + (1 - \frac{1}{X_n})\phi_2 + \chi\, \phi_2^2] \qquad 5.45$$

Expanding the logarithm term in a series expansion

$$(\mu_1 - \mu_1^\circ)_E = -\,RT[-\frac{\phi_2}{X_n} - (\frac{1}{2} - \chi)\,\phi_2^2 + \phi_2^2/3 + \ldots] \qquad 5.46$$

Within the random coil, there are many solvent molecules per polymer segment, so that φ_2 is small. The higher order terms in ϕ_2 can be neglected.

$$(\mu_1 - \mu_1^\circ)_E = RT \ln a_1 \cong -RT\left(-\frac{\phi_2}{X_n} - (\frac{1}{2} - \chi)\,\phi_2^2\right) \qquad 5.47$$

Criterion for Complete Miscibility in Dilute Solution. Equation 5.47 defines a criterion for complete miscibility of a polymer molecule in dilute solution. The solvent activity should be close to one in dilute solution. When the molecular weight of the polymer is large, the first term in the right-hand side of Equation 5.47 is small. The second term is small when the interaction parameter is less than or equal to one-half. Solvents meeting this criterion are considered good solvents, since phase separation should not occur over the entire volume fraction range. Equation 5.47 can be decomposed into an excess partial molar enthalpy and an excess partial molar entropy,

$$(\mu_1 - \mu_1^\circ)_E = RT\,(K_1 - F_1)\,\phi_2^2 \qquad 5.48$$

where K_1 is the enthalpy coefficient and F_1/T is the entropy coefficient. When the polymer solution is ideal, the excess chemical potential will be zero. A temperature, θ, can be defined so that the enthalpy and entropy contributions will exactly balance. This temperature is the ideal, Flory, or theta temperature,

$$\theta = K_1\,T/F_1 \qquad 5.49$$

Substituting Equation 5.49 into Equation 5.48,

$$(\mu_1 - \mu_1^\circ)_E = -\,RT\,(1 - \theta/T)\,\phi_2^2 \qquad 5.50$$

When $T = \theta$, the deviations from ideality vanish. K_1 and F_1 should have the same sign for this analysis. For most nonpolar polymer solutions, K_1 and F_1 are both positive. The free energy of mixing is negative because the entropy term is much larger than the enthalpy term.

The expansion factor of a dissolved polymer molecule can be calculated as a function of the theta temperature. a is defined as the ratio of the end-to-end distances of the actual and ideal polymer chains and is related to the theta temperature by

$$a^5 - a^3 = 2\,C_\infty\,F_1\,(1 - \theta/T)\,M^{1/2} \qquad 5.51$$

a increases with increasing molecular weight but decreases as V_1 increases. At the theta temperature, $a = 1$. At temperatures above the theta temperature, the polymer expands relative to its minimum dimensions. At temperatures below the theta temperatures, the polymer will aggregate to form a precipitate phase. Figure 5.11 shows data for the system, cyclohexane/polystyrene, taken by various techniques.

5.4.5 Extension to Polydisperse Systems

The Flory–Huggins theory was developed for monodisperse polymer systems, but can be extended to polymer samples with a range of molecular weights. The relation analogous to Equation 5.31 is

$$\ln a_1 = \ln \phi_1 + \left(1 - \frac{1}{X_n}\right)\phi_2 + \chi\,\phi_2^2 \qquad 5.52$$

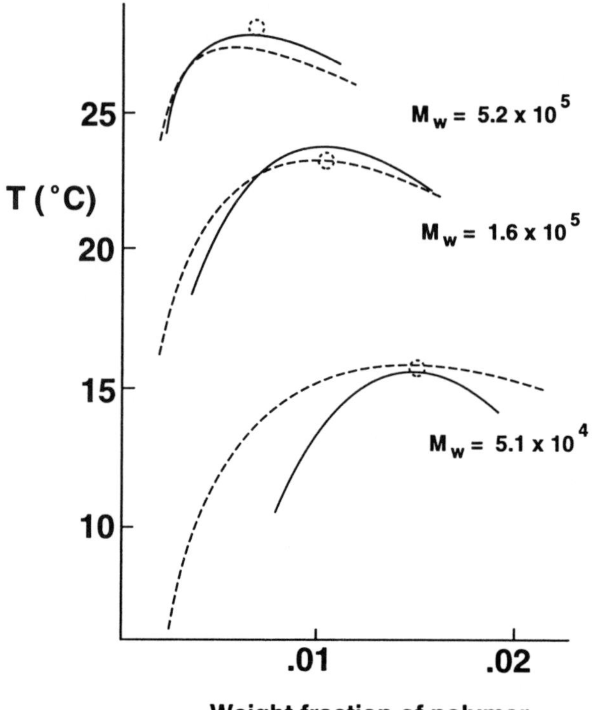

Figure 5.11 Spinodals and critical miscibility data for cyclohexane/polystyrene samples. The open circles show the measured critical points. Light-scattering data for the spinodals are the dashed curves. The soid curves are the Koningsveld and Kleintjens model (1971).

5.5 POLYMER PHASE EQUILIBRIA

5.5.1 Analysis of Phase Diagrams

Determination of phase equilibria is very important to developing sound polymer products and processes. For example, it is important to find plasticizers that do not phase separate at use temperatures. Polymerization conditions can be set such that the polymer will precipitate from solution as it gets to some critical molecular weight.

The free energy of mixing must be less than zero for the solution of one material in another. The entropy of mixing will always be greater than zero since the two pure components would be going to less ordered states. Therefore, the entropic term will always contribute a negative quantity to the free energy of mixing. There are two cases to consider for the enthalpy of mixing term.

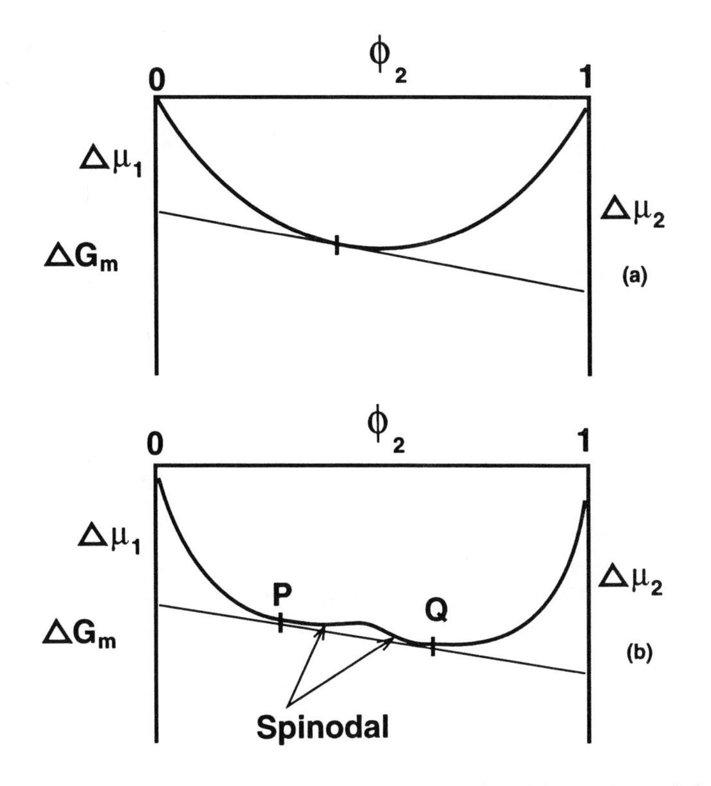

Figure 5.12 Free energy versus polymer mole fraction: **(a)** negative enthalpy of mixing; **(b)** positive enthalpy of mixing.

Case 1: Negative Enthalpy of Mixing.

A solvent-polymer system is always miscible if the enthalpy of mixing is negative for all mixture concentrations. This case is sketched in Figure 5.12a. The left vertical axis represents the chemical potential change for the solvent and the right vertical axis represents the chemical potential change for the polymer. At a given value of x_2, the free energy of mixing for the system is

$$\Delta G_m = x_1 \Delta\mu_1 + x_2 \Delta\mu_2 \tag{5.53}$$

where $\Delta\mu_1$ and $\Delta\mu_2$ are partial molal properties and are given by Equation 5.9c. Equation 5.53 can be rearranged to give

$$\Delta G_m = \Delta\mu_1 + x_2 (\Delta\mu_2 - \Delta\mu_1) \tag{5.54}$$

This is the equation of the tangent having the intercept, $\Delta\mu_1$, at $x_2 = 0$ and the slope of $(\Delta\mu_2 - \Delta\mu_1)$. Each point along the free energy curve has a unique set of values for

$\Delta\mu_1$ and $\Delta\mu_2$. There are not two points with the same values for the chemical potential, and only one phase will exist.

Case 2: Positive Enthalpy of Mixing.
The free energy of mixing is not always less than zero when the enthalpy of mixing term is positive. Figure 5.12b shows two points, P and Q, that have the same tangent line. This means that the chemical potential differences for these two polymer mole fractions are the same, and that they are in equilibrium with each other. Between P and Q, the curve goes through a maximum. The free energy changes on mixing for these compositions are greater than the minimum free energy of mixing for the system. These solutions separate into two solutions with the compositions, P and Q. The tie line rule determines the relative amounts of the P and Q phases.

Upper and Lower Critical Solution Temperatures. Figure 5.13 shows a phase diagram for a polymer mixture. The range of complete miscibility lies between the upper and lower curves. Upper critical solution temperatures (UCST) are known for a variety of solvent-polymer systems. Lower critical solution temperatures (LCST) have been discovered more recently. They often occur for systems with strong hydrogen bonding between the solvent and the polymer. Polymer solubility decreases with increasing temperatures. Many polymer systems showing UCST's also have LCST's at temperatures approaching the critical point of the solvent. The cause is thought to be the much larger expansion coefficient of the solvent as temperature changes.

Binodals, Spinodals, and the Critical Point. Figure 5.14 sketches the free energy of mixing versus polymer mole fraction for the UCST and two values below it. The lowest curve contains the upper critical solution temperature and is the lowest temperature for complete miscibility for this system. The highest curve represents two

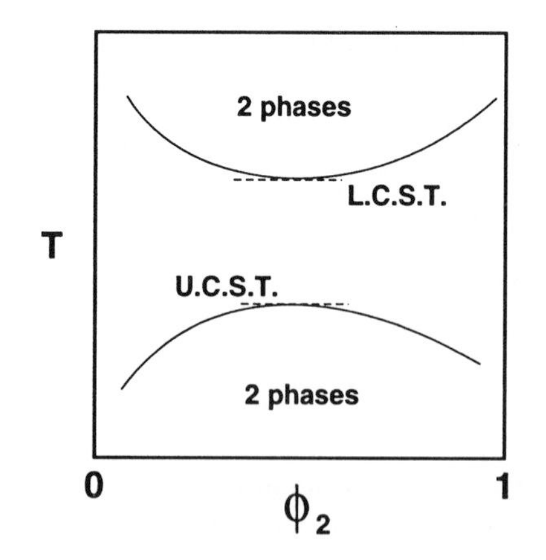

Figure 5.13 Phase diagram for a polymer mixture. The one- and two-phase regions are shown as functions of temperature and volume fraction. The range of miscibility lies between the upper critical solution temperature (UCST) and the lower critical solution temperature (LCST).

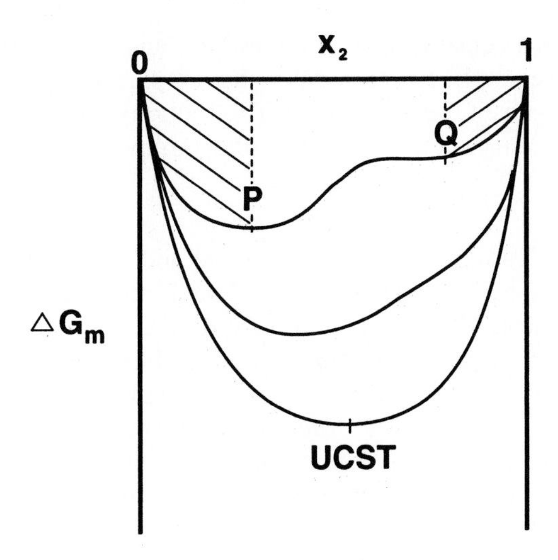

Figure 5.14 Free energy of mixing versus polymer mole fraction as a function of temperature. Mixtures with $0 < X_2 < P$ and $Q < X_2 < 1$ are miscible.

phases, Phase P and Phase Q. Phase P can have all polymer mole fractions up to x_P at point P. Phase Q can have all polymer mole fractions from 1 to x_Q at point Q. Maxima and minima along this curve can be obtained by setting the first derivative of the free energy of mixing to zero.

$$\frac{\partial \Delta G_m}{\partial x_2} = 0 \qquad 5.55$$

The phase envelope, or binodal curve, can be calculated by finding the compositions of the solvent-rich and polymer-rich phases that satisfy Equation 5.55.

There are two inflection, or spinodal, points between points P and Q. Spinodals are defined as having the second derivative of the free energy of mixing with respect to mole fraction equal to zero.

$$\frac{\partial^2 \Delta G_m}{\partial x_2^2} = 0 \qquad 5.56$$

The critical point represents the displacement of an inflection point and is defined by

$$\frac{\partial^3 \Delta G_m}{\partial x_2^3} = 0 \qquad 5.57$$

Mixtures with compositions between those of the spinodal points are unstable and will spontaneously phase separate. Two phases should be observed over a wide range of experimental times. Mixture having compositions between the binodal and spinodal (between point P and its associated spinodal point) are unstable from a thermodynamic point of view. However, the kinetics of the phase separation process may be slow compared to the time scale of the experiment. This section of the phase diagram is called metastable. Small changes in the mole fraction or temperature will cause the system to go unstable.

Prediction of Phase Separation. The spinodal and critical points can be calculated by the Flory–Huggins model. Equation 5.55 is the first derivative of the free energy of mixing. Since the second and third derivatives will be set equal to zero, either x_1 or x_2 can be used for differentiation. Furthermore, since the temperature is constant, either the volume fraction of solvent or the volume fraction of the polymer can be used for differentiation, since the derivative of volume fraction with respect to mole fraction will be a constant. It is easiest to differentiate the solvent chemical potential difference by the polymer volume fraction, giving, for the spinodals,

$$\frac{\partial \Delta\mu_1}{\partial \phi_2} = 0 = -\frac{1}{1-\phi_2} + (1+\frac{1}{N}) + 2\chi\phi_2 \quad 5.58$$

The condition for the critical point is that

$$\frac{\partial^2 \Delta\mu_1}{\partial \phi_2^2} = -\left[\frac{1}{1-\phi_2}\right]^2 + 2\chi \quad 5.59$$

Both Equations 5.58 and 5.59 hold at the critical point. The interaction parameter at the critical point is

$$\chi_c = \frac{1}{2(1-\phi_{2c})^2} = \frac{1}{2\phi_{1c}^2} \quad 5.60$$

The solvent volume fraction at the critical point is

$$\phi_{1c} = \frac{1}{1+1/\sqrt{N}} \quad 5.61$$

For large values of n, the interaction parameter should approach one-half and the critical solvent volume fraction should approach one. This theory predicts the critical interaction parameter much better than it predicts the critical solvent volume fraction. The phase envelope is defined by

$$\mu_i(\phi_2^{1}) = \mu_i(\phi_2^{11}) \quad 5.62$$

where the phase compositions, φ_2^{I} and φ_2^{II}, describe the polymer volume fractions in the two phases. The phase equilibrium equations become (Eqs. 5.29a and 5.29b, with g = constant):

$$\ln(1 - \phi_2^{\mathrm{I}}) + (1 - \frac{1}{N})\,\phi_2^{\mathrm{I}} + \chi(\phi_2^{\mathrm{I}})^2 =$$

$$\ln(1 - \phi_2^{\mathrm{II}}) + (1 - \frac{1}{N})\,\phi_2^{\mathrm{II}} + \chi(\phi_2^{\mathrm{II}})^2 \qquad 5.63$$

and

$$\ln(\phi_2^{\mathrm{I}}) - (N - 1)(1 - \phi_2^{\mathrm{I}}) + N\chi(1 - \phi_2^{\mathrm{I}})^2 =$$

$$\ln(\phi_2^{\mathrm{II}}) - (N - 1)(1 - \phi_2^{\mathrm{II}}) + N\chi(1 - \phi_2^{\mathrm{II}})^2 \qquad 5.64$$

These equations can be solved numerically to find the polymer and solvent composition in each phase.

Figure 5.15 shows the calculation of binodals and spinodal curves and the critical points for several values of N. The binodals are tangent to the spinodals at the critical point. For $\chi < \chi_c$ for a given value of N, the polymer and solvent are miscible in all proportions. For $\chi > \chi_c$, the solution separates into two phases. The molar volume ratio, N, is often assumed to be the degree of polymerization. N values of 10 and 30 correspond to oligomer solutions. These systems form homogeneous solutions for a wide range of interaction parameter values. Most commodity polymer systems have

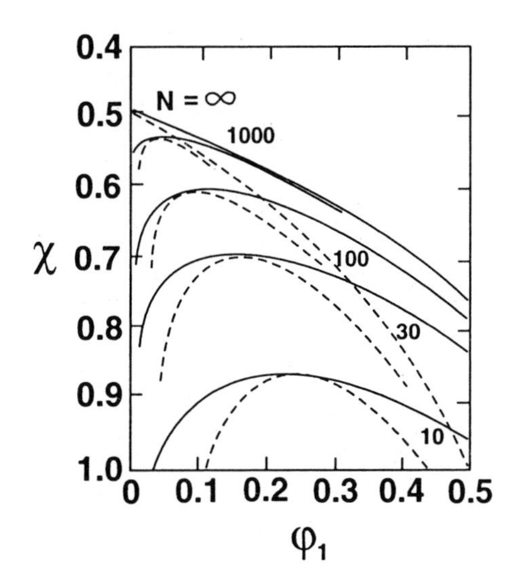

Figure 5.15 Calculated binodals (solid lines) and spinodals (dashed lines) of binary solutions of a polymer as a function of relative chain length P1. Critical points occur at the points of tangency between the binodal and spinodal curves. Note the inverted scale for χ (Calculations of Shultz and Flory, 1959).

degrees of polymerization greater than 1000, and are approximated by the $N = 1000$ curve. The critical point for this case is at a low value of φ_1. The polymer phase contains a very low volume fraction of solvent, while the solvent phase contains significant volume fractions of both. The binodal and spinodal curves can be generated for concentration dependent values of χ (either the power series expansion or the lattice theory model).

5.5.2 Polymer Fractionation

There are two basic ways in which to fractionate polymer samples. Fractional precipitation recovers polymer from a polymer-rich phase. Coacervate extraction recovers polymer from a dilute solution phase. In either case, precipitation is induced by adding or subtracting solvent, or by changing the temperature. Equation 5.64 can be rearranged to give the ratio of the volume fractions of an n-mer in Phases 1 and 11:

$$\ln \frac{\phi_2^1}{\phi_2^{11}} = n\left\{\frac{1-n}{n}(\phi_1^{11} - \phi_1^1) + \chi\left[(\phi_1^{11})^2 - (\phi_1^1)^2\right]\right\} \qquad 5.65$$

One of the phases will be dilute in polymer compared to the other phase. If the volume fraction of this phase is kept large, the polymer recovery is maximized. Also, as the chain length (or molar volume) of the polymer increases, the phase separation increases.

5.5.3 Experimental Methods for Phase Diagrams

Solvent-Polymer Miscibility. Light-scattering techniques are usually used for binary systems of a solvent and a polymer. At a given temperature, the compositions of the phases that first appear cloudy (either by adding or taking away solvent) are plotted as the boundary of the two-phase envelope. Solutions or thin films can be observed under a microscope.

Temperature effects phase equilibria by changing the interaction parameter. The Flory–Krigbaum description of χ as a function of temperature can be combined with χ_c to determine the critical temperature, T_c.

$$\frac{1}{T_c} = \frac{1}{\theta} + \frac{1}{f_1\theta}\left(\frac{1}{\sqrt{N}} + \frac{1}{2N}\right) \qquad 5.66$$

Figure 5.15 shows T versus ϕ_1, cloud point curves for various molecular weights of polystyrene in cyclohexane (Schultz and Flory, 1952). The solid curves describe the cloud point data curves and the dashed lines show the calculated binodal curves using a constant value of χ. For monodisperse samples, the binodal curve and the cloud point curve are identical. The effect of chain length dependence on T_c is shown in Figure 5.16. The Flory–Huggins theory describes the temperature dependence of the critical points well, but does not represent the entire cloud point curve. The computed critical solvent concentration, ϕ_{1c}, is lower than the measured value, and the calculated difference between phase compositions is smaller than observed.

Polymer-Polymer Miscibility. Polymer-polymer miscibility tests can be done by electron microscopy, dynamic mechanical analysis, or glass transition temperature measurements. Using electron microscopy, the sample would be stained to establish contrast between the phases and then examined to determine whether one or two phases exist. A miscible blend of two polymers will exhibit only one T_g, while an immiscible blend will give 2 T_g's. The T_g of the sample can be determined using mechanical methods, dilatometry, calorimetry, mutual solvents, or inverse-phase gas chromatography.

The molecular weights of the homopolymers, branching, and cross-linking affect blend miscibility by affecting the free energy of the polymer mixing process. Furthermore, the size of the homopolymer phases and the method of blending affect the rate at which the phases are interdispersed. The goal of the blending operation is to make as fine a dispersion of one polymer in other as practical. Fine dispersions generate high interfacial areas for strong bonding between the two polymer phases. Figure 3.9 shows the interface between two homopolymers. If the polymers are miscible, chains of one type of homopolymer can move through the interface into the second polymer phase. The areas of "homopolymer" near the interface are called the interphase region, describing the different local composition from the interface to the bulk phase.

Blending operations that rely on physical means may not produce finely dispersed products. The mixer may not be able to get enough energy into the melt to

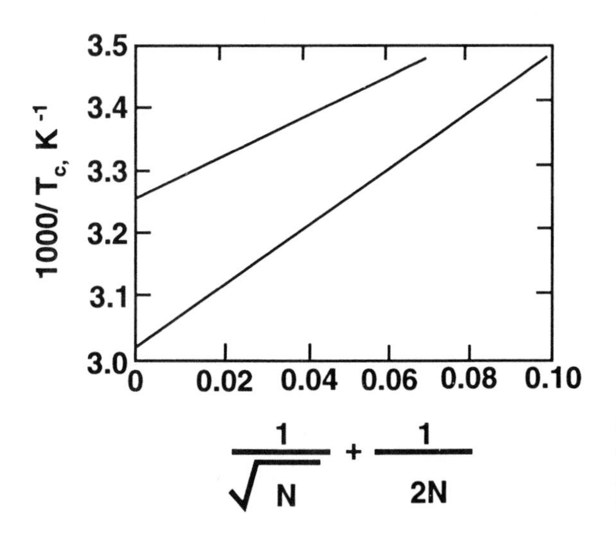

Figure 5.16 Chain length dependence of critical point T_c (Threshold temperature for T_t) two systems (Shultz and Flory, 1959).

disperse the phases before degradation of one or both polymers occurs. A typical commercial blending operation might include powder blending followed by extrusion or injection molding to form the product. The extruder alone might be used as both the powder blending and polymer melting device. Polymer in pellet form may be more difficult to blend because more energy would be needed to reduce the homopolymer phase size and disperse it in the other polymer.

Solution blending is an alternative to physical blending, but it has other drawbacks. The first problem is choosing a solvent in which both polymers are miscible over the entire ternary system composition range. Some solvents will cause phase separation between homopolymer solutions that are miscible by themselves. Figure 5.17

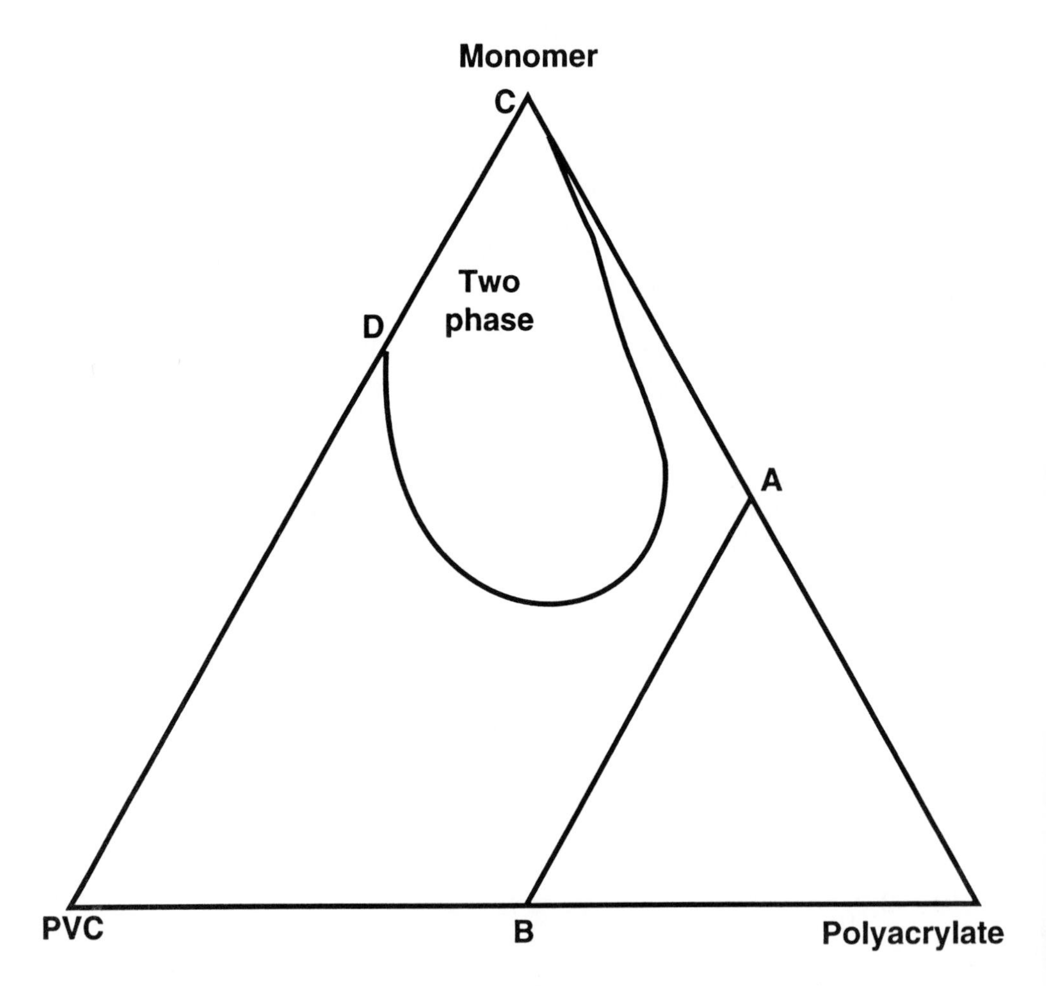

Figure 5.17 Phase diagram for the vinyl chloride/PVC/poly(butyl acrylate) ternary system (Walsh and Cheng, 1989). Typical graft polymerizations start at point **A** and follow the curve, **AB**. The points **C** and **D** show the phase separation of vinyl chloride and poly(vinyl chloride).

shows this for the vinyl chloride/PVC/poly(butyl acrylate) ternary system. The two homopolymers are miscible, but the addition of vinyl chloride to the blend can cause phase separation. The path A-B represents a typical polymerization pathway starting with vinyl chloride monomer and polyacrylate and grafting PVC to the PA chains. Other polyacrylates show similar phase separation with vinyl chloride and PVC. A second problem associated with solution blending is the removal of the solvent. Except for adhesive products, the solvent is normally removed and recycled.

In situ polymerization is a very good method for making blends when ternary system phase separation can be controlled under reaction conditions. High impact polystyrene is an example. Butadiene is polymerized, and styrene is added to the reactor at the end of the homopolymerization. Styrene is soluble in the polybutadiene, and the polystyrene forms an intimate mixture with a separate phase. This is a preferred method for making blends, as the sizes of the two homopolymer phases, the choice of the continuous phase, and the molecular weights can be controlled.

EXAMPLE 5.5 POLYMER BLEND MISCIBILITY: PVC WITH POLYACRYLATES AND POLYMETHACRYLATES

These example systems (Walsh and Cheng, 1984) show some of the characteristic difficulties associated with making polymer blends. The T_g data presented here are for blends made by in situ polymerization of vinyl chloride on particles of the other homopolymers. The copolymers were analyzed for glass transition temperature by dynamic mechanical analysis (Fig. 5.18 and 5.19). The quantity, tan δ, is a measure of energy dissipation under a cyclic stress. When the energy loss per cycle goes through a maximum, that temperature should be associated with the onset of molecular mobility. The presence of two peaks suggests immiscibility, while one peak suggests a compatible blend. Poly(vinyl chloride) forms a miscible blend with poly(ethyl acrylate). Poly(methyl acrylate) may blend slightly with PVC, but the presence of two peaks suggests a two-phase blend. Other characteristics of two-phase blends include opacity and brittleness. The T_g's of these in situ blends are higher than those made by solvent casting. It seems likely that residual solvent affected the T_g measurements for these other samples.

5.5.4 Models for Miscible Polymer Blend Properties

There are several models for describing the effects of blending on the properties of miscible blends. The properties of immiscible blends usually are not predictable. The

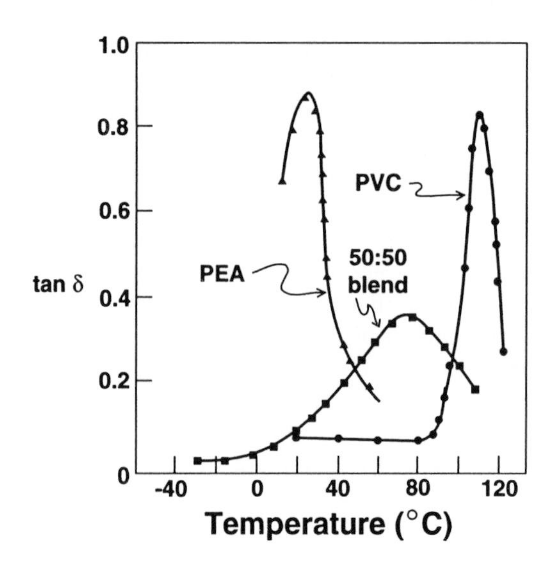

Figure 5.18 Dynamic mechanical analysis measurement of T_g. Miscible blends (Walsh and Cheng, 1989). Plots of tan δ (not normalized against temperature for PVC(●) and poly(ethyl acrylate) (▲) and a 50:50 in situ polymerized blend (■), showing a single intermediate glass gransition indicative of a one-phase miscible blend.

phases may not adhere together, so that morphology and properties vary from sample to sample. Mixing rules vary depending on the system and usually are empirical. A simple linear mixing rule for a mixture property is Equation 5.8.

The glass transition temperature is an important property of miscible polymer blends. Since T_g is not a thermodynamic quantity, the models used to describe it are considered empirical. Equation 5.8 can be applied to the T_g of blends by using the polymer weight fractions as the weighting factors.

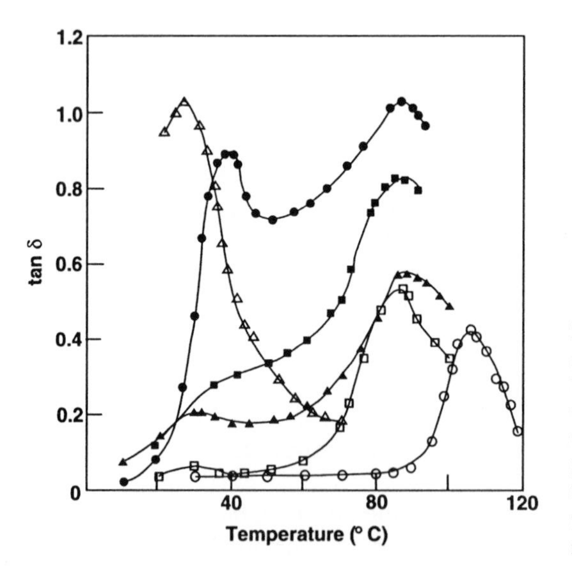

Figure 5.19 Dynamic mechanical analysis measurement of T_g. Immiscible blends (Walsh and Cheng, 1989). Plots of tan δ (not normalized against temperature for PVC(○) and poly(methyl acrylate) (△) and PMA/PVC blends. The blends show the presence of two glass transitions indicative of a two-phase immiscible blend.

$$T_{g,\,blend} = w_1\, T_{g1} + w_2\, T_{g2} \qquad 5.67$$

An alternative model is the Fox equation (Fox, 1956).

$$\frac{1}{T_{g,\,blend}} = \frac{w_1}{T_{g1}} + \frac{w_2}{T_{g2}} \qquad 5.68$$

The glass transition temperature should be expressed in absolute units. The Gordon–Taylor equation has a constant that indicates the strength of the interaction between the polymers (Gordon and Taylor, 1952).

$$T_{g,blend} = \frac{w_1\, T_{g1} + K\, w_2\, T_{g2}}{w_1 + K\, w_2} \qquad 5.69$$

When $K > 1$, the T_g is greater than that predicted by the linear model. When $0 < K < 1$, the T_g is lower than the linear model prediction. Negative values of K are needed to describe blends for which the T_g is higher than those of the pure homopolymers. Walsh and Cheng (1984) fit some data with $K < 0$, but this may not describe the T_g versus weight fraction curve over the entire composition range. When $w_1 = -K/(1 - K)$, the $T_{g,\,blend}$ is undefined and equation 5.69 is discontinuous.

The Redlich–Kister expansion for excess thermodynamic properties can be applied to polymer systems. An excess property for a polymer blend, Z^{ex}, can be defined

$$Z^{ex} = Z_{mix} - (x_1\, Z_1 + x_2\, Z_2) \qquad 5.70$$

Z^{ex} is zero for pure homopolymers, and may be nonzero at other compositions. A Redlich-Kister–type expansion would be

$$Z^{ex} = x_1\, x_2\, [A + B\,(x_1 - x_2) + C\,(x_1 - x_2)^2 + \ldots] \qquad 5.71$$

This model can provide an empirical model for properties of binary polymer blends that can have values above or below the linear model (Eq. 5.67). Equations 5.70 and 5.71 applied to the T_g of a blend are

$$T_g^{ex} = T_{g,blend} - (w_1\, T_{g1} + w_2\, T_{g2}) \qquad 5.72$$

$$T_g^{ex} = w_1\, w_2\, [A + B\,(w_1 - w_2) + C(w_1 - w_2)^2 + \ldots] \qquad 5.73$$

As with the Gordon–Taylor models, a single point can be used to determine the model constant, A. If several points are known, then other coefficients can be determined, depending on the degrees of freedom of the data set.

EXAMPLE 5.6 MODELS FOR T_G, BLEND

The data in Table 5.5 was taken for a 50:50 weight fraction blend. The Gordon–Taylor and Redlich–Kister coefficients were found for each data point. Figure 5.20 shows the predictions of several models for the PVC/poly(ethyl acrylate) blend. The Gordon–Taylor model has been fit to go through the data point. The Fox and the linear models have no adjustable parameters and do not fit the data point.

An advantage for the Redlich–Kister model can be seen with the polymethacrylate series. It can predict a continuous curve for T_g even when the blend value is above that of the two homopolymers (Fig. 5.21). Both models describe systems in which there are moderate deviations (positive or negative) from a linear model. Figure 5.22 shows the predictions of the Gordon–Taylor model for the polymethacrylate series. Clearly, the Fox model would be a poor choice for three of these four blends, as it would give the incorrect curvature to the plots.

EXAMPLE 5.7 PRECIPITATION DURING COPOLYMERIZATION

Polymer Film, Inc., is producing a copolymer of vinyl chloride and vinylidene chloride containing 75wt% vinylidene chloride by a batch process. The polymer composition will change during a batch polymerization. The variable copolymer compositions could cause phase separation in the film product. Compute whether copolymer produced later in the reaction might precipitate from polymer produced at the beginning. Assume that the reaction takes place at 75°C, that the system can be characterized as a two-polymer system (monomer does not affect the equilibrium), and that the solubility parameter of the copolymer is given by

$$\delta_{copolymer},\ (cal/cm^3)^{1/2} = 9.7\ \phi_{PVC} + 12.2\ \phi_{PVCl_2}$$

Data:	vinyl chloride (1)	vinylidene chloride (2)
r_i	0.20	0.36
ρ_i	1.40 g/cm^3	1.70 g/cm^3
M_i	62.5	97
N	800	800
ϕ_i	0.288	0.712

The polymer composition is located on a copolymer composition diagram (Section 4.2.1) to determine which monomer is preferentially added. This polymer preferentially adds vinylidene chloride. Precipitation is assumed to occur if the

(continued)

interaction parameter for the polymer pair is greater than 0.5. Equation 5.44 can be used to calculate the solubility parameter difference for copolymer precipitation. The partial molar volume of the first copolymer is

$$V_1 = (.75 \cdot \frac{97}{1.7} + .25 \cdot \frac{62.5}{1.4}) \cdot 800$$

$$= 43{,}200 \frac{cm^3}{gmol}$$

The solubility parameter difference for precipitation is

$$\delta_1 - \delta_2 = \sqrt{(0.5 - 0.34)\left(\frac{RT}{V_i}\right)} = \sqrt{(0.5 - 0.34)\left(\frac{1.987 \cdot 348}{43{,}200}\right)} = 0.051$$

The solubility parameter for the starting polymer is

$$\delta_1 = 9.7 \cdot 0.288 + 12.2 \cdot 0.712 = 11.48$$

and the volume fraction of vinylidene chloride in the precipitation copolymer will be 0.732 or greater.

TABLE 5.5 Glass Transitions Temperatures of PVC Blends

Polymer	T_g, C	T_g	T	K*	A†
Poly(vinyl chloride)	103	(50 : 50 blend)			
Poly(methyl acrylate)		two T_g's			
Poly(ethyl acrylate)	15	71	12	.571	48
Poly(butyl acrylate)	−25	54	15	1.6	60
Poly(octyl acrylate)		two T_g's			
Poly(methyl methacrylate)	128	118	2.5	1.5	10
Poly(ethyl methacrylate)	85	105	11	−10‡	44
Poly(butyl methacrylate)	40	82.5	11	2.1	44
Poly(hexyl methacrylate)	15	45	−10	.48	−56

* = Gordon–Taylor coefficient
† = Redlich–Kister coefficient
‡The recommended K value is 10. This value underpredicts the T_g of the 50 : 50 blend but gives a continuous prediction for all w_1 values.
SOURCE: Data of D.J. Walsh and G.L. Cheng, *Polymer 25* (4), 495–498 (1984).

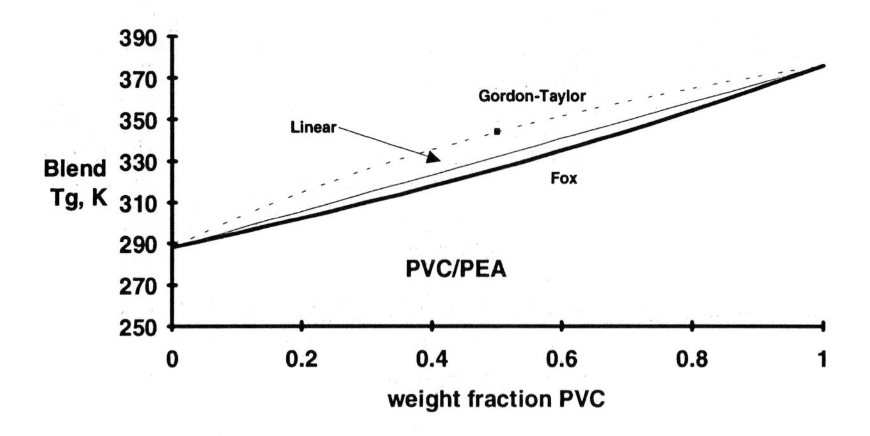

Figure 5.20 Comparison of Fox, linear, and Gordon–Taylor models for data in Table 5.5.

Polydisperse Polymer Solutions. If χ is assumed to be a phenomenological description of the solution, a mutual interaction parameter, then polydisperse systems can be modeled. The critical point parameters, ϕ_{2c} and χ_c, are

$$\phi_c^2 = \frac{1}{1 + \overline{M_w}\ \overline{M_z}^{-1/2}} \tag{5.74a}$$

$$\chi_c = \frac{1}{2}\left(\frac{1 + \overline{M_z}^{1/2}}{\overline{M_w}}\right)\left(1 + \overline{M_z}^{-1/2}\right) \tag{5.74b}$$

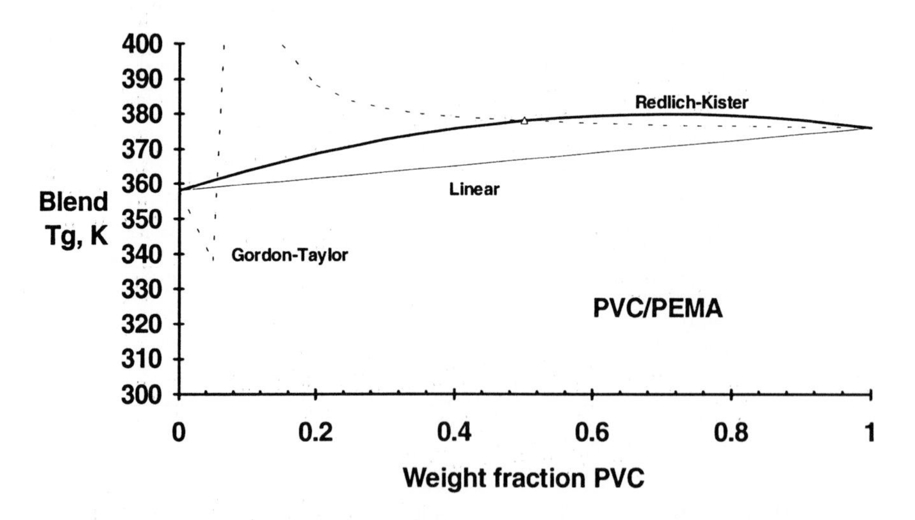

Figure 5.21 $T_{g,blend}$ for PVC and PEMA.

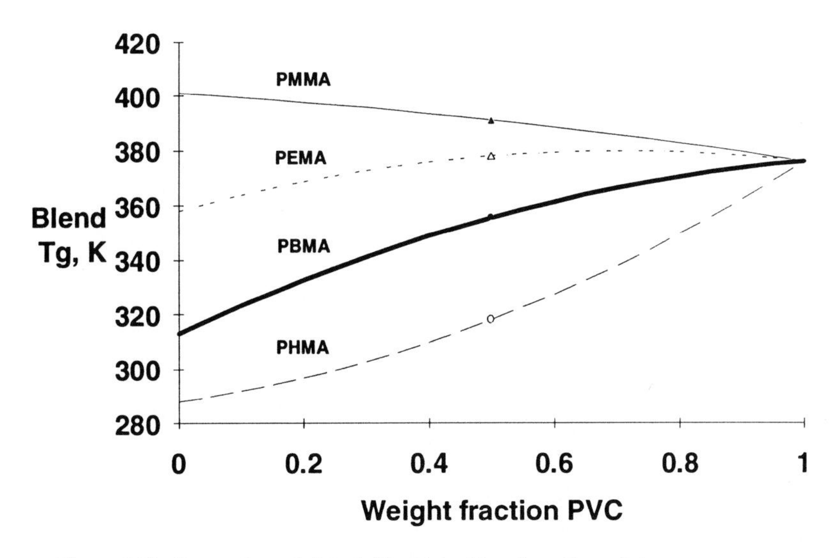

Figure 5.22 $T_{g,blend}$ for poly(vinyl chloride) with polymethacrylate.

where $\overline{M_w}$ is the weight average molecular weight, $\overline{M_z}$ is the z average molecular weight, and ϕ_{2c} is the total volume fraction of polymer.

5.5.6 Typical Phase Equilibria for Ternary Systems

Case 1: *Two Monodisperse Polymers in One Solvent.*

Some polymer systems are bidisperse; that is, there is a bimodal distribution of molecular weights. In such a system, the two polymer fractions might be miscible in each other, but immiscible in some proportions in the presence of a solvent. Two key factors affecting phase separation are the system temperature and the system composition. Variations in temperature and composition are typical during the processing and use of polymer systems.

Figure 5.23 shows a ternary diagram for a solvent (0), a low molecular weight polymer (1) and a high molecular weight polymer (2). The two polymers vary in chain length but not configuration and are assumed miscible in all proportions (there are no separate phases along line 1-2). The high molecular weight polymer is assumed to phase-separate with solvent over a range of temperatures, T_1 to T_{c2}. In general, solubility decreases with increasing chain length. Above the upper critical solution temperature of the mixture, T_{c2}, the mixture is miscible at all compositions.

The curve for T_1 describes the two-phase region. The curve of P''_1 - P'_1 - critical point represents the polymer-rich phase (P''_1) in equilibrium with the solvent-rich phase (P'_1). Lines connecting points such as Q'_1 and Q''_1 are tie lines between phases

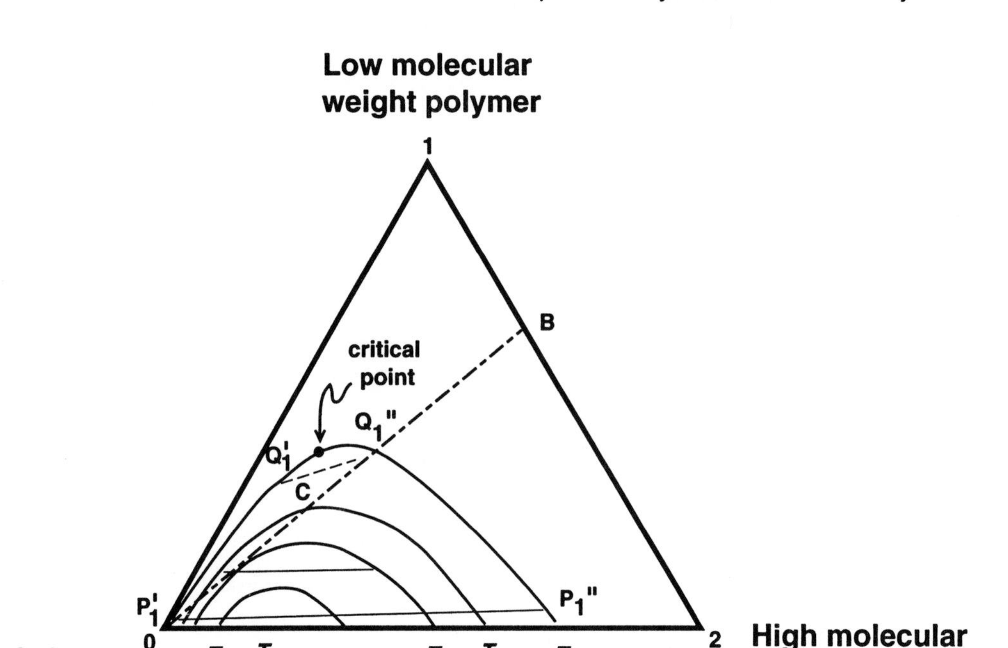

Figure 5.23 Phase diagram of a ternary sytem (solvent 0 + polymer 1 + polymer 2). Thick solid lines: binodals; thin solid lines: tie lines; dot-dash line: composition line (Kurata, 1982).

in equilibrium. Some polymer systems might have a lower critical solution temperature as well, but this phenomenon is not demonstrated on the diagram.

This diagram could represent a polymerizing system in which higher molecular weight polymer precipitates from solution at the polymerization temperature, T_1. Suppose that the polymerization mechanism forms high and low molecular weight polymer in a constant ratio (B) throughout the reaction and that the solvent, 0, is taken to be monomer. The composition of the reaction mixture would start at 100% monomer, or the left point of the diagram. The composition follows the line, 0-B, as the monomer is converted to polymer. Between the initial composition (0 = 100%) and composition P_1', the mixture is one phase. Between compositions P_1' and Q_1'', the mixture splits into a two-phase mixture, with compositions established by the tie lines and volumes described by the lever rule. At Q_1'', the last of the solvent-rich phase disappears and the polymerization continues in the polymer-rich phase. If the polymerization rate were different between the two phases, then two rate expressions would be required to model the process.

A polymer solution method for dissolving B in pure solvent would follow the reverse process. The solvent would swell the polymer initially (if added slowly enough). At high compositions, two liquid phases would form; and at high solvent concentrations, only one phase would exist.

Case 2: Two Different Homopolymers in One Solvent

Solvent casting is often used as a way to blend two polymers with different chemical structure. Different polymer species generally are incompatible. Three interaction parameters would be needed; χ_{o1}, for solvent (0) and polymer (1); χ_{o2} for solvent (0) and polymer (2); and χ_{12}, for the polymer mixture. A simple case would be one in which the polymer-polymer interaction parameter controlled the phase equilibrium.

If the solvent interacts with each polymer about the same, $\chi_{01} = \chi_{02} = \chi$, and the molar volume ratios are the same, $N_1 = N_2 = N$, the conditions for phase equilibrium are

Solvent 0

$$\begin{aligned} &\ln\phi_0' + \left(1 - \frac{1}{N}\right)(1 - \phi_0') + \chi(1 - \phi_1')^2 - \chi_{12}\phi_1'\phi_2' = \\ &\ln\phi_0'' + \left(1 - \frac{1}{N}\right)(1 - \phi_0'') + \chi(1 - \phi_0'')^2 - \chi_{12}\phi_1''\phi_2'' \end{aligned} \qquad 5.75$$

Polymer 1

$$\begin{aligned} &\ln\phi_1' + (1 - N)\phi_0' + N\chi\phi_0'^2 + N\chi_{12}(1 - \phi_1')\phi_2' = \\ &\ln\phi_1'' + (1 - N)\phi_0'' + N\chi\phi_0''^2 + N\chi_{12}(1 - \phi_1'')\phi_2'' \end{aligned} \qquad 5.76$$

Polymer 2

$$\begin{aligned} &\ln\phi_2' + (1 - N)\phi_0' + N\chi\phi_0'^2 + N\chi_{12}(1 - \phi_2')\phi_1' = \\ &\ln\phi_2'' + (1 - N)\phi_0'' + N\chi\phi_0''^2 + N\chi_{12}(1 - \phi_2'')\phi_1'' \end{aligned} \qquad 5.77$$

The binodal curve is independent of χ and

$$\begin{aligned} \phi_{0c} &= 1 - \frac{2}{N\chi_{12}} \\ \phi_{1c} &= \phi_{2c} = \frac{1}{N\chi_{12}} \end{aligned} \qquad 5.78$$

If a solution of solvent and polymer 2 is mixed with a solution of solvent and polymer 1, the mixture separates into two phases: Q′ containing mostly polymer 1 and Q″ containing mostly polymer 2. This phenomenon is called demixing and occurs at very low polymer fractions. Two-phase aqueous separations are one example of this phenomenon. Figure 5.24 shows the demixing of poly(vinyl acetate) and polystyrene in benzene. In general, the effect of the solvent is to lower the value of χ_{12}.

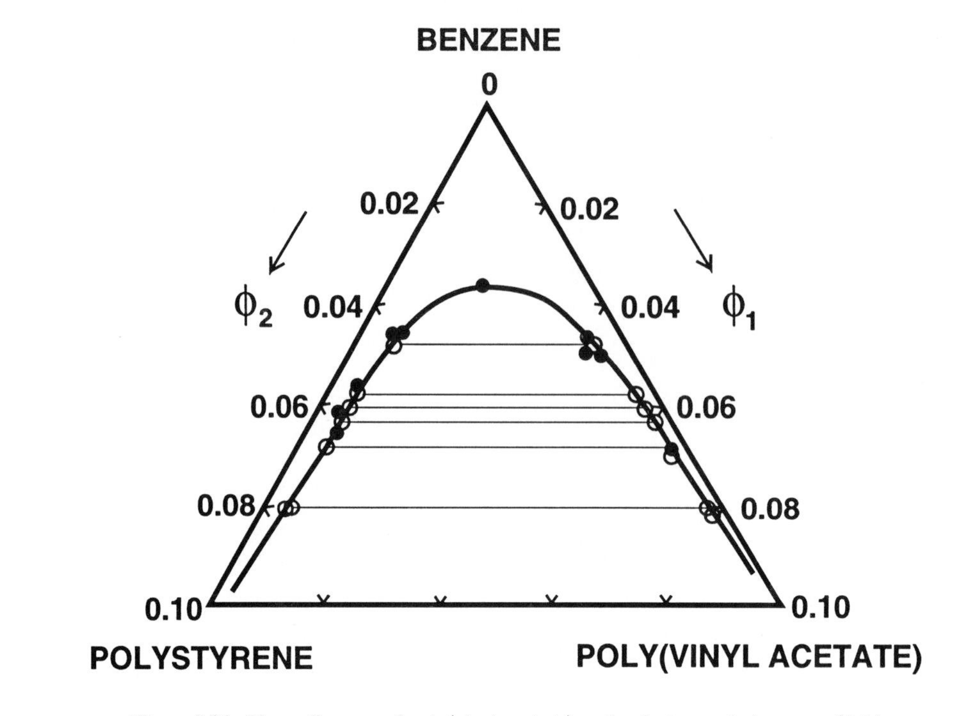

Figure 5.24 Phase diagram of poly(vinyl acetate) and polystyrene in benzene (Seki and Sakurada, 1969).

Case 3: One Polymer in Two Solvents

A. Poor-Solvent (0); Good-Solvent (1); Polymer (2)

The phase diagram for this case is similar to that for solvent and two homopolymers (Fig. 5.23). The low molecular weight polymer can be considered a good solvent for the high molecular weight polymer. The poor solvent can precipitate the high molecular weight polymer over some composition ranges. Three phases can exist near the critical point of the mixtures; however, these are not shown on the diagram. The good solvent is said to be selectively absorbed by the polymer.

The usual process goal is to optimize the separation between the good solvent and the polymer: The polymer phase should be very lean in solvent, and the solvent phase should contain very little polymer. Residual solvent in the polymer would need to be removed by further drying steps. Polymer remaining in the solvent phase would be difficult to recover, would be harmful to solvent recovery and recycle operations, and would be very difficult to recover for sale (it probably would represent a yield loss to the polymerization process). Precipitating solvents usually are chosen so that there is a large two-phase region that is, $\chi_{12} > 0.5$.

B. Poor Solvent (0); Poor-Solvent (1); Polymer (2)

The two solvents will precipitate the polymer when $\chi_{02} > 0.5$ and $\chi_{12} > 0.5$. Figure 5.25 shows a phase diagram that happens to have two critical points on the 0-1 side of the

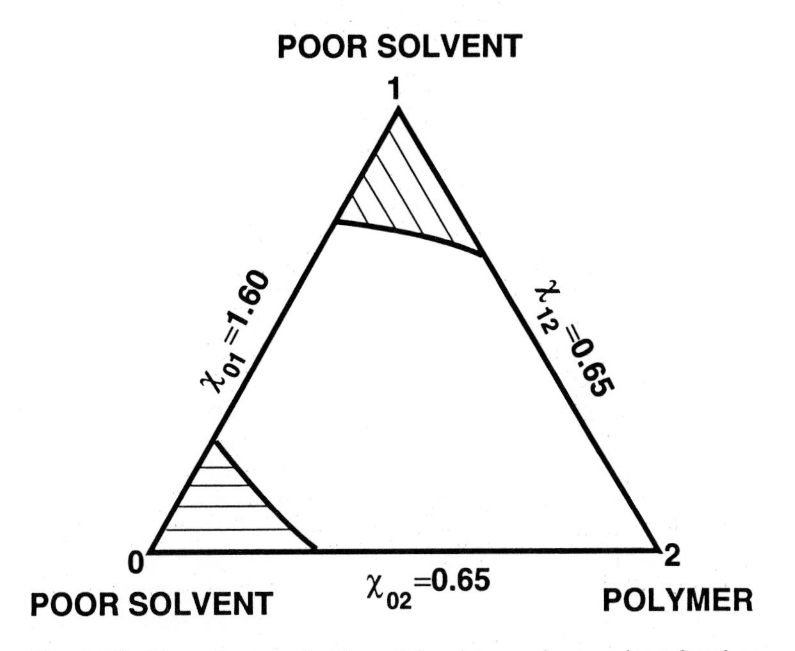

Figure 5.25 Phase diagram of polymer 2 in mixtures of poor solvent 0 and poor solvent 1 (Kurata, 1982).

triangle ($N_2 = \infty$, the polymer has an infinite molecular weight). For $N_2 < \infty$ but within the range of many commodity polymers, the critical points would move away from the 0-1 side slightly.

If the solubility parameter of the polymer, δ_2, is intermediate between those of the two solvents, the mean value of the solvent phase solubility parameter would approach that of the polymer, and theory would suggest that the solvent mixture could dissolve the polymer. This phenomenon has been observed experimentally for chloroform-ethanol-cellulose acetate (Bamford and Tompa, 1950). However, rigorous application of theory suggests some inconsistencies and does not completely describe the phenomenon.

C. Good Solvent (0); Good Solvent (1); Polymer (2)
For this case, no phase separation would occur along the sides of the phase diagram as all the binary pairs are miscible. For some systems, there may be phase separation inside the triangular diagram. The binodal curve forms a loop and has two critical points (Tompa, 1956).

5.6 PHASE EQUILIBRIA PROBLEMS

Two-phase equilibria problems associated with polymerization are presented in this section. These problems are typical of those found in industry.

EXAMPLE 5.8 PHASE EQUILIBRIA IN POLYETHYLENE POLYMERIZATION

The high pressure process for making polyethylene is shown in Figure 5.26. Ethylene is compressed from its storage conditions to 2000 atm and 200°C in a two-stage operation. Catalytic amounts of oxygen are added before the compressor. The polymerization occurs in a long tubular reactor, while the pressure falls to about 300 atm and the temperature drops to about 100°C. Ethylene conversion per pass ranges from 10 to 30%, and the monomer must be recycled. The reaction mixture is thought to form a single phase in the tube. The polymerization rate probably depends mostly on temperature, slightly on pressure, and should be low at the exit of the reactor. At this point, the reactor effluent would be sent to a flash operation, with the monomer being recycled and the polymer being sent to an extruder for forming into pellets.

The goal of this problem is to estimate the composition of the gas and polymer streams leaving the flash operation. These compositions could be used in process design and process modeling. Application of the Gibbs phase rule to a two-component, two-phase system gives two degrees of freedom. Therefore, if T and P are known, and equilibrium models are assumed, the phase compositions can be calculated. The Flory–Huggins model will be used to describe the phase equilibrium of ethylene. The following table shows some of the assumptions:

Component	Flash Feed, wt%	Vapor Composition, y	Polymer Melt, ϕ
Ethylene	70	1	$\phi_1 = ?$
Polyethylene	30	0	$\phi_2 = ?$

Basis: 1 k_g of feed to the flash; 300 g of polymer, 700 g of monomer
Feed = Vapor + Polymer, mass units

1. Calculate χ. If the values for χ cannot be found, it can be estimated using solubility parameters. Solubility parameters for polyethylene range from 7.9 to 8.2 $(cal/cm^3)^{1/2}$. The solubility parameter for ethylene has been listed as 5.8 $(cal/cm^3)^{1/2}$. The interaction parameter will be estimated by using the solubility parameters at room temperature, and then using Equation 5.42 to scale χ with temperature and molar volume of the solvent. The effect of pressure on the interaction parameter is assumed to be modest at 300 atm.

   The molar volume of ethylene can be estimated using the Peng–Robinson equation of state. The value at 100°C and 300 bar is 82.01 $cm^3/gmol$.

$$\chi = 0.34 + \frac{82.01\ cm^3/gmol}{1.987 \dfrac{cal}{gmol\,°K} \cdot 373°K}(5.8 - 8.2)^2$$

(continued)

$$\chi = 0.977$$

The large value is due to the high molar volume of ethylene at supercritical conditions. Notice that the solubility parameter model includes the effect of pressure via the solvent molar volume. This effect should be small except at reduced pressures near or greater than one.

2. Calculate the solvent volume fraction in the polymer phase. The gas phase is assumed to be pure ethylene and $a_1 = 1$. The effects of gaseous contaminants such as oxygen have been ignored, since they should be present only in very low amounts, if at all. If the gas phase were not pure ethylene, a reference state would need to be chosen in order to calculate its activity. Equation 5.31 becomes

$$0 = \ln\phi_1 + \phi_2 + \chi\phi_2^2$$

and

$$\phi_1 = 0.33$$

3. Calculate the flash unit material balance. All the polyethylene is assumed to be in the molten polymer phase (if it weren't, the compressor would foul quickly). The density of polyethylene at room temperature is 0.92 g/cm^3, so the total volume of polyethylene in the polymer phase is

$$300g\ PE\ \frac{1\ cm^3}{.92\ g} = 326\ cm^3\ PE$$

It would be preferable to use a polymer density at 300 bar and 100°C if it were known. The total volume of the polymer phase is

$$\frac{326\ cm^3\ PE}{.67\ volume\ fraction} = 486.7\ cm^3$$

The volume of ethylene in this phase is

$$486.7\ cm^3\ polymer - 326\ cm^3\ PE = 160.7\ cm^3\ ethylene$$

The weight of ethylene in the polymer phase is

$$\frac{160.7\ cm^3\ ethylene}{82.01\ cm^3/gmol} = 1.96\ gmol = 54.9\ g\ ethylene$$

The overall material balance for the flash unit is

	Feed,g	*Vapor,g*	*Melt,g*
Ethylene	700	645.1	54.9
Polyethylene	300	0	300

(continued)

4. Estimate of χ from literature data. The above calculation suffers from many simplifications, the most important of which is the estimate for the interaction parameter. Figure 5.27 shows the phase envelope for this system at 130°C, with the upper critical solution pressure of 1760 atm (Ehrlich, 1965). The solvent-rich phase lies along the right-hand pressure axis. This information will be used to get a better estimate for χ at the conditions of the problem. χ will be extrapolated with temperature and pressure to obtain a value for our problem.

   The polymer has a range of molecular weights (Fig. 5.28 shows the critical loci for two fractions of this polymer); but, since the distribution is unknown, the polymer will be treated as a monodisperse sample. Each point along the curve can be used to estimate χ. The four data points give

P, atm	ϕ_1	χ	*V, cm*^{3}*/gmol*
1760	.957	.515	51.0
1750	.936	.525	50.9
1390	.679	.642	51.5
1290	.536	.741	52.4
300	—	—	93.1

The last column of the table above gives the Peng–Robinson calculation for the molar volume of ethylene for various pressures at 130°C. χ can be extrapolated using power series expansions either in ϕ_2 or in P. The pressure dependence of the interaction parameter can be estimated as

$$\chi = \chi_o + \frac{d\chi}{dP} P$$

For the data in Figure 5.27, the equation is

$$\chi = 1.27 - .00043\, P(bar)$$

The interaction parameter decreases as pressure increases, which is typical of many polymer systems. However, large changes in pressure are required to have significant changes in the phase equilibrium. This model gives a value of $\chi = 1.14$ and $\phi_1 = 0.25$ at 130°C and 300 bar. The value of χ at 100°C and 300 bar can be scaled using the solubility parameter model for χ.

   χ can be used to solve for the square of the solubility parameter difference.

$$1.14 = 0.34 + \frac{93.1\ cm^3/gmol}{1.987 \cdot 403°K}(\delta_1 - \delta_2)^2$$

$$(\delta_1 - \delta_2)^2 = 6.92$$

(continued)

The value of χ at 100°C and 300 bar should be

$$\chi = 0.34 + \frac{82.02\ cm^3/gmol}{1.987 \cdot 373°K}(6.92) = 1.10$$

and the volume fraction of ethylene in the polymer melt should be 0.27. The flash unit material balance based on literature data is

	Feed,g	*Vapor,g*	*Melt,g*
ethylene	700	658.8	41.2
polyethylene	300	0	300

The difference between the two material balances is significant. Equilibrium data near the flash unit conditions should be obtained in order to improve the model coefficients.

The general effect of increasing system pressure is to increase polymer solubility. Figure 5.29 shows a three-dimensional phase diagram for ethylene/polyethylene. The two-phase region becomes smaller as temperature and pressure are increased. Isotherms A and C show upper critical solution temperatures (UCST). The pressure at which the UCST occurs decreases as temperature is raised. Except at very low pressures, the ethylene phase in equilibrium with the polymer melt contains little or no polymer (the polymer vapor pressure is essentially zero).

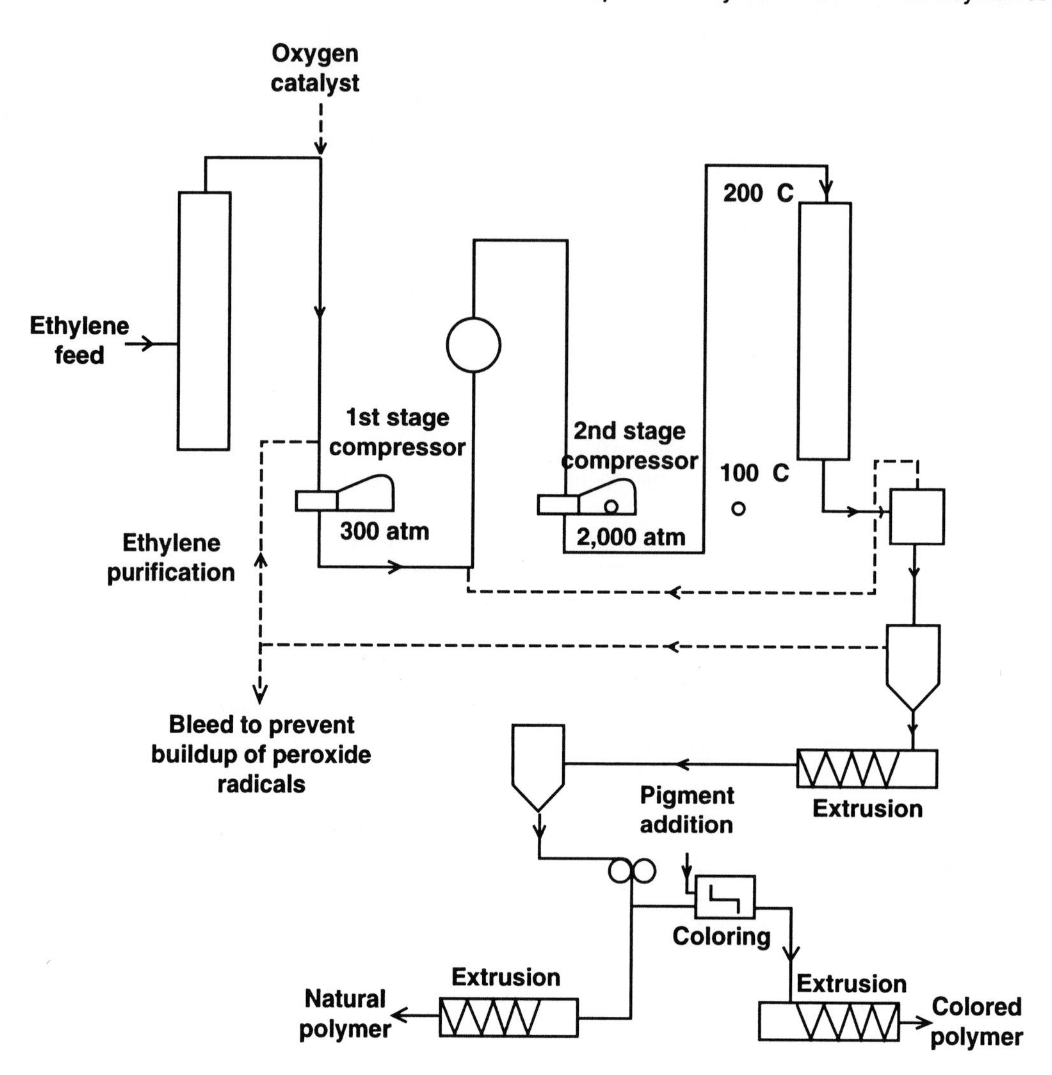

Figure 5.26 Simplified flow scheme for the high pressure polymerization of ethylene (Miles and Briston, 1965).

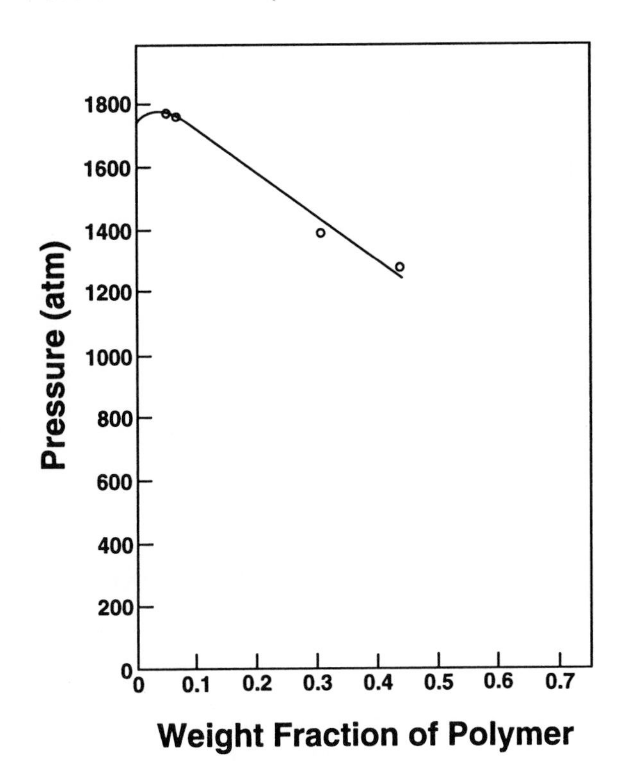

Figure 5.27 130° isotherm for polyethylene with ethylene (whole polymer) (Ehrlich, 1965).

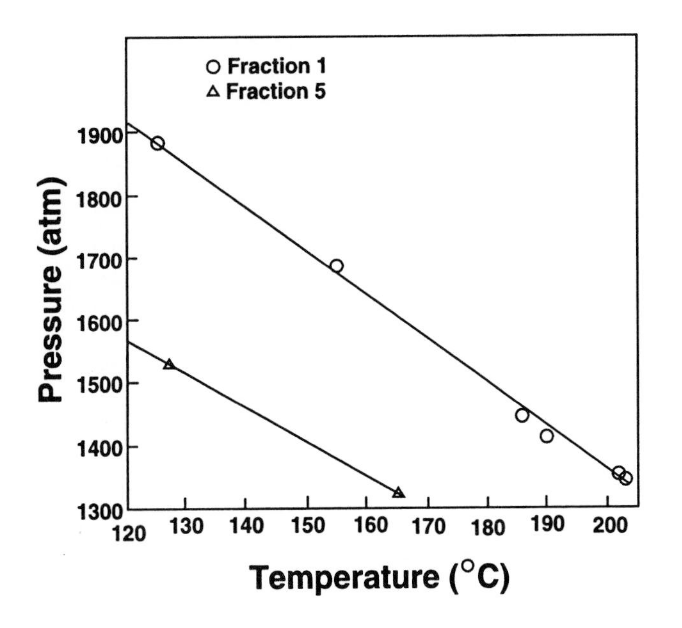

Figure 5-28 Critical locus for two polyethylene fractions with ethylene (Ehrlich, 1965).

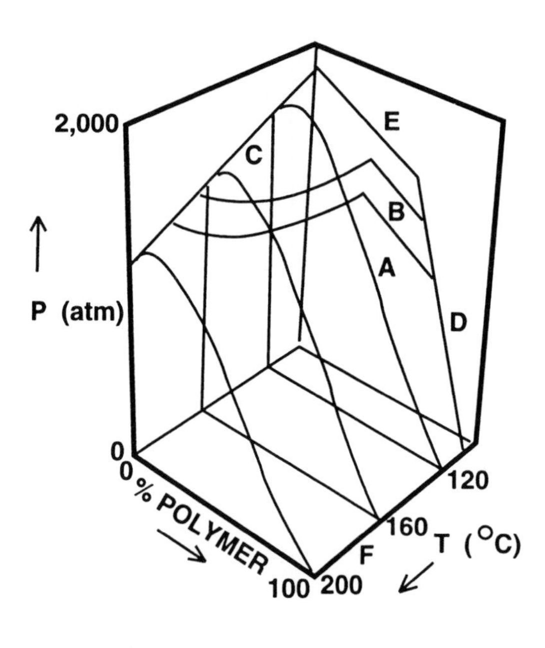

Figure 5.29 Three dimensional (P,T, composition) model for polyethylene-ethylene: isotherm, **A**; isobar, **B**; critical locus, **C**; melting point of pure polymer as *f*(P), **D**; melting point as *f*(diluent), **E**; vapor pressure of pure polymer as *f*(T), **F** (Ehrlich, 1965).

EXAMPLE 5.9 PHASE EQUILIBRIA IN SUSPENSION POLYMERIZATION

A number of commercial polymers are manufactured using a suspension of the monomer in a nonsolvent, water. Polymer forms in the monomer droplets. The water continuous phase improves heat transfer by reducing mixture viscosity, and it provides a large sink for the heat of polymerization. In a number of systems, monomer is not very soluble in the polymer, and it partitions between the water, polymer, and gas phases.

Poly(vinylidene fluoride) (PVDF) is being produced in a suspension process. The problem objective is to estimate the extent of conversion based on the reactor conditions at blow-down: $T = 180°F$, $p = 40$ psia, reactor volume $= 100$ ft^3, and 90vol% is filled with reaction slurry. The reactor contents are considered to be at equilibrium. Knowing the liquid and vapor volumes makes the calculations easier. In practice, this could be done using level measurements or a weigh cell to determine the total weight in the reactor.

The table below shows the composition descriptions of the three phases.

Component	Vapor	Liquid	Polymer
VF2	y_{vf2}	x_{vf2}	ϕ_{vf2}
Water	y_{H_2O}	x_{H_2O}	0
PVDF	0	0	$\phi_{polymer}$
	$\Sigma y_i = 1$	$\Sigma x_i = 1$	$\Sigma \phi_i = 1$

(continued)

The polymer has no vapor pressure and is essentially insoluble in the water phase. Water is essentially insoluble in the polymer phase. Monomer exists in all three phases. The phase equilibria are

VF2: $a^{v}_{vf2} = a^{l}_{vf2} = a^{p}_{vf2}$

Water: $a^{v}_{H_2O} = a^{l}_{H_2O}$

Polymer: not needed as it appears only in one phase

At the temperature and pressure of the reactor, VF_2 exists as a vapor and water exists as a liquid in their pure component states. The phase equilibria relations are

Polymer Phase Model: Equation 5.31 with 1 = monomer, 2 = polymer, $\chi = 3.2$.
Liquid Phase Model: $f^{v}_{vf2} = f^{l}_{vf2} = x_{vf2} H_{vf2}, H_{vf2} = 25000$ bar.

There are several ways to model the liquid phase. Vinylidene fluoride is a polar organic that is sparingly soluble in water. While activity coefficients could be used, it is easier to use Henry's law solubility data if they are available.

Gas

$$f_i^v = y_i P \left(\frac{f}{p}\right)_i \cong y_i P$$

A check of $(f/P)_i$ at reactor conditions for water and vinylidene fluoride shows that this ratio is near one for each component, and the gas phase is ideal. The physical properties of the components are

Component	ρ(180°F, 40 psia)	Weight Charged, lb
VF_2	.416 g/cm^3	1685
H_2O	0.9706 g/cm^3	4500 lb
PVDF	1.73 g/cm^3	—

If all the vinylidene fluoride were converted to polymer, the product would be a 27wt% slurry, typical of industrial practice.

1. Compute the partial pressure of water in the vapor phase.

$$f^L\, H_2O(180°F,\, 40\; psia) = x\, H_2O \gamma\; H_2O P^{vap}(180°F)$$

$$\cong 1 \cdot 1 \cdot 6.0 \text{ psia} \equiv 0.4083\; bar$$

$$f^V\, H_2O = f^L\, H_2O = 0.4083\; bar = y_i P = y\, H_2O 2.722\; bar$$

$$y_{H_2O} = \frac{0.4083}{2.722} = 0.15$$

2. Compute the partial pressure of VF_2 and its fugacity

$$y_{vf2} = 1 - y_{H_2O} = 0.85$$

$$f_{vf2} = 0.85 \cdot 2.7218\ bar = 2.314\ bar$$

3. Compute the mole fraction of VF_2 in the liquid phase.

$$x'_{vf_2} = \frac{2.314\ bar}{25{,}000\ bar} = 0.00012;\ x'_{H_2O} = 0.99988$$

4. The activity of monomer in the vapor phase will be defined as equal to its mole fraction

$$a_{vf2} = 0.85.$$

5. Compute the monomer volume fraction in the polymer phase.

$$\ln a_{vf_2} = \ln 0.85 = \ln \phi_1 + \phi_2 + \chi \phi_2^2$$

$$\chi = 3.2,\ \phi_1 = 0.014,\ \phi_2 = 0.986$$

6. Calculate the material balances.

The vapor phase composition is calculated using the vapor phase volume and the ideal gas law. Then, the total mass of water charged, the water density, and the liquid phase volume are used to calculate the volume of polymer. Finally, monomer conversion is calculated using the monomer volume fraction.

Phase Component	Vapor	Liquid	Solid
lbmoles			
VF_2	.04951	.0300	.09
H_2O	.00874	249.99	0
lbs			
VF_2	3.171	2.00	5.7
H_2O	.1573	4499.84	0
Kynar	—	—	1661

This calculation shows that significant amounts of monomer remain in each of the three phases. The overall conversion is 99.3wt%. It probably would be economical to recover some of the unreacted monomer and recycle it.

NOMENCLATURE

a_i	activity of component *i*
A	constant for the lattice theory model
b	length of a random flight chain segment
B	constant for the lattice theory model
C^{∞}	ratio of real chain to random flight chain expansion
D	constant for the lattice theory model
ΔE_i^v	energy of vaporization of *i*
f_i	fugacity of component *i*
f_i^0	standard state fugacity of component *i*
F_i	molar attraction constant for group *i*
F_1	entropy corrector for polymer solutions
g	polymer-solvent interaction parameter
ΔG_m	free energy of mixing
h	end-to-end distance
$\langle h^2 \rangle$	average of the end-to-end distance squared
ΔH_i^v	enthalpy of vaporization
ΔH_m	enthalpy of mixing
k	Boltzmann constant
K	constant to correct heat of mixing of regular solutions
K_1	enthalpy corrector for polymer solutions
L	contour length
M	molecular weight of *i*
n	number of repeating units in an n-mer
N	ratio of molar volumes, V_2/V_1
n_i	number of molecules of *i*
p_i	partial pressure of component *i*
P	pressure
P_i^{sat}	saturated vapor pressure of component *i*
r_a	bond rotation angle
R	ideal gas constant
s	radius of gyration of a polymer molecule
ΔS_m	change in entropy on mixing
T	temperature
T_g	glass transition temperature
T_m	melting temperature
v_a	valence angle
V	system volume
V_i	partial molar volume of component *i* in mixture
V_i^0	partial molar volume of pure component *i*
ΔV_m	change in volume on mixing
$W_{(i\text{-}j)}$	energy of breaking an i-j bond

$\Delta W_{(i\text{-}j)}$ change in energy on mixing due to breaking i-j bond
$W(x_i)$ probability that a chain assumes a length between x and x + dx in the x direction
x_i mole fraction of component i in the liquid phase
y_i mole fraction of component i in the vapor phase
z total number of segments in chain model
z_s total number of segments in real chain model
Z thermodynamic property
Z_i molar property of component i
Z_i^o molar property of pure component i
ΔZ_m change in thermodynamic property on mixing

Greek Symbols

a ratio of end-to-end distances of actual and ideal chains
γ_i activity coefficient for component i
δ solubility parameter
δ_i term of three-dimensional solubility parameter
μ_i chemical potential of component i
$\mu^{\cdot}_i$ standard state chemical potential of component i
ξ polymer interaction parameter
φ_i volume fraction of component i
$\varphi_{i,c}$ component i volume fraction at the critical point
θ theta temperature
χ solvent interaction parameter
χ_c interaction parameter value at the critical point
P_i density of component i

PROBLEMS

5.1 One significant use of the solubility parameter is in the selection of solvents in the paint industry. Use the data in the table below to choose an appropriate solvent for polystyrene.

Component	δ	δ_d	δ_p	δ_h
Acetone	9.77	7.58	4.4	2.5
MEK	9.27	7.49	5.0	2.0
Benzene	9.15	8.95	0.5	1.0
Chloroform	9.57	9.28	2.1	1.0
Polystyrene	9.8	8.6	5.0	2.0

5.2. Cellulose acetate is not soluble in ethanol or chloroform alone, but is soluble in a mixture of these two solvents. Using the data given below, rationalize these phenomena.

Component	δ	δ_d	δ_p	δ_h
Cellulose acetate	12.7	9.5	6.0	6.0
Ethanol	15.0	7.73	4.0	9.7
Chloroform	9.2	8.65	1.2	5.0

5.3. Describe a lower critical solution temperature and explain why the phenomenon occurs.

5.4. Based on general criteria, methyl ethyl ketone, acetone, benzene, and chloroform might all be solvents for PVC. Benzene and chloroform swell PVC but are not particularly good solvents. Explain this phenomena using the data below.

Component	δ	δ_d	δ_p	δ_h
MEK	9.27	7.49	5.0	2.0
Acetone	9.77	7.58	4.4	2.5
Benzene	9.15	8.95	0.5	1.0
Chloroform	9.23	8.65	1.2	5.0
PVC	9.54	8.16	5.5	5.5

5.5. Compute the interaction parameter for the system, polystyrene-benzene, by using Equations 5.41, 5.42, and 5.44. Compare the computed solubility parameters and the interaction parameter to literature values.

5.6. The research department has proposed hexane as a solvent for the polymerization of propylene. If the reaction mixture was to contain 50:50 hexane:propylene, estimate whether polypropylene would phase-separate.

5.7. A plastic cup inspector opened a sealed box containing polystyrene cups and thought she detected the monomer. If her nose was sensitive to 10 ppm styrene, estimate how much residual monomer is in the cups.

5.8. Polymer demixing has been proposed as a method for purifying two polymers in a mixed recycle stream. Suppose that the polymers are poly(vinyl acetate) and polystyrene with benzene as the solvent. Using Equations 5.75 through 78, calculate the phase compositions for $P = 1000$ and a dilute mixture. Comment on the advantages and disadvantages of using this method to purify mixed plastic waste streams.

5.9. Plasticized poly(vinyl chloride) is a good barrier material for vegetable oil but is not used for liquids containing alcohol. Rationalize this performance using phase equilibria arguments.

5.10. Bulk polymerizations are done with monomer and its homopolymer in solution. Estimate whether any of the following bulk polymerizations will phase-separate: polyethylene, polystyrene, poly(methyl methacrylate), and poly(vinyl chloride). If they phase-separate, what will be the maximum molecular weight of polymer chains soluble in the monomer?

REFERENCES

C.H. BAKER, W.B. BROWN, G.GEE, J.S. ROWLINSON, D. STUBLEN, and Y.E. YEADAN, *Polymer 3*, 215 (1962).

C.H. BAMFORD and H. TOMPA, *Trans. Faraday Soc. 46*, 310 (1950).

A.F.M. BARTON, *CRC Handbook of Solubility Parameters and Other Cohesion Parameters*, CRC Press, Boca Raton, FL, 1983.

R.S. CHAHAL, W.-P. KAO, and D. PATTERSON, *J. Chem. Soc.*, Faraday Trans., 1, (*69*), 1834 (1973).

F. CHEN, A. FREDENSLUND, and P. RASMUSSEN, *Ind. Eng. Chem. Res. 29*, 875–882 (1990).

F. CHEN, A. FREDENSLUND and P. RASMUSSEN, *Ind. Eng. Chem. Res. 30*, 2506 (1991).

E.A. COLLINS, J. BARES, and F.W. BILLMEYER, *Experiment in Polymer Science*, Wiley-Interscience, NY, 1973.

E.L. DERR and C.H. DEAL, *Inst. Chem. Eng. Symp. Ser. 32*, 40 (1969).

P. EHRLICH, *J. Polymer Science: Part A 3*, 131 (1965).

B.E. EICHINGER and P.J. FLORY, *Trans. Faraday Soc. 64*, 2053 (1968a).

B.E. EICHINGER and P.J. FLORY, *Trans. Faraday Soc. 64* 2061 (1968b).

B.E. EICHINGER and P.J. FLORY, *Trans. Faraday Soc. 64*, 2066 (1968c).

P.J. FLORY, *J. Am. Chem. Soc. 65*, 374 (1943).

P.J. FLORY, *Principles of Polymer Chemistry*, Cornell University Press, Ithaca, NY, 1953.

P.J. FLORY and H. DAOUST, *J. Polym. Sci. 25*, 429 (1957).

T.G. FOX, *Bull. Am. Phys. Soc. 1*, 123 (1956).

M. GORDON and J.S. TAYLOR, *J. Appl. Chem. 2*, 493 (1952).

E.A. GRULKE, "Solubility Parameter Values", in *Polymer Handbook* (B.J. Brandrup and E.H. Immergut, Eds.), Wiley, NY, 1989.

F. GUNDERT and B.A. WOLF, "Polymer-Solvent Interaction Parameters," in *Polymer Handbook,* (B.J. Brandrup and E.H. Immergut, Eds.), Wiley, NY, 1989.

C.M. HANSEN, *Ind. Eng. Chem. Prod. Res. Dev. 8*, 2 (1969).

J.H. HILDEBRAND and R.L. SCOTT, *The Solubility of Nonelectrolytes*, 3rd ed., Reinhold, NY, 1950.

K.L. HOY, *J. Paint Technol. 42*, 76 (1970).

M.L. Huggins, *Physical Chemistry of High Polymers*, John Wiley, NY, 1958.

R. KONINGSVELD and L.A. KLEINTJENS, *Macromolecules 4*, 637 (1971).

W.R. KRIGBAUM, and D.O. GEYMER, *J. Am. Chem. Soc. 81* 1859 (1959).

M. KURATA, *Thermodynamics of Polymer Solutions*, Harwood Academic Publishers, NY, 1982.

D.C. MILES and J.H. BRISTON, *Polymer Technology*, Chemical Publishing, NY, 1965.

M.J. MISOVICH, E.A. GRULKE, and R.F. BLANKS, *I&EC Proc. Des. Dev. 24*, 1036 (1985).

R.D. NEWMAN and J.M. PRAUSNITZ, *J. Phys. Chem. 76*, 1492 (1972).

I. NODA, Y. HIGO, N. UENO, and T. FUJIMOTO, *Macromolecules 17*, 1055 (1984).

T. OISHI and J.M. PRAUSNITZ, *Ind. Eng. Chem. Proc. Des. Dev. 17*, 333 (1978).

O. OLABISI, L.M. ROBESON, and M.T. Shaw, *Polymer–Polymer Miscibility*, Academic Press, NY, 1979.

J.M. PRAUSNITZ, *Molecular Thermodynamics of Fluid-Phase Equilibria*, Prentice-Hall, NY, 1969.

I. SAKURADA and K. SEKI in *Molecular Properties of Polymers*, Part 2 (A. Nakajima and M. Hosono, Eds.), Kagaku-dohin, Kyoto, 1969.

S.I. SANDLER, *Chemical and Engineering Thermodynamics, 2nd ed.*, Wiley, NY, 1989.

G. SCATCHARD, *Chem. Rev. 44*, 7 (1949).

A.R. SCHULTZ and P.J. FLORY, *J. Am. Chem. Soc. 74*, 4760 (1952).

P.A. SMALL, *J. Appl. Chem. 3* 71 (1953).

H. TOMPA, *Polymer Solutions*, Butterworth, London, 1956.

D.W. VAN KREVELEN, *Fuel 44*, 229 (1965).

D.W. VAN KREVELEN, *Properties of Polymers, Correlations with Chemical Structures*, Elsevier, NY, 1990.

D.J. WALSH and G.L. CHENG, *Polymer 25*(4), 495–498 (1984).

H. YAMAKAWA, *Modern Theory of Polymer Solutions*, Harper and Row, NY, 1971.

6

Molecular Weight Analyses

There are many tests that characterize polymers with respect to their chemical and physical properties. Four polymer properties critical to the engineer are the molecular weight, thermal transitions, flow properties, and mechanical properties. This property set helps define how the material will process and perform in its applications.

The molecular weight distribution shows the frequency with which molecules of a specific chain length, or molecular weight, occur in the sample. The frequency distribution can be analyzed to determine an average molecular weight of the material. For some applications, the average value is sufficient to compare different polymer samples and predict their properties. For example, polymer lots from one supplier usually are polymerized in similar equipment with similar recipes and operating procedures. Because the chain growth and chain termination steps are similar, the complete molecular weight frequencies follow the same distribution function even when the average molecular weights vary. In this case, an average molecular weight value is adequate for describing a sample.

Molecular weight distributions and physical properties are linked. This concept is illustrated by comparing molecular weight distributions and thermal transitions for samples of a polyurethane that have been degraded by high temperature treatments. Figure 6.1 shows thermal transitions measured by differential scanning calorimetry (DSC) and molecular weight distributions measured by gel permeation chromatography (GPC) on polyurethane samples having different thermal histories (Yang et al., 1986). The DSC curves appear above the GPC curves for the samples. Sample A was precipitated, Sample B was annealed at 170°C for two hours, and Sample C was heated in a DSC from 50°C to 250°C at 10°C/min. The DSC peaks are typical of exothermic reaction events. The GPC

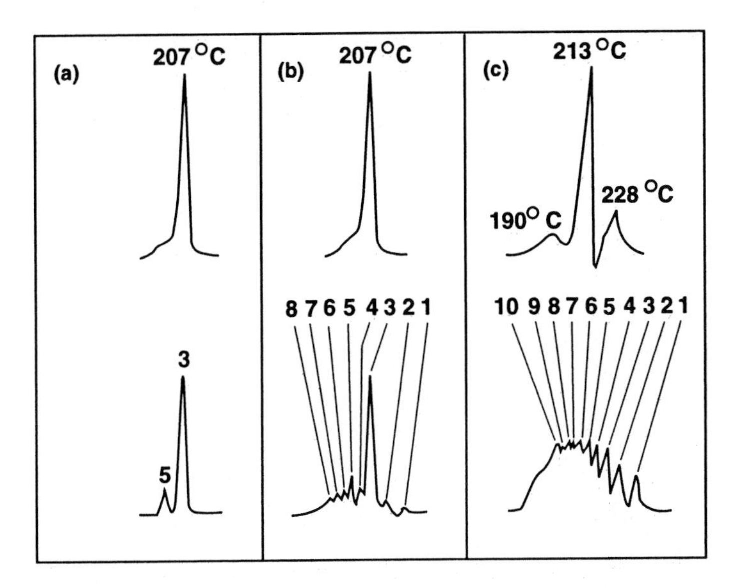

Figure 6.1 Differential scanning calorimetry (DSC) upper curves and gel permeation chromatography (GPC) lower curves for analysis of polyurethane samples (Yang et al., 1986): (**a**) precipitated polymer, (**b**) 2 hour annealing at 170°C, (**c**) 10°C/min ramp from 50 to 250°C.

curve of the original sample shows a narrow molecular weight distribution. The samples with higher temperature histories show changes in the reaction events as well as broader molecular weight distributions. These data are consistent with the hypothesis that high temperatures cause degradation followed by repolymerization for this material. These materials will have different end-use properties and illustrate the fundamental importance of molecular weight and thermal property tests.

6.1 MOLECULAR WEIGHT DISTRIBUTION

In contrast with low molecular weight compounds, polymers do not have a unique molecular weight. Polymers with the same constitution and configuration but different molecular weights can have different properties. Polystyrene of low molecular weight is a viscous liquid; polystyrene of 10,000 molecular weight is a brittle plastic; and polystyrene of 250,000 molecular weight is a hard plastic. The breadth of the distribution will affect the performance of the polymer in commercial products. For this reason, much attention has been paid to measuring and understanding molecular weight distributions.

6.2 STATISTICAL ANALYSIS OF DISTRIBUTIONS

There are two different methods for reporting molecular weight distributions: discrete and continuous.

6.2.1 Discrete Distributions

Discrete distributions report frequencies for a finite number of degrees of polymerization. If the frequency of the distribution is known, then the discrete probability density function, $g(x)$, is known for each x. The probability density function has the property that

$$\sum_{A} g(x) = 1 \tag{6.1}$$

where the summation is over the one-dimensional, A, space of the random variable, x, the degree of polymerization of the molecules. The frequency, $g(x^*)$, is defined as the probability that a randomly chosen molecule has a degree of polymerization equal to x^*.

Sometimes the cumulative frequency is reported rather than the frequency. The cumulative frequency, $G(x^*)$, is defined as the probability that a randomly chosen molecule has a degree of polymerization less than or equal to x^*. The cumulative frequency is related to the frequency by

$$G(x*) = \sum_{w \leq x*}^{x*} g(w) \tag{6.2}$$

where the ith value of the frequency occurs at $x = x^*$. The frequency can be recovered from the cumulative frequency by

$$g(i) = G(i) - G(i - 1) \tag{6.3}$$

Figures 6.2a and 6.2b show a discrete and cumulative distribution for a short chain oligomer homopolymer. The discrete distribution shows that no monomer is present [$f(1) = 0$] and that all the chains are between two and ten monomer units long. A number frequency distribution shows the number of molecules having a specific molecular weight. A weight frequency distribution also could have been chosen. The cumulative distribution is generated from the discrete distribution by Equations 6.1 and 6.2.

Table 6.1 shows the raw data for this sample. The first column is the degree of polymerization of the fraction. The second column is the number of chains for each fraction. The quantities in this column are not $f(x)$ values since the sum of the column

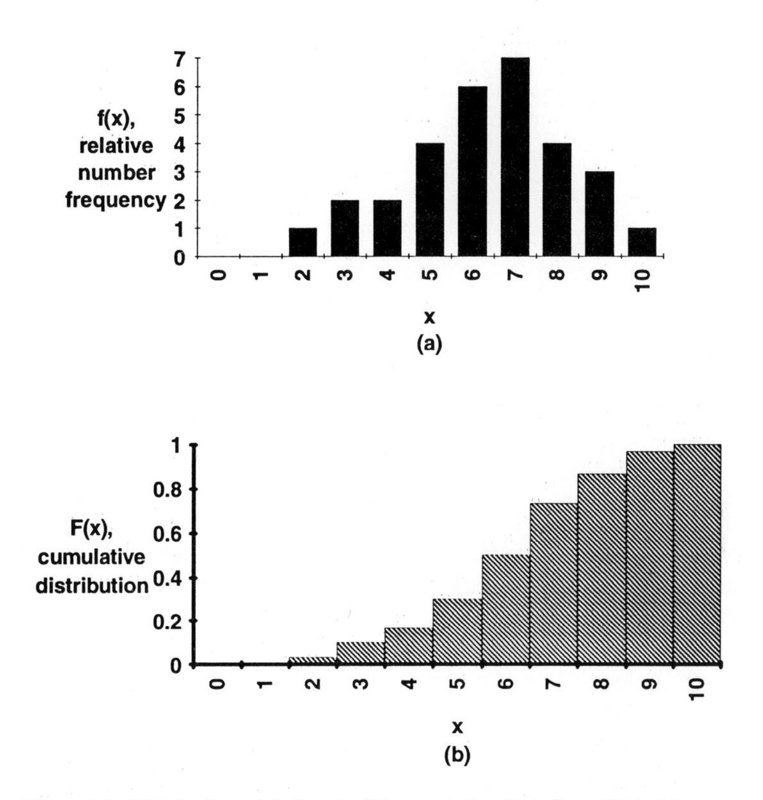

Figure 6-2 Distributions: (a) discrete, (b) cumulative. Data from Table 6.1.

does not equal one (Eq. 6.1). The third column represents the cumulative fraction of chains. These numbers were determined by applying Equation 6.2 to the data in the first two columns.

The data show that the relative number of chains with degree of polymerization equal to 6(x = 6) is six. This corresponds to a $g(x)$ value of 6/30 or 0.20. The cumulative distribution shows that 50% of all chains have a degree of polymerization less than or equal to six.

The number average degree of polymerization, of this sample is 6.3, and the weight average degree of polymerization is 6.9. This oligomer has a small polydispersity value of 6.9/6.3 = 1.1. Most (70%) of the molecules in this sample have chain lengths between five and eight, inclusive. This is easy to determine by using the cumulative distribution. The number of molecules having chain lengths of eight or less is 0.87. The number fraction of molecules having chain lengths of four or less is 0.17. The fraction of molecules having chain lengths greater than four but less than or equal to eight is 0.87 − 0.17 = 0.70.

TABLE 6.1 Discrete Distribution Data for a Homopolymer Oligomer

X	Number of chains	Cumulative fraction of chains
0	0	0.00
1	0	0.00
2	1	0.03
3	2	0.10
4	2	0.17
5	4	0.30
6	6	0.50
7	7	0.73
8	4	0.87
9	3	0.97
10	1	1.00
Total number of chains	30	

This discrete distribution is a hypothetical one. It would be difficult to develop such data for most polymer samples because the analytical methods would have to distinguish between dimers, trimers, tetramers, etc. This is not feasible for many polymers.

There are several practical methods for determining molecular weights. An average molecular weight can be measured for the sample. A second method is to separate the sample into several fractions, measure the average molecular weight of each fraction, and generate a discrete distribution. Finally, a continuous distribution can be measured by chromatographic methods.

6.2.2 Continuous Distributions

Much molecular weight analysis now is done by gel permeation chromatography, which yields a continuous distribution. Continuous distributions can be reported as either differential or integral (analogous to frequency or cumulative frequency). The probability density function, $f(x)$, is defined as the probability that a given molecule has a degree of polymerization between x and $x + dx$. It has the property that

$$\int_A f(x)\,dx = 1 \tag{6.4}$$

The integral distribution, $F(x^*)$, is defined as the probability that a molecule has degree of polymerization less than or equal to x^*.

$$F(x^*) = \int_o^{x^*} f(x)\,dx \tag{6.5}$$

The differential distribution can be recovered from the integral distribution by noting that

$$f(x) = \frac{d\,F(x)}{dx} \tag{6.6}$$

For both discrete and continuous distributions, the frequency, $f(x)$, may be plotted as either the fractional number of molecules or the fractional weight of the molecules.

Figures 6.3a and 6.3b show differential and integral distributions of a thermoplastic homopolymer. The degree of polymerization ranges to 2000. The differential distribution shows a maximum at a degree of polymerization less than 200. The integral distribution shows that 80% of all the molecules have molecular weights less than 300. The number average degree of polymerization of this sample is 700, and the weight average degree of polymerization is 2100. It has a polydispersity value of 3.

6.2.3 Statistical Tools

Several analytical tools of distributions of random variables will be useful in interpreting molecular weight distributions. They are the expected values and the moments of distributions.

Expected Value. Suppose that x is a random variable with a probability density function, $f(x)$. Also, suppose $u(x)$ is a weighting function such that the weighting function concept can be applied to continuous or discrete distributions. Then the mathematical expectation or expected value of $u(x)$ is

$$E[u(x)] = \int_{-\infty}^{\infty} u(x)f(x)dx \tag{6.7a}$$

or

$$E[u(x)] = \sum u(x)\, g(x) \tag{6.7b}$$

where the first equation is for continuous distributions and the second is for discrete. The expected value, $E[u(x)]$, requires that the integral or sum converges absolutely. Some special mathematical expectations are associated with distributions having one random variable. Suppose that the function, $u(x)$, equals x. Then

$$E[u(x)] = \int_{-\infty}^{\infty} xf(x)dx \tag{6.8a}$$

and

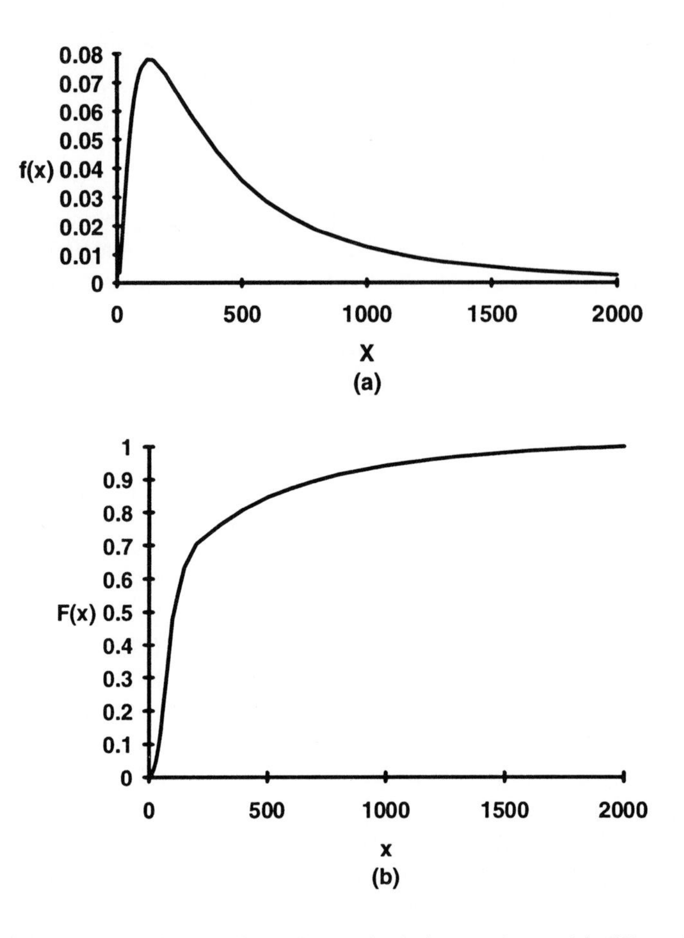

Figure 6.3 Distributions for a thermoplastic homopolymer: (a) differential, (b) integral.

$$E[u(x)] = \Sigma\, xg(x). \qquad 6.8b$$

The integral and sum are simply the average of all values of f(x) using x as a weighting factor. E(x) is the arithmetic mean of the values of x, or the mean value of x, or the mean value of the distribution, μ.

Consider the expected value for the function, $(x - \mu)^2$. The quantity, $x - \mu$, is the difference between a specific value of x and the average value of x for the distribution, μ. The expected values will be

$$E[(x - \mu)^2] = \int_{-\infty}^{\infty} (x - \mu)^2 f(x)dx \qquad 6.9a$$

$$= \Sigma(x - \mu)^2\, g(x). \qquad 6.9b$$

This integral or sum is the weighted average of the squares of the deviations of the x's from their mean value, μ, and is called the variance of x, s^2. The positive root of the variance is the standard deviation of x, $\sqrt{s}$.

Now consider the function $u(x) = x^m$. The expected values will be

$$E[x^m] = \int_{-\infty}^{\infty} x^m f(x) dx \tag{6.10a}$$

$$= \sum x^m g(x). \tag{6.10b}$$

Moments of Distribution. The integral and sum of x^m are called the moments of the distribution. When $m = 0$, the expected value is the area under the distribution curve. This area should be one if the frequencies, $f(x)$, have been normalized ($m = 0$ is the zeroth moment of the curve). The moments of the distribution will have characteristic values for different types of probability density functions and can be used to describe the complete curve with just a few numbers.

6.2.4 Mole Fraction and Weight Fraction Distributions

Discrete distributions of molecular weight come from several types of experiments. Polymer samples may be separated by fractional precipitation: A portion of the sample is precipitated from solvent solution by raising or lowering the temperature. Each fraction can be analyzed for molecular weight. The method of analysis will determine what type of molecular weight distribution is measured. For example, colligative property measurements yield number average molecular weight distributions.

Probability density functions based on the number average molecular weight can be generated directly from the reaction mechanisms, as shown in Chapter 4. Continuous weight average molecular weight distributions can be obtained from gel permeation chromatography. Once a distribution is available, it can be converted to any form desired.

Table 6.2 shows the moments of continuous distributions. The number average molecular weight can be computed from the zeroth and first moments of the mole fraction distribution, or from the zeroth and minus one moment of the weight fraction distribution. Similarly, the weight average molecular weight is the first moment of the weight fraction curve divided by the zeroth moment (the area under the curve). Since the number and weight distributions are simply connected, $n(x) = w(x)/M(x)$, it is easy to compute the desired average from either set of moments.

The meaning of average molecular weights is easy to interpret using the moments of the distributions. The number average molecular weight is the expected value of the number frequency distribution. The weight average molecular weight is the expected value of the weight frequency distribution. In Table 6.2, both the number frequency and the weight frequency are assumed to be normalized so that their

TABLE 6.2 Moments of Continuous Distributions

Definitions		
Mole Fraction Distribution		Weight Fraction Distributions
η_i	Statistical Moments	ω_i
$n(x)$	Distribution Function, $f(x)$	$w(x) = n(x)M(x)$
[number frequency]		[weight frequency]
	Moments	
$\eta_0 = \int n(M)\,dM = 1$		$\omega_{-1} = \int \frac{w(M)}{M}\,dM$
$\eta_1 = \int Mn(M)\,dM$		$\omega_0 = \int w(M)\,dM = 1$
$\eta_2 = \int M^2 n(M)\,dM$		$\omega_1 = \int Mw(M)\,dM$
$\eta_i = \int M^i n(M)\,dM$		$\omega_2 = \int M^2 w(M)\,dM$
		$\omega_i = \int M^i w(M)\,dM$
	Molecular Weights	
η_1/η_0	$\overline{M}_n$	ω_0/ω_{-1}
η_2/η_1	$\overline{M}_w$	ω_1/ω_0
η_3/η_2	$\overline{M}_z$	ω_2/ω_1
η_4/η_3	$\overline{M}_{z+1}$	ω_3/ω_2
⋮	⋮	⋮

integral over the complete variable range equals one. However, the use of moment ratios to determine the averages means that even raw data can be used as the distribution function.

Four molecular weight averages can be measured experimentally: $\overline{M}_n$, $\overline{M}_w$, $\overline{M}_z$, and $\overline{M}_{z+1}$. The number average molecular weight, $\overline{M}_n$, can be measured using colligative properties, which depend on the number of molecules in solution. $\overline{M}_w$ can be determined by light-scattering measurements or sedimentation equilibrium methods, which are sensitive to the weight of the molecule. Sedimentation experiments can be done to measure $\overline{M}_z$ and $\overline{M}_{z+1}$.

Viscosity experiments give molecular weight averages in between $\overline{M}_n$ and $\overline{M}_w$. Using number frequency distributions as a basis, viscosity measurements give an average for $\overline{M}_{c+1}$, where $0.5 \leq c \leq 0.8$.

Discrete distribution averages can be represented by an equation developed by Meyerhoff,

$$\overline{M_c} = \frac{\sum n_i M_i^{c+1}}{\sum n_i M_i^c} \tag{6.11}$$

where n_i is the mole fraction of molecules of molecular weight, M_i, and c is a variable. When c equals zero, one, and two, Equation 6.11 computes the number average, weight average, and z-average molecular weights. Equation 6.11 is analogous to moment ratios shown in Table 6.2. The number and weight fractions are simply connected.

$$w_i = n_i M_i \tag{6.12}$$

The number average molecular weight is the total weight of the polymer sample divided by the total number of moles of the sample.

$$\overline{M_n} = \frac{\sum n_i M_i^1}{\sum n_i M_i^0} \tag{6.13}$$

The weight average molecular weight is the average based on the weight frequency rather than the number frequency of the distribution.

$$\overline{M_w} = \frac{\sum w_i M_i}{\sum w_i} = \frac{\sum n_i M_i^2}{\sum n_i M_i^1} \tag{6.14}$$

The ratio of the weight average molecular weight to the number average molecular weight is the polydispersity of the sample. If the polydispersity is one, the sample is monodisperse and has one molecular weight. Only proteins and other polypeptides are considered to be monodisperse. Most commercial polymers have polydispersities in the range of 5 to 20. Different methods of measuring the molecular weight give different molecular weight averages. Table 6.3 connects the measurement technique to the appropriate molecular weight averages.

TABLE 6.3 Molecular Weight Determination Methods

Determination Method	Average	Power = c*
End group	$\overline{M_n}$	0
Osmotic Pressure	$\overline{M_n}$	0
Light Scattering	$\overline{M_w}$	1
Sedimentation (ultracentrifuge)	$\approx \overline{M_w}$	0.9
Sedimentation equilibrium	$\overline{M_z}$	2
Viscosity	$\approx \overline{M_w}$	$0.8 \rightarrow 1.0$

*Equation 6.11.

EXAMPLE 6.1 MOLECULAR WEIGHT AVERAGES

This example should help you interpret the number fraction, n_i, and the weight fraction, w_i, distribution frequencies. Suppose that you have 100 *g* of spheres consisting of ten total spheres of three different sizes. Compute the number average and weight average size based on the following data:

Fraction	n_i (no. of spheres)	M_i (wt./sphere)	w_i (fraction weight)
A	4	6.25	25
B	5	10	50
C	1	25	25
	10 spheres		100 g

Notice that the number of spheres, the weight per sphere, and the weight of each sphere type are related by Equation 6.12. The number average size is given by Equation 6.13.

$$\overline{M_n} = \frac{25 + 50 + 25}{4 + 5 + 1} = \frac{100}{10} = 10$$

The weight average size is given by Equation 6.14.

$$\overline{M_w} = \frac{156.25 + 500 + 625}{25 + 50 + 25} = \frac{1281.25}{100} = 12.8$$

The weight average size is larger than the number average size, and the polydispersity of this sample is 1.28.

EXAMPLE 6.2 MOLECULAR WEIGHT AVERAGES OF BLENDS

Two polymer samples with different molecular weights are being powder-blended to produce material with intermediate properties. Estimate the number and weight average molecular weights of the blends. The values of $\overline{M_n}$ and $\overline{M_w}$ for each sample are given in the list that follows.

(continued)

Sample	$\overline{M_n}$	$\overline{M_w}$
A	70,000	210,000
B	120,000	360,000

The three blends have 25, 50, and 75wt% of sample A.

Discrete distribution equations can be used to solve this problem. Since the blend compositions are given in weight fractions, it is simple to begin with the weight averages. Equation 6.14 can be used directly for the 25A:75B blend,

$$\overline{M_w} = \frac{156.25 + 500 + 625}{25 + 50 + 25} = \frac{1281.25}{100} = 12.8$$

The number fraction for each blend composition should be calculated using Equation 6.12. For the 25A : 75B blend,

$$n_A = \frac{w_A}{M_A} = \frac{.25}{70{,}000} = 3.57 \times 10^{-6}$$

$$n_B = \frac{w_B}{M_B} = \frac{.75}{120{,}000} = 6.25 \times 10^{-6}$$

and Equation 6.13 can be used to calculate $\overline{M_n}$;

$$\overline{M_n} = \frac{3.57 \times 10^{-6} \cdot 70{,}000 + 6.25 \times 10^{-6} \cdot 120{,}000}{3.57 \times 10^{-6} + 6.25 \times 10^{-6}} = 101{,}800$$

The calculated averages for all these blends are

Blend	$\overline{M_n}$	$\overline{M_w}$
25A : 75B	101,800	322,500
50A : 50B	88,400	285,000
75A : 25B	78,200	247,500

Figure 6.4 shows the average molecular weights of these blends as a function of weight fraction of Sample A. Because the blends are based on weight fraction measures, $\overline{M_w}$ should be linear with w_B, and it is. However, $\overline{M_n}$ should not be linear with w_B because equal weight fractions do not represent equal mole fractions for two samples with different molecular weights. The $\overline{M_n}$ of the blend can be estimated using weight fraction "weightings." The relative error for using a weight fraction weighting to average $\overline{M_n}$'s would be greatest for the 50 : 50 blend. $\overline{M_n}$ for this sample based on weight fraction weightings is

$$\overline{M_n}\,(50:50) = .50 \times 70{,}000 + .50 \times 120{,}000 = 95{,}000$$

(continued)

This value is high by 7%. This difference might not be significant for normal processing, but it could be critical for phase separation phenomona.

6.3 SELECTED DISTRIBUTION FUNCTIONS

The Gaussian distribution, or the normal distribution, provides an important comparison to distribution functions describing polymer systems. It can be used for polymer systems, but it permits negative values of the independent variable to have nonzero frequencies. In other words, a polymer sample following a Gaussian distribution could have negative chain lengths, which is not possible. The mole fraction distribution is

$$n(M) = \left(\frac{1}{2\pi}\right)^{1/2} \frac{1}{\sigma_n} \exp\left(\frac{-(M-\overline{M}_n)^2}{2\sigma_n^2}\right) \qquad 6.15$$

where σ_n is the standard deviation of the distribution and $\overline{M}_n$ is the median value of the distribution.

$$\int_{-\infty}^{\overline{M}_n} n(M)\, dM = 0.5 \qquad 6.16$$

The Gaussian distribution is symmetric about the mean. The standard deviation can be related to $\overline{M}_n$ and $\overline{M}_w$ by

$$\sigma_n = (\overline{M}_w \overline{M}_n - \overline{M}_n^2)^{1/2} \qquad 6.17$$

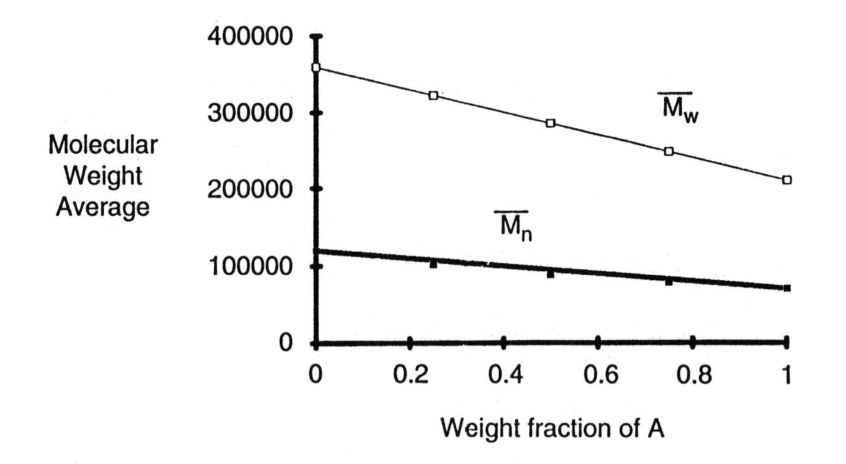

Figure 6.4 Average molecular weights for Example 6.2.

If a Gaussian distribution described a polymer sample, the standard deviation could be determined from the number and weight average molecular weights. For this distribution only, the standard deviation is an accurate measure of its width about the mean. The weight fraction distribution is

$$w(M) = \left(\frac{1}{2\pi}\right)^{1/2} \frac{1}{\sigma_w} \exp\left(\frac{-(M - \overline{M_w})^2}{2\,\sigma_w^2}\right) \qquad 6.18$$

where σ_w is the standard deviation for the weight fraction distribution, w(M) is the weight frequency of the material with molecular weight, M, and $\sigma_w = (\overline{M_z\,M_w} - \overline{M_w^2})^{1/2}$. Several of the distribution functions used to model polymer systems have forms related to the Gaussian distribution. The integrals of these functions converge absolutely over the range $0 < M < \infty$, so that they meet the criteria of expected value functions.

Three distribution functions are commonly used to describe molecular weight distributions: the Schulz–Flory, the log normal (Wesslau), and the Poisson distributions. This section describes these models and demonstrates their use.

Schulz–Flory Distribution. The Schulz–Flory distribution is based on a model for chain polymerization and is sometimes called the most probable distribution. Sections 4.1.5 and 4.3.5 showed how to compute probabilities of finding molecules of given degrees of polymerization for chain and step mechanisms. It will be useful to convert these distributions into continuous forms. Equation 4.26 computes the probability of finding a molecule of a specific molecular weight. When the extent of reaction is high, the extent of reaction, p, can be replaced by $1 - e$, where e is small:

$$n(x) = (1 - e)^{(x-1)}e \rightarrow e\,exp(-ex) \qquad 6.19$$

The term, exp(-ex), can be shown to approximate the power term by expansion in a power series and consideration of small e. The mole and weight fraction probability densities based on the degree of polymerization are

$$n(x) = e\,exp(-ex) \qquad 6.20a$$

$$w(x) = x\,n(x) = e\,x\,exp(-ex) \qquad 6.20b$$

The mole and weight fraction probability density functions are

$$n(M) = \frac{\beta^{k+1} M^{k-1} \exp(-\beta M)}{\Gamma(k+1)} \qquad 6.21a$$

$$w(M) = \frac{\beta^{k+1} M^{k} \exp(-\beta M)}{\Gamma(k+1)} \qquad 6.21b$$

where $\beta = k/\overline{M}_n$ and $\Gamma(k + 1)$ is the gamma function of k + 1. Values for the gamma function are given in Table 6.4. The value for k is 1 for disproportionation termination and 2 for termination by coupling. This distribution has the property that the polydispersity equals 2 for disproportionation. The number, weight, and z averages are connected by

$$\frac{\overline{M}_n}{k} = \frac{\overline{M}_w}{k+1} = \frac{\overline{M}_z}{k+2} \tag{6.22}$$

Log Normal Distribution. Log normal distributions can be generalized

$$w(M) = \frac{1}{(2\pi)^{1/2}} \frac{1}{\sigma_w^*} \frac{M^A}{B\overline{M}_m^{A+1}} \exp\left[\frac{-(\ln M - \ln \overline{M}_m)^2}{2\sigma_w^{*2}}\right] \tag{6.23}$$

where

$$B = \exp\left[\frac{\sigma_w^{*2}}{2}(A+1)^2\right],$$

σ_w^* is the standard deviation for this distribution, $\overline{M}_m$ is the median value, and A is a constant. There are two empirical variations of this distribution:

Lansing-Kraemer: $$A = 0, B = \exp\left[\frac{\sigma_w^{*2}}{2}\right],$$

TABLE 6.4 Selected Values of the Gamma Function

x	$\Gamma(x)$
1	1.000
1.1	0.951
1.25	0.906
1.50	0.886
1.75	0.919
1.90	0.962
2.00	1.000
Integer Values	
$\Gamma(n + 1) = n!$	

SOURCE: M. Abramowitz and T.A. Stegun, *Handbook of Mathematical Functions*, Dover Publications, NY, 1970.

and

Wesslau: $A = -1, B = 1$

The molecular weight averages are related to the median value by

$$\overline{M_n} = \overline{M_m} \exp\left[\frac{(2A+1)\sigma_w^{*2}}{2}\right] \qquad 6.24a$$

$$\overline{M_w} = \overline{M_m} \exp\left[\frac{(2A+3)\sigma_w^{*2}}{2}\right] \qquad 6.24b$$

$$\overline{M_z} = \overline{M_m} \exp\left[\frac{(2A+5)\sigma_w^{*2}}{2}\right] \qquad 6.24c$$

The mole fraction distribution is similar to Equation 6.23 except that the variance becomes σ_n^2, the variance of the mole fraction distribution. The mean value of the log normal distribution occurs between the number average and the weight average degree of polymerization.

$$\overline{M_n} < \overline{M_m} < \overline{M_w}$$

The polydispersity is given by the ratio of Equations 6.24b to Equation 6.24a.

$$\frac{\overline{M_w}}{\overline{M_n}} = \exp(\sigma_w^{*2}) \qquad 6.25$$

The log normal sample would have a greater fraction of very high molecular weight material than the Schulz–Flory sample. Normally, this difference would not be significant unless the polymer exhibited large viscoelastic effects. If the materials were processed at very high shear rates, the log normal sample might be more difficult to work. The mechanical properties of these samples could be similar, but the thermal properties of the log normal sample might be lowered due to the higher fraction of low molecular weight material.

Poisson Distribution. The Poisson distribution applies to living polymer systems where there is no termination mechanism. Due to the lack of a termination step, the polymer is almost monodisperse and the mole fraction distribution is

EXAMPLE 6.3 COMPARISON OF SCHULZ–FLORY AND LOG NORMAL DISTRIBUTIONS

Polymer samples can have the same average molecular weights and polydispersities, but quite different processing and performance properties. This example illustrates differences between samples having similar average properties but different *distributions*. Each sample has the same number average degree of polymerization, $\overline{X_n} = 5000$, and polydispersity, $\overline{X_w}/\overline{X_n} = 2$. One sample follows the Schulz–Flory distribution and the other follows the log normal distribution.

For the polymer sample that follows the Schulz–Flory distribution, $\overline{X_n}$ and $\overline{X_w}$ are related by Equation 6.22. The polydispersity is

$$\frac{\overline{X_w}}{\overline{X_n}} = \frac{k+1}{k} = 2$$

The value of k is one and $\beta = k/\overline{X_n} = 2*10^{-4}$. The frequency distributions, $n(x)$ and $w(x)$, can be calculated using Equations 6.21a and 6.21b. Each equation can be converted from a molecular weight to a chain length basis.

$$n(x) = \frac{\beta^{k+1} x^{k-1} \exp(-\beta x)}{\Gamma(k+1)}$$

$$w(x) = \frac{\beta^{k+1} x^{k} \exp(-\beta x)}{\Gamma(k+1)}$$

The Wesslau log normal distribution also is a two-parameter model. The variance is (Eq. 6.25)

$$\frac{\overline{M_w}}{\overline{M_n}} = \frac{\overline{X_w}}{\overline{X_n}} = 2 = \exp(\sigma_w^{*2})$$

and $\sigma_w^* = .8326$. The median number fraction, $\overline{X_m}$, is determined using Equation 6.24a.

$$5000 = \overline{X_m} \exp\left(\frac{-\sigma_w^{*2}}{2}\right)$$

$$\overline{X_m} = \frac{5000}{.707} = 7070$$

(continued)

The Schulz–Flory distribution is represented by the solid curve in Figures 6.5a and 6.5b. The dashed lines in Figures 6.5a and 6.5b show the Wesslau distribution. The log normal distribution has higher fractions of low and very high molecular weight material than the Schulz–Flory. Because low molecular weight material often acts as plasticizer, the log normal material might be easier to process under low to moderate shear rates.

EXAMPLE 6.4 COMPARISON AT SAMPLES WITH DIFFERENT AVERAGE PROPERTIES

The previous example compared two samples with the same average properties but different differential distributions. This example compares two samples following the same distribution model but having different average properties. The same sample characteristics as Example 6.3 are used and the Wesslau distribution is the model.

The sample parameters are

	Sample A	Sample B
$\overline{M}_n$	70,000	120,000
$\overline{M}_w$	210,000	360,000
σ_w^2	1.099	1.099
$\overline{M}_m$	123,000	208,000

Both samples have the same variance because the polydispersities are the same. Figures 6.6a through 6.6c show the number frequency, weight frequency, and integral distributions. Both differential distributions show that the sample with the lower average molecular weight has more small chains. The differences between the integral distributions are evident, but not as easy to interpret.

Figure 6.5b shows samples with identical average properties but following different distributions, while Figure 6.6b shows samples with different average properties but following the same distribution model. For critical sample comparisons, it is best to know both the average properties as well as the model of the differential distribution. Both types of data may be needed to identify such problems as chain scission during processing and side reactions such as branching and chain transfer in the polymerization step.

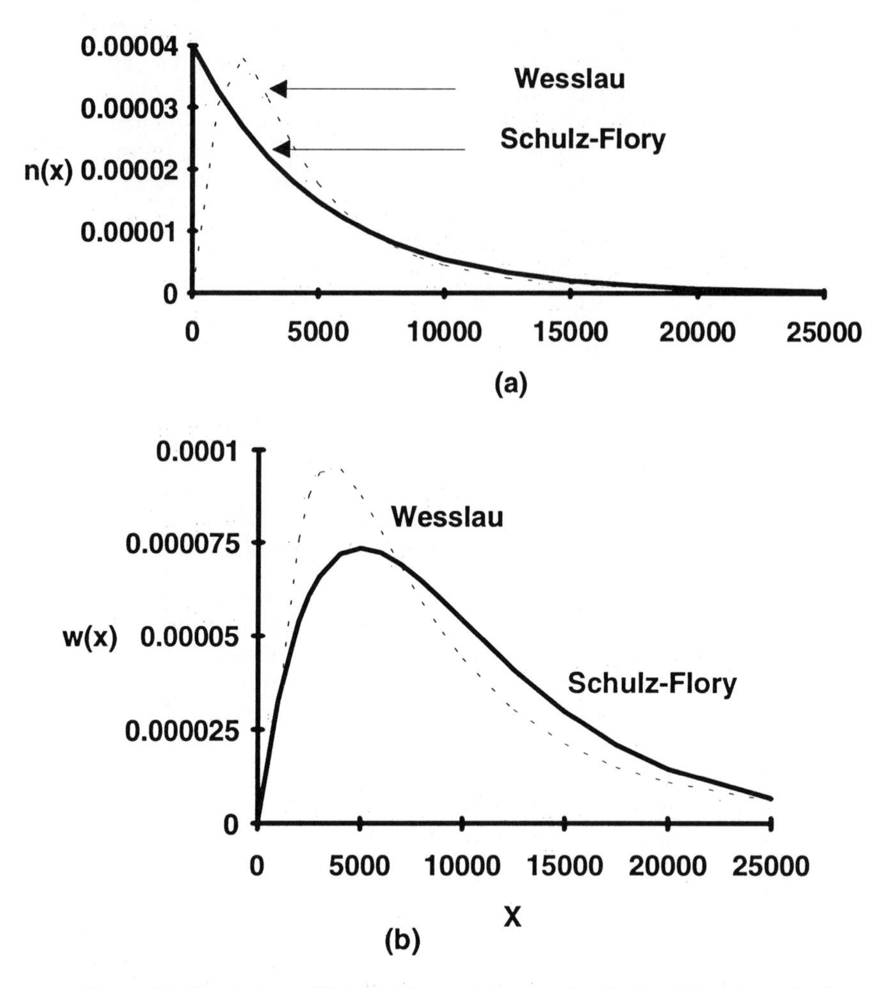

Figure 6.5 Comparison of Schultz–Flory and Wesslau distribution: **(a)** $n(x)$, number frequency; **(b)** $w(x)$, weight frequency.

$$n(x) = \frac{v^{(x-1)}}{\Gamma(x)} \exp(-v); \; v = \overline{X_n} - 1 \tag{6.26}$$

The polydispersity is

$$\frac{\overline{X_w}}{\overline{X_n}} = 1 + \frac{1}{\overline{X_n}} - \left(\frac{1}{\overline{X_n}}\right)^2 \tag{6.27}$$

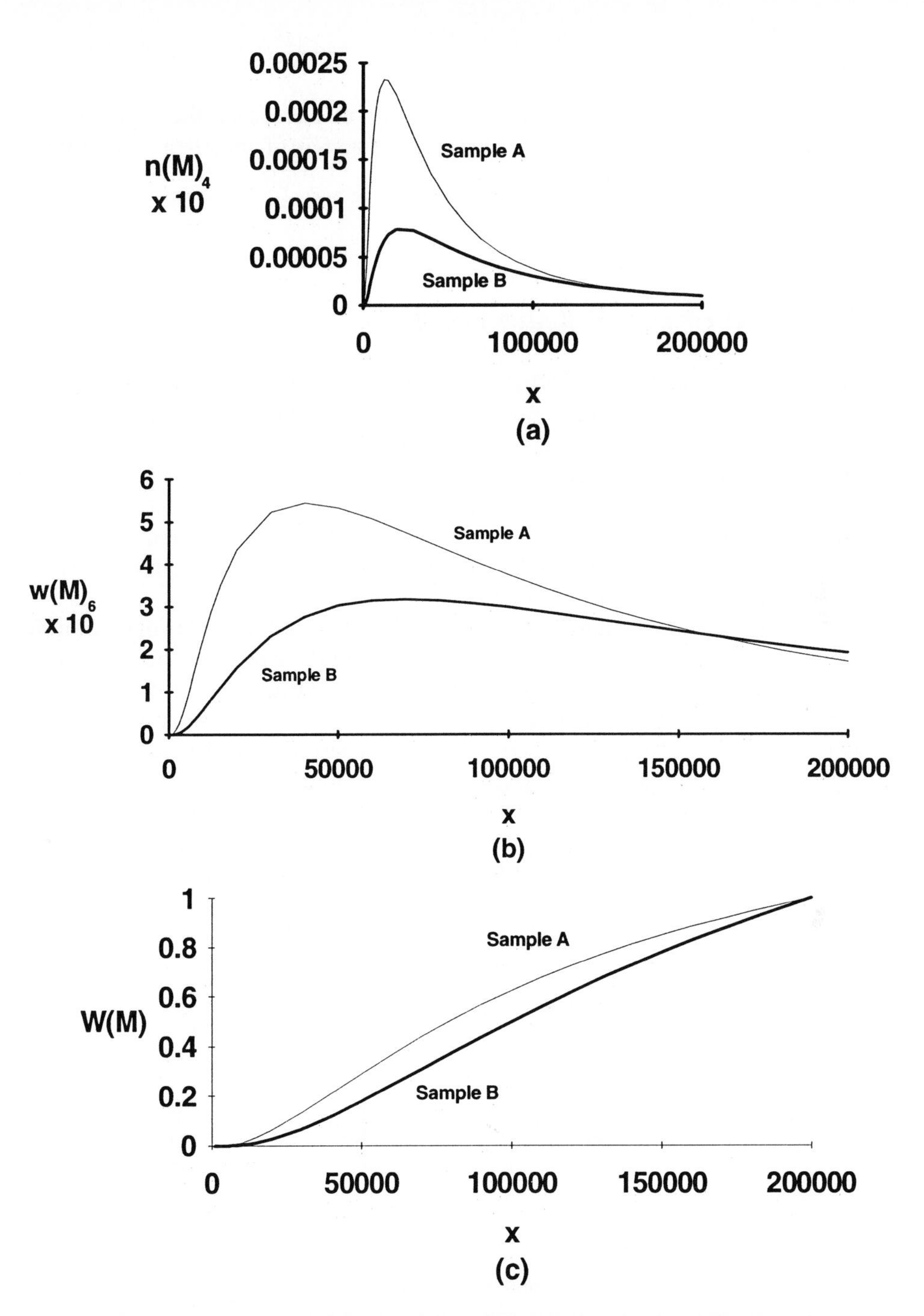

Figure 6.6 Distributions for two samples following the Wesslau distribution. Data from Example 6.4. **(a)** number frequency distributions; **(b)** weight frequency distributions; **(c)** integral distributions.

Equation 6.27 suggests that the polydispersity will decrease as the degree of polymerization increases. The polydispersity is only a function of number average molecular weight.

6.4 MEASUREMENT OF MOLECULAR WEIGHT

There are two types of methods for measuring polymer molecular weight. Single value methods were developed first. They represent significant and creative applications of physical and chemical principles to polymer systems. Gel permeation chromatography (GPC) provides complete molecular weight distributions. Once a GPC method has been calibrated using the other single value techniques, it is faster and easier to use. Some discussion of the other methods is given here for completeness. More detailed descriptions of the classical methods are given by Billingham (1977), Elias (1984), and Collins, Bares, and Billmeyer (1973).

Table 6.3 shows a number of methods for determining single values of polymer sample molecular weights. End-group analysis and the colligative methods of membrane osmometry and vapor pressure depression give the number average molecular weight. Sedimentation, light scattering, and capillary viscometry give molecular weights that relate to the size and shape of the macromolecules in solution.

All of the methods giving single values of the molecular weight can be used to determine sample molecular weight distributions. This is done by separating the sample into fractions containing narrow ranges of molecular weights. Partial fractionation of polymers from solution can be done by adding small amounts of a nonsolvent in order to precipitate the highest molecular weight material (Fig. 6.7). The precipitate is resuspended for analysis, and the polymer solution is recycled for the next fractionation.

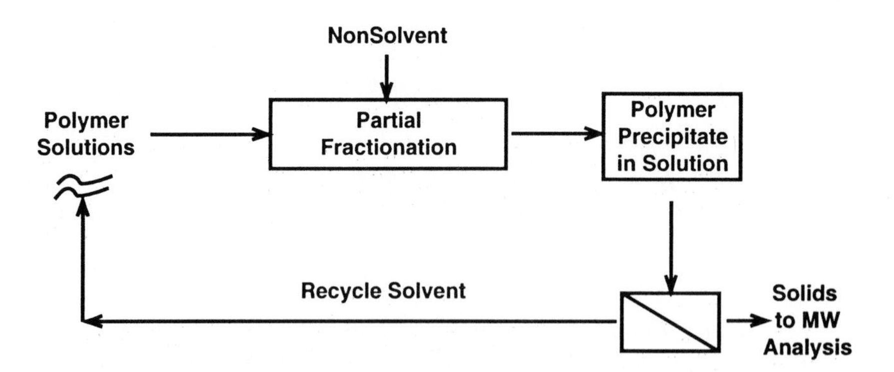

Figure 6.7 Method for determining molecular weight distribution by partial fractionation.

6.4.1 End-Group Analysis

End-group analysis is usually used to determine the molecular weight of step reaction polymers with molecular weights less than 10,000. These polymers must have reactive chain end groups. If each polymer molecule has b end groups, then the total number of moles of end groups in a polymer sample is

$$\frac{b\sum n_i}{\sum n_i} \tag{6.28}$$

Since the sample weight, W, equals

$$W = \frac{\sum n_i M_i}{\sum n_i} \tag{6.29}$$

the concentration of end groups per unit weight, E, is

$$E = \frac{b\sum n_i}{\sum n_i}\frac{\sum n_i}{\sum n_i M_i} = \frac{b}{\overline{M}_n} \tag{6.30}$$

Since E is inversely proportional to molecular weight, the accuracy of the method decreases with increasing molecular weight.

The analysis must take into account the presence of different end groups and the reaction of initiators, catalysts, chain transfer agents, and chain termination agents used in the reaction mixture. Analysis is usually done in dilute solution. Solvent purity is critical to accurate results. The chemical reaction chosen must go to completion. Bulky neighboring groups, high solution viscosity and microgel formation may prevent all the available group from reacting. Side reactions with the polymer are obviously undesirable. End-group analysis is most often used for polyesters, polyamides, polyurethanes, and polyacetals. It is also used in "living polymer" systems, initiated by butyl lithium or sodium naphthenate.

6.4.2 Colligative Methods

Thermodynamic Basis. Colligative properties are those that depend on the total number of molecules in a system. The four standard methods are vapor pressure lowering, boiling point elevation (ebulliometry), freezing point depression (cryoscopy), and membrane osmometry.

Consider a polymer solution phase and a pure solvent phase, in equilibrium with a vapor phase. The solvent in the solution is not in equilibrium with the pure solvent at the same temperature and pressure because the chemical potential of the solvent in solution, μ_i, differs from that of the pure solvent, μ_i^o. At equilibrium, $\mu_i^o = \mu_i^v$ and $\mu_i = \mu_i^v$. Equilibrium can be restored if μ_i is made equal to μ_i^o, by making a small change in the temperature of the polymer solution relative to the pure solvent. The amount of change required to achieve equilibrium gives a direct measure of the activity of the solvent in the polymer solution.

If the solution is ideal, it is possible to estimate the magnitude of colligative property measurements for solutes of different sizes. Table 6.5 shows such estimates for four methods: vapor pressure lowering, boiling point elevation, freezing point depression, and solvent osmotic pressure. The precision of the methods depends on how well the changes in the colligative property can be measured. The first three techniques will be most accurate for low molecular weight materials. Osmotic pressure experiments will be difficult to do for large changes in solvent height and are used for materials with molecular weights greater than 10,000. An additional experiment difficulty with osmotic pressure measurements is finding membranes that are not permeable to oligomers.

Membrane Osmometry. In membrane osmometry, pure solvent and a polymer solution are separated by a reverse osmosis membrane. The membrane permits the passage of solvent but not of polymer. Figure 6.8 shows a typical arrangement. Using the relations just developed, a small change in a chemical potential of the solvent in the solution can be related to a change in the activity by

$$d\mu_1(B) = RT\, d \ln a_1 \tag{6.31}$$

The solvent's chemical potential in the polymer solution can be made equal to that of the pure solvent by applying pressure to the solution. In practice, pressure is applied to the polymer solution side of the membrane by a height difference in the two rise tubes. The change in the chemical potential can be related to the pressure change by

TABLE 6.5 Magnitude of Colligative Properties—5 g/l Solutes in Benzene at 0°C.

Molar mass, g/mol	A^*	$B^\dagger$	$C^\ddagger$	$D^\S$
10^2	3.9×10^{-4}	0.0126	0.0253	1426
10^3	3.9×10^{-5}	1.26×10^{-3}	2.53×10^{-3}	142.6
10^4	3.9×10^{-6}	1.26×10^{-4}	2.53×10^{-4}	16.26
10^5	3.9×10^{-7}	1.26×10^{-5}	2.53×10^{-5}	1.43
10^6	3.9×10^{-8}	1.26×10^{-6}	2.53×10^{-6}	0.143

*A-Relative lowering of vapor pressure, $[P - P_0]/P_0$
†B-Boiling point elevation, ΔT_b, °K.
‡C-Freezing point depression, ΔT_f, °K.
§D-Osmotic pressure, cm solvent.

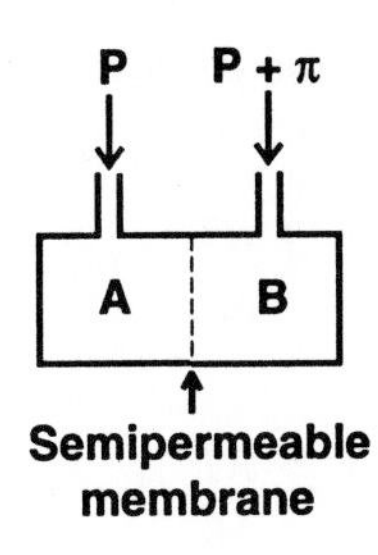

Figure 6.8 Schematic of a membrane osmometer. A = pure solvent. B = polymer solution.

$$d\mu_1(B) = V_1\, dP \tag{6.32}$$

Since equilibrium has been restored with the pure solvent, the overall change in μ_1(B) must be zero and

$$d\mu_1(B) = 0 = RT\, d\ln \mathrm{a}_1 + \mathrm{V}_1\, dP \tag{6.33}$$

This equation can be integrated to give a relationship between the activity and the applied pressure.

$$RT \int_1^{a_1} d \ln a_1 = -\int_0^{\pi} V_1\, dP \tag{6.34a}$$

$$RT \ln a_1 = -V_1 \pi \tag{6.34b}$$

For an ideal solution, the activity is related to the solvent molar volume. The substitution of this relation gives the relationship between the applied pressure, π = osmotic pressure, and the number average molecular weight.

$$\frac{\pi}{C} = \frac{RT}{\overline{M}_n} \tag{6.35}$$

The above discussion would be sufficient if polymer solutions were ideal. However, the solvent activity is related to the polymer's volume fraction.

$$\ln a_1 = -\frac{\phi_2}{N} - A_2\, \phi_2^2 - A_3\, \phi_2^3 \ldots \tag{6.36}$$

where N is the ratio of molar volumes of the polymer and solvents, ϕ_2 is the polymer volume fraction, and A_i's are the virial coefficients. The combination of Equations 6.34b and 6.36 give

$$\pi = \frac{RT}{V_1}\left(\frac{\phi_2}{N} + A_2\, \phi_2^2 + A_3\, \phi_2^3 \ldots\right) \tag{6.37}$$

The polymer solutions are made on a concentration basis, so it is convenient to develop Equation 6.37 in an expansion using concentration as the independent variable.

$$\frac{\pi}{C} = \frac{RT}{\overline{M_n}} + DC + EC^2 + \ldots \tag{6.38a}$$

$$\frac{\pi}{C} = \frac{RT}{\overline{M_n}}(1 + FC + GC^2 + \ldots) \tag{6.38b}$$

The above equations can be applied to dilute polymer solutions with no reference to a specific model. If the Flory–Huggins or Flory–Krigbaum theories are applied, specific relations for the second virial coefficient, A_2, can be derived. Flory–Huggins:

$$A_2 = (1/2 - \chi)\frac{V_2^2}{V_1} = F_1(1 - \frac{\theta}{T})\frac{V_2^2}{V_1} \tag{6.39}$$

Flory–Krigbaum:

$$A_2 = F_1(1 - \frac{\theta}{T})\frac{V_2^2}{V_1} f(x) \tag{6.40}$$

Both of these relations show A_2 should be temperature-dependent and should be zero at the theta temperature, θ.

Osmotic pressure measurements are best made when the solution is close to its theta temperature. At these conditions, the second virial coefficient is small and a plot of π versus C is linear, with a slope equal to the second virial coefficient and an intercept inversely proportional to the number average molecular weight. Osmotic pressureo measurements are usually reported in terms of centimeters of solvent head. A convenient value for the ideal gas constant, R, is $86.76/\rho$($\ell cm\ solvent/mol°K$) where ρ is the solvent density at the measurement temperature.

EXAMPLE 6.5 DETERMINATION OF $\overline{M_n}$ FROM OSMOTIC PRESSURE MEASUREMENTS FOR POLYSTYRENE IN METHYL ETHYL KETONE AT 37°C.

Table 6.6 shows typical osmotic pressure data for a polystyrene sample in methyl ethyl ketone. The polymer concentration is given in g/dm^3 (g/ℓ) and the osmotic pressure has been recorded as cm solvent. In Figure 6.8, this would be the difference between the height of the liquid columns between sides A and B.

(continued)

The membranes are not 100% efficient in rejecting small solutes, so osmotic pressure measurements may give low values of $\overline{M_n}$. Experimental times to accomplish equilibrium can be long (greater than 24 hours), so accurate measurements require patience. If the polystyrene solution were ideal over this concentration range, then the osmotic pressure measurement should vary with concentration to the first power and a plot of π/C versus C should be a horizontal line with an intercept value related to $1/\overline{M_n}$ (Eq. 6.38). Figure 6.9a shows that the osmotic pressure does not vary linearly with concentration. Figure 6.9b shows that Equation 6.38c with a second virial coefficient should model the data well. The plot will be used to estimate $\overline{M_n}$.

The intercept of the linear model is 0.187 cm solvent/(g/dm^3). The molecular weight will be

$$\overline{M_n} = \frac{RT}{\pi / C}$$

An appropriate value for the ideal gas constant is

$$R = (84.76/\rho)\frac{\ell - cm\ solvent}{mol^\circ K}$$

with the liquid density given in g/ℓ. For methyl ethyl ketone, $\rho = .805\ g/cm^3$, and the number average molecular weight is

$$\overline{M_n} = \left(\frac{84.76}{0.805}\right)\frac{\ell - cm\ solvent}{gmol - {}^\circ K} \cdot \frac{310^\circ K}{0.187\ cm\ solvent/(g/dm^3)} = 175{,}000 \frac{g}{gmol}$$

The nonideal behavior of polymer solutions can be used to determine virial coefficients and other model parameters. In this example problem, the osmotic pressure data will be used to determine the second virial coefficient and the Flory–Huggins interaction parameter.

The concentration factor is expanded in a power series to determine the virial coefficients (Eq. 6.38b), where A_2 is the second virial coefficient and A_3 is the third virial coefficient. The interaction parameter can be related to the second virial coefficient using Equation 6.39.

$$A_2 = (1/2 - \chi)\frac{V_2}{V_1}$$

The slope of the line in Figure 6.9b is

$$7.08 * 10^{-3} \frac{cm\ solvent/(g/\ell)}{(g/\ell)} = \frac{RT}{\overline{M_n}} \cdot F = B$$

(continued)

The second virial coefficient, A_2, is

$$A_2 = \frac{F}{\overline{M_n}} = \frac{B}{RT} = \frac{\dfrac{7.08 * 10^{-3}\ cm\ solvent/(g/\ell)}{(g/\ell)}}{\dfrac{105\ \ell-cm\ solvent \cdot 310^\circ K}{mol^\circ K}} = 2.17 x 10^{-7} \frac{l-mol}{g^2}$$

The partial specific volume of the polymer is

$$V_2 = \frac{1\,cm^3}{1.02g} = \frac{.98\ cm^3}{g} \equiv \frac{.00098\ell}{g}$$

The partial molar volume of methyl ethyl ketone is

$$V_1 = \frac{1\,cm^3}{.805g} * \frac{72.1g}{gmol} \equiv .0896 \frac{\ell}{gmol}$$

The Flory–Huggins interaction parameter can be calculated

$$\chi = 1/2 - A_2 \frac{V_1}{V_2^2}$$

$$= 1/2 - 2.17 * 10^{-7} \cdot \frac{.0896}{9.60 * 10^{-7}}$$

$$\chi = .500 - .020$$

$$\chi = .480$$

This calculation suggests that methyl ethyl ketone is a good solvent for polystyrene.

Vapor Pressure Lowering. The vapor pressure of solvent from a solution may be equated with the partial pressure of solvent in the vapor phase. For an ideal gas, the activity equals the vapor pressure ratio

$$a_1 = \frac{P_1}{P_1^{sat}} = \gamma_1 x_1 = 1 - x_2 \qquad 6.41$$

TABLE 6.6 Osmotic Pressure of a Polystyrene Sample in Methyl Ethyl Ketone at 37°C

Concentration, g/dm^3	Osmotic Pressure, cm solvent
0	0
1.3	0.2548
3	0.624
4.8	1.104
6.2	1.457
7.8	1.8876
8.7	2.2272

$$\frac{P_1^{sat} - P_1}{P_1^{sat}} = x_2 \tag{6.42}$$

The measurement of the relative lowering of the vapor pressure for a dilute solution allows the measurement of the number average molecular weight. If a drop of solution is exposed to the vapor of pure solvent, its lower vapor pressure will result in condensation of solvent from the vapor phase. The heat of condensation will cause the solution drop to increase its temperature. Equilibrium will be attained when the temperature increase raises the solution pressure to equal that of pure solvent. The technique is rapid and is not useful for low molecular weight samples. Figure 6.10 shows a schematic diagram of the equipment.

6.4.3 Sedimentation

In sedimentation experiments, polymer solutions are subjected to high gravitational forces, and individual polymer molecules will migrate with constant velocities, dependent on their size. If polymer chains were free-draining, or linear, there would be little dependence of the sedimentation velocity on molecular weight (for rod-like molecules, sedimentation velocity is independent of length). Most polymers exist in solution in the form of coils, with considerable solvent trapped between chain segments. Sedimentation experiments yield the z average molecular weight. Sedimentation experiments are greatly affected by solvent choice. Poor solvents are usually chosen so that the effective coil size is as uniform as possible.

6.4.4 Capillary Viscometry

Staudinger introduced the viscosity method for determining molecular weight. Viscosity measurements are very sensitive to changes in molecular weight, which affects the space occupied by molecules in solution. The basis for using viscosity to measure

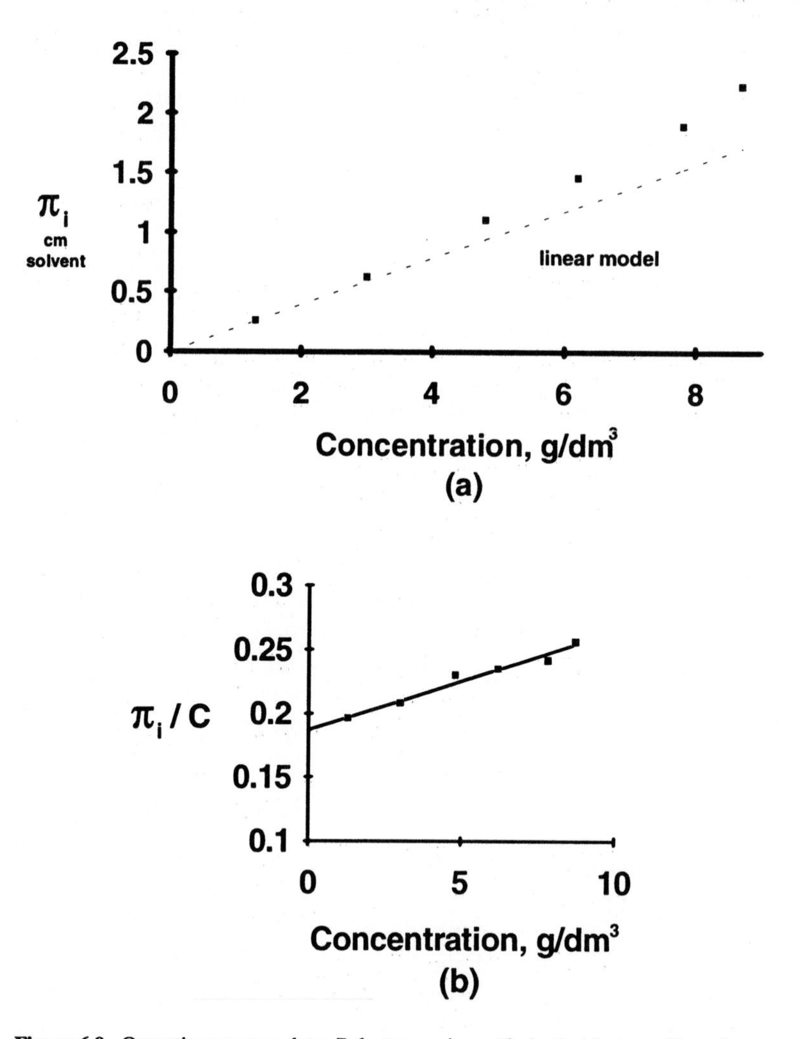

Figure 6.9 Osmotic pressure data. Polystyrene in methyl ethyl ketone. Data from Table 6.6. **(a)** π vs c; **(b)** π/c vs. c.

molecular weight is the Stokes–Einstein equation, which relates the relative viscosity of a dispersion of spherical particles to their volume fraction as the dispersed phase.

Table 6.7 shows some definitions of different viscosities. One of the more commonly reported measurements is that of intrinsic viscosity, which is independent of concentration (actually the limiting value for inherent viscosity as polymer concentration goes to zero). Intrinsic viscosity is affected by the choice of solvent. An important correlating tool is the Mark–Houwink relationship,

$$[\eta] = K\,M^{a} \tag{6.43}$$

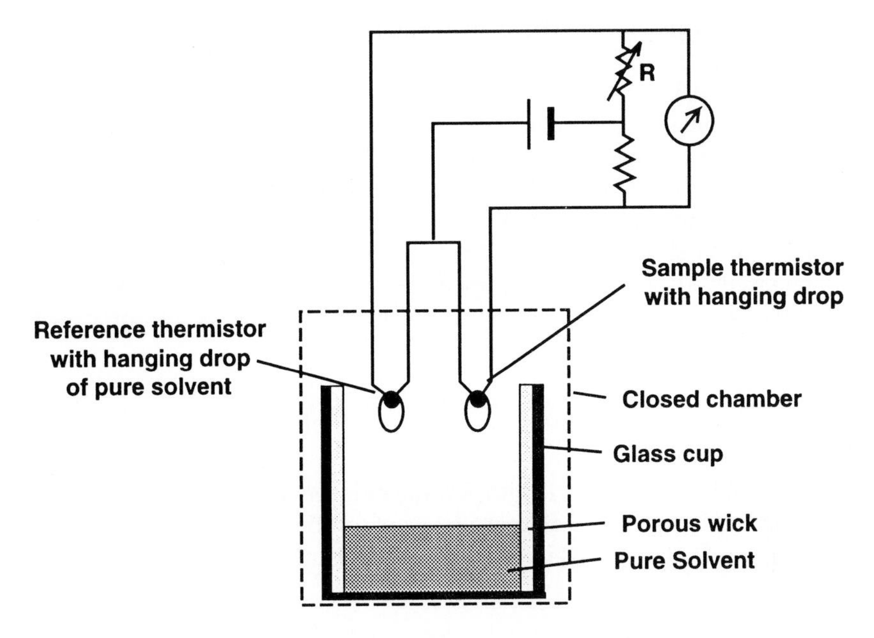

Figure 6.10 Schematic diagram of vapor point depression equipment.

where $[\eta]$ is the intrinsic viscosity, K is a constant, and a is the power exponent on molecular weight. The power exponent, a, typically has values in the range of 3.4 for high polymers, and much theoretical effort has been expended to predict the value of a from first principles. The constants for Equation 6.43 change depending on the polymer-solvent system considered and should be obtained for each case. Table 1 in Appendix F contains a table of Mark–Houwink parameters for several solvent systems. The average molecular weight obtained by viscosity

TABLE 6.7 Definitions of Different Viscosity Types

Symbol	Type of Viscosity	Equation	Units	IUPAC
η	Solution	—	Poise	
η_s	Solvent	—	Poise	
η_r	Relative	η/η_s	Dimensionless	Viscosity ratio
η_{sp}	Specific	$\left(\dfrac{\eta - \eta_s}{\eta_s}\right)$	Dimensionless	
η_{in}	Inherent	$(\ln \eta_r)/c$	Deciliter/g	
$[\eta]$	Intrinsic	$(\eta_{sp}/c)_{c=0} = (\ln \eta_r)/c_{c=0}$	Deciliter/g	Limiting viscosity number
η_{red}	Reduced	η_{sp}/c	Deciliter/g	

measurements lies between the number and weight average molecular weights. Experimental results give the exponent for the Meyerhoff relation (Eq. 6.11) between $0.5 < c < 0.8$.

Capillary viscometers are easy to use and are often used in polymer laboratory experiments. Figure 6.11 shows several types of these devices. The measurement is based on the laminar flow of polymer solution through a tube of constant radius. Polymer solution is drawn into the reservoir above the capillary tube (on the right-hand side of each device) and the flow is started. The time for the solution meniscus to pass from the upper reservoir mark to the lower reservoir mark is directly related to the viscosity. The time required for the process to occur is

$$t = \frac{8\eta L}{\pi g \rho R^4} \int_{h_1}^{h_2} \frac{dQ}{h} \tag{6.44}$$

where L is the capillary length, R is the radius, h_1 and h_2 are the upper and lower liquid head heights, Q is the volumetric flow of solution, ρ is the density, and η is the solution viscosity. Equation 6.44 is derived by solving the equation of motion for constant flow

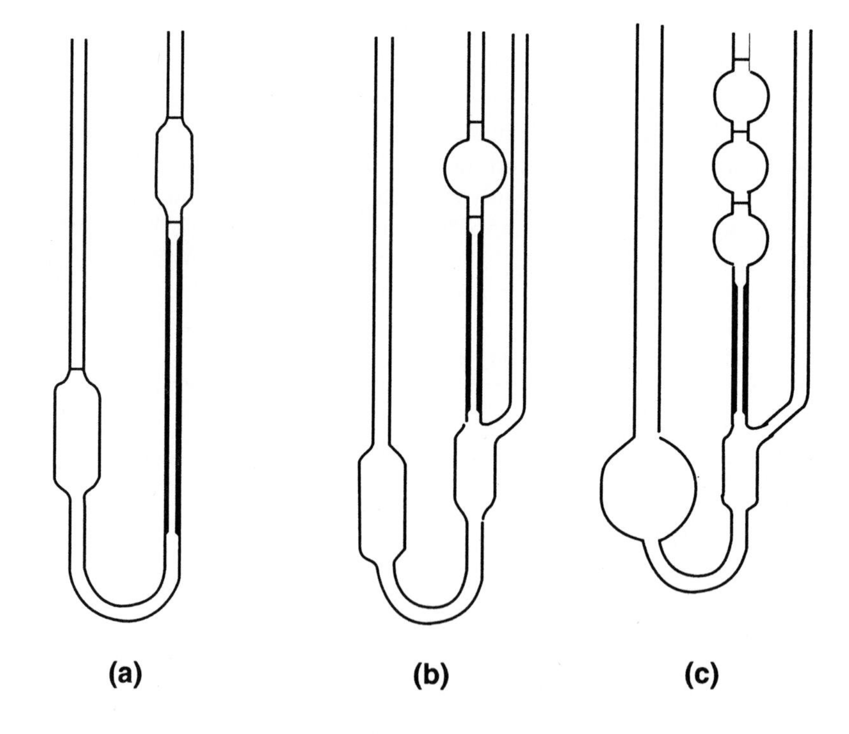

Figure 6.11 Several types of capillary viscometers.

and constant velocity. The integral and several of the parameters are constant for a specific viscometer, so the flow time is

$$t = \frac{\eta}{A\rho} \qquad 6.45$$

where A is the viscometer constant. The quantity, η/ρ, is the dynamic viscosity of the fluid. The viscometer constant is determined by calibrating it with standard solutions.

The ratio of the solution viscosity to the solvent viscosity is the reduced viscosity,

$$\eta_r = \frac{\eta}{\eta_o} = \frac{t\rho}{t_o\rho_o} \qquad 6.46$$

where η_r is the reduced viscosity and the o subscripts refer to solvent values. The density ratio is often one for dilute solutions, so that the reduced viscosity is the ratio of flow times. The reduced viscosity approaches one as a limit when the polymer concentration approaches zero. The specific viscosity describes the fractional increase in viscosity caused by the polymer,

$$\eta_{sp} = \eta_r - 1 = \frac{t - t_o}{t_o} \qquad 6.47$$

The specific viscosity is nonlinear with polymer concentration and has been modeled using a power series expansion.

$$\eta_{red} = \frac{\eta_{sp}}{C} = [\eta] + \alpha C + \beta C^2 + \ldots \qquad 6.48$$

The reduced viscosity, η_{red}, is the ratio of the specific viscosity to the concentration. It has a limiting value as concentration approaches zero of $[\eta]$, the intrinsic viscosity. The intrinsic viscosity is a measure of the polymer's ability to increase solution viscosity in the absence of intermolecular effects (in the limit as C goes to zero).

It has been difficult to relate the coefficients a and β of Equation 6.48 to the properties of the polymer solution. There are two popular empirical models for solution viscosities. The Huggins model assumes that the a coefficient is proportional to $[\eta]^2$.

$$\frac{\eta_{sp}}{C} = [\eta] + k[\eta]^2 C \qquad 6.49$$

A plot of η_{sp}/c versus C should be linear with an intercept of $[\eta]$ and a slope of $k[\eta]^2$. The Huggins coefficient, k, usually is a constant for polymer solvent and often has a value between 0.3 and 0.5.

The inherent viscosity, ln η_r/c, often approaches a limiting value as concentration goes to zero. The Kraemer model uses this property to estimate [η].

$$\eta_i = \frac{\ln \eta_r}{C} = [\eta] + k'[\eta]^2 C \qquad 6.50$$

EXAMPLE 6.6 INTRINSIC VISCOSITY OF POLY(METHYL METHACRYLATE) IN CHLOROFORM AT 35°C.

A student has taken some capillary rheology data and needs to determine the intrinsic viscosity of the sample. The data are shown in the following table.

Concentration, *g* polystyrene/cm^3	Flow Time, *s*
0 (pure solvent)	25
.0015	31.1
.0033	39.6
.0047	47.2
.0062	56.1
.0095	79.1

This problem can be solved by using Equations 6.46 and 6.47 to calculate η_{sp}, plotting η_{sp}/C versus C and then determining the slope and intercept. The conversion of the raw data is shown in Table 6.8. Figure 6.12 shows the data plotted using Equation 6.48. The solid curve is the Huggins model using $[\eta] = 152$ and $k = 0.35$. The model fits this data set well.

Both models are plotted on the same graph to give a value for the intrinsic viscosity. There are alternative models for determining the intrinsic viscosity from capillary rheometer measurements. Table 6.9 lists some of them.

Relation between Intrinsic Viscosity and Molecular Size. The intrinsic viscosity has importance beyond simple viscosity measurements: It can be related to the size of the polymer molecule in solution as well as to its molecular weight. Table 6.10 gives some of these methods. The Einstein–Simha model is based on the viscosity of a solution of large particles. Its application depends on defining V_n, the hydrodynamic volume.

The Flory–Fox model is based on the partially permeable polymer coil model. It leads to an equation relating the intrinsic viscosity to the radius of the polymer in solution and the molecular weight. However, the polymer radius is a function of a molecular weight, so the result is

TABLE 6.8 Conversion of Flow Data to η_{sp}/c

c, g/cm^3	t, s	η_r	η_{sp}	η_{sp}/c, cm^3/g
0	25	1	—	—
.0015	31.1	1.244	.244	163
.0033	39.6	1.584	.584	177
.0047	47.2	1.888	.888	189
.0062	56.1	2.244	1.24	201
.0095	79.1	3.164	2.16	228

$$[\eta] = \Phi_o \left(\frac{<r^2>_o}{M} \right)^{3/2} \alpha^3 M^{1/2} \tag{6.51}$$

where Φ_o is a constant (2.5×10^{23} for many flexible polymers in good solvents).

6.4.5 Gel Permeation Chromatography

GPC measurements have become popular recently because they yield continuous distributions and can be automated. They are based on the mechanical separation effected by flowing a pulse of polymer solution through a packed bed of porous particles. At slow flow rates, small molecules will diffuse into the pores of the bed particles, and their net velocity through the bed will be retarded. Large particles will not diffuse as quickly and will flow through the interstitial space (Fig. 6.13). An absorbance detector monitors the effluent stream to detect polymer concentration as a function of time. GPC curves have the large polymer molecules exiting

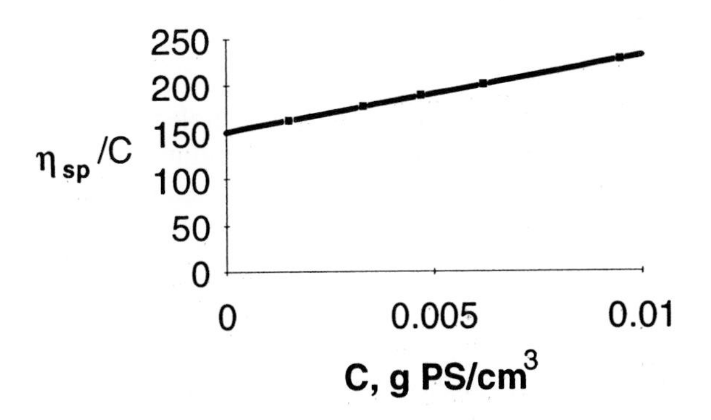

Figure 6.12 Plot of reduced viscosity versus concentration. Poly(methyl methacrylate) in chloroform at 35°C.

TABLE 6.9 Equations for Estimating Intrinsic Viscosity

Equations for Extrapolation of Viscosity Data

1. Huggins equation:

$$\eta_{sp}/c = [\eta] + k[\eta]^2 c$$

The most generally useful equation.

2. Kraemer equation:

$$(\ln \eta_r)/c = [\eta] + k'[\eta]^2 c$$

Usually used in combination with Huggins equation; may cause difficulty if $k' \geq 0.5$.

3. Martin equation:

$$\log(\eta_{sp}/c) = \log[\eta] + k'[\eta]c$$

Useful if the Huggins plots shows upward curvature; validity at infinite dilution not fully established. Sometimes used to check linearity of a Huggins plot.

4. Schulz–Blaschke equation:

$$(\eta_{sp}/c) = [\eta] + \gamma\eta_{sp}[\eta]$$

Generally linear over a wider concentration range than the Huggins equation. Not often used.

5. Heller equations:

$$(c/\eta_{sp}) = (1/[\eta]) - \gamma c$$
$$c/\ln \eta_r = (1/[\eta]) - \gamma' c$$

Proposed as a pair of equations to replace the Huggins and Kraemer equations

the column at short times and are the reverse of normal distribution curves until they are inverted.

A polymer sample in solvent at a concentration of 0.1 to 1wt% is injected into the solvent stream. The retardation of a given molecule depends on its size and the distribution of pore sizes in the gel. The solvent volume needed to elute a specific sized molecule is the retention volume (V_r) and is related to the excluded solvent volume (V_m) and the volume of the stationary solvent (V_s) by

$$V_r = V_m + EV_s \tag{6.52}$$

where $O < E < 1$. Figure 6.14 shows the relationship between the molecular size and the retention volume for the size exclusion mechanism. This assumes that there are no interactions between the solutes and the packaging material. Solvent is passed

TABLE 6.10 Models Relating the Intrinsic Viscosity to Molecular Weight and Molecular Size

Einstein–Simha:

$$[\eta] = \gamma N_A V_1 / M$$

Flory–Fox:

$$[\eta] = \phi_0 [\langle r^2 \rangle]^{3/2} / M$$

$$= \phi_0 \left[\frac{\langle r^2 \rangle_0}{M} \right]^{3/2} \alpha^3 M^{1/2}$$

Ptitsyn–Eizner:

$$[\eta] = \phi(\varepsilon) [\langle r^2 \rangle]^{3/2} / M$$

$$\phi(\varepsilon) = \phi_0 (1 - 2.63\varepsilon + 2.86\varepsilon^2)$$

$$\varepsilon = (2a - 1)/3$$

Mark–Houwink:

$$[\eta] = KM^a.$$

through both a test column and a reference column. The difference in the monitored response of both these streams is used to determine the concentration of polymer passing the test column detector at a specific time or solvent volume (Fig. 6.15). The detector response is usually proportional to the weight concentration of polymer in the solvent. (This measurement is often not dependent on molecular weight except for low molecular weight materials.)

If GPC separations depend only on the size of the molecules, then different solvent-polymer choices should lead to different calibration curves. Calibration procedures include using monodisperse polymers, using polymer fractions or whole polymer samples, and using the universal calibration procedure. Figures 6.16a and 6.16b show universal calibrations. This universal technique depends on the fact that the intrinsic viscosity of a polymer in solution is a measure of its size. The GPC can be calibrated with polystyrene. A different polymer should have about the same size as polystyrene if similar Mark–Houwink constants exist for the polymer at specific temperature and solvent conditions. Solvents that have similar Mark–Houwink exponents are chosen, and intrinsic viscosity data are taken on the recovered GPC fractions. The corrected GPC curve can be calculated from these data.

Figure 6.17 shows a typical gel permeation chromatogram. The marks correspond to a syphon counter, which measures total volume flow through the column. The high molecular weight material is eluted first. The small peaks at high counts (>26) correspond to low molecular weight impurities. These usually occur below the calibration range of the column and are excluded from the analysis.

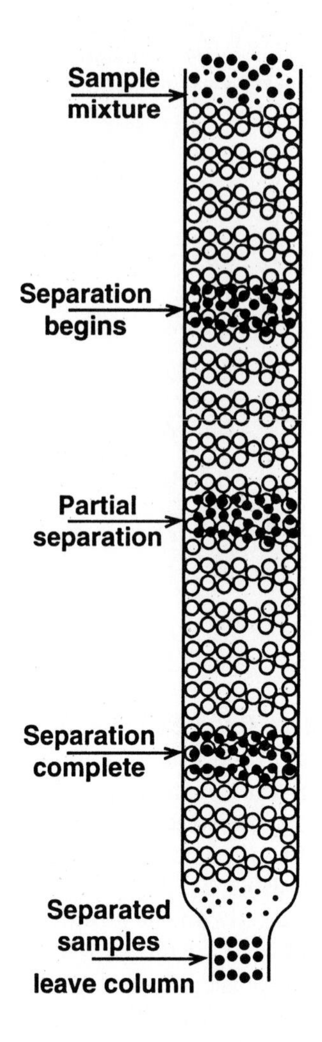

Figure 6.13 Sketch of small and large molecules interacting with porous resins.

The chromatogram of Figure 6.17 needs to be converted into a molecular weight distribution for analysis. This can be done by using the universal calibration curve. Figure 6.18 shows a schematic representation of the conversion process.

Nomenclature

Symbol	*Definition*
a	molecular weight exponent, Mark–Houwink equation
a_1	activity of component 1, or solvent

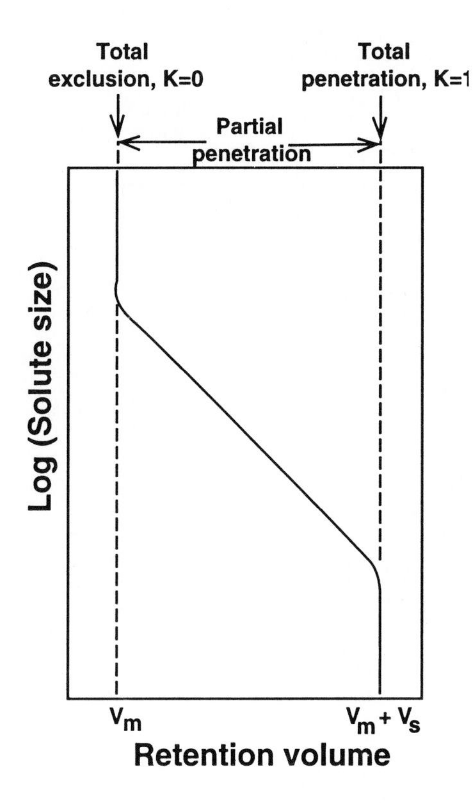

Figure 6.14 Relationship between solute molecular size and retention volume. Theoretical relationship between solute size and retention volume for the steric exclusion mechanism.

Symbol	*Definition*
A	molecular weight exponent, log normal distribution
A_i	virial coefficients
b	number of end groups
c	molecular weight exponent, Meyerhoff equation
C	mass concentration, g/ℓ
D_o	diffusion coefficient
e	unpolymerized monomer
E	concentration of end groups
$E[u(x)]$	expected value of $u(x)$
$f(x)$	continuous probability density function
$F(x)$	integral of continuous probability density function
F_1	enthalpy corrector for polymer solutions
$g(x)$	discrete probability density function
$G(x)$	cumulative density function

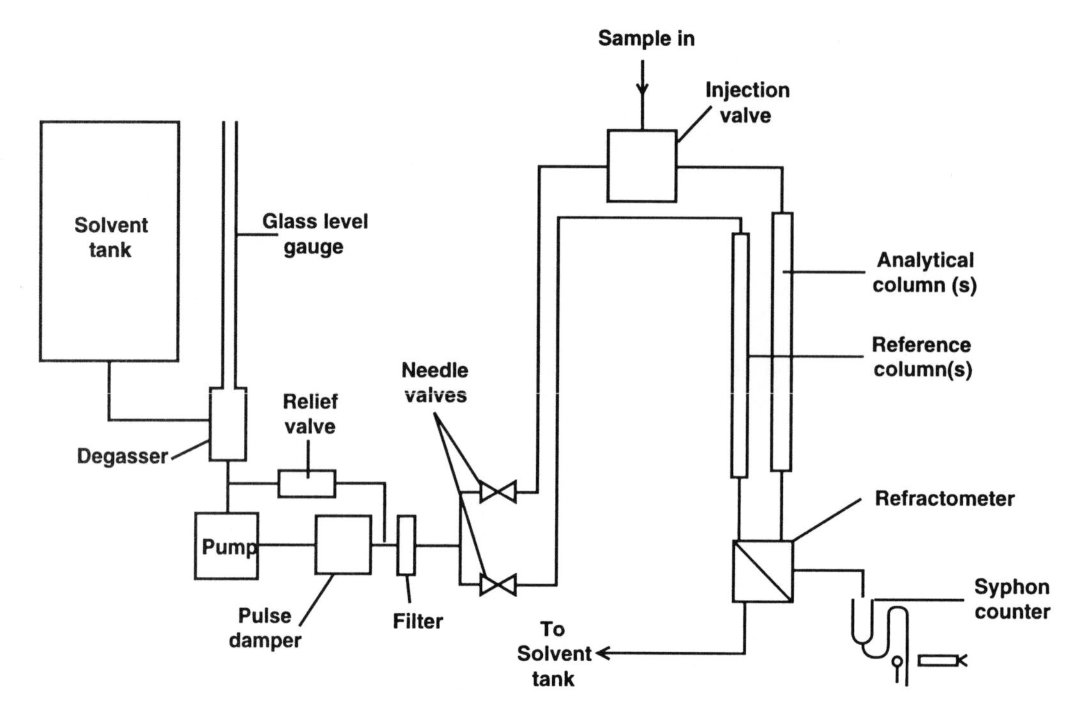

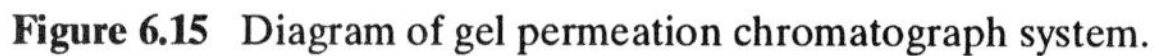

Figure 6.15 Diagram of gel permeation chromatograph system.

Symbol	*Definition*
k	Schulz–Flory constant relating to termination mechanism
K	proportionality constant
M_i	molecular weight
$\overline{M_c}$	an average molecular weight for the cth weighting factor
$\overline{M_m}$	median value of the log normal distribution
$\overline{M_n}$	average number molecular weight
$\overline{M_w}$	average weight molecular weight
$\overline{M_v}$	molecular weight average from viscometry
$\overline{M_z}$	z average molecular weight
$n(M)$	mole fraction probability function based on molecular weight
$n(x)$	mole fraction probability function based on degree of polymerization
n_i	number fraction of component i

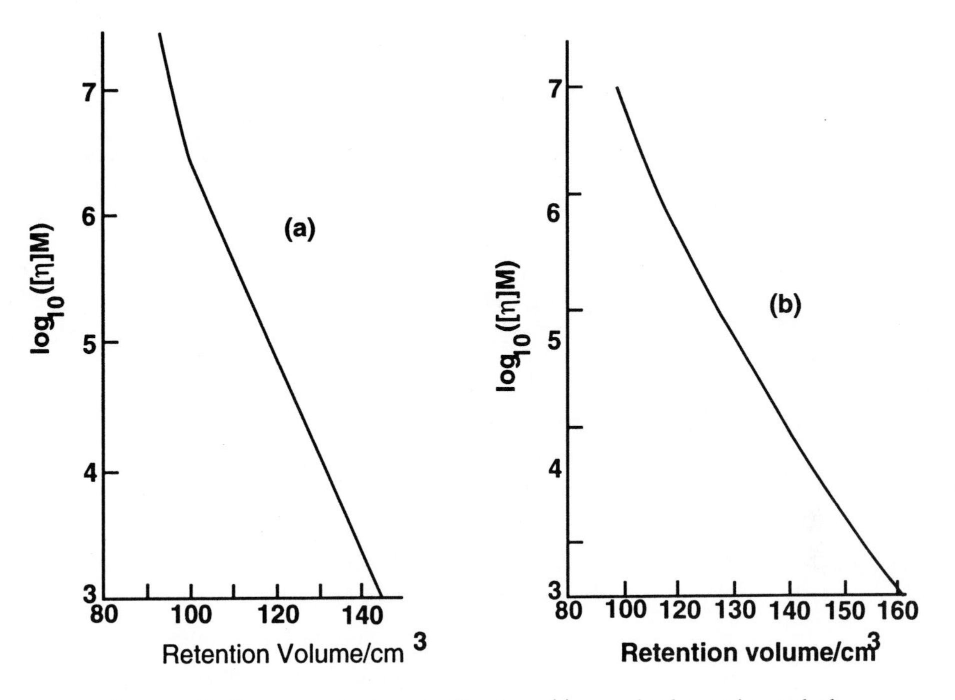

Figure 6.16 Examples of universal calibrations: (a) several polymers in tetrahydrofuram (Benoit et al., 1966): (b) several types of polyethylene in o-dichlorobenzene (Wild and Guliana, 1967).

Symbol	*Definition*
N	ratio of polymer and solvent molar volumes
N_x	probability of an x-mer
N_o	initial monomer concentration
p	probability that a monomer has reacted
P	pressure
P_i	partial pressure of component i
P_i^{sat}	saturated vapor pressure of component i
R	ideal gas constant
S	standard deviation
S^2	variance
T	temperature
$u(x)$	weighting function
V	solution volume
V_m	excluded solvent volume

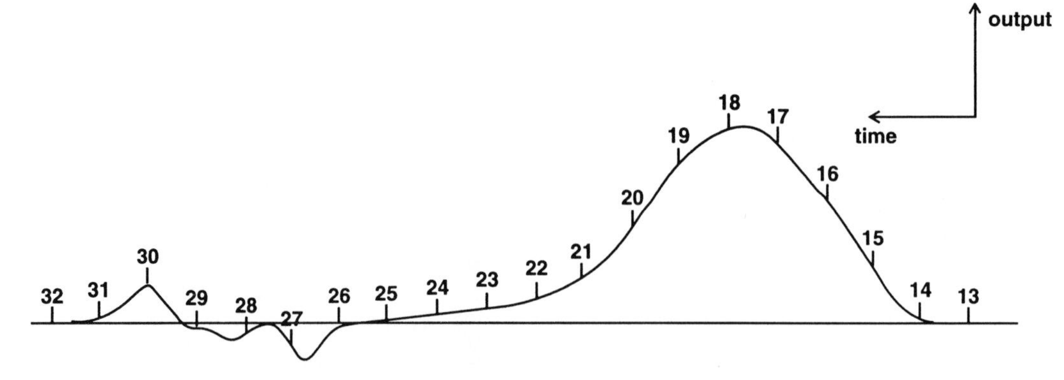

Figure 6.17 A typical gel permeation chromatograph. Vertical ticks show the time intervals.

Symbol	*Definition*
V_r	retention volume
V_s	volume of stationary solvent
V_i	partial molar volume of component i
w_i	weight fraction of component i
$w(M)$	weight fraction probability function based on molecular weight
$w(x)$	weight fraction probability function based on degree of polymerization
x	degree of polymerization
$x(M)$	mole fraction probability function
$\overline{X_m}$	median value of a log normal distribution based on degree of polymerization
$\overline{X_n}$	number average degree of polymerization
$\overline{X_w}$	weight average degree of polymerization
$\overline{X_z}$	z average degree of polymerization
Greek Symbols	
β	$k/\overline{X_n}$, constant in Schulz–Flory distribution
$\Gamma(i)$	gamma function for i
π	osmotic pressure
ϕ_2	polymer volume fraction
$[\eta]$	intrinsic viscosity
μ_1	chemical potential
μ_1°	standard state chemical potential

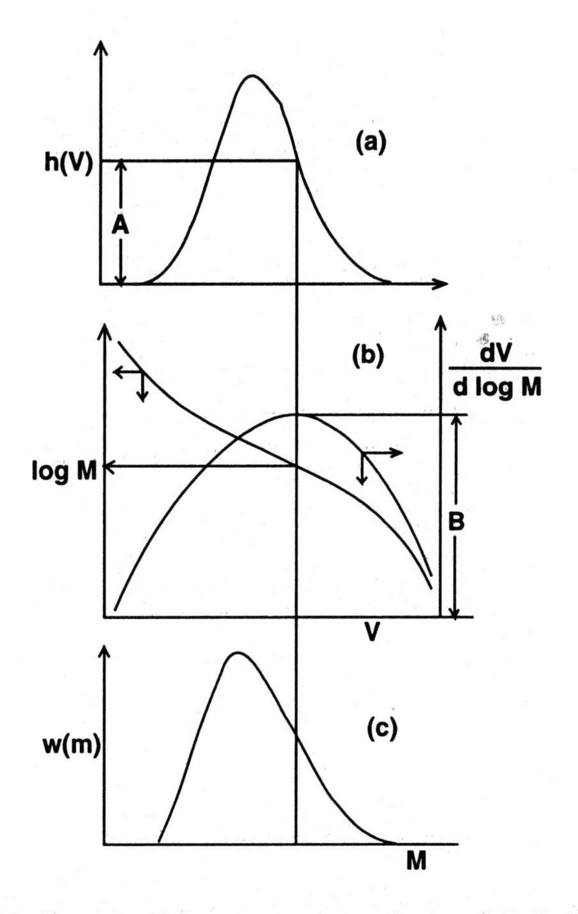

Figure 6.18 Conversion of detector response to molecular weight distribution: **(a)** detector response curve; **(b)** molecular weight calibration curve; **(c)** converted molecular weight distribution.

Symbol	*Definition*
σ_n	standard deviation of a mole fraction Gaussian distribution
σ_w^*	standard deviation of the weight fraction log normal distribution
σ_w	standard deviation of a weight fraction Gaussian distribution
θ	theta temperature
ω_i	moment of a weight frequency distribution
η_i	moment of a number frequency distribution
ν	$\overline{X_n} - 1$

Problems

6.1 Chemical analysis of poly(methyl methacrylate) was used to determine the following distribution function:

x	$w(x)$
100	0.10
300	0.15
400	0.25
500	0.15
600	0.15
700	0.10

(a.) Calculate $\overline{X_n}$ and $\overline{X_w}$. Is the polydispersity larger or smaller than that predicted by the Schulz–Flory distribution?

(b.) A light-scattering experiment was run on the same polymer sample and measured a molecular weight of 33,000. Is this result consistent with the above data?

6.2 The degree of polymerization of several GPC fraction of poly(vinyl acetate) were determined by analysis of the carbonyl content. Compute the polydispersity of the sample.

Fraction	x	$W(x)$	$w(x)$
1	100	0.10	.3125
2	300	0.40	.125
3	600	0.75	.2344
4	1000	0.95	.2969
5	1200	1.00	.3125

6.3 Osmotic pressure data are given below for the sample in Problem 6.2. Use these data to decide whether the number average molecular weight calculated in 6.2 is a good or poor estimate and explain your answer.

Solvent	C, g/dl	$(\pi/C)^{1/2}$, $cm^{1/2}$
Methyl isopropyl ketone	1	28
	2	28
	3	27
	4	28
	5	29
	6	29
Methyl ethyl ketone	2.1	33
	2.5	40

Solvent	C, *g/dl*	$(\pi/C)^{1/2}$, $cm^{1/2}$
	3.0	47
	3.3	55
	3.6	62

6.4 The intrinsic viscosity of a polymer in solution can be used to determine the size of the molecule in solution. If you wanted to determine the minimum size of poly(vinyl acetate) in solution, which of the two solvents mentioned in Problem 6.3 would you use and why?

6.5 Polymer B is generated from polymer A by carrying out a postpolymerization reaction. Polymer A has a polydispersity of 1.22. If the functional groups added to A were uniformly distributed to each polymer chain, B would have the same polydispersity as A. On the other hand, if the functional group reacted with only half the A molecules, the polydispersity might change. Determine which of these two possibilities has occurred using the following data:

Polymer	Molecular Weight	Weight Fraction
A	45,000	0.30
	105,000	0.40
	135,000	0.30
B	69,000	0.30
	162,000	0.40
	208,000	0.30

6.6 Gel permeation chromatography has been used to fractionate a polymer sample into three fractions. Fractions A, B, and C have been tested in a membrane osmometer to determine the molecular weights. Assuming that there is an equal weight of polymer in each fraction, compute the polydispersity of the original polymer sample. The data was taken at 25°C.

Fraction	C, g/ml × 10	$\pi_i/(RTC)$ × 10
A	1.0	1.0
	2.0	1.1
	3.0	1.1
	6.0	1.2
B	1.0	1.8
	2.0	2.2
	2.5	2.4
	3.5	2.8
C	1.0	2.3
	2.0	2.7
	2.5	2.9
	3.5	3.3

6.7 The fractions in Problem 6.6 have also been run in an Ostwald–Fenske capillary viscometer to determine intrinsic viscosities. Rank the fractions with regard to their intrinsic viscosities and explain your answer.

6.8 For the data given for poly(methyl methacrylate) in acetone at 30°C, find the intrinsic viscosity and the Huggins constant k. (dl = deciliter)

Reduced Viscosity, *dl/g*	Concentration, *g/dl*
0.61	.275
0.62	.375
0.63	.375
0.70	.90
0.74	1.20
0.83	1.60
0.89	1.80

6.9 Some molecular weight averages can be additive. Derive an equation for the weight average molecular weight of a mixture of homopolymer samples A and B having $\overline{M}_w$'s of M_a and M_b, with weight fractions W_a and W_b. Determine whether the $\overline{M}_w$ of the blend will be linear with w_a.

6.10 Poly(vinyl chloride) samples often have log normal distributions. Fitting a distribution to a sample can help identify the presence of excess amounts of high molecular weight polymer, which can make processing difficult. Compute the amount of polymer with molecular weight greater than 250,000 for two PVC samples with the number average molecular weight 80,000 and polydispersities of 2.0 and 2.2.

6.11 Sketch the Schultz–Flory distribution for combination termination and a number average molecular weight of 125,000.

6.12 A condensation polymer has been fractionated to determine the molecular weight distribution. The polymerization engineer believes that the reaction has proceeded to 99.7% conversion (he made an enthalpy balance around the reactor and also did end-group analysis on the product).

(a) Determine the number average and weight average degree of polymerization of this material based on the following data.

X_i, degree of Polymerization	W_i, weight fraction
100	0.18
200	0.22
300	0.29
400	0.15
500	0.08
600	0.08

(b) Does this material conform to the expected distribution for a product of its conversion?

(c) What effect(s) might the small number of samples for analysis have had on your answer to part b?

6.13 The Schultz–Flory distribution can be applied to free radical polymerizations. Consider two polymer samples having the same number average degree of polymerization (1000), which were made by disproportionation and coupling. Compute the weight average degree of polymerization for each of these fractions. Compare the weight fraction density functions at $X = 2000$.

6.14 A polymer sample has been fractionated and its discrete distribution is

M_w	W_i
2500	0.1
7500	0.3
12500	0.4
17500	0.2

(a) Compute $\overline{M_n}$, $\overline{M_w}$, *and* $\overline{M_z}$.

(b) Which analytical techniques would you choose to verify your values for $\overline{M_n}$? $\overline{M_w}$? $\overline{M_z}$?

6.15 Compute the number average molecular weight, weight average molecular weight, and the polydispersity for this sample:

M_i	N_i
5,000	500
25,000	200
50,000	150
80,000	70
120,000	30
175,000	10

6.16 The degree of polymerization of several GPC fractions of poly(vinyl acetate) were determined by analysis of carbonyl content. Compute the polydispersity of the sample.

Fraction	w_i	X_i	W_i
1	.10	100	.10
2	.30	300	.40
3	.35	600	.75
4	.20	1000	.95
5	.05	1200	1.00

6.17 A variety of methods are available for determining molecular weight. Match the statements to the techniques and explain why the method measures the specific average.

____Gel permeation chromatography	(a) Gives $\overline{M_w}$
____Osmometry	(b) Determines a molecular weight distribution
____Viscometry	(c) Used with condensation polymers
____End-group analysis	(d) Gives $\overline{M_n}$

6.18 A sample of polystyrene was fractionated by precipitation into four fractions. The weight average molecular weight was determined by viscometry for each fraction. Compute $\overline{M_w}$ for the sample based on the following data:

Fraction	Total Weight of Fraction, *g*	Molecular Wt. $\overline{M_w}$
1	7.1	27,000
2	24.1	58,000
3	37.4	115,000
4	7.5	230,000

6.19 Taylor (1947) fractionated a sample of nylon 6,6 and obtained the following weight fraction of n-mers. Determine the number average, weight average, z average molecular weights and the polydispersity for this material.

n	weight	*n*	weight
12	6.5	311	15.2
35	19.6	334	14.1
58	29.4	357	13.0
81	33.0	380	11.5
104	35.4	403	11.0
127	36.5	426	9.1
150	33.0	449	7.2
173	27.6	472	6.5
196	25.2	495	4.9
219	22.9	518	4.3
242	19.4	541	3.9
265	18.5	564	3.3
288	16.8		

6.20 Calcium stearate ($M_w = 606$) can be used as a plasticizer for PVC. Several weight percent of a plasticizer will reduce the molecular weight of a compound and change its flow and mechanical properties. Determine M_n for a PVC compound containing 2wt% calcium stearate and PVC with $M_n = 25{,}000$. How would the melt viscosity of this compound compare to that of the homopolymer?

References

M. ABRAMOWITZ and I.A. STEGUN, *Handbook of Mathematical Functions*, Dover Publications, Inc., NY, 1970.

H. BENOIT, Z. GRUBISIC, P. REMPP, D. DECKER, and J.G. ZILLIOX, *J. Chem. Phys. 63*, 1507 (1966).

N.C. BILLINGHAM, "Molar Mass Measurements" in *Polymer Science*, Wiley, NY, 1977.

E.A. COLLINS, J. BARES, and F.W. BILLMEYER, *Experiments in Polymer Science*, Wiley-Interscience, NY, 1973.

A. EINSTEIN, *Ann. Physik 19*, 289 (1906). A. Einstein *Ann. Physik 34*, 591 (1911).

H.G. ELIAS, *Macromolecules*, Vol. 1 and 2, Plenum Press, NY, 1984.

P.J FLORY and T.G. FOX, *J. Am. Chem. Soc. 73*, 1904 (1951).

M.L. HUGGINS, *J. Am. Chem. Soc. 64*, 2716 (1942).

E.O. KRAEMER, *Ind. Eng. Chem. 30*, 1200 (1938).

O.B. PTITSYN and Y.E. EIZNER, *J. Phys. Chem.* USSR *32*, 2464 (1958).

O.B. PTITSYN, and Y.E. Eizner, *J. Tech. Phys.* USSR, *29* 1117 (1959).

R. SIMHA, *J. Phys. Chem. 44*, 25 (1940).

R. SIMHA, *Science, 92*, 132 (1940).

E.B. TAYLOR, *J. Am. Chem. Soc. 69*, 638 (1947).

L. WILD and R. GULIANA, *J. Polymer Sci. A2,, 5*, 1087 (1967).

W.P. YANG, C.W. MACOSKO, and S.T. WELLINGHOFF, *Polymer 27*, 1235 (1986).

7

Polymer Analyses

This chapter covers several areas of polymer analysis: qualitative evaluations, thermal analysis, chain micro structure, polymer macrostructure, and thermophysical properties. The qualitative evaluation section provides an overview for the complete description of a polymer sample. Five property groups—thermal, solution, stability, mechanical, and composition—are used for rapid comparisons of polymeric materials. The practicing polymer engineer should have a series of practical tests to perform on his/her polymers, such as those described here.

Thermal testing has become easier over the last two decades with the appearance of automated equipment for testing small samples. These systems speed the characterization of new and modified polymers, allowing efficient research and development as well as accurate production monitoring. The chemical composition tests are important also, although the chemical compositions of most materials sold commercially are well-known. These tests are useful in determining the composition of unknown samples or verifying that the expected composition was achieved. They also can be very helpful in identifying impurities, side reactions, and additives that can have a big effect on performance properties.

7.1 QUALITATIVE EVALUATIONS

Simple, qualitative evaluations of polymer properties and characteristics are very useful. The polymer scientist synthesizing a new material, the plant engineer sampling a new production run, the designer selecting a material for an application, and the student in the polymer laboratory can benefit from simple tests of a polymer's characteristics. The systematic approach described here has been developed by Sherr (1965), Saunders (1966) and Collins, Bares, and Billmeyer (1973).

There is no substitute for first-hand information about a particular polymer sample or product. Direct manipulation of the material (pressing a film, melting a sample, pulling a molten filament, or performing solubility tests) helps give an intuitive feel for how the material will perform during processing and in its end use. Simple evaluations can provide a basis for deciding on further tests and can help select which tests will be used. The engineer will use her judgment in choosing tests to include and to eliminate from the protocol. It would be unwise to reject a material based on poor qualitative tests, but they usually indicate a deficiency in the material. These tests help focus future testing, engineering, and research work, as well as directing the match between the material and some market segment need.

Table 7.1 shows a number of important polymer properties. The five property groups included in this protocol are: thermal properties, solubility, stability, mechanical behavior, and composition and molecular architecture. Qualitative techniques are discussed for each property group. Molecular weight analyses are not included in this method. In order to perform well in these five property areas, the molecular weight will be adequate. This evaluation procedure should be applied to samples that will be used without extensive compounding, blending, or further reaction.

TABLE 7.1 Important Polymer Properties

Type of Property	Quantities
Thermophysical	Volume—density, molar volume, thermal expansion, isothermal compression
	Calorimetric—heat capacity, enthalpy, entropy, enthalpy and entropy of fusion
	Transitions—T_g, T_m, others
	Solutions—solubility parameters, interaction parameters, other equilibria constants
	Interfacial—surface energy, interfacial tension
Other physical	Chemical structure
	Optical—refractive index, reflection, birefringence, light scattering
	Electrical—dielectric constant, conductivity
	Magnetic—magnetic resonance
	Acoustic—sound propagation, sound absorption
Transport	Rheological—shear, extension, elasticity
	Thermal conductivity
	Mass transfer—diffusion, permeation
	Crystallization
Thermochemical	Polymerization rate coefficients
	Reaction equilibria—free energy of reaction
	Thermal degradation
	Chemical degradation
Processing	Extrusion—melt flow index, melt flow rate, die swell
	Molding—spiral flow test, molding area diagram, moldability index
	Spinning—melt strength
	Calendaring—draw strength
Product	Mechanical—deformation, toughnesss, hardness, friction, wear
	Environmental—thermal, flammability, solvent effects, pollutant attack, recycling/reuse

7.1.1 Thermal Properties

Thermal properties probably are the most important characteristic of a polymer sample. They determine whether the material will perform as a solid, an elastomer, or a fluid in the end-use application. They affect the processing methods used to convert the reactor product into finished parts. There are three thermal performance properties that should be considered: an evaluation of the material's melting or flowing characteristics, flammability of the melt, and thermal degradation. Each topic includes some suggestions on how to manipulate the sample in order to determine some of its bulk characteristics.

Melting/Softening/Flow Analysis. The temperatures at which a sample melts, softens, or flows can be determined by several tests. A *hot block* (Fisher–Johns melting point apparatus) has a gradient of temperature along a metal strip. Slowly pressing the sample against the strip and moving it from the cold end to the hot end will determine the temperature at which the material softens, melts, and begins to flow. A hot stage microscope or Kofler hot bench can also be used for this task.

Taking the sample through several heating/cooling cycles can be used to distinguish between thermoplastic and thermoset materials. When there is no degradation, a thermoplastic material can be cycled repeatedly without much change in its viscosity at the high temperature. A thermoset material may become fluid at the high temperature of the first cycle, but additional cross-linking also will take place. On subsequent heats, the sample will no longer be pliable and should be hard.

The softening temperature can be estimated by probing the sample with a spatula or metal rod while the sample is being heated slowly (1–5°C/min). The softening point is the temperature at which a small force on the sample causes a deformation. The melt rheology may vary from highly viscous (a firm force causes a permanent deformation) to highly elastic (the polymer recovers nearly all of its original shape after the force is removed).

Crystalline polymers have narrower melting ranges over which the thermal characteristics change than amorphous polymers going through a softening point. At temperatures slightly above T_m, crystalline polymers show large reductions in melt viscosity. Amorphous polymers have a more gradual change in viscosity as temperature is changed. Experimentation with known samples can help demonstrate this phenomenon.

Crystalline polymers that are cycled slowly through the melting point can grow larger crystals than those originally present. A crystalline sample that has been cycled several times may have a higher melting temperature than that measured originally. The larger crystals grown during the temperature cycles will melt at a slightly higher temperature. Crystallinity sometimes can be confirmed by quenching and reheating the sample. A crystalline sample heated above its melting point and then cooled very quickly will be highly amorphous. The amorphous regions can crystallize when the sample is heated above T_g.

Polymer just above T_g or T_m does not "flow." The flow temperature usually is higher and is defined as the temperature at which the sample wets a hot metal surface and leaves a thin film after being drawn across it. A polymer must be above its flow temperature to be processed easily. A melt flow indexer (ASTM D 1238) gives good data on flow properties.

The optical characteristics of the material help show thermal instability. Polymer degradation is often accompanied by formation of double bonds (leading to off colors), loss of functional groups (leading to large reductions or increases in viscosity), cross-linking (leading to hardening), or off-gassing of low molecular weight volatile products. Impurities and contamination often can be detected in polymers that become transparent or translucent at flow temperatures.

High molecular weight materials can be formed into fibers by pinching a small amount of the melt with a forceps and pulling. The fibers can be allowed to cool below the crystallization temperature and then further oriented by cold drawing. Material that can be cold-drawn usually has high molecular weight and good mechanical properties.

The tack point is defined as the temperature at which two pieces will adhere when pressed together. It may be related to the temperature at which the melt viscosity is near 10^8 poise (McLaren, et al. 1951). Adhesive properties can be tested by allowing molten material to coat a clean metal spatula and then trying to remove it after cooling to room temperature. A shiny spatula has surface defects less than 10 microns in size, and a sample that adheres well has good adhesive properties.

Copolymers and polymer blends may show phase separation during heating. Plasticizers and other compounding ingredients also can phase-separate. Samples that have nonequilibrium structures because they have been quenched or precipitated quickly may show anomalous behavior during a first heat. In these cases, direct observation is preferable to standard analytical tests in which the sample is not easily visible during the changes in temperature.

Flammability. Polymers, like other materials, can be burned by heating them to a high enough temperature so that they degrade. In order to continue burning after ignition, polymers must generate volatile fuels. A material is considered nonflammable either when the volatiles generated during heat degradation are not flammable or when the material will not burn. Poly(vinyl chloride) and poly(acrylic) acid are nonflammable. PVC generates hydrogen chloride as the primary volatile combustion product, and poly(acrylic acid) generates carbon dioxide and water. Both materials will extinguish themselves after ignition; combustion will continue only if there is another source of heat or flame.

Most polymers burn after ignition and may be compounded with flame retardant agents for applications in which there are high temperatures or the possibility of ignition. The major mechanism for flame spreading is heat conduction through the melt. If a polymer melts below 200°C, it usually is flammable. Polymers that degrade to aliphatic, aromatic, or alcoholic compounds usually are flammable. A flammable material produces little char when heated above 400°C in an oxidative environment.

Polymers that form dense chars when heated in an oxidative environment, or do not degrade below 400°C, are flame retardant.

Thermal Degradation (pyrolysis). Thermal degradation can be used to establish polymer composition of unknown samples. Under controlled pyrolysis conditions, polymers degrade by specific mechanisms, releasing specific volatile products that can be used to identify the sample. Some polymers release significant fractions of monomer when they are degraded. Examples include polystyrene, poly(methyl methacrylate), polyisobutylene, and poly(methacrylic esters). Others decompose to fragments of the monomer molecule (poly[vinyl chloride] and poly[vinyl acetate]). Others decompose by random chain scission, leaving decomposition products over a range of molecular weights. Polyethylene, polyurethanes, polyamides, and poly(acrylic esters) degrade by random chain scission.

Pyrolysis systems linked to gas chromatographs, or GC/MS systems, can be used to verify composition of polymers and copolymers and determine the composition of unknown samples. There are two common types of pyrolysis units: resistively heated devices (filaments, coils, and ribbons) and inductively heated devices (Curie point units) (Fig. 7.1). The resistively heated units are very versatile, but they have the disadvantages that the heated material (often platinum) may be catalytic to polymer degradation and that the material temperature may be hard to measure. The inductively heated devices use a wire coated with polymer in a radio frequency electromagnetic field. The wire absorbs energy and heats to its Curie point temperature, where the energy absorption slows and the temperature stabilizes. The wires are coated with polymer by dipping them in a polymer solution.

Figure 7.2 shows the effect of sample size on the time-temperature profile of a filament pyrolyzer. Heating curves for three levels of polyethylene are compared to that of an uncoated filament. Increasing the film thickness slows the heating rate of the sample. Fragmentation of the sample occurs quickly and reproducibly only when the time-temperature profile is matched exactly.

Polymers pyrolyzed at specific conditions degrade reproducibly. Figure 7.3 shows the off-gases of a Kraton 1107 sample, a copolymer of styrene and isoprene. The isoprene units degrade to isoprene and dipentene, while the styrene units release styrene. The mechanisms for the formation of isoprene and dipentene are different, leading to different ratios of these products as the pyrolysis temperature changes.

7.1.2 Solution Properties (Solvency Testing)

The most important factor in determining polymer solubility is its chemical structure. For instance, the group contribution solubility parameter model for predicting polymer solubility is based on the chemical groups that make up the polymer. Within one polymer type, other factors have secondary effects on solubility. In general, increasing the molecular weight decreases the polymer's solubility. Crystalline segments of a polymer may be insoluble, while amorphous regions of the same polymer may

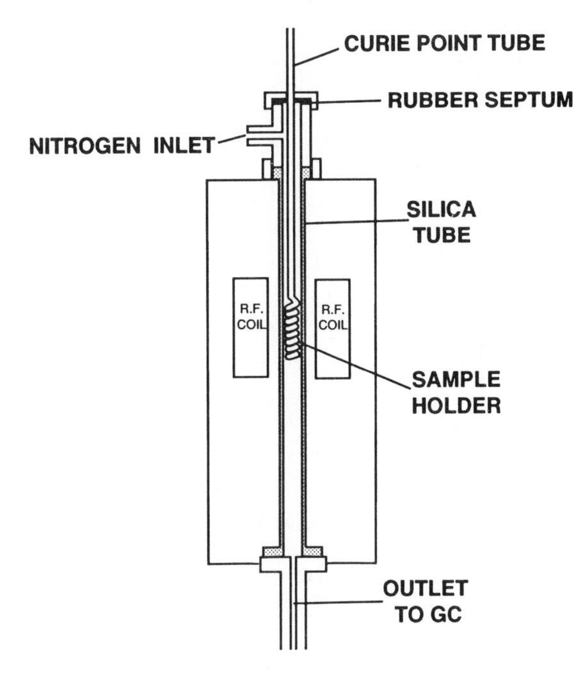

Figure 7.1 Typical Curie point pyrolyzer and filament wire (Berezkin et al., 1977).

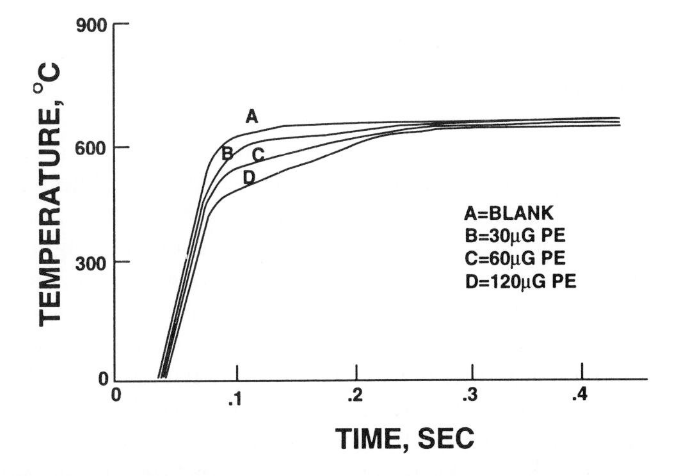

Figure 7.2 Time-temperature profile of a filament pyrolyzer (Levy et al., 1972).

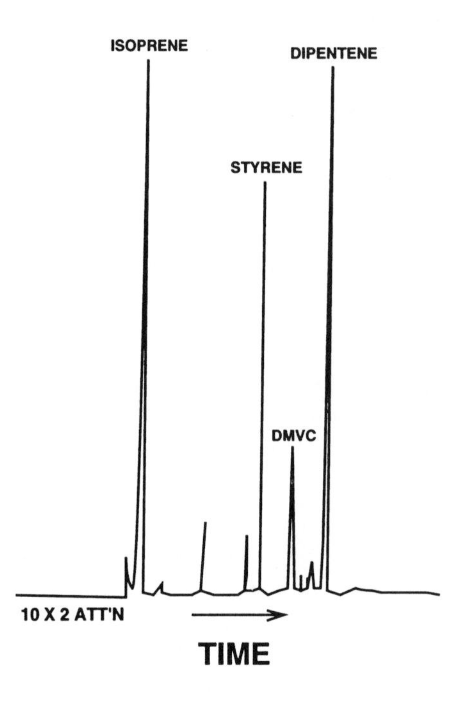

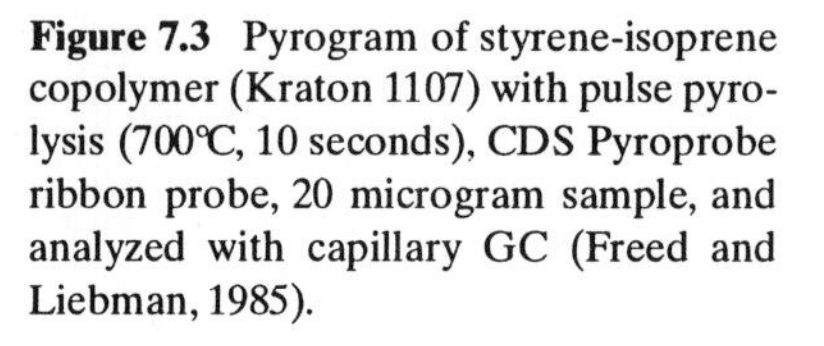
Figure 7.3 Pyrogram of styrene-isoprene copolymer (Kraton 1107) with pulse pyrolysis (700°C, 10 seconds), CDS Pyroprobe ribbon probe, 20 microgram sample, and analyzed with capillary GC (Freed and Liebman, 1985).

dissolve. Many crystalline, nonpolar polymers only dissolve when heated to temperatures near their melting temperature. The presence of cross-links decreases solubility by increasing the molecular weight. The swelling of cross-linked polymers can be related to their solubility parameter. A solvency testing procedure is discussed in Example 5.3.

7.1.3 Environmental Stability

The environmental factors of heat/cold, chemical attack, and radiation can degrade polymer properties over long periods of time. Predicting the long-term effects of environmental attack on polymer performance is difficult. A number of properties can be used to track environmental stability: total sample weight, solution viscosity, color, melting point, cross-link density, modulus, elongation, dimensional stability, etc.

Stability tests usually are done on small, compression-molded films or plaques. The effects of aggressive chemicals—such as sulfuric acid, caustic solution, boiling water, and salt solutions—is determined by immersing the sample in the liquid and measuring property changes with time. The effects of these chemicals often can be detected by weight and tensile strength changes in 24-hour tests.

Thermal stability can be evaluated by short-term heating tests. Keeping polymer films at 120°C for several hours is sufficient to determine initial thermal stability.

Samples that discolor or degrade have few uses. Samples that can withstand temperatures greater than 120°C can be used in a variety of applications involving boiling water, for example, kitchen appliances, major appliances, and automotive applications. Materials that can withstand temperatures greater than 200°C are very desirable (see Table 1.3).

Determining the effects of exposure to radiation, humidity, and acid rain are best done by long-term environmental tests. It is very difficult to use accelerated weathering tests to predict sample performance accurately. Effects such as cross-linking, surface finish changes, color changes, or mechanical property changes are good indicators of environmental attack.

7.1.4 Mechanical Properties

Stress-strain measurements are a good method for rapid evaluation of mechanical properties. Typical stress-strain curves for several different types of polymers are shown in Figure 7.4. ASTM methods (ASTM D 638 and ASTM D 882) are recommended because they offer consistent methods for testing polymer samples. Materials may range from rubbery to glassy. Manipulating the sample (pulling, stretching, flexing, and twisting) can give some indication of the mechanical properties when standard tests are not available. Yield and fracture of brittle (Figure 7.4b) and ductile (Figure 7.4e) polymers are discussed in Section 1.8.

Figure 7.4c shows a generalized stress-strain curve for a polymer from which six parameters can be determined. These parameters help define the mechanical properties of the material and are very valuable for initial comparison of samples. Most samples show a region of a linear change in stress with strain. The slope of this portion of the curve is the tensile modulus and is an indication of the stiffness of the material. Strains imposed on the sample in this region are reversible; that is, the material is elastic and can recover to its original dimensions. At higher strains, the material yields and the stress-strain curve goes through a maximum. This is the maximum stress the material can withstand for elastic performance. Further straining of the sample after the yield stress will not be reversible. Above the yield stress, some samples can be elongated by pulling at nearly constant stresses (the drawing stress). Finally, the sample breaks. The area under the entire curve is the impact strength (the energy to break). Table 7.2 relates these factors to general properties of polymers. Further description of these curves is given in Section 8.4.

7.1.5 Composition and Molecular Architecture

Spectroscopy is important for verifying chemical composition. Ultraviolet and visible spectra usually are done in solution. Infrared spectra can be obtained with powdered or film samples. IR analysis can identify the major groups along the chain, distinguish between isomers, detect crystallinity, determine copolymer composition, detect

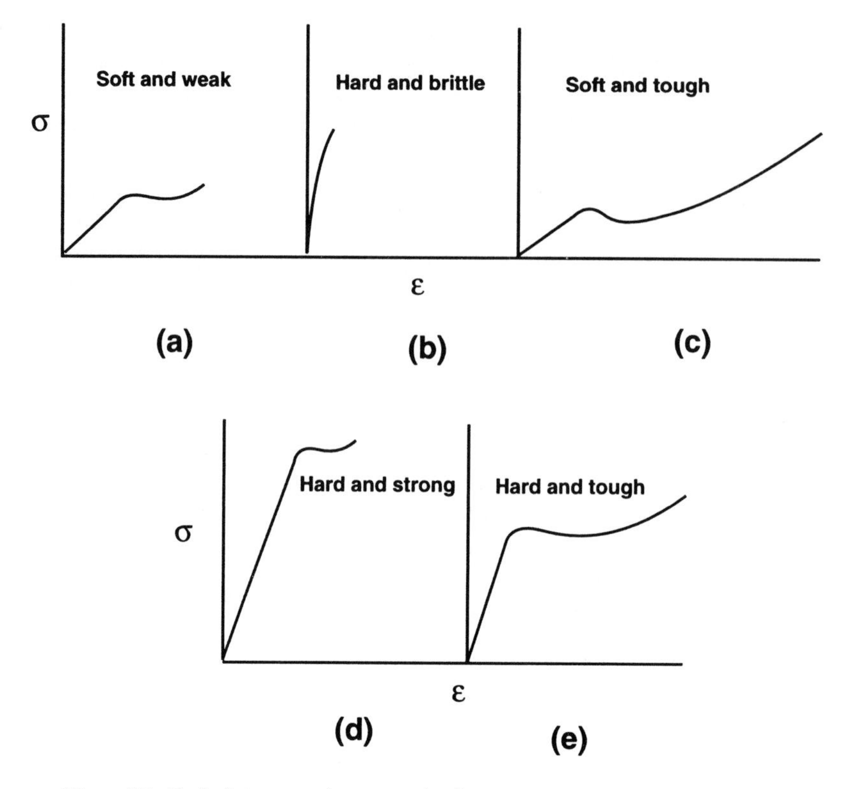

Figure 7.4 Typical stress-strain curves of polymers.

chemical attack (such as oxidation), and give sequence information. Further descriptions of these methods are given in Section 7.3.

7.2 THERMAL ANALYSIS

Thermal analysis of polymers is done by measuring physical properties of the polymer as it is subjected to controlled temperature changes. Thermal analysis is performed on condensed matter, specifically solids, glasses, liquids, and solutions. Table 7.3 contains a list of the more popular methods for thermal analysis along with their abbreviations. Four methods—thermogravimetric analysis, differential scanning calorimetry, thermomechanical analysis, and dynamic mechanical analysis—are widely practiced in the polymer and composites industries. Thermal properties are very important to end-use applications and to the processing methods used to make polymer products. There are a number of important thermal transitions that relate to processing, including T_g and T_m. Many polymers degrade or depolymerize when they are heated 100°C or more above their processing temperature. Except in a few cases,

TABLE 7.2 Characteristic Features of Stress-Strain Curves as Related to Polymer Properties

Description of Polymer	Characteristics of Stress-Strain Curve			
	Modulus	Yield Stress	Ultimate Strength	Elongation at Break
Soft, weak	Low	Low	Low	Moderate
Soft, tough	Low	Low	Yield stress	High
Hard, brittle	High	None	Moderate	Low
Hard, strong	High	High	High	Moderate
Hard, tough	High	High	High	High

SOURCE: C.W. Winding and G.D. Hiatt, *Polymeric Materials*, McGraw-Hill, NY, 1961.

polymers do not degrade to reform monomer units, but react with oxygen or with themselves to form a wide variety of volatile and nonvolatile products.

7.2.1 Thermogravimetric Analysis

Thermogravimetric analysis (TGA or TG) is used to measure a variety of polymeric phenomena involving weight change. Typical phenomena include rate of sorption of gases; desorption of volatile contaminants (monomers, solvents, plasticizers, and other additives); diffusion and permeation of gases; and polymer degradations in oxidative, inert, and vacuum environments. When gaseous materials evolve from the sample, TGA is often used in tandem with gas chromatography (GC) or mass spectroscopy (MS) to identify the lost materials; for example, it can be used to determine the carbon content of rubbers and the composition of polymer composites. TGA/GC/MS pyrolysis of samples is used for identification and characterization of homopolymers, copolymers, blends, and mixtures. In some cases, branching and tacticity can be detected.

Figure 7.5 shows a sketch of a TGA device. The sample is placed on a balance

TABLE 7.3 Thermal Analysis Methods

Method		Use
Thermogravimetric Analysic	TG	Change in gain or loss of weight
Differential scanning calorimetry	DSC	Change in specific heat
Differential thermal analysis	DTA	Heat capacity (rate of enthalpy change)
Thermomechanical analysis	TMA	Change in dimensions
Dynamic mechanical analysis	DMA	Loss moduli
Thermodilatometry		Change in volume
Dielectric thermal analysis	DETA	Change in dielectric constant
Evolved gas detection	EGD	Pyrolysis or degradation products
Evolved gas analysis	EGA	Solvent loss

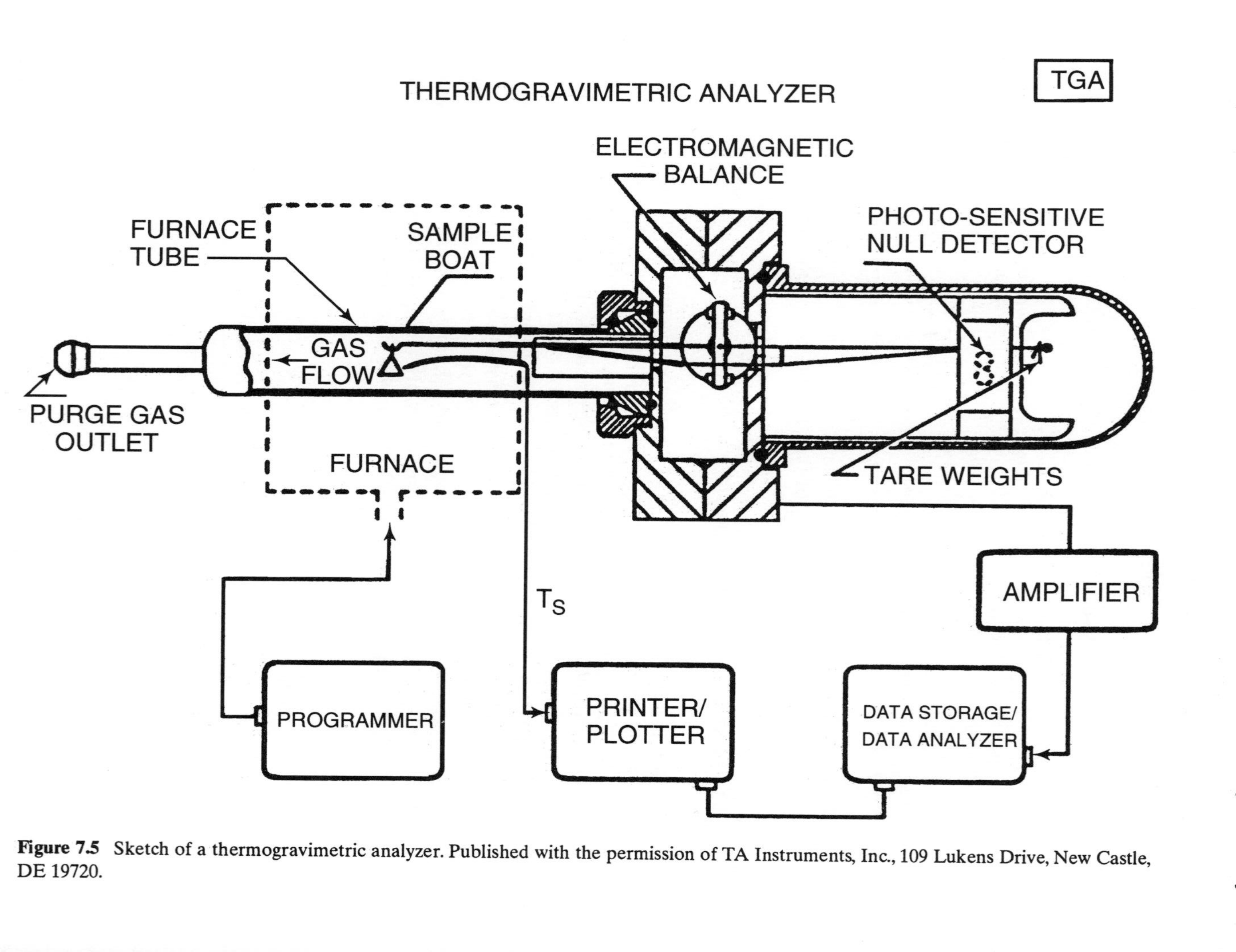

Figure 7.5 Sketch of a thermogravimetric analyzer. Published with the permission of TA Instruments, Inc., 109 Lukens Drive, New Castle, DE 19720.

beam in an oven. The position of the beam is detected by an electro-optical system that maintains the beam at a constant position by applying torque. The torque needed to maintain the position is related to the sample weight. Temperature is measured near the sample; however, the sensor may lag the actual sample temperature. The oven is controlled to provide the needed temperature profile as a function of time.

Thermogravimetric curves for several polymers are shown in Figure 7.6. Polyethylene, poly(methyl methacrylate) and polytetrafluoroethylene are unusual in that they lose weight at uniform rates throughout the degradation process. The curves for poly(vinyl chloride) and polypyromellitimide are typical of most other polymers, including blends, composites, and copolymers. Poly(vinyl chloride) degrades in several steps. Hydrogen chloride is released from chains in an "unzipping" mechanism, leaving a polyene structure behind. The polyene can react with oxygen at higher temperatures, although some residue char is left. Polypyromellitimide still has half of its original weight remaining at 800°C.

Thermal gradients in the material can exist because of the low thermal conductivity and the possibility of local "hot spots," where reactions are taking place. Rapid weight loss of solvent can lead to unsteady weight changes as it boils off. Heating rates should be kept low.

Degradation Kinetics. TGA is one of several methods used for determining degradation kinetics. Its advantages are that it may show new aspects of the degradation process, including cycling, jump temperatures, pressure effects, and purge gas effects. Because the conditions of the experiment are not well controlled, TGA may not be the best method for a first-principles analysis of a degradation process. However, it may be adequate for quality control, oxidative stability, and thermal breakdown applications.

The curve for poly(vinyl chloride) in Figure 7.6 shows several plateaus, each of which corresponds to a different process. The degree of conversion for each process is

$$p = \frac{(x_i - x)}{(x_i - x_f)} \tag{7.1}$$

where x_i is the value of the variable (sample weight) before the process, x_f is the value of the variable after the complete process, and x is the value at some intermediate time. The extent of reaction is

$$(1 - p) = \frac{(x - x_f)}{(x_i - x_f)} \tag{7.2}$$

Assuming an nth order reaction for the process,

$$\frac{dp}{dt} = k_d(1 - p)^n = k\, f(p) \tag{7.3}$$

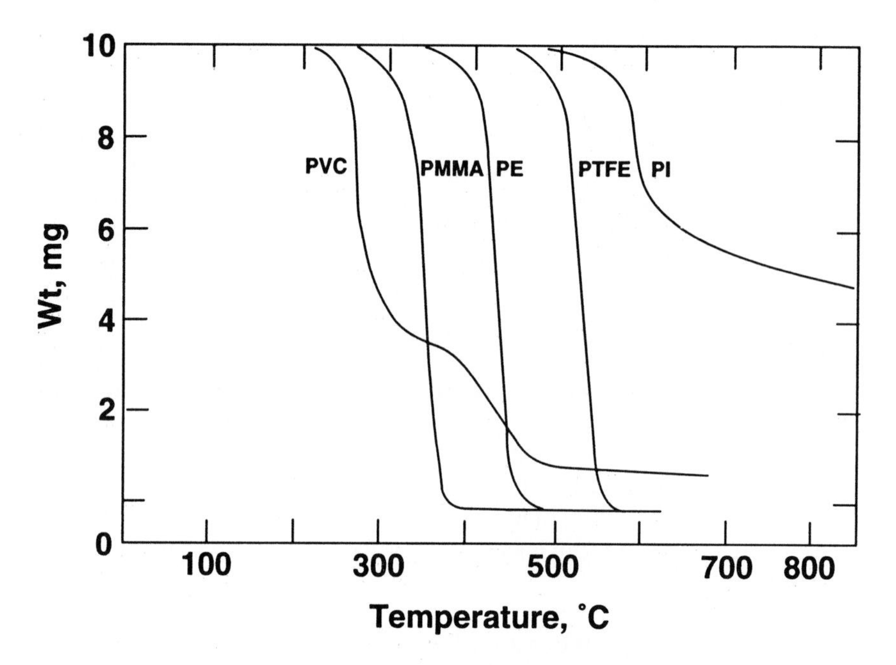

Figure 7.6 Thermogravimetric curves for several polymers. PVC: Poly(vinyl chloride); PMMA: Poly(methyl methacrylate); PE: Polyethylene; PTFE: Polytetrafluoroethylene; and PI: polypyromellitimide. Conditions: 10 mg heated at 5°C/min in nitrogen (Chin, 1966). Published with permission of John Wiley & Sons, Ltd.

where k is the reaction rate constant (which scales by an Arrhenius function with temperature), n is the reaction rate order, and $f(p)$ is the rate model.

Estimating the effect of temperature on polymer degradation is important for predicting, controlling, and preventing degradation. Determining the constants for the Arrhenius model will help predict this change. The Arrhenius model can be substituted for k in Equation 7.3.

$$\frac{dp}{dt} = A_d f(p) \exp(-E_d / RT) \quad 7.4$$

Taking the natural logarithm of each side of Equation 7.4,

$$\ln \frac{dp}{dt} = \ln A_d f(p) - E_d / RT \quad 7.5$$

A plot of the logarithm of the process rate versus inverse temperature should give the activation energy for the process.

TGA tests have been applied to predicting long-term stability of polymer samples in various thermal and gaseous environments. However, these predictions depend on the reaction mechanism being similar over the testing period as well as the end use. Few such extrapolations have been successful. TGA has been applied to composition analysis of elastomers, as it detects volatiles, plasticizers, and oils. Oxidative induction times of polyolefin wire coatings can be determined by heating samples to a fixed temperature and measuring the time required to get to maximum oxygen uptake.

Stepwise polymerizations often have volatile leaving groups. TGA can be used to determine the extent of cure by measuring the amount of additional leaving groups in the sample after the original cure has taken place. Figure 7.7 shows TGA applied to the cure of several thermoset samples. The weight loss curves are the solid lines and the dashed lines are the rates of weight loss for samples A, B, and C. Sample A is an uncured phenolic resin. At temperatures above 140°C, the cure begins and water starts to leave the sample. At 300°C, the reaction is essentially completed and 3% of the material has been lost. The ratio of water loss from cured samples to this uncured sample can be used to determine the extent of cure.

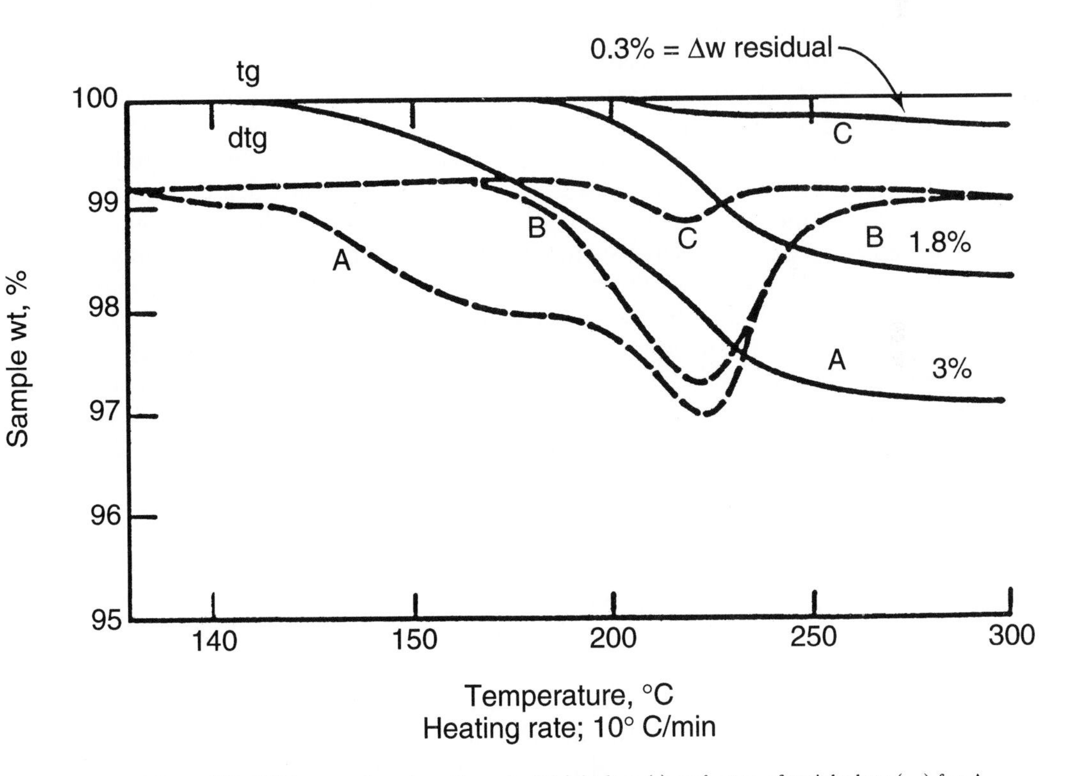

Figure 7.7 TGA curve for phenolic resin. Weight loss (-) and rates of weight loss (—) for **A** (uncured phenolic bonding resin), **B** (cured 1 min. at 160°C), **C** (cured 1 min. at 180°C). Relative degrees of cure: **B**, 40%; **C**, 40%; **C**, 90% (Day et al,.). Published with the permission of the Society of Plastics Engineers.

Sample B has been cured for one minute at 160°C. As expected, no additional curing occurs until the sample is hotter than the original curing temperature. At 180°C, the weight of this sample begins to drop. It reaches a stable weight loss of 1.8%, so its original extent of cure was 100 * (1 − 1.8/3.0), or 40%. Sample C has been cured for 1 minute at 180°C. It begins to lose weight at temperatures greater than 200°C. Its stable weight loss is 0.3%, so its original extent of cure was 100 * (1 − .3/3.0), or 90%.

7.2.2 Differential Scanning Calorimetry

Differential scanning calorimetry (DSC) measures the difference between a reference and a sample during a controlled temperature change. Changes in the heating rate of the sample relative to the reference can be converted into heat capacity and enthalpy changes. DSC experiments can be run at isothermal conditions or, usually, with temperature ramped with time.

There are three methods used for DSC: classic DTA, heat flux DSC, and power-compensated DSC. The classic DTA method has the sample and reference materials in a large heated block. A thermocouple is placed in each material, and the temperature difference is recorded as the block is heated through a specific temperature profile. In DTA, the heat flow depends on the sample material and temperature, so calibration can be difficult. The technique now is used for very high temperature or high pressure studies for which DSC instruments have not been modified. Heat flow DSC has temperature sensors placed in heat sinks just below the sample. The sample and the reference material are placed in small pans that can be hermetically sealed to retain any volatile materials. Most of the heat flow to the sample passes through the heat sink, and the instrument can be calibrated. The power-compensated DSC has heaters for the sample and the reference (Figure 7.8). The instrument maintains the same temperature in each cell by applying power to each heater as needed. The difference between the power flow to each heater is a direct measure of the rate of energy change of the sample. The DTA method is used with moderate heating rates in order to detect temperature differences, while the two DSC methods can be used isothermally or with very low heating rates.

DSC responds to all thermal events, including chemical reactions, physical transitions, release of strains, loss of volatiles, and decompositions. Mass loss during analysis changes the total heat capacity of the sample and changes the baseline. Therefore, interpretation of DSC data should be done with care. However, it has the advantage of measuring thermal transitions and chemical reactions quickly using small sample sizes. This technique has been used to determine thermodynamic constants; cure rates of resins and monomers; rates of thermal oxidative and radiative degradation; rates of physical and chemical changes; and effects of preparation, annealing, molecular weight distribution, and composition. Sample size, sample morphology, and heating rates can affect the results and interpretation of DSC data. Users should consult analysis reviews (Flynn, 1989) and literature specific to the polymer family they are investigating.

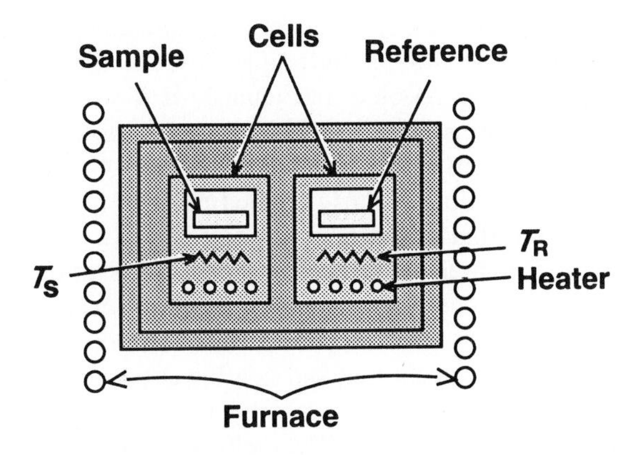

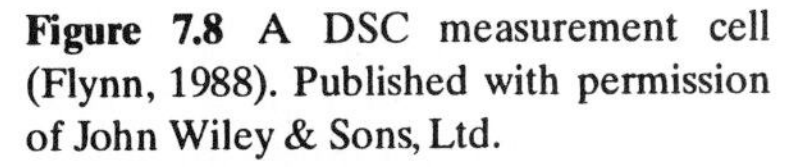
Figure 7.8 A DSC measurement cell (Flynn, 1988). Published with permission of John Wiley & Sons, Ltd.

Typical DSC Scan. Figure 7.9 shows typical heating and cooling scans for a polymer sample. For this scan, exothermic events appear as peaks and endothermic events appear as valleys. At low temperatures, there is little change in the energy flow into the sample relative to the reference, and the heat flux curve is horizontal. Most polymers contain enough amorphous material to exhibit a glass transition. At T_g, the heat capacity of the material changes and the curve drops to a lower value of heat flux. T_g is an endothermic event as heat flows into the sample to cause chain segments

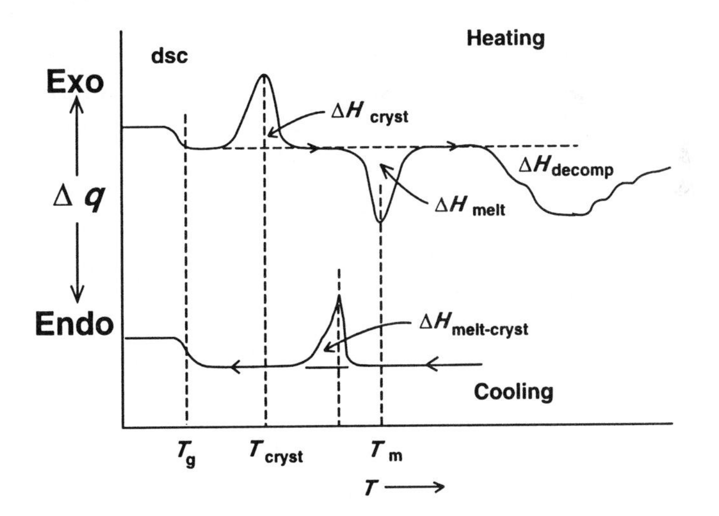

Figure 7.9 Heating and cooling curves for a partially crystallized polymer. On heating, the sample exhibits at T_g. Above T_g but below T_m, additional material can crystallize, releasing the heat of crystallization, ΔH_{cryst}. At the melting temperature, all crystallites melt and the sample absorbs energy. At higher temperature, the sample decomposes, releasing energy. On cooling, the sample crystallizes below T_m and exhibits T_g. ΔH_{cryst} = heat of crystallization, ΔH_{melt} = heat of melting, ΔH_{decomp} = heat of degradation, $\Delta H_{melt\text{-}cryst}$ = heat of crystallization from the melt. (Flynn, 1988). Published with permission of John Wiley & Sons, Ltd.

to move. There are several criteria for choosing the sample T_g, including using the point of inflection of the downward slope and the midpoint of the change.

Above T_g, the sample may be supercooled, particularly if it was quenched quickly from a high temperature or precipitated quickly. Either process can give internal strains in the material and prevent equilibrium crystallization. Samples with internal stresses may crystallize above T_g, particularly on the first DSC heating cycle. Crystallization is an exothermic event that releases heat; it appears as a peak on the diagram. The area under the curve is related to the enthalpy of crystallization for the portion of the sample that crystallized. Samples that are near equilibrium crystallinity may not show any crystallization event. Cooling curves can be used to identify thermal events, but there may be hysteresis between cooling and heating curves.

Near T_m, the sample begins to melt. Energy needs to be added to the sample to melt crystallites, and the T_m event appears as a valley. The melting range may be broad because crystallites can melt at different temperatures and because the heating rate and sample morphology can affect the melting process. Above the melting temperature, most polymers undergo degradation, particularly if oxygen is present. The degradation reactions may be endothermic or exothermic. Many of them result in sample weight loss and volatile gas production. Weight loss makes interpretation of the thermogram difficult, and degradations are usually avoided if possible. Some polymers degrade before or near their melting temperature (crystalline cellulose is one example).

Because previous thermal history affects the sample's response to temperature changes, samples are often heated, cooled, and then heated to "erase" the polymer's memory and provide better data on the transitions. This operating procedure is typical for thermoplastics, which do not undergo further polymerizations during temperature cycles. Thermosets often react further when their temperature is raised above the polymerization temperature and they are studied on the first heat.

Measurement of Specific Heat Capacity. When the temperature is changed at constant heating rates, the baseline for empty sample and reference pans shows the balance of heat losses between the two sides. When the sample and the reference are the same, the steady-state baseline should be a straight line. When the sample and reference sides are different, the steady-state baseline is displaced from the isothermal baseline by an amount equal to the heating rate times the difference in heat capacities between the two sides.

Measurement of the Glass Transition Temperature. T_g is a second-order transition and appears as a change in slope of the heating curve, which is endothermic. As with the melting temperature, heating rates will affect the measured T_g. In addition, prior thermal history of the sample also affects the results. Heat/cool cycles and different heating rates are used to identify these problems.

Measurement of the Melting Temperature. Polymer melting peaks tend to be broad relative to the melting peaks of pure compounds. The polymer crystals contain materials with different molecular weights, which melt at slightly different temperatures. Slow cooling of the sample often results in large, well-ordered crystals that melt at higher temperatures. T_m is affected by molecular weight distributions, chain branching, and tacticity as well. The area in the melting event relative to the baseline is a measure of the total enthalpy change of the sample. If the enthalpy of fusion is known, the percent crystallinity can be found by dividing the enthalpy per gram determined by DSC by the enthalpy of fusion. Conversely, if the percent crystallinity of a sample is known from other measurements, such as x-ray diffraction, the enthalpy of fusion can be determined using a DSC experiment.

Measurement of Process Kinetics. Physical changes in polymer systems are slower than for lower molecular weight organic liquids. DSC can be used to study changes that take place over minutes, such as melting, crystallization, glass transition, and chemical reactions. Curing of stepwise polymerizations often is done using differential scanning calorimetry. Typical heats of reaction are 20 kcal/mol (85 kJ/mol), which are easy to detect. Figure 7.10 shows the rate of heat released versus time for a curing reaction that is induced by oxygen. After a lag time, the curing process starts, goes through a maximum curing rate at T_m, and then slows with time. The area under the curve is the enthalpy of reaction for the entire event. The curing rate is represented by the distance between the curve and the baseline. In this case, the curing rate probably is dependent on the diffusion of oxygen through the sample. The observed rate is not the kinetic rate of the curing reaction, but includes transport dynamics as well.

Curing reactions often are limited by gelation or by the glass transition temperature of the system. In either case, the reactions between functional groups can decrease dramatically because the local viscosity is very high. At T_g of the system, the

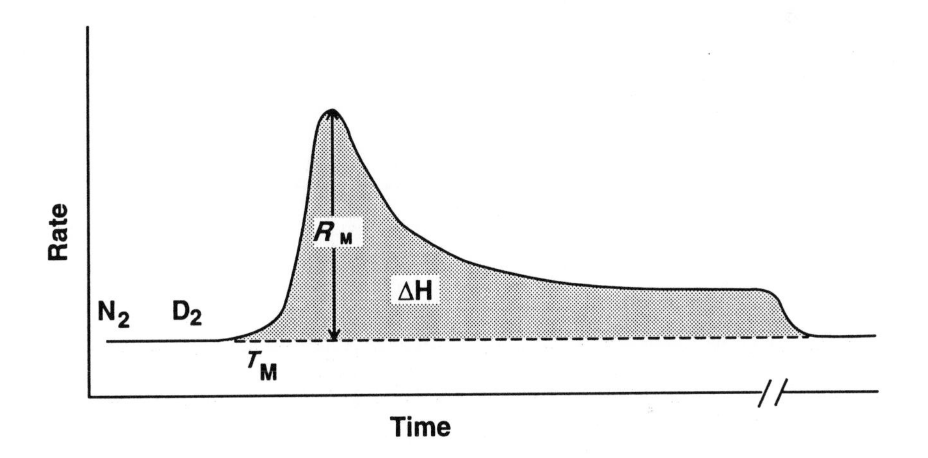

Figure 7.10 DSC of a curing reaction. R_m: maximum rate, T_M: time to rate (incipiency of cure), ΔH: total area (Flynn, 1989).

reaction appears to stop. Raising the temperature will allow the cure to continue. A major problem with DSC analysis of many curing reactions is a change in the baseline due to volatile loss.

Determination of Phase Diagrams. DSC can be used to determine phase diagrams. Figure 7.11 shows a phase diagram for polyethylene and a high melting solvent, 1,2,4,5-tetrachlorobenzene, determined using DSC. The corresponding thermograms are shown in Figure 7.12. Pure 1,2,4,5-tetrachlorobenzene melts between 141°C and 143°C. Pure polyethylene (Marlex 6009) melts between 120°C and 137°C. The broad melting range is typical of polymeric solids. The samples were prepared by

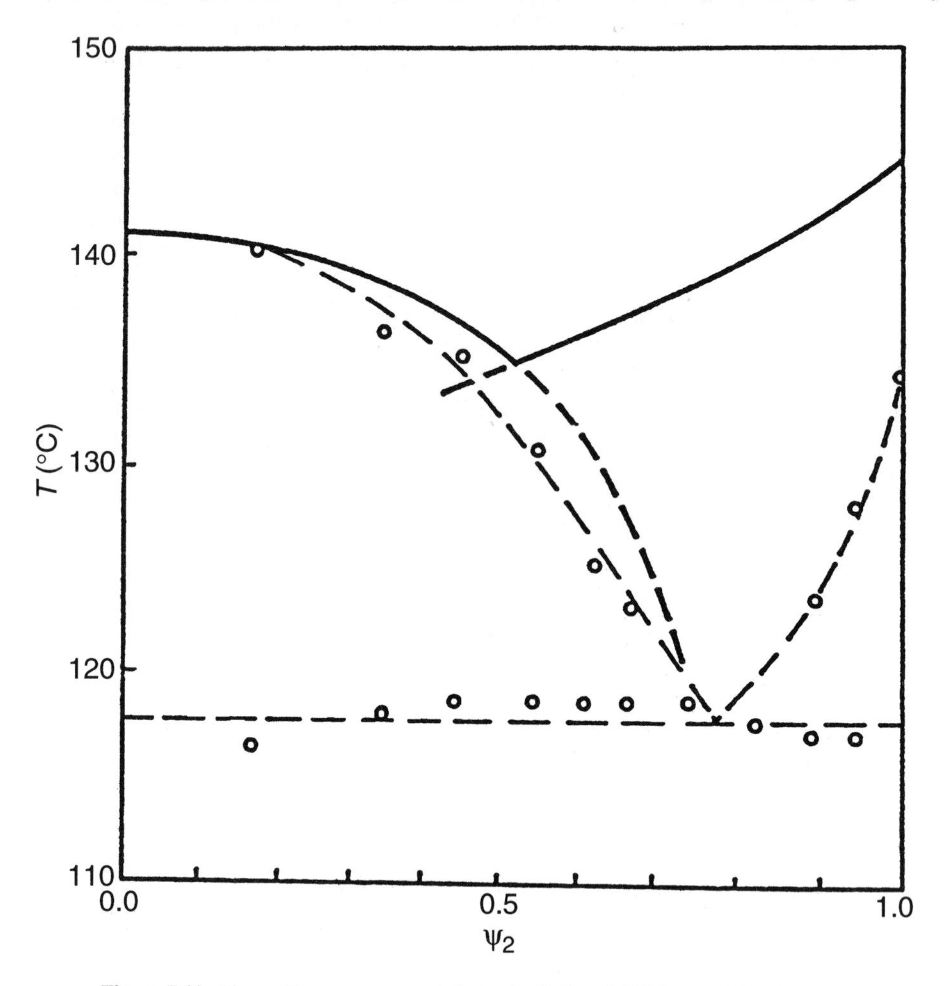

Figure 7.11 Phase diagram prepared for polyethylene/1,2,4,5-tetrachlorobenzene using a DSC. (Smith and Pennings, 1974). By permission of the publishers, Butterworth Heinemann, Ltd.©

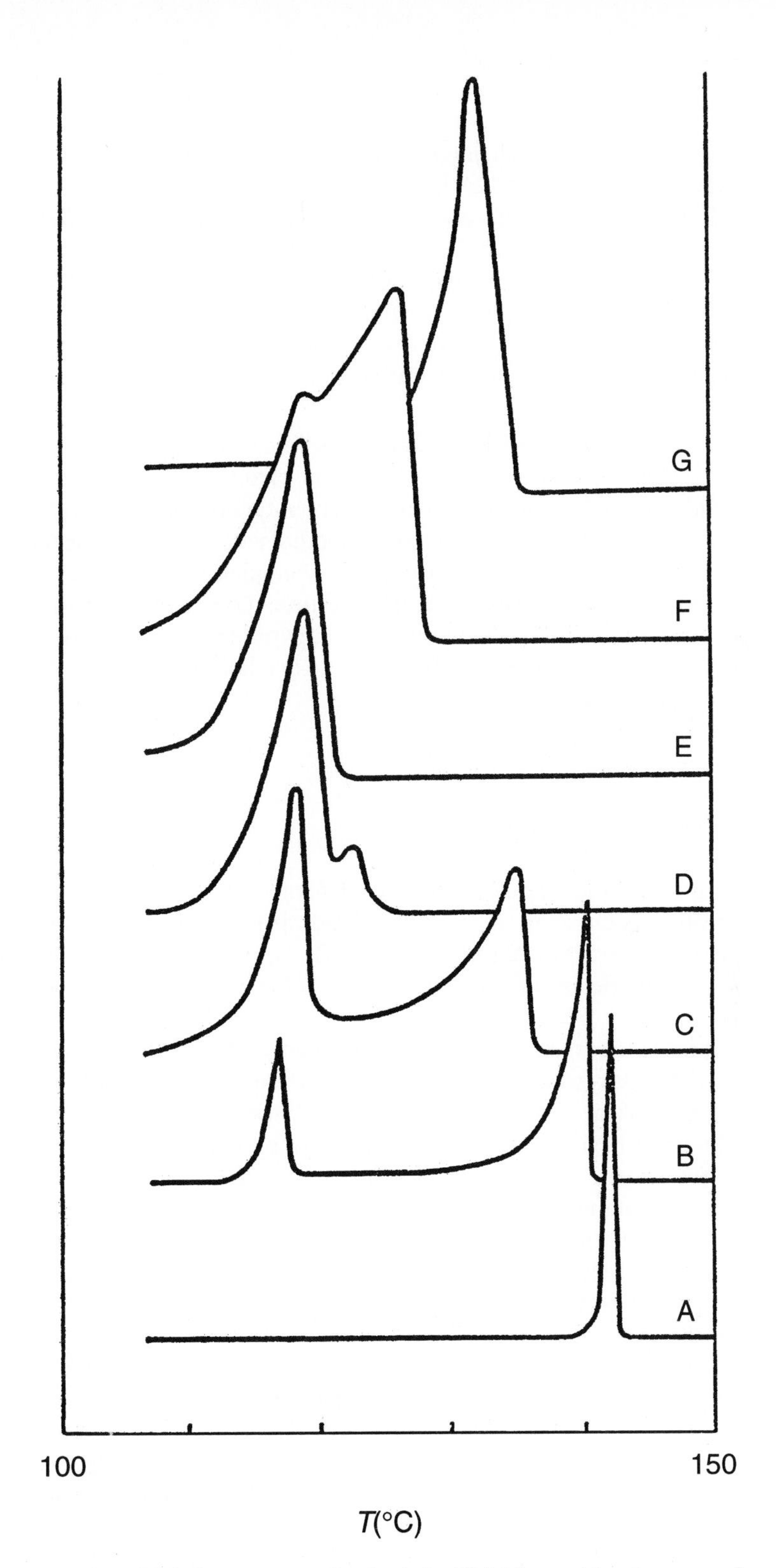

Figure 7-12 DSC thermograms of polyethylene/1,2,4,5-tetrachloro-benzene mixtures. (Smith and Pennings, 1974). By permission of the publishers, Butterworth Heinemann, Ltd.©

blending at 200°C and then quenching to 80°C. This produced small crystal phases that melted readily during the heating cycle.

The pure components and the blend with .8 polyethylene/.2 solvent (Curve E) show one melting peak each. The double peaks present in the rest of the thermograms suggest that the blend is incompatible. The compatible blend of .8 PE/.2 solvent is taken as a eutectic composition, which should have the same melting point in all thermograms. Figure 7.12 shows that it does. The peaks at higher temperatures are either pure polymer or pure solvent crystals in the polymer solution.

Determination of Blend Compatibility. Differential scanning calorimetry is an ideal tool for probing the structure of blends and copolymers. For example, a polyurethane can be made from a mixture of butanediol and a polyester. The polyester soft segments will have a T_g, and the butanediol will have a melting temperature. DSC can be used to measure both thermal events and determine the effects of chain length on them. The thermal sensitivity of these transitions will affect the mechanical and rheological properties of the material. Compatible polymer blends have glass transition temperatures intermediate between the two homopolymers. Incompatible blends have two distinct transition temperatures.

7.2.3 Thermal Mechanical Analysis

Thermal mechanical analysis (TMA) deforms a sample under a static load as its temperature is changed. At very low loads, it measures the volume change of the sample with temperature (dilatometry). The applied load can be in compression, in tension, or in flexure. Oscillating load studies are dynamic mechanical analyses (DMA).

Tension loads are used to study fibers and films. Compression loads are used to measure the glass transition temperature and softening points. Flexural loads correlate with the Vicat softening point and the elastic modulus.

Figure 7.13 shows TMA equipment. The equipment measures the linear expansion of the polymer and works with solid samples only. The volume change of a sample is an integral change, while the change in heat capacity is a differential change. T_g and the softening temperature can be determined using TMA. Figure 7.14 shows this method for a sample of poly(vinyl chloride). At T_g, the sample contracts. At the softening temperature, the tip penetrates into the sample and the instrument records a large change in dimension.

Oriented polymers often are analyzed using TMA. TMA can measure the amount of orientation, which can be used in turn to optimize the spinning or drawing process. Figure 7.15 shows fiber shrinkage for a series of poly(ethylene terephthalate) samples that experienced different wind-up speeds. The T_g of PET is about 80°C, and the melting temperature is about 240°C. As wind-up speed is increased from 2000 to 3500 m/min, fiber shrinkage increases due to noncrystalline orientation, which de-

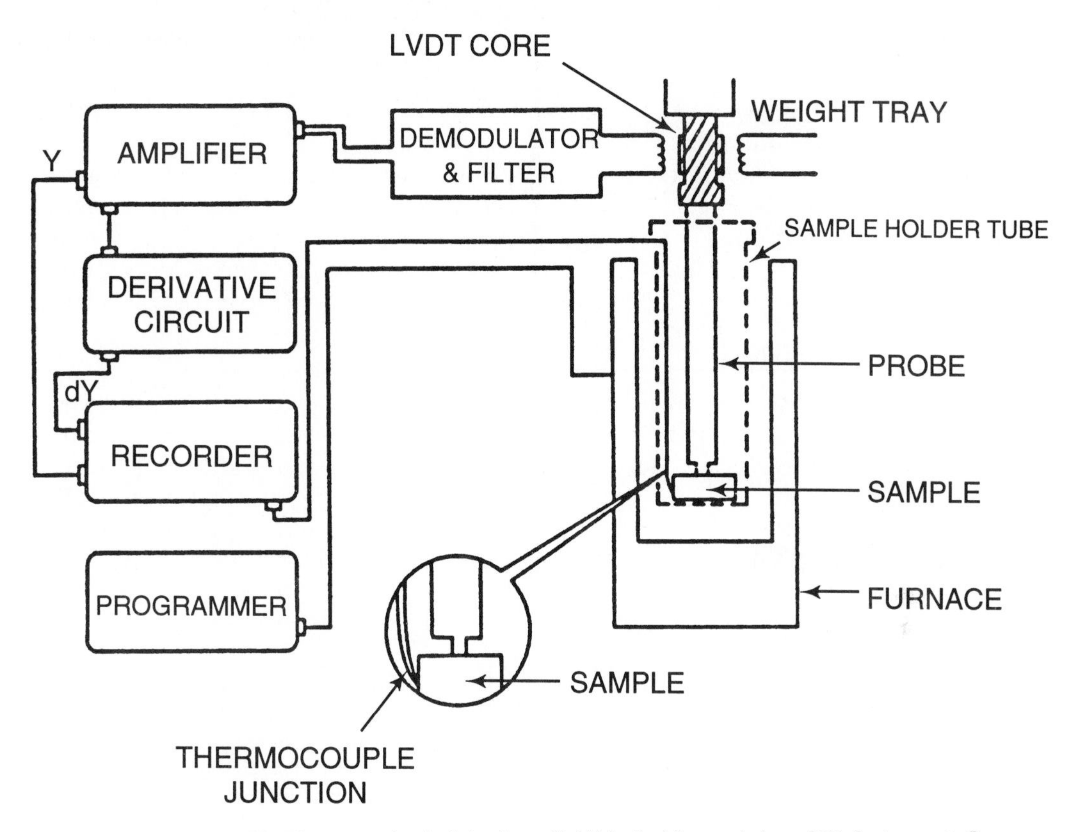

Figure 7.13 Thermomechanical Analyzer. Published with permission of TA Instruments, Inc., 109 Lukens Drive, New Castle, DE 19720.

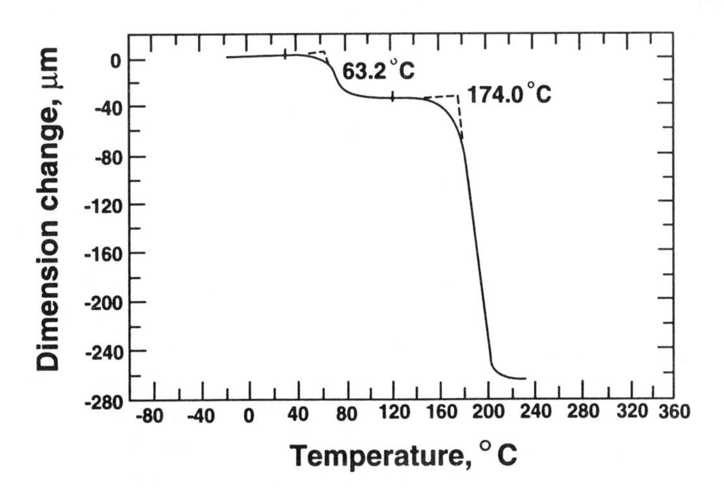

Figure 7.14 Softening point of a thermoplastic: T_g is 63.2°C; softening point temperature is 174.0°C. (De Francis).

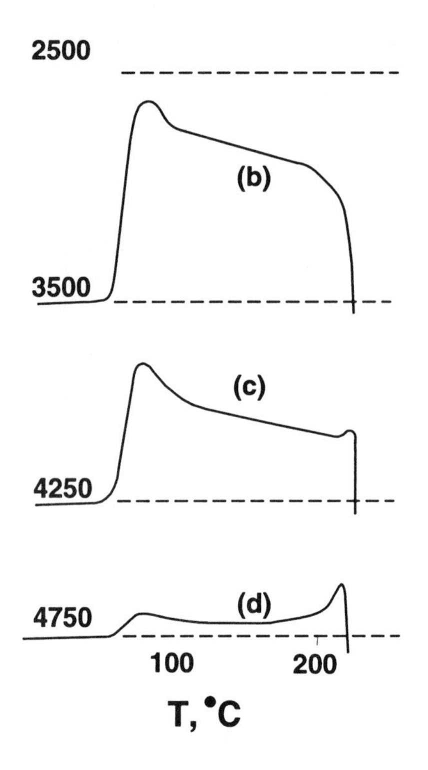

Figure 7.15 Shrinkage determination by TMA of as-spun PET yarns. Dimensional change versus temperature. Yarn wind-up speeds are listed by the curves (Heuval and Huisman, 1978).

creases crystallization during the cooling process. Above wind-up speeds of 4000 m/min, crystallization during spinning increases and fiber shrinkage decreases. At very high wind-up rates, fiber shrinkage is very low until the fiber reaches the PET melting temperature.

7.2.4 Dynamic Mechanical Analysis

Dynamic mechanical analysis (DMA) (Fig. 7.16) measures the response of the material to an oscillatory load during a temperature cycle. The measurement of the material response to the imposed load gives a direct measure of the material modulus. The measurement of the lag of the material response to the mechanical forcing function gives a measure of the damping, or loss factor, of the material. Only solid materials can be measured, but liquids can be studied as well by coating them on the surfaces of solid substrates (fibers or films) and comparing the response of the coated sample to the response of the substrate. Dielectric thermal analysis is done by using an oscillating electric field as the forcing function.

Small oscillations are used to perturb the sample, so the major variable is the

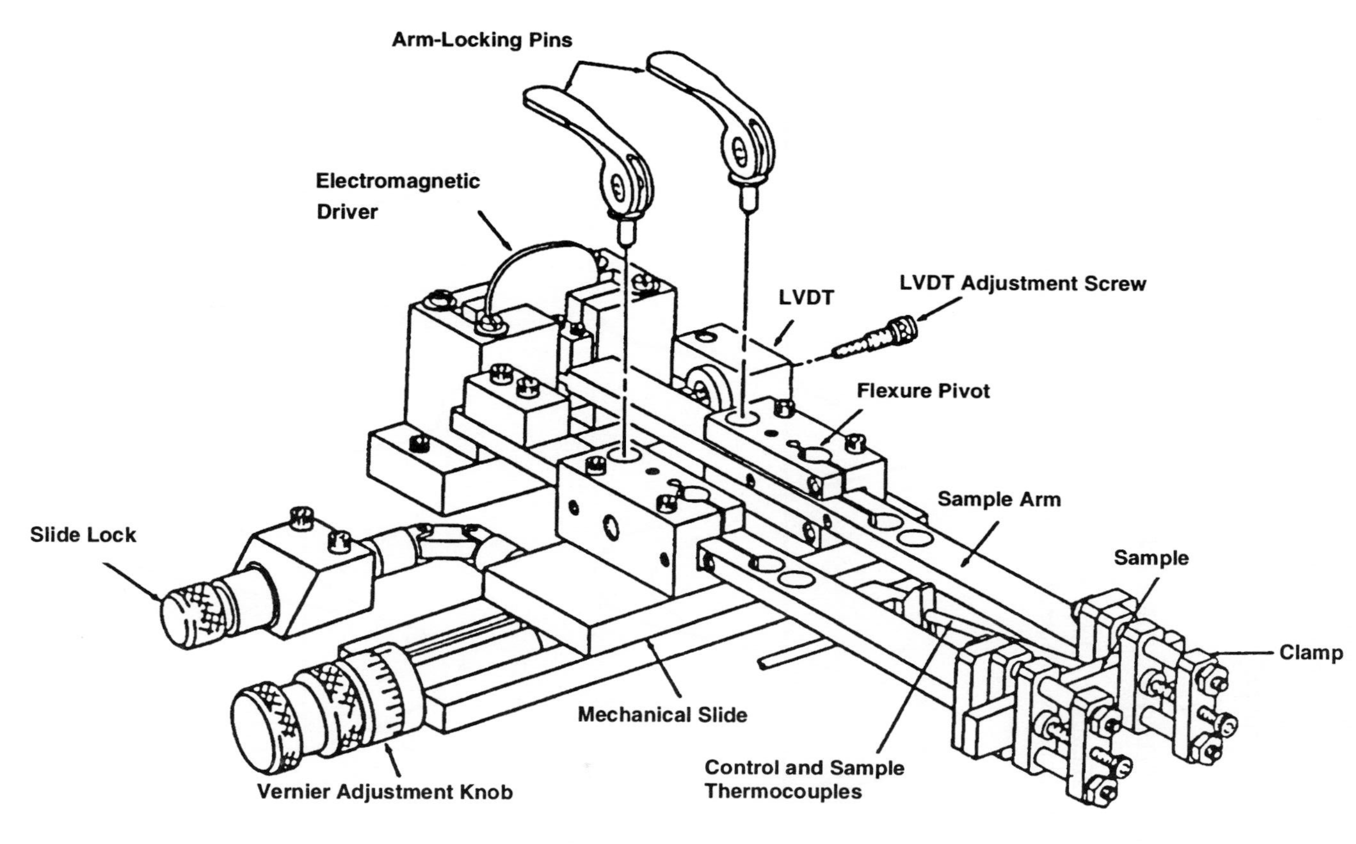

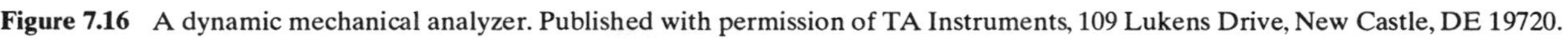

Figure 7.16 A dynamic mechanical analyzer. Published with permission of TA Instruments, 109 Lukens Drive, New Castle, DE 19720.

frequency of the oscillations (Fig. 7.17). The measurement of T_g and other thermal events depends on the frequency of the test. Usually, the peak of a thermal event decreases as forcing frequency decreases. Impact strength, tensile strength, toughness, creep rate, the coefficient of friction, and tackiness have been related to modulus (E′) and loss factor (E'') measurements.

7.3 CHAIN MICROSTRUCTURE ANALYSIS

The evaluation of chain microstructure includes determining the configuration and conformation of the units of the chains. Both chemical (Table 7.4) and physical tests

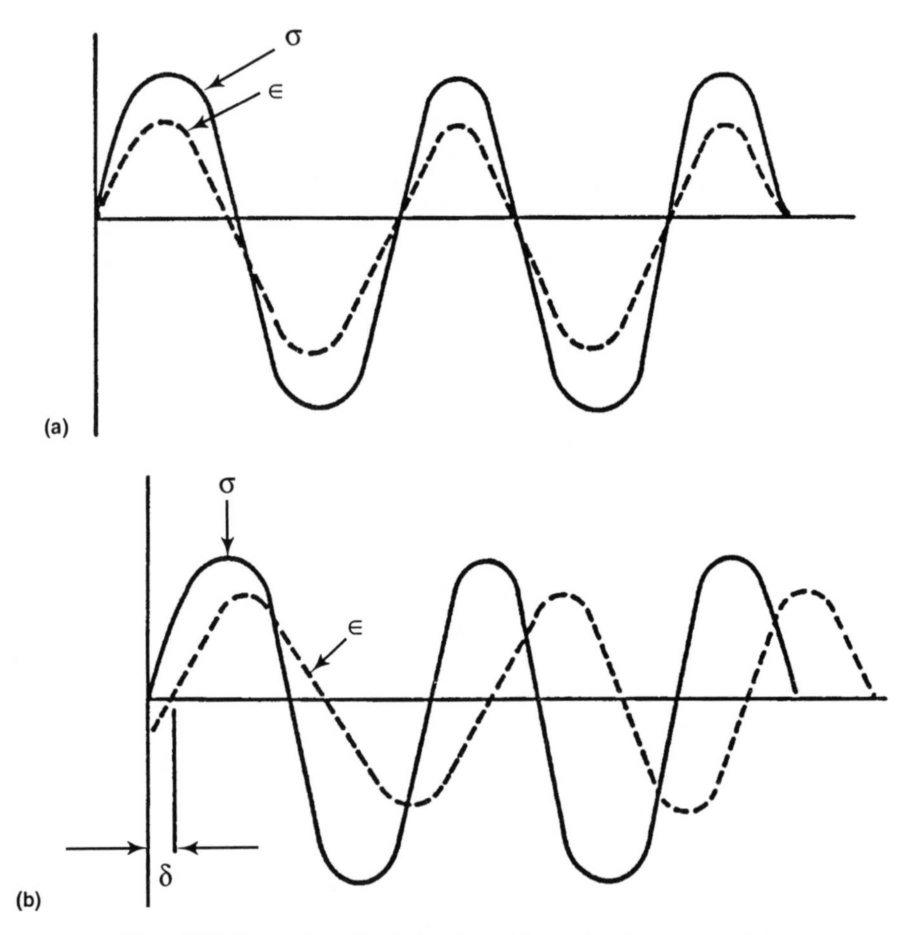

Figure 7.17 Dynamic mechanical analysis: **(a)** stress/strain response of elastic materials, $\delta = 0$. **(b)** stress/strain response of viscoelastic material, $\delta \geq 0$. Published with permission of TA Instruments, Inc., 109 Lukens Drive, New Castle, DE 19720.

TABLE 7.4 Chemical Methods of Determining Polymer Chain Microstructure

Method	Application	Reference
Elemental analysis	Gross composition, yielding the percent composition of each element; C, H, N, O, S, ...	(a)
Functional group analysis	Reaction of a specific group with a known reagent. Acids, bases, and oxidizing and reducing agents are common. Example: titration of carboxyl groups.	(b), (c)
Selective degradation	Selective scissions of particular bonds, frequently by oxidation or hydrolysis. Example: ozonalysis of polymers containing double bonds.	(d)
Cyclization reactions	Sequence analysis through formation of lactones, lactams, imides, α-tetralenes, and endone rings.	(e)
Cooperative reactions	Sequence analysis using reactions of one group with a neighboring group.	(f)

a) Critchfield and Johnson, 1961.
b) Siggia, 1963.
c) Bikales, 1971.
d) Hill, Lewis, and Simonsen, 1939.
e) Tanaka, Nishimura, and Shono, 1975.
f) Gonzales and Hammer, 1976.

(Table 7.5) are used. These methods are used for polymers and for many other materials; however, there often are special techniques needed for the analysis of polymers.

Chemical Methods. The chemical methods can help address configuration and conformation questions. Elemental analysis involves the controlled degradation of a small sample of polymer. This method helps identify unknown materials and can aid in confirming the results of new syntheses when only small amounts of polymer are available. Polymer purity also can be measured, either by demonstrating that the ratios of the elements are near their expected values or by showing that the ash contains the expected metal contaminants.

One important chemical method is determining the molecular weight of step polymerization polymers, in which specific functional groups remain at the chain ends of the molecules. The method requires dissolving the polymer in a solvent and performing specific reactions on accessible functional groups.

Selective degradation helps determine the quantities of specific chemical groups. An example is the determination of residual double bonds in cross-linked elastomers by ozonation. The last two methods of Table 7.4 determine the sequence of the units in the chains.

TABLE 7.5 Physical Methods of Determining Polymer Chain Microstructure

Method	Application	Reference
Nuclear magnetic resonance	Determination of steric configuration in homopolymers; composition of copolymers and proteins; chemical functionality; oxidation products; structural, geometric and substitutional isomerism, conformation, and microstructure.	(a–c)
Infrared and Raman spectroscopy	Molecular identification; determination of functionality; chain and sequence length; quantitative analysis; stereochemical configuration; chain conformation.	(d, e)
Ultraviolet and visible light spectroscopy	Identification and analysis; sequence length.	(f)
Mass spectroscopy	Polymer degradation mechanisms; order of block copolymers.	(g)
Electron spectroscopy (ESCA)	Microstructure of polymers, particularly surfaces	(h)
X-Ray and electron diffraction	Crystal unit cell; inter- and intramolecular spacings; chain conformation and configuration.	(i)

a) Bovey, 1972.
b) McDonald, Phillips, and Glickson, 1971.
c) Randall, 1975.
d) Haslam, Willis, and Squirrell, 1972.
e) Koenig, 1971.
f) Winston and Wichackeewa, 1973.
g) Lee and Sedgwick, 1978.
h) Clark and Feast, 1975.
i) Natta, 1960.

Physical Methods. Table 7.5 contains a list of physical methods for determining polymer chain microstructure. Nuclear magnetic resonance can be used to identify configurations of homopolymers and copolymers, as well as chemical functionality. Nuclei with spin quantum numbers of ½ or greater have magnetic moments. Protons (^1H), deuterium (^2H), fluorine (^{19}F), carbon-13 (^{13}C), nitrogen-14 (^{14}N), nitrogen-15 (^{15}N), and phosphorous-31 (^{31}P) have spins of ½ or greater. These nuclei can be incorporated into polymer chains during synthesis or by exchange reactions in some cases.

When these nuclei are in a magnetic field, they can occupy either of two energy levels, corresponding to their magnetic moments being aligned with or against the field. Energy is absorbed when the nuclei are raised from the lower energy state to the higher state, and it is emitted when the nuclei revert to the lower state. The energy field on a specific nuclei is altered by the nearby presence of other nuclei. Energy changes are used to identify local neighbors and, thereby, to deduce the molecular structure. Structured units of polymers are identified from a combination of chemical shift data and spin-spin splitting. These permit analysis of polymer stereochemistry and monomer sequencing.

NMR studies require good sample preparation and careful analysis of the spectra obtained, usually in comparison to well-chosen reference samples. An important feature of ^{13}C NMR is its ability to study molecular motion via measurement of the relaxation parameters of different carbon atoms in the chain. This example shows the application of ^{13}C NMR to the analysis of microphase separation in low molecular weight styrene-isoprene diblock copolymers (Morèse-Séguéla et al., 1980).

Styrene-isoprene diblock copolymers have hard (styrene) and soft (isoprene) microphases. Theories of phase separation for AB block copolymers consider regions of pure A, pure B, and a region of mixed A and B. If a region of mixed A and B occurs at the interphase between two pure A and B regions, its volume fraction should increase with decreasing molecular weight due to greater dispersion. This effect should be easier to observe at low molecular weights of the diblock chains. Polyisoprene and polystyrene have different glass transition temperatures: T_{g1} is 209°K for polyisoprene and T_{g2} is 375°K for high molecular weight polystyrene. Both T_g's change with chain molecular weight, as shown in Figure 7.18. However, the T_g of the polystyrene hard block phase is more sensitive to molecular weight than the T_g of the polyisoprene block.

Electron spin resonance (ESR) or electron paramagnetic resonance (EPR) works on the same principle as NMR, but using microwave frequencies to detect the spin transitions of unpaired electrons. ESR is useful for analyzing free radical reactions in polymerizations, oxidations, or degradations.

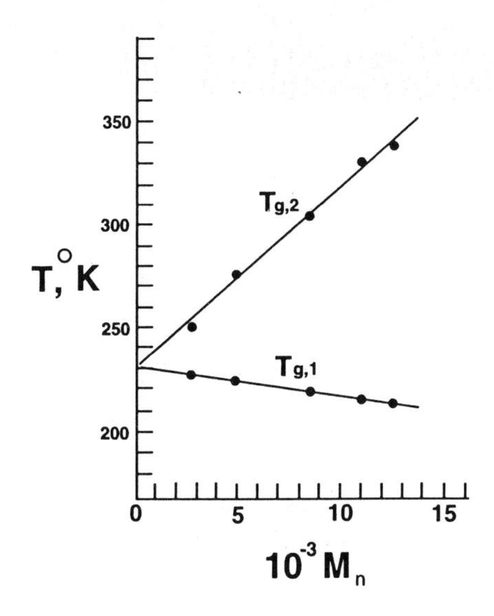

Figure 7.18 Glass transition temperatures of the soft ($T_{g,1}$) and hard ($T_{g,2}$) phases of block copolymers as a function of molecular weight. (Morèse-Séguéla et al., 1980). Published with permission of the American Chemical Society.

Infrared measurements can be used to identify changes in the rotational and vibrational spin energies of atoms in the chain. Infrared radiation is passed through the sample and the amount absorbed is measured as a function of wavelength. Absorption at specific wavelengths then is related to specific molecular motions, such as C=O stretching. Fourier transform infrared spectroscopy (FTIR) allows a wide range of wavelengths to be scanned quickly. Infrared spectra of many polymers are known (Haslam et al., 1972; Siesler and Holland-Moritz, 1980), so unknown samples can be identified by comparison. Polymers with different stereochemistries and monomer sequence distributions usually have different spectra. Polymer structure may be deduced by identifying functional group absorption bands or by comparing the polymer's absorption spectra to those of monomers or low molecular weight analogs.

FTIR can be used to distinguish between miscible and immiscible polymer blends. Immiscible blends have spectra that are the superposition of the two materials. Miscible blends show spectra of three components; the two homopolymers and the interacting homopolymer chains. Noryl is a commercial blend of styrene and poly(phenylene oxide). Subtraction of the homopolymer spectra from that of the blend shows an interaction spectra. The spectrum arises from conformational changes caused by dipole-dipole interactions between the benzene rings of the homopolymers. The spectrum is strongest for a blend of 30:70 poly(phenylene oxide)/polystyrene, the concentration giving the best mechanical properties.

Raman spectroscopy uses a light-scattering technique to probe molecular structure. The incoming light exchanges energy with the molecule, causing light of slightly changed wavelength to be scattered (the Raman effect). The differences in scattering wavelengths can be related to specific motions along the chain.

IR spectroscopy reveals unsymmetric band stretching and bending, while Raman spectroscopy shows symmetric vibrations. The scattering intensities from Raman experiments are very low, so strong incident light sources are needed (lasers). The technique can be applied to conformational studies of carbon chains by comparison to model compounds of about 20–60 carbon atoms. Raman is very useful for studying biological molecule shapes at various pH's, because the scattering of water is negligible.

Ultraviolet and visible light spectroscopy can detect simple changes in structure and often correlates well with visible changes in the sample. For example, a number of polymers form conjugated double bonds during degradation. These can lead to slight color changes, from white to slightly yellow or slightly brown. Mass spectroscopy can be used to determine the mechanism of polymer degradation. Gases from oxidations, degradations, or pyrolysis experiments can be analyzed for chemical content. Electron spectroscopy (ESCA or XPS) can be used to determine the composition of polymer surfaces. The technique works by reflecting x-rays from the surface of the sample and analyzing the spectra. X-ray diffraction is used to measure crystallinity and determine the morphology of crystalline repeating units. It also can be used to determine conformation of proteins and other biopolymers.

7.4 POLYMER MACROSTRUCTURE ANALYSIS

As discussed in Chapter 2 on polymer morphology, polymer chains can have large-scale, three-dimensional structures that are important to the performance of the material. The smallest scale is that of chain segments that associate and pack in specific ways. These atomic level structures can be probed using the techniques discussed in the previous section. There are longer range structures, with sizes between 10 nm to over 10's of microns, that contribute to polymer performance as well. These include blend phase sizes, fillers, and fibers. Optical and electron microscopy are useful in detecting such structures. Optical microscopy can be used to show structures about the size of the wavelength of visible light, near one micron. Electron microscopy has resolutions to several nanometers.

Transmission electron microscopy (TEM) is done on ultrathin samples. Hydrocarbon materials often need to be less than 100 nm thick and less than 3 mm in diameter. The material is evaluated by the electrons that are transmitted through the sample.

Slightly thicker samples will diffract electrons. Crystalline materials can be analyzed for crystal size, orientation, and polymorphism. Small areas of the sample, down to several hundred nm's, can be analyzed to determine the local morphology.

Scanning electron microscopy (SEM) is a surface technique that can be done on samples of any thickness. It analyzes electrons that are scattered from the sample's surface. The major drawback of TEM and SEM analysis is the radiation damage of the electrons on the material. The intensity of the electron beam on the area being analyzed is quite high, and local heating and cross-linking of the sample occur. Heating can reduce or destroy crystallinity. Cross-linking and degradation reactions can change the sample, as well as harm the instrument. Figure 7.19 has optical and SEM photographs of a layered composite material. Figure 7.19a and 7.19b show the compression and tension failure of the specimen. Figure 7.19c shows an SEM of the edge of the failed tension specimen. Cracks occur in the matrix material and extend long distances from the failure surface (as shown by the arrows).

Table 7.6 shows several methods for analyzing the surface of polymeric materials. Applications include coatings, fibers, polymer-catalyst interfaces, polymer-filler interfaces, surface oxidation, and surface morphology.

7.5 THERMOPHYSICAL PROPERTIES

Energy and material balances are fundamental to process design. This section describes some models and correlations for the thermophysical properties needed for these balances. They are the thermal transitions, volumetric properties and thermal properties (including enthalpy and entropy of fusion and thermal conductivity). Viscosity is needed for models of polymer processing and is covered in Chapter 8. Diffusivity is needed for mass transfer models and is discussed in Chapter 10.

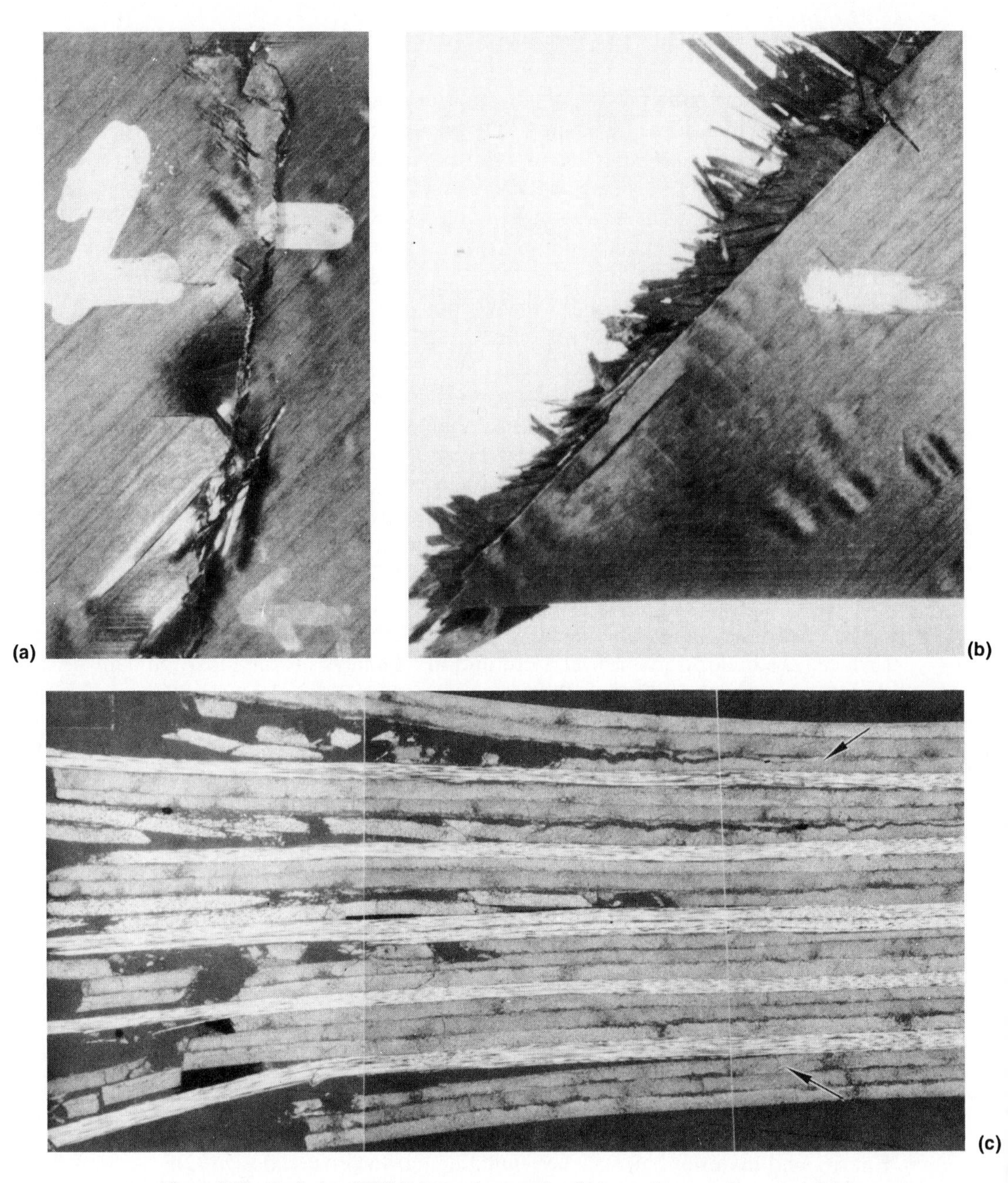

Figure 7.19 Optical and SEM photomicrographs of a layered composite material: **(a)** compression failure; **(b)** tensile failure; **(c)** SEM of a cross section of the failed tensile specimen. By permission of A. Lee, Michigan State University.

TABLE 7.6 Surface Composition Analysis

Method	Basis
Auger electron spectroscopy (AES)	An electron beam excites surface atoms and results in the emission of electrons with energies related to the emitting atom. The emitted electrons are counted and analyzed for energy to determine the atomic composition of the surface. Spatial resolution can be 25 μm.
X-ray photoelectron spectroscopy (XPS, ESCA)	X-rays irradiate the surface and cause the emission of electrons. The mean free path of the electrons is short, and only the first few atomic layers can be detected. Spatial resolution can be 250 μm.
Sputtered ion mass spectroscopy (SIMS)	Ions are aimed at the surface and are reflected from the surface. Surface atoms may be sputtered off during the collision. If a single ion source is used, the sputtered ions can be analyzed by mass spectroscopy.

7.5.1 Thermal Transitions

There are a number of thermal transition events that can occur in polymer systems. A thermal transition is defined as a change in macroscopic properties as a small temperature change is made. Two well-known transitions are the glass-liquid transition temperature, T_g, and the crystal-liquid transition temperature, T_m. The measurement of both these quantities is only moderately sensitive to the time scale of the experiment. As an example, the T_g of polystyrene as measured by the change in the elastic modulus, E, at different heating rates only varies by 6°C (Fig. 7.20).

Except for specially prepared single crystals, all polymer solids contain amorphous material and have T_g's. The T_g's and T_m's for a number of commercial polymers are listed in Table 1 of Appendix B and Table 4 of Appendix E. In addition to T_g and T_m, there are other important molecular relaxations that lead to smaller property changes. Table 7.7 lists five thermal transitions of ordinary polymers. T_β, $T_{l,l}$, and T_δ are often called secondary transitions because they are more difficult to detect and quantify. They involve changes in the local segmental degrees of freedom and do not lead to overall motion of the polymer backbone atoms.

Crystal-Liquid Transition, T_m. T_m represents the transition from crystalline material to a viscous liquid and can be analyzed using thermodynamics. The melting point depression caused by impurities can be modeled as

$$\frac{1}{T_m} - \frac{1}{T_m^\infty} = \frac{R}{\Delta H_m} \ln a_2 \qquad 7.6$$

where a_2 is the activity of the polymer when it contains the impurity, T_m is the melting point of pure crystallites with infinite molecular weight, T_m is the melting point of the

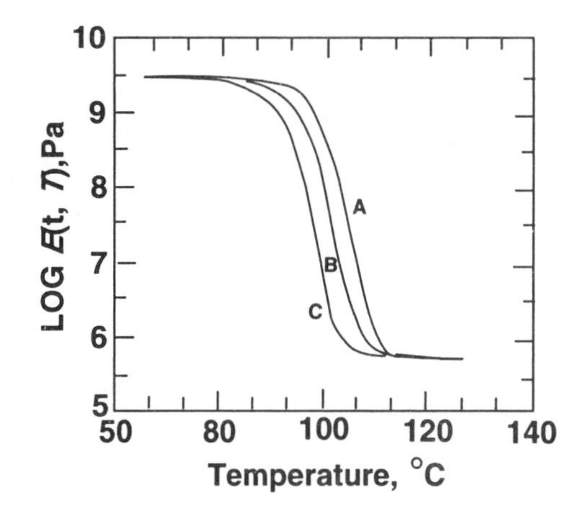

Figure 7.20 Tensile modulus of atactic polystyrene at different times. (**A**, 10s; **B**, 100s; **C**, 1000s) as a function of temperature near the glass transition (Bendler, 1989).

mixture, ΔH_m is the enthalpy of fusion, and R is the ideal gas constant. There are a number of possible impurities. Chain ends are an impurity because they will not fit easily into the crystal lattice. Solvents and plasticizers affect the melting point as well.

When the mole fraction of the impurity, x_1, is low, then there are some approximations that can be used to simplify Equation 7.6. The activity of the polymer can be approximated by its mole fraction, x_2. When the impurity level is low, then

$$-\ln x_2 = -\ln(1 - x_1) \cong x_1 \qquad 7.7$$

and

$$\frac{1}{T_m} - \frac{1}{T_m^\infty} = \frac{R}{\Delta H_m} x_1 \qquad 7.8$$

TABLE 7.7 Transition Temperatures of Polymers without Mesophases

Transition	Symbol	Definition
Melting temperature	T_m	Crystalline-rubber transition
Premelting transition	T_{ac}	Hindered rotation of polymer chains inside crystallites, ~ 0.9 T_m
Liquid-liquid transition	T_{ll}	Transition from viscoelastic to viscous state, ~ 1.2 T_g
Glass transition temperature	T_g	Segmental motion of 20 to 50 C-atom segments, ~ 0.5 T_m for symmetric polymers, ~ 0.66 T_m for unsymmetric polymers
Local relaxation mode	T_β	Segmental motion of 4 to 10 C-atom segments, ~ 0.75 T_g. Polymers are brittle below T_β.

The chain ends have a different chemical structure and can be treated as an impurity related to the molecular weight of the polymer. If both ends are identical and have the group molecular weight, M_o, the mole fraction of chain ends is 2 M_o/ Mn, and the melting point depression is

$$\frac{1}{T_m} - \frac{1}{T_m^{\infty}} = \frac{R}{\Delta H_m} \frac{2M_o}{\overline{M}_n} \tag{7.9}$$

As expected, the highest melting crystallites are those with the highest molecular weight. Equation 7.9 suggests that the melting temperature of a sample approaches that of pure crystallites as its molecular weight increases.

The depression of T_m by solvent and plasticizer can be modeled using the Flory–Huggins equation (Mandelkern and Flory, 1951). The result is

$$\frac{1}{T_m} - \frac{1}{T_m^{\infty}} = \frac{R}{\Delta H_m} \frac{V_s}{V_1} (\phi_1 - \chi \phi_1^2) \tag{7.10}$$

where V_s is the volume of a chain repeating unit, and V_1, is the solvent molar volume.

Pressure also affects the crystalline melting point and is modeled by (Karasz and Jones, 1967)

$$P_m - P_m^o = \frac{RT_m}{\Delta V_m} \frac{V_s}{V_1} (\phi_1 - \chi \phi_1^2) \tag{7.11}$$

where ΔV_m is the change in volume on melting, P_m^o is the reference pressure, and P_m is the system pressure.

Effects of Chemical Structure on T_m. Melting temperatures can be predicted directly from chemical structures. However, side group composition and chain segment length between polar function groups have nonlinear effects on T_m values. The poly(α-olefin) series allows comparison of the effects of increasing side chain length on T_m. Figure 7.21 shows T_m's versus side chain lengths. When $N = 0$, the molecule is polyethylene, which has a zig-zag molecular conformation and crystallizes as a folded chain into lamellae or spherulites. $N = 1$ is polypropylene, which forms an α-helix and has the highest melting point of the series. As the side chain length is increased, the helix widens and the fusion temperature drops. It reaches a minimum between five and seven methylene groups long. The crystal unit cell is larger, and there are lower interactions between chain segments. Side chains with longer end groups increase the melting point, probably due to steric hindrance effects for rotation and movement about the chain backbone.

Cohesive energy density can model the solution properties of polymers, but it may not predict crystal-liquid transitions well. Figure 7.22 compares a series of

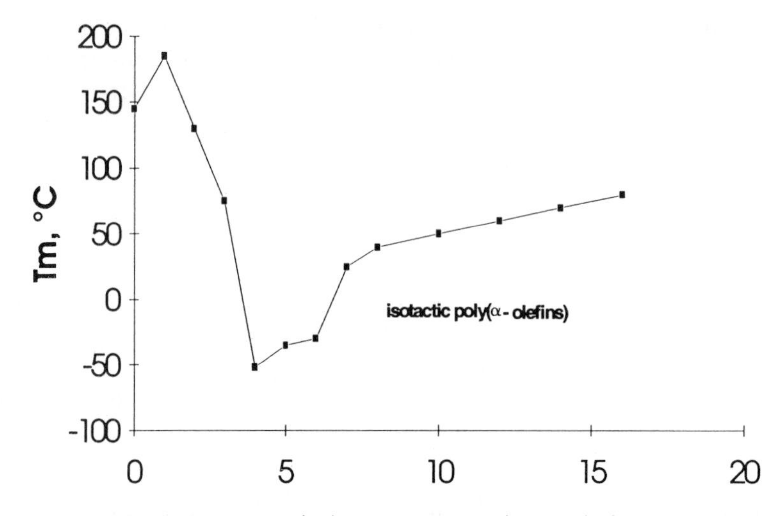

Figure 7.21 Melting point (T_m) of isotactic poly(α-olefins), $(\text{-CH}_2\text{-CHR-})_n$ as a function of the number of methylene groups, N, in the unit R = $(\text{-CH2})_N\text{H}$ (Elias, 1984).

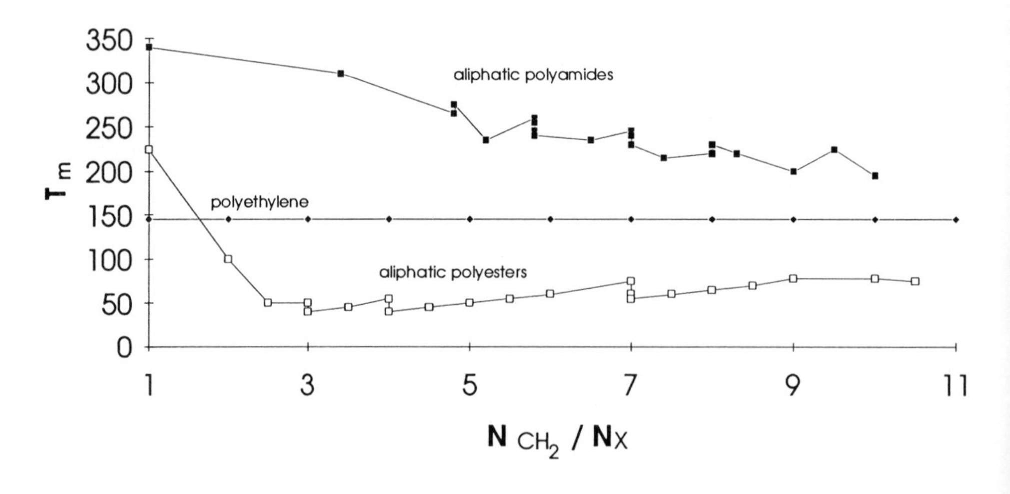

Figure 7.22 Dependence of the melting point (T_m) of aliphatic polyamides (PA) and aliphatic polyesters (PES) with X = amide or ester groups as a function of the group content. Polyethylene is given by the dashed line (Elias, 1984).

polyamide and polyester melting points to that of polyethylene. The cohesive energy density of the methylene unit is 2.85 kJ/mol, of the ester group is 12.1 kJ/mol, and of the amide group is 35.6 kJ/mol. If cohesive energy density alone determined the melting points, then the melting points of polyesters and polyamides should increase as the methylene chain lengths decrease. Polyamides show monotomically decreasing melting points as methylene chain length increases. Polyesters show melting points below that of polyethylene except for the shortest methylene chain unit, and the melting points then increase with methylene chain length. The ester groups have a lower potential energy barrier for rotation about the backbone than do methylene and amide groups. Therefore, molecule flexibility is a primary factor in establishing the melting point.

Premelting Transition, T_{ac}. Some polymers show a change in mechanical properties just below their melting temperatures. The transition has been related to mechanical loss peaks observed using DMA. T_{ac} is thought to be related to hindered rotation of chains in folded crystallites. NMR studies of polyethylene indicate that oscillation about chain axes within chain-folded crystals takes place at this temperature. For a number of polymers, $T_{ac} = 0.9\ T_m$.

Glass-Liquid Transition, T_g. The glass transition process marks the freezing (on cooling) or the unfreezing (on heating) of micro-Brownian motion of chain segments that are 20–50 carbon atoms in length. This motion includes oscillation or rotation about backbone bonds in a given chain, as well as in neighboring chains and possibly side group rotation. The measurement of T_g seems to be a time-dependent relaxation phenomena and not a thermodynamic transition in the sense that the melting point is. However, some theories of T_g describe it as being related to a thermodynamic transition that should be measured at very long times.

All the physical properties of amorphous polymers change at T_g. Figure 7.23 shows the changes in specific volume; its first derivative, α; the change in enthalpy, H; its first derivative, C_p; the change in shear modulus, G'; and its first derivative, G''. Because there is a discontinuity in these derivative properties at T_g, the glass transition temperature is often called a second-order transition (not to be confused with a "secondary" transition).

A number of glassy polymers show quantitatively similar property changes around the glass transition temperature. These include

$$\Delta C_p = 11.3\ J/°\text{K-mol} \quad (7.12a)$$

$$\Delta \alpha = 0.113 \quad (7.12b)$$

$$\Delta \kappa = (V_s/RT_g) \times 7.62 \times 10^{-3}\ \text{Pa}^{-1} \quad (7.12c)$$

where α is the thermal expansion coefficient, κ is the compressibility coefficient, and V_s is the molar volume of a repeating unit of the polymer.

T_g's can be detected by over 20 methods, including dilatometry, DSC, IR, x-ray

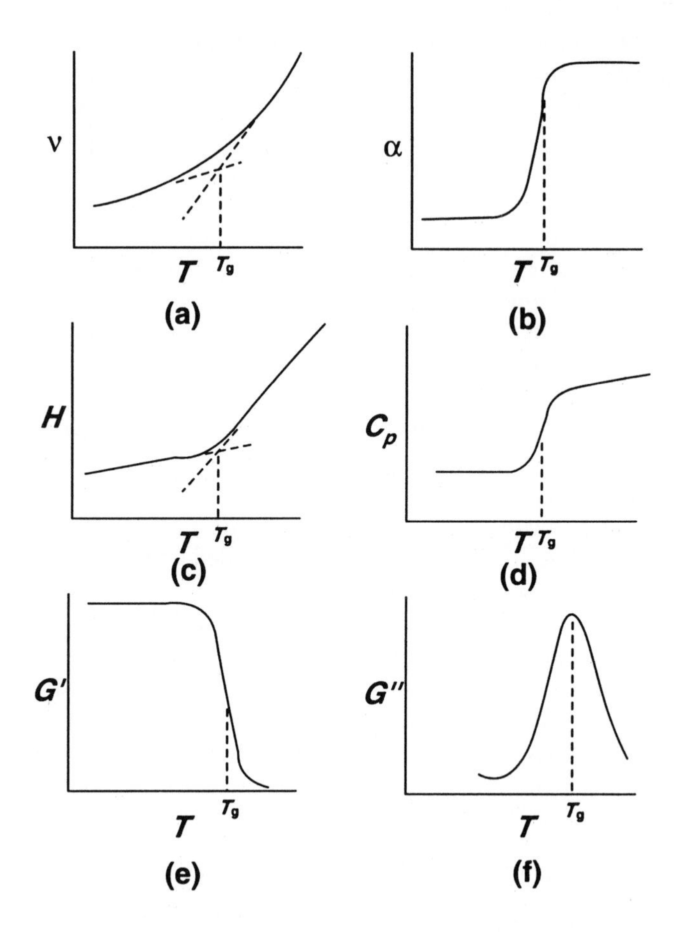

Figure 7.23 Idealized variations in volume (v), enthalpy (H) and their derivatives, the volume coefficient of expansion (α) and heat capacity (Cp) as functions of temperature. Also shown are the storage shear modulus (G′) and the loss shear modulus (G″).

diffraction, NMR, and the torsional pendulum. The methods give different values for T_g. The range of T_g's is usually small for atactic polymers and can be attributed to differences in sample preparation and impurities. Semicrystalline polymers, 30% to 60% crystalline, also usually have a small range of T_g values. The measurement of T_g is quite variable for highly crystalline polymers. Polyethylene and polytetrafluoroethylene have fractional crystallinities of greater than 80% and have wide ranges of reported T_g's (150°C for PE). The ambiguity of these results makes it difficult for the polymer engineer to depend on a specific temperature at which the amorphous regions of these polymers will solidify.

T_g is not a thermodynamic transition in the sense that the melting point is, and most measurements of T_g are time-dependent relaxation phenomena. Even though

T_g does not represent a thermodynamic phase change, the effect of molecular weight on T_g has been modeled similar to Equation 7.10.

$$\frac{1}{T_g} - \frac{1}{T_g^{\infty}} = \frac{A}{\overline{M}_n} \qquad 7.13$$

Relation between T_g and T_m. There is a relationship between T_g and T_m for polymer series, but there is not one ratio that describes all samples. The most general relation is (Boyer, 1975)

$$T_g/T_m = 2/3 \qquad 7.14$$

This ratio describes about 55% of all polymer systems and can be used to estimate transition values for unknown samples. The ratio is ½ for polymers that are symmetrical and contain small side groups (such as polyethylene, poly[oxymethylene] and poly[vinylidene fluoride]). Most vinyl and condensation polymers are adequately modeled by Equation 7.14. Some polymers have a higher ratio of T_g/T_m. This set includes the poly(α-olefins) with bulky side groups and materials such as poly(2,6-dimethylphenylene oxide), which has a ratio of 0.93.

T_g and Vitrification. A common phenomenon in the production of thermoset parts is the onset of vitrification or the glass transition temperature. Figure 7.24 is a sketch of the typical gelling time as a function of curing temperature for a cross-linking thermoset system. During a condensation polymerization, the molecular weight is continuously increasing. As suggested by Equation 7.13, the T_g of the forming polymer should be continuously increasing. There will be a T_g^{∞} for high polymer. Also, there may be a point at which vitrification (the onset of the glass transition) and gelation (cross-linking to form infinite molecular weight polymer) occur simultaneously, ${}_{gel}T_g$. The dashed curve leaving this point suggests that gelation occurs at shorter times when the cure temperature is raised. However, the gelled polymer may be at the vitrification point before all possible reactions have been completed. This will cause the polymerization to stop before it reaches high conversion. If the sample is heated to a higher temperature, the reaction will continue. Epoxy reactions are done with postcure temperatures above T_g^{∞}, so that the reaction is 100% completed. Otherwise, the part would not develop full properties until after it had been heated above T_g^{∞} while in service.

Effects of Crystallinity on T_g. The variation of T_g with crystalline fraction seems to be polymer-specific. The crystal morphology will affect the size and defects in crystalline regions, which will in turn affect the size of micro-Brownian motion that occurs at T_g. Sample temperature and processing history will obviously have an impact.

Effects of Diluents on T_g. Low molecular weight solvents or polymer can depress T_g. For example, residual styrene of 1% in polystyrene will depress the T_g by

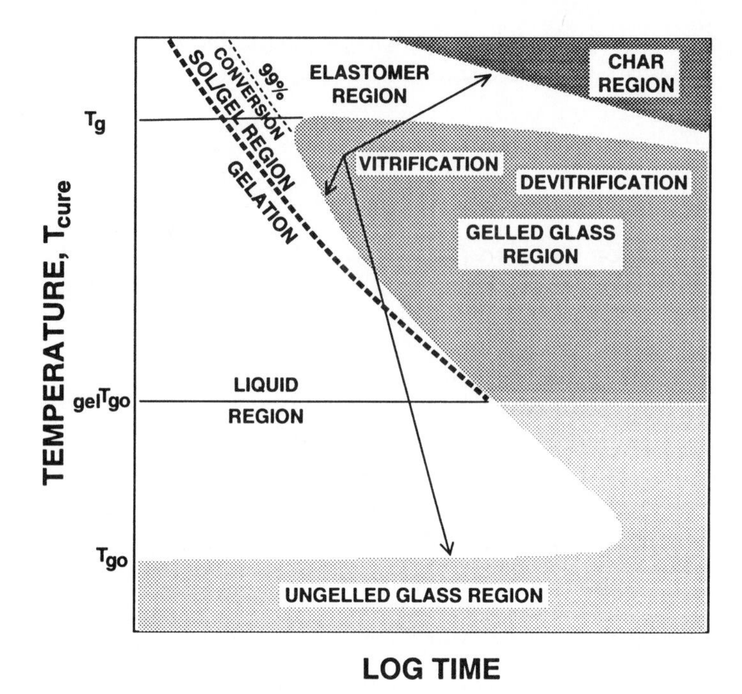

Figure 7.24 The thermosetting process as illustrated by the time-temperature transform reaction diagram (Sperling, 1986).

5°C. Plasticizers are added to deliberately lower T_g to improve end-use performance. The change in T_g with plasticizer addition is quite variable and different polymer-plasticizer systems follow quite different models (see Section 5.4). In general, diluents lower T_g in proportion to the difference between the T_g's of the polymer and the diluent. Diluents increase free volume above T_g but they tend to decrease free volume in the glassy state.

Effect of Chemical Structure on T_g. The effects of chemical structure on T_g are very difficult to generalize. It is possible to find consistent trends within families of polymers, but extrapolating trends from one family to another is misleading and often wrong. High intermolecular forces (as characterized by the cohesive energy density), intrachain steric hindrance, and bulky stiff side groups tend to increase T_g. Intrachain groups increasing flexibility, flexible side groups and symmetric substitution tend to decrease T_g.

There are many factors that can affect segmental motion. For the polystyrene family, bulky side groups tend to hinder motion and lead to higher T_g's. The poly(methyl methacrylate) family is unusual because complexes tend to form between its isotactic and syndiotactic stereoisomers. The glass transition temperature of nylons is largely determined by the hydrogen bonding of the amide groups. Annealing nylon samples has a major effect on the measured T_g, since it helps establish hydrogen bonding.

Processing and T_g. Extrusion and thermoforming processes depend on cooling the polymer so that it crystallizes or becomes a glassy solid. A number of manufacturing operations take advantage of the changes in properties around T_g and T_m to shape parts. Because mold filling is unsteady state and has variable velocities, pressures, temperatures, and cycle times, the properties of the polymer can vary from point to point in the finished part. The kinetics of crystallization and glass formation are important to the production processes as well as the performance of the product.

When a polymer sample is cooled quickly through the glass transition temperature, the polymer chains "freeze" in place at T_g. This results in higher enthalpy in the cooled part than expected if the polymer could relax during the cooling process. These residual stresses can cause deformations when the sample is reheated or can lower the effective strength of the part under load.

Other Relaxations. There are a number of important thermal transitions other than T_g and T_m. The glass transition is called the α transition by convention. Transitions at lower temperatures than T_g are labeled β, γ, etc. in succession. All transitions have not been observed in all polymers. Table 7.8 shows transitions for polystyrene. The melting point of isotactic polystyrene is 240°C (513°K), although stereoregular polymer is not made in most commercial processes. Its T_g is 373°K, so the ratio of T_g to T_m is 0.73. The β transition occurs at about 325°C. It has been related to the torsional vibrations of the phenyl groups. The γ transition occurs at 130°K and is due to motions involving about four carbon backbone atoms. The δ transition occurs at about 48°K and is associated with the oscillation of phenyl groups. Different polymers have different molecular motions at their transitions.

Dynamic mechanical loss peaks can show the T_β, T_g, and T_{ll} transitions. Figure 7.25 shows such data for polystyrene. The low temperature event, at 0.75 T_g, is the T_β transition. This can be defined as the freezing in of local chain motion of 2–10 carbon atoms in length. T_β is also called the local mode relaxation peak. The event at 1.2 T_g is the liquid state loss peak. It is the point at which the motion of entire polymer chains

TABLE 7.8 Multiple Transitions in Amorphous Polystyrene

Temperature	Transitions	Suggested Mechanisms
433°K (160°C)	T_{11}	Liquid$_1$ → liquid$_2$
373°K (100°C)	T_g	Long-range chain motions
325 ± 4°K (~ 50°C)	β	Torsional vibrations of phenyl groups
130°K	γ	Motions due to four-carbon backbone segments
38–48°K	δ	Oscillation or wagging of phenyl groups

SOURCE: R. Boyer, "Transitions and Relaxations," in *Encyclopedia of Polymer Science and Technology*, Vol. 1, Wiley-Interscience, NY, 1975.

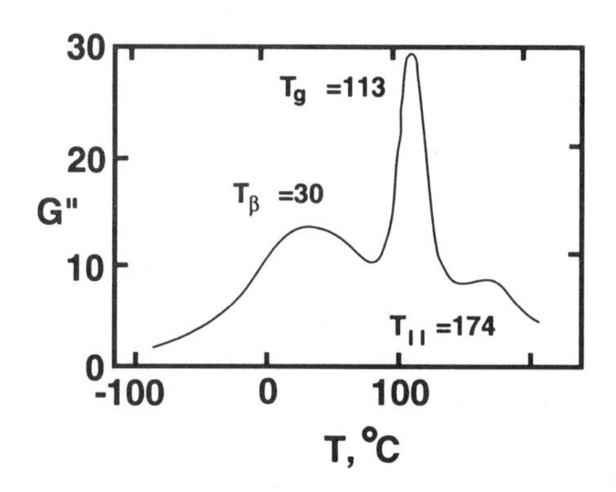

Figure 7.25 Fluid-Solid transitions for polystyrene identified using the loss shear modulus.

is frozen in. In crystalline polymers, crystallinity will freeze in the motion of entire molecules. Micro-Brownian motion and local mode relaxations still can occur.

T_β is not present in all polymers. It is missing in some symmetrically substituted polymers, such as poly(vinylidene chloride). The local mode relaxation process is sensitive to the thermal history of the polymer and can be suppressed by annealing. Many elastomers become brittle below T_β.

Since T_g is generally regarded as the onset of liquid properties, T_{ll} can be considered as a transition from one liquid state to another. T_{ll} has been observed for some polymer systems and is known to vary with molecular weight and added plasticizers. The fact that a second phenomenon exists above T_g for polymer melts indicates that T_g may not be an equilibrium state. T_{ll} is thought to represent the beginning of the motion of the entire molecule and may have applications in polymer processing.

7.5.2 Volumetric Properties

The presence of two solid polymer phases, amorphous and crystalline, means that the volumetric properties of polymers are different from simple organic liquids. Simha and Boyer (1955) developed a simple model of polymer molar volume as a function of temperature (Fig. 7.26). Molar volume usually decreases with temperature.

The cooling curves for a highly crystalline polymer and for a glassy polymer of the same composition are different. Both materials follow the liquid curve above the melting temperature. At T_m, the crystalline polymer has an abrupt change in volume, ΔV_m and contracts to a lower molar volume. The crystalline solid follows a curve with a slope different from that of the liquid as the temperature is decreased further.

The glassy sample would not experience an abrupt change in volume below T_m and would continue to follow the liquid curve to T_g. At this temperature, the slope of the volume-temperature curve changes to that of the crystalline solid. The molar thermal expansivities are similar for both crystalline and glassy solids because they are highly packed. If the liquid sample could continue to contract, it would reach $V_{0,1}$,

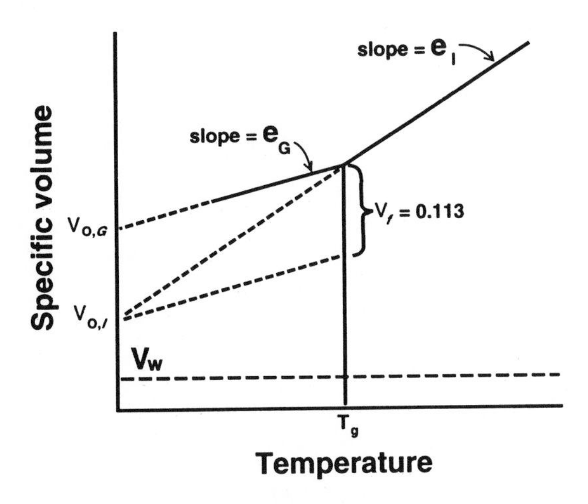

Figure 7.26 Free volume as a function of temperature as modeled by Simha and Boyer. At T_g, the sample specific thermal expansivity changes from that of the liquid to that of the glass. The extrapolations of the slopes of the liquid and glass curves give the specific volumes of the liquid and the glass at 0°K. The difference between these two volumes is the excess free volume of the glassy state. The van der Waals volume is shown for comparison.

the extrapolation of the liquid specific volume to 0°K. The extrapolation of the glassy volume to 0°K ($V_{0,g}$) accounts for the actual sample contraction. The actual sample has two kinds of volume: the effective volume of the atoms and free volume. The difference between these two extrapolations is an estimate for the free volume of the sample. The van der Waals volume of the samples also is shown in Figure 7.26. This is the space occupied by the molecule that cannot be penetrated by other atoms.

Several parameters are needed to model thermal expansion of polymers: the transition temperatures T_g and T_m, $V_l(T)$, $V_c(T)$, $V_g(T)$, and ΔV_m. The subscripts l, c, and g refer to the liquid (rubbery), crystalline, and glassy phases. These quantities are defined in Table 7.9. Definitions of thermal expansion properties are given in Table 7.10.

Correlations for some of these parameters have been developed by analyzing molar volume data for many polymers (van Krevelen, 1989). The three molar thermal expansivities are

TABLE 7.9 Volumetric Properties

Quantity	Symbol	Definition*
Density	ρ	$\frac{\text{Mass}}{\text{volume}}$
Van der Waals Volume	V_w	Space occupied by the molecule
Molar Volume	$\bar{V}$	$\frac{\text{Volume}}{\text{mole}}$
	V_r	Molar volume, rubbery amorphous polymer
	V_g	Molar volume, glassy polymer
	V_c	Molar volume, crystalline polymer
	V_f	Free volume ($V_g(0) - V_r(0)$)

*Can be defined per molecule or per structural unit.

TABLE 7.10 Thermal Expansion Properties

Quantity	Symbol	Definition	Units
Specific thermal expansivity	e	$\left(\frac{\delta V}{\delta T}\right)_p$	$\frac{m^3}{kg \cdot K}, \frac{cm^3}{g \cdot K}$
Temperature coefficient of density	g	$\left(\frac{\delta \rho}{\delta T}\right)_p$	$\frac{kg}{m^3 \cdot K}, \frac{g}{cm^3 \cdot K}$
Coefficient of thermal expansion	β	$\frac{1}{V}\left(\frac{\delta V}{\delta T}\right)_p$	K^{-1}
Linear coefficient of thermal expansion	α	$\frac{1}{L}\left(\frac{\delta L}{\delta T}\right)_p$	K^{-1}
Molar thermal expansivity	E	$\left(\frac{\delta \bar{V}}{\delta T}\right)_p$	$\frac{m^3}{kmol \cdot K}, \frac{cm^3}{mol \cdot K}$

$$E_l = 1.00 * 10^{-3} V_w \qquad 7.15a$$

$$E_g = E_c = 0.45 * 10^{-3} V_w \qquad 7.15b$$

The excess volume in the glassy state is

$$\Delta V_g(0) = \Delta V_g(T_g) = 0.55 * 10^{-3} V_w T_g \qquad 7.16$$

The melting expansion is

$$\Delta V_m = 0.55 * 10^{-3} V_w T_m \qquad 7.17$$

The molar volumes can be related to the van der Waals volume by

$$V_l(298) = 1.60\ V_w \qquad 7.18a$$

$$V_c(298) = 1.435 V_w \qquad 7.18b$$

$$V_g(298) = (1.43 + 0.55 * 10^{-4} T_g) V_w \qquad 7.18c$$

Equations. 7.15 and 7.18 can be combined to give the temperature dependence of the molar volumes.

$$V_l(T) = V_w(1.30 + 10^{-3} T) \qquad 7.19a$$

$$V_g(T) = V_w(1.30 + 0.55 * 10^{-3} T_g + 0.45 * 10^{-3} T) \qquad 7.19b$$

$$V_c(T) = V^w(1.30 + 0.45 * 10^{-3} T) \qquad 7.19c$$

Equations. 7.15 through 7.19 can be used to estimate the volumetric properties of polymers. Group contributions also can be used.

Liquid crystal polymers have complicated volumetric properties because several crystal morphologies can exist. Finkelman and Rehage (1984) show some examples.

7.5.3 Thermal Properties

The thermophysical properties discussed in this section are heat capacity, enthalpy, entropy, heat of reaction, and entropy of reaction. Thermoconductivity is included with this group. As with volumetric properties, the possibility of two solid phases requires different models from simple organic liquids. Several definitions for heat capacity are given in Table 7.11. Constant pressure heat capacities are available most often. Enthalpy and entropy balances are shown in Table 7.12. These balances are based on a reference temperature, heat capacities for the solid and liquid phases, T_g and T_m, and the enthalpy and entropy of fusion. They include the pressure correction to the enthalpy and entropy balances. Pressure corrections become important at the high pressures used in polymer processing and in some polymerization systems.

Figure 7.27 shows the molar heat capacity of a semicrystalline propylene sample. For many polymers, the heat capacity of the liquid and solid phases are linear with temperature over wide ranges. Crystalline and solid amorphous polymer phases have the same heat capacities. The heat capacity of the sample can be modeled as that of

TABLE 7.11 Heat Capacities*

Quantity	Symbol	Definition	Units
Constant volume heat capacity	C_v	$\left(\frac{\delta U}{\delta T}\right)_v$	$\frac{J}{kg \cdot K}$
Constant pressure heat capacity	C_p	$\left(\frac{\delta H}{\delta T}\right)_p$	$\frac{J}{kg \cdot K}$
Constant volume molar heat capacity	$\overline{C_v}$	$M\,C_v$	$\frac{J}{mol \cdot K}$
Constant pressure molar heat capacity	$\overline{C_p}$	$\left(\frac{\delta \overline{H}}{\delta T}\right)_p$	$\frac{J}{mol \cdot K}$

*Heat capacities can be defined for liquid, solid (glassy), and crystalline polymer.

TABLE 7.12 Enthalpy and Entropy

ENTHALPY
Crystalline polymers
$T < T_m$

$$\overline{H}_c(T, P) = \overline{H}_c(O) + \int_0^T \overline{C}_p^s \, dT + \int_0^P \left(\overline{V}_c - T\left(\frac{\partial \overline{V}_c}{\partial T}\right)_P \right) dP$$

$T > T_m$

$$\overline{H}_l(T) = \overline{H}_c(O) + \int_0^{T_m} \overline{C}_p^s \, dT + \int_{T_m}^T \overline{C}_p^l \, dT + \Delta \overline{H}_m + \int_0^P \left(\overline{V}_1 - T\left(\frac{\partial \overline{V}_1}{\partial T}\right)_P \right) dP$$

Liquid polymers

$$\overline{H}_l(T) = \overline{H}_c(O) + \int_0^{T_g} \overline{C}_p^s \, dT + \int_{T_g}^T \overline{C}_p^l \, dT + \int_0^P \left(\overline{V}_1 - T\left(\frac{\partial \overline{V}_1}{\partial T}\right)_P \right) dP$$

ENTROPY

Crystalline polymers
$T < T_m$

$$\overline{S}_c(T) = \overline{S}_c(O) + \int_0^T \frac{\overline{C}_p^s}{T} \, dT - \int_0^P \left(\frac{\partial \overline{V}_c}{\partial T}\right)_P dP$$

$T > T_m$

$$\overline{S}_l(T) = \overline{S}_c(O) + \int_0^{T_m} \frac{\overline{C}_p^s}{T} \, dT + \int_{T_m}^T \frac{\overline{C}_p^l}{T} \, dT + \Delta \overline{S}_m - \int_0^P \left(\frac{\partial \overline{V}_1}{\partial T}\right)_p dP$$

Liquid polymers

$$\overline{S}_l(T) = \overline{S}_l(O) + \int_0^{T_g} \overline{C}_p^s \, dT + \int_{T_g}^T \overline{C}_p^l \, dT - \int_0^P \left(\frac{\partial \overline{V}_1}{\partial T}\right)_p dP$$

the liquid above $T_m(C_p^l)$, that of the solid below $T_g(C_p^s)$, and a combination of the liquid and solid for $T_g < T < T_m$ based on the percent crystallinity.

Heat capacities are nearly linear with temperature. Van Krevelen (1989) has shown that the values for a number of common polymers correlate as

$$\frac{1}{C_p^s(298)} \frac{dC_p^s}{dT} = 3 \pm 0.15 * 10^{-3} \,{}^\circ K^{-1} \qquad 7.20a$$

$$\frac{1}{C_p^l(298)} \frac{dC_p^l}{dT} = 1.2 \pm .36 * 10^{-3} \,{}^\circ K^{-1} \qquad 7.20b$$

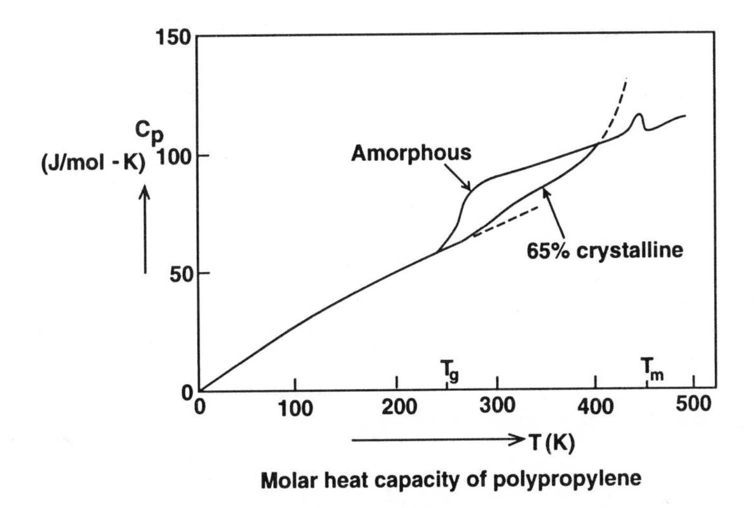

Figure 7.27 Molar heat capacity of polypropylene. 100% amorphous material shows an abrupt change in heat capacity at T_g. 100% crystalline material would show a change in heat capacity at the crystallization temperature (Figure 7.12). This typical sample exhibits a response in between these limits (van Krevelen, 1989).

These models can be used to predict heat capacities at different temperatures.

$$C_p^s(T) = C_p^s(298)\,(0.106 + 3 * 10^{-3}T) \quad 7.21a$$

$$C_p^l(T) = C_p^l(298)(0.64 + 1.2 * 10^{-3}T) \quad 7.21b$$

Enthalpy and Entropy of Fusion. The enthalpy of fusion is needed for energy balances in polymer processes. While few entropy balances are done for polymer processes, the entropy of fusion is useful in calculating the free energy of the polymer. The enthalpy of fusion can be difficult to measure since the percent crystallinity of the sample must be known. The entropy of fusion can be calculated from the enthalpy of fusion and the melting temperature by applying Equation 3.1 to the phase change.

$$\Delta S_m = \frac{\Delta H_m}{T_m} \quad 7.22$$

Figure 7.28 shows the enthalpy of crystalline and amorphous polypropylene. The difference between the liquid and amorphous curves at T_m is the enthalpy of fusion. Below the glass transition temperature, the curves are nearly parallel. The equations shown in Table 7.12 can be used to calculate enthalpy as a function of temperature.

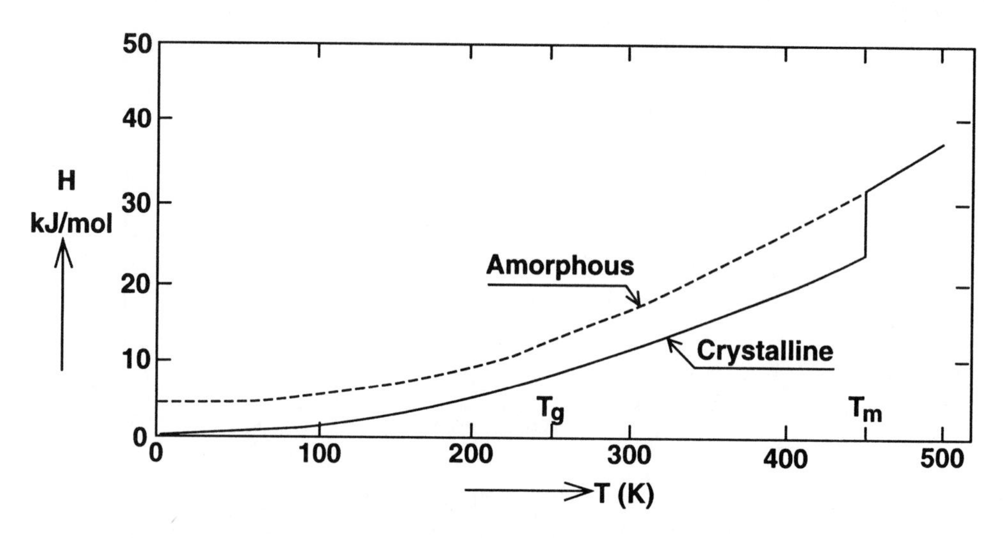

Figure 7.28 Enthalpy of polypropylene (van Krevelen, 1989).

Thermal Conductivity. The thermal conductivity of polymers is needed for calculating heat transfer rates. The Debye theory of thermal conductivity has been applied to polymer,

$$k \propto C_v \rho u_s \tag{7.23}$$

where C_v is the constant volume heat capacity, ρ is the density, and u_s is the speed of sound in the medium. The heat capacity has a discontinuity upward and a smaller slope above T_g. The density (the inverse of the molar volume) has a change in slope above T_g. The speed of sound is reduced above T_g because the chains are not as tightly packed. The overall effect is a maximum value of thermal conductivity at T_g.

Many amorphous polymers follow this general trend. Figure 7.29 shows a plot of the thermal conductivity ratio, $k(T)/k(T_g)$, versus the temperature ratio, T/T_g. This empirical plot can be used to estimate the thermal conductivity at any temperature when T_g and the value of k at another temperature are known.

Crystalline polymers have higher thermal conductivities because of their increase in molecular order. Several correlations and empirical relations relate the thermal conductivities of amorphous and crystalline polymers. These require knowledge of the density ratio between the crystalline and amorphous materials. Eiermann (1965) developed the relation

$$\frac{k_c}{k_a} - 1 = 5.8\left(\frac{\rho_c}{\rho_a} - 1\right) \tag{7.24}$$

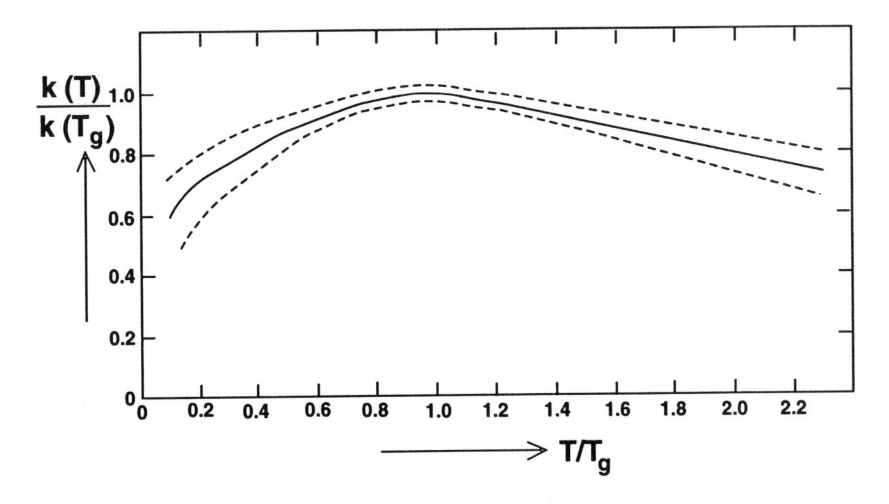

Figure 7.29 Generalized correlation for the thermal conductivity of amorphous polymers based on the thermal conductivity at T_g. Solid line shows the data average and the dashed line-represent the range of the data (van Krevelen, 1989).

The correlations for molar volume can be used to estimate the crystalline and amorphous polymer densities when they are not available.

Free Energy of Formation. Polymerizations are reversible, and free energies of formation can be used to calculate the reaction equilibrium. Reaction equilibria calculations often are applied to stepwise polymerizations, since high conversions are required to achieve high molecular weight materials. These calculations can be used to calculate the maximum reaction temperature at which polymer can be formed. The polymerization process can be analyzed using chemical equilibrium. For a reaction,

$$aA + bB \rightleftarrows cC + dD \tag{7.25}$$

the equilibrium constant describes the ratio,

$$K_{eq} = \frac{C^c D^d}{A^a B^b} \tag{7.26}$$

where the capital letters of Equations 7.25 and 7.26 denote concentrations. Activities also can be used to define K_{eq}. The equilibrium constant is related to the free energy of the process.

$$\Delta G_{rxn} = -RT \ln K_{eq} \tag{7.27}$$

When $K_{eq} > 1$, the right-hand side of Equation 7.27 is negative, ΔG_{rxn} is negative, and the reaction described by Equation 7.25 is shifted to the right. When $K_{eq} < 1$, the reaction is shifted to the left. The effect of temperature on the reaction rate constant is

$$\frac{d \ln K_{eq}}{dT} = \frac{\Delta H_{rxn}}{RT^2} \qquad 7.28$$

For many polymerizations, ΔH_{rxn} is negative, and the equilibrium constant decreases as temperature increases. When $K_{eq} = 1$, the reaction is equally likely in either direction. The temperature at which this occurs is the ceiling temperature, T_c. This temperature represents the upper bound for the polymer, since depolymerization is favored thermodynamically at higher temperatures.

Although the reaction equations for chain reactions usually are written irreversibly, monomer can be formed from polymer at high temperatures. Ring-opening polymerizations are reversible and the reversion of polymer to monomer at high temperatures can be a major processing problem. One commercial polyurethane takes advantage of depolymerization at high temperatures to lower melt viscosity during forming operations. High polymer reforms as the part cools.

Equilibrium constant can be calculated using free energies of formation for the individual reactants. Group contribution methods can be used to estimate free energies of formation. The free energy of reaction is

$$\Delta G_{rxn} = c\,\Delta G^o_{fC} + d\,G^o_{fD} - a\,\Delta G^o_{fA} - b\,G^o_{fB} \qquad 7.29$$

where ΔG^o_{fC} is the standard free energy of C.

NOMENCLATURE

a_i	activity of component i
A_d	frequency factor
C_i	heat capacity of type i for phase j defined in Table 7.11
ΔC_p	change in the heat capacity at T_g
e	specific thermal expansivity
E_i	molar thermal expansivity for phase i
E_d	activation energy
g	temperature coefficient of density
G_{rxn}	free energy of reaction
ΔH_m	enthalpy of melting
k_i	thermal conductivity of phase i
K_{eg}	reaction equilibrium constant
k_d	degradation rate constant

M_o	group molecular weight
M_n	number average molecular weight
p	degree of conversion
P_m	melting pressure
R	ideal gas constant
ΔS_m	entropy of melting
T	temperature, Kelvin
T_i	thermal events defined in Table 7.7
T_g	glass transition temperature
T_g^∞	glass transition temperature for an infinite chain
T_m	melting temperature
T_m^∞	melting temperature for an infinite chain
u_s	speed of sound in the polymer
x_i	variable for the rate analysis of degradation
V_i	molar volume of type i defined in Table 7.9
V_1	solvent molar volume
V_s	chain segment repeating unit molar volume
Greek Symbols	
α	linear coefficient of thermal expansion
$\Delta\alpha$	change in the thermal expansion coefficient at T_g
β	coefficient of thermal expansion
ϕ_i	volume fraction of i
ρ	density
χ	solvent interaction parameter ΔV_m volume change on melting
κ	compressibility coefficient
$\Delta\kappa$	change in the compressibility coefficient at T_g

Problems

7.1 Using the definitions of Table 7.10, show the following:

(a) $e = -V_g^2$

(b) $g = -e\rho^2$

(c) $e = \beta V = \beta/\rho$

(d) $g = -\beta\rho = \beta/V$

(e) $\beta = 3a$

(f) $E = Me = \beta V = \beta M/\rho$

7.2 For the polymers below, estimate the heat capacities at 298°C, the thermal transitions, and the thermal conductivities using the correlations given in Chapter 7. Compare your estimates to literature values.

(a) Polyethylene (d) *cis*-Polyisoprene
(b) Polystyrene (e) Polytetrafluorethylene
(c) Polybutadiene (f) Nylon 6

7.3 Estimate the molar volumes, for the polymers below. Compare your estimates to literature values.

(a) Polyethylene (d) *cis*-Polyisoprene
(b) Polystyrene (e) Polytetrafluoroethylene
(c) Polybutadiene (f) Nylon 6

7.4 Determine the Arrhenius constant for the degradation of polyethylene in Figure 7.6. What assumptions need to be made to do this calculation?

7.5 The melting temperature for a polymer of 100,000 molecular weight is 10°C less than that of an infinite chain. Determine what increase in T_m would be expected if the molecular weight could be increase to 250,000.

References

J.T. BENDLER "Transitions and Relaxations", in *Encyclopedia of Polymer Science and Technology*, Vol. 17, Wiley-Interscience, NY, 1989.

V.G. BEREZKIN, V.R. ALISHOYEV, and I.B. NEMIROUSKAYA, *Gas Chromatography of Polymers*, Elsevier, Amsterdam, 1977.

N. BIKALES, "Characterization of Polymers" in *Encyclopedia of Polymer Science and Technology*, Wiley-Interscience, NY, 1971.

F. BOVEY, *High Resolution NMR of Macromolecules*, Academic Press, NY, 1972.

R. BOYER, "Transitions and Relaxations," in *Encyclopedia of Polymer Science and Technology*, Vol. 17, Wiley-Interscience, NY, 1975.

J. CHIN *Appl. Polym. Symp 2*, 25 (1966).

D.T. CLARK, and W.J. FEAST, "Application of Electron Spectroscopy for Chemical Applications (ESCA) to Studies of Structure and Bonding in Polymeric Systems," *Journal of Macromolecular Science-Chemistry C12*, 191 (1975).

E.A. COLLINS, J. BARES, and F.W. BILLMEYER, *Experiments in Polymer Science*, Wiley-Interscience, NY, 1973.

F.E. CRITCHFIELD, and D.P. JOHNSON, "Chemical Analysis of Polymers," *Anal. Chem. 33*, 1834 (1961).

M. DAY, J. COONEY, and D. WILES, *SPE Proceedings of 46th Conf. ANTEC88, Atlanta, GA*, Apr. 18–21, 1988, 933–936, Society of Plastics Engineers, Brookfield, CT.

J.H. DEFRANCIS, *Inst. Systems*, E.I. duPont de Nemours & Co., Wilmington, DE.

K. EIERMANN, *Kunststoffe 55*, 335 (1965).

H.G. ELIAS, *Macromolecules*, Plenum Press, NY, 1984.

P.G. FAIR, *Inst. Systems*, E.I. duPont de Nemours & Co., Wilmington, DE.

H. FINKELMANN and G. REHAGE, "Liquid Crystal Polymers" in *Advances in Polymer Science*, Vol. 60 (M. Gordon and N.A. Plate, Eds.), NY, 1984.

J.H. FLYNN, "Thermal Analysis" in *Encyclopedia of Polymer Science and Technology*, supplemental volume (Mark, Bikales, Overberger, and Menges, Eds.), Wiley-Interscience, NY, 1989.

J.H. FLYNN, *J. Therm. Anal. 34*, 367 (1988).

P.J. FREED, and S.A. LIEBMAN, "Basic Analytical Pyrolysis Instrumentation," in *Pyrolysis and GC in Polymer Analysis*, (S.A. Liebman and E.J. Levy, Eds.), Chromatographic Science Series, 29, Marcel Dekker, NY, 1985.

J.J. GONZALES, and P.C. HAMMER, "Sequential Analysis by Cooperative Reactions on Copolymers. I. A Theoretical Treatment," *J. Polym. Sci., Polymer Letters Ed. 14*, 645 (1976).

S.M. Gumbrell, L. Mullins, and R.S. Riulin, *Trans. Faradoy Soc.*, *49*, 1495 (1953).

J. HASLAM, H.A. WILLIS, and M. SQUIRRELL, *Identification and Analysis of Plastics*, 2nd ed., Ileffe, London, 1972.

M. HEUVAL, and R. HUISMAN, "Effect of Winding Speed on the Physical Structure of As-Spun Poly(ethylene Terephthalate) Fibers, Including Orientation-Induced Crystallization," *J. Appl. Polym. Sci. 22*, 2229–2243 (1978).

R. HILL, J.R. LEWIS, and J. SIMONSEN, "Butadiene Co-Polymers: Elucidation of Structure by Ozonolysis," *Trans. Soc. 35*, 1073 (1939).

F.E. KARASZ, and L.D. JONES, *J. Phys. Chem. 71*, 2234 (1967).

KOENIG, J.L., *Appl. Spectroscopy Rev. 4*, 233 (1971).

A.K. LEE, and R.D. SEDGWICK, "Application of Mass Spectrometry to Copolymers and Ethylenes and Propylene Oxides. I. Structural Analysis," *J. Polymer Sci. Polym. Chem. Ed. 16*, 685 (1978).

R.L. LEVY, D.L. FANTER, and C.J. WOLF, "Temperature Rise Time and Time Pyrolysis Temperature in Pulse Mode Pyrolysis Gas Chromatography," *Analy. Chem. 44*, 38C (1972).

L. MANDELKERN, and P.J. FLORY, *J. Am. Chem. Soc. 73*, 3206 (1951).

C.C. McDONALD, W.D. PHILLIPS, and J.D. GLICKSON, "Nuclear Magnetic Resonance Study of the Mechanism of Reversible Denaturation of Lysozyme," *J. Am. Chem. Soc.*, *93*, 235 (1971).

A.D. McLAREN, G.G. LI, R. ROGER, and H. MARK, *J. Polymer Sci. 7*, 463 (1951).

B. MORESE-SÉGUÉLA, M. ST. JACQUES, J.M. RENAUD, and J. PRUD'HOMME, "Microphase Separation in Low Molecular Weight Styrene-Isoprene Diblock Copolymers Studied by DSC and ^{13}C NMR", *Macromolecules 13* (1), 100C (1980).

NATTA, G., *Makromoleculare Chemie 35*, 94 (1960).

J.C. RANDALL, "The Distribution of Stereochemical Configurations in Polystyrene as Observed with ^{13}C NMR", *J. Polymer Sci., Polym. Phys. Ed. 13*, 889 (1975).

K.J. SAUNDERS, *The Identification of Plastics and Rubbers*, Chapman and Hall, London, 1966.

A.E. SHERR, *Soc. Plast. Eng. 21* (1), 1965.

H.W. SIESLER and K. HOLLAND-MORITZ, *Infrared and Raman Spectroscopy of Polymers*, Marcel Dekker, NY, 1980.

S. SIGGIA, *Quantitative Organic Analysis via Functional Groups*, 3rd ed., Wiley, NY, 1963.

R. SIMHA and R.F. BOYER, *J. Chem. Phys.*, *37*, 1003 (1962).

P. SMITH, and A.J. PENNINGS, "Enterctic Crystallization of Psuedobinary Systems of Polyethylene and High Meltings Diluents," *Polymer 15*, 413–419 (1974).

L.H. SPERLING, *Introduction to Physical Polymer Science*, Wiley-Interscience, NY, 1986.

M. TANAKA, F. NISHIMURA, and T. SHONO, *Anal. Chim. Acta 74*, 119 (1975).

D.W. VAN KREVELEN, *Properties of Polymers*, 3rd ed., Elseview NY, 1989.

R.A. WEISS, in *Proc. 17th NATAS Conf. 2*, Lake Buena Vista, FL, Oct. 9–12, 1988 (C.M. Earnest, ed.) NATAS, Scotch Plains, NJ.

A. WINSTON, and P. WICHACKEEWA, "Sequence Distribution in 1-Chloro-1,3-butadiene Styrene Copolymers," *Macromolecules 6,* 200 (1973).

C.W. WINDING and G.D. HIATT, *Polymeric Materials*, McGraw-Hill, NY, 1961.

8

Flow and Mechanical Properties of Polymers

This chapter presents some of the flow and mechanical properties of polymers. A variety of coefficients and definitions are used to describe the deformations of fluids and solids. This introduction reviews several of these parameters.

Fluid Shear. When a fluid element is sheared due to a velocity gradient (Fig. 8.1), the shear stress is related to the velocity gradient by

$$\tau_{yx} = -\mu \frac{dU_x}{dy} \qquad 8.1$$

where τ_{yx} is the shear stress, μ is the fluid viscosity, and $\frac{dU_x}{dy}$ is the gradient of the x component of velocity in the y direction. An equivalent description is

$$\tau_{yx} = -\mu \, \dot{\gamma}_{xy} \qquad 8.2$$

where $\dot{\gamma}_{xy}$ describes the shear rate. Fluids with high viscosities will have lower velocity gradients (lower shear rates) for the same shear stress. Either the shear stress or the shear rate can be the forcing function in an experiment to measure the viscosity.

Volume Changes on Deformations. Some fluids and solids deform isochorically (constant volume), and some do not. Many fluids are considered to have

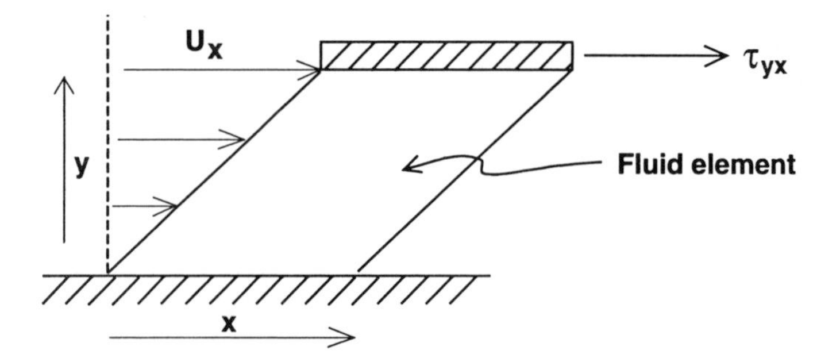

Figure 8.1 Velocity gradient of U_x in the y direction. Flow is caused by one wall moving relative to the other, which is caused by the shear stress, τ_{yx}.

constant density under shear. Elastic solids, and especially cross-linked elastomers, deform with constant volume. Figure 8.2 shows the normal stresses on a solid that is loaded by a uniaxial tensile stress, σ_x. Since the volume of the solid before and after deformation is the same,

$$V = L_x^o\, L_y^o\, L_z^o = L_x\, L_y\, L_z$$

where L_i^o is the original length in the i direction and L_i is the deformed length. When the material is isotropic, the deformations and stresses in the y and z directions are identical. The relationship between the stresses for uniaxial loading is

$$\upsilon\, \sigma_x = \sigma_y = \sigma_z \qquad 8.3$$

where υ is Poisson's ratio. When $\upsilon = 0.50$, the volume of the solid after deformation is unchanged. Poisson's ratios vary from 0.5 to zero, which represents no change in

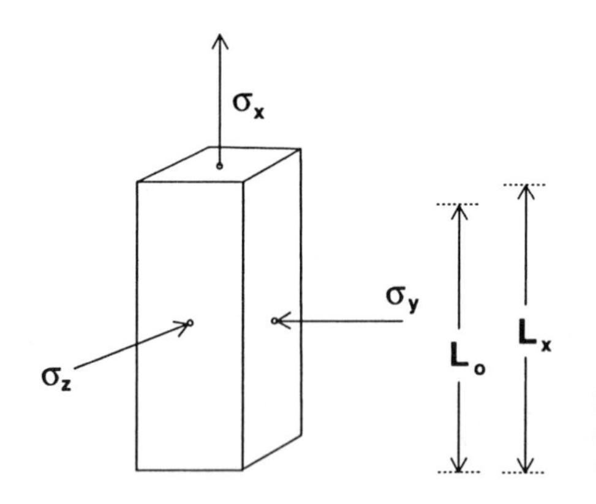

Figure 8.2 Uniaxial load on a solid. L_o is the initial length in the i direction and L is the loaded length.

the y and z directions due to a change in the x direction. Table 4 in Appendix F shows Poisson's ratios for several materials.

Modulus of Elasticity (Young's Modulus). The elongation, or strain, of the solid in Figure 8.2, $\epsilon = (L_x - L_o)/L_o$, is related to the load by the modulus of elasticity, E,

$$\sigma = E\epsilon \qquad 8.4$$

where σ is the load. Equation 8.4 is Hooke's law and is appropriate for strains less than the yield strain while the sample is elastic. Either the stress or the strain can be the forcing function in an experiment to measure Young's modulus (E).

Shear Modulus. Figure 8.1 shows a fluid element undergoing shear. The material between the plates could be a solid under shear. For this case, shear strain is

$$\tau = Gs \qquad 8.5$$

where τ is the shear stress, s is the shear strain, and G is the shear modulus.

The elastic and shear modulus can be related using the bulk modulus and Poisson's ratio. The bulk modulus, B, is a measure of how the sample's volume changes with pressure. The equations are

$$E = 3B(1 - 2\upsilon) \qquad 8.6a$$

$$= 2(1 + \upsilon)G \qquad 8.6b$$

When the solid is perfectly elastic ($\upsilon = 0.50$), the tensile modulus is exactly three times the shear modulus ($E = 3G$).

Compliance. The compliance of a material is the inverse of its elastic modulus,

$$J = \frac{1}{E} \qquad 8.7$$

where J is the compliance.

The above definitions of the elastic, shear, and bulk moduli are for static measurements of fluid and solid properties (the flow, load, or strain is constant). Mechanical properties may be measured using dynamic experiments in which the load or strain is varied in an oscillatory manner. Moduli can be defined for these experiments, but they contain real and imaginary components. For example, the dynamic modulus, E^*, is

$$E^* = E' + iE'' \qquad 8.8$$

where E^{I} is the storage modulus (the elastic response) and E^{II} is the loss modulus (the damping response). Two coefficients are needed to describe the response function of the solid. The other moduli and the viscosity can be defined in similar ways.

8.1 FLOW PROPERTIES AND RHEOLOGY

Rheology is the study of the deformation and the flow of material. Polymer rheology is important to the process designer from the polymerization vessel to the final fabrication step. Many polymer processing steps depend on how shear affects the flow of polymer solutions and melts. The goal of the rheologist is to find a constitutive equation for the particular polymer in a specific application. The constitutive equation is analogous to an equation of state: It relates molecular properties to the polymer's flow response. This section on flow properties will review Newtonian and non-Newtonian rheology.

There are a wide range of uses for rheological flow curves. These include: raw material characterization and control, prediction/evaluation of performance, product development, screw and die design, process simulation and control, compounding, scrap utilization, process improvement, and down-time reduction.

8.1.1 Newtonian Fluids

Equation 8.1 defines a Newtonian fluid: The shear stress always is proportional to the shear rate times a constant. The constant of proportionality is called the viscosity, μ. Figure 8.3 shows a plot of shear rate versus shear stress for a Newtonian fluid. Many authors plot such graphs with the axis labels reversed. The viscosity of a simple Newtonian fluid is independent of both shear stress and shear rate. Newtonian fluids have molecular relaxations that are much faster than the time scale of the imposed shear rates and shear forces. Water and glycerin are Newtonian fluids. In steady wide-channel or capillary flows, the velocity profile is constant through the device and does not vary with time. In oscillatory flows, the fluid responds instantaneously to changes in the forcing function. The viscosities of these fluids vary with temperature and pressure.

8.1.2 Non-Newtonian Fluids

Some fluids have viscosities that change with shear rate. An apparent viscosity, μ_a, is often used to describe the flow behavior of these cases. The defining equation is similar to Equation 8.1.

$$\tau_{yx} = -\mu_a \frac{dU_x}{dy} \qquad 8.9$$

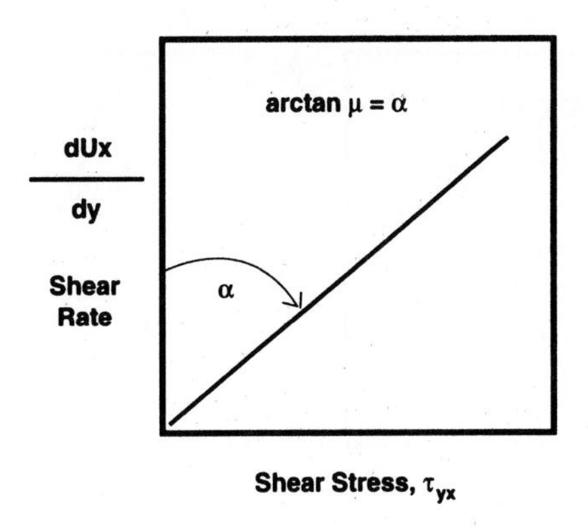

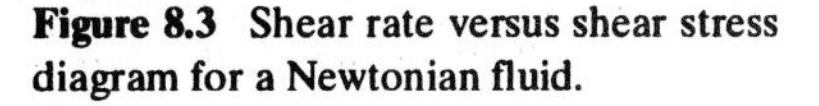

Figure 8.3 Shear rate versus shear stress diagram for a Newtonian fluid.

The apparent viscosity can be defined for any point on the shear diagram, and μ_a is not necessarily constant (Fig. 8.4). The apparent viscosity can be a function of shear rate and can be time-dependent.

Shear Thinning. Shear-thinning materials have apparent viscosities that decrease with increasing shear. Bingham plastic is one example of a shear-thinning material. Figure 8.5 shows the shear stress–shear rate and viscosity–shear stress diagrams for these materials. The fluid can be defined by

$$|\tau_{yx}| < \tau_o; \mu_a = \infty \tag{8.10a}$$

$$|\tau_{yx}| > \tau_o; |\tau_{yx}| - \tau_o = -\mu_a \frac{dU_x}{dy} \tag{8.10b}$$

where τ_o is the critical shear stress for the material and μ_a is the viscosity that relates the shear rate to the shear stress. These fluids exhibit no flow below their critical shear stress, τ_o. At shear stresses above τ_o, the material flows. A convenient definition for the shear stress of the flowing material is the actual shear stress minus the shear stress

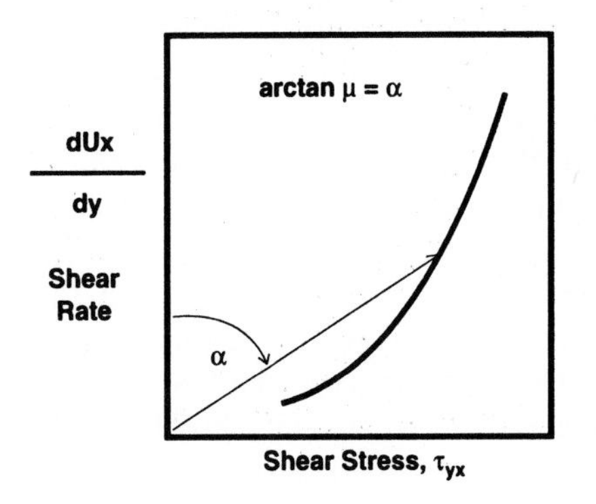

Figure 8.4 Definition of apparent viscosity on a shear rate versus shear stress diagram.

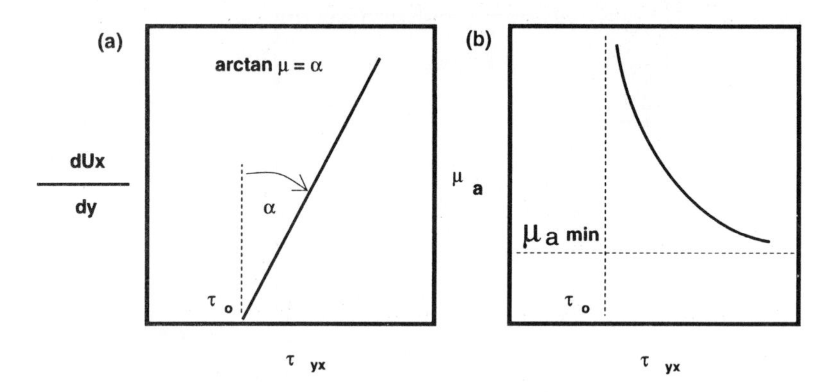

Figure 8.5 Bingham plastic fluid: **(a)** shear rate–shear stress diagram showing minimum shear stress for flow; **(b)** viscosity as a function of shear stress showing limiting viscosity at high shear.

required to initiate flow. Suspensions of clay, drawing ink, paint, toothpaste, and drilling muds are Bingham plastics. When a gel toothpaste tube is held upside down, no toothpaste flows from it (the gravitation forces on the fluid are less than the critical shear stress). As pressure is applied to the tube, the critical shear stress is exceeded and the material flows out. Careful observation of the gel shows that the material very near the tube wall has a velocity profile, while the toothpaste near the center is moving as a plug. The velocity profile is quite different from that of a Newtonian fluid, which would show a continuous gradient in velocity all the way to the center of the tube in laminar flow.

Pseudoplastic is another name for shear-thinning materials and is based on the observation that most thermoplastic polymers show shear-thinning behavior. Pseudoplastics have no specific yield stress value but do show a decreasing apparent viscosity with increasing shear rate. The viscosity of a pseudoplastic shows no time-dependent effects: The apparent viscosity is the same for any measurement period. Melts of high molecular weight polymers, solutions of these polymers, and some emulsion systems are pseudoplastic fluids.

Most pseudoplastics show more complex behavior than Figure 8.6 suggests. Figure 8.6 shows that the viscosity continuously decreases as the shear stress or shear rate increases. Many pseudoplastic fluids exhibit Newtonian behavior at low shear rates, show shear-thinning behavior at moderate shear rates, and then approach Newtonian behavior again at high shear rates. Such materials follow an Ostwald curve, shown in Figure 8.7.

The flow of Ostwald fluids is consistent with the random coil picture of polymer solutions and melts (Fig. 8.8). At low shear rates (Fig. 8.8a), flow occurs by random coil molecules slipping past one another. The whole molecule moves as a unit and is not deformed by the velocity gradient. The large arrow shows the movement of one coil. Figure 8.8b shows the coil conformations at an

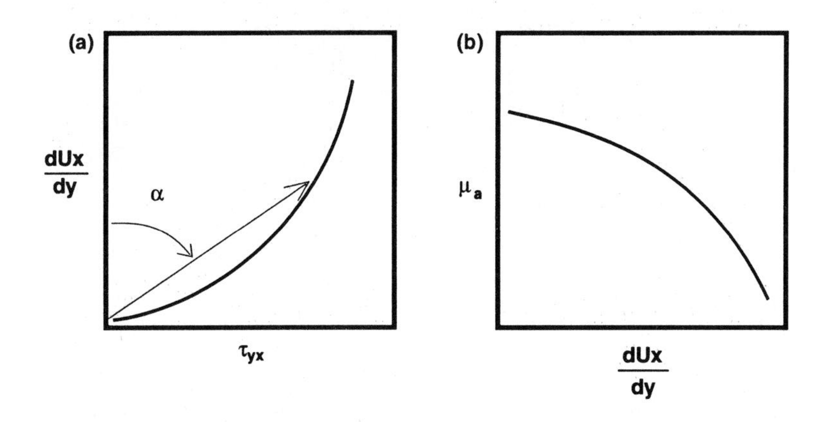

Figure 8.6 Shear-thinning fluid: **(a)** shear rate as a function of shear stress; **(b)** apparent viscosity as a function of shear rate.

intermediate shear rate. Under these conditions, the coils are deformed by the velocity gradient and slip past each other more easily than at low shear rates. Some energy is stored in the coil because it has been deformed from its random coil conformation. At very high shear rates (Fig. 8.8c), the coils have been distorted as much as possible by the flow field and offer low resistance to flow. The viscosity does not go to zero at very high shear rates because of entanglements between chains. This explanation is consistent with the concept of structural viscosity. The reptation model of molecular movement in concentrated solutions also could be applied to explain shear-thinning polymer melts. The "tubes" defining the molecule could be partially aligned with the flow field, reducing resistance of the molecule to shear-induced flow (Fig. 1.11).

Shear-thickening materials have increasing apparent viscosity with shear rate (Fig. 8.9). Dilatant fluids are one set of shear-thickening fluids that contain suspended solids (Fig. 8.10a). The solids can become close-packed under shear, and thereby increase the rigidity of the material. Figure 8.10b shows the solids of Figure 8.10a under shear. The movement of particles has resulted in a packed

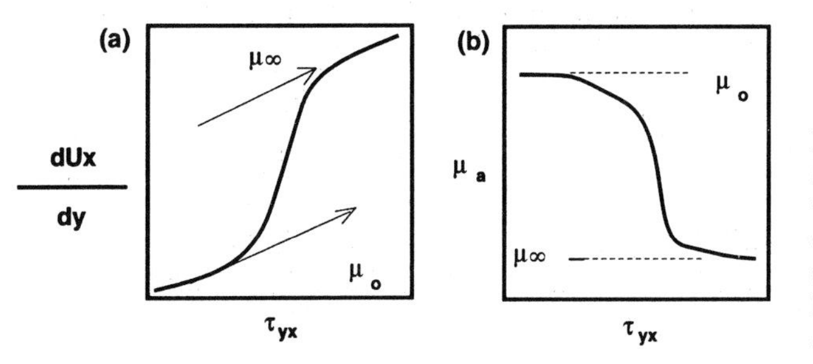

Figure 8.7 Ostwald fluid: **(a)** shear rate as a function of shear stress with the limiting viscosities shown as the tangents to the curve at low and high shear rates; **(b)** viscosity as a function of shear stress with the limiting viscosities shown as the maximum and minimum values.

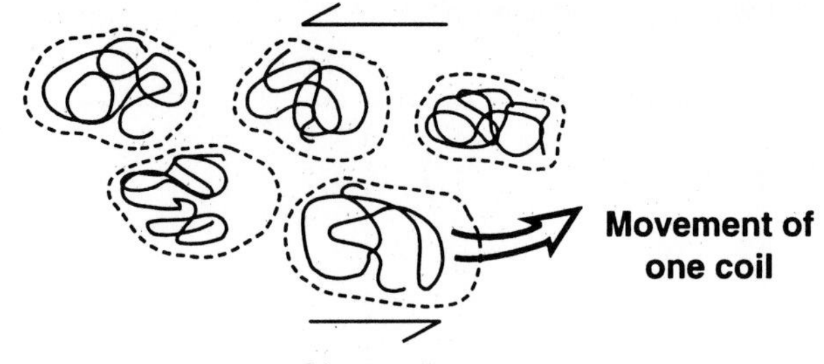

(a)

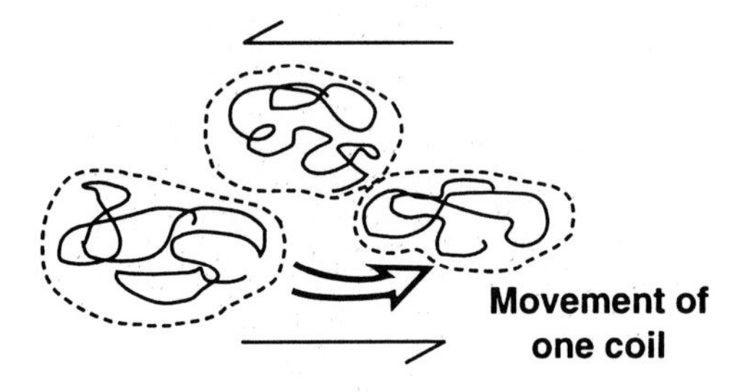

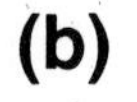

(b)

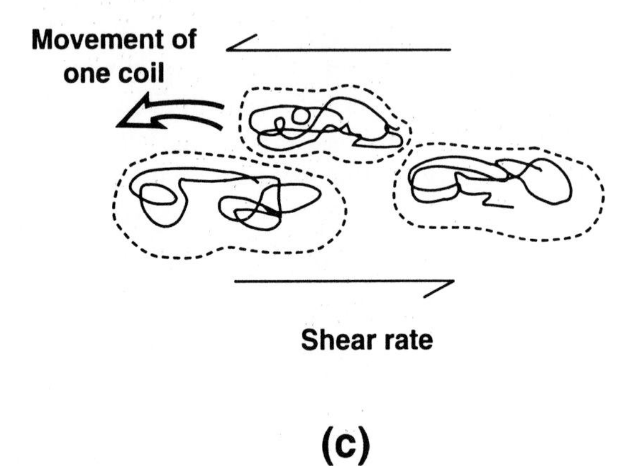

(c)

Figure 8.8 Movement of random coils in shear fields: **(a)** low shear rate, $\mu = \mu_o$, fluid motion does not distort the coils; **(b)** intermediate shear rate, $\mu_0 > \mu > \mu_\infty$, velocity gradient elongates the coils in the shear direction; **(c)** high shear rate, $\mu = \mu_\infty$, coil distortion has reached a dynamic equilibrium between the elongations caused by the shear and the recoil of the molecule.

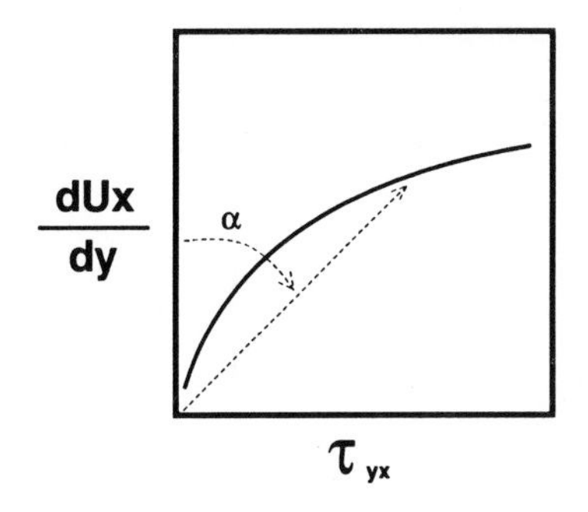

Figure 8.9 Shear rate versus shear stress for a shear-thickening fluid.

aggregate indicated by the dashed curve. The aggregate could resist further particle movement and would increase the local viscosity.

8.1.3 Time-Dependent Behavior

Many polymer fluids exhibit time-dependent behavior; that is, the viscosity measured by a rheometer or another method will change with the time period of the observation. The time-temperature superposition concept was introduced in Chapter 1. Time-dependent behavior can be classified as either thinning with time or thickening with time. Thixotropic fluids are shear-thinning with time. However, the original fluid structure or viscosity can be recovered when the stress is removed. The time-dependent behavior is consistent with the orientation of the random coil

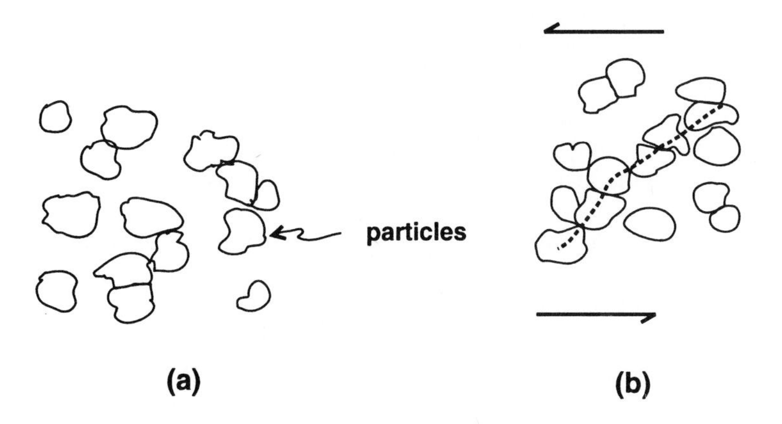

Figure 8.10 Dilatant fluid: **(a)** fluid at rest, **(b)** fluid under shear. Velocity gradient causes particles to pack into larger agglomerates that have increased resistance to flow. Dashed line suggests the effective diameter of the particle unit.

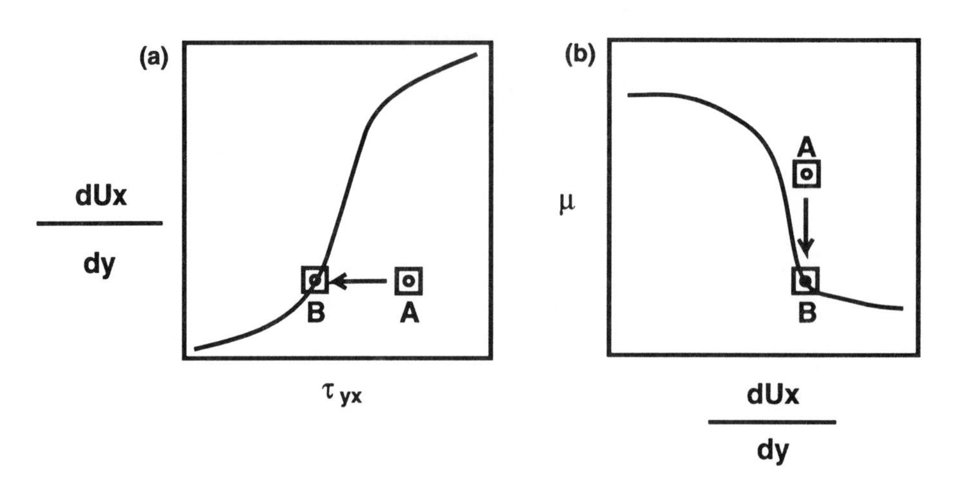

Figure 8.11 Thixotropic fluid in a constant shear rate experiment: **(a)** shear rate versus shear stress, **(b)** viscosity versus shear rate. The initial measurement gives the viscosity shown at point A. After long flow times, the viscosity approaches point B as the "equilibrium" value.

in shear fields. Because of their long chain length, polymer molecules do not reach a statistical distribution of shapes in a shear field immediately. Some time is required for the shear field to deform the coils. If the time needed to deform the coils to their average conformation is long compared to the experimental time, then the measured viscosity will change with time. In a steady flow experiment (melt index measurement by capillary flow), the polymer flow rate will increase with time at a constant extrusion pressure. Figure 8.11 shows the shear stress–shear rate diagram for a thixotropic fluid. The initial viscosity measurement gives the value shown as point A. As the fluid continues to flow, the viscosity decreases and approaches that shown as point B. The "equilibrium" should be considered to be dynamic; that is, after long times under steady shear, the chain conformations have reached a steady-state distribution. In an oscillatory flow experiment, the measured viscosity will depend on the oscillation frequency. Higher frequencies will give higher viscosities, as the material will have less time to deform with each cycle.

Thixotropic fluids thin without structural loss. This means that the fluid can recover its original high value for viscosity (point A of Fig. 8.11) after the shear has been removed and the solution allowed to rest. Some shear-thinning fluids have been reported to undergo structure loss. This would require either breaking carbon-carbon bonds or cross-links between chain segments. When chain scission occurs, the material has changed and has lower molecular weight. Some shear-thickening fluids can thin with time (dilatant fluids).

Materials that increase their apparent viscosity with time are called antithixotropic. Ammonium oleate suspensions behave in this manner. Rheopectic materials are thixotropic materials whose original structure or viscosity can be restored by low shear motions after shear thickening has been achieved by high shear rates.

Equation 8.9 can be applied to shear-thinning or shear-thickening fluids by modeling the viscosity as a function of the absolute value of the shear rate.

$$\tau_{yx} = -\mu \frac{dU_x}{dy}; \quad \mu = f\left(\frac{dU_x}{dy} \; or \; \tau_{yx}\right) \tag{8.11}$$

Equation 8.11 can be generalized to relate the stress tensor to the deformation tensor, where the viscosity is a function of the scalar invariants of the stress or deformation tensor,

$$\underline{\underline{\tau}} = -\mu \, \underline{\underline{\dot{\gamma}}}; \; \mu = f(scalar \; invariants \; of \; \underline{\underline{\tau}} \; or \; \underline{\underline{\dot{\gamma}}}) \tag{8.12}$$

where $\underline{\underline{\tau}}$ is the shear stress tensor and $\underline{\underline{\dot{\gamma}}}$ is the shear rate tensor. A scalar invariant of a tensor should be used as the independent variable in a model for the viscosity of a fluid. Three independent scalar quantities can be determined by finding the traces of $\underline{\underline{\dot{\gamma}}}$, $\underline{\underline{\dot{\gamma}}}^2$, and $\underline{\underline{\dot{\gamma}}}^3$. The trace is found by summing the diagonal elements of these quantities. The scalar invariants of the shear rate tensor are

$$I_{\dot{\gamma}} = \sum_i \dot{\gamma}_{ii} \tag{8.13a}$$

$$II_{\dot{\gamma}} = \sum_i \sum_j \dot{\gamma}_{ij} \dot{\gamma}_{ji} \tag{8.13b}$$

$$III_{\dot{\gamma}} = \sum_i \sum_j \sum_k \dot{\gamma}_{ij} \dot{\gamma}_{jk} \dot{\gamma}_{ki} \tag{8.13c}$$

$I_{\dot{\gamma}}$, $II_{\dot{\gamma}}$, and $III_{\dot{\gamma}}$ values are independent of the coordinate system used to determine them. Other scalar invariants can be formed, but they will be combinations of these three. For incompressible fluids, $I_{\dot{\gamma}} = 2(\nabla \cdot \upsilon) = 0$. For shearing flows, $III_{\dot{\gamma}}$ is zero. Therefore, the viscosity can be modeled as a function of $II_{\dot{\gamma}}$.

The generalized Newtonian viscosity is a scalar. Since it is dependent on the deformation tensor, it must depend only on those components of the tensor that are scalar quantities independent of the coordinate system. Generalized Newtonian models are appropriate for the following steady flows: tubular, axial annular, tangential annular, helical annular, parallel plates, rotating disks, and cone-and-plate flows. The capillary, the Couette, and the cone-and-plate viscometers are used to determine the viscosity of polymer solutions.

Example 8.1 Viscosity Modeling for Simple Shear Flows

The scalar invariants of the shear rate tensor need to be determined in order to develop models for polymer fluid viscosities. The shear rate tensor for planar shear flow (Fig. 8.1) is:

$$\underline{\underline{\dot{\gamma}}} = \begin{vmatrix} \dot{\gamma}_{xx} & \dot{\gamma}_{xy} & \dot{\gamma}_{xz} \\ \dot{\gamma}_{yx} & \dot{\gamma}_{yy} & \dot{\gamma}_{yz} \\ \dot{\gamma}_{zx} & \dot{\gamma}_{zy} & \dot{\gamma}_{zz} \end{vmatrix}$$

Since shear occurs only by the gradient of U_x in the y direction

$$\underline{\underline{\dot{\gamma}}} = \begin{vmatrix} 0 & \dot{\gamma}_{xy} & 0 \\ \dot{\gamma}_{yx} & 0 & 0 \\ 0 & 0 & 0 \end{vmatrix}$$

By symmetry, $\dot{\gamma}_{xy} = \dot{\gamma}_{yx}$. The value of $II_{\dot{\gamma}}$ is

$$II_{\dot{\gamma}} = \sum_i \sum_j \dot{\gamma}_{ij}\dot{\gamma}_{ji} = \dot{\gamma}^2_{yx}$$

Therefore, the viscosity can be modeled as $\mu = f(\dot{\gamma}^2_{yx})$. A simpler choice is the absolute magnitude of the shear rate, $|\dot{\gamma}_{yx}|$. Many of the empirical viscosity models describe the viscosity as $\mu = f(|\dot{\gamma}_{yx}|$ or $|\tau_{yx}|)$.

8.1.4 Generalized Newtonian Models

Generalized Newtonian models for viscosity have been widely used to describe polymer rheology, despite their limitations. These models assume instant and full response to any deforming force. As a result, they do not predict phenomena such as normal forces or viscoelasticity and cannot be used (at least on theoretical grounds) when polymers exhibit the following properties: recoil, stress relaxation, or die swell. They also cannot be used in special flows, such as elongational flows. Of course, these models can be used when the time-dependent effects are small or occur on a long time scale.

Power Law Model. Most commercial processing is done in regions of the viscosity versus shear rate curve where the viscosity is decreasing with shear rate. Here, the viscosity can be well-represented by a power law fit,

$$\tau_{yx} = -k|\dot{\gamma}_{yx}|^n \tag{8.14a}$$

$$= -k|\dot{\gamma}_{yx}|^{n-1}\dot{\gamma}_{yx} \tag{8.14b}$$

where k is a proportionality constant relating the shear stress to the nth power of the shear rate (Ostwald, 1925). Equation 8.14b is written in the form of Equation 8.1, so that the power law viscosity is

$$\mu(\dot{\gamma}) = k|\dot{\gamma}|^{n-1} \tag{8.15}$$

Figure 8.12 shows a logarithmic plot of viscosity versus shear rate for a polystyrene sample at 180°C. Equation 8.15 will model the melt viscosity well over the shear rate range, 10° to 10^2 sec^{-1}. The constants, k and n need to be determined by fitting the equation to data. The model's advantages are that it is easily integrable and differentiable—characteristics that make it very useful in modeling of process equipment. When $n = 1$, the fluid is Newtonian. $n < 1$ implies that the fluid is shear-thinning, and $n > 1$ implies that the fluid is shear-thickening. Table 8.1 shows typical power law constants for several common materials.

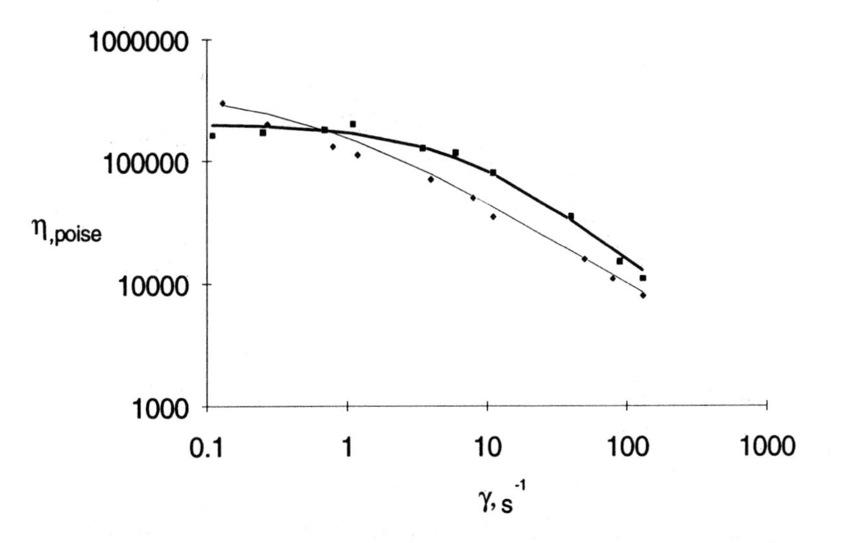

Figure 8.12 Plot of the shear rate dependent viscosity of a narrow molecular weight distribution (PS)(■) at 180°C, showing the "Newtonian plateau" and the "power law" regions and a broad distribution PS (◆) (Graessley et al., 1970).

TABLE 8.1 Typical Power Law Parameters for Well-known Materials

Material	k (Pa · s^n)	n	Shear rate range (s^{-1})
Ball-point pen ink	10	0.85	10^0–10^3
Fabric conditioner	10	0.6	10^0–10^2
Polymer melt	10000	0.6	10^2–10^4
Molten chocolate	50	0.5	10^{-1}–10
Synovial fluid	0.5	0.4	10^{-1}–10^2
Toothpaste	300	0.3	10^0–10^3
Skin cream	250	0.1	10^0–10^2
Lubricating grease	1000	0.1	10^{-1}–10^2

Truncated Power Law. Much polymer melt data can be represented by a bent line articulated at some critical shear rate. The data in Figure 8.12 shows that viscosity is constant at low shear rates and then starts to decrease at shear rates greater than 1 sec^{-1}. This two-line fit of the data is the basis for the truncated power law model. The three constants are the zero shear rate viscosity, μ_o, the critical shear rate, $\dot{\gamma}_{yx,o}$, and the exponent n.

$$\mu = \mu_o; \dot{\gamma}_{yx} < \dot{\gamma}_{yx,o} \qquad 8.16a$$

$$\mu = u_o(\dot{\gamma}_{yx}/\dot{\gamma}_{yx,o})^{n-1}; \dot{\gamma}_{yx} > \dot{\gamma}_{yx,o} \qquad 8.16b$$

At low shear rates, there is a constant viscosity, μ_o, for the fluid; this region is the Newtonian region. At higher shear rates, the material is shear-thinning and viscosity decreases as shear rate increases. At very high shear rates, the viscosity has a limiting value, μ_∞ A monodisperse sample will exhibit a sharper break in its viscosity versus shear rate curve than a sample with a broad molecular weight distribution (Fig. 8.12). In the polydisperse sample, the small molecules act as lubricants for the larger molecules, and the shear-thinning response is not as noticeable. At a given screw speed or power input to an extruder, the broad molecular weight distribution materials flow faster and result in a higher output. However, the broad molecular weight distribution does affect mechanical properties. In general, as the distribution increases, the tensile strength of a finished part will decrease and the elongation to break will increase. Flow curves help predict processing behavior and can also be used as a characterizing tool for molecular weight and molecular weight distribution.

The limiting viscosity as shear rate approaches zero is proportional to the molecular weight of the sample. Therefore, viscometry offers an easy way to compare molecular weights. Most polymers approach their limiting values of viscosity at shear rates of less than 0.01 reciprocal seconds. Nylon and PET are exceptions to this rule. The Newtonian region becomes smaller as the molecular weight of the sample increases (Fig. 8.13) and the molecular weight distribution broadens. A similar phenomenon occurs in solutions of polymers.

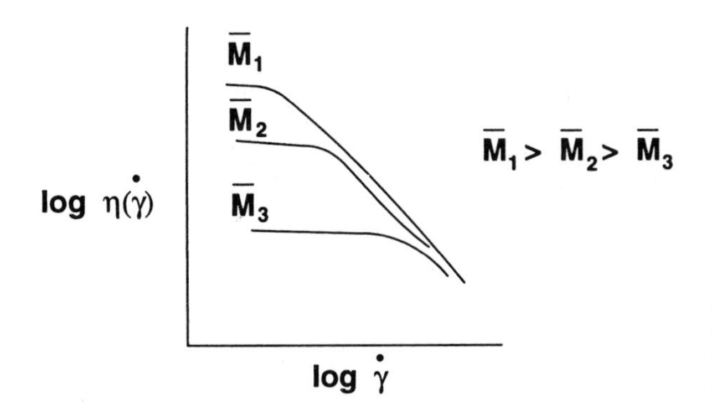

Figure 8.13 Effect of molecular weight on melt viscosity. Samples with higher molecular weights start shear-thinning at lower shear rates.

Polymers can exhibit properties ranging from liquid-like to solid-like, depending on the conditions under which they are tested. The Deborah number attempts to describe the relationship between the material response time and the experimental time. It is defined as

$$N_{Deb} = \frac{Material\,Response\ Time}{Deformation\,Time} \qquad 8.17$$

If the material responds faster than the forcing function, then its Deborah number will be less than one. Such materials should show a Newtonian type response and flexible, uncrosslinked polymer systems should behave as liquids. If the Deborah number is much greater than one, then a flexible polymer should behave like a solid. The Deborah number equals one at about the shear rate at which the power law region begins.

Ellis Model. The viscosity can be modeled based on either the shear rate or the shear stress. This model describes the changes in viscosity as a function of the shear stress,

$$\frac{\mu_o}{\mu} = 1 + \left(\frac{\tau_{yx}}{\tau_{1/2}}\right)^{n-1} \qquad 8.18$$

where μ_o is the zero shear rate viscosity, $\tau_{½}$ is the value of the shear stress when the viscosity is one-half the zero shear viscosity and n is the power exponent. The Ellis model is easy to fit to experimental data and predicts a Newtonian plateau at low shear rates.

Example 8.2 Fitting the Ellis Model to Polystyrene Data

Determine values of the zero shear viscosity, $\tau_{½}$, and n for the data of polystyrene at 178°C.

(continued)

$\dot{\gamma}_{yx}$, sec^{-1}	τ_{yx}, *MPa*	μ, *MPa-s*
1.5	103	68.9
5	241	48.3
8	345	43.1
15	414	27.6
50	552	11.0
80	655	8.19
150	827	5.52
500	1034	2.07
800	1379	1.72

One method for fitting the data to the Ellis model would be to use a multiple regression analysis. Another method would be to estimate the values of the constants from limiting cases and continue iterating until an appropriate fit occurred. (This second method might be used to get reasonable initial guesses for the values of the constants for a regression analysis.) As an initial trial, assume that the zero shear viscosity is near the highest reported value. Then, by picking two well-separated points, solve for n and $\tau_{1/2}$. The first and last point yield the following values for the constants: $\mu_o = 75$ *MPa-s*, $n = 2.9$, and $\tau_{1/2} = 148$ *MPa*. These constants are not a perfect fit to the data since $\tau_{1/2}$ is supposed to be the value of the shear stress at $\mu = \mu_o/2$. An improved fit with $\mu_o = 75$ *MPa-s*, $a =$ 3.6. $\tau_{1/2} = 290$ is shown in Figure 8.14.

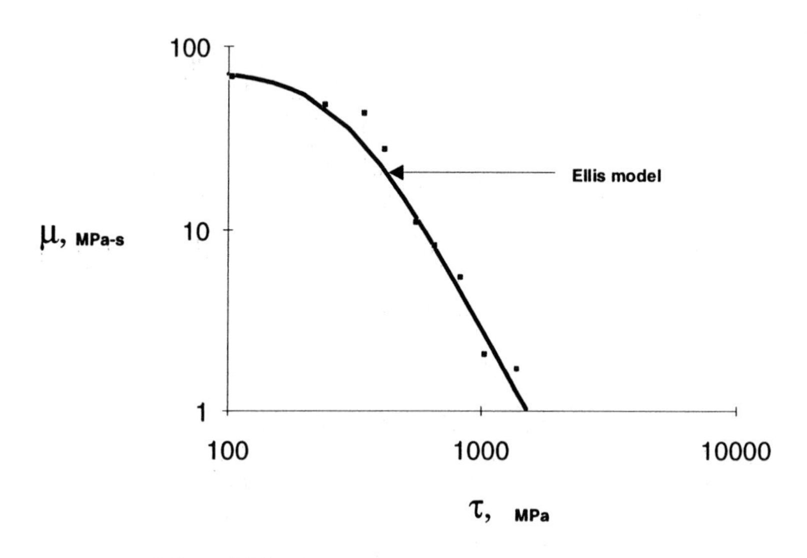

Figure 8.14 Ellis model fitted to polystyrene data.

Carreau Model. Recall that many shear-thinning materials could be plotted on the Ostwald curve and that many had zero shear rate and infinite shear rate limiting viscosities. Such a fluid can be fairly well represented by a four-constant model,

$$\frac{\mu - \mu_\infty}{\mu_o - \mu_\infty} = \left[1 + (\lambda \dot{\gamma})^2\right]^{\frac{n-1}{2}} \quad 8.19$$

where μ_o is the zero shear viscosity, μ_∞ is the infinite shear viscosity, λ is a time constant, and n is an exponent.

8.1.5 Free Volume Model of Polymer Melt Viscosity

The generalized Newtonian models discussed above do not attempt to relate the polymer viscosity to its chemical and physical properties. Physicochemical models can be very useful since polymer melt viscosity is known to be a function of the following properties: molecular weight, temperature, pressure, polydispersity, branching, stereoisomerism, unsaturation, chain stiffness, T_g, and polarity.

Williams, Landel, and Ferry conjectured that the mobility of the polymer at temperatures above T_g should be a function of the amount of free volume in the melt. This concept had been used to describe the viscosity of simple liquids. The logarithm of the viscosity is taken to be proportional to the fractional change in the free volume. Other theories of the liquid state propose a temperature dependence of viscosity that is an Arrhenius-type formulation. This seems to be a good physical picture of polymer liquids at temperatures much higher than T_g.

The temperature variation of viscosity is primarily due to the fluid's thermal expansion. Simple hydrocarbon liquids such as the n-alkanes show exponential changes in viscosity as the free volume changes. A better model for polymers is

$$\mu = A\exp(BV_o/V_f) = A\exp(B/f) \quad 8.20$$

where V_o is the occupied volume, and V_f is the free volume, f is approximately the volume fraction of free volume when the free volume is small compared to the occupied volume. Equation 8.20 is divided by a reference viscosity at a reference condition, giving

$$\ln(\mu/\mu_o) = B(1/f - 1/f_o) \quad 8.21$$

It is convenient to use the glass transition temperature as the reference condition. The free volume of the fluid is modeled as a function of the thermal expansion coefficient, α.

$$T > T_g: f = f_g + a(T - T_g) T < T_g: f = f_g \quad 8.22$$

When Equation 8.22 is substituted into Equation 8.21 and the result rearranged, the equation is

TABLE 8.2 WLF Coefficients for the Temperature Dependence of the Shift Factor

$$\log a_T = \frac{-c_1^0(T - T_0)}{(c_2^0 + T - T_0)}$$

Polymer	T_0, K	c_1^0	c_2^0	T_g, K
ELASTOMERS				
Hevea rubber	248	8.86	101.6	200
Polybutadiene	298	3.44	196.6	161
Polybutadiene	298	6.23	72.5	261
Butyl rubber	298	9.03	201.6	205
Styrene-butadiene	298	4.57	113.6	210
Ethylene-propylene	298	5.52	96.7	242
Polyurethane	283	8.86	101.6	238
THERMOPLASTICS				
Polyisobutylene	298	8.61	200.4	205
Poly(vinyl acetate)	349	8.86	101.6	305
Polystyrene	373	12.7	49.8	370
Poly(1-hexene)	218	17.4	51.6	218
Poly(dimethyl siloxane)	303	1.90	222	150
Poly(propylene oxide)	198	16.2	24	198
Poly(methyl methacrylate)	381	34.0	80	381

SOURCE: J.D. Ferry, *Viscoelastic Properties of Polymers*, 3rd ed., Wiley, NY, 1980.

$$\ln(\mu/\mu_o) = \frac{-A(T - T_g)}{B + (T - T_g)} = \frac{-17.44\ (T - T_g)}{51.6 + (T - T_g)} \qquad 8.23$$

The universal constants of Equation 8.23 apply to many polymer systems and can be used when more specific values are unknown. This leads to the conclusion that the free volume of a polymer at the glass temperature is constant, $f_g = 0.025$ volume fraction. Table 8.2 lists Williams–Landel–Ferry (WLF) constants for specific polymers. The equation for these constants is based on an arbitrary reference temperature, T_o.

8.1.6 Dependence of Viscosity on Molecular Weight, Temperature, and Pressure

An equation similar to the Mark–Houwink equation correlates the molecular weight with the melt viscosity. Below a critical molecular weight, the intrinsic viscosity is proportional to molecular weight to the first power. The critical molecular weight, M_c, is thought to be the point when molecular entanglements begin to restrict the slippage of polymer molecules past each other.

$$M < M_c,\ \mu = K\,M^1;\ M > M_c,\ \mu = K\,M^{3.4} \qquad 8.24$$

Polymer viscosity is also a function of the temperature. An Arrhenius-type equation

Example 8.3 Molecular Weight Dependence of Viscosity

Processing equipment can be designed to handle materials within specific viscosity ranges. The temperature and the molecular weight of the polymer are two variables used to change the melt properties. For molten polystyrene at 217°C, calculate how much change in molecular weight is needed to halve the viscosity above and below the critical molecular weight. A plot of viscosity data for polystyrene at 217°C yielded the following models for the melt viscosity dependence on molecular weight:

$$\mu = 1.7 \times 10^{-5}\, M^{1.4};\ M < 50{,}000$$
$$\mu = 3.0 \times 10^{-13}\, M^{3.1};\ M > 50{,}000$$

Below the critical molecular weight, the molecular weight of the polymer must be reduced by 39% to halve the viscosity. Above the critical molecular weight, M must be reduced by 20% to halve the viscosity. Clearly, control of molecular weight will have a large impact on the processing performance of a polymer. Since polystyrene is pseudoplastic, the effect of molecular weight changes on viscosity will not be as dramatic under actual shearing conditions.

can be used to relate viscosity to an activation energy. However, viscosity also is a function of the shear stress or shear rate. Viscosity can be modeled as

$$\mu = A \exp(-E_\mu/RT)$$
$$\mu = B \exp(E(\dot{\gamma})/RT)$$
$$\mu = C \exp(E(\tau)/RT) \qquad 8.25$$

where E_μ is the activation energy for viscous flow, and $E(\dot{\gamma})$ and $E(\tau)$ give the activation energy for flow at constant shear rate or shear stress. Using an activation energy for flow to describe the temperature dependence works well for narrow temperature ranges. The molecular weight distribution does have an effect on the temperature sensitivity of viscosity. A polymer with a broad molecular weight distribution is less

Example 8.4 Shear Dependence of Flow Properties

For three different shear rates, compute how much the temperature must be changed to halve the polymer viscosity from its value at 215°C. Data for polystyrene are given in the following table:

(continued)

T,°C	$\dot{\gamma}$, sec^{-1}	μ	$\dot{\gamma}$, sec^{-1}	μ	$\dot{\gamma}$, sec^{-1}	μ
178	10	5,000	100	1,000	1,000	250
205		2,500		600		125
232		900		400		80
260		400		200		60
288		100		70		30

The data sets are fitted to Equation 8.25 to get the shear rate activation energy. Then the temperature change for 50% viscosity decrease can be calculated. The results are given in the following table:

$\dot{\gamma}$	$E(\dot{\gamma})/R$	ΔT, °C for 50% Decrease in Viscosity
10	8830	+ 19 C (234 C)
100	5840	+ 30 C (245 C)
1000	4580	+ 39 C (254 C)

The activation energy decreases with increasing shear rate. There will be an upper bound for the temperature increase, since excessive temperatures can result in increases in the degradation rate. Polystyrene degradation rates become high above 250°C.

sensitive to temperature. Also, the temperature sensitivity of viscosity decreases as the shear rate increases.

Pressure changes will also result in changes in the melt viscosity. In general, high pressures will decrease the free volume and thereby increase the viscosity of the melt. Pressure effects on viscosity can be correlated using

$$\mu = A \exp (B P) \tag{8.26}$$

where μ is the viscosity, P is pressure, and A and B are constants at a given temperature. The pressure dependence of viscosity is not large for many systems.

Branched Polymers. The previous discussion assumed that the polymer was linear and amorphous with few branch points. Branched polymers have slightly different melt rheology. The melt viscosity of low molecular weight branched polymers is actually smaller than that of linear polymers. The volume occupied by a branched molecule is smaller than the volume occupied by linear molecules of the same molecular weight. A lower effective size means lower drag forces as molecules pass one another. The zero shear rate viscosity is also less for a branched molecule of low molecular weight. When the polymer molecular weight is greater than 30,000, this effect is reversed. Figure 8.15 compares linear and branched molecules of high molecular weight. Since most forming processes occur at high shear rates, linear polymers usually have lower melt viscosities and are preferred.

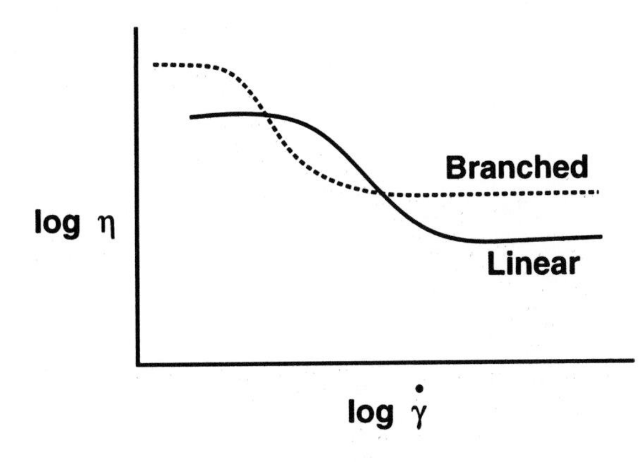

Figure 8.15 Effect of chain branching on melt viscosity. Branched chains start shear-thinning at lower shear rates than linear chains. Branched chains tend to have higher limiting viscosities because branches tend to "stiffen" the macromolecule.

Compound Additives. Many commercial polymers contain compounding additives either to improve their processing or end-use performance. Figure 8.16 illustrates how processing variables affect the viscosity of the melt, assuming that all other properties are constant. The directions of the arrows suggest the way the curve should shift if the property is increased. The magnitude of the effect depends on the polymer family.

Viscometers. There are a number of viscometers that can be used to determine fluid viscosity as a function of shear rate in steady flow fields. These devices are simple to operate and are fairly reliable. Table 8.3 shows the equations for shear stress, shear rate, and viscosity for several viscometers with power law fluids. Figures 8.17

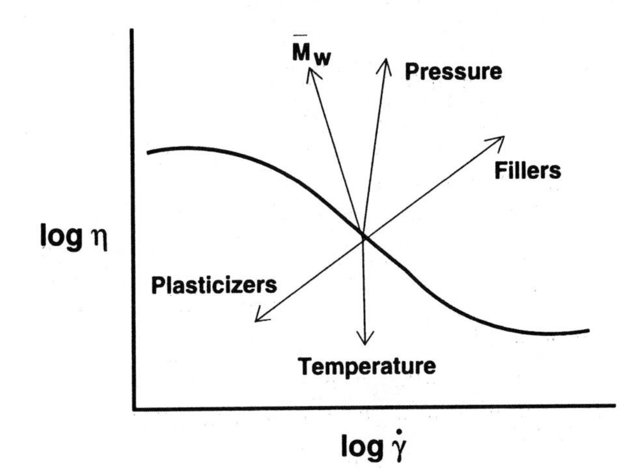

Figure 8.16 Effects of processing variables on melt viscosity. The arrows show the direction the viscosity curve will be shifted by an increase in the variable.

TABLE 8.3 Shear Rate, Shear Stress, and Viscosity for Several Viscometers in Steady Flow[a]

Viscometer	Shear Rate, $\dot{\gamma}$	Shear Stress, τ	Viscosity, η	Comment
Narrow gap concentric cylinder $r_1/r_0 > .97$	$\frac{r_0 \Omega}{r_0 - r}$	$\frac{T}{2\pi r_0^2 L}$	$\frac{T(r_0 - r_1)}{2\pi r_0^3 \Omega_1 L}$	
Wide-gap concentric cylinder	$\frac{2\Omega}{n(1 - b^{2/n})}$	$\frac{T}{2\pi r_1^2 L}$	$\frac{T_n(1 - b^{2/n})}{4\pi r_1^2 L \Omega_1}$	n refers to power law exponent
Rotating cylinder in large volume	$\frac{2\Omega_1}{n}$	$\frac{T}{2\pi r_1^2 L}$	$\frac{T_n}{4\pi r_1^2 \Omega_1 L}$	.1 to 10 s^{-1}
Cone-and-plate viscometer	$\frac{\Omega_1}{\theta_0}$	$\frac{3T}{2\pi a^3}$	$\frac{3T\theta_0}{2\pi a^3 \Omega_1}$	
Parallel plate viscometer	$\frac{a\Omega_1}{n}$	$\frac{3T}{2\pi a^3\left(1 + \frac{n}{3}\right)}$	$\frac{3T_n}{2\pi a^4 \Omega_1\left(1 + \frac{n}{3}\right)}$	
Capillary viscometer	$\frac{4Q_f}{\pi a^3}\left[\frac{3}{4} + \frac{1}{4}\frac{d \ln Q_f}{d \ln \tau_w}\right]$	$\frac{a}{2}\frac{dP}{dL}$	$\frac{\pi a^4\, dP/dL}{8Q\left[\frac{3}{4} + \frac{1}{4}\frac{d \ln Q_f}{d \ln \tau_w}\right]}$	Q_f = flow

[a] a = radius, Ω = rotation speed, T = torque

and 8.18 show cone-and-plate and Couette viscometers. Capillary rheometers were discussed in Chapter 6. Except for the capillary flow method, all the viscometers in Table 8.3 use rotation to shear the fluid. One surface is rotated and the torque is measured on the other, although some devices have the rotation and torque measurements on the same surface.

The capillary viscometer is a practical way for measuring viscosities under flow conditions similar to those experienced during processing. It is sensitive to nonuniform flows before the capillary entrance (entrance effects), the distance required to establish the steady-state velocity profile, and exit effects. The flow of a power law fluid in a capillary of radius R gives the following expression for pressure drop per unit length for a power law model:

$$\frac{dP}{dL} = \frac{2k}{R}\left[\frac{(3n + 1)Q_f}{\pi n R^3}\right]^n \tag{8.27}$$

When $n = 1$, the fluid is Newtonian, and the pressure gradient is a function of R^{-4} for a fixed flow rate. If a shear-thinning power law fluid is in the capillary, the dependence

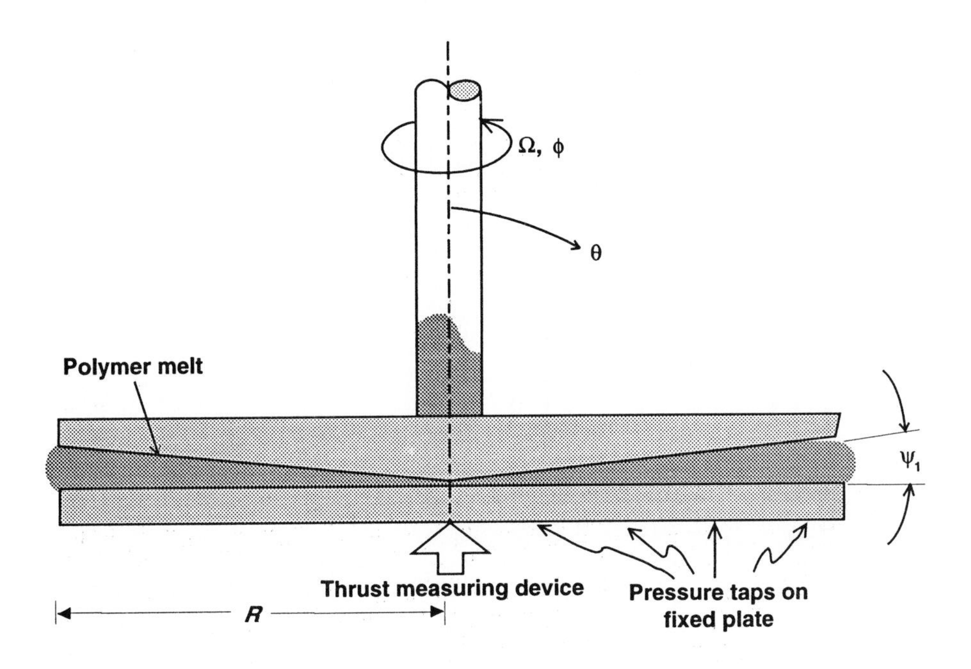

Figure 8.17 Schematic representation of the cone-and-plate viscometer. The cone angle is only several degrees or less and has been exaggerated in this sketch. In this set-up, the cone is rotated with speed, Ω, and the bottom plate is stationary. The torque on the cone and its speed can be used to determine viscosity. Pressure taps and the thrust measurement can be used to determine normal forces.

on radius is reduced. For example, molten chocolate would give a pressure drop proportional to $R^{-2.5}$. The velocity profile is given by

$$v(r) = \frac{Q_f(3n+1)}{\pi R^2(n+1)}\left(1-\frac{r}{R}\right)^{\frac{n+1}{n}} \qquad 8.28$$

Low values of n lead to "plug flow" velocity profiles. For a given pressure drop, a shear-thinning fluid will have a higher flow rate than a Newtonian fluid. Because the velocity profile is steeper near the wall, heat transfer rates are improved. Shear-thickening fluids give the opposite effects.

Melt Index. Despite the rather sophisticated models and measurement techniques that are available to the polymer engineer, simpler measurements are usually

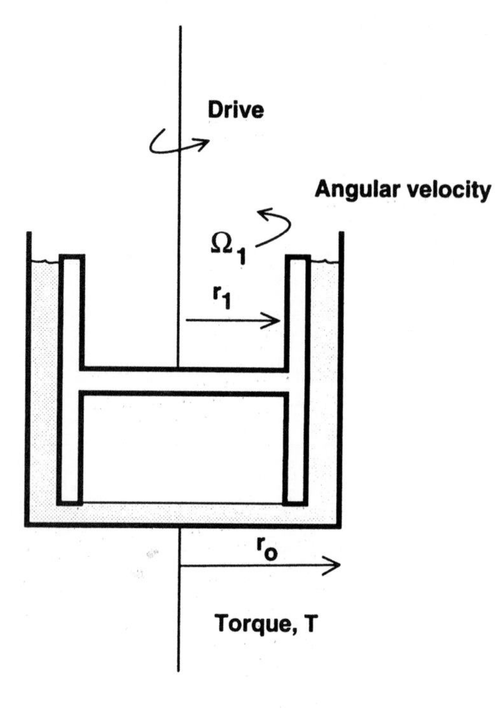

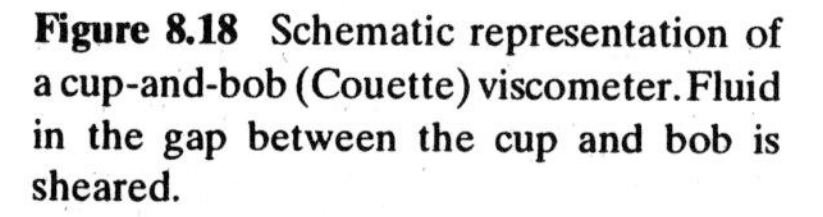
Figure 8.18 Schematic representation of a cup-and-bob (Couette) viscometer. Fluid in the gap between the cup and bob is sheared.

used to characterize polymer melt flow and to design equipment for plant operations. This is due to the difficulty of modeling processing equipment even when the constitutive equation for the polymer melt is known. The melt index is often used to compare different polymer samples. It is defined as the amount of polymer that will flow through a die in a given time and is usually obtained in a capillary rheometer. It is important because it does show the response of the material at the temperature, pressure, and flow conditions under which it will be processed. High melt index material has a high flow rate and low viscosity. An "easy flowing" plastic has a high melt index, low crystallinity, and low molecular weight. The melt index is used to set an upper limit in the weight average molecular weight of polymer for a specific purpose. The lower limit on the number average molecular weight is usually set by mechanical properties.

8.1.7 Linear Viscoelastic Models

Viscoelasticity can be modeled using linear or nonlinear models. Viscoelastic material exhibits a partial elastic response similar to ideal solids at low deformations and a partial viscous response similar to the flow of a liquid at low shear rates. Linear models are based on the superposition principle: The response of the fluid is proportional to the deforming force, and forces applied at different times are additive. Linear viscoelastic models have linear differential equations with constant coefficients. The time derivatives are ordinary partial derivatives. As a result, these models can be applied to small deformations only. A general model for linear viscoelastic fluids is

$$\left(1+\alpha_1\frac{d}{dt}+\alpha_2\frac{d^2}{dt^2}+\ldots+\alpha_n\frac{d^n}{dt^n}\right)\sigma=\left(\beta_0+\beta_1\frac{d}{dt}+\beta_2\frac{d^2}{dt^2}+\ldots+\beta_m\frac{d^m}{dt^m}\right)\gamma \qquad 8.29$$

where $n = m$ or $n = m - 1$. Equation 8.29 relates stress and strain directly, rather than through the strain rate. The scalar variables, σ and γ, could be replaced by their tensor representations if desired.

There are some special cases of Equation 8.29 that the student should recognize.

Hooke's Law. When β_0 is the only nonzero coefficient, we have an equation describing linear elastic solid behavior,

$$\sigma = \beta_0\,\gamma \qquad 8.30$$

where β_0 is the shear or rigidity modulus, and the material responds like a spring.

Newton's Law. When β_1 is the only nonzero coefficient, the equation is similar to Equation 8.1 and represents Newtonian viscous flow,

$$\sigma = \beta_1\frac{d\gamma}{dt} = \beta_1\,\dot{\gamma} = \eta\,\dot{\gamma} \qquad 8.31$$

where β_1 is the dynamic viscosity, and the material responds like a dashpot.

Voigt (Kelvin) Model. When β_0, the shear modulus, and β_1, the viscosity, are nonzero, the general equation reduces to one of two simple linear viscoelastic models,

$$\sigma = \beta_0\,\gamma + \beta_1\,\dot{\gamma} = G\gamma + \eta\,\dot{\gamma} \qquad 8.32$$

where β_0 equals the shear modulus, G, and β_1 equals the viscosity. When a change in stress, σ, is made at $t = 0$, the strain will change as

$$\gamma = \frac{\sigma}{G}\left[1 - \exp\left(\frac{-tG}{\eta}\right)\right] \qquad 8.33$$

Comparison of Equations 8.30 and 8.33 suggests a time-dependent response by the Voigt model to attain the Hooke model modulus. The group η/G is analogous to a relaxation time constant; τ_M.

Maxwell Model. The Maxwell model is formulated with α_1 and β_1 as nonzero coefficients,

$$\sigma + a_1 \dot{\sigma} = \sigma + \tau_M \dot{\sigma} + \eta \dot{\gamma} = \beta_1 \dot{\gamma} \quad 8.34$$

where $a_1 = \tau_M$ and $\beta_1 = \eta$. When there is a step change in strain, γ, the stress in the material at constant strain decreases with time.

$$\sigma = \eta \gamma [1 - \exp(-t/\tau_M)] \quad 8.35$$

There are mechanical analogies for these models using springs to describe the elastic response and dashpots to describe the viscous response of the fluid. These are derived in the Appendix at the end of this chapter. These classic descriptions are useful for introducing viscoelastic concepts, but do not describe most systems well. More complex models can be generated by increasing the number of nonzero coefficients. For example, the Jeffreys model has nonzero a_1, β_1, and β_2; the Burgers model has four nonzero coefficients and is analogous to the Voigt and Maxwell models in series.

The linear models can replicate some of the characteristics of the viscoelastic response of polymer solutions and solids. However, two, three, or four coefficient models usually are not sufficient to give good predictions of polymer behavior. A series of Voigt or Maxwell elements gives a better description, leading to the concept of a continuous distribution of relaxation times. For a distribution of Maxwell models in series, the result is

$$\sigma(t) = \int_0^{\infty} \frac{N(\tau)}{\tau} \int_{-\infty}^{t} \exp\left(\frac{-(t - t^1)}{\tau}\right) \dot{\gamma}(t^1)\, dt^1\, d\tau \quad 8.36$$

where the first integration describes the time dependency of the shear rate and the second integration accounts for the distribution of relaxation times, $N(\tau)$. Equation 8.36 represents a generalization of Equation 8.35. Another generalization would be to consider a distribution of elastic moduli.

There are several limitations to the linear viscoelastic models. They are most accurate for small deformations, and do not give good predictions for large deformations for which the material response often is nonlinear. The spring and dashpot analogy is helpful because it is easy to visualize the material response. However, it is difficult to relate the model coefficients to molecular events and characteristics of polymers. The concept of a distribution of response times is valuable because these can be measured, even if it may be difficult to connect them with polymer properties.

Material Response to Oscillating Deformations. It is possible to deform the fluids with small amplitude oscillations in the rotating viscometers in Table 8.3. Suppose that the strain of the sample is given by

$$\gamma = \gamma_0 \cos \omega t \quad 8.37$$

where ω is the frequency and γ_0 is the strain amplitude. The strain rate is the differential of Equation 8.37 with respect to time.

$$\dot{\gamma} = -\gamma_0 \omega \sin \omega t \tag{8.38}$$

This forcing function is substituted into Equation 8.32 to solve for the stress as a function of time.

$$\sigma = \frac{\eta \omega \gamma_0}{(1 + \omega^2 \tau^2)} (\omega \tau \cos \omega\tau - \sin \omega\tau) \tag{8.39a}$$

$$= G^{I} \cos \omega t - G^{II} \sin \omega t \tag{8.39b}$$

There are two components to the stress response. The coefficient of the cosine term is the elastic response of the material and is in phase with the cosine forcing function. The coefficient of the sine function is the relaxation modulus of the material. Figure 8.19 shows the responses of a Newtonian and a viscoelastic fluid to an oscillatory

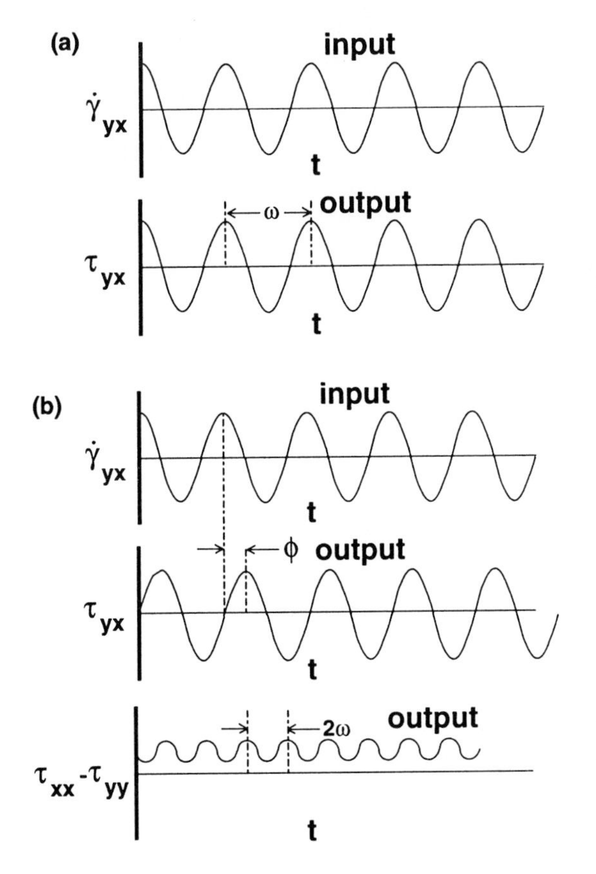

Figure 8.19a Time-dependent response of a Newtonian fluid. For a sinusoidal shear rate, there is no lag time for the shear stress response.

Figure 8.19b Time-dependent response of a viscoelastic fluid. For a sinusoidal shear rate input, the shear rate output lags by ϕ. There is a measurable, nonzero normal shear stress difference with a nonzero mean and frequency 2ω.

strain. The Newtonian material has a shear stress response that is in phase with the forcing function ($G^{II} = 0$). The viscoelastic fluid has a response that lags the input function by a constant amount ($G^{II} \neq 0$). The model for the system is

$$\sigma(t) = G^*(\omega)\,\gamma\,(t) \qquad 8.40$$

where G^* (ω) is the complex shear modulus. G^* is modeled as

$$G^* = G^I + iG^{II} \qquad 8.41$$

where $i = \sqrt{-1}$. A complex viscosity, η^*, can be defined

$$\eta^* = \eta^I + i\eta^{II} \qquad 8.42$$

where η^I is the dynamic viscosity and η^{II} represents energy stored by the fluid. Analogous equations can be written for other material properties, such as the compliance.

An alternate method for describing the shear modulus is to use the elastic modulus, G^I, and the loss angle, ϕ, which is the ratio of G^{II}/G^I. Figure 8.20 shows how the normalized dynamic viscosity, η^1/η, normalized storage modulus, G^1/G, and normalized loss modulus, G^{II}/G, vary with $\omega\tau$ for a Maxwell model. The dynamic viscosity decreases and the storage modulus rises over a narrow, two-decade change in $\omega\tau$. This range of $\omega\tau$ represents the viscoelastic region for the Maxwell material. At low frequencies, the material has a viscous response ($G^I \cong 0$). At high frequencies, the material has an elastic response ($\eta^I = 0$).

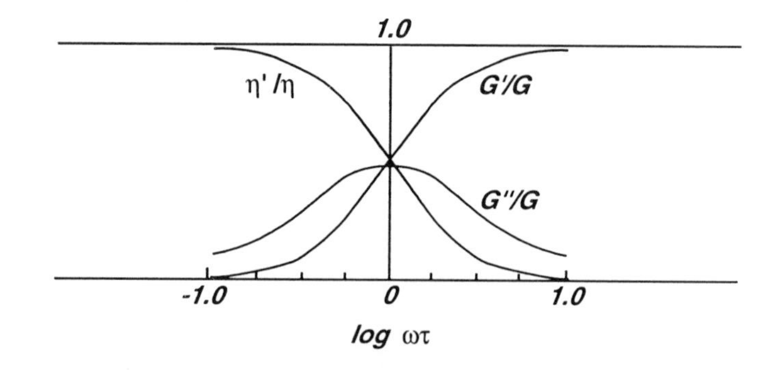

Figure 8.20 The Maxwell model in oscillatory shear. Variation of the normalized moduli and viscosity with normalized frequency ($\tau = \eta/G$). The superscripts indicate storage (′) and the loss (″) moduli. Published with permission from the publisher from, *An Introduction in Rheology*, H.A. Barnes, J.F. Hutton, and K. Walters. Copyright 1989 by Elsevier Science Publishing Co., Inc.

8.1.8 Normal Stresses and Stress Relaxation

Stresses and shear rates in polymer processing equipment are not usually steady. Two examples are the melt flow past a screw flight as it turns past the barrel and banding of polymer on a two-roll mill. Newtonian fluids would respond instantly to the deforming force and would do so only in the direction of the applied shear force. Non-Newtonian fluids, which includes most polymers and polymer solutions, exhibit time-dependent responses to deforming forces and respond in directions other than the shear direction. Therefore, it is important to have models for viscosity that can predict polymer performance in unsteady shear flows.

Consider an experiment in which a polymer film is subjected to a small amplitude oscillatory shear. The bottom plate is fixed and the top plate oscillates. The plate velocity can be represented by

$$U_x = (\gamma_0 \cos \omega t)\tau \qquad 8.43$$

A Newtonian fluid will have the shear stress in phase with the shear rate (Fig. 8.19a). The vertical pressure on the top plate will be zero, the normal stress difference, $\tau_{xx} - \tau_{yy}$, is zero, and its mean is zero. The shear stress tensor for a Newtonian fluid in this shear flow is

$$\underline{\underline{\tau}} = \begin{vmatrix} p & \tau_{yx} & 0 \\ \tau_{yx} & p & 0 \\ 0 & 0 & p \end{vmatrix} \qquad 8.44$$

where $\tau_{xy} = \tau_{xy}$ by symmetry. A non-Newtonian fluid with normal stresses will have a shear stress out of phase with the shear rate (Fig. 8.19b). There will be a measurable pressure on the top plate due to the normal forces in the y direction, and the normal force difference will have a nonzero mean. The shear stress tensor for this flow is

$$\underline{\underline{\tau}} = \begin{vmatrix} p + \tau_{xx} & \tau_{yx} & 0 \\ \tau_{yx} & p + \tau_{yy} & 0 \\ 0 & 0 & p + \tau_{zz} \end{vmatrix} \qquad 8.45$$

The normal stress components contain the total system pressure as a common factor, so that an easy way to determine the magnitude of the normal stress components (τ_{ii}) is to subtract the normal stresses. Direction 1 is the direction of the velocity component, direction 2 is the direction of the velocity component gradient, and direction 3 is the neutral direction.

Primary Normal Stress Difference:

$$(p + \tau_{11}) - (p + \tau_{22}) = \tau_{11} - \tau_{22} = \tau_{xx} - \tau_{yy} \qquad 8.46$$

Secondary Normal Stress Difference:

$$(p + \tau_{22}) - (p + \tau_{33}) = \tau_{22} - \tau_{33} = \tau_{yy} - \tau_{zz} \tag{8.47}$$

A number of important practical problems in polymer processing arise out of normal stress responses. One common example is die swell. Shear stresses are imposed on the polymer solution inside the capillary. Upon exiting the die and forming a jet, the polymer is free to respond to its own internal forces and swells in the radial direction.

The Weissenberg effect is another example. A polymer solution with normal stresses will climb a stirring rod. You might try to draw the forces on a small volumetric element of polymer solution and decide whether this response makes physical sense. The primary normal stress difference is usually larger than the secondary normal stress difference by about an order of magnitude.

8.1.9 Nonlinear Viscoelastic (Dynamic) Models

Dynamic models have been developed to describe the time-dependent response of polymer melts and solutions to shear. These models can also be applied to deformations of solid polymers in the gel or rubber states. A number of investigators have tried to relate the characteristics of polymer coils to their time-dependent phenomena. The Rouse model is one of the more successful ones. The macromolecule is represented by a bead and spring model, which is to replicate the motions of subunits of the chain. The shear storage modulus and the shear loss modulus models are

$$G'(\omega) = NkT\,(\omega\tau)^2/[1 + (\omega\tau)^2] \tag{8.48a}$$

$$G''(\omega) = \omega\,\gamma_0 + NkT\,\omega\tau/[1 + (\omega\tau)^2] \tag{8.48b}$$

where k is the Boltzman constant and N is the number of segments. These equations relate fairly well to the dependence of the storage and loss moduli on perturbation frequency for real polymer samples. The Rouse model is just one of many that have been developed. Further discussion of dynamic rheology is found in graduate texts on polymer rheology and polymer processing.

8.1.10 Elongational Flows

The models discussed above for viscosity were for shear flow fields. There are a number of important polymer flows that do not involve shear gradients—these flows are elongational, extensional, or shear free flows. Fiber drawing (also known as fiber spinning), film forming, parison blowing, and converging channel flow are all examples of elongational flow. In practice, these types of flows are nonisothermal and often involve a change in phase. In the case of fiber spinning, the polymer is changing from a melt to an oriented solid.

Figure 8.21 shows a fiber being formed by pulling a polymer melt in the 1 direction, causing an excess normal stress in that direction. The total stress in the 1 direction is

$$\tau_{ll} + p = F_1/A_1 \tag{8.49}$$

where F_1 is the total force needed to cause the flow and A_1 is the cross-sectional area normal to the flow. The normal stresses in the other directions equal the static pressure. Such flows have the general flow field,

$$U_i = a_i x_i;\ 2a_i = \gamma_{ii} \tag{8.50}$$

where U_i's are the velocities in the normal directions, and the coefficients, a_i, are constants. The velocities sum to zero for incompressible fluids

$$\Sigma\, U_i = 0;\ \Sigma\, a_i = 0 \tag{8.51}$$

The velocity of the fiber in the machine direction is $U_1 = \dot{\varepsilon}\, x_1$. Since the sum of a_i's equals zero, and by symmetry the change in a_2 and a_3 should be identical,

$$a_2 = a_3 = -\dot{\varepsilon}/2 \tag{8.52}$$

The diagonal terms in the shear rate tensor are

$$\gamma_{11} = 2\dot{\varepsilon};\ \dot{\gamma}_{22} = \dot{\gamma}_{33} = \dot{\varepsilon} \tag{8.53}$$

and the stress tensor is

$$\dot{\underline{\underline{\gamma}}} = \begin{vmatrix} 2\,\varepsilon & 0 & 0 \\ 0 & -\dot{\varepsilon} & 0 \\ 0 & 0 & -\dot{\varepsilon} \end{vmatrix} \tag{8.54}$$

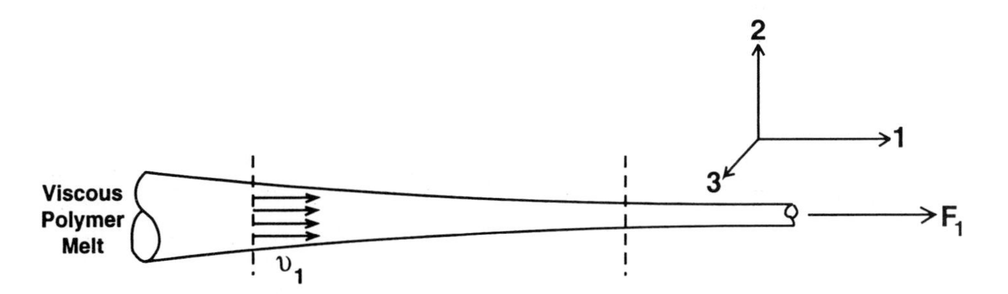

Figure 8.21 One-dimensional extension/elongation fiber drawing.

As with generalized Newtonian fluids, the elongational viscosity should be related to scalar invariants of the shear rate or shear stress tensor. For simple elongational flow, the scalar invariants are

$$I_{\dot{\gamma}} = 0; II_{\dot{\gamma}} = 6\,\dot{\gamma}^2; III_{\dot{\gamma}} = 2\,\dot{\varepsilon}^2 \qquad 8.55$$

Elongational viscosities are very difficult to measure. The typical experiment is to pull the sample at an exponential rate with time so that the rate of elongation, $\dot{\varepsilon}$, is a constant. An elongational viscosity can be defined by dividing the excess stress needed to cause the flow (the force above atmospheric pressure) by the elongation rate.

$$\eta(\dot{\varepsilon}) = \frac{-(\tau_{11} - \tau_{33})}{\dot{\varepsilon}} \qquad 8.56$$

One relation that does seem to hold is the Trouton relationship, which states that the elongational viscosity taken at zero elongation rate is three times the zero shear rate viscosity.

$$\eta(\dot{\varepsilon}) = 3\mu \qquad 8.57$$

Elongational flows are an important area for study because they occur in many polymer forming processes.

8.2 THE ELASTIC STATE

Mechanical Model of Elastomers. The width of the rubber-elastic region along the temperature axis in a modulus-temperature diagram (Fig. 1.15) is dependent on the molecular weight. The higher the molecular weight, the wider the region. Elastomers are used above their glass transition temperatures. Covalent cross-links between elastomer molecules are normally formed in commercial products in order for the material to retain its shape.

Under very low stress, there will be no visible elongation of the elastomer (a minimum amount of energy is needed to start strain). Crystallization can occur in the stretched state and will increase the tensile strength of the material. Think of stretching a rubber band; it is harder to stretch it near the breaking point than to deform it from its original shape. In the absence of crystallization, the energy stored while stretching an elastomer is in the changed conformation of the chains. This means that the elastic modulus increases with temperature, rather than decreasing as would happen for energy stored by deforming primary bonds (enthalpic).

8.2.1 Thermodynamic Analysis of Polymeric Solids

Many students are familiar with models of the thermodynamic state properties of internal energy, enthalpy and entropy based on the volume change of the system. These models form the basis for gas equations of state. Thermodynamic analysis of polymer volume can lead to equations of state for polymers. Similar analyses of polymer extension and surface tension lead to mechanical and surface equations of state for polymers.

Polymer Volume as a Function of Temperature and Pressure. Estimating solid volume as a function of temperature and pressure is fundamental to a wide variety of polymer engineering problems. It also leads to models analogous to equations of state for gases. The total derivative of volume with respect to temperature and pressure is

$$dV = \left(\frac{\partial V}{\partial T}\right)_P dT + \left(\frac{\partial V}{\partial P}\right)_T dP \qquad 8.58$$

where the two partial derivatives are determined while the other property is held constant. It is convenient to relate the thermodynamic coefficients to the percent or fractional change in volume when another property is varied. The fractional volume change is given by dividing all factors in Equation 8.58 by the volume, V.

$$\frac{dV}{V} = \frac{1}{V}\left(\frac{\partial V}{\partial T}\right)_P dT + \frac{1}{V}\left(\frac{\partial V}{\partial P}\right)_T dP \qquad 8.59$$

The multiplier for differential temperature is the volume expansivity, β. See Table 8.4 for a listing of this and several of the following coefficients and definitions. The multiplier for differential pressure is the isothermal compressibility, κ. The inverse of κ often is reported as the bulk modulus, B. The fractional volume change as a function of temperature and pressure is

$$\frac{dV}{V} = \beta\, dT - \kappa\, dP \qquad 8.60$$

The coefficients, β and κ, are nearly independent of temperature and pressure, respectively, for modest changes in these variables. An important goal of polymer equations of state is to predict β and κ as functions of system conditions and polymer properties.

Some problems require models for pressure changes with temperature when the volume of the solid is held constant. This partial derivative is

$$-\left(\frac{\partial P}{\partial T}\right)_V = \left(\frac{\partial P}{\partial V}\right)_T \left(\frac{\partial V}{\partial T}\right)_P = \frac{-\beta}{\kappa} \qquad 8.61$$

TABLE 8.4 Definitions of Solid Properties and Coefficients

Property/Coefficient	Symbol	Definition
VOLUME		
Thermal expansivity, T^{-1}	β	$\frac{1}{V}\left(\frac{\partial V}{\partial T}\right)_P$
Isothermal compressibility, P^{-1}	κ	$\frac{-1}{V}\left(\frac{\partial V}{\partial P}\right)_T$
Bulk modulus, P	B	$-V\left(\frac{\partial P}{\partial V}\right)_T$
	$\frac{\beta}{\kappa}$	$\left(\frac{\partial P}{\partial T}\right)_V$
LENGTH		
Linear expansivity, T^{-1}	α	$\frac{1}{L}\left(\frac{\partial L}{\partial T}\right)_F$
Young's modulus, F	E	$\frac{L}{A}\left(\frac{\partial F}{\partial L}\right)_T$, A = area
	$-\alpha AE$	$\left(\frac{\partial F}{\partial T}\right)_L$
SURFACE TENSION		
Isothermal extensibility, $\dot{\varepsilon}^{-1}$	ε	$\frac{1}{\sigma_0}\left(\frac{\partial \sigma}{\partial \dot{\varepsilon}}\right)_T$
Thermal thinning coefficient, T^{-1}	τ	$\frac{-1}{\sigma_0}\left(\frac{\partial \sigma}{\partial T}\right)_{\dot{\varepsilon}}$

Elongation with Temperature and Force. Mechanical properties involving dimensional changes and imposed forces can be modeled as well. The strain in a sample can be related to the imposed force, F, and the temperature, T. The change in sample length is

$$dL = \left(\frac{\partial L}{\partial T}\right)_F dT + \left(\frac{\partial L}{\partial F}\right)_T dF \qquad 8.62$$

The fractional change in length is

$$\varepsilon = \frac{dL}{L} = \frac{1}{L}\left(\frac{\partial L}{\partial T}\right)_F dT + \frac{1}{L}\left(\frac{\partial L}{\partial F}\right)_T dF = \alpha\, dT + \frac{A}{E}\, dF \qquad 8.63$$

Equation 8.63 is a mechanical equation of state for a polymer solid and defines the linear expansivity and the Young's modulus (Table 8.4). The change in force at constant extension with variable temperature is the ratio of the linear expansivity to the Young's modulus.

Surface Tension with Rate of Elongation and Temperature. Surface tension can be defined as an equation of state. Often, surface tension is a function of temperature and the rate of elongation of the surface, $\dot{\varepsilon}$. Two coefficients—an isothermal extensibility and a thinning coefficient—come from this analysis (Table 8.4).

Internal Energy and Heat Capacities. The internal energy of the polymer can be used to define heat capacities. The equation for internal energy is

$$dU = dQ - dW \tag{8.64}$$

where U is internal energy, Q is heat, and W is work. The definition of work depends on the equation of state choice.

Volume	$dW = PdV$	8.65a
Length	$dW = -FdL$	8.65b
Surface Tension	$dW = -\sigma dA$	8.65c

The heat capacities of the solid are defined based on the system choice. Heat capacity for most polymers is a strong function of temperature and a weak function of pressure. The pressure dependence is not important except when this variable changes by order of magnitude during processing. A case in point is the change in pressure during extrusion or injection molding. In these cases, the polymer heat capacity may change by 10% to 20% during processing (Kamel and Levan, 1972).

Enthalpy and Entropy. Enthalpy and entropy can be related to heat capacity, thermal expansivity, and isothermal compressibility. The derivative of enthalpy is

$$dH = dU + PdV + VdP \tag{8.66}$$

Using the volume form of different work, differential enthalpy is

$$dH = dQ + VdP \tag{8.67}$$

Differentiating the enthalpy with respect to temperature at constant pressure gives

$$\left(\frac{\partial H}{\partial T}\right)_P = \left(\frac{\partial Q}{\partial T}\right)_P = C_P \tag{8.68}$$

The entropy is

$$dS = C_P \frac{dT}{T} - \left(\frac{\partial V}{\partial T}\right)_P dP = C_P \frac{dT}{T} - V\,\beta\,dP \qquad 8.69$$

Using other definitions of work, enthalpy and entropy can be related to the Young's modulus, the linear expansivity, and the constant length or constant force heat capacity. The thermodynamic analysis of rubber elasticity is an example of a case in which such results are very useful.

8.2.2 Deformation of Elastomer Chains

The analysis of the extension of an individual chain will help show how elastomers deform elastically. The results apply best to systems that are *lightly cross-linked*, having several cross-linking sites per one hundred repeating units. These systems could permit chain motion under deforming forces. The analysis is done on an individual chain segment between branching points. In a real elastomer, other chain segments could be imagined in the space between the chains being analyzed. Choosing $|\overline{AB}|$ as the end-to-end distance of the polymer chain, the entropy of this chain segment is

$$S = S_o + k \ln P(r) \qquad 8.70$$

where S_o is a zero state entropy, k is the Boltzmann constant, and $P(r)$ is the probability of having the end-to-end distance, $|\overline{AB}|$. The end-to-end distance is related to the density for Gaussian chains,

$$|\overline{AB}| = \left\langle r_0^2 \right\rangle = \frac{3}{2}\rho^2 \qquad 8.71$$

and the probability that a chain has length, r, is

$$P(r) = \frac{\exp\left[-(r/\rho)^2\right]}{(\rho\sqrt{\pi})^3} \qquad 8.72$$

The model for entropy becomes

$$S = S_0 - k\left[3\ln\left(\sqrt{\pi}\rho\right) + (r/\rho)^2\right] \qquad 8.73$$

The change in entropy as length changes is

$$\left(\frac{\partial S}{\partial r}\right) = \frac{-2kr}{\rho^2} \qquad 8.74$$

Entropy can be linked to the tensile force by the Helmholtz free energy. This state function is used because elastomers deform with essentially no volume change (their Poisson's ratio is near 0.5). The Helmholtz free energy, A, is defined as

$$A = U - TS \tag{8.75}$$

$$dA = dU - TdS - SdT \tag{8.76}$$

The change in internal energy of the chain is

$$dU = TdS + dW \tag{8.78}$$

and

$$dA = dW - SdT \tag{8.79}$$

Most chain extension will take place isothermally, so the change in the Helmholtz free energy is directly related to the work of extension.

$$dA = dW = FdL \tag{8.80}$$

Equation 8.79 can be written as

$$F = \left(\frac{\partial A}{\partial L}\right)_{T,V} = \left(\frac{\partial U}{\partial L}\right)_{T,V} - T\left(\frac{\partial S}{\partial L}\right)_{T,V} \tag{8.81}$$

Metal and ceramic materials have large changes in internal energy with extension since primary bonds are being deformed. Their entropy changes little with extension because individual atoms do not move past one another. These materials are elastic with high moduli. The opposite conditions describe rubber elasticity. The internal energy does not change much with extension, but there are large changes in the chain segment conformations. Entropy decreases with extension because there are fewer conformations for the larger end-to-end distance. Most of the force can be related to the second factor on the right-hand side of Equation 8.81.

$$F \cong -T\left(\frac{\partial S}{\partial L}\right)_{T,V} \tag{8.82}$$

Substituting Equation 8.74 into Equation 8.82 gives

$$F = \frac{2kTr}{\rho^2} = Kr \tag{8.83}$$

The force required to extend or compress the chain segment is directly proportional to the distance, r, and is directly proportional to temperature. The chain acts like a spring (elastically), and its modulus increases with temperature.

8.2.3 Deformation of Elastomer Networks

Suppose that a elastomer volume with dimensions X_0, Y_0, and Z_0 is deformed by extensional stress to a new volume defined by X, Y, and Z. There is no change in volume on this extension, and the material is isotropic. No change in volume implies that

$$X_0 Y_0 Z_0 = XYZ \tag{8.84}$$

The extension ratios in each direction are

$$\lambda_x = \frac{X}{X_0}, \lambda_y = \frac{Y}{Y_0}, \lambda_z = \frac{Z}{Z_0} \tag{8.85}$$

By definition, the product of the three extension ratios is one. The deformation is assumed to be affine, which means that displacements are uniform across the solid. All chain displacements can be represented by a single deformation vector. The undeformed end-to-end vector is

$$r_0 = x_0 i + y_0 j + z_0 k \tag{8.86}$$

The deformed vector will be

$$r = \lambda_x x_0 \, i + \lambda_y i_0 \, j + \lambda_z z_0 \, k \tag{8.87}$$

The forces for each vector are

$$f_0 = kr_0; \; f = K r \tag{8.88}$$

The deformed vector is for a uniaxial extension. If the material is isotropic, $\lambda_x \lambda_y \lambda_z = 1$ and the extension ratio in the z direction equals 4.0.

$$\lambda_x = \lambda_y = \frac{1}{\sqrt{\lambda_z}} \tag{8.89}$$

$$\lambda_z = 4, \lambda_x = \lambda_y = \frac{1}{2} \tag{8.90}$$

The individual force components are

$$f_z = K(4z_0 - z_0) = 3Kz_0 \tag{8.91a}$$

$$f_y = K\left(\frac{y_0}{2} - y_0\right) = -Ky_0/2 \tag{8.91b}$$

$$f_x = K\left(\frac{x_0}{2} - x_0\right) = -Kx_0/2 \tag{8.91c}$$

The stress-strain properties of this material are computed by equating the external deforming forces to the resulting internal forces. The work in the z direction is

$$w_z = \int_{z_0}^{\lambda_z z_0} f_z d_z \qquad 8.92a$$

$$= \frac{2kT}{\rho^2} \int_{z_0}^{\lambda_z z_0} z d_z \qquad 8.92b$$

$$= \frac{kT}{\rho^2} (\lambda_z^2 - 1) z_0^2 \qquad 8.92c$$

where Equation 8.83 is used to describe the force on the chain. Equation 8.92c shows the work for one chain with end-to-end distance in the z direction, z_0. The sum of w_z over all chains, N, is the total work for the sample. Since the deformation is affine and the density is constant,

$$\sum_1^N w_z = \frac{kT}{\rho^2} (\lambda_z^2 - 1) N \langle z_0^2 \rangle \qquad 8.93$$

where $\langle z_0^2 \rangle$ is the mean squared value of z_0. Since the undeformed state is isotropic,

$$\langle z_0^2 \rangle = \langle y_0^2 \rangle = \langle x_0^2 \rangle = \frac{\langle r_0^2 \rangle}{3} \qquad 8.94$$

The density can be related to the end-to-end distance of a chain that is not cross-linked, $\langle r_1^2 \rangle$. The final equation for the z component of work is

$$\sum_1^N w_z = \frac{NkT}{2} \frac{\langle r^2 \rangle_0}{\langle r^2 \rangle_l} (\lambda_z^2 - 1) \qquad 8.95$$

The equations for the other components are similar, so

$$w = \frac{NkT}{2} \frac{\langle r_0^2 \rangle}{\langle r_l^2 \rangle} (\lambda_x^2 + \lambda_y^2 + \lambda_z^2 - 3) \qquad 8.96$$

When the number of chains per unit volume, $\upsilon = N/V$, is known,

$$w = \frac{VG}{2} (\lambda_x^2 + \lambda_y^2 + \lambda_z^2 - 3); \; G = \nu k T \frac{\langle r_0^2 \rangle}{\langle r_l^2 \rangle} \qquad 8.97$$

Example 8.5 Stress-Strain Relationship for Uniaxial Stress

A uniaxial stress is a simple case to model. The model is appropriate for compression or extension. The extension ratios are given by Equation 8.90. The total work done is

$$w = \frac{VG}{2}(\lambda_x^2 + \lambda_y^2 + \lambda_z^2 - 3)$$

$$= \frac{VG}{2}\left(\frac{1}{\lambda_z} + \frac{1}{\lambda_z} + \lambda_z^2 - 3\right)$$

The force causing the deformation is related to the work by Equation 8.80. Therefore,

$$F = \frac{dw}{dL} = \frac{dw}{d\lambda_z} \cdot \frac{d\lambda_z}{dL}$$

The first derivative is

$$\frac{dw}{d\lambda_z} = \frac{VG}{2}\left(2\lambda_z - \frac{2}{\lambda_z^2}\right)$$

The second derivative is given by the definition of the extension ratio.

$$\frac{d\lambda_z}{dL} = \frac{1}{L_0}$$

Therefore,

$$F = \frac{VG}{L_0}\left(\lambda_z - \frac{1}{\lambda_z^2}\right) = A_oG\left(\lambda_z - \frac{1}{\lambda_z^2}\right)$$

For the extension ratio of four,

$$F = \frac{63}{16}A_oG$$

The results of Example 8.5 can be used to define the engineering stress.

$$\sigma = \frac{F}{A_o} = G\left(\lambda - \frac{1}{\lambda^2}\right) \tag{8.98}$$

Equation 8.98 is tested in Figure 8.22 for a natural rubber sample. The theory is excellent for compressions but fails for tensions in which $\lambda > 1.2$. For high extensions, natural rubber crystallizes, so the density changes as does the enthalpy of retraction. The simple model assumptions do not hold; however, the model gives a reasonable qualitative description of stress-strain behavior of elastomers.

The true stress of the solid is the load divided by the actual area of the sample. Area can be related to the sample volume divided by the length, so the engineering tensile stress is

$$\sigma_t = G\left(\lambda^2 - \frac{2}{\lambda}\right) \tag{8.99}$$

8.2.4 Other Models for Rubber Elasticity

The Mooney–Rivlin equation is similar to Equation 8.99. It is derived based on symmetry arguments and gives the work of deformation as

$$w = C_1\,(\lambda_x^2 + \lambda_y^2 + \lambda_z^2 - 3) + C_2\left(\frac{1}{\lambda_x^2} + \frac{1}{\lambda_y^2} + \frac{1}{\lambda_z^2} - 3\right) \tag{8.100}$$

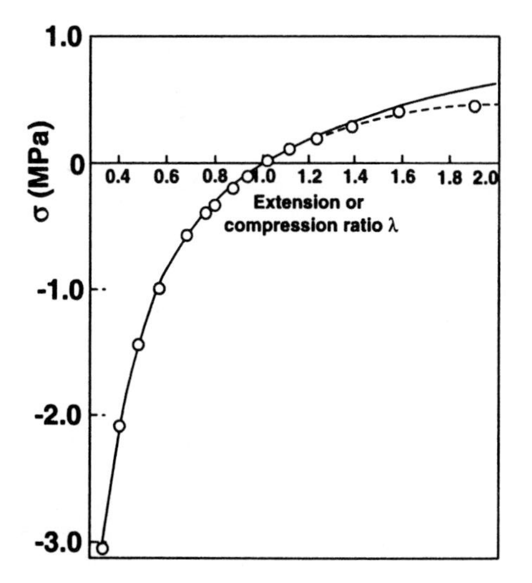

Figure 8.22 A comparison of theory and experiment for the tensile elongation and compression of a rubber prism. McCrum et al., 1988.

This equation is equivalent to the Gaussian network theory (Eq. 8.96) when $C_2 = 0$ and $C_1 = G/2$. The equation for the stress is

$$\sigma = 2C_1\left(\lambda - \frac{1}{\lambda^2}\right) + 2C_2\left(1 - \frac{1}{\lambda^3}\right) = \left(2C_1 + \frac{2C_2}{\lambda}\right)\left(\lambda - \frac{1}{\lambda^2}\right) \qquad 8.101$$

Gaussian network theory predicts that the group, $\sigma/(\lambda - 1/\lambda^2)$, is constant. The continuum theory suggests that this group varies with λ.

$$\frac{\sigma}{\lambda - \frac{1}{\lambda^2}} = 2C_1 + \frac{2C_2}{\lambda} \qquad 8.102$$

Figure 8.23 shows data for vulcanized rubbers plotted according to Equation 8.102. The value of C_2 for these data is about 2 kg/cm². The value for C_1 ranges from 2 to 6 kg/cm². Physical interpretation of these constants is difficult. Swelling of rubber samples with solvent decreases the value of C_2, which approaches zero as the volume fraction of the polymer approaches 0.2. The data in Figure 8.24 show this behavior.

An improved model for rubber elasticity can be developed by writing general forms for the strain-energy functions.

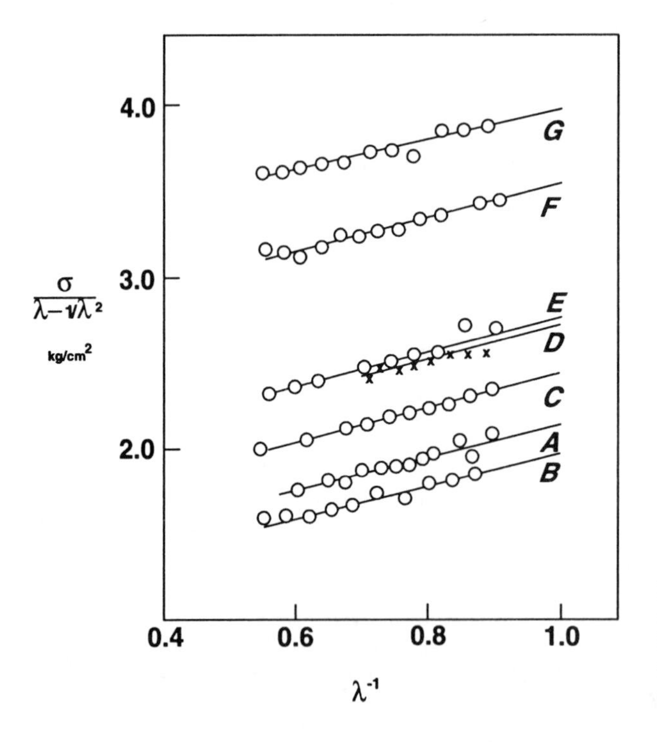

Figure 8.23 Plot of $\sigma/(\lambda - 1/\lambda^2)$ versus λ^{-1} for a range of natural rubber vulcanizates. Sulfur content increases from 3% to 4%, with time of vulcanization and other quantities as variables (Gumbrell et al., 1953).

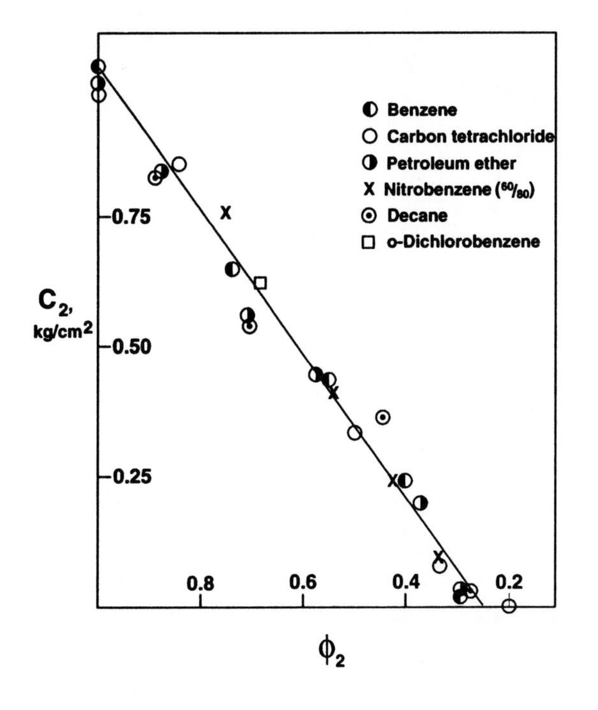

Figure 8.24 Dependence of C_2 on ϕ_2 for synthetic rubber vulcanizates. Open circle, butadiene-styrene, (95/5); circle shaded right, butadiene-styrene, (90/10); circle shaded left, butadiene-styrene, (85/15); solid circle, butadiene-styrene, (75/25); circle with dot, butadiene-styrene, (70/30); X, butadiene-acrylonitrile, 75/25) (Gumbrell et al., 1953).

$$\sigma = \left(C + \frac{C^{I}}{\lambda} + C^{II}\lambda^2 \right)\left(\lambda - \frac{1}{\lambda^2} \right) \qquad 8.103$$

This empirical model fits data very well; however, the constants do not have molecular interpretations. An example of a good fit is shown in Figure 8.25.

Solvent swelling of cross-linked elastomers leads to a different type of problem. Swelling is modeled as affine (deformation is the same in all directions) but not isochoric. The sample volume increases as the solvent fraction increases, so the product of the deformation ratios does not equal one,

$$\lambda_x \lambda_y \lambda_z = \frac{1}{\phi_2} \qquad 8.104$$

where ϕ_2 is the polymer volume fraction. The volume change affects two terms in the shear modulus (Eq. 8.97): the ratio of the radii and the concentration of chain segments. The radii ratio is increased by a factor of $1/\phi_2^{2/3}$ relative to the base case. The number of chain segments per unit volume is reduced by a factor of $1/\phi_2$. The net result is

$$\sigma = \phi_2^{1/3}\, G\left(\lambda - \frac{1}{\lambda^2} \right) \qquad 8.105$$

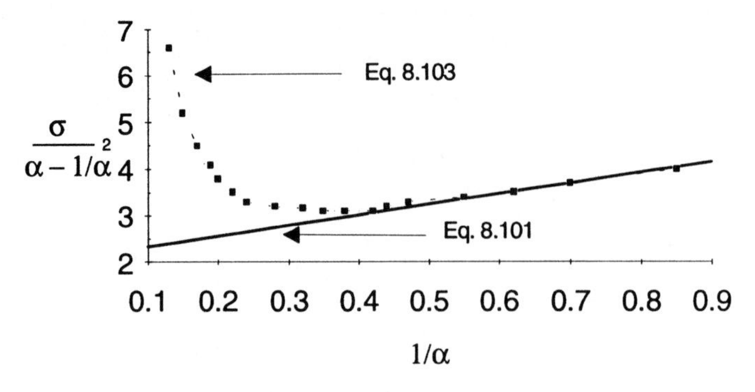

Figure 8.25 Mooney–Rivlin plot for sulfur-vulcanized natural rubber. Solid line: Equation 8.101; dotted line: Equation 8.103 (Sperling, 1986).

The stress required to achieve a given elongation ratio, λ, is lowered by the factor, $\phi_2^{1/3}$, as a combination of these two effects.

8.3 FLUID-SOLID TRANSITIONS AND RELAXATIONS

Significant flow of polymers does not occur below the glass transition temperature. Above T_g, viscosity decreases quickly as temperature increases. As discussed in Section 8.1.5 free volume theory and the WLF equation can be used to estimate melt viscosity between 50°C and 100°C above T_g. This range is near processing temperatures for many thermoplastics. Arrehnius equations can be used for scaling viscosities above these temperatures. The glass transition temperature can be changed by adding diluents or by changing the system pressure. The effects of both methods can be analyzed using free volume theory.

8.3.1 Effects of Diluents on T_g

One result of the theory is that free volume of many polymers at T_g is a constant: $f_g = 0.025$. This estimate for f_g has practical and theoretical significance. If f_g is a constant, then the glass transition temperature of a polymer can be changed by altering its free volume fraction. There are several ways to do this. One is to mix in other molecules that have larger free volume fractions at a given temperature. A second is to change the molecular weight of the polymer, which changes the number of chain ends. Chain ends have larger free volumes than the rest of the molecule because they are attached by one, not two, covalent bonds to the chain. Suppose that the extra free volume associated with chain ends is defined as θ, then the additional free volume per cm^3 due to chain ends is

$$\frac{2\rho N_A \theta}{M} \tag{8.106}$$

where ρ is the polymer density, N_A is Avogadro's number, and M is the molecular weight. Polymer having no chain ends (infinite molecular weight) can be used as a reference, giving the maximum T_g possible for the material. The group in Equation 8.106 should be equal to the thermal expansivity of the melt (see Table 7.10) times the difference between the T_g of the sample and T_g^{∞}, the glass transition temperature of an infinite molecular weight polymer.

$$T_g = T_g^{\infty} - \frac{K}{M};\ K = \frac{2\rho\, N_A\, \theta}{\alpha} \qquad 8.107$$

A plot of T_g versus inverse molecular weight should be linear, and its slope should be related to the free volume associated with chain ends. Figure 8.26 shows data for polystyrene samples treated in this fashion. The value of θ for styrene polymers is about 0.08 nm^3. The volume of a repeating unit in polystyrene is about 0.166 nm^3, so the volume ratio of chain ends to chain segments is 0.5, making chain ends significant contributors to free volume.

This concept can be used to estimate the effect of a plasticizer on the glass transition temperature. The Kelly–Bueche equation represents the T_g of a number of plasticized systems,

$$T_g = \frac{\alpha_p \phi_p T_{g,p} + \alpha_d \phi_d T_{g,d}}{\alpha_p \phi_p + \alpha_d \phi_d} \qquad 8.108$$

where the subscripts describe either the polymer (p) or diluent (d), and ϕ_i is the volume fraction of the component. Other effects are found in random copolymer systems, in which chain ends of one monomer may not interact well with chain segments of the second monomer type.

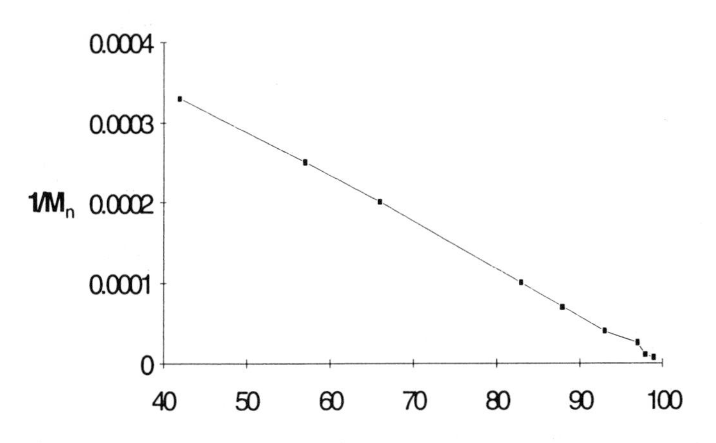

Figure 8.26 Glass-transition temperatures of polystyrene fractions plotted versus 1/M$_n$ (Fox and Flory, 1950).

8.3.2 Effects of Pressure on T_g

Increasing the system pressure decreases the free volume in the polymer and reduces T_g. One method for estimating the change in T_g with pressure is based on Equation 8.22 for the free volume. This is modified to include a compressibility term,

$$f(T, P) = f_g + \alpha(T - T_g) - \kappa P \qquad 8.109$$

where the glass transition temperature at zero pressure is used and κ is the free volume compressibility. If T_g is an iso-free volume event, then $f(T,P) = f_g$ at a given temperature and pressure, and we can equate the last two terms on the right hand side of Equation 8.109. This leads to

$$\frac{\partial T_g}{\partial P} = \frac{\Delta \kappa}{\Delta \alpha} \qquad 8.110$$

This relation applies to polystyrene.

8.3.3 Time-Temperature Superposition Principle

Polymer properties often are functions of the time period over which the measurement is made. It would be possible to plot elastic moduli curves, similar to those of Figure 1.14, that would have the time period of the measurement as the independent variable. A shifting procedure has been established experimentally that allows the construction of a master curve (complete modulus-time behavior at constant temperature). The free volume concept can be used to calculate the effect of temperature on properties other than viscosity. Rather than performing experiments across the complete time range at one temperature, experiments are performed across fairly limited time ranges, and temperature is changed from series to series as a way of slowing or speeding up the response. Example 8.6 shows how to shift an elastic modulus.

Example 8.6 Time-Temperature Superposition

The Deborah number of an experiment determines how a polymer will behave. Temperature can affect the material response time and thereby change the Deborah number. The elastic modulus (E) of polyisobutylene (PIB) has been measured after ten hours of stress at 25°C and found to be 10^6 dynes/cm^2. Suppose that this is a particularly accurate experimental set-up, and a measure-

(continued)

ment of the same modulus is needed at -20°C. Determine the amount of time needed before the readings should be taken.

The T_g for PIB is −70°C. The universal constants will be used for this example problem, but it would be more accurate to use constants for PIB (see Table 8.2). This problem will be solved by estimating the time needed for the event to occur at the reference temperature and then scaling it to −20°C.

Time for Event at T_g:

$$\log \frac{t(25^\circ C)}{t(-70^\circ C)} = \frac{-17.44(25-(-70))}{51.6+(25-(-70))}$$

$$t(-70) = 2 \times 10^{12} \text{ hrs}$$

Time for Event at −20°C:

$$\log \frac{t(-20^\circ C)}{t(-70^\circ C)} = \frac{-17.44(-20-(-70))}{51.6+(-20-(-70))}$$

$$t(-20) = (2 \times 10^{12}) \times (2.57 \times 10^{-9}) = 5140 \text{ hrs}$$

Time-temperature superposition is used for predicting long time performance. For example, accelerated tests can be done at high temperatures, and the results can be extrapolated to use temperature as a way of estimating performance properties at long times. This method can be used for evaluating plastic pipe, which is subjected to hydraulic pressure over long time periods and can exhibit creep.

8.3.4 Polymer Properties and Polymer Performance: Engineering Use Temperatures

General guidelines for using polymers based on the master curve for viscoelastic response were given in Chapter 1. Elastomers are used well above the glass transition temperature. There must be high segment mobility in order to have elastic responses. Amorphous structural polymers, which rely on rigidity, should be used well below T_g. Tough, leather-like polymers, such as plasticized PVC, can be used near T_g. Highly crystalline and oriented polymers should be used well below T_m. Semicrystalline polymers, such as polyethylene, are used in between T_g and T_m.

There are often unobvious relationships between polymer structure as it is measured in the laboratory and the mechanical properties of a finished part. There are extensive data on some crystalline polymers, and relationships can be established between processing and performance. Amorphous polymers are less characterized, and the relationships between processing and final properties are not as well known. Small changes in reactor conditions, catalysts, and impurities can make significant changes in performance. For these reasons, polymers are often characterized with respect to processing characteristics, even though these characteristics must be a result of the underlying polymer physics. Even if the polymer scientist has completely characterized a material with respect to T_g, viscosity, molecular weight, etc., the polymer engineer still wants to know the melt index.

Crystallinity and glass transition temperature are highly related to a polymer's processing characteristics. During the cooling process, there is a period of time required for polymer molecules to crystallize. This induction period is significant with regards to "machine time" and many crystalline polymers can be made in an amorphous form by quenching their melt to below T_g. The crystallization rate depends on the type of crystals being formed and the number of nuclei present.

Crystallization can be difficult to relate to the properties of a final product. Processing occurs under extreme conditions, and crystal growth tends to be sporadic. Crystals may develop during postprocessing treatment or during aging. Compounded resin may have new nucleation sites. Selective crystallization by molecular weight may also occur. The time-temperature superposition principle combined with the notion of linear viscoelasticity suggests that all prior thermal and stress history affects the state of the polymer at any given time. Therefore, the actual properties of a sample at the time of use by the customer are difficult to predict. In most cases, the customer is not sensitive to small changes in physical properties. However, when the material is being worked near the boundaries of its performance envelope, prior thermal and mechanical stresses may be critical and probably need to be controlled.

For similar reasons, the relationship between T_g and polymer properties is critical for amorphous polymers. For example, polymer blends or copolymers typically have T_g's in between those of the constituents. Tough, pliable polymers can be made by carefully choosing the components so that the blend has good characteristics. Cooling rates and conditions are very important for amorphous polymers. Supercooling can occur between T_m and T_g. Different glasses can be made by changing cooling rates; these glasses will have different properties.

8.4 MECHANICAL PROPERTIES OF POLYMER SYSTEMS

A wide variety of mechanical tests are available for evaluating end-use properties. These can be categorized with respect to short- and long-term responses (Table 8.5). Mechanical deformation tests measure material properties in the elastic and nonelastic regions. Durability tests evaluate ultimate properties of the material at failure. Surface tests include measures of hardness and wear. Discussions of two specific tests—tensile and impact strength—are given here as examples, although there are many others that could be included as well. Tensile tests are done to obtain the fundamental stress-strain curve of the material and are used in many applications.

TABLE 8.5 Mechanical End-use Properties

Type	Response	
	Short-Term	Long-Term
Deformation	*Stiffness*	*Creep*
	Stress-strain	Uniaxial
	Modulus	Flexural
	Yield stress	Creep
Durability	*Toughness*	*Endurance*
	Ductile and brittle fracture	Creep rupture
	Stress and elongation at break	Crazing, cracking
	Impact strength	Flex resistance
		Fatigue failure
Surface	*Hardness*	*Friction, Wear*
	Scratch resistance	Coefficient of friction
	Indentation hardness	Abrasion resistance

Impact strength testing is specific to parts that must withstand sudden forces. It is a typical example of a failure test.

8.4.1 Tensile Tests

The value of studying tensile test results is that much work has been done to relate failure mechanisms and test results to material properties on the molecular and conformation levels. This discussion of tensile testing should provide a good background for interpreting other mechanical tests. This test applies a uniform load in one direction to a dumbbell-shaped sample. The load on the sample and strain of the sample are measured while the specimen is stretched to fracture failure. The dumbbell shape ensures that the failure will occur in known cross-sectional area so that the stress can be calculated. The elongation can be determined either by determining the relative position of the jaws to each other or by strain gauges placed on the sample.

Definitions. Tensile test parameters have been described in Section 1.8 and 7.1.4 and are defined here for the gauge section of a tensile bar. Figure 8.27 shows a bar sample with initial cross-sectional area, A_o, and initial length, L. When a force, F, is applied to the sample, the sample elongates to the length, L, and has a new cross-sectional area, A. Elastic deformation is a constant volume process for small deformations, so that

$$A_o L_o = AL = constant \quad 8.111$$

There are two generally accepted sets of definitions of stress and strain. The engineering practice definitions are

$$\sigma_{eng} = engineering\ stress = \frac{load}{initial\ area} = \frac{F}{A_o} \tag{8.112a}$$

$$\varepsilon_{eng} = engineering\ strain = \frac{length\ change}{initial\ length} = \frac{L - L_o}{L_o} \tag{8.112b}$$

where the load, area, and lengths are shown in Figure 8.27. Stress and strain can be evaluated at the actual deformation of the sample.

$$\sigma_{true} = true\ stress = \frac{load}{area\ at\ load} = \frac{F}{A} \tag{8.113a}$$

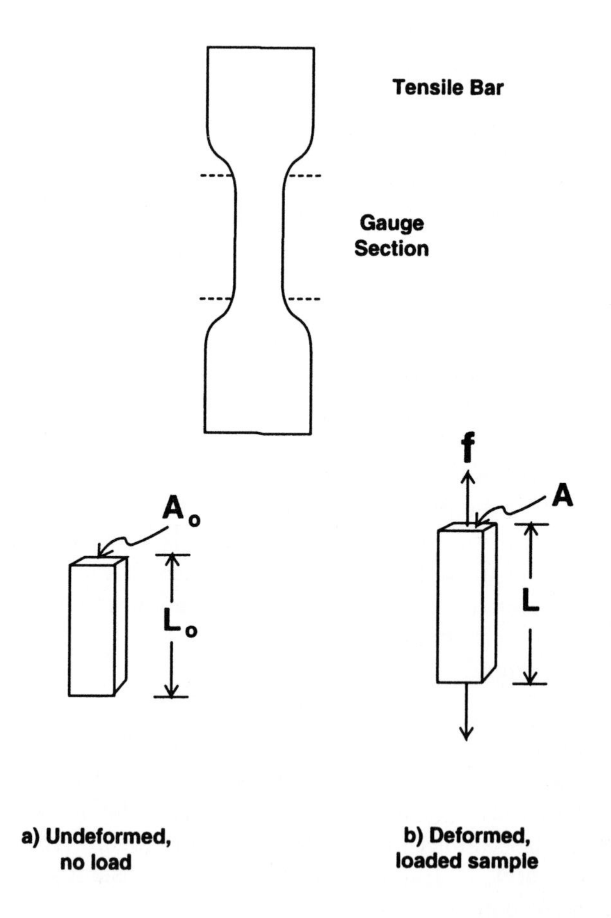

Figure 8.27 Tensile bar sample under unidirectional load: **(a)** no load; **(b)** loaded sample.

$$\varepsilon_{true} = true\ strain = \ln\left(\frac{length}{initial\ length}\right) = \ln\left(\frac{L}{L_o}\right) \quad 8.113b$$

Many polymers will become oriented during high strains, with portions of the material undergoing crystallization. In these cases, Equation 8.111 is not valid and the following analysis is incorrect. Under the constant volume restriction, the above equations can be combined to show that

$$\sigma_{true} = \frac{F(L_o/L)}{A_o} = \sigma_{eng}(L/L_o) = \sigma_{eng}(1 + \varepsilon_{eng}) \quad 8.114$$

and

$$\varepsilon_{true} = \ln(\varepsilon_{eng} + 1) \quad 8.115$$

The true stress and true strain are always larger than the engineering stress and strain values. When there are density changes in the sample during strain, the true stress and strain are determined by direct measurement.

When a material is deformed, it absorbs energy because the force has acted over the deformation distance, $L - L_o$. The sample's strength is the stress required to make it fail. Ductility describes the amount of permanent strain prior to fracture failure (the breakage of the sample). Toughness is the amount of energy absorbed by the material during fracture failure (the area under the stress-strain curve). The initial yield, or yield point, defines the stress or strain up to which deformations are elastic and the material can recover its initial dimensions. The maximum tensile strength is related to the highest load the material can take prior to fracture. Table 8.6 shows standard definitions and units for mechanical properties of materials.

Elastic deformations occur below the yield point, and the stress-strain curve is linear with the slope, *E*, Young's modulus. The modulus of elasticity is defined by Equation 8.4. Stresses below the yield stress are absorbed elastically, and the material recovers its original dimensions once the stress has been removed. Resiliency is a measure of the amount of energy that can be absorbed under elastic loading and that is completely recovered when the sample is unloaded. It is the area under the stress-strain curve to the yield point (resilience $= 1/2\ \sigma_{max}\ \varepsilon_{max}$).

At higher stresses, permanent strain occurs in the sample. When the stress is unloaded, the sample does not recover its full original length, L_0. In cold working operations, such as the cold drawing of fibers, stress above the yield stress is desired because it will help crystallize the material. For some applications, exceeding the yield strength of the part results in product failure, because the original dimensions are not recovered. In other applications, the definition of failure is the breakage of the part. The total strain prior to fracture (ductility or percent elongation) is important. The tensile strength is determined by dividing the maximum force at fracture by the original cross-sectional area.

TABLE 8.6 Mechanical Properties of Materials

Property, or Characteristic	Symbol	Definition (or comments)	Common units	
			SI	English
Stress	σ	Force/unit area (F/A)	pascal* ($N^{\dagger}/m^2$)	psi* $lb_f/in.^2$
Strain	ε	Fractional deformation $(\Delta L/L)$	—	—
Elastic modulus	E	Stress/elastic strain	pascal	psi
Strength				
yield	σ_y	Resistance to initial plastic deformation	pascal	psi
tensile	σ_t	Maximum strength (based on original dimensions)	pascal	psi
Ductility				
elongation	ε_f	$(L_f - L_0)/L_0$	§	§
reduction of area	R of A	$(A_0 - A_f)/A_0$	§	§
Toughness		Energy for failure by fracture	joules	ft-lb
Hardness‡		Resistance to plastic indentation	Empirical units	

*1 pascal (Pa) = 1 newton/m^2 = 0.145×10^{-3} psi; 1000 psi = 6.894 Mpa.

†A load of 1 kg mass produces a force F of 9.8 newtons (N) by gravity.

‡Three different procedures are commonly used to determine hardness values:

Brinell(BHN): A large indenter is used. The hardness is related to the diameter (1 to 4 mm) of the indentation.

Rockwell(R): A small indenter is used. The hardness is related to the penetration depth. Several different scales are available, based on the indenter size and the applied load.

Vickers(DPH): A diamond pyramid is used. A very light load may be used to measure the hardness in a microscopic area.

§Dimensionless. Based on original (o) and final or fractured (f) measurements (usually expressed as percent).

Example 8.7 Elastic Modulus of a Glassy Polymer as a Function of Temperature

The data in Figure 8.28 can be used to calculate moduli for poly(methyl methacrylate) as a function of temperature. One method for calculating the elastic moduli is choosing the yield point for each curve and applying Equation 8.4. There are several ways to define the yield point. The proportional limit is the stress below which the stress is proportional to the strain. The elastic limit is the stress level below which the strain is fully reversible. The proportional and elastic limits are identical for most engineering materials. The yield strength is the stress level at which plastic (irreversible) flow begins. This is more difficult to determine, as very sensitive strain gages can detect plastic deformation at stresses far below the proportional limit. A standard method has been adopted for estimating the yield strength (ASTM E 8-69). A line parallel to the proportional segment of the curve is drawn with an offset of 0.2% strain (Fig. 8.29). The intersection of the parallel

(continued)

line with the stress-strain curve is defined as the yield point. The results of this method applied to the data of Figure 8.28 are shown in the following table.

T, °K	Stress, kpsi	Strain, %	E, 10^6 psi
78	17	2	.85
213	14.3	2.8	.51
253	8	1.8	.44
295	4.4	1.0	.44
313	2.1	.5	.42
373	.8	.4	.20

The moduli decrease with temperature as expected, based on the previous discussions of the regions of viscoelastic behavior (Section 1.7).

The toughness of a sample is a measure of the energy required to break the material. Energy is the product of force times distance and can be estimated as the area under the stress-strain curve for unnotched tensile bar samples.

$$Energy/Volume = \int_{o}^{\varepsilon_{max}} \sigma d\varepsilon \tag{8.116}$$

Optimal toughness is obtained by a combination of strength and ductility. High strength and high ductility alone are not sufficient to high toughness.

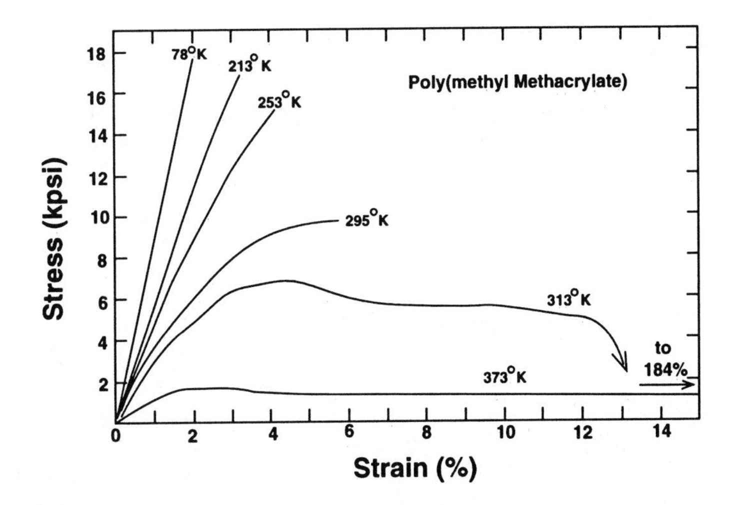

Figure 8.28 Tensile stress-strain curves for poly(methyl methacrylate) (Rabinowitz and Beardmore, 1972).

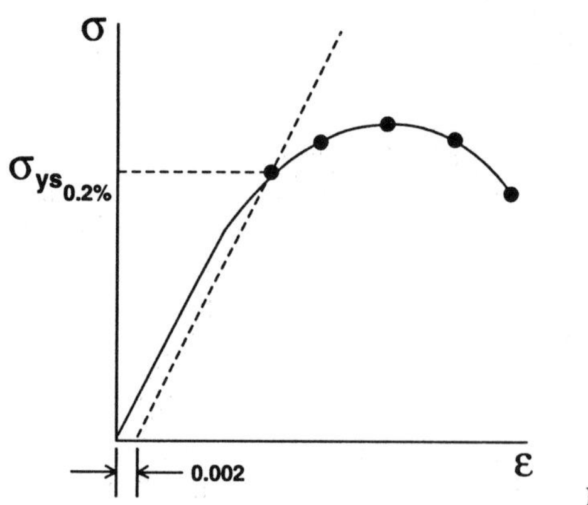

Figure 8.29 Estimation of the yield stress.

Example 8.8 Toughness of Poly(methyl Methacrylate) as a Function of Temperature

The toughness of PMMA at various temperatures can be determined by estimating the area under the stress-strain curves in Figure 8.28. The results are shown.

T, °K	Energy/Volume, kpsi
73	0.17
213	0.35
253	0.40
295	0.42
313	0.72
373	2.39*

The asterisk by the value at 373°K is due to the fact that the stress at high strains is not known. At lower temperatures, PMMA is very brittle, as demonstrated by the low toughness values.

8.4.2 Failure Mechanisms for Polymers

A number of different models have been developed to help interpret failure data. Several response regimes have been identified for engineering polymers: purely elastic deformation, brittle fracture initiated by shear banding or crazing, plasticity terminating in ductile fracture, cold drawing, rubbery and viscous flow, and adiabatic heating (Ahmad and Ashby, 1988). The stress-strain curves for poly(methyl methacrylate) (Fig. 8.28) are typical of a number of engineering thermoplastics.

Brittle Fracture. At temperatures less than 0.8 T_g, the material fails by brittle fracture and the stress-strain curve is nearly linear up to the break point (the T_g of PMMA is 380 °K). The curves below 295 °K show elongations of less than several percent, which is typical of brittle materials. High strain rates can lead to brittle failure as well. Catastrophic failure of brittle materials may be initiated by localized shear yielding or crazing.

Failure in tension usually is initiated at cracks or flaws in the sample, and the critical tensile stress for sample failure can be related to the inverse square root of the crack radius (Griffith, 1921). Polymers show a limiting critical flaw size, below which fracture stress is independent of flaws that have been introduced into the sample artificially. This finding has been interpreted to mean that polymers have internal defects that can initiate failure. The critical flaw size for PMMA at room temperature is about 0.05 mm (Berry, 1972). The internal defects are thought to be crazes or shear bands that have cracklike stress fields associated with them. Figure 8.30 is a sketch of a crack perpendicular to the applied stress.

Failure strength in compression may be an order of magnitude larger than the tension value. Crack growth under compression is thought to be much more difficult than under tension, and a plasticity mechanism may be operating in which failure occurs by plastic flow of the material.

Crazes become important near 0.8 T_g and higher. Crazes are cracks that are filled with oriented, load-bearing material. Crazes usually are initiated at a free surface of the sample. Craze propagation is thought to be a microdrawing process, resulting in fibrillation of the polymer in the craze. The fibrils have similar diameters, and crazes are thought to thicken by pulling more material into the fibrils. The thickening process stops when the local stress has decreased due to the deformation process.

Plasticity/Ductile Fracture. Plasticity (nonlinear, nonelastic deformation) becomes significant at temperatures above 0.8 T_g under tension. Under compression, plasticity occurs at lower temperatures and is associated with shear banding. Shear

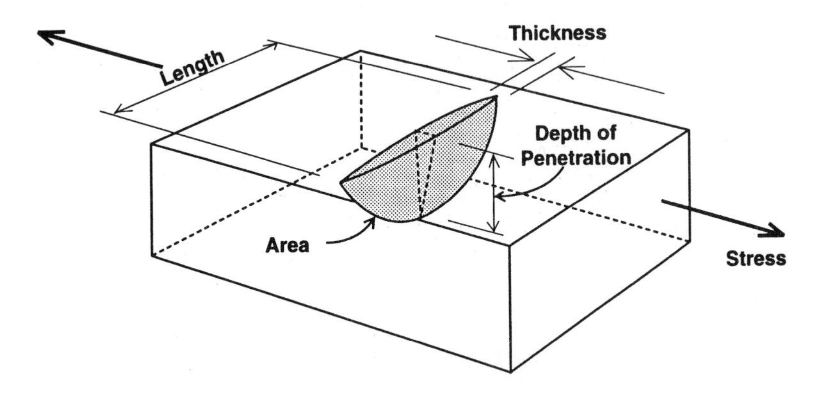

Figure 8.30 Craze morphology (Rabinowitz and Beardmore, 1972).

yielding or banding is observed as "kink bands," local changes in orientation of the polymer often at an angle to the tensile or compressive force. Shear yielding modes are most common under compression or at conditions near pure shear of the sample. The curve in Figure 8.28 at 373°K shows plastic deformation followed by ductile fracture. The sample has stretched at constant stress to very high deformation followed by breaking of the filament.

Cold Drawing. Semicrystalline polymers exhibit a different behavior from glassy polymers under tension. Figure 1.16 shows a response curve for a semicrystalline polymer that can be cold drawn. In contrast to curves that show a yield point followed by a region of constant stress and then stress loss, semicrystalline thermoplastics (examples are nylon 6,6 and polypropylene) show a yield point followed by a load drop. At higher elongations, the material starts to neck and the polymer starts to crystallize as chain segments are mechanically moved past each other. The neck is nearly constant in cross-sectional area and grows at each end of the sample as more material becomes oriented. The alignment of polymer chains in the solid state gives long-term orientation to the sample and much higher tensile strength. The increase in the engineering stress at high elongations corresponds to a large increase in the true stress, as the necking process reduces the cross-sectional area of the sample. Unloading the sample does not result in a return to the original dimensions; the material is permanently deformed.

Viscous Flow. At temperatures above 1.1 T_g, polymers deform via viscous flow. The WLF equation can be used to correlate the temperature dependence of the viscosity. The upper temperature limit of material failure usually is the decomposition or degradation temperature of the material.

Failure of Cross-linked Elastomers. Cross-linked elastomers have stress-strain curves somewhat similar to semicrystalline polymers. Because they are cross-linked, there is a proportional region at small strains with no yield point. Above the proportional region, the chains are being aligned and crystallized, so that the engineering stress increases. If the strain is removed before the sample breaks, the original dimensions of the sample can be recovered, although there may be some hysteresis in the curve.

Adiabatic Heating. Plastic deformation is accompanied by local heating in the craze and shear-banding regions. Polymers have low thermal conductivities and may have local temperature increases near shear bands. The higher temperature reduces the local elastic modulus and can result in strain softening. The sample may begin to neck, which tends to favor cold drawing in semicrystalline materials.

Other Factors Affecting Failure. There are a variety of external factors that affect the tensile behavior of polymers. These include system temperature and pressure, strain rate, annealing, and cold drawing. The effects of temperature have already been shown in Figure 8.28. Modulus decreases as temperature increases, and the elongation at break increases with temperature. Because the modulus decreases with temperature, it is controlled by the internal energy of the polymer, not its entropy. In general, increases in system pressure increase the Young's modulus and the yield stress. This response may be caused by a denser packing of amorphous regions of the material, increasing the difficulty for stretching of the chains.

The strain rate of the sample can make a difference in the recorded stress-strain curve. Figure 8.31 shows the tenacity of nylon samples as a function of strain rate of the tensile test. The tenacity of the sample, in grams per denier, can be converted to tensile stress in MPa by multiplying by the sample density (g/cm^3) and by 88.3. At high strain rates, the samples have higher yield stresses and appear to be more brittle. This is consistent with the time-temperature superposition principle discussed in Section 8.3.

Sample annealing and cold forming have large effects on tensile behavior. Annealing is done by raising the temperature of the sample, holding it there for a specified period of time, and returning it to room temperature. Annealing permits more crystallization to take place, allows the crystallites to grow in size, and increases the density (reduces the free volume) in the amorphous phase. All of these effects contribute to the tensile behavior: the modulus increases, the ultimate tensile strength increases, and the ultimate elongation decreases.

Deformations done below the melt temperature have the greatest effects on the properties of the sample. Figure 8.32 shows the effect of draw ratio (L/L_o) on the

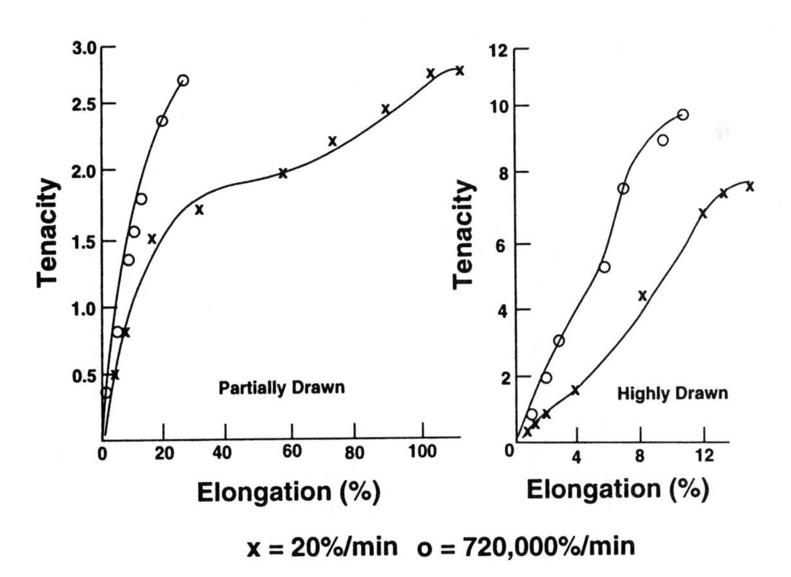

Figure 8.31 Effect of drawing rate on tensile strength of nylon yarn. Tenacity (grams per denier) plotted versus percent elongation (Butterworth and Abbott, 1967).

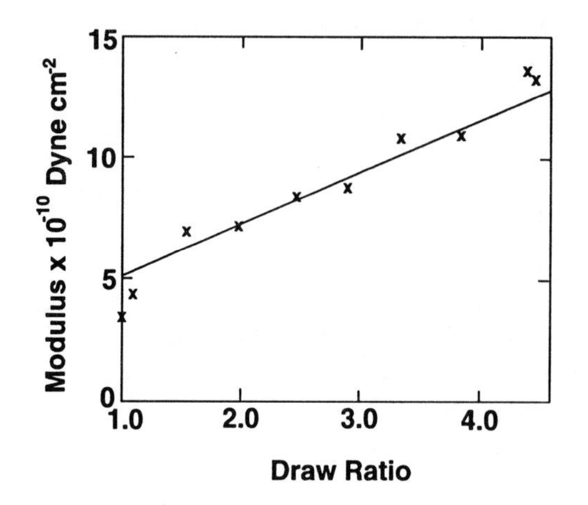

Figure 8.32 Effect of draw ratio on tensile modulus of polypropylene fibers (Samuels, 1967).

modulus of polypropylene. Drawing is quite complex and probably involves mechanical working of the material as well as reorientation of crystalline regions.

8.4.3 Tensile Properties of Structured Composites

As discussed in Section 3.10, composite materials provide very high elastic moduli at low weights and moderate costs compared to metals. Figure 8.33 shows a composite plaque under load. The composite is composed of continuous fibers parallel with the tensile direction. The matrix is well-bonded to the fibers, and the strain is small enough so that the material in both phases behave elastically. The total load, F, is the sum of the load carried by the fibers, F_f, and the load carried by the matrix, F_m.

$$F = F_f + F_m \tag{8.117}$$

This simple analysis does not account for the functions of the interphase material between the fiber surface and the bulk matrix, but it does give models that can bound the performance of the composite.

For a uniform strain on the plaque, the strain on each phase must be identical.

$$\varepsilon = \varepsilon_f = \varepsilon_m \tag{8.118}$$

From the definition of stress (Eq. 8.112) and load (Eq. 8.4),

$$\sigma A = \sigma_f A_f + \sigma_m A_m \tag{8.119}$$

By the geometry of the sample, the ratio of the fiber area to the total area equals the fiber volume fraction. Dividing each side of Equation 8.119 by the composite area

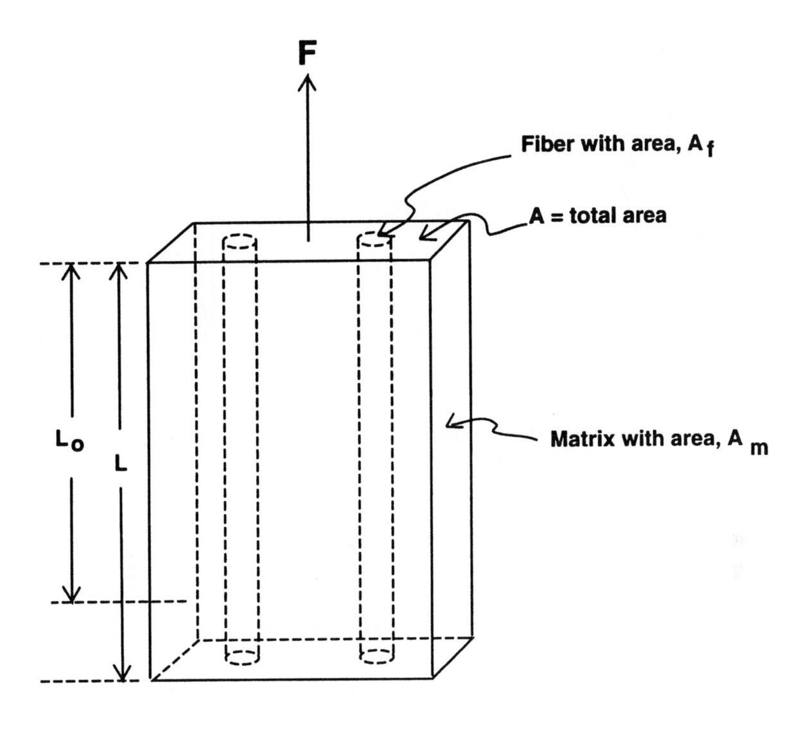

Figure 8.33 Analysis of tensile behavior on oriented composite plaque.

and using the elastic modulus definitions, we can solve for the total stress on the sample,

$$\sigma = E_f \varepsilon_f V_f + E_m \varepsilon_m V_m \tag{8.120}$$

and the modulus of the composite,

$$E = E_f V_f + E_m V_m \tag{8.121}$$

There are two limiting cases: The first is the fiber reaching its yield stress before the matrix material does; the second is the matrix material reaching its yield stress first. The cases are sketched in Figure 8.34. In each case, the composite has tensile properties intermediate between those of the fiber and matrix phases. In the first case, the material fails similarly to the fibers but at a lower modulus value (the value of the composite). When the matrix fails first, Equation 8.120 is incorrect, and the sample stress can be approximated by

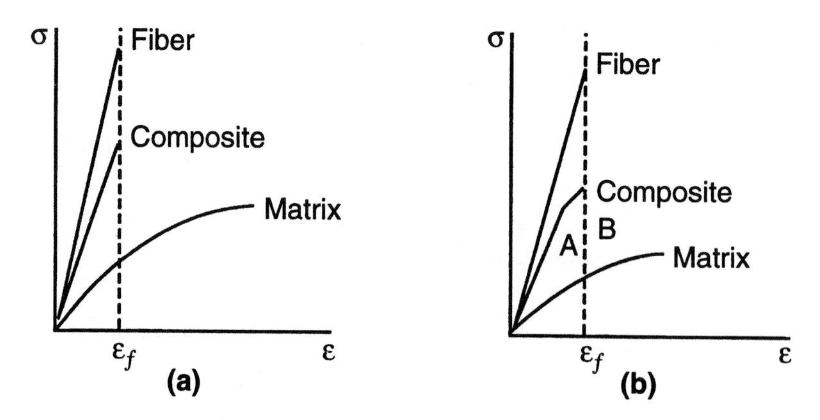

Figure 8.34 Stress-strain response of fiber, matrix, and composite with **(a)** both phases elastic and **(b)** elastic fiber with plastically deformed matrix at fracture.

$$\sigma = \sigma_f V_f + \sigma_m^1(1 - V_f) \quad 8.122$$

where σ_m^1 is the stress acting on the matrix at the fracture stress. The composite acts elastically up to point A in Figure 8.34b, and then deforms plastically to point B.

The fiber volume fraction does not increase the modulus linearly. At low fiber volume fractions, the strength of the composite can be less than the matrix. For many systems, there is a minimum value for the fiber volume fraction needed for reinforcement of the composite.

There are two major factors that reduce the modulus of a chopped fiber composite relative to an ideal, continuous fiber sample. The applied load is transmitted to the fibers by shear stresses acting along the interface between the fiber and the matrix. The shear stresses produce a normal axial stress in the fibers that is dependent on the fiber radius and length. The presence of the fiber ends reduces the load they can carry because the axial force must be zero at the end of the fiber. There will be a minimum length, L_c, for reinforcement in order for the fiber to reach its maximum load prior to failure. The effective tensile strength of chopped fibers is a function of their length relative to the critical reinforcing length.

The second factor affecting the effective strength of the fibers is their orientation in the sample relative to the tensile load. Different orientations lead to different failure mechanisms. For example, fibers aligned near the tensile load direction will experience shear fracture, while those aligned transverse to the load will experience transverse fracture. In this second case, the load is transmitted across the fiber radius, and the composite's strength is much lower.

8.4.4 Impact Resistance

Impact resistance is a typical failure test. It measures the strength of the material near failing. Failure tests are very sensitive to defects. In fact, careful examination of parts

Example 8.9 Composite Tensile Strength

A composite material has 30wt% glass fibers dispersed in polycarbonate. Estimate its modulus. The modulus of polycarbonate is 2345 MPa (Appendix C). The modules of glass fibers can be determined by knowing their breaking tenacity and the elongation at break. The breaking tenacity of glass fibers is 9.9 g/denier, which is equivalent to 2190 Mpa tensile strength. The elongation of glass at the fracture point is 3.1%, so the modulus of the fibers is σ/ε or 70,500 MPa. The specific gravity of the glass fibers is about 2.5 g/cm^3, while that of the matrix is 1.2 g/cm^3. The volume fraction of the fibers is 17%. The modulus of the composites is

$$E = E_f V_f + E_m V_m$$

$$= 70{,}500 \text{ MPa} \cdot 0.17 + 2345 \text{ MPa} \cdot .83 = 14{,}000 \text{ MPa}$$

This value compares to an actual material value of 8500 MPa for a 30wt% chopped fiber composite.

that have failed through impact, abrasion, tearing, or flexing often reveals a contaminant, air bubble, copolymer composition variation, fiber deficiency in a composition, or some other defect that acted as a site for the failure to initiate.

There are at least four elements to selecting and designing failure tests: reproducibility; mechanism for failure; correlation between the test and the end-use; and relation between the test measurements and polymer properties, including configuration, molecular weight distribution, and processing. Because the failure of the sample is often associated with a defect, the test reproducibility can vary greatly. Therefore, sample preparation is tightly controlled, and the tests are fairly well specified with respect to the equipment used and the testing protocol. ASTM test methods should be used because they have been reviewed by a number of experts. Even these methods may have deficiencies for a specific application. Engineers developing new materials or new applications need to apply standard tests with caution.

The failure test usually is established to evaluate the material in one mode of failure. The manufacturer wants to relate the test results to the material performance in a specific application, in which the part will be most susceptible to one or maybe two types of failure. The manufacturer also wants to use the results of the failure tests to help optimize the polymerization process, the product formulation, and the processing and fabrication of the material.

Impact Resistance Test Devices. Impact resistance tests are intended to test the brittle failure of the sample. There are a number of tests that measure this property (Table 8.7). The impact can be imposed by a free-falling weight or by a swinging arm. In each test, it is important to verify that the sample exhibits brittle failure and that contaminants or other defects did not contribute to the premature initiation of the crack.

The test choice depends on the polymer type and its applications. For example, plastic pipe can be subjected to impacts from falling tools, or from being dropped during transportation, or at the work site. These types of abuse can be modeled as a rapid force being applied at a well-defined location on the pipe. Impact tests that apply force rapidly to a small area should replicate field problems well.

For this application, it is preferable to test the material as a finished product, a piece of pipe, because the samples will have some internal stresses left from the fabrication process as well as contamination from the production processes. Both the internal stresses and the contamination will tend to lower impact resistance. Therefore, a pipe manufacturer might prefer to use ASTM D 2444 to test finished pipe at the plant prior to shipping. This impact test uses a falling weight (called a tup) from various heights to measure the impact resistance of the sample. The pipe would be placed so that the tup would strike a direct blow on the vertical radius (a glancing blow would not direct all the energy of the tup to breaking the sample).

In the research and development lab, impact resistance is one of a series of tests used to characterize new and modified materials. When there is no specific application in mind, the impact resistance of a bar of material should be adequate, and the Izod impact test, ASTM D 256, might be chosen.

TABLE 8.7 Tests for Impact Resistance

Test	ASTM Number	Description
Brittle temperature	D 746	Temperature at which plastics and elastomers exhibit brittle failure after impact
Falling weight	D 3029	Energy to break or crack rigid plastics by a falling weight (tup); constant height and variable weights used.
Falling weight	D 1709	Impact resistance of polyethylene film using a free-falling dart
Falling weight	D 2444	Impact resistance of thermoplastic pipe and fittings by a falling weight (tup).
Fracture toughness	E 399	Plane-strain fracture toughness
High rate stress/ strain (tension)	D2289	Area under stress-strain curve relates to impact resistance at testing speeds up to 254 m/min.
Izod impact	D 256	Energy to break a notched specimen as a cantilever beam by impact from a pendulum. Notch tends to promote brittle failure.
Tensile impact	D 1822	For materials too flexible, thin or rigid to be tested by ASTM D 256. Measures energy to break in tension imparted by pendulum.

The Izod (Fig. 8.35) and Charpy impact testers are in common use. These devices break sample bars that are ¼″ thick. The Charpy test breaks the sample by applying the impact directly opposite the notch. The Izod impact is applied to the face of the sample having the notch a fixed distance away. The striker velocity is 2.4 m/s for the Izod test and 3.4 m/s for the Charpy test. Different thickness can be used, but the results should not be reported as Izod tests. The specimens have notches machined in them to act as sites for initiating cracks. The intent is to cause the crack to start at the same location on the specimen and travel in a straight line to the opposite side. Cracks that do this will break the same cross-sectional area of the sample and should take the same amount of energy to cause the failure.

The notch makes a stress concentration that produces a brittle fracture instead of a ductile fracture. Sample preparation and equipment set-up can have a significant effect on the measurements. The cutting of the notch requires special attention. Small changes in the notch geometry can make big changes in the impact values. Samples with notches molded in are preferred since they are smoother than cut notches. Some polymers will heat and distort under cutting conditions, so the sample should be cooled using water to prevent ductile flow of the polymer and to reduce thermal stresses. Notch cutting is not recommended for crystalline polymers, since the heating and cooling changes the polymer

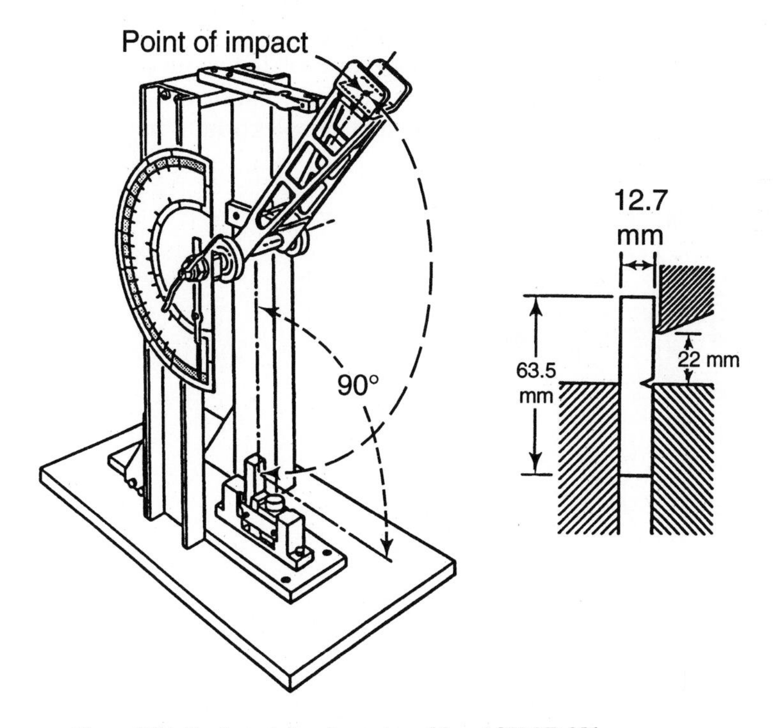

Figure 8.35 Izod pendulum impact machine. ASTM D 256.

morphology near the notch compared to the rest of the sample. Changes in the notch shape can change the impact values by 10%, while clamping pressure and the striker alignment can change values by 15%.

Analysis of Impact Resistance Tests. The effects of sample preparation and equipment set-up on test values are direct results of the failure mechanisms. The Izod test measures the total energy of the following steps: fracture initiation, fracture propagation across the specimen, plastic deformation (yield), separation of the fragment from the sample, and losses due to vibration.

The notch geometry has a big effect on fracture initiation and propagation. Figure 8.36 shows the impact strength as a function of notch tip radius for several polymers. Different ASTM methods are shown with arrows. PVC results are particularly sensitive to notch radius because it is brittle. ABS and PMMA results are fairly insensitive to notch size.

Some materials show plastic deformation, or ductile failure, even when they are notched. The separation of the fragment from the sample is a small amount of the total energy except for high density materials having low impact, such as filled phenolic resin systems. The energy associated with fragmentation sometimes can be estimated by striking the fragment again to measure the energy needed to put it in motion.

The time-temperature superposition principle suggests that the speed of the impact testing device and the temperature of the sample will affect the results. Figure 8.37 shows the effect of sample temperature on the impact strength of PVC for several notch radii. Notice that samples with 1 mm notches will be very

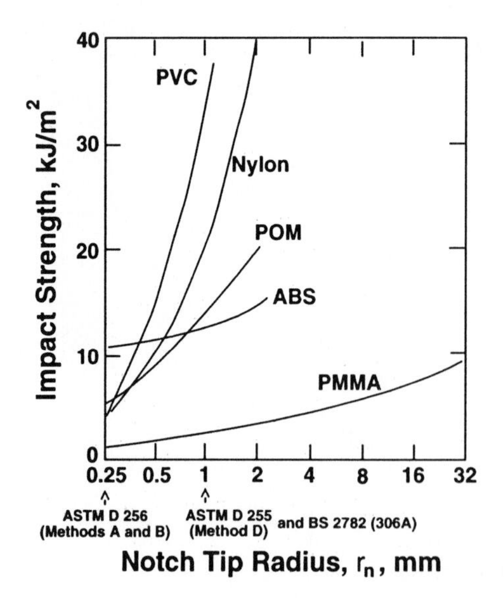

Figure 8.36 Effect of notch tip radius on impact strength for several polymers (Vincent, 1971).

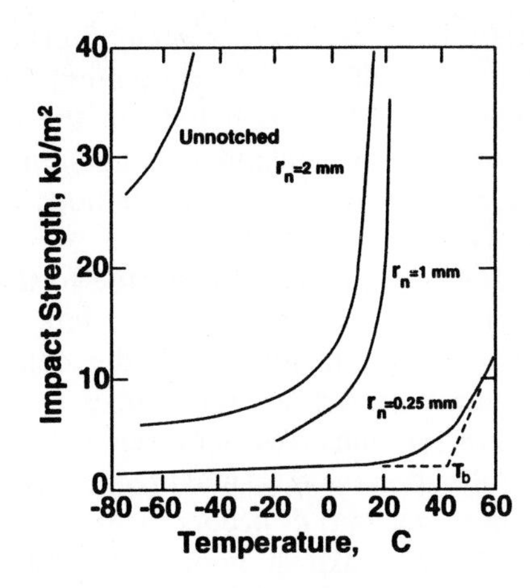

Figure 8.37 Effect of temperature on the impact strength of PVC for several notch tip radii. T_b: brittle temperature (Vincent, 1971).

sensitive to testing temperatures near 20°C, the ambient conditions in most labs. Smaller notch radii cause great reductions in impact strength. The effects of test speed and sample thickness are shown in Figure 8.38. Thin samples (3.2 mm) of polycarbonate are ductile and show high fracture energies over a wide range of testing speeds. As the samples become thick (6.4 mm), they become brittle. The samples of intermediate size go through brittle-to-ductile transitions in failure mechanisms at speeds between 0.1 and 1.0 cm/s.

The impact test is a measure of the total energy needed to break the sample. Because the sample cross-sectional area may change during the test due to necking and other factors, it is difficult to convert impact values to true stress–true strain diagrams. However, if samples fail by the same mechanism, it is possible to rank their impact performance using these tests. Relative rankings can be very valuable in optimizing material and processing conditions for applications.

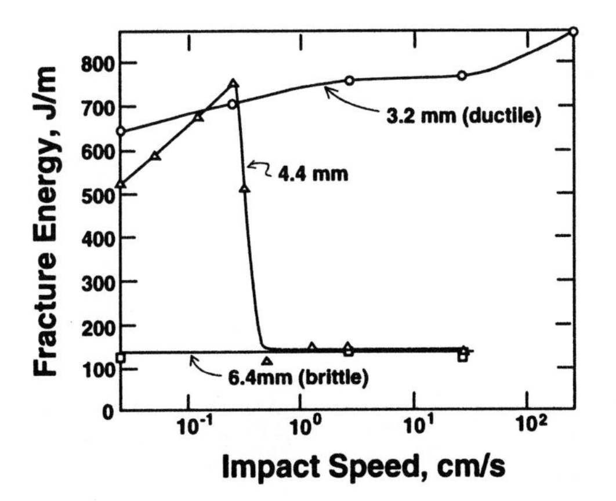

Figure 8.38 Effect of impact speed on the fracture energy of polycarbonate using the Charpy test. Specimen thickness indicated. 4.4 mm sample undergoes brittle to ductile transition at 0.3 cm/s (Yee, 1977).

It is important to relate failure tests to fundamental property measurements. This correlation can be used to modify the sample material in order to improve its performance on the failure tests, and hopefully in the end-use application. There are a number of ways to do this, including changing molecular weight and molecular weight distribution, altering processing methods and conditions, and changing the product formula.

Impact strength of thermoplastics has been related to the dynamic modulus as well as to mechanical losses (Vincent, 1974). Test data for twenty different thermoplastics were correlated. Careful analysis of the data showed that, in a number of cases, plots of Charpy impact strength versus temperature had peaks close to peaks in tan δ, the loss modulus. Two representative examples are shown in Figures 8.39 and 8.40. Polysulfone has a maximum Charpy impact strength near −40°C, and a maximum in tan δ occurs near the same temperature. Polytetrafluoroethylene has three peaks in impact strength: −90°C, 40°C, and 140°C—which relate closely to peaks in tan δ.

Appendix: MECHANICAL ANALOGIES TO VISCOELASTICITY

The Maxwell and Voigt models shown in Figure 8.A1 are often used to describe the performance of polymers in mechanical tests. In the Maxwell model, the spring is in series with the dashpot, so that the applied stress is the same in each element. The

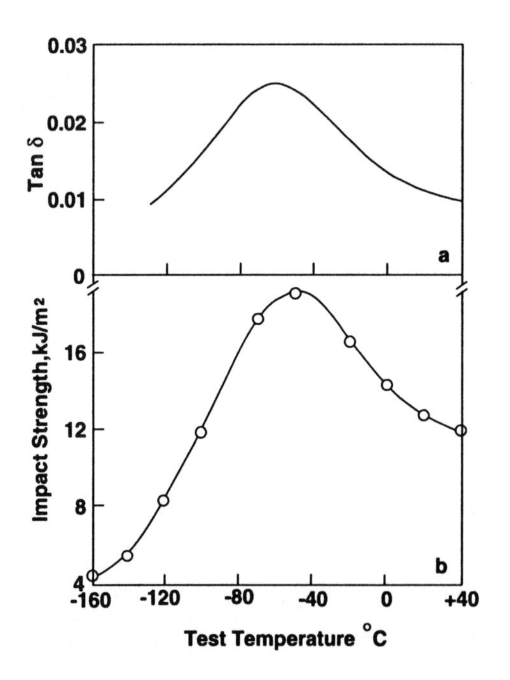

Figure 8.39 Relationships among test temperatures, loss modulus, and impact strength for polysulfone (Vincent, 1974).

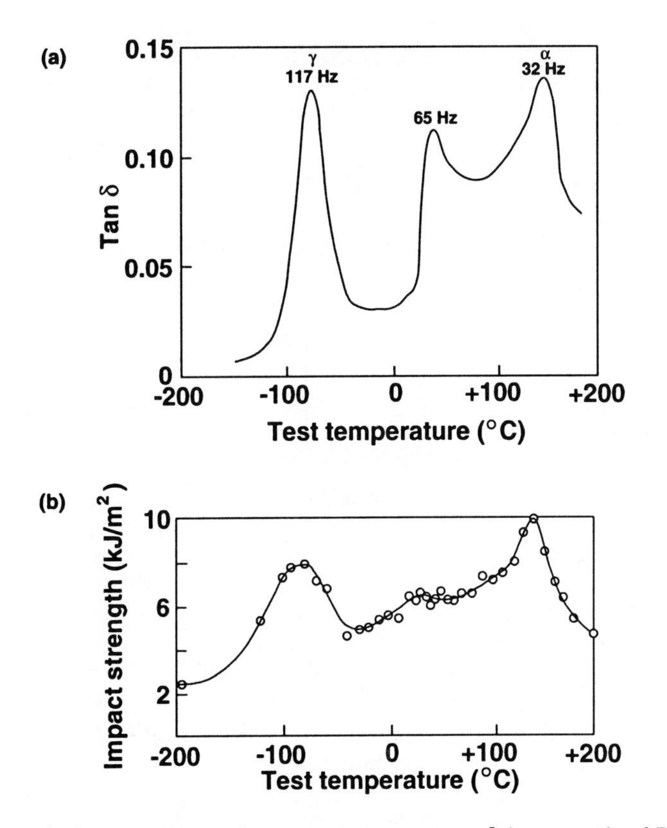

Figure 8.40 **(a)** Effect of test temperature on tan δ for a sample of PTFE. The frequencies at the peaks were as marked. **(b)** Effect of test temperature on impact strength with very sharp notches for a sample of PTFE similar to that of the figure above (Vincent, 1974).

total elongation of the sample is the sum of the elongation of each element. These equations are

Spring: $$\sigma_1 = E\varepsilon_1 \qquad \text{8.A1}$$

Dashpot: $$\sigma_2 = \eta \frac{d\varepsilon}{dt} \qquad \text{8.A2}$$

Total Stress: $$\sigma = \sigma_1 = \sigma_2 \qquad \text{8.A3}$$

Total Strain: $$\varepsilon = \varepsilon_1 + \varepsilon_2 \qquad \text{8.A4}$$

Creep Test: Maxwell Model

In a creep test, a constant load is applied to a sample and the deformation with time is observed. The change in the elongation (strain) with time is

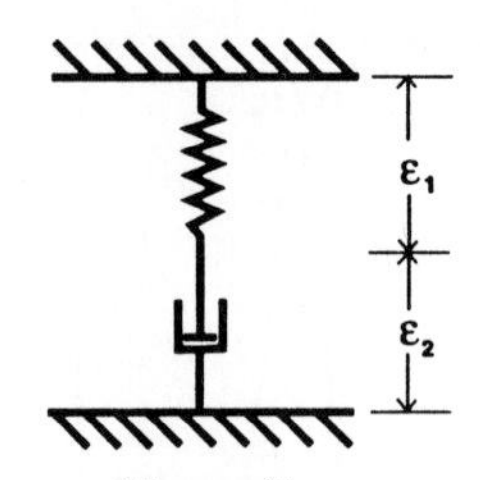

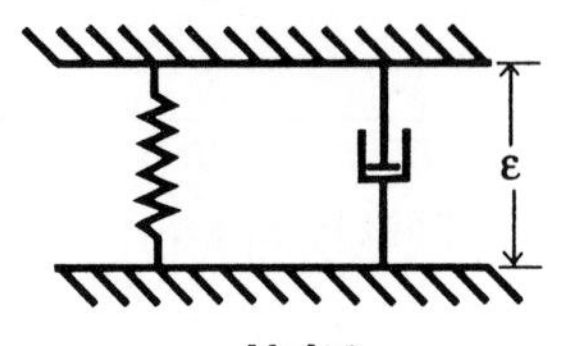

Figure 8.A1 Maxwell and Voigt models.

$$\frac{d\varepsilon}{dt} = \frac{d\varepsilon_1}{dt} + \frac{d\varepsilon_2}{dt} = \frac{1}{E}\frac{d\sigma}{dt} + \frac{\sigma}{\eta} \qquad 8.A5$$

The derivative of the spring strain is zero (after application of the initial stress), so we can integrate the following equation:

$$\int_{\varepsilon_o}^{\varepsilon(t)} d\varepsilon = \frac{\sigma}{\eta}\int_{o}^{t} dt \qquad 8.A6$$

and get

$$\varepsilon(t) = \sigma\left(\frac{1}{E_o} + \frac{t}{\eta}\right) \qquad 8.A7$$

The sketch in Figure 8.A2 shows a typical creep response for an uncrosslinked polymer.

The Maxwell model deforms immediately on the application of the load. The initial strain, ε_o, is set by the load, σ, divided by the Young's modulus of the spring. The dashpot creeps linearly with time, with slope σ/η. When the load is removed, the spring element returns to its initial elongation. However, the dashpot has undergone elongation that is not recovered when the load is removed. In a polymer sample, this would be described as permanent set.

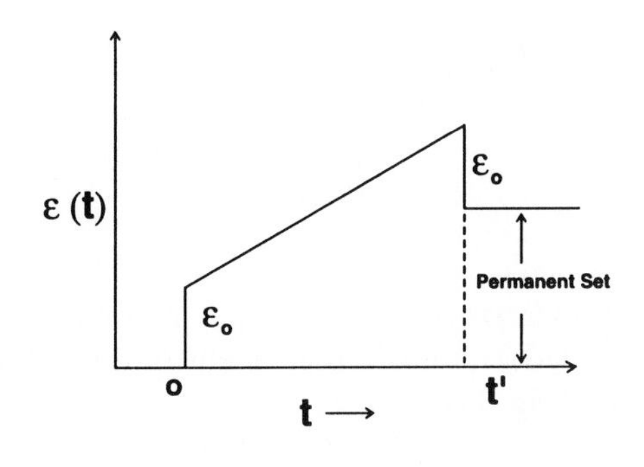

Figure 8.A2 Creep response of the Maxwell model. Element unloaded at $t = t^1$.

Creep Test: Voigt Model

The Voigt model has a different creep response.

Spring: $\sigma_1 = E\varepsilon_1$ 8.A8

Dashpot: $\sigma_2 = \frac{d\varepsilon_2}{dt}$ 8.A9

Total Stress: $\sigma = \sigma_1 + \sigma_2$ 8.A10

Total Strain: $\varepsilon = \varepsilon_1 = \varepsilon_2$ 8.A11

The total stress on the element is constant and the two strains are equal. We can use the total stress balance to solve the problem.

$$\sigma = \sigma_1 + \sigma_2 = E\varepsilon + \frac{\eta\, d\varepsilon}{dt} \tag{8.A12}$$

$$\sigma - E\varepsilon = \frac{\eta\, d\varepsilon}{dt} \tag{8.A13}$$

We can separate variables

$$\int_o^{t^1} \frac{1}{\eta} dt = \int_o^{\varepsilon^1} \frac{d\varepsilon}{\sigma - E\varepsilon} \tag{8.A14}$$

$$\frac{t}{\eta} = -\ln \frac{(\sigma - E\varepsilon)|_{\varepsilon^1}}{(\sigma - E\varepsilon)|_o} \tag{8.A15}$$

$$\frac{t}{\eta} = \ln\frac{\sigma}{(\sigma - E\varepsilon)} \qquad 8.A16$$

$$\varepsilon = \frac{\sigma}{E}[1 - \exp(-t/\eta)] \qquad 8.A17$$

The Voigt element has an exponential rise in strain to a limiting value based on the material modulus and the imposed strain. When the element is unloaded, the sample returns to its initial dimensions (Fig. 8.A3).

Stress Relaxation: Maxwell Model

In a stress relaxation experiment, a constant strain is placed on a sample and the stress change with time is measured. For the Maxwell model, the following equations apply:

Total strain: $\varepsilon = \varepsilon_1 + \varepsilon_2 = constant$ 8.A18

Total stress(Eq. 8.A12): $\sigma = \sigma_1 = \sigma_2$

Since the total strain is constant, its derivative is zero.

$$\frac{d\varepsilon}{dt} = 0 = \frac{d\varepsilon_1}{dt} + \frac{d\varepsilon_2}{dt} = \frac{1}{E}\frac{d\sigma}{dt} + \frac{\sigma}{\eta} \qquad 8.A19$$

This can be integrated to

$$\sigma(t) = \sigma_o \exp\left(\frac{-Et}{\eta}\right) \qquad 8.A20$$

Figure 8.A3 shows the response of a Maxwell element to a stress relaxation experiment. The stress quickly rises to σ_o, based on the spring modulus and load. The load then decreases as the dashpot extends (Fig. 8.A4).

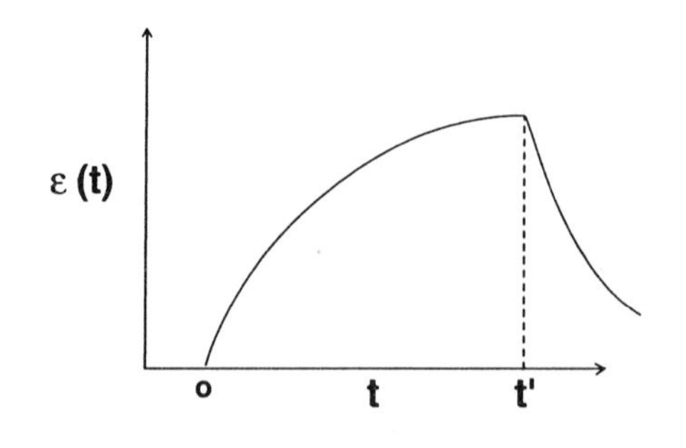

Figure 8.A3 Creep response of the Voigt (Kelvin) element. Element unloaded at $t = t^1$.

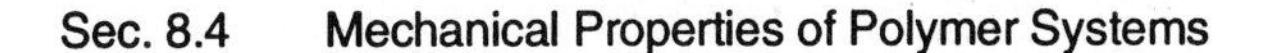

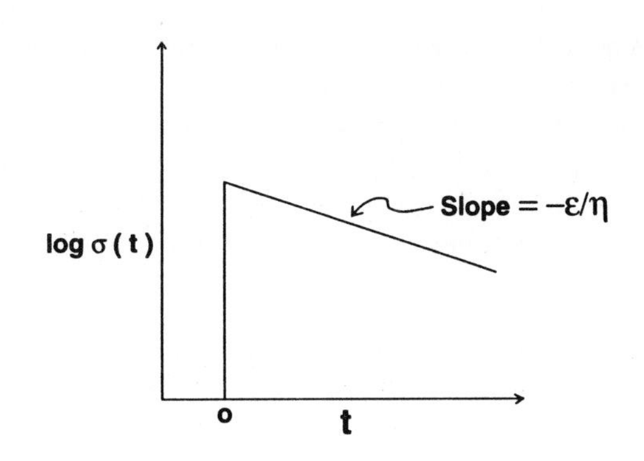

Figure 8.A4 Stress relaxation of the Maxwell element.

A combination of Maxwell and Voigt models captures the major viscoelastic phenomena. Figure 8.A5 shows the four-element model and its response in a creep experiment. A combination of Maxwell and Voigt elements will exhibit typical phenomena found in most creep experiments: instantaneous elastic strain and recovery, retarded elastic strain and recovery, and equilibrium flow and permanent set. The dashpot can be considered analogous to molecular slippage, while the spring can be

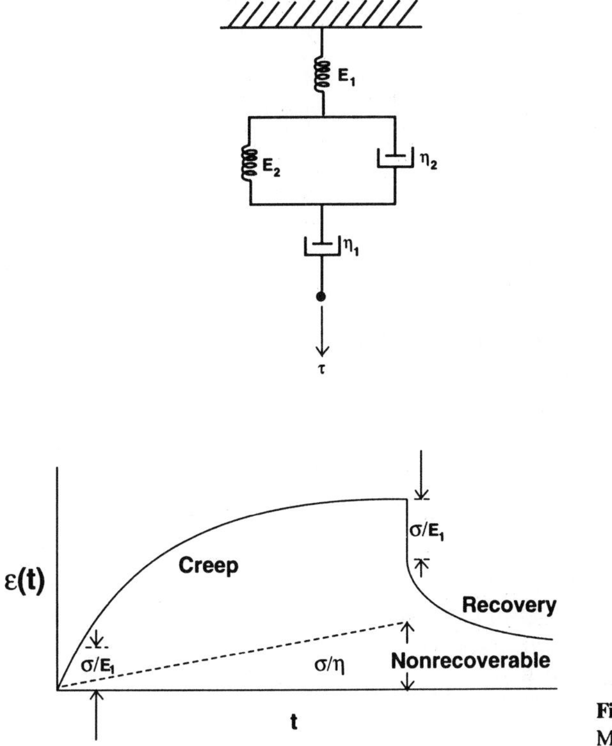

Figure 8.A5 Stress relaxation of a Voigt–Maxwell element.

considered analogous to elastic straining of the bond angles. Entanglements might contribute to retarded elasticity, as might entropy.

More complicated models consisting of many springs and dashpots can be developed. In general, the stresses are linearly additive (Boltzmann superposition principle), and the stress in a material at a given time depends on the entire past strain history. As with rheological testing, time-dependent (dynamic) mechanical tests are used to measure complex moduli.

NOMENCLATURE

a	exponent in Mark–Houwink equation
A	Helmholtz free energy
a_i	scalar coefficient of the normal velocity in the i direction
B	bulk modulus
C_i	heat capacity at constant property i
E	Young's modulus (E^*, dynamic; E', storage; E'', loss)
E_a	activation energy (may be measured at constant stress or shear rate)
F	load
f	fraction of free volume
f_g	fraction of free volume in the glass
G	shear modulus (G^*, dynamic; G', storage; G'', loss)
H	enthalpy
i	square root of -1
J	compliance
k	proportionality constant, power law fluid
K	Mark–Houwink constant
L	length
L_i^o	original length in i direction
L_i	deformed length in i direction
M	molecular weight
M_c	critical molecular weight
$N(\tau)$	distribution of relaxation times
N_{Deb}	Deborah number
n	exponent
N	number of chains
P	pressure

Q	heat
Q_f	flow rate in a capillary rheometer
r	end-to-end distance
$<r_o^2>$	chain end-to-end distance at rest
$<r^2>$	chain end-to-end distance in deformed state
R	ideal gas constant
S	entropy
t	time
T	temperature
T_g	glass transition temperature
T_m	melting temperature
U	internal energy
U_i	velocity in the i direction
V	volume
V_f	free volume
$V(r)$	fluid velocity as a function of radial position
W	work

Greek symbols

α	linear expansivity
α_i	stress coefficients, linear viscoelastic model
β	coefficient of thermal expansion
β_i	shear coefficients, linear viscoelastic model
γ	shear
$\dot{\gamma}_{ij}$	component of rate of strain tensor
$\dot{\gamma}_{ij}^o$	critical shear rate for the truncated power law
ε	strain
$\dot{\varepsilon}$	strain rate
$[\eta]$	intrinsic viscosity
κ	isothermal compressibility
η	viscosity (η^*, complex; η', dynamic; η'', imaginary)
λ_i	extension ratio in the i direction
μ_a	apparent viscosity
μ	viscosity
μ_0	zero shear viscosity
μ_∞	infinite shear viscosity
υ	Poisson's ratio

ρ	density
σ_i	uniaxial tensile stress in the i direction
τ_r	relaxation time constant
τ_{xy}	shear stress
τ_o	critical shear stress for a Bingham plastic
$\tau_{1/2}$	Ellis model constant
ϕ	loss angle
ω	frequency

PROBLEMS

8.1 A lab technician is measuring the modulus of poly(methyl methacrylate) as a function of time and temperature. He has measured a modulus of 10^{10} dynes/cm at a measurement time of 0.10 hours at 105°C (T_g), but he thinks that his data point may be inaccurate due to the equipment response time. Compute the temperature at which he should run the experiment to determine the same modulus overnight (16 hrs).

8.2 For the polystyrene data given below, determine the constants for the Mark–Houwink equation and predict the melt viscosity for a polymer sample with molecular weight of 600,000.

Melt Viscosity, poise	Molecular Weight
10	10,200
20	16,000
100	50,000
1,000	130,000
10,000	200,000
100,000	500,000

8.3 Use the WLF equation to estimate the viscosity of polystyrene with a molecular weight of 200,000 at 217°C and compare to the answer for Problem 8.2. Explain the basis of the WLF theory and why melt viscosities of polymers might be related to T_g.

8.4 Derive expressions for a simple Voigt element in the following experiments:

(a) creep experiment—time-dependent compliance,

(b) stress relaxation experiment—time-dependent modulus,

(c) constant extension rate experiment—change in stress.

8.5 Samples of an amorphous polymer are strained 1% at a range of temperatures. Draw a diagram of the stress ten seconds into the experiment versus temperature. Identify T_g and rubbery plateau region on the diagram. Show how and explain why the diagram changes if the following parameters are varied:

(a) the material is degraded to 90% of the original molecular weight,

(b) the material is loosely cross-linked,

(c) the diagram is plotted at an experimental time of ten hours.

8.6 One weekend at your home, you are trying to convince your parents of the practical value of your polymer engineering classes. Your mother decides to make a cup of hot tea and before you can stop her, she sticks a tea bag and a plastic spoon into a cup of boiling water. The plastic spoon is distorted after it is removed from the cup. This is your big chance to impress her with the value of your college education.

(a) From which of the following polymers is the spoon made, and how do you know? Polyethylene, polypropylene, or polystyrene.

(b) Your mother asks, "why is it that plastic spoons melt without even getting hot?" Explain what happened to the spoon.

8.7 A short rod of Silly Putty (unvulcanized silicone gum rubber) will fracture when extended very rapidly. Does this experiment have a high or low Deborah number? Explain why.

8.8 The Mooney–Rivlin equation relates the stress in a deformed rubber sample to the volume deformation and the chain end-to-end distance. An alternate form of this equation is

$$\sigma = NRT\left(1 - \frac{2M_c}{M}\right)\frac{\overline{r_i^2}}{\overline{r_o^2}}\left(\alpha - \frac{1}{\alpha^2}\right)$$

where M_c is the molecular weight between cross-links, M is the primary chain molecular weight, and N is the number of active chain segments per unit volume.

(a) How could this equation be applied to cross-linked rubber samples?

(b) How could this equation be used to determine the molecular weight of the chains?

8.9 Experimental data on the shear rate versus shear stress for two samples of polystyrene are given in the following table:

Sample A		Sample B	
Shear Rate	**Shear Stress**	**Shear Rate**	**Shear Stress**
0.2	63,000	80	127,000
1.	251,000	100	157,000
2.	400,000	200	251,000
10.	500,000	1000	1,000,000
60.	950,000		

Shear rate is given in 1/sec, and shear stress is given in dynes/cm2.

(a) Categorize these two samples with respect to their rheology.

(b) You would like to interpolate the data for Sample A at shear rates greater than 2 sec^{-1}. Fit a power law model to the data and determine the constants.

(c) If these two samples differed with respect to measurement temperature, which would have been tested at the higher temperature? Explain why, using the concept of a Deborah number.

(d) If the weight average molecular weight ratio of Sample A to Sample B is 5, estimate the Mark–Houwink exponent.

8.10 Plot the relationship between the elastic and shear moduli for Poisson's ratios over the range, 0.30 to 0.50. Indicate where polyethylene, natural rubber, and polyesters should occur. Do moduli data for these polymers agree with your predictions?

8.11 Squash balls do not have a high bounce until they have been "warmed up" by hitting them against the wall. Explain this phenomenon.

8.12 Bingham plastics exhibit non-Newtonian fluid behavior. Describe the velocity profile of a Bingham plastic in pipe flow, in a cup-and-bob viscosity, and in a plate-and-cone viscometer. Make drawings or graphs of the profile.

8.13 Assuming that the Ellis model describes the polymer solution, calculate the flow rate in gallons per hour through a long, vertical pipe with three bar pressure drop. $\eta_0 = 100$ poises, $\tau_{1/2} = 3120$ dynes/cm^2, $a = 1.75$, $p = 1.0$ g/cm^3, $r = 1.0$ cm. Calculate a Reynolds number using the apparent viscosity.

8.14 Estimate the effect of pressure on the T_g of polystyrene over the pressure range, 1 to 1000 bar.

8.15 Calculate the tensile modulus of polycarbonate/ glass composites over the fiber volume fraction range, 5 vol% to 40 vol%. How would you expect the predictions to relate to tensile measurements?

8.16 Analyze the internal energy of a polymeric solid under tension to define a constant length heat capacity and a constant force heat capacity. Relate the constant force heat capacity to the thermal expansion coefficient, α.

8.17 Use the Kelly–Bueche equation to estimate the T_g of plasticized poly(vinyl chloride) over the entire composition range. The T_g of the plasticizer is 25°C and its thermal expansion coefficient is 1.5 times that of the polymer.

8.18 The WLF equation can be used to relate viscosity measurements taken at different temperatures. If the viscosity of a polymer is 10^{-3} poise at 120°C, what would be the viscosity of the sample at 150°C? $T_g = 102$°C.

8.19 Fit either a power law or truncated power law model to the data shown in Figure 8.12. Determine all constants and justify your choice of model.

8.20 A polymer melt is being extruded from a capillary die. Write the shear stress tensor for this flow, define the primary and secondary normal stress difference, and describe and explain the expected flow profile of the melt leaving the die.

8.21 Your plant is forming polyisobutylene tubing at an extrusion temperature of

210°C. Because of polymer stability problems, production would like to drop the processing temperature to 195°C. Estimate the expected change in viscosity of the polymer melt caused by this change.

8.22 Production is considering using a polyisobutylene of higher degree of polymerization for extra strength tubing. The degree of polymerization of the material now is 1500 and the plant wishes to switch to material with a degree of polymerization of 2000. Estimate the melt viscosity of the new material at 210°C.

8.23 Sketch the differences that might be observed between a shear-thickening antithixotropic fluid and a pseudoplastic thixotropic fluid in pipe flow.

8.24 A polymer solution can be represented by this equation relating stress, τ, to the rate of shear $\dot{\gamma}$.

$$\tau = 2.70 \times 10^3 \, dyne/(cm^2)(\sec^{0.635}) \, \dot{\gamma}^{0.635}$$

What is the viscosity in centipoises at a stress of 10,000 dynes/cm^2? Why is this equation likely to be misleading at very low shear rates? Is this solution dilatant, thixotropic, or rheopectic?

8.25 For many polymers, the melt viscosity is proportional to molecular weight to the 3.4 power above some critical molecular weight.

$$\eta = K \, M_w^{3.4}$$

Using the data below, evaluate the constant K for polystyrene at 217°C and predict the melt viscosity for a polymer with a molecular weight of 600,000.

Melt Viscosity, Poise	Molecular Weight
10	10,200
20	16,000
100	50,000
1,000	130,000
10,000	200,000
100,000	500,000

8.26 Polyacrylamide dispersed in glycerine exhibits normal stresses. Discuss the response of this fluid to rotational shear in terms of the stress tensor. Draw a picture.

8.27 The Ellis model can be used to describe non-Newtonian behavior. Can it be used to describe the behavior of a thixotropic Bingham plastic? If so, what are the restrictions, if any? HINT: Draw a thixotropic fluid + a Bingham plastic on the same graph.

8.28 A rectangular bar of a natural rubber has original dimensions of $L_x = 5$ cm, L_y

= 1 cm, and L_z = 1 cm. It is stretched 50% in the x direction, and the material has a Poisson's ratio of 0.50. Calculate the new dimensions and the change in volume.

8.29 Provide explanations for the frequency responses of the moduli shown in Figure 8.19b, based on the molecular characteristics of polymers. Speculate on the effects of chain length on these responses.

REFERENCES

Z.V. AHMAD and M.F. ASHBY, *J. Mat. Sci., 23,* 2037–2050 (1988).

H.A. BARNES, J.F. HUTTON, and K. Walters, *An Introduction to Rheology.* Elsevier, Amsterdam, 1989.

P. BEAHAN, M. BEVIS, and D. HULL, *Phil. Mag. 24,* 1267 (1971).

J.T. BENDLER, "Transitions and Relaxations" in *Encyclopedia of Polymer Science and Technology,* Vol. 17, Wiley-Interscience, NY, 1989.

J.P. BERRY, *Fracture,* Vol. 7 (H. Liebowitz, Ed.), Academic Press, NY. 1972.

R. BOYER, "Transitions and Relaxations" in *Encyclopedia of Polymer Science and Technology,* Wiley-Interscience, NY, Vol. 17, 1989.

N. BROWN and S.K. BHATTACHARYA, *J. Mat. Sci. 20,* 4553 (1985).

G.A.M. BUTTERWORTH and N.J. ABBOT, *J. Materials 2,* 487 (1967).

J.D. FERRY, *Viscoelastic Properties of Polymers,* 3rd ed., Wiley, NY, 1980.

W.W. GRAESSLEY, et al, *Trans. Soc. Rheol. 14,* 519 (1970).

A.A. GRIFFITH, *Phil. Trans. R. Soc. A221,* 163 (1921).

S.M. GUMBRELL, L. MULLINS, and R.S. RIVLIN, *Trans. Faraday Soc. 49,* 1495 (1953).

R.W. HERTZBERG, *Deformation and Fracture Mechanics of Engineering Materials,* 3rd ed., J Wiley NY, 1989.

M.R. KAMEL and N.T. LEVAN, *SPE ANTEC Tech. Pap. 18,* 367 (1972).

F.E. KARASZ and L.D. JONES, *J. Phys. Chem. 71,* 2234 (1967).

L. MANDELKERN and P.J. FLORY, *J. Am. Chem. Soc. 73,* 3206 (1951).

L. MANDELKERN, R.R. GARRETT, and P.J. FLORY, *J. Am. Chem. Soc. 74,* 3949 (1952).

W. OSTWALD, *Kolloid-Z 36,* 99 (1925).

S. RABINOWITZ and P. BEARDMORE, *CRC Crit. Rev. in Macromol. Sci. 1,* 1 (1972).

R.J. SAMUELS, *J. Polymer Sci. 20C* 253, (1967).

N.G. MCCRUM, C.P. BUCKLEY, and C.B. BUCKNELL, *Principles of Polymer Engineering,* Oxford University Press, NY, 1988.

M.C. SHEN and A. EISENBERG, *Prog. Solid State Chem. 3,* 407 (1966); *Rubber Chem. Technol. 43,* 95 (1970); *Rubber Chem. Technol. 43,* 156 (1970).

L.H. SPERLING, *Introduction to Physical Polymer Science,* Wiley-Interscience, NY, 1986.

J.L. THRONE, *Plastics Process Engineering,* Marcel Dekker, NY, 1979.

L.H. VAN VLACK, *Elements of Materials Science and Engineering,* 4th ed., Addison Wesley, Reading, MA, 1980.

P.I. VINCENT, *Impact Tests and Service Performance of Thermoplastics,* Plastics Institute, London, 1971.

P.I. VINCENT, *Polymer, 15,* 111–116 (1974).

A.F. YEE, *J. Mat. Sci. 12,* 757 (1977).

A.F. YEE, "Impact Resistance" in Encyclopedia of Polymer Science and Technology, Vol. 8, 1986.

9

Polymer Production Systems

The final two chapters of this text describe some practical problems associated with making and processing polymers. This chapter reviews polymerization process from monomer synthesis to devolatilization of the polymer product. Chapter 10 reviews processing from the bulk storage of polymers and oligomers to parts and products. There are generic process schemes for each area that show typical operations and systems needed to convert the raw materials into products.

9.1 POLYMER PRODUCTION SYSTEMS

Figure 9.1 shows a block flow diagram for polymer production systems. There are four major steps for the production of polymer from raw material feedstocks: monomer synthesis and purification, monomer and storage transport, polymerization, and recovery. Some companies are vertically integrated and perform all the steps themselves. Others may specialize in monomer synthesis or polymerization. In either case, it usually is convenient to separate the monomer synthesis function from the polymerization function. The plant is designed by setting the monomer production capacity and then applying operating time percentages and conversion losses. Because of these losses, the flow of material decreases gradually through the process, as does the design capacity of the equipment. A process for making liquid monomer can be imagined as a large funnel with the wide mouth at the raw material end and the narrower outlet at the monomer end of the process. The equipment capacity will stay the same or decrease from inlet to outlet. By contrast, processes that make solid, particulate polymer often have increases in equipment handling capacity downstream of the solid

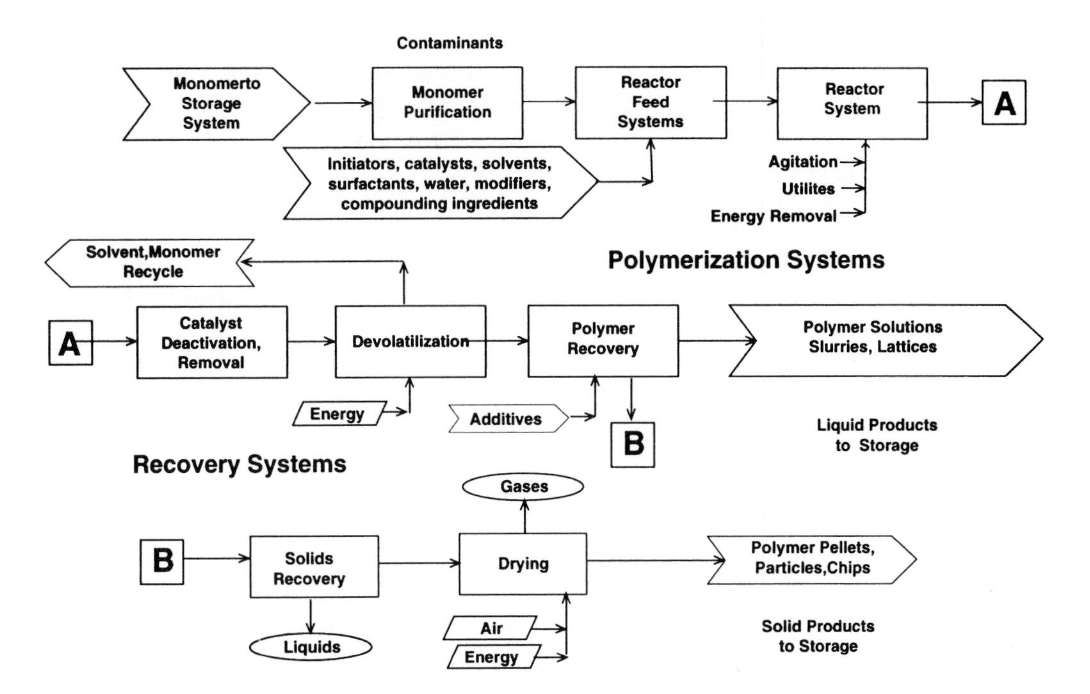

Figure 9.1 Block flow diagram for polymer production systems.

formation step. This design permits the system to remove material from the polymerization sequence rapidly, which often is the bottleneck of the process.

Monomer Synthesis and Purification. A monomer synthesis production unit converts chemical feedstocks into monomers using catalysts, energy, and other utilities. There may be several commercial processes for making a specific monomer. This is particularly true for the commodity polymers. Each process will have its own "fingerprint" of by-products, which must be removed from the crude monomer before shipping. Contaminants can affect the polymerization process and product by poisoning catalysts, depleting initiator from the reaction solution, acting as chain transfer agents, and acting as branching sites or unstable defects in the chains. The inability to make high purity monomers slowed the commercial development of the polymer industry. The major improvements in distillation technology that occurred in the 1930s and 1940s made the production of high molecular weight polymers possible. Emissions from the process into the air, into water effluents, and as solids for incineration and disposal are part of the overall material balance and process economics.

Monomer Transportation and Storage. Monomer is moved by pipelines, barges, railcars, tanker trucks, and drums. These methods are listed in increasing order of tendency for contamination. One important concern in transportation and storage of hydrocarbons is reducing evaporative losses and emissions. Many monomers will be stored as compressed liquids, so the possibilities of contamination by leakages from the environment into the storage container are remote. However, contamination of storage facilities by water can be a major concern. The presence of particulate solids in some batches of monomer can lead to long-term contamination problems. The ability to separate and clean up material that is out of specification is very desirable. In fact, in areas that have high volumes of monomer production and shipping, there are a number of small, custom plants that purify barge-load quantities of hydrocarbons.

A second concern in transportation and storage is preventing the polymerization or degradation of the monomer. A number of monomers that undergo free radical reactions are stabilized prior to shipping. Styrene is one such monomer stabilized during shipping. Reaction of volatile monomers during storage can be disastrous, because the storage containers are designed to hold large volumes of high vapor pressure liquids with the minimum of surface area (the lowest cost tank). These vessels are not designed to remove large quantities of heat.

Reaction Systems. The generic reaction system flow diagram (Fig. 9.1) is applicable to the production of a wide variety of polymers. It includes a number of operations that might be used to make polymer products. Obviously, the process flow diagram for a specific polymer may not include all these operations or have the operations in the same order.

The first step in the sequence is purifying incoming monomer. This step is necessary for monomers that have been stabilized during storage and shipping. Other raw materials also may be purified (solvents, water, and other large volume chemicals). Off-specification lots would be rejected.

Reactor feed systems should be considered as a special operation because of the mixing problems. Except for the solvent and water streams in some systems, most of the additives make up less than 1 wt% of the reactor volume. Usually, the goal of the reactor feed systems is to meter precise amounts of material into the reaction system in such a manner that the reactor agitation system can disperse the material uniformly. Each polymerization process type—bulk (solid and liquid phase), solution, suspension, and emulsion—will have its own agitation requirements.

The reactor often is the most important section of the process with respect to capital and operating costs. The major task of the reaction system is to develop the desired molecular weight distribution at reproducible residence or batch times. Agitation is a key factor for dispersing ingredients, maintaining residence times for continuous systems, and removing the heat of reaction. In most commercial systems, the rate-limiting step of the system is the rate of heat removal.

Recovery Systems. There are two types of recovery systems: volatiles recovery and polymer recovery. Initiator destruction or catalyst inactivation may be required for stable products. In general, catalysts and initiators are not recovered and recycled. Few commercial chain polymerizations are carried out to conversions higher than 90%. At these levels, the reaction rate slows and the reaction mixture can become quite viscous. It usually is economical to recover monomer and recycle it. Step polymerizations often are done in two stages. The first stage would be carried out to 90–95% conversion, making low molecular weight oligomers. Monomer and/or the leaving groups would be recovered from this prepolymer before sending it to a second reaction stage. The two-stage process allows the stoichiometry to be balanced prior to polymerizing to very high conversions (> 99.9%) to obtain high molecular weight materials (see Section 4.3.3 for a discussion of these effects).

Bulk and solution polymerizations have viscous polymeric fluids leaving the reactor. These fluids can be devolatilized above T_g, where the diffusivities of monomers and solvents are high. Suspension and emulsion polymerizations make solid particles, which are stripped of monomers. The stripping often is done while the solids are suspended in their original water slurry. Monomers and solvents from any volatile recovery system are separated and purified for recycle or resale.

Polymer recovery systems are used to make dry powder or pellet products from suspension and emulsion processes. Suspension systems can be dewatered using centrifuges. Drying is required to remove surface and interstitial water. Emulsion polymer systems may be used as the latex or may be coagulated. Coagulation is the process by which the small emulsion particles (diameters of less than 1 micron) are agglomerated into larger particles. The coagulation product may have a size as small as 10 microns, which is suitable for spray drying, or may be a crumb of several centimeters or larger, which can be recovered by screening. Coagulation can be done by shear, temperature, changes in ion type, ionic strength, or pH, or a combination of these techniques. It is preferrable to remove monomer and solvents prior to polymer recovery, since they often affect the surface properties of the polymeric solids during coagulation. In bulk and solution polymerizations, the volatile recovery process can serve as the polymer recovery process. The stripped monomer phase usually will be a viscous melt, since devolatilization usually is done at conditions between $T/T_g = 1.07$ (slurry stripping) to $T/T_g = 1.3$ (melt devolatilization). The melt can be sent to extrusion or other melt processing equipment for pelletizing or chipping.

When the polymerization product is an oligomer or a polymer solution, only volatile recovery is needed. The production of a stable liquid or solid product constitutes the end of the polymerization process. Solid products would be processed by methods discussed in Chapter 10. Liquid oligomers intended for thermoset applications or composites would be reacted further in the shape of the part. This will be discussed in part in this Chapter.

9.2 MONOMER SYNTHESIS

The economics of petrochemicals is tied very closely to the cost of the raw material, and much infrastructure has been built around transporting and using the products. Monomers made from sources other than petroleum—such as the silicones, cellulose-containing monomers, and monomers incorporating fluorine and chlorine—are less affected by changes in the crude oil price structure. The cost of the monomer is often the most significant cost portion of the commodity polymer cost. Therefore, commodity polymer prices tend to track the costs of their monomers and the cost of petroleum. Commodity polymers are made at the billion-pound-per-year level, and their markets are very competitive. Since monomer costs are a major portion of the polymer price, many businesses try to become vertically integrated so that they can control their costs at each step of the overall process. Monomers from different sources often have different trace impurities, which can lead to different defects in the polymer product.

Propylene. Propylene is typical of the α-olefins and is a product of alkylate gasoline refining. Thermodynamic and physical properties are shown in Table 1 of Appendix B. Other engineering properties useful for design are the upper and lower explosive limits in air (2.4 and 11.1 volume percent), the autoignition temperature (224 °C), and the solubility in water (44.6 ml gas/100 ml water at 20 °C, 101 kPa). Propylene is used to make acrylonitrile, propylene oxide, cumene, isopropyl alcohol, and butyraldehyde. It is used to make the homopolymer, polypropylene, as well as several important commercial copolymers, particularly those with ethylene. Its price in the United States is largely determined by its value in alkylating feedstocks for gasoline (Lieberman and Barbe, 1988).

The major commercial process for propylene production is its recovery as a by-product of ethylene production by steam cracking. The process is emphasized in Europe and Japan, where there are higher molecular weight feedstocks used for the steam-cracking process. Steam cracking is the pyrolysis of hydrocarbons in the presence of steam over the temperature range of 650°C to 950°C. The process residence time is on the order of 0.1 to 0.6 seconds with a steam to hydrocarbon ratio of 0.3:1.0 (Kniel et al., 1980). The pyrolysis occurs via free radical mechanisms and produces a number of unsaturated products. Ethylene is the primary product, and the ratio of ethylene to propylene is enhanced by raising the temperature and the residence time or by increasing the light hydrocarbons in the feed. A crude C_3 stream (the stream contains several components having three carbon atoms) is distilled to give propylene overhead containing propane, allene, and methylacetylene impurities. The polymerization grade is 99.8% pure (refinery, 50%, and chemical, 93%, grades also are produced).

Catalytic cracking is an alternative route to propylene. Petroleum fractions are passed over a zeolite catalyst at moderate temperatures (380°C to 530°C), low pressures (2.5 to 4.0 bar) with residence times between 5 and 120 seconds. The final product contains CO_2, COS, and H_2S as impurities. Polymerization grade monomer is used for gas phase processes, while chemical grade monomer can be used for liquid slurry processes. The lower quality monomer can be used for liquid phase reactions because of the high molar concentration of monomer (moles per unit volume) in the liquid phase.

Coordination complex catalysts for the polymerization of the poly(α-olefins) are becoming more efficient, so the presence of even trace amounts of catalyst poisons can be detrimental to polymer quality (Hahn et al., 1975). There are three types of impurities in polymerization grade propylene: inerts, copolymerizable monomers, and catalyst poisons. The inerts are primarily propane, but also include ethane, methane, nitrogen, and other saturated hydrocarbons. These components act as diluents, reducing the propylene concentration in some polymerization processes. Copolymerizable monomers, such as ethylene and butene, make elastic copolymers that are undesirable. The catalyst poisons include acetylene, dienes, CO, CO_2, oxygen, water, alcohols, H_2S, COS, and other sulfur-containing compounds. Different catalysts have different sensitivities to these poisons. Many polypropylene polymerization processes include molecular sieves to remove polar compounds such as CO_2 and water, metal oxides to absorb H_2S, oxidation systems for CO, and hydrogenation systems for unsaturated compounds. Table 9.1 shows some typical specifications for polymerization grade propylene for two different catalyst systems.

9.3 REACTANT STORAGE, TRANSPORT, AND FEED SYSTEMS

The goal of storage, transport, and feed systems from the point of view of the polymerization process is to get monomer, solvents, and other reactants from their

TABLE 9.1 Specifications for Propylene Purity
($TiCl_3$, gas phase[a]; $MgCl_2$ supported[b]—99.5 mol % propylene)

Components	Maximum	
	$TiCl_3$	$MgCl_2$
CONTAMINANTS, mol %		
Propane	0.5	
Methane and ethane	0.04	
TRACES	wt ppm	vol ppm
Ethylene	45	
Acetylene, propadiene	6	5
Butadiene, allene	10	
C_4	10	5
CO	5	0.3
CO_2	5	5
O_2	5	2
H_2	10	
H_2O	10	
S	2 (as H2S)	1 (total)
CH_3OH	5	5
COS		0.03
C6–C12 hydrocarbons		20

SOURCE: [a]Hahn et al., 1975.
[b]Ross and Bowles, 1985.

suppliers into the reaction system with the minimum of contamination and losses. There are a variety of methods for shipping and storing reactive organic liquids, and each has specific contamination problems. The most common contaminants are dust and dirt from air-borne sources; iron and other metal oxides from tank, pipe, valve, and fitting scale; water from the normal "breathing" of atmospheric storage tanks leading to condensation of water vapor on the walls; and oxygen and nitrogen absorption from air in contact with the liquid.

The different polymerization mechanisms are affected in different ways by these contaminants. Coordination complex catalysts are particularly sensitive to water. In fact, water or steam often are used to deactivate residual catalyst in poly(α-olefin)s after they leave the reactor. In these water-sensitive systems, large quantities of water contamination ($>1\%$) would be removed from reactants by distillation. Their reactor feed systems include molecular sieves to remove water from monomer and solvent.

Metal oxides can have two types of effects on polymer processes. They can catalyze the polymerization or can form a second product with an undesirable molecular weight distribution. If they pass through the process and are incorporated into the polymer, they can reduce its thermal stability. Some oxides catalyze dehydrohalogenation and depolymerization reactions, leading to products that discolor or degrade at temperatures much lower than those of the pure polymer. Metal oxides often occur in the form of scale and can be removed with filters having moderate particle size exclusions (10 microns would be typical). Since metal scale can be formed in most sections of the storage and transport system, large storage vessels should have filtering systems on all their inlets to prevent incoming particulates from contaminating large quantities of reactants. In most cases, filters should be used just ahead of the reactor in the reactor feed system as a final purification step.

Oxygen, nitrogen, and other gases dissolved in reactants may have subtle effects on polymerization systems. Oxygen often is a poison or inhibitor for free radical reactions because it can react with initiators to form peroxides that do not initiate chains. However, in some free radical reactions (LDPE for example), oxygen is a catalyst. A reasonable goal is to achieve reproducibly low levels of oxygen, rather than eliminate it completely. The polymerization system can be adjusted using extra initiator. Nitrogen and other gaseous inerts may come into the process when inert gas "blankets" are used to reduce oxygen contamination of organic liquids stored at atmospheric pressure. Noncondensible gases add to the total pressure of the gas phase, reducing the efficiency of reflux condenser cooling systems, lowering the partial pressure of reactants in the gas phase and building up in recycle systems. Inert vent systems would be part of the reactant system design.

9.4 TYPES OF POLYMERIZATION SYSTEMS

Polymerization systems must address typical design problems—such as conversion, residence time, agitation, and heat transfer—that are affected by changes in the physical properties of the system during reaction. Liquids with high viscosities may be the product from the reactor. High viscosities can impair heat transfer, so that methods for enhancing heat transfer may need to be devised. Typical solutions to the heat transfer problem include inserting cooling tubes, diluting the reaction media with organic solvent to make a polymer slurry, or designing suspension systems using water as a continuous phase. Suspended solid systems need a balance of agitation, surfactants, and particle size to function properly.

This section will consider the basic kinds of polymerization schemes and the situations in which they might find use. There are basically four types of continuous or batch polymerization systems: bulk, solution, suspension and emulsion. Table 9.2 compares these systems.

9.4.1 Bulk Polymerization Systems

Bulk polymerizations have been characterized as the polymerization from pure monomer. This definition would include systems with low amounts of solvents, catalysts, initiators, and chain transfer agents, as well as the new fluidized bed technology for preparing solid poly (α-olefins). However, the reactor types and design problems are quite different among these systems. The fluid viscosity of liquid phase bulk polymerizations can increase by a factor of 10^4 or more during the process. A revised definition is the polymerization from solutions high in monomer concentration (Nauman, 1982). This definition groups together systems in which the solution viscosities are very high during polymerization. Most bulk polymerization systems require a reactor system linked with a devolatilization system. The reactor systems will be described in this section, and devolatilizers will be discussed in Section 9.6.

TABLE 9.2 Commercial Polymerization Systems

Type	Advantages	Disadvantages
Bulk: batch	Simple procedure, composite lay-up and casting of unusual shapes.	Exothermic, wide molecular weight range.
Bulk: continuous	Temperature control, controlled MWD.	Low conversion per pass, monomer-polymer separation, high temperature, pressure may be needed.
Solution	Temperature control, solution form may be direct product.	Solvent may be expensive, polymer drying costly.
Suspension	Temperature control, dried particles may be a direct product.	Agitation and scale-up difficult, additives needed.
Emulsion	Temperature control, fast reaction rates.	High surfactant concentrations, poor stability.

General Design Factors. Bulk systems usually have two advantages over other types of polymerization schemes. They yield purer product in most cases, as the major contaminant is the catalyst (if one is used). No large inventories of solvents, emulsifiers, suspending agents, or their recovery equipment are required. However, they are subject to some very important limitations. Batch, stirred tank, and tubular reactors can be used, but are not practical for all systems. Factors such as the method of heat removal, stoichiometry control, and monomer removal, rather than volumetric efficiency, dominate the reactor selection.

Continuous bulk polymerizations of vinyl monomers tend to have limited conversions. This is especially the case in tubular reactors. As the conversion approaches 20% to 30% (or less depending on the conditions and the polymer), the viscosity of the stream becomes so large that mechanical flow becomes very difficult. Pumping, by pumps or imposed pressure drops, becomes costly, and the tube tends to plug with polymer. Mixing within the tube is suppressed, and catalysts may not be fully utilized. In stirred tanks or autoclaves, mixing can often be provided, but there may still be upper limits on conversion.

Figure 9.2 shows the increase in melt viscosity of polystyrene with temperature and polystyrene wt% in styrene solutions (Nauman, 1982). At conversion above 10 wt%, the solutions are viscous enough that tube flow will be laminar. Turbulent heat transfer is not practical, and natural convection would be very slow. Therefore,

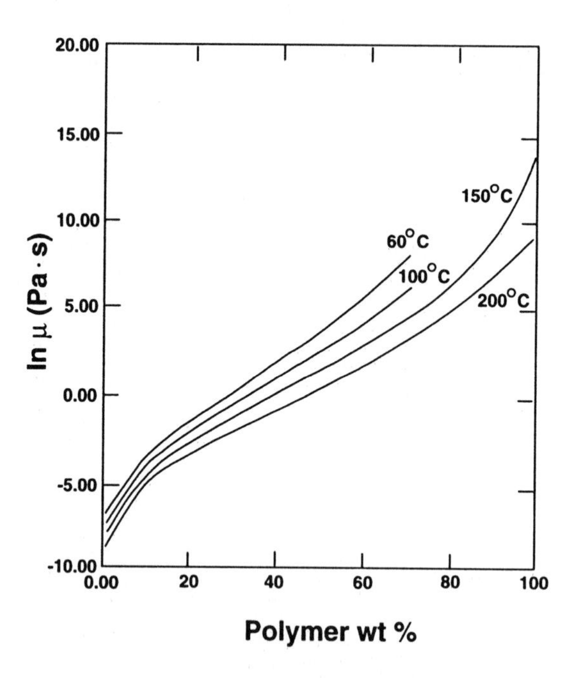

Figure 9.2 Viscosities in the styrene-polystyrene system: $\overline{M}_w = 259{,}000$, $\overline{M}_n = 110{,}000$ (Nauman, 1982).

heat transfer coefficients between reactor surfaces and the melt are several orders of magnitude below those for low viscosity systems.

Viscosity increases reduce heat transfer rates and change mechanical properties of the system. The heat of polymerization commonly is approximately −20 kcal/mol for a vinyl polymerization. Unless this heat is removed, the reaction will tend to autoaccelerate. As conversion increases, the heat transfer rate falls low enough that the heat liberated in the polymerization can no longer be removed. Excessive high temperatures may result in polymer depolymerization or decomposition, which would obviously affect polymer properties.

Cooling and heating systems are needed for bulk batch vinyl polymerizations. Heat input is required in the initial stages of reaction to bring the monomer to reaction temperatures. After reaction is initiated, heat rejection is necessary to control the distribution of molecular weights and to forestall the onset of the Trommsdorff effect (which results in high molecular weight polymer, Section 4.1.2). As the viscosity increases, the termination rate slows. Each initiated chain tends to react for a longer time before termination, resulting in increased molecular weights. The increased molecular weight, in turn, further slows termination due to its higher inherent viscosity.

Continuous bulk polymerizations of condensation polymers have rather different characteristics. The heat of reaction tends to be much smaller, so that temperature control is not so crucial. However, very high conversions are necessary to obtain high molecular weights. This requires very careful control of stoichiometry in a single reactor. In practice, most processes do not use a single-stage reaction from monomer to final polymer. The monomer may be prepolymerized to a degree of polymerization of 10 to 20. The material is separated from nonreacted monomer and further polymerized to give high molecular weight polymer. The second stage polmerization may be batch since the resultant viscosity is very high, and continuous processes do not seem to be able to handle the materials as well. In the case of multifunctional condensation polymer, the prepolymer may be polymerized at the same time it is molded into final product (cross-linking due to multifunctionality causes the material to be a thermoset).

The volatility of the monomer can be used to control temperature in batch and continuous vinyl polymerizations. Running the reaction at the boiling point of the monomer allows a reflux condenser to remove the heat of polymerization and to control the temperature. The adiabatic temperature rise can be calculated by knowing the heat of polymerization and the monomer density. Table 9.3 lists the adiabatic temperature rise for several bulk systems, ranked in order of their heat transfer requirements.

Batch Reactors. Casting is a traditional method for making small parts and shapes of some polymers. Poly(methyl methacrylate) and nylon 6 often are polymerized in the mold. The casting process is nearly adiabatic, and heat transfer limits part size. Larger parts can be made of nylon 6 than PMMA because of nylon's lower heat of polymerization. Figure 9.3 shows the temperature as a function of time for a PMMA

TABLE 9.3 Typical Heats of Polymerization and Adiabatic Temperature Rises

Polymer	ΔH, kJ/mol*	ΔT_{adiab},[†] °C
Polyethylene	95.0	1610
Poly(vinyl chloride)	95.8	730
Polystyrene	69.9	320
Poly(methyl methacrylate)	56.5	270
Polycaprolactam	15.9	68
Polysulfone	25.1	24
Polycarbonate	0	0
Poly(butylene terephthalate)	0	0

*Per mole of bonds formed.
[†]Assumes constant heat capacity of 2J/g ·°C).

casting. Although the adiabatic temperature rise can be 270°C for methyl methacrylate monomer, the temperature rise has been reduced to 80°C by using prepolymerized material.

Agitated batch reactors usually use ribbon agitators or very large diameter pitched turbine agitators. These provide top-to-bottom mixing and only modest heat removal capacity. Chain addition polymerizations usually are discharged at conversion less than 90%, and devolatilization is done in other systems. Stepwise polymerizations are done batchwise by getting the correct stoichiometry for complete conversion by careful reactant metering, and then stopping the polymerization at the desired molecular weight. The major mechanism for heat removal in either case is vaporizing monomer or solvent and using reflux condensation to maintain stoichiometry.

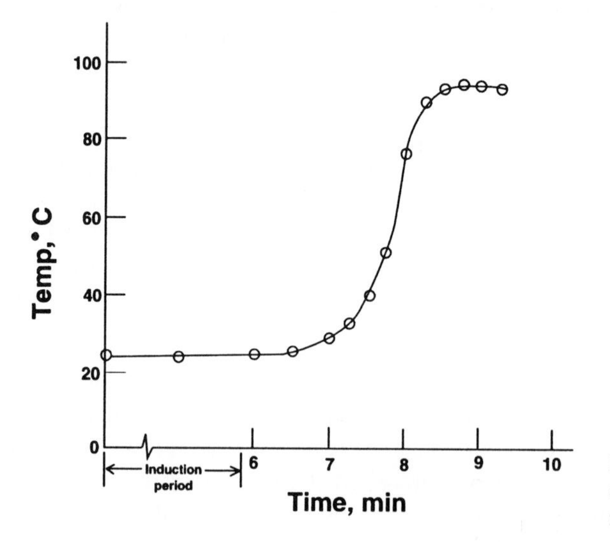

Figure 9.3 Reaction exotherm for a methyl methacrylate casting system (Nauman, 1982).

Continuous Stirred Tank Reactors. CSTR's often are used for bulk polymerizations having free radical mechanisms. High circulation rates can be maintained by using the agitator types mentioned above. Uniform temperatures are fairly easy to accomplish with good agitation and reflux cooling. Large reactors may use reflux cooling. Due to the pressure change caused by the liquid height, boiling occurs in the top portion of the reactor. The boiling process helps to agitate the top portion of the vessel, but may not have much affect on the mixing in the bottom of the tank where product usually is drawn off. Uniform polymer composition depends on uniform mixing of the reactor contents and may be hard to achieve.

The distribution of residence times from a CSTR has a moderate effect on the molecular weight distribution from chain addition processes, because the individual chains form very quickly. Stepwise polymerizations do not form high molecular weight polymer until high conversions are reached. Any fluid elements "short-circuiting" the reactor would have low molecular weight oligomers. Therefore, stepwise polymerizations are very sensitive to residence time distributions (Tadmor and Biesenberger, 1966), and CSTR's normally are not selected for these systems.

Because the degree of polymerization of chain reactions is directly related to the rates of polymerization and termination, the CSTR can produce precise molecular weight distributions as compared to batch reactors, in which the average molecular weight changes slowly with conversion (Example 4.4). CSTR's can be used to make uniform composition copolymers, reducing the problem of precipitation of varied composition copolymers (Example 5.5). Neither of these advantages can be realized if the agitation system cannot maintain uniform composition.

Plug Flow Reactors. Tubular reactors can be used when the viscosity of the fluid in the tube is uniform. Thermal runaways, hot spots, and flow instabilities are major operating problems. Constant fluid viscosities usually require that the polymer concentration throughout the tube be high and the temperature gradients be low. Tubular reactors can be used as finishing reactors for polystyrene. When the conversion through the reactor is small, temperature changes are modest and viscosity fluctuations are not as critical to polymer properties. Tubular reactors may be used to make polymers from liquid monomers in the pilot scale, but often fail on scale-up. For example, polystyrene can be made from monomer in small tubular reactors. In larger tubes, the polymerization tends to occur at the wall, increasing the local viscosity and reaction rate, and resulting in jets of unreacted monomer exiting the centerline of the flow (Lynn, 1977). Parallel multitube reactors suffer from wall polymerizations and from the probability that pressure drop differences between tubes will lead to uncontrolled flow rates. In-line static mixers have been used to improve the performance of tubular reactors.

Examples of Bulk Polymerization Systems. Table 9.4 shows five polymers that are made commercially via bulk polymerization systems. Some aspects of each process are discussed below. The production of low density polyethylene has been done in tubular reactors for many years. Recent developments in fluidized bed

TABLE 9.4 Some Commercial Bulk Polymerizations

Polymer	Reactor	Comments
Polyethylene (LDPE)*	Tubular reactor	250 MPa, 15% conversion per pass. Major commercial process for LDPE, now being replaced.
Poly(ethylene terephthalate) (PET)†		
1. Ester Exchange	Multistage reactor	Ethylene glycol (slight excess) plus dimethyl terephthalate reacts to diglycol terephthate and methanol moderate viscosity product. Multistage system gives products with low methanol content, which is desirable.
2. Poly condensation of diglycol terephthalate (DGT)	Multistage reactor	DGT is self-condensing with correct stoichiometry. Ethylene glycol is removed and recycled to ester exchange reactor.
3. Polymerization to high molecular weight	Devolatilizing twin screw extruder	High temperatures to permit viscous melt to be pumped. Vent(s) for removal of ethylene glycol under vacuum.
Poly(methyl methacrylate)‡	Sheet, rod, and tube molds	Casting process with 20% polymer dissolved in monomer. Heat removal and shrinkage are design problems. Cycle times include 45 minute reaction and 15 minute annealing.
Polystyrene§	Two stirred tanks in series, tubular reactor and devolatizer	Temperature and conversion increase through the system. Flash devolatilization at high temperatures (240–260°C), with water-induced foaming in second stage.
Polysulfone‖	Batch reactor	Bisphenol A reacted with NaOH to form sodium salt. Water removed by refluxing with a solvent. Dichlorodiphenyl sulfone is added. Methylchloride is used to end cap the polymer.

*Donati et al., 1982.
†Ravindranath and Machelkar, 1984.
‡Kine and Novak, 1981.
§Amos, 1974.
‖Johnson et al., 1967.

technology for olefin polymerizations will probably cause these systems to be replaced by the new designs. The conversion per pass is low to reduce thermal runaways and hydrodynamic instabilities. Wall fouling can be a problem but is reduced by periodic purges. Devolatilization may be done in two or three stages.

Poly(ethylene terephthalate) is made in a sequence of three different bulk polymerizations. The first stage is the ester interchange between excess ethylene glycol and dimethyl terephthalate, giving diglycol terephthalate and methanol as reaction products. A multistage reactor is used with methanol removal at each stage. The second step is the formation of prepolymers of diglycol terephthalate. Ethylene glycol is devolatilized throughout a second multistage reactor. The viscosities are low because of the modest molecular weight of the fluid. The final step is carried out in a devolatilizing twin screw extruder that can pump the viscous polymer solution while removing the remaining ethylene glycol to force the reaction to the right.

Poly(methyl methacrylate) is made into sheet, rod, and tube parts by direct casting. The process is improved by starting with 20% polymer in monomer and adding some initiators. Removing the heat of polymerization is a major problem, as is the shrinkage of the part on cooling due to the density differences between the monomer and polymer. Hot spots often develop due to nonhomogeneous polymerization and lead to internal stresses in the parts. Annealing is used to relieve the stresses.

Polystyrene processes often use two CSTR's in series to make 40% conversion material (130–140°C) and 60% conversion material (150–160°C), followed by a tubular reactor, which has an exit temperature of 200°C. Several devolatilizer designs have been used, but foaming devolatilizers based on water injection work well.

Polysulfone is an example of a condensation polymer that can be made by sequenced batch operations. Batch weights can be measured accurately, so Bisphenol A and sodium hydroxide are added stoichiometrically to form the sodium salt. Water is removed by refluxing with a solvent. Dichlorodiphenylsulfone is added, and high polymer is made with sodium chloride as a leaving group. There is no equilibrium limitation in this reaction mechanism. Sodium chloride precipitates from the reaction solution and forms a discontinuous phase. The reaction is quenched by cold solvent, and the polymer is end-capped with methyl chloride. The polymer is washed with water to remove the second phase. The maximum polymer concentration in the solvent is 25%, due to the high viscosity of the polymer phase during washing decreasing the efficiency of salt removal. Devolatilization requirements are high, and three stages may be needed.

9.4.2 Solution Polymerization Systems

Solution polymerizations are done with a solvent that will dissolve both the monomer and its polymer. Organic solvents and water can be used as the solvent. Free radical, coordination complex, and ionic mechanism systems use solution polymerization. Table 9.5 shows some solution polymers and their 1986 production levels. Some classifications of solution polymerization include the LDPE process as a solution process. Batch solution polymerizations have an advantage over suspension polymerizations in that additional initiators and catalysts can be readily dispersed in the reaction medium at any time during the process. This capability can lead to precise control of molecular weight and precise control of copolymer compositions.

There are a number of alternative solution processes for some of the commodity

TABLE 9.5 Commercial Solution Polymerizations†, 10^3 Metric Tons

Polymer	United States	Total
NONAQUEOUS PROCESS		
High density polyethylene	440	499
Linear low density polyethylene	313	400
Acrylic resins	68	113
Vinyl chloride–vinyl acetate copolymers	36	59
Fluoropolymers	14	25
AQUEOUS PROCESS		
Poly(acrylic acids)*	21	

*Includes poly(methacrylic acid) and copolymers.
†1986 Production.
SOURCE: Swift and Hughes, 1989

polymers. Linear low density polyethylene (LLDPE) provides one example. Table 9.6 shows four different solution routes to this product using slurry catalysts. Figure 9.4 shows the Phillips double loop process flow diagram. The monomer, comonomer, and isobutane are fed to a purification system. The product of this step is combined with fresh catalyst and recycled solvent as the reactor feed. The polymer actually phase-separates into discrete particles in the reactor, which makes this process not quite fit the formal definition of a solution process. The slurry of particles, catalyst, monomers, and solvent is circulated through the loop reactor. The polymer settles in the bottom of the loop as a way of concentrating and separating the solid from the unreacted material. Intermittent discharges of the bottoms content into a flash drum provides the method for removing polymer from the system. The solids concentration of this slurry ranges from 50% to 60%. A two-stage flash unit removes the monomer and solvent from the solid polymer. Solvent is recovered from the overhead stream of the flash and recycled. Ethylene is separated from the comonomer and each is recycled. The polyethylene solid is compounded and extruded into pellets for storage and shipping.

TABLE 9.6 Solution Polymerization Routes to Linear Low Density Polyethylene (LLDPE)

Reactor Type	Solvent	Supplier
Continuous path double loop	Isobutane	Phillips
Loop reactor	Hexane	National Distillers/USI Solvay
Stirred tank	Hexane, heptane	Hoechst Montedison Asahi
Liquid-pool process	Propane, butane	El Paso

SOURCE: James, 1986.

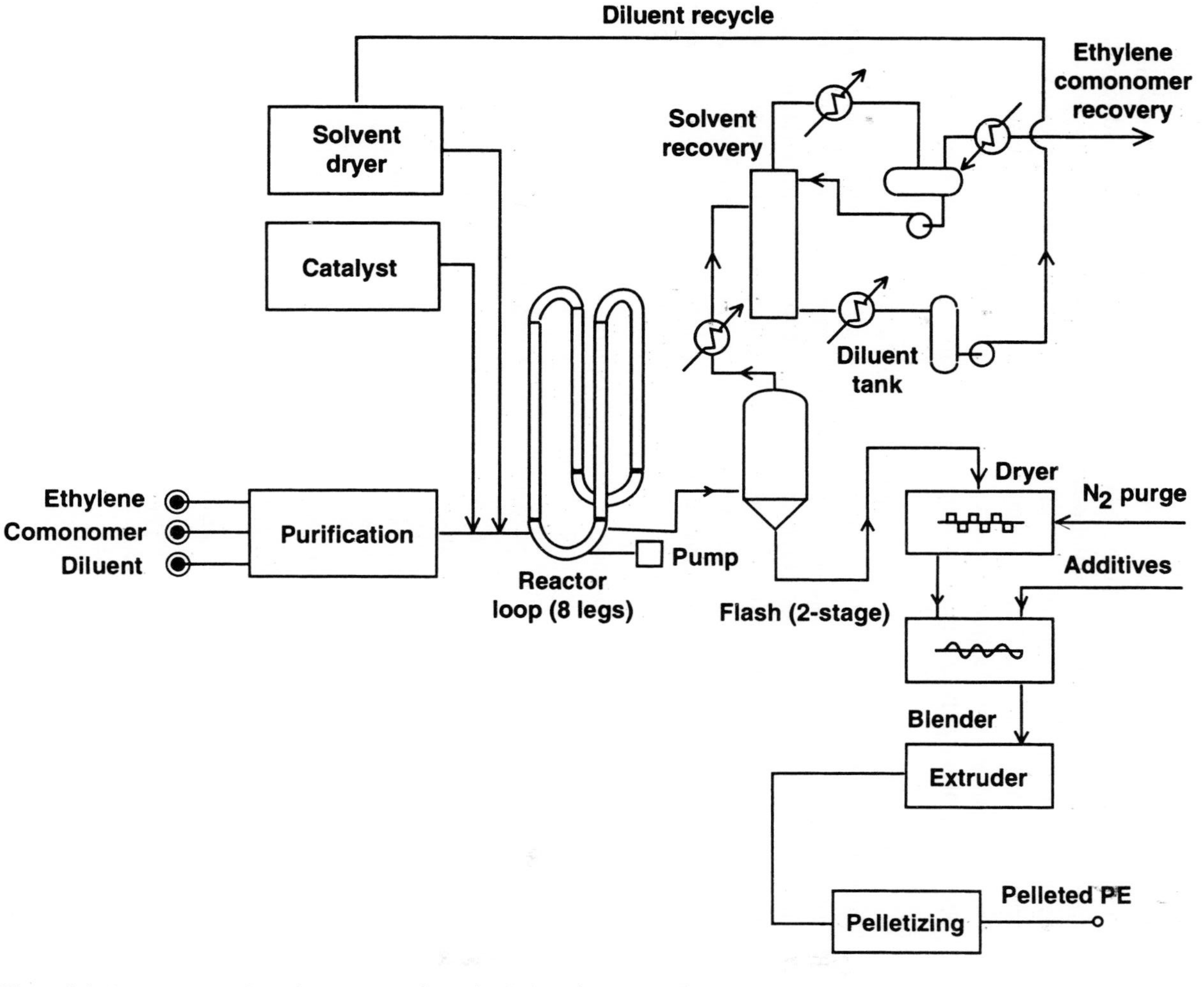

Figure 9.4 Low pressure slurry loop process for polyethylene (James, 1986).

Many of the solvents used for polyethylene solution processes have relatively low boiling points. While they may dissolve polyethylene near its melt temperature, they are not very soluble in the polymer at storage conditions. As a result, there are low levels of these solvents in the pellet product and even less in finished parts. Many of the other organic solvents that might be chosen for other polymerization systems have higher boiling points and are more difficult to remove from polymers. While the economic recovery level might have been 1wt% at one time, the new and important concerns about environmental pollution have resulted in more regulations on volatile hydrocarbons. The regulations on volatile organic content and worker exposure resulting from these concerns are putting pressure on the use of many organic solvents. Now, solution polymerization processes using organic solvents might only be developed after other polymerization systems having fewer volatile recovery difficulties have been tried.

The polymers made in aqueous solution do not have the same type of regulatory pressure. Table 9.7 lists some free radical solution systems. Monomers that are not soluble in water are candidates for suspension and emulsion systems. Water is not a concern as a contaminant for polymers with respect to health and environmental issues. Furthermore, the presence of water reduces the activity of the monomer in the system, making it easier to remove. Water-soluble polymers are used in detergents, dispersants, thickeners, and water treatment applications. They include poly(acrylic

TABLE 9.7 Monomers Suitable for Free Radical Solution Polymerization

Monomer	Manufacturer	Organic	Solvent Water
Acrylic acid	Rohm & Haas	+	+
	BASF		
	Hoechst-Celanese		
Methacrylic acid	Rohm & Haas	+	+
Methyl methacrylate	Rohm & Haas	+	−
Methyl acrylate	Rohm & Haas	+	+
Ethyl acrylate	Rohm & Haas	+	−
Butyl acrylate	Rohm & Haas	+	−
Maleic acid	Monsanto	+	+
Itaconic acid	Pfizer	+	+
Acrylonitrile	American Cyanamid	+	+
Styrene	Dow	+	−
	USX Corporation		
Vinyl acetate	Hoechst-Celanese	+	+
Methyl vinyl ether	GAF	+	+
Butadiene	Exxon	+	−
α-methylstyrene	USX Corporation	+	−
Vinyl pyrrolidinone	GAF	+	+
Vinyl chloride	Dow	+	−
Vinylidene chloride	Dow	+	−

SOURCE: Swift and Hughes, 1989.

acid), poly(methacrylic acid), poly(maleic acid), polyacrylamide, polymethacrylamide, poly(vinyl alcohol), poly(N-vinylpyrolidone), and their copolymers. The rate mechanisms follow the chain polymerization discussion in Chapter 4, with the exception that redox initiation systems using transition metals and hydrogen peroxide can be used, as these materials are soluble in water.

As in suspension and emulsion processes, the large heat capacity of the solvent greatly improves temperature control. Reflux cooling can be used. The solvent also serves to lower the viscosity of the reacting mixture as long as the polymer remains dissolved. Most commercial polymerizations have limits on polymer solubility between 5% and 15% by weight. Above this point, polymer either crystallizes or separates as a second liquid phase. The first alternative may be handled either by filtration of reactor effluent or by heating the effluent to melt the polymer. The formation of a second liquid phase is usually not desirable since it will have a high viscosity and is difficult to recover.

Solution processes allow intimate mixing of initiators and monomer. Catalysts can be prepared and added to the reaction mixture as a solution or suspension in the solvent (depending on the nature of the catalyst). This improves catalyst effectiveness and product uniformity. Continuous solution processes are common since they are easy to interface with continuous solvent recovery systems. Solvents and monomers are recovered, purified, and recycled. Suspending agents, catalysts, and other contaminants must be removed. Solvents and monomers are often removed by flashing or steam stripping. Catalysts can be removed by filtration or centrifugation. In some cases, specifically Ziegler-Natta–type catalysts, the catalysts are so efficient that the residuum is not removed from the polymer.

9.4.3 Suspension Polymerization Systems

Suspension polymerizations are systems in which monomer(s) are suspended as the discontinuous phase of droplets in a continuous phase (usually water) and polymerized. The reactor product is a slurry of suspended polymer particles. The monomers suitable for suspension polymerization are those that can be polymerized by free radical mechanisms and that are sparingly soluble in water. The terms pearl polymerization and bead polymerization refer to the smooth, spherical particles made in some suspension systems.

Suspension particle morphology has technical value for downstream processing, affecting how the solid performs in mixing operations and in applications. Suspension particles of poly(vinyl chloride) are 50 to 200 microns in diameter, with the size distribution depending on the monomer type, surfactant system, and agitation. Some products with sizes slightly smaller and larger than this range are made. Microsuspension processes can be used to make particles in the range of 20 to 30 microns, while dispersion processes make particles in the range of 400 microns.

The description of suspension particle sizes depends on the specific monomer system and what levels of surfactant are required. Emulsion systems make thermodynamically stable droplets that usually do not require agitation to prevent agglomeration once they are formed. Suspension droplets are not thermodynamically stable and tend to coalesce even with agitation. The particle size distribution achieved is a dynamic balance maintained by the surfactant and the agitation system. Suspension particles are much larger than those made in emulsion systems (20–1,000 nm), move as free-flowing powders, and are relatively easy to remove from water by mechanical means. Table 9.8 shows some polymers made by suspension polymerizations and their applications.

Several characteristics of suspension polymerization are common to most systems. The weight ratio of the continuous water phase to the discontinuous monomer phase varies from 1:1 to 4:1. The bulk viscosity of the slurry is nearly that of water for most of the polymerization. Agitation is required to prevent particle agglomeration during polymerization. The engineering advantages of suspension polymerization include low conversion cost when water is the suspending agent, excellent heat transfer due to the high heat capacity of water and the low slurry viscosity, particle morphology control, and low levels of additives in the polymer.

There are some disadvantages to suspension processes. The suspending agents can become incorporated in the surface of the particles, affecting properties such as particle fusion. The water may contain contaminants, such as metal ions, that adversely affect thermal stability. Monomer soluble in the water phase can polymerize and the resulting polymer will have properties different from that made in the particles. Most suspension processes are batch because of the problems of coalescence and polymer build-up on reactor surfaces.

Process Description. The steps of a typical suspension process include preparation of the suspension system (suspension agents, buffers), formation of the monomer suspension with initiator in the water continuous phase, initiation and polymerization, monomer recovery, dewatering, and powder storage. Each particle can be considered a mini–"bulk" reactor for modeling purposes. Figure 9.5 shows the

TABLE 9.8 Examples of Suspension Polymerization Systems

Polymer	Suspension System	Typical Application
Poly(vinyl acetate)	Poly(vinyl alcohol) in water	Hydrolysis to poly(vinyl alcohol) in methanol solution with sodium methoxide catalyst.
Poly(methyl methacrylate)	Sodium polyacrylate in water	Particles and pellets for a variety of extruded applications.
Poly(vinyl chloride)	Poly(vinyl alcohol), methyl cellulose in water	Particles for compounding into injection molding and profile extrusion formulations. Copolymers for bottle applications.
Polystyrene		Ion exchange resins after postpolymerization reactions; beads for foamed products.
Poly(vinylidene chloride)		Copolymers to make packaging films.

conversion versus time for a suspension system. Except near the completion of polymerization, most of the monomer is in droplets or polymerizing particles because it is sparingly soluble in the water phase. Most systems use initiators that are soluble in the monomer phase rather than the water phase. Reactions are carried out above room temperature, but below the boiling point of the water phase. Reactor pressure is related to the vapor pressure of the monomer, and it often decreases near the end of the polymerization. In some copolymer processes, the pressure actually increases toward the end of the polymerization.

Most polymerizations show a lag time after initiation, during which oxygen and other free radical traps are consumed. At high conversions, the polymerization rate may be lowered because of the slow diffusion of monomer to the free radical chain ends. When the reactor pressure is reduced to atmospheric pressure, the rates are too low to be economical and the reactor is discharged.

At the end of the polymerization, the resins still contain monomer and may be near their T_g. Free radicals may still exist due to the very slow reaction rates. These need to be destroyed for safe storage and transport of the polymer. Residual free radicals can be destroyed by adding free radical traps, adding terminating agents, or raising the temperature. The last technique is the most common as most suspension resins are stripped of residual monomer by raising the temperature of the slurry. The slurry is dewatered by centrifugation and dried. The resin is transported to storage silos by mechanical or pneumatic means. Figure 9.6 shows the process flow diagram for a poly(vinyl chloride) suspension process.

Particle Formation. The formation of solid particles during the polymerization is an advantage of suspension polymerization. Key problems are forming of a

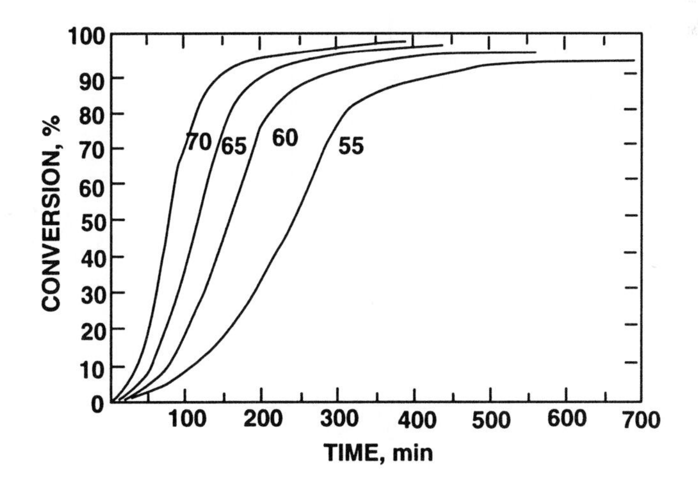

Figure 9.5 Suspension polymerization of vinylchloride monomer at different temperature, with AIBN as an initiator. $[I]_o$ = 0.25 wt%; temperature in °C (Xie et al., 1991).

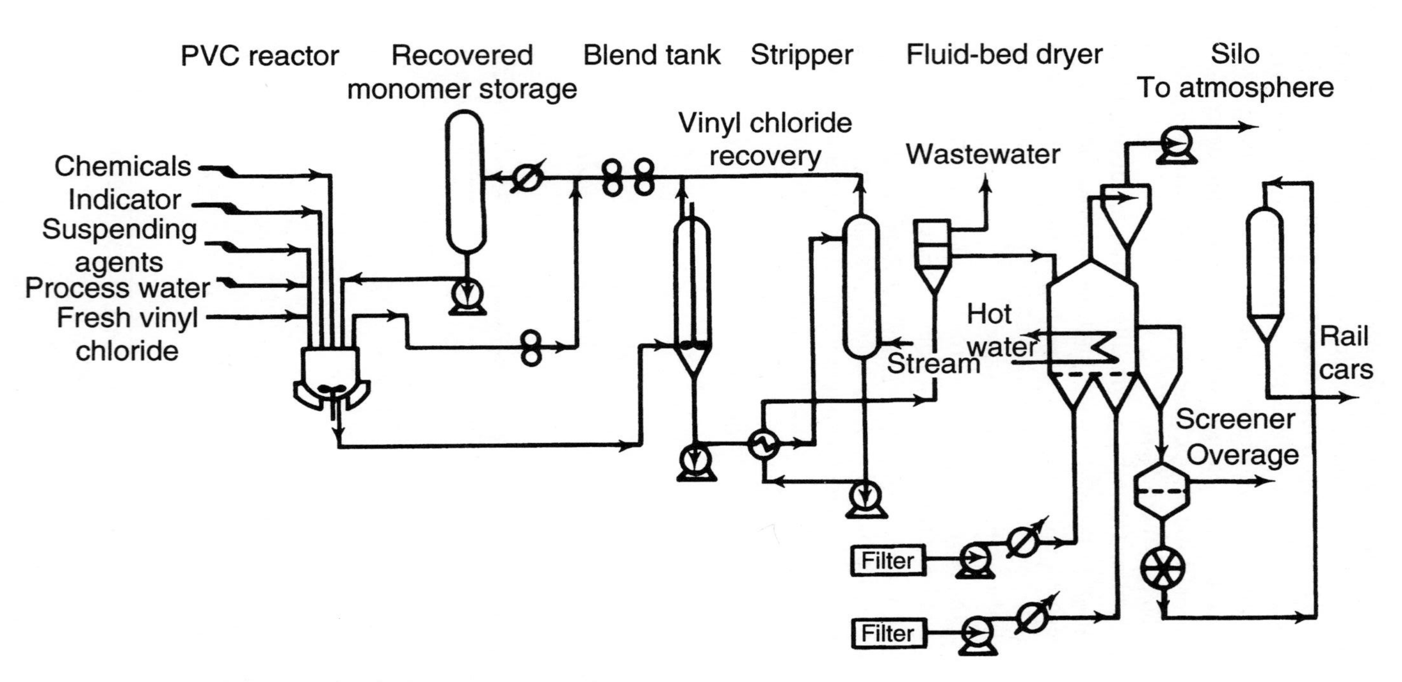

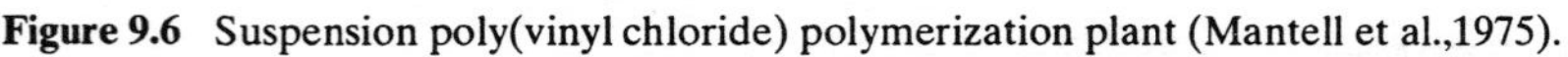
Figure 9.6 Suspension poly(vinyl chloride) polymerization plant (Mantell et al.,1975).

"uniform" suspension of monomer droplets and preventing the formation of large agglomerates. Large agglomerates have several undesirable properties. They settle at faster rates than the rest of the particles and are harder to suspend with agitation; they can become sites for runaway reactions; they can adhere to reactor surfaces; and they do not take up stabilizer and other compounding ingredients well.

The monomer must be relatively insoluble in water for droplets to form. The typical organic monomer has a lower surface tension than water. When these monomers are mixed into water with no surfactants present, an unstable dispersion forms due to the continuous breakup and coalescence of monomer droplets. If agitation stops, the dispersion will settle into two phases. Particle coalescence is not prevented, but it is controlled by a combination of surfactant and agitation system design.

The surfactant affects particle size, porosity, and shape, as well as the clarity, transparency, and film-forming properties of the resin. Most suspending agents function by the mechanism of adsorption at the monomer-water interface. In some systems, the surfactant polymerizes with monomer to form a thin skin. The surfactant maintains droplet dispersion from the start of the reaction through the stage in which the polymerizing droplets are tacky. After this point, the tendency to agglomerate is reduced, and the agitation system is sufficient to suspend the particles.

The agitation system is designed to balance the requirements for preventing coalescence, providing good top-to-bottom mixing, and providing good velocity past heat transfer surfaces. The heat transfer rate is related to slurry viscosity. The rheology of suspensions is discussed in Section 9.5.2. It is preferable to design for heat transfer by using correlations based on concentrated, particulate slurries rather than isotropic fluids. The particles cause subtle differences in the flow fields in agitated vessels, and the impact of particles on the vessel walls will affect heat transfer.

The scale-up of agitation systems from the bench to commercial scales is not trivial, and much of the information is proprietary. The impeller Weber number has been used to correlate particle sizes for vinyl chloride (Johnson, 1980) and styrene (Leng and Quarderer, 1982) polymerizations. The mean PVC particle size is not linear with Weber number, as would be expected for oil-in-water dispersions, but is parabolic with a minimum (Fig. 9.7). The impeller Weber number is defined as

$$W_e = \frac{\rho N^2 D^3}{\sigma g_c} \qquad 9.1$$

where ρ is the total density of the charge, N is the impeller speed, D is the impeller diameter, σ is the interfacial tension, and g_c is the gravitational constant. At low Weber numbers, the average particle size is high. Monomer droplets decrease in size as the Weber number increases, increasing the droplet surface area (which affects the coalescence rate). The droplet size goes through a minimum and continues to rise with Weber number as agglomeration dominates.

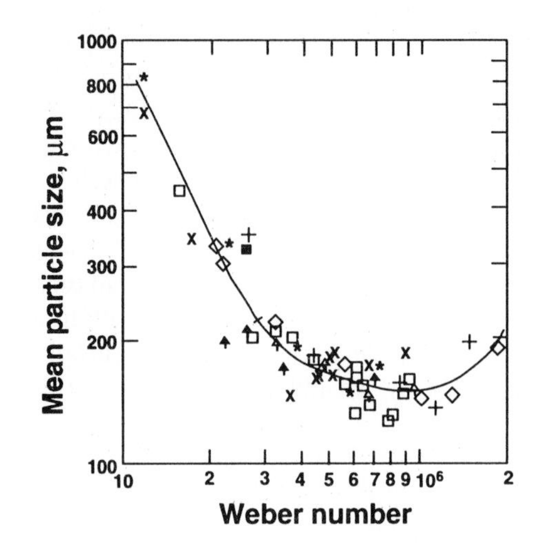

Figure 9.7 Particle size versus Weber number for suspension poly (vinyl chloride) (Johnson, 1980).

Example 9.1 *Controlling Suspension Particle Size* on Scale-up

A research engineer prepared a PVC resin having an average particle size of 200 microns in a bench scale batch polymerization vessel. What conditions should be used in a commercial scale unit to produce the same average particle size product?

Basis:

Variable	Bench Scale	Commercial Scale
D_T, tank diameter	0.105 m	2.25 m
H_T, tank height	0.315 m	6.75 m
N, impeller speed	1000 rpm	?
D_I/D_T, impeller/tank	0.75	0.67
ρ, slurry	1.15	1.15

Solution:

The two reaction vessels are assumed to be similar geometrically to each other with similar standard internals. The Weber number correlation of Figure 9.7 should apply, and the scale-up goal is to match the Weber number of the bench unit with that of the commercial unit. While there are four parameters in Equation 9.1, only two typically would be used to control N_{We}. The slurry density would be kept constant, as high product loading is desirable. The surface tension would

(continued)

be kept constant, since it would be normal to use the same surfactant system in each unit. Therefore, impeller rpm and impeller diameter might be varied in the plant. Assuming that it is easier to change rpm (using a variable motor drive or a change in the gear box), the commercial unit's rpm is

$$(N^2 D_I^3)_{bench} = (N^2 D_I^3)_{commercial}$$

or

$$N_{commercial} = N_{bench}\left[\frac{D_{I,bench}}{D_{I,\,commercial}}\right]^{\frac{3}{2}}$$

$$= 1000\left(\frac{.105}{2.25}\right)^{1.5}$$

$$= 12.1\ rpm$$

Comments:

The difference in size between these vessels is large. Unless the scale-up model had been carefully established between these two vessels, intermediate size vessels might be used to verify the scale-up model. The curve shown in Figure 9.7 may be specific to the polymerization system and vessel internals studied. For best results, the correlation would be tested carefully by the user.

Polymerization Rates Polymerization rates can affect the performance properties of suspension products, including molecular weight distributions, side reactions, residual monomer, and particle morphology. The kinetics are similar to those for bulk polymerizations. However, there are three different phase behaviors for suspension polymerizations: polymer soluble in its monomer, polymer swollen in its monomer, and polymer insoluble in its monomer. The first system has a single monomer-polymer phase during the entire polymerization. The free radical polymerization equations developed in Chapter 4 can be applied directly to these systems.

Polymers that are swollen by monomer can have two phases present during the polymerization, a monomer-rich and a polymer-rich phase. Polymerization can take place in either phase. The kinetics have been modeled by using a set of polymerization equations for each phase having different coefficients (Davidson and Gardner, 1983; Johnston, 1976; Ugelsted at al., 1981; Xie et al., 1991). At moderate conversions, the polymerization rate in the monomer-rich phase is lower than that in the polymer-rich phase, primarily because the termination rate constant is much lower in the polymer-rich phase.

Polymers that are insoluble in their monomers often crystallize as they polymerize. An example is the poly(vinylidene chloride) system. These systems have been difficult to model, and some data suggest that the polymerization and crystallization processes are coupled (Abdel-Alim and Hamielec, 1972; Burnett and Melville, 1950).

Particle Morphology. Suspension particle morphology depends on the phase separation phenomena, the surfactant, and the agitation. In general, polymer has a higher density than monomer. As the monomer reacts, the volume of the droplet decreases. If polymer forms near the surface of the droplet, the reduction in volume can cause depressions in the particle surface. This phenomenon has been called density collapse. Polymer soluble in its monomer often makes spherical beads (pearl polymerization). Figure 9.8 shows some polystyrene ion exchange beads that are nearly spherical. The beads show no sign of density collapse since they are cross-linked. Figure 9.9 shows some poly(vinyl chloride) suspension particles. A cross

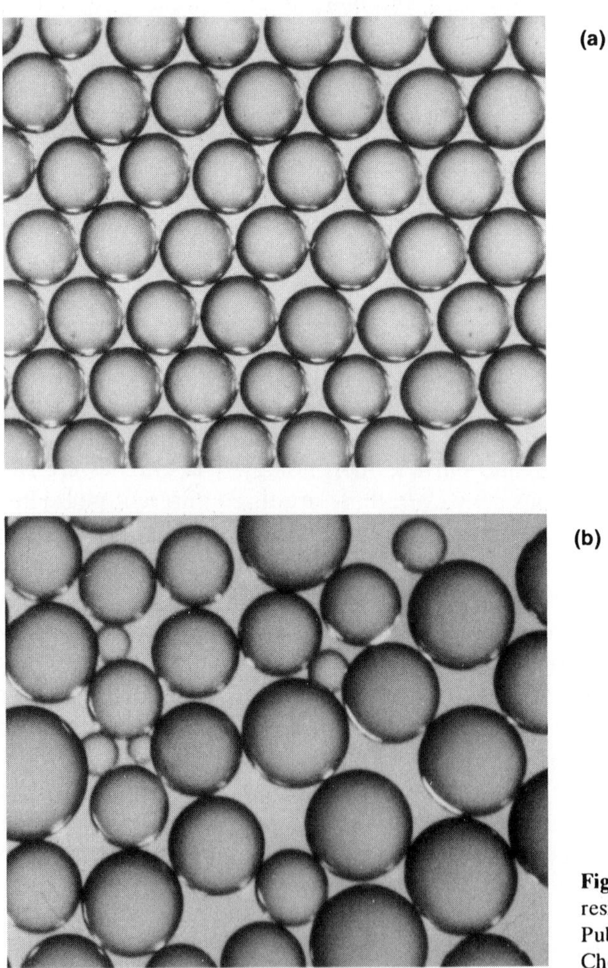

Figure 9.8 Polystyrene ion exchange resin: **(a)** monodisperse, **(b)** polydisperse. Published with the permission of the Dow Chemical Company.

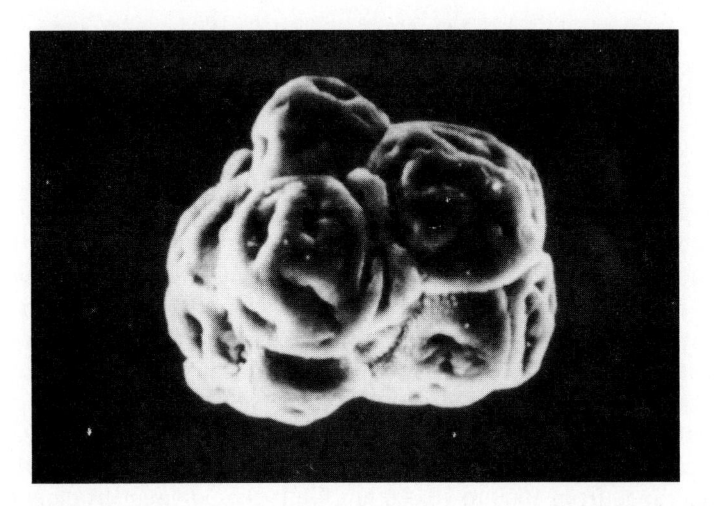

Figure 9.9 Poly(vinyl chloride) resin agglomerate. Published with the permission of the Greon Company.

section of the particle shows primary particles that have precipitated from the monomer phase inside the large monomer droplet. The polymerization has continued inside the primary particles, and they agglomerate to form a network that becomes the large particle. The final product particles are agglomerates of several large particles that presumably came together during the tacky stage of the polymerization.

9.4.4 Emulsion Polymerization Systems

In an emulsion polymerization process, monomers are dispersed as fine droplets into a continuous phase water using high levels of surfactants and polymerized by the free radical method. The continuous phase often is water, and the droplets are colloidal in size, usually $<< 1$ micron. The product slurry is called a latex. In contrast to suspension polymerizations, emulsion dispersions tend to be thermodynamically stable. Once the dispersion is formed, it will not coagulate immediately in the absence of agitation. The small particle size leads to some unique polymerization models.

Acrylic monomers, acrylamide, acrylonitrile, butadiene, chloroprene, ethylene, styrene, vinyl acetate, and vinyl chloride can be made into lattices via emulsion polymerization. Latex paints are one example of consumer products made by this technique. The small particles made in the process have some desirable properties. Their small diameter makes nonuniformities in packing undetectable to the eye. Therefore, they form smooth, glossy coatings when applied from the liquid or dry forms. Blended or block copolymers based on emulsion products with the diameter of the dispersed phase less than one micron in size tend to have improved mechanical properties. The typical size range possible in emulsion polymerizations goes from 10 nm to over one micron. Most commercial products are based on oil-in-water emulsions, but water-in-oil emulsions can also be made for hydrophilic monomers such as acrylic acid and acrylamide.

Emulsion Formation and Polymerization Mechanism. This discussion is based on systems with water as the continuous phase. Emulsion systems require a number of additives (given in order of their volume fraction) a water phase, a monomer phase (monomer, comonomer, and solvent), emulsifiers (surfactants plus water phase buffers to control the ionic levels and pH), initiators (usually soluble in the water phase rather than the monomer phase), and possibly a seed latex, swelling agents, or other solvents. The order in which these components are mixed, the agitation system used to create the dispersion, changes in the components during reaction (adding monomer for example), and the postpolymerization processing all affect the latex characteristics and properties. Water dispersions of monomer droplets often range from 1000 to 10,000 nm, while the polymer droplets formed range from 100 to 400 nm. This size change is too large to be accounted for by the change in density between monomer and polymer. In contrast to suspension polymerization, an emulsion process is not simply the polymerization inside monomer droplets.

Harkins (1947, 1950) developed a model for the emulsion polymerization of styrene and butadiene that is a good description of the process. There are three steps in this mechanism. The first step is polymer particle nucleation, the second step is monomer droplet depletion, and the third step is the polymerization to high conversions.

Step 1. Polymer Particle Nucleation. Monomer is added to water containing the surfactants, buffers, and the initiator. Agitation is used to increase the surface area, allowing the surfactant system to make small monomer droplets in the 1 to 10 micron range (see Fig. 9.10). The surfactant system must be above its critical micelle formation level in order to stabilize monomer droplets. The surfactant is present in the water, in the micelles, and at the monomer-water interface. Monomer is present in the monomer droplets, is dissolved in the water phase, and also swells the micelles. Polymerization is thought to start when water-soluble initiator diffuses to the surface of a micelle and decomposes. The polymerization of monomer in the micelle depletes its local concentration, and more monomer diffuses to the micelle from the surrounding water phase. The water-soluble monomer is replaced by monomer diffusing from the monomer droplets. The probability that an initiator will begin a chain in a micelle is high compared to the probability that a chain will be initiated in a monomer droplet, due to the big difference in surface area per unit volume. This mechanism suggests that changes in the surfactant system that lead to different micelle sizes and numbers of micelles would affect the polymerization.

The polymer droplet grows quickly and takes up monomer and surfactant from the water phase. Polymer droplets keep on forming (nucleating) until all the emulsifier is either on the surface of a monomer droplet or on the surface of a polymer particle (there are no more micelles). This occurs at conversions between 2% and 10% for a

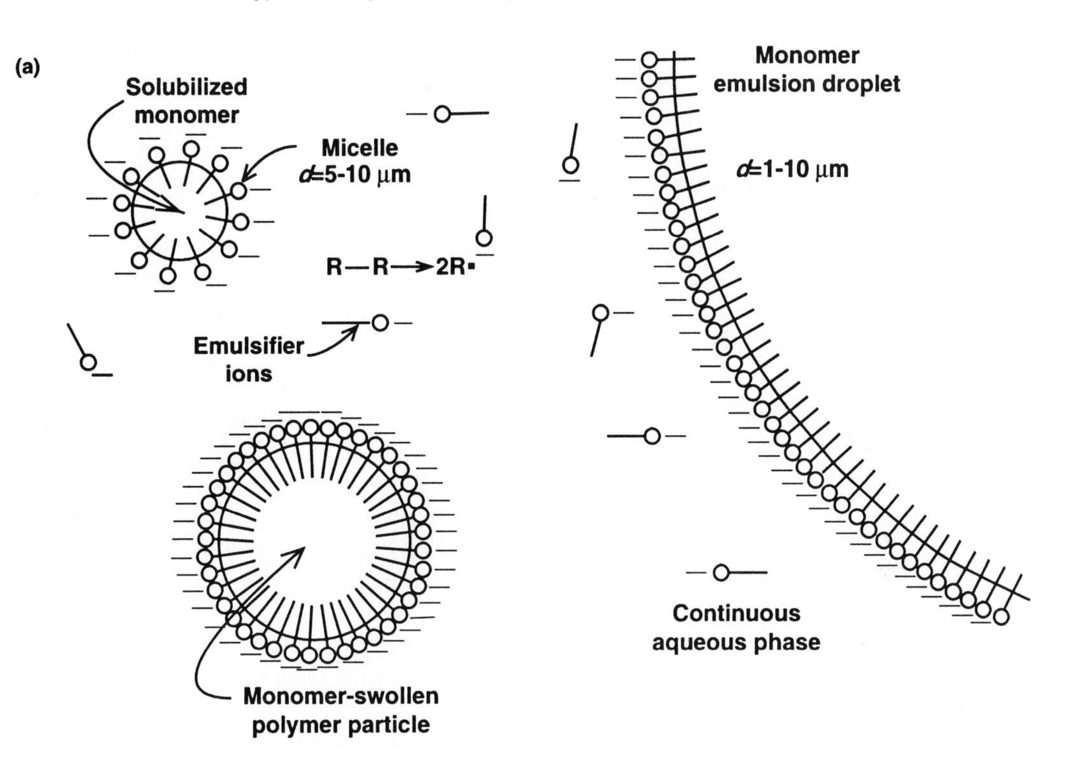

Figure 9.10 Emulsion polymerization: **(a)** schematic showing solution during Interval 1—the number of micelles, particles, and droplets are arbitrary; d is diameter (Vanderhoff, 1969); **(b)** scanning force photomicrograph (Topometrix TMX 2000) of finished latex particles. Published with the permission of F.F. Lin and D.J. Meier, Michigan Molecular Institute.

number of emulsion systems. If desired, nucleation could be continued by adding more surfactant.

Poehlein (1986) has suggested several other phenomena that can affect the polymer particles nucleated in the first step. Water-soluble initiators can leave ionic end groups on the polymer chains that should remain at the water interface in the polymer particle. These groups will act as surfactants to stabilize the particles and supplement the original surfactant. This mechanism helps explain the fact that stable polymer lattices can be formed without emulsifiers. Depending on the sizes of the micelles and monomer droplets, initiation in the monomer droplets may be important. Other factors include polymerization of monomer in the water phase and particle coagulation.

Step 2. Monomer Droplet Depletion. Once all the polymer particles have been nucleated, monomer droplets decrease in size as the monomer transfers to the polymer particle phase. The rate of the diffusion processes usually is high compared to the rate of polymerization in the particles. The surface area of the particles increases as they grow and the emulsifier content at their surface decreases. At some point (40% conversion for typical styrene emulsion systems), all the monomer droplets disappear and the monomer is present in the particle and water phases only.

The concentration of monomer in the polymer particle is nearly constant during the second step of the process. As long as monomer droplets exist, the activity of the monomer in each phase (monomer droplets, water, and polymer particles) should be nearly one. The Flory–Huggins equation could be used to estimate the monomer concentration in the polymer particle. However, because of the small particle size, the surface energy should be considered in the free energy balance. This correction will be discussed later in this section.

Step 3. Polymerization to the Endpoint. Monomer in the particles continues to polymerize until the reaction is stopped. The monomer concentration continuously decreases during this step, leading to side reactions such as chain branching and to the Trommsdorff effect. Side reactions are particularly important in elastomer emulsions and can be controlled by lowering the polymerization temperature, adding chain transfer agents, and stopping the reaction at lower conversions.

There are a number of other mechanisms that can affect emulsion polymerizations. These include the termination of growing chains at the polymer particle surface, diffusion of small free radicals back into the water phase, the low probability that two free radicals exist in one particle, the possibility that mass transfer can be rate limiting for monomers that are sparingly soluble in water, and coagulation of the latex particles caused in part by the different surface chemistries of monomer and polymer interfaces.

There are many emulsification systems available. One method for selecting surfactants is the HLB (hydrophilic-lyophilic balance) system (Griffin, 1949). Numbers between 1 and 20 are assigned to surfactants based on the type of systems they stabilize. Table 9.9 lists the applications for various HLB ranges. Most emulsion polymerizations are designed around oil-in-water emulsions, with the monomer

TABLE 9.9 HLB Ranges and Their Applications

HLB Range	Applications
4–6	Water-in-oil emulsions
7–9	Wetting agents
8–18	Oil-in-water emulsions
13–15	Detergents
10–18	Solubilizers

suspended as discrete particles in a continuous phase of water. There are ranges of HLB values that give the best emulsion stability for specific polymer systems, so several surfactants should be tried for new systems. The stability of the polymer particles will depend in part on initiator fragments bound to the chains and concentrating at the particle-water interface. Latex stability can be changed by changing initiator systems.

Reactor System Operation. Batch and continuous reactors are used for the commercial production of lattices. In contrast to suspension systems in which coagulation is often a problem, emulsion systems tend to have greater stability. The advantages of CSTR's include constant polymerization rates, constant and heat transfer duties in each vessel, control of copolymer compositions, and control of molecular weights.

Batch systems usually are used in the initial stages of process research and development because large numbers of experiments can be carried out quickly. Batch systems are preferred to continuous systems for commercial production when a wide variety of products needs to be made. These systems are more flexible for handling varied cycle times, product formulations, and for segregating lots and batches.

Emulsion chemistry and operating conditions can be used to vary the properties of the latex. Emulsion chemistry is complex and is reviewed by Poehlein (1986). Reaction temperature affects reaction rates as well as side reactions such as branching and cross-linking. Pressure usually is determined by the vapor pressures of monomers and water at the reaction temperatures. A separate monomer phase is present through Steps 1 and 2. Reactors making homopolymers at constant temperature during these steps will have constant pressure. In Step 3, when the monomer phase disappears, the monomer vapor pressure will decrease as will the total reactor pressure. Copolymer systems may exhibit a pressure change during Steps 1 and 2 if one monomer is reacted preferentially.

Heat removal often is the rate-limiting problem for commercial processes. Jacketed reactors, heat exchanger baffles, and condenser reflux cooling are used to remove the heat of polymerization. Scale-up of heat transfer systems can be difficult. Mixing is at the core of good heat transfer and other phenomena in the reactor. Heat transfer depends on the fluid hydrodynamics near heat transfer surfaces. Efficient top-to-bottom mixing is required for CSTR systems in order to maintain constant residence times with narrow distributions. Vessel geometry, fill level, and internals all affect flow behavior. Poor mixing near walls, baffles, corners, and gas-liquid interfaces can lead to wall build-up. Agitation scale-up and models for Newtonian fluids are fairly reliable as long as one mechanism (particle suspension, droplet dispersion, pumping) can be chosen as the design basis. The rheology of suspensions and lattices is more complex (see Section 9.5), and many commercial reactors operate near nonideal regions with respect to the effect of dispersed phase volume on viscosity.

Semibatch operation is common for several reasons. Polymers are more dense than their monomers, so a batch reactor will have a reduction of slurry volume during the polymerization. Semibatch production with continuous monomer feed can maintain the reactor fill level, resulting in an increase in productivity over the batch case. Semibatch operation is used for copolymer production when one monomer is preferentially consumed. The faster–reacting monomer will be fed to maintain constant copolymer composition.

Continuous polymerization systems are used when high volumes of one product are needed or when the differences between recipes is small. In this second case, the "changeover" material usually is blended away into one or both products. Figure 9.11 shows a train of CSTR's for the production of SBR. All the reactors are equal in size. The product is shortstopped with free radical traps to stop the reaction prior to latex stripping.

One of the challenges of operating CSTR's for emulsion polymerizations is controlling the conversion and particle concentration. Particle formation and growth

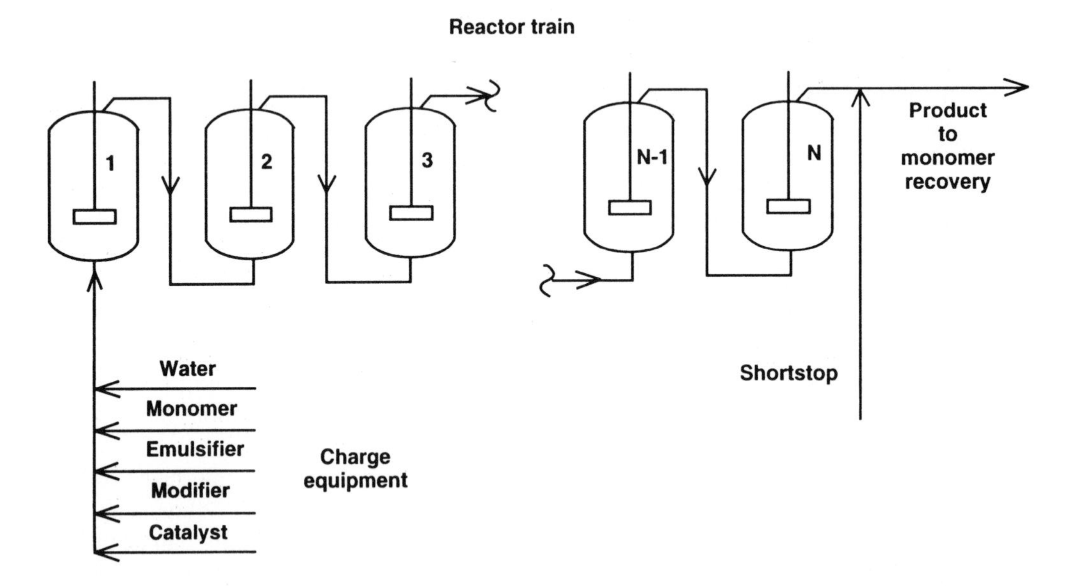

Figure 9.11 Flow diagram of typical continuous reactor system (Poehlein and Dougherty, 1977).

processes can cause conversion oscillations in the first CSTR, which may propagate through the system. When these oscillations are not damped, the product can have variable conversion and particle size.

Characteristics of Emulsion Polymers. Commercial lattices have 30 to 50 vol% solids. Lower solids volume fractions rarely are economical, and higher volume fractions may be difficult to pump and process. Broad or narrow particle size distributions can be made. Broad distributions can be made by extending the nucleation period so that many different particle ages exist (recall that the size of particle is related to its polymerization time). Broad distributions can be processed at higher volume fractions than narrow distributions. Narrow distributions are made with narrow nucleation time ranges.

A key advantage of emulsion polymerizations is that the reaction rate is proportional to the number of particles rather than the initiator concentration. In bulk, solution, and suspension free radical processes, the addition of more initiator to the monomer phase in order to increase the reaction rate also decreases the molecular weight of the polymer. Reaction rates in emulsion systems can be increased by increasing the number of nucleating particles, but this does not change the molecular weight distribution of the polymer. The kinetic model for emulsion polymerization presented below section captures some of the key mechanisms for these systems. Downstream processing includes monomer stripping, blending with additives and other emulsions, coagulation, and drying. Coagulation needs to be prevented during polymerization and stripping. In some systems, stabilizers are added to prevent coagulation during shipping and storage.

Kinetic Model for Emulsion Polymerization. The emulsion process leads to a different initiation mechanism, which affects the overall polymerization rate. This kinetic model is specific to free radical emulsion polymerizations. Figure 9.10 shows that monomer is present in micelles, in monomer-swollen polymer particles, in the continuous aqueous phase, and in monomer emulsion droplets. Based on the previous process description, the reaction is thought to occur in the monomer-swollen polymer particles, and the material balances are written on this phase. Water phase polymerization also occurs, but usually produces much less polymer than that made in the polymer particles.

The loss of monomer during a free radical reaction is given by

$$\frac{-dM}{dt} = k_p M M_o \tag{9.2}$$

where k_p is the propagation rate constant, M is the monomer concentration, and M_o is the concentration of growing free radical chains. Equation 9.2 is equivalent to Equation 4.4. The difference between free radical emulsion kinetic models and the chain additions models of Chapter 4 is the term used to describe the concentration of growing free radical chains, M_o. The number of free radicals per particle, n, is expected to be small and should vary with the size of the particle. For small particles, the instantaneous values of n should be zero or one. For larger particles, the instantaneous values of n would range from zero to integer values higher than one. A value for the average number of radicals per particle, $\bar{n}$, will be observed in an emulsion polymerization. The rate equation for particle size i becomes

$$\frac{-dM}{dt}\Big|_{size\ i} = k_p M <\bar{n}>_i \frac{N_i}{N_A} \tag{9.3}$$

where $< >$ denotes a time average over all particles of size i, N_i is the total number of particles of size i per unit volume, and N_A is Avogadro's number. The reaction rate is integrated over all particle sizes to give

$$\frac{-dM}{dt} = k_p M\left(\frac{<\bar{n}>\,N}{N_A}\right) \tag{9.4}$$

When the nucleation process is controlled to make a narrow particle size distribution, $<\bar{n}>$ approaches $\bar{n}$ for the average particle size. Equation 9.4 suggests that the polymerization rate is directly proportional to the number of particles, N. The propagation rate constant is that of ordinary free radical polymerizations (Table 1 in Appendix D). The monomer concentration can be calculated using the Flory–Huggins equation modified by a term accounting for the interfacial free energy of the emulsion drops (Morton et al. 1954):

$$\ln a_1 = \ln \phi_1 + \left(1 - \frac{1}{X_n}\right)\phi_2 + \chi\,\phi_2^2 + \frac{2\,V_1\gamma}{r_1\,RT} \tag{9.5}$$

where V_1 is the molar volume of the solvent, r_1 is the droplet radius, and γ is the interfacial tension between the polymer particles and the water phase. When monomer droplets exists $a_1 = 1$. Equation 9.5 predicts modest changes in the monomer concentration in the polymer particles as they grow during Steps 1 and 2. During Step 3, the monomer concentration is determined by a material balance. Swellings agents and surfactants can be used to modify the chemistry and monomer concentration in the polymer particles so as to produce large emulsion particles. The analysis of CSTR systems results in different kinetic models (Poehlein, 1986). The major goal of most models is to predict the particle concentration, N, as well as the average number of free radicals per particle, $\bar{n}$.

9.5 MIXING AND POLYMERIZATION

There are a number of important mixing problems in polymerization reactor designs. There are significant differences between agitation in single phase and multiphase systems. Multiphase systems can be separated into liquid slurries (suspensions and emulsions) and gas fluidized beds (the UNIPOL process for polyethylene). There are common objectives for the mixing system: ensuring good heat transfer, mixing ingredients efficiently and uniformly, keeping particles suspended, and nucleating particles reproducibly. The rheology of single phase systems are characterized by high viscosity and the rheology of multiphase systems is dominated by the presence of particles. Some of the mixing problems encountered in single phase reaction systems are similar to those encountered during melt processing of polymers.

9.5.1 Mixing Concepts

Table 9.10 summarizes some of the design objectives for mixing systems. In low viscosity fluids, three mechanisms are important for the mixing of two streams of similar viscosity and density: molecular diffusion, mixing caused by turbulent eddies, and convective mixing (Brodkey, 1966). In viscous polymer solutions and melts, molecular diffusion of polymer molecules is very low, and the high viscosity suppresses

TABLE 9.10 Agitation/Mixing Objectives

Reaction System	Design Objectives
SINGLE PHASE	
Bulk and solution polymerizations	1. Steady flow through equipment 2. Uniform dispersion of catalysts, initiators, and additives 3. Low or control dispersion of residence times in continuous systems 4. Controlled heat transfer
Reaction injection molding	1. Rapid and complete mixing of reactants (polyol and isocyanates)
MULTIPHASE	
Suspension polymerization	1. Particle suspension 2. Monomer droplet formation 3. Steady flow past heat transfer surfaces 4. Uniform monomer distribution during semibatch operation and reflux cooling 5. Controlled coalescence
Emulsion polymerization	1. Uniform dispersion of emulsifier ingredients 2. Uniform reactor flow patterns 3. Uniform monomer distribution during semibatch operations and reflux cooling
Gas fluidized beds	1. Uniform heat transfer 2. Rapid catalyst dispersion and particle nucleation 3. Uniform product stream

turbulent eddies. The primary mixing mechanism is convection. Solids mixing also occurs by the mechanism of convective mixing; molecular diffusion of low molecular weight species (plasticizer, monomer, or solvent) may be rate limiting for stripping and devolatilization processes. Convection can result in mixing of liquid-liquid systems when the interfacial area between the two phases increases. In solid-liquid and solid-solid mixing, convection distributes the dispersed phase through the continuous phase. Convective mixing can be random, the type of mixing expected in an intermittent device such as a V-type solid mixer, or ordered, the type of mixing expected in a motionless mixer. Liquid-liquid and liquid-solid mixing occurs in shear, elongation or extension, and squeezing flows. Elongation occurs when the fluid is stretched, and squeezing occurs when the fluid is kneaded. The most important of these flows in reactive polymer processing is shear flow. Squeezing flows become important in calendering operations.

Effect of Shear on Striation Thickness. One goal of the deformation of fluid elements is to increase the surface area between two phases. If the initial area, A_o, is increased to A and the fluid is incompressible, then the increase in interfacial area is directly proportional to the total strain γ, (Tadmor and Gogos, 1979).

$$\frac{A}{A_o} \propto \gamma \qquad 9.6$$

The mean change in interfacial area for elements randomly aligned relative to the shear direction is

$$\frac{A}{A_o} = \frac{\gamma}{2} \qquad 9.7$$

The thickness of layers, or striations, resulting from such a mixing process is

$$r = \frac{2}{A/V} \qquad 9.8$$

where r is the striation thickness, A is the new area, and V is the volume of the sample. For large strains, the striation thickness can be related to the strain

$$r = \frac{2r_o}{\gamma} = \frac{2}{3}\frac{L}{\gamma\phi_d} \qquad 9.9$$

where r_o is the initial striation thickness, L is the diameter of the dispersed phase, and ϕ_d is its volume fraction. These results have a simple interpretation. A larger particle of the discontinuous phase will have a larger striation thickness after a shearing process. The final striation thickness is inversely proportional to the strain and to the ratio, ϕ_d/L.

Residence Time Distribution in Newtonian Pipe Flow. The residence time distribution of a polymer solution through processing equipment can cause spatial variations in the properties of the liquid. Figure 9.12 shows the velocity profile for a Newtonian fluid in laminar pipe flow. The velocity down the pipe is

$$v_z(r) = \frac{\Delta p}{4L\mu}(R^2 - r^2) = C(R^2 - r^2) \tag{9.10}$$

where Δp is the pressure drop, L is the tube length, μ is the viscosity, R is the inside radius of the pipe, and r is the radius of interest. The flow has a maximum at r = 0 and is zero at the pipe wall. The residence time along a streamline is

$$t = \frac{L}{v_z(r)} = \frac{L}{C\left(R^2 - r^2\right)} \tag{9.11}$$

The minimum residence time is experienced by the fluid following the centerline of the pipe:

$$t_{\min} = \frac{L}{CR^2} \tag{9.12}$$

Residence times range from $t_{\min}$ at the centerline to infinity at the wall. The external residence time distribution of the system, $f(t)$, is the fraction of the flow exiting between t and $t + dt$. The average residence time, $\overline{t}$, is the integral of $f(t)$ from $t_{\min}$ to infinity,

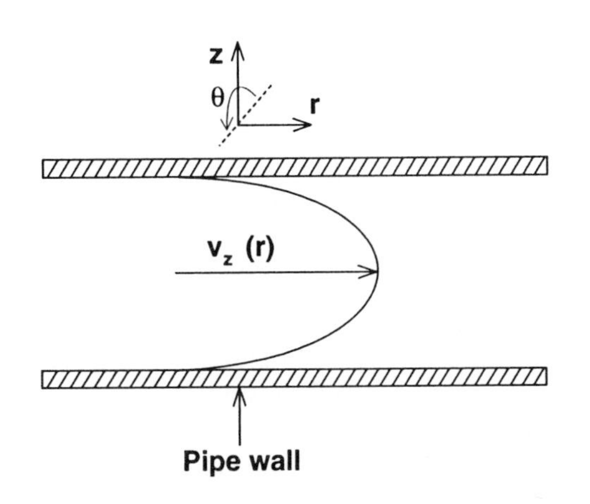

Figure 9.12 Laminar flow of a Newtonian fluid in a pipe.

$$\bar{t} = \int_{t_{min}}^{\infty} tf(t)\, dt = \frac{V}{Q_f} \qquad 9.13$$

and it must equal the system volume, V_z divided by its flow rate, Q_f. The external residence time distribution is related to the differential flow rate by

$$f(t)\, dt = \frac{dQ_f}{Q_f} = \frac{2\pi r\, v_z(r)\, dr}{\int_o^R 2\pi r\, v_z(r)\, dr} = \frac{2r v_z(r)\, dr}{R^2 L/2\, t_{min}} \qquad 9.14$$

Equation 9.14 is integrated by substituting Equations 9.11 and 9.12 for expressions in t, giving

$$f(t)\, dt = \frac{2 t_{min}^2 dt}{t^3},\ t \geq t_{min} \qquad 9.15$$

The mean residence time, $\bar{t} = 2\, t_{min}$. This means that half of the flow stays in the pipe longer than $2\, t_{min}$.

Example 9.2 Problem: *Polymer Conversion in Laminar Flow Tube*

Polystyrene can be made in a series of CSTR's. In this problem, styrene monomer and benzoyl peroxide are mixed in-line and sent to a CSTR. The feed concentrations are 9 mol/l monomer and 0.005 mol/l initiator. The efficiency factor for initiation is ideal (f = 1). The monomer reacts to 30% conversion in the first CSTR and flows through a pipe to the second CSTR. Polymerization continues in the pipe, so that the second CSTR receives feed with a higher conversion than that in CSTR 1. The objective of this problem is to calculate the additional conversion taking place in the pipe as a function of residence time. This information can be used to choose a desired residence time and the pipe size.

Solution:

The polymerization model for monomer concentration as a function of reaction time will be combined with residence time distributions in the pipe to determine the monomer concentrations exiting the tube. Equation 9.15 gives the residence time distribution function for laminar flow in a pipe. The concept of expected value (Section 6.2.3, Eq. 6.7a & 6.7b) can be applied to this problem. Equation 9.15 is the probability density function, and the integrated equation for monomer concentration is the weighting function.

(continued)

$$M(t) = M_o \exp\left[-k_p \left(\frac{k_d f I}{k_{tc}}\right)^{\frac{1}{2}} t\right]$$

CONSTANT INITIATOR CONCENTRATION

$$M(t) = M_o \exp\left\{-k_p \left(\frac{k_d f I_o}{k_{tc}}\right)^{\frac{1}{2}} \left(\frac{k_d}{2}\right)\left[1 - \exp\left(\frac{-k_d t}{2}\right)\right]\right\}$$

VARIABLE INITIATOR CONCENTRATION

The expected value for the average monomer concentration at any exit time, $t_{ave} = 2t_{min}$, is:

$$\overline{M} = \int_{t_{min}}^{\infty} M(t) \cdot f(t) dt$$

The lower limit of integration is t_{min}, since no fluid element exits the pipe prior to this time. The integral can be evaluated analytically for this combination of simple distribution and kinetic functions. Numerical integrations are useful for complicated kinetic models or experimental residence time distributions. The following table shows the results of the calculations using a simple numerical integration technique.

		I = constant		**T = variable**	
t_{ave}, s	**t_{min}, s**	**M**	**% conv.**	**M**	**% conv.**
60	30	5.85	2.6	5.85	2.5
120	60	5.76	4.1	5.76	3.9
240	120	5.58	6.9	5.61	6.5
600	300	5.10	15.0	5.24	12.6

The assumption that the initiator concentration is constant would not be accurate in practice, but it gives an upper bound on the conversion (and a lower bound on the monomer concentration). Figure 9.13a shows the molar concentrations of monomer and initiator for the two initiator assumptions for an average time of 240 *s*. Even for this short residence time, there are differences between the cases. The material in the center of the pipe has a modest conversion (Figure 9.13b). Material near the wall has higher conversion and much lower initiator concentrations. For a residence time four times this base case (960 *s*), the initiator has been consumed in the fluid elements near the wall.

Comments

A complete analysis of this problem would include the material balance shown above, an energy balance, and a momentum balance. A residence time of four

(continued)

minutes in the tube converts 6.5% of the monomer to polymer. Although the conversion near the tube wall is the highest, there still is a significant conversion of 3% along the centerline of the flow, as shown in Figure 9.13c. It is not likely that the fluid temperature would be isothermal across the tube, as has been assumed. The temperature gradient would alter the reaction rate as well as the viscosity of the reaction mixture along streamlines. The velocity profile would not fit the laminar flow profile and would need to be determined as a function of axial distance. In addition, entrance and exit effects on the velocity profiles would be more important as the tube length becomes shorter. Finally, the design of a commercial reactor would include consideration of side reactions that lead to undesirable properties in the product.

A CSTR is another mixing device that provides back mixing. Its external residence time distribution function is

$$f(t)\,dt = \frac{1}{t}\exp\left(-t/\bar{t}\right) \tag{9.16}$$

In contrast to pipe flow, in which no material exits before $t_{min,}$ a CSTR has some material that exits immediately after entering the tank.

9.5.2 Rheology of Slurries

There are many polymer systems that can be characterized as suspensions. Not only suspension and emulsion products, but paints, printing inks, SMC's, RRIM's, and many foods are suspensions. The general viscosity–shear rate curve for all suspensions is shown in Figure 9.14. At low shear rates, there is a Newtonian plateau. At higher shear rates, the material is shear-thinning and reaches a constant viscosity plateau at higher shear rates. At very high shear rates, the viscosity may increase again due to flow-induced packing of solids. Figure 9.15 shows a viscosity–shear stress curve for latex particles suspended in several solvents. A key independent variable for modeling suspension rheology is the solid volume fraction, ϕ_s, in the continuous liquid phase. Solid phase volume affects the hydrodynamic forces on the surface of a particle or an aggregate of particles.

Three different types of forces act on suspended particles: colloidal forces, Brownian motion, and viscous forces. Colloidal forces are the attractive and repulsive forces between particles resulting from van der Waals or electrostatic interactions. They are most important for very small particles (< one micron) of materials with high dielectric constants. Brownian motion also is size dependent and is most important for small particles (< 5 microns). Viscous forces are proportional to the velocity

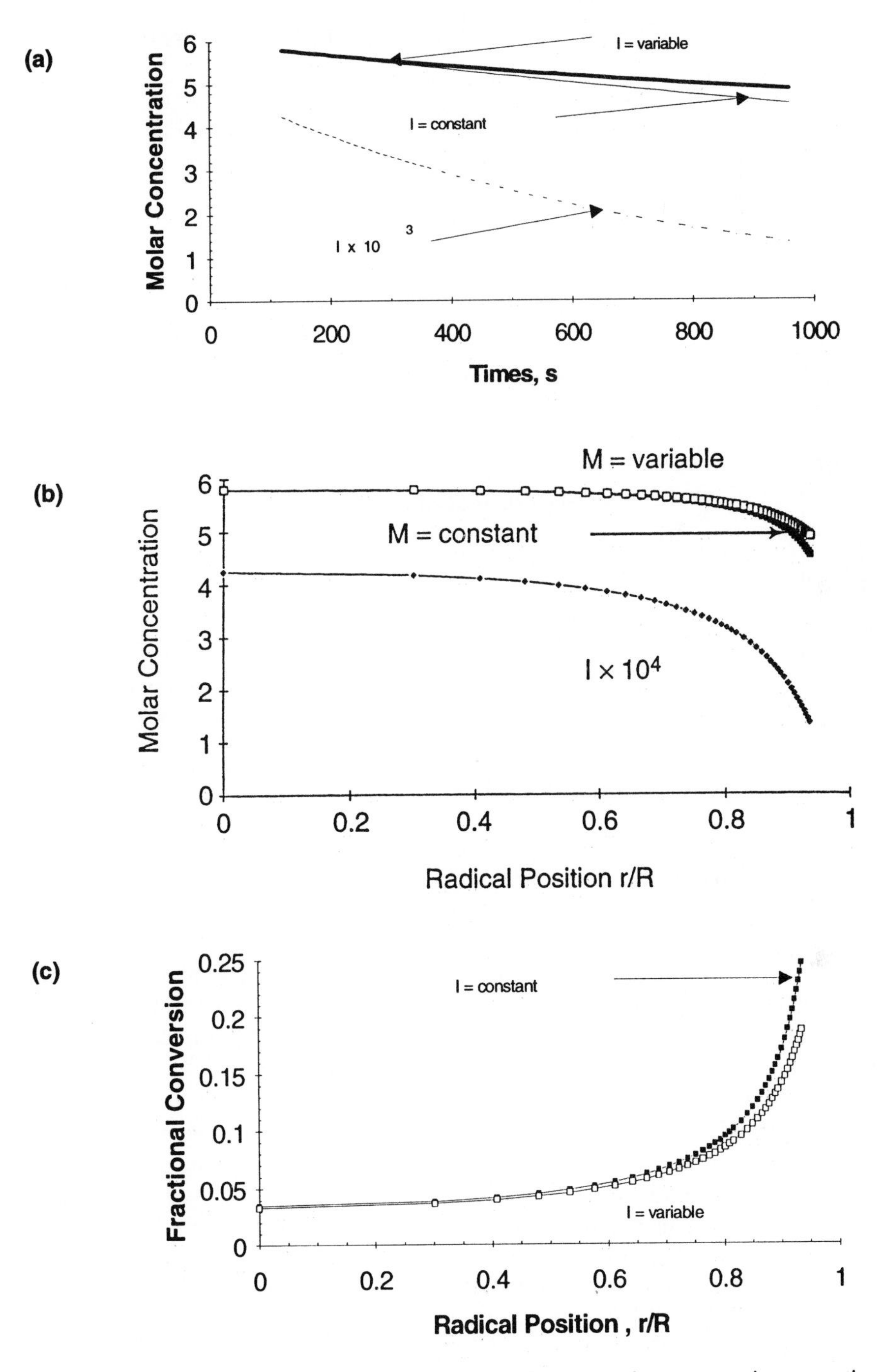

Figure 9.13 Polymerization during laminar pipe flow: **(a)** molar concentration versus time, **(b)** molar concentration versus radial position, **(c)** fractional conversion versus radial position.

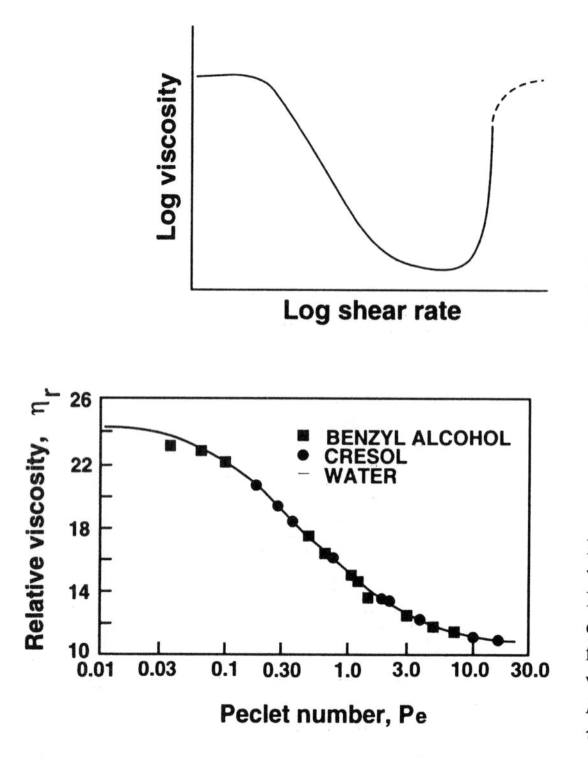

Figure 9.14 Schematic representation of the flow curve of a concentrated suspension.

Figure 9.15 Composite curve of relative viscosity versus modified Péclet number. Points are for 0.1–0.5 μm latex particles dispersed in two solvents; the solid line is for the same sized particles dispersed in water; the volume fraction is 0.50. $P_e = 6\tau a^3/kT$, where a is the particle radius, τ the shear stress (Krieger, 1972).

difference between the particles and the continuous phase. Suspension rheology models often calculate the viscosity relative to that of the continuous phase. The microstructure of the fluid affects its viscosity. When particles are first added to a fluid, they disrupt the fluid streamlines and increase the viscosity. At higher concentrations, the particles add to viscous dissipation by moving past each other. Particles that form flows or other aggregates provide more resistance to flow because the effective volume of the solid is higher than its simple volume fraction.

Spherical particles have different effects than irregularly shaped ellipsoids, plates, and rods. The surface treatment of the solid also can have an effect. The steric forces caused by sorbed surfactants and macromolecules and the electrostatic effects with low salt levels in water affect the viscosity of concentrated suspensions. Flow induces structure in macromolecules and also can induce structure in suspended solids. Moderate shear rates can induce structure in solids that can not be restored at the experimental time scale by Brownian motion. Higher shear may lead to more orientation. Asymmetric solids—such as clays, soaps, and short fibers—tend to align in the direction of flow.

Dilute Suspensions: Einstein's Equation. The prediction of viscosity in dilute systems ($\phi_s << 0.10$) is done using Einstein's equation,

$$\eta = \eta_s (1 + 2.5\,\phi_s) \qquad 9.17$$

where η is the suspension viscosity, η_s is the solvent viscosity, and ϕ_s is the solid volume fraction. This equation will not account for differences in particle size or position. The interactions caused by neighboring particles can be accounted for with higher order terms in ϕ_s. Batchelor's model (1977), which models extensional viscosity, gives improved fits to data.

$$\eta = \eta_s (1 + 2.5\phi_s + 6.2\phi_s^2) \qquad 9.18$$

Correlations using the form of Equation 9.18 have been fit to shear viscosity data, but the coefficients of ϕ_s^2 range from 5 to 15. Most commercial suspensions cannot be considered dilute, and the two equations described above do not apply.

Concentrated Suspensions. There are maximum packing fractions for solids in liquids; that is, at higher solid fractions, the viscosity is essentially infinite. The value of the maximum packing fraction, $\phi_{s,max}$, depends on the packing array. Table 9.11 lists some values of $\phi_{s,max}$ for monodisperse spheres. The $\phi_{s,max}$ for broad particle size distributions can have higher values than 0.75 because the small particles can pack in between the large ones. Nonspherical particles and flocculation can lead to lower values of $\phi_{s,max}$.

Newtonian Suspensions. Ball and Richmond (1980) assumed that a differential form of Equation 9.17 is correct, that is, the effects of all particles are related to the effect of particles added sequentially (differentially) to the suspension.

$$d\eta = \frac{5\eta}{2}\, d\phi_s \qquad 9.19$$

The viscosity of the final suspension is obtained by integrating Equation 9.19 between 0 and ϕ_s:

$$\eta = \eta_s \exp\left(\frac{5\phi_s}{2}\right) \qquad 9.20$$

TABLE 9.11 The Maximum Packing Fraction of Various Arrangements of Monodisperse Spheres

Arrangement	Maximum Packing Fraction
Simple cubic	0.52
Minimum thermodynamically stable configuration	0.548
Hexagonally packed sheets just touching	0.605
Random close packing	0.637
Body-centered cubic packing	0.68
Face-centered cubic/hexagonal close packed	0.74

This integration neglects the finite size of the spheres, so that the differential change in volume fraction should be

$$\frac{d\phi}{1 - k\phi}$$

The improved model is

$$\eta = \eta_s(1 - k\phi_s)^{-5/2k} \tag{9.21}$$

where k describes the crowding effect. When $k = 1/\phi_s$, the viscosity is infinite, so $k = 1/\phi_{s,\max}$. The 5/2 factor can be replaced by $[\eta]$ for asymmetric particles. The final result is

$$\eta = \eta_s\left(1 - \frac{\phi_s}{\phi_{s,\max}}\right)^{-[\eta]\,\phi_{s,\max}} \tag{9.22}$$

Figure 9.16 shows a plot of relative viscosity versus fraction of large particles for various volume fractions. At high volume fractions, there are minima in the viscosity as a function of the particle size distribution. The fitting parameter, $\phi_{s,\max}$, is best determined by empirical application of Equation 9.22 to data. The intrinsic viscosity parameter can be determined in the normal way.

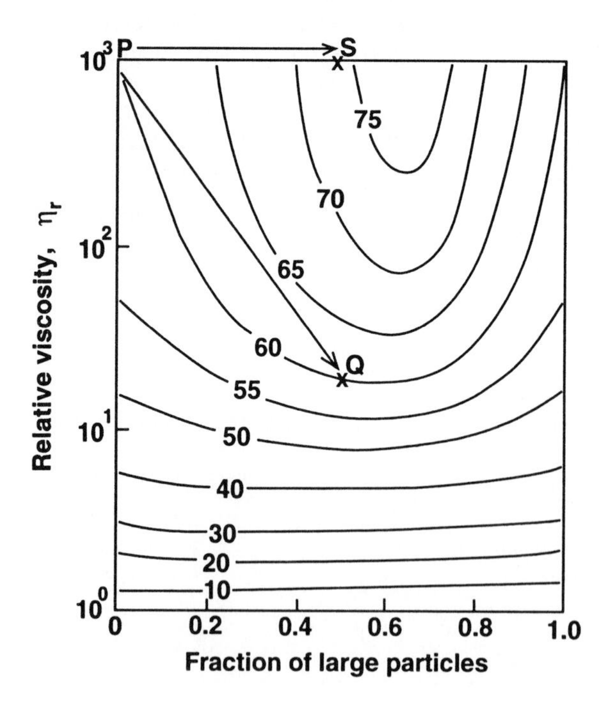

Figure 9.16 Effect of bimodel particle-size fraction on suspension viscosity, with total percent phase volume as parameter. The particle-size ratio is 5:1. P→Q illustrates the fiftyfold reduction in viscosity when a 60% v/v suspension is changed from a mono- to a bimodal (50/50) mixture. P→S illustrates the 15% increase in phase volume possible for the same viscosity when a suspension is changed from mono to bimodal (Barnes et al., 1989).

TABLE 9.12 The Values of $[\eta]$ and $\phi_{s,\max}$ for a Number of Suspensions of Asymmetric Particles

System	$[\eta]$	$\phi_{s,\max}$	$[\eta]\phi_{s,\max}$	Reference
Spheres (submicron)	2.7	0.71	1.92	de Kruif et al. (1985)
Spheres (40 μm)	3.28	0.61	2.00	Giesekus (1983)
Ground gypsum	3.25	0.69	2.24	Turian and Yuan (1977)
Titanium dioxide	5.0	0.55	2.77	Turian and Yuan (1977)
Laterite	9.0	0.35	3.15	Turian and Yuan (1977)
Glass rods (30 × 700 μm)	9.25	0.268	2.48	Clarke (1967)
Glass plates (100 × 400 μm)	9.87	0.382	3.77	Clarke (1967)
Quartz grains (53–76 μm)	5.8	0.371	2.15	Clarke (1967)
Glass fibers:				
axial ratio-7	3.8	0.374	1.42	Giesekus (1983)
axial ratio-14	5.03	0.26	1.31	Giesekus (1983)
axial ratio-21	6.0	0.233	1.40	Giesekus (1983)

Table 9.12 shows some values of $[\eta]$ and $\phi_{s,max}$ for some asymmetric particles. The L/D ratio of fibers affects the relative viscosity. The higher the value of L/D, the higher the relative viscosity at a constant ϕ_s. This effect has implications for sheet-molding compound (SMC) and reinforced reaction injection-molding (RRIM) applications. Both $[\eta]$ and $\phi_{s,max}$ can vary with shear rate.

The suspension viscosity is very sensitive to solid volume fraction. Weighing errors can cause major differences between samples. Wall effects (the depletion of particles near the wall) and local particle packing also cause anomalous effects. Figure 9.14 shows that suspensions can be shear-thickening at high solids loadings. This is particularly common at volume fractions above 0.50. The effect is sensitive to particle size (Fig. 9.17). Suspensions have viscoelastic behavior that can be measured in dynamic experiments.

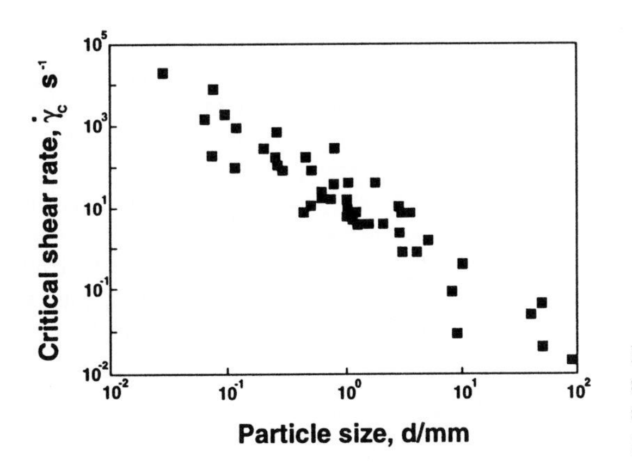

Figure 9.17 Shear rate for the onset of shear thickening γ_c, versus particle size for suspensions with phase volumes of 0.5 (Barnes et al., 1989).

Example 9.3 Problem: *Effects of Solids on Polymer System Viscosities*

Equation 9.22 can be applied to a number of systems to illustrate the effect of solids on the viscosity of the system. Three examples will be calculated: suspensions of spheres, fillers, and glass fibers. The constants for Equation 9.22 are taken from Table 9.12.

Spheres. Latex and suspension resins are important products made in water slurries. Despite careful process control, there can be changes in solid volume fraction and in particle size for different batches. The reduced viscosity of the suspension is used for comparison. Figure 9.18a shows the reduced viscosity as a function of solid volume fraction for ideal spheres, submicron spheres, and 40 micron spheres. As the spheres increase in size, the maximum packing fraction decreases. The maximum fraction, $\phi_{s,max}$, represents the point at which the suspension is solid-like and cannot be stirred effectively.

Many suspension systems are operated at solids volume fractions between 0.30 and 0.40, while latex systems may have slightly higher solids. The curves in Figure 9.18a show that the reduced viscosities are only an order of magnitude above that of the solvent. Such viscosity changes would be handled easily by conventional pumps and agitation systems. Notice that the slurry reduced viscosity increases as the particle size increases.

Fillers. Particle shapes and particle size distributions can have a great effect on slurry viscosities. Titanium dioxide is a filler used in many polymer products to make them white and to extend the polymer. T_iO_2 particles are irregular and have a range of sizes. The amount of fillers used varies widely but can be in the range of 0.25 to 0.40 volume fraction. At these levels, the reduced viscosity of the polymer system may range from 10 to 100 and would affect the polymer flow in forming operations. The reduced viscosity of ideal spheres is shown for comparison (Fig. 9.18b).

Glass Fibers. Fibers are used as fillers in a number of polymer products. As shown in Table 1, of Appendix B they increase the tensile strength of the polymer and can increase the heat deflection temperature (see poly[aryl ether ketone] poly[ether ether ketone] and polycarbonate as examples). The effects of glass fibers on reduced viscosity depend on their aspect ratio (L/D). The fibers shown in Figure 9.18c are fairly short compared to fibers used in sheet-molding compounds but have a great effect on reduced viscosity at moderate solid volume fractions. Fiber length often is reduced due to mechanical breakage during blending, processing, and forming operations. Notice that the reduced viscosity is very sensitive to aspect ratio, so that small changes in fiber comminution make big changes in reduced viscosity. This sensitivity has important implications for recycling fiber-filled systems. Regrinding waste or recycled parts will break fibers and change the processing characteristics (viscosity) and performance properties (tensile strength, heat deflection temperature).

TABLE 9.13 Typical Reaction Injection Molding Cycle

Time, s	Mold/Mixhead Operation	Reactant Feed Operation
0	Close mold	Fill metering piston
11	Prepressurize	Achieve high pressure
18	Fill mold	Fill mold
20	Empty cylinders	Low pressure recirculation
22	Cure part	Low pressure recirculation
50	Open mold	Low pressure recirculation
60	Remove part and flashing	Low pressure recirculation
90	End of cycle	Low pressure recirculation

9.5.3 Reaction Injection Molding

Reaction injection molding (RIM) is used to produce panels, gaskets, and machine and appliance housings quickly from low viscosity monomers and oligomers. Two reactive liquids are combined by impingement mixing just before they enter the mold (Fig. 1.10). The polymer forms a solid phase by cross-linking or by phase separation. Cycle times on the process are on the order of a minute. Table 9.13 shows an example cycle time for 5 kg part of urea-urethane used for automobile fascia. The mold-filling portion of the cycle is two seconds long, and the monomers must be mixed immediately on entering the mold system. In contrast to standard injection molding, reactive monomers and oligomers fill the mold and polymerize rather than filling the mold

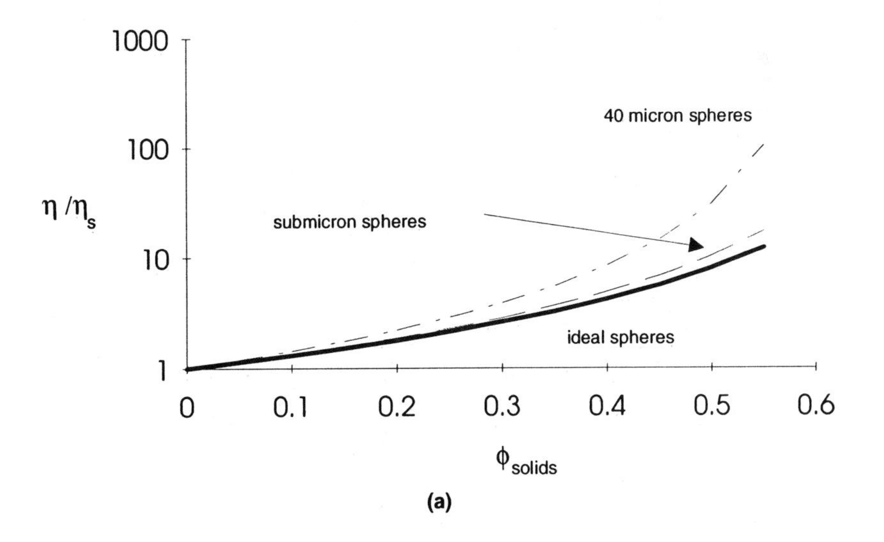

Figure 9.18 Viscosity versus particle volume fraction for several filled polymers: **(a)** spherical particles, **(b)** commercial fillers, and **(c)** fibers.

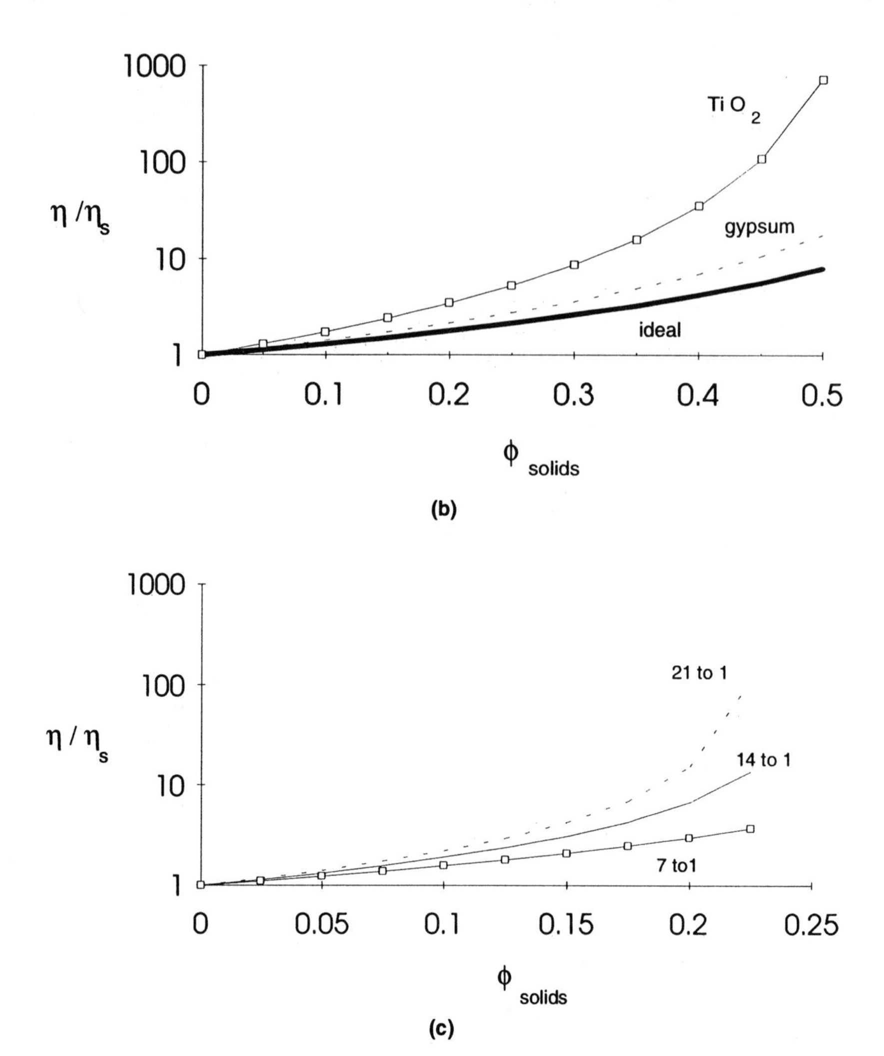

Figure 9.18 *(continued)*

with molten polymer that is cooled into a shape. The RIM mold temperature is about the same as the polymer, and the reaction is initiated by turbulent mixing of the liquid streams. Table 9.14 shows some of the chemical systems used in RIM applications.

The low viscosity mold filling has advantages over the injection molding of thermoplastic materials. The mold shape is filled by expansion of the gas so that low mold-clamping pressures are required. Table 9.15 compares the process requirements of RIM and IM systems. Macosko (1989) has a good review of RIM engineering science and technology.

TABLE 9.14 Reaction Injection Molding Systems

Chemicals/Polymer	Temperature, °C T_0	Temperature, °C T_{mold}	ΔT, exotherm
Urethane	40	70	130
Urea	40	60	110
Nylon 6	90	135	40
Dicyclopentadiene	35	60	170
Polyester	25	120	100
Acrylamate	25	100	100
Epoxy	60	120	150

SOURCE: Macosko, 1989.

TABLE 9.15 Comparison between Typical RIM and Thermoplastic Injection Molding

	RIM	TIM
Temperature		
reactants	40°C	200°C
mold	70°C	25°C
Material viscosity	0.1–1 Pa · s	10^2–10^5 Pa · s
Injection pressure	100 bar	1000 bar
Clamping force (for 1 m^2 surface part)	50 ton	3000 ton

SOURCE: Lee, 1980.

Mixing prior to mold filling is one of the critical parts of the process, as it affects the reaction results. Many mixhead designs are based on "T" geometry, with two streams impinging on each other (Fig. 1.10). Fluids mixed in "T" mixers may be uniform in just a few pipe diameters downstream. In the field, qualitative measures of mixing are used to adjust the flow geometries, flow velocities, pressures, and residence times. An alternate approach is to scale mixing with the Reynolds number of the flow. This dimensionless group relates the kinetic energy of the flow to the viscous dissipation of the flow. High ratios lead to high mixing relative to energy dissipation. The nozzle Reynolds number is

$$Re_{nozzle} = \frac{4\,\rho_1 Q_1}{\pi\,\eta_1 D_{nozzle}} \qquad 9.23$$

where Q_1 is the flow rate of component 1, ρ_1 is its density, η_1 is its viscosity, and D_{nozzle} is the nozzle diameter. Because nozzle systems are complex, the choice of D_{nozzle} may be arbitrary and is a major problem on scale-up. Alternatives to Equation 9.23 use a hydraulic diameter or an effective diameter based on the pressure drop through the flow contraction. Equation 9.24 is one description that uses a drag coefficient (C_D).

$$\Delta P = \frac{8\,C_D\,\rho_1\,Q_1^2}{\pi^2\,D_{nozzle}^4} \qquad 9.24$$

For conical nozzles at Reynolds numbers greater than 500, the drag coefficient is near one.

Equation 9.24 can be used to calculate the Reynolds number for each stream. In the polyurethane system (isocyanate + polyol), the polyol system has the highest viscosity and the lowest Reynolds number at stoichiometric flow rates; therefore, it tends to control the mixing process. Uniform mixing should be seen in the part by the surface finish, distribution of bubble sizes, streaks in the color, and low measurement distributions for mechanical tests. Table 9.16 shows some qualitative mixing evaluations and the Reynolds numbers of the monomer flows. The polyol stream has the lower nozzle Reynolds numbers, and changes in this value affect the qualitative observations. The optimum Reynolds number for mixing probably depends on the system used, the difference between the Reynolds numbers of the two streams, the mold shape, and the failure mode of the part.

Another method for demonstrating uniform mixing is following the adiabatic temperature rise of the part. Rapid temperature rises suggest better mixing than slow temperature rises. Figure 9.19 shows that the time to adiabatic temperature rise correlates well with the Reynolds number of the flow. In this case, the Reynolds number was increased by increasing the flow rate. The higher values correspond to high shear rates, which should reduce the striation thickness (see Section 9.5.1). Thinner striations should permit faster diffusion and reaction of the ingredients.

9.6 POLYMER DEVOLATILIZATION: MONOMER AND SOLVENT RECOVERY

Most polymerizations are not taken to 100% conversion, which means that much monomer needs to be removed. Solution polymerizations and many bulk polymeri-

TABLE 9.16 Mixing Quality Correlated to Reynolds Number

Isocyanate Stream					Polyol Stream					
T °C	η mPa · s	ρQ g/s	d mm	Re	T °C	η mPa · s	ρQ g/s	d mm	Re	Appearance
46	(low)	100	0.5	800	46	351	150	1	240	Sticky
46		100	0.5	800	50	290	150	1	280	Streaks
46		100	0.5	800	52	264	150	1	300	Good
46		100	0.5	800	54	240	150	1	320	Good

SOURCE: Macosko, 1989.

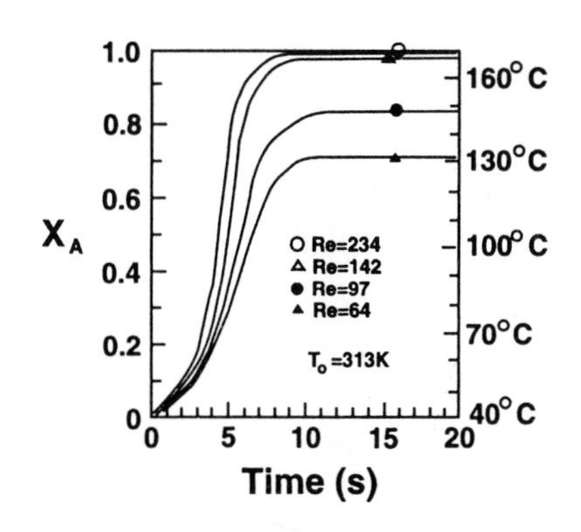

Figure 9.19 Adiabatic temperature rise for a cross-linking polyrurethane RIM formulation F-1. Reynold's increased by increasing flow rate (Chella and Ottino, 1983).

zations have solvents and diluents to remove. In general, this is done as close to the reactor as possible, while the polymer is molten and above its T_g. In this text, devolatilization refers to the removal of solvents above T_g, while stripping and drying refer to solute removal below T_g when the polymer is a solid. Devolatilization is described here, while stripping is described in Chapter 10. It is most convenient to recover solvents from a polymer molten phase to a gas phase (rather than a second liquid phase). The term "monomer" is used in this section to describe the organic vapor being removed. Remember that solvents and inerts also may be in the mixture.

The flux of solvent from a polymer depends on the activity driving force between the gas and polymer melt phases, the diffusion coefficient of the solvent in the melt, the interfacial area, and the contact time between the two phases. Phase equilibria can be modeled using Henry's law coefficients, Flory–Huggins interaction parameters, or group contribution techniques. The discussion of these models in Chapter 5 should be applicable to most problems. The high viscosities of polymer melts make solvent removal difficult compared to conventional liquids. Diffusion coefficients of low molecular weight solutes are several orders of magnitude below those in low viscosity liquids. The high viscosity means that turbulent flow does not occur and there are no turbulent mixing mechanisms operating. Surface area can be difficult to create, and the contact time between the two phases may be long.

These problems have generated the need for specialized equipment for removing solvents (Table 9.17). This equipment generates new surface area either by shear or by generating bubbles in the polymer melt. Often the equipment needs to be staged, since the monomer concentrations leaving the reactor may be 10% to 20%, while the residual monomer content of the product may be specified as 100 ppm.

9.6.1 Devolatilization Theory

The driving force for equilibrium is the activity difference between the gas and polymer phases. The phase equilibria relations developed in Chapter 5 will be used

TABLE 9.17 Devolatization Equipment

Devolatilization Method	
Shear	Single screw extruders, vented; twin screw extruders, vented; wiped film evaporator; high and low viscosity disk processor
Low shear, no shear	Flash systems; falling strand devolatilization

SOURCE: Mehta et al, 1984.

here. Solvent activity in the gas phase controls the driving force. This solvent activity can be reduced by raising the temperature (which raises the saturated vapor pressure of the solvent) or by diluting the gas phase with an inert gas or water. The second method makes some changes in solvent activity, but also can cause the formation of bubbles in the polymer melt. The increased surface area between the gas bubbles and melt phases greatly increases the flux of solvent.

Rate-Limiting Steps. Devolatilization is an unsteady-state process occurring in a polymer melt. The unsteady-state material balance across a thin film melt is

$$\frac{\partial C}{\partial t} = \frac{\partial\left(D\frac{\partial C}{\partial x_i}\right)}{\partial x_i} \qquad 9.25$$

where C is the concentration of monomer in the melt, x represents a spatial coordinate, and D is the diffusion coefficient of monomer. The diffusion coefficient in Equation 9.25 has been left inside the partial differential intentionally to suggest that it is a variable for some devolatilization problems. At temperatures between T_g and T_g + 50°C, the diffusion coefficients on many monomers and solvents are very dependent on their melt concentrations. The best theory for describing this dependence is the free volume theory developed by Duda and Vrentas (Zielinski and Duda, 1992). It relates the monomer diffusion coefficient to estimates of the free volume in the polymer as modeled by the WLF equation. Although a number of coefficients are required, this theory gives good agreement with data, particularly when accurate thermodynamic models are used (Mossner and Grulke, 1988).

The diffusion coefficients of monomers in polymer melts are low, in the range of 10^{-14} to 10^{-18} m²/s. At these small values, the depth of the polymer melt across which the concentration changes is very thin, and most problems can be modeled as diffusion from a semi-infinite slab. The solution to Equation 9.25 is

$$\frac{C(t, x) - C_o}{C_\infty - C_o} = erf\left(\frac{x}{(4\,tD)^{1/2}}\right) \qquad 9.26$$

where the diffusion coefficient has been assumed to be constant, C_o is the initial monomer concentration, and C_∞ is the monomer concentration at infinite time.

Example 9.4 Problem: *Diffusion in a Thin Molten Polymer Film*

The importance of mixing is illustrated by the following example. Suppose that the diffusion coefficient of the monomer is 1×10^{-14} m^2/s and that a thin film of molten polymer has been exposed to the gas phase for five minutes. How far into the sample will the monomer concentration be 90% of the total that can be lost? (The right-hand side of Eq. 9.26 is 0.10.)

An error function table can be used to show that at a depth of 1 micron the concentration will be 10% of the surface value. At lower diffusivities, the depth will be less. Residence times in devolatizers are in the range of minutes, and typical changes in monomer concentration per stage are between 1/3 to 1/10. Therefore, the continuous generation of surface area and mixing of the continuous phase are essential to efficient devolatilization.

This calculation suggests the need for high surface areas between the polymer and the gas phases as well as mixing of the melt for rapid devolatilization. Surface area can be generated through shearing and elongational flows or by causing bubbles to form.

Bubble Formation. The pressure in a bubble containing monomer vapor would be its equilibrium vapor pressure at the system temperature and the weight fraction of monomer in the melt. If the system pressure is above this level, bubbles will be unstable and collapse. If the system pressure is below this level, then monomer bubbles will be stable. Any higher system pressure will cause no bubbles to form. It is convenient to use a weight fraction Henry's law constant, K_w, to model the equilibrium,

$$P_1 = K_w W_1 \quad 9.27$$

where P_1 is the equilibrium vapor pressure of monomer in the gas phase and w_1 is the weight fraction of monomer in the polymer melt.

Example 9.5 Problem: *Equilibrium Vapor Pressure of Monomer in a Polymer Melt*

A comparison of the equilibrium vapor pressure calculations using Henry's law and the Flory–Huggins equation is given in Table 9.18. These calculations are

(continued)

for various weight fractions of styrene in polystyrene at 260°C. Both models give similar styrene pressures below $w_1 = 0.005$. System pressures below the listed values will cause monomer vapor bubbles to form. Henry's law is appropriate for ideal systems with low weight fractions of solvent or monomer. Above 0.005 weight fraction of styrene in polystyrene, the Flory–Huggins equation should be used.

Bubble Size. The pressure inside the bubble will exceed the hydrostatic pressure in the surrounding polymer. This phenomenon was modeled as a change in free energy of the bubble in Section 9.4.4 on emulsion polymerization. An alternate model calculates the pressure increase in the bubble. Figure 9.20 shows two gas bubbles in a continuous polymer melt phase. Large bubbles will have pressures near that of the continuous phase, while small bubbles can maintain significant overpressures due to the curved surface.

The pressure of the bubble can be calculated from the mechanical force balance demonstrated on the larger bubble in Figure 9.20. The polymer melt exerts a pressure, P, on the top hemisphere. The total force exerted by the melt is the pressure times the cross-sectional area of the bubble in the vertical direction. In addition, the surface tension results in a downward force around the perimeter of the bubble. The gas in the bubble exerts a pressure against the polymer. The bubble expansion force, P_i, is balanced by the compression force of the melt and surface tension force,

$$\pi r^2_i P_i = \pi r^2_i P + 2\pi r_i \sigma \qquad 9.28$$

where r_i is the bubble radius, P_i is the internal pressure of bubbles with radius r_i, and σ is the surface tension. Dividing by the cross-sectional area gives the Laplace–Kelvin equation.

TABLE 9.18 Comparison of Equilibrium Monomer Pressures. Styrene in Polystyrene at 160°C

		Equilibrium Model	
		P, atm	
Weight Fraction Styrene	ppm	Henry's Law	Flory–Huggins
.10	100,000	3.9	3.29
.05	50,000	1.95	1.79
.01	10,000	.393	.387
.005	5,000	.195	.194
.001	1,000	.0392	.0393
.0005	500	.0196	.0196
.0001	100	.00392	.00393
.00005	50	.00194	.00195
.00001	10	.00039	.00039

$^*P^{sat}_{styrene} = 10$ atm, $\chi = 0.23$, $K_w = 39$ atm.

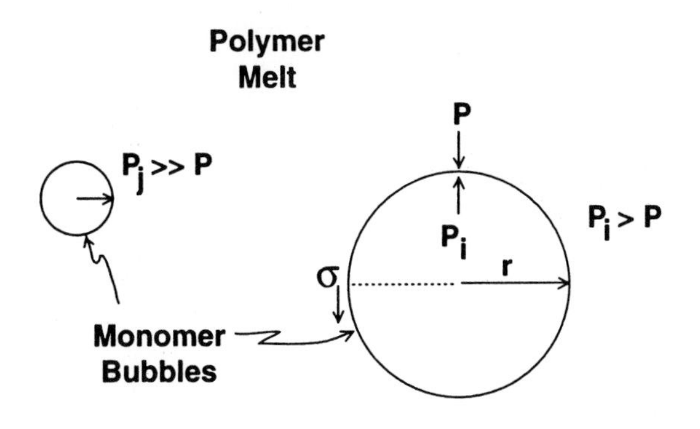

Figure 9.20 Force balance in a monomer bubble.

$$P_i = P + \frac{2\sigma}{r_i} \qquad 9.29$$

A similar result can be derived using the Helmholtz free energy of the bubble and expanding it with respect to volume and surface area.

Example 9.6 Problem: *Bubble Radius as a Function of Polymer Melt Pressure and Monomer Weight Fraction*

The pressure difference across the bubble surface is inversely proportional to bubble radius. Rearranging Equation 9.29 gives the bubble radius as a function of the pressure difference. This permits the calculation of the bubble size as a function of the pressure difference.

$$r_i = \frac{2\sigma}{P_i - P}$$

The styrene/polystyrene system will be used to illustrate these calculations. A value of 25 erg/cm^2 (dynes/cm) will be used for the interfacial tension. Two external variables control the bubble radius: the equilibrium pressure of the monomer in the melt and the external pressure on the system. When the external pressure of the system is greater than the equilibrium vapor pressure, no bubbles will form. If the melt has 1wt% monomer, the system pressure must be lower than 0.387 atm to form an infinitely large bubble. As the pressure is reduced, smaller bubbles can form because the surface tension forces will balance the pressure difference across the bubble surface.

However, there is a limit on the size of the bubble, dependent on the equilibrium vapor pressure. If the system pressure is nearly zero, the vapor

(continued)

pressure in the bubble is balanced only by the surface tension. Therefore, the smallest bubbles will be formed at the highest weight fraction of solvent as shown in Figure 9.21. Small bubbles have the advantage of high surface area and the disadvantage of low rise velocities. Because the transfer of monomer from the melt phase to the gas phase is a molecular diffusion process, smaller bubbles are preferred. Devolatilization is an unsteady-state process, so the monomer concentration in the melt is continuously decreasing as bubbles rise to a bulk surface of the melt and rupture, releasing their contents into the gas phase where an inert gas or a vacuum stream removes them.

Nucleation. There are two types of nucleation: homogeneous and heterogeneous. Homogeneous nucleation occurs when local density fluctuations in the fluid permit a second phase to be initiated. This mechanism is enhanced by agitation and mixing. Heterogeneous nucleation is caused by surfaces in the fluid phase. These could be solid particulates, impurities, the walls and surfaces of the device, or a second gas phase. An example of the latter is the use of steam or inert gases as devolatilization aids. When such surfaces are present, heterogeneous nucleation is the dominant mechanism. For either mechanism, the rate of nucleation can be related to the supersaturation of the system, $P_i - P$. The standard model is

$$r_{nucleation} = A \exp\left(\frac{-B\sigma^3}{T(P_i - P)^2}\right) \tag{9.30}$$

where A and B are constants for a specific monomer/polymer system.

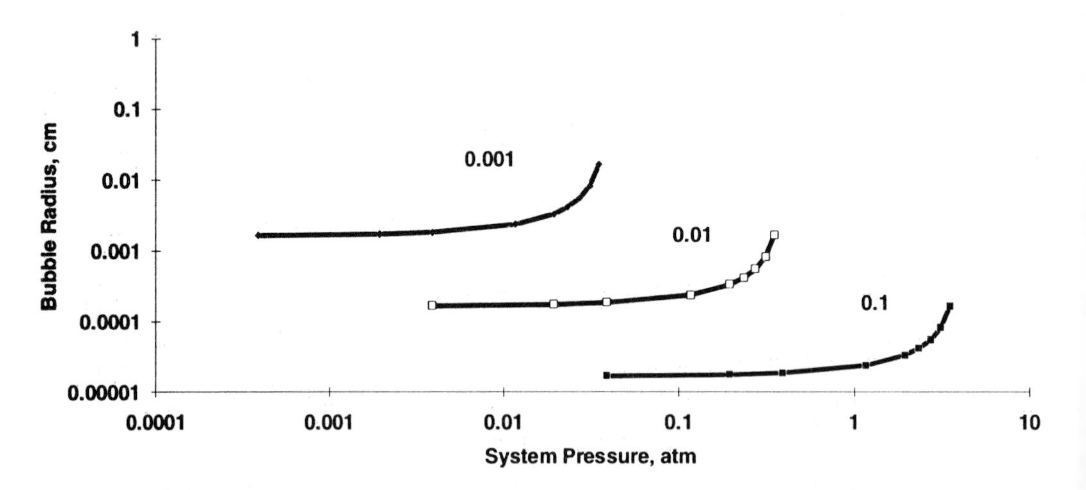

Figure 9.21 Effect of system pressure and monomer weight fraction on bubble radius for styrene in polystyrene, 160°C.

Growth, Migration and Rupture. After nucleation, the bubbles must grow, migrate to a free surface, and rupture. The growth process will be fairly independent of the mixing; however, migration to a surface can be significantly accelerated by proper flow patterns. In the case of crystalline polymers near their melting temperature, bubble growth may be accompanied by the orientation of the melt at the bubble surface. When dense swarms of bubbles are formed, foams may result. The foams may be unstable (quick to rupture) or may be stabilized by dilute quantities of impurities that act as surfactants. Therefore, foam and bubble disruption can be part of the equipment design.

9.6.2 Devolatilization Equipment

Most devolatilization equipment is designed to be continuous for lower cost. When big changes in monomer concentration are required, the equipment will be staged. This results in higher efficiency per unit (because of reduced backmixing) and smaller equipment size and cost. Equipment choice is dependent on many factors. At high monomer weight fractions, the melt viscosity can be fairly low, and simple flash systems that generate melt gas surface area can be effective. At lower monomer weight fractions, the viscosity can be much higher and agitated systems are required.

Falling Strand Devolatilization. The falling strand devolatilizer is an example of a device in which the monomer is removed without mechanical agitation. Figure 9.22 shows a schematic of such a device. Polymer melt is pumped at a high pressure through a spaghetti die into a large vessel. The falling strands have high surface area, and the large quantities of vapor can be removed by vacuum systems fitted with condensers to recover monomer. Most of the devolatilization occurs during bubble formation and expansion in the strands, rather than in the melt pool at the bottom of the vessel.

At high weight fractions of monomer, the pressure difference between the system and the equilibrium pressure may be great, and the devolatilization rate will be limited by bubble rupture and coalescence (boiling). The heat of vaporization is important for systems in which a large fraction of monomer is removed. For example, if 10% monomer is to be reduced to 1% monomer, most of the heat of vaporization must be supplied by the melt. This can cause local cooling and much lower viscosities than expected, lowering the rates of bubble growth, migration, and collapse. In falling strand designs, the bubbles must be able to migrate to the surface of the strands before they fall into the melt pool.

At lower values of the pressure difference, the process rate may be controlled by bubble initiation and growth. The transition from boiling to bubble growth limitation occurs between 10% and 20% monomer. The boiling mechanism can be enhanced by increasing the vacuum, increasing the temperature, and injecting a lower boiling condensible or noncondensible gas into the melt prior to extrusion through the die. Steam is a popular choice for a condensible gas. Flash systems have several advantages over mechanical system. They are cheaper because there are very few moving parts and thermal energy is used to drive the process. The melt pool can rise

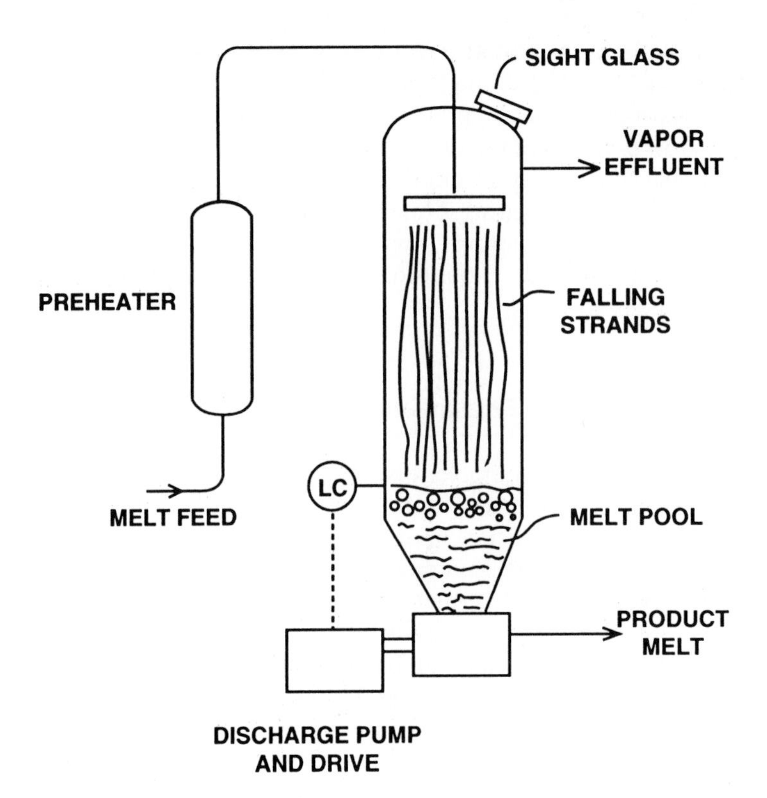

Figure 9.22 Falling strand devolatizer.

and fall, or the melt feed can be recycled, permitting the system to provide surge capacity. They are preferred for shear-sensitive polymers.

Extruders (Single and Twin Screw). Extruders are natural choices for devolatilization. Many polymer melts will be processed into pellets using extruders immediately after the reactor system. The extruders can be designed with vent systems to remove volatile gases. The screw is designed to build up pressure in front of the vent (thick radius screw) and release the pressure under the vent (thin radius screw). The screw must maintain a seal of polymer along its flights ahead of and behind the vent, but must be operated starved (the space between the barrel and the screw must not be completely filled with melt) in the venting region to allow the vapors to escape from the melt. The conveying capacity of the screw after the vent must be higher than that before the vent in order for the vent region to operate starved. Foaming melts can be generated using high vacuums or inert gases mixed with the melt. However, these systems can be difficult to control, since molten polymer can be entrained with the gases leaving the vent and can solidify in downstream processing equipment.

9.6.3 Staged Systems

Staged devolatilization systems are preferred when there are several orders of magnitude reduction in the monomer concentration. Suppose that the monomer concentration in the polymer melt exiting the reactor is 20 wt% and the target level of monomer in the pellet product is 100 ppm. It might be convenient to remove the monomer in four stages with an equal reduction occurring across each stage.

Example 9.7. Problem: *Staged Devolatilization Process*

The styrene-polystyrene system can be used to illustrate the design of a staged devolatilization process. A sketch of the process is shown in Figure 9.23. The associated material balance table (Table 9.19) shows the flows of each stream. No polymer leaves with the vapor streams. The monomer in the first stage could be removed by flashing (stream 3). The heat of vaporization requirements of each stage are listed in Table 9.20. The adiabatic temperature drop for the first stage is 54°C, which would be rather large for most processes; therefore, it would be wiser to choose a lower monomer recovery for the first stage. The first stage conditions might be set at 0.03 weight fraction monomer, which would give a pressure greater than atmospheric in the exit gas and an adiabatic temperature drop of 38°C. The first stage could be done as a flash operation, and the other stages could be wiped film evaporators and devolatilizing extruders.

9.7 POLYMER RECOVERY

There are two different types of recovery systems, which depend on the form of the polymer being produced. Polymers that are made in solution or bulk processes usually are devolatilized and then extruded into pellet form. No further work is needed prior to storage, shipping, and fabricating into parts and products.

The production of solid polymer in suspension or emulsion processes may require a dewatering step. In the cases where the emulsion is the product, additional stabilizers may be added after polymerization to keep the product emulsion stable during storage, transportation, and use. The other emulsion products are coagulated, dewatered, dried, and stored. Emulsions can be coagulated by high shear, high or low temperatures, changes in pH, changes in ionic strength, or a combination of these methods. A key challenge is to control the size of the coagulated particles to permit water to permeate through the structure while making a solid large enough to recover using screening, centrifuging, or other conventional techniques. Most suspension products are dewatered, dried, and stored.

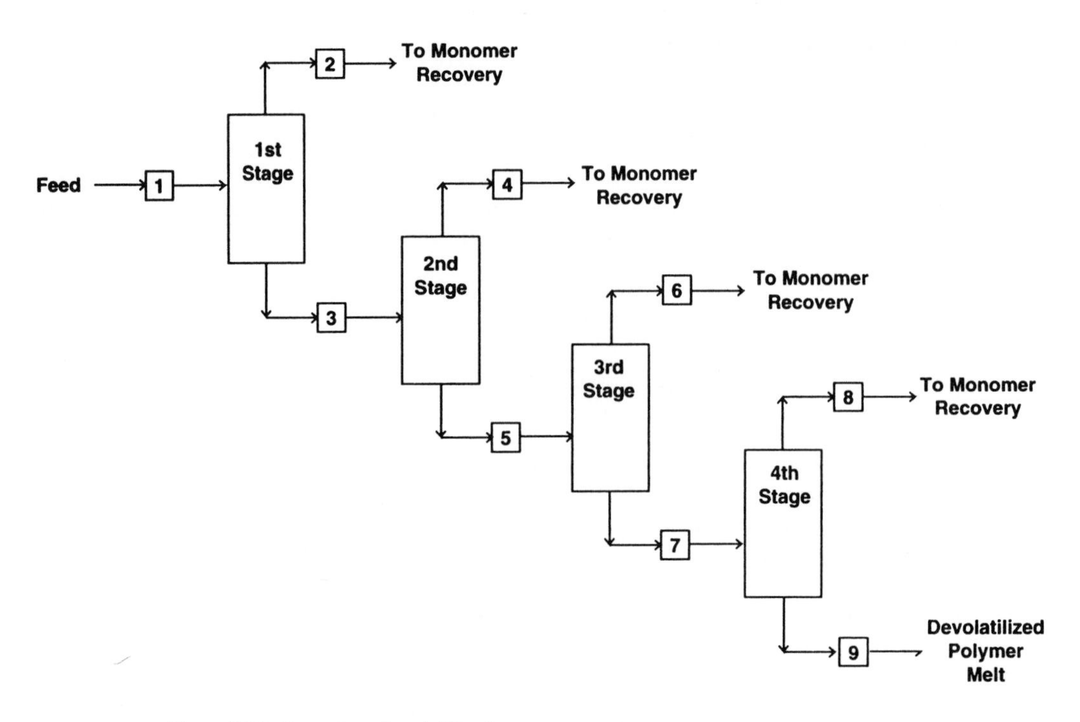

Figure 9.23 Four-stage devolatilization process.

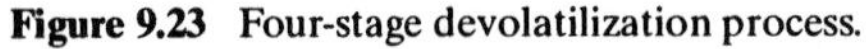

Even when the reaction system is continuous and solid polymer is not produced until the polymer recovery step, the presence of a solid almost always means batchwise handling. Unless the solids-handling system is designed with care, it can become the rate-limiting step in the production sequence. When an interruption occurs in a solids-handling unit, the flow to the unit is cut off until the problem is corrected. If the solids-handling portion of the plant has the same capacity as the reaction system, disruptions in the ability to take material away from the reactors will cause slowdowns or shutdowns. This means that, from the first appearance of solids in the process to solid storage, the process equipment capacity should be expanding. This intentional

TABLE 9.19 Staged Devolatilization Process Material Balance

	Stream kg/hr								
Component	1	2	3	4	5	6	7	8	9
Inert	0.00								
Monomer	200.00	179.49	20.51	18.00	2.51	2.20	0.31	0.27	0.04
Polymer	800.00		800.00		800.00		800.00		800.00
Total	1000.00	179.49	820.51	18.00	802.51	2.20	800.31	0.27	800.04
M/P	0.2		0.025		0.003125		3.91E-04		4.88E-05

TABLE 9.20 Heat Load for Each Stage

	Stage			
Heat Load	1	2	3	4
kg monomer	179.49	18.00	2.20	0.27
kcal/hr	15597.44	1564.63	190.76	23.77
Adiabatic ΔT	54.09	5.57	0.68	0.08
$C_{p,\, polymer}$	0.35			
$C_{p,\, monomer}$	0.407			

increase in capacity permits the solids-handling equipment to lag behind and then catch up to the reactor capacity.

9.8 POSTPOLYMERIZATION REACTIONS

There are a variety of postpolymerization reactions done on polymers (Table 9.21). There are several major types: copolymer preparation, cross-linking, surface modification, modification of the bulk polymer, and the addition of functional groups intended as binding sites. Some of these processes already have been discussed. Graft and block copolymers are prepared after one chain has been made. The major cross-linking reaction is the vulcanization of elastomers. Radiation-sensitive polymers can be cross-linked in place on computer chips, providing a mask over which to vapor deposit semiconductor materials. Surface modification is one way to vary the barrier properties or fouling properties of polymeric solids without extensive reactions. Chlorinated polyethylene and chlorinate poly(vinyl chloride) have properties that are significantly different from a homopolymer with the same bulk composition. The addition of functional groups on a variety of resins is the basis of many ion exchange resins. Only a few of these techniques will be discussed (vulcanization, polystyrene-based ion exchange resins, and chlorinated PVC).

There are two important aspects to postpolymerization reactions on polymers. The first is the diffusion of the reacting group to a site on the polymer. Diffusion coefficients for gases and low molecular weight materials in glassy polymers range from 10^{-14} m²/s to 10^{-18} m²/s. Similar solutes diffusing in swollen polymers or polymer solutions have diffusion coefficients several orders of magnitude higher. In order to get fast processes, either the polymer should be intimately mixed with the reactants before the process begins (as in vulcanization), the polymer should be expanded with a swelling agent or the reactant itself (sulfonation of ion exchange resins is an example), or small particles and thin films should be treated (fluorination of polyethylene gasoline tanks). When two reactants are needed to functionalize the polymer, the diffusion paths must be large enough so that both have access to polymer sites.

The second factor is the effects of nearby chains and functional groups on the reactivity of chain sites. Many solid polymer reactions are fundamentally different

TABLE 9.21 Postpolymerization Reactions

Type	Examples
Copolymer preparation	Graft and block copolymers
Cross-linking	Vulcanization of elastomers Cross-linking of phenol-formaldehyde for ion exchange resins Radiation cross-linking of polymers for high resolution electron beam microlithography
OTHER MODIFICATIONS	
Surfaces	Heparinization to reduce interactions with blood components Fluorination to improve barrier properties (surface-fluorinated PE gasoline tanks) Sulfonation to change wettability
Bulk properties	Chlorinated polyethylene Chlorinated poly(vinyl chloride)
Functional groups	Controlled hydrolysis of nylons for improved dyeing Alcohol exchange to convert poly(vinyl acetate) to poly(vinyl alcohol) Addition of various ligands to sulfonated polystyrene to make ion exchange resins

from those done on model compounds in dilute solution. Even in swollen systems, the diffusion rates of reactants and by-products are modest. The reactants tend to react where they are, leading to substitutions and by-products that may be very difficult to observe in dilute solution. In the chain-dense polymer phase, steric hindrance by large groups can be important. As an example, extensive model compound studies of sulfonation of liquid organic molecules do not relate well to the sulfonation of polystyrene to prepare ion exchange resins with respect to substitution rates or substitution sites. Changes in reactivity can be due to solubility of the reactant in the polymer, electrostatic effects, changes in the activation energy, and neighboring group effects.

Vulcanization. Vulcanization is a cross-linking process specific to elastomers. Cross-links can be made by several techniques, including a group of sulfur atoms in a short chain, a single sulfur atom, a carbon-to-carbon bond, polyvalent organic radical, an ionic cluster, or a polyvalent metal ion. The Goodyear–Hancock vulcanization process based on sulfur was developed in 1841. The elastomer molecules have high molecular weight, often between 100,000 and 500,000. The vulcanization process results in chain segments of 4,000 to 10,000 molecular weight between cross-links. The vulcanized elastomer is insoluble in all solvents and takes on the shape of the mold. Any changes in shape require breaking backbone bonds. Vulcanization is used for a variety of elastomers containing double bonds: natural rubber, styrene/butadiene rubber, butadiene rubber, ethylene-propylene-diene monomer (EPDM), butyl rubber, halobutyl rubbers, and nitrile rubber.

The process depends on the rate of cross-link formation and the extent of cross-linking. Scorch resistance (the ability to resist early cross-linking) is important to prevent resistance to flow in the mold. Rheometry is a good tool for studying the vulcanization process. Sulfur vulcanization alone is no longer practiced because it is slow. A typical recipe was to add elemental sulfur to natural rubber (8 parts per 100 parts) and heat for five hours at 140°C. The recipe ingredients are masticated into the rubber by moderate temperature, high shear mixing devices. The mastication has been a batch process because of the many different types of rubber compounds that are used in the tire business. Zinc oxide added to this recipe reduced the cross-linking time to three hours. Accelerators have reduced the time to one to three minutes. A typical vulcanization recipe is

Item	PHR (parts per hundred ratio)
Elastomer	100
Zinc oxide	10 to 2
Fatty acid (stearic)	4 to 1
Sulfur	4 to 0.5
Accelerator	2 to 0.5

The accelerated process is thought to work as follows (Fig. 9.24): (a) The accelerator, Ac, reacts with the sulfur, S, to give polysulfides of the structure, $Ac - S_x - Ac$; (b) one end of the polysulfide reacts with the allylic position near the double bond on the elastomer to give rubber, $- S_x - Ac$, (c) one accelerator end group reacts to form a mercaptan; and (d) the other end of the cross-link then reacts with another double bond or with an intermediate to form the cross-link. The process is carried out via free radical mechanisms. Zinc oxide may help the process by forming ligand complexes with the polysulfides.

Increasing the sulfur and accelerator content will increase the cross-link densities and give high moduli, stiffness, and hardness. High accelerator to sulfur ratios lead to more unreacted polysulfide end groups on the chain. Low ratios lead to more polysulfide cross-links and heterocyclic ring compounds. Because of the batch-to-batch variation in some elastomer stocks, the complex chemistry, and the difficulty of obtaining uniform mixing with highly viscoelastic polymer, statistical analysis and optimization of rubber recipes has become an important development tool. Fatigue life is an important test for elastomers. Low values of fatigue life often occur for low sulfur-to-accelerator ratios due to the appended groups and high concentrations of monosulfidic cross-links. These cross-links are not able to exchange, rearrange, or break to relieve local stresses without breaking the main polymer chain. Polysulfidic cross-links can rearrange under stress, and generally improve flex life. There are other curing systems for elastomers, including phenolic, benzoquinone, and bimaleimide systems. Figure 9.25 shows the general property trends for rubbers as a function of cross-link density.

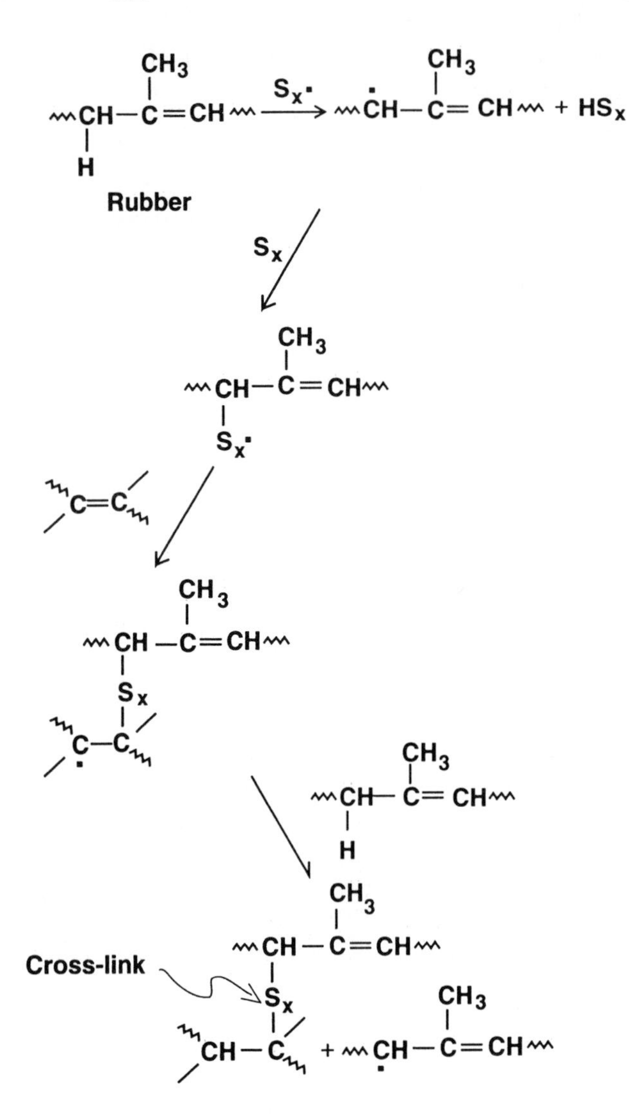

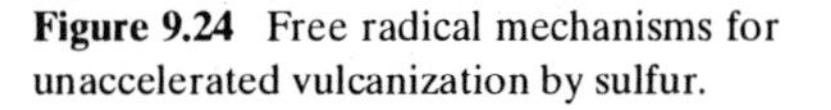
Figure 9.24 Free radical mechanisms for unaccelerated vulcanization by sulfur.

Although the mechanism by which they are elastic is quite different, the saturated hydrocarbon elastomers such as ethylene-propylene rubber also can be cross-linked. Peroxides are used to abstract hydrogen from neighboring carbon atoms on different chains. The free radicals recombine to form a cross-link. Silicone rubbers can be cross-linked by peroxides also.

Ion Exchange Resins. The sulfonation of lightly cross-linked styrene-divinylbenzene copolymer spheres to functionalize them for ion exchange resins (Fig. 9.26) is an example of the bulk modification of a swollen polymer. In this case, the

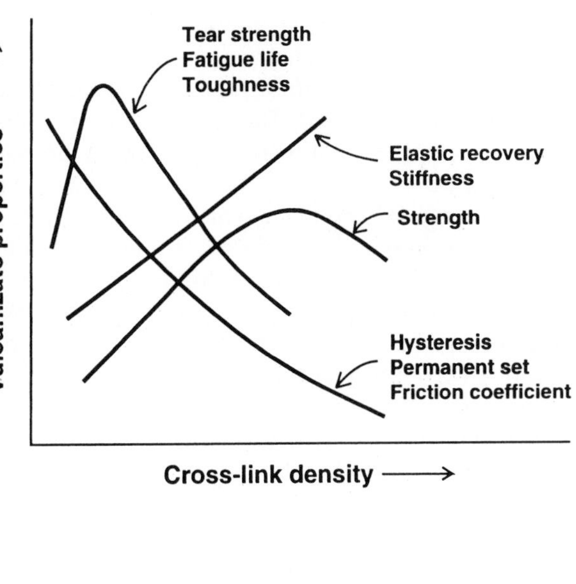

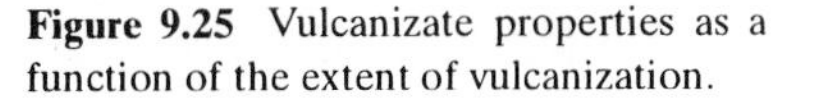

Figure 9.25 Vulcanizate properties as a function of the extent of vulcanization.

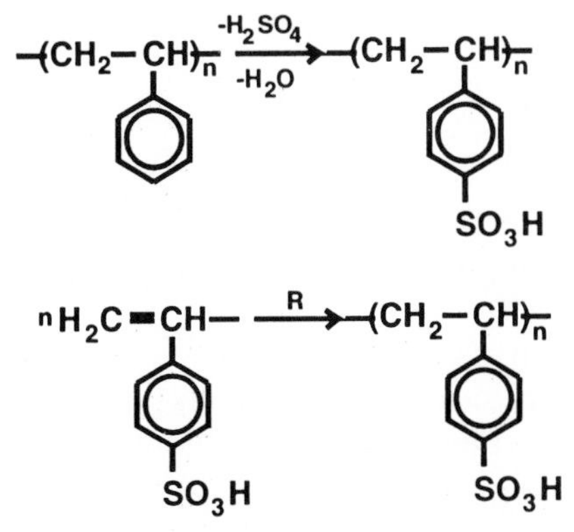

Figure 9.26 Polystyrene with sulfonic acid groups can be made by sulfonation of polystyrene or polymerization of 4-vinylbenzene sulfonic acid.

sulfonic acid group tends to add in the para position on the aromatic ring. A similar material can be made by polymerizing 4-vinylbenzene sulfonic acid. The polymers made by these two routes have different properties. The homopolymer has a uniform distribution of groups throughout its volume. The functionalized polymer may not have a uniform distribution of sulfonic acid groups, but those available should be readily accessible to ionic species.

Strong acid ion exchange resins are made by sulfonating solid polystyrene beads in sulfuric acid (Millar et al., 1963). The polymer is swollen by using solvents or by using high temperatures with sulfuric acid alone. The polarity of the resin changes during the reaction, and the flux of reactant into the beads seems to increase. The resin becomes water-wettable.

Chlorinated Poly(vinyl Chloride). Poly(vinyl chloride) can be chlorinated by redispersing suspension resin in water and adding chlorine gas at modest pressures. The chlorine gas dissolves in the water phase and diffuses to the surface of the solid polymer particles, where it is soluble to some extent. The reaction can be initiated thermally by raising the reaction temperature so that there is sufficient by-product of the hydrogen abstraction from the chains.

Old processes used swelling agents to expand the polymer particles. These may have increased the chlorine solubility as well as its diffusivity in the solid phase. Similar results can be obtained by increasing the partial pressure of chlorine, which also acts as a swelling agent. The distribution of chlorine can vary through the particle, as can the substitution site. The preferred reaction is that on the unchlorinated carbon of each repeating unit. The 1,1-substituted material has some undesirable properties. The optimal reactor balances the diffusive flux of chlorine into the resin with the reaction rate to provide as uniform a chlorination as possible. The rate-limiting step can vary from the flux of chlorine into the water phase, to the flux of chlorine to the particle, to the flux of chlorine into the particles, to the chemical reaction rate. The hydrogen chloride needs to be stripped from the product prior to compounding and extrusion. The chlorinated PVC has a glass transition temperature above 100°C, making it suitable for commercial and residential hot water pipe.

PROBLEMS

Some of the problems for Chapter 9 can be answered best by referring to descriptions of commercial processes or their process flow diagrams.

9.1 During the polymerization of vinyl monomers by emulsion polymerization, the reaction kinetics are not the same as those for bulk polymerization. Explain why, and relate your answer to a physical picture for this type of commercial process.

9.2 You have been called into the plant at 3 a.m. to solve a quality problem on the ABS compounding line. The operator complains of nonuniform fluxing on the mill. Because of a low inventory of ABS, the polymerization line has been feeding product directly to the compounding hopper. After checking the polymerization production logs, you have noticed three unusual operating conditions:

(a) The evening shift operator has been adding short-stop too soon.

(b) The pressure in the stripper has been higher than normal.

(c) The styrene flow to the monomer mixing tank has been higher than desired.

What could have caused the milling problem and why?

9.3 Find a process flow diagram for polypropylene and answer the following questions:

(a) What type of polymerization is this?

(b) Will oxygen or water affect the reaction rate?

(c) What would you expect to be the rate-limiting part of the process?

9.4 Methyl methacrylate can be polymerized by a bulk polymerization in which only monomer, initiator, and polymer are present. Reaction kinetics for the process are predicted to be first-order in monomer, with initiation concentration high enough so that it can be assumed constant. However, as monomer is converted to polymer, the viscosity of the reaction solution increases dramatically. This has little effect on small monomer molecules, which can still diffuse easily to growing chains; however, large growing chains cannot diffuse easily toward each other.

(a) What effect will the viscosity increase have on the rate constants k_i, k_p, and k_t?

(b) The overall rate of reaction is given by:

$$R_p = k_p\left(\frac{k_1}{k_t}\right)^{0.5} [M]\,[1]^{0.5}$$

Plot the percent polymerization versus time.

(c) How will the molecular weight of polymer formed after the viscosity increase compare to that formed early in the reaction?

(d) Besides the obvious problem of material transfer for a high viscosity material, what other reactor design problem might arise in this situation? (Think about the consequences of your answer to (b)).

9.5 The kinetic rate equations for polymerization developed in Chapter 4 can be used with materials balances to design reactor systems. The goal of this problem is to compare different reactor systems for the production of 1000 kg/hr of poly(methyl methacrylate). Assume that all reactors should reach 65% conversion and that heat transfer systems can be designed to keep the contents isothermal. Determine the volume of a batch, a CSTR, and plug flow reactor for this duty (consult reaction engineering texts for the design equations). Propose your preferred reactor system and explain why.

9.6 The purpose of this problem is assemble up-to-date information on a specific polymer with regards to the following items:

(a) Monomer synthesis route(s)

(b) Monomer synthesis process flow diagrams

(c) Polymerization reactor design

(d) Polymerization process design with emphasis on pollution control and monomer recovery

Look through *Hydrocarbon Processing* and *Polymer Technology* journals to find the most recent monomer synthesis and polymerization process. Whenever possible, include process flow diagrams. Discuss the operating improvements, product property improvements, and/or cost reductions realized by these new processes.

Your specific assignment is taken from the following table.

Polymer
Polyethylene—Union Carbide fluidized bed
Polybutylene—Witco, Shell
Polysulfone
Cellulose Acetate—Kodak, coal-derived monomers
Polyaramide—Dupont, liquid crystalline polymer
Urethane Foams—Reaction Injection molding
Phenolic Resins—Novolac
Polystyrene
Polyesters
Poly (methyl methacrylate)

9.7 Polyesters such as poly(ethylene terephthalate) are often made by ester-interchange. In the ester-interchange reactor, DMT (dimethyl terephthalate) is reacted with ethylene glycol and methanol is removed overhead. In the polycondensation reactors, the glycol is removed under vacuum and reaction to high molecular weight polymer continues

(a) Is this a bulk, suspension solution or an emulsion process?

(b) Your PET operating plant has been experiencing difficulties in spinning the polymer. Dye swell is excessive, spinning rates are low, and the consensus of opinion is that viscosity is too high at the spinning manifold. Someone suggests adding a solvent (such as ethylene glycol) in the second polycondensation reactor. List advantages and disadvantages of this "solution" method for solving the problem.

(c) During the discussion of the above proposal, an operator mentions that steam supply to the polycondensation reactors has been erratic for a week and that he has observed lower-than-normal temperatures in the process streams out of these vessels. Suggest an alternate way to solve the spinning problem and explain why it might work.

9.8 Nylon 6,12 is known in the fabric industry as Qiana. Suppose that cyclododecane was a by-product of the production of 1,5,9-cyclododecatriene and that it is not separated from the triene before oxidation to the diacid. What property changes, if any, might be expected from the polymer made from the contaminated triene versus the pure triene?

9.9 The Mobil—Witco process to produce poly-1-butene uses a Ziegler-type catalyst to produce isotactic polymer.

(a) Draw the structure of this polymer.

(b) The polymerization is continuous and includes a recycle loop at the reactor for heat removal purposes. Temperature and pressure are maintained so that the polymer remains dissolved in 1-butene during reaction. Besides catalyst addition, what means does the operator have for controlling molecular weight? Use mechanisms to reinforce your arguments.

(c) Crystalline poly-1-butene exists in three crystal habits. It initially crystallizes as Form II from the melt and begins a slow transition to Form I. The change to Form I includes a change in the unit cell shape and size; a change in density and strength is reversible and requires several days. Describe how you would perform quality control tests on this polymer.

(d) How would you handle distribution and inventory on this material?

9.10 Latex is the preferred form for a number of elastomers.

(a) Describe two methods for coagulating latex.

(b) Monomers need to be removed from latices prior to their use. Sketch a downstream processing system for an elastomer and suggest the locations of the coagulation and devolatilization steps. Explain your process choices.

9.11 Contrast the water-removal steps in the production of nylon 6,6 & nylon 6 with those for PET.

9.12 Find a process flow sketch for the manufacture of poly(methyl methacrylate) by a suspension process. Should the suspending agent be removed from the final product, and if so, where? (What effects might the suspending agent have on the polymer's properties?)

9.13 Compute the adiabatic temperature rise for the following polystyrene polymerization systems: bulk, suspension with a 1:4 styrene to water weight ratio, and solution with a 1:4 styrene to toluene weight ratio. The initial temperature of the system is assumed to be 90°C.

9.14 The first CSTR in a polystyrene process operates at 40% conversion and 140°C. If the feed is pure styrene at 80°C, calculate the residence time needed for this step of the polymerization. What will the heating or cooling duty be for this reactor?

9.15 Calculate the pressures in the first and second polystyrene CSTR's if the conditions are:

CSTR 1: $T = 140°C, X = .40$
CSTR 2: $T = 160°C, X = .60$

Suggest a system to get the outlet stream of the first CSTR to the inlet port of the second.

9.16 The adiabatic temperature rise gives an estimate of the temperature in the reactor after a runaway reaction. Estimate the pressure in each of the reaction systems of problem 9.15 after such an event. What safety systems and controls should be considered for managing styrene polymerization systems?

9.17 The ester interchange reaction between ethylene glycol and dimethylterephthalate is done in a multistage reactor rather than a CSTR. Do some engineering calculations that show the advantage of the multistage system. What might be the target concentration of methanol from the final stage of the multistage reactor?

9.18 Gypsum and titanium dioxide are being considered as fillers and whiteners for a thermoplastic compound. Compare the viscosity increase ratios (compound viscosity over homopolymer viscosity) for these fillers over the range, 1 to 20vol%.

REFERENCES

A.H. ABDEL-ALIM and A.E. HAMIELEC, *J. Appl. Polym. Sci. 16,* 783 (1972).

J.L. AMOS., *Polym. Eng. Sci. 14,* 1 (1974).

R. BALL and P. RICHMOND, *J. Phys. Chem. Liquids 9,* 99–116 (1980).

H.A. BARNES, J.F. HUTTON, and K. WALTERS, *An Introduction to Rheology*, Elsevier, Amsterdam, 1989.

G.K. BATCHELOR, *J. Fluid Mechanics 83*, 97–117 (1977).

R.S. BRODKEY, "Fluid Motion and Mixing," in *Mixing*, Vol. 1 (H. Uhl and J.B. Gray, Eds.), Academic Press, NY, 1966.

J.D. BURNETT and H.W. MELVILLE, *Trans. Faraday Soc. 46*, 976 (1950).

R. CHELLA and J.M. OTTINO, *AIChEJ 29,* 373 (1983).

B. CLARKE, *Trans. Inst. Chem. Eng. 45*, 251–256 (1967).

J.A. DAVIDSON and K.L. GARDNER, "Vinyl Polymers, Poly(vinyl Chloride)" in *Kirk-Othmer Encyclopedia of Chemical Technology*, 3rd ed., Vol. 23, (M. Grayson, Ed.), Wiley-Interscience, NY, 1983.

C.G. DEKRUIF, E.M.F. VAN IEVSEL, A. VRIJ, and W.B. RUSSEL, *J. Chem. Phys. 83*, 4717–4725 (1985).

G. DONATI et al, *ACS Symp. Ser. 196*, 579 (1982).

H. GIESEKUS, "Disperse Systems," in *Physical Properties of Foods* (R. Jowitt et al, Eds.), Applied Science Publishers, 1983.

W.C. GRIFFIN, *J. Soc. Cosmet. Chem. 1*(5), 311 (1949).

A. HAHN, A. CHAPTAL, and J. SIALELLI, *Hydrocarbon Proc. 54*(2), 89 (1975).

W.D. HARKINS, *J. Am. Chem. Soc. 69*, 1428 (1947), W.D. HARKINS, *J. Polym. Sci. 5*, 217 (1950).

K. RAVINDRANATH and R.A. MACHELKAR, *AIChE Jr. 30*, 415 (1984).

D.E. JAMES, "Linear Polyethylene," in *Encyclopedia of Polymer Science and Engineering*, Vol. 6, Wiley-Interscience, NY, 1986.

R.N. JOHNSON A.G. FARMHAM, R.A. CLENDINNING, W.F. HALE, and C.N. MERRIAN, *J. Polym. Sci*, Part A-1, *5*, 2375 (1967).

G.R. JOHNSON, *J. Vinyl Technol. 2*, 138 (1980).

C.W. JOHNSTON in *Encyclopedia of PVC*, Vol. 1, Chap. 3 (L.I. Ness, Ed.,) Marcel Dekker, NY, 1976.

B.B. KINE and R.W. NOVAK, "Methacrylate Polymers" in (M. Grayson, Ed.), *Kirk-Othmer Encyclopedia of Chemical Technology*, 3rd ed., Vol. 15, Wiley-Interscience, NY, 1981.

L. KNIEL, O. WINTER, and C.H. TSAI, in "Ethylene" *Kirk-Othmer Encyclopedia of Chemical Technology,* 3rd Ed., Vol. 9, p. 400 1980.

I.M. KRIEGER, *Adv. Colloid Interface Sci. 3*, 111–136 (1972).

L. J. LEE, *Rubber Chem. Tech. 53,* 542 (1980).

D.E. LENG and G.J. QUARDERER, *Chem. Eng. Commun. 14,* 177 (1982).

R.B. LIEBERMAN and P.C. BARBE, "Propylene Polymers," in *Encyclopedia of Polymer Science and Engineering*, Vol. 13, Wiley-Interscience, 1988.

S. LYNN, *AIChE J. 23,* 387 (1977).

C.W. MACOSKO, *RIM Fundamentals of Reaction Injection Molding*, Hanser, NY, 1989.

G.J. MANTELL, J.T. BARR, and R.K.S. CHAN, *Chem. Eng. Prog. 71*(9), 54 (1975).

P.S. MEHTA, L.N. VALSAMIS, Z. TADMOR, *Polym. Proc. Eng. 2,* 103–128 (1984).

J.R. MILLAR, D.G. SMITH, W.E. MARR, and T.R.E. KRESSMAN, *J. Chem. Soc. 38,* 218 (1963).

M. MORTON, S. KAIZERMAN, and M.W. ALTIER, *J. Colloid Sco. 9,* 300 (1954).

L.S. MOSSNER and E.A. GRULKE, *J. Appl. Polym. Sci. 35,* 923–936 (1988).

H. MULLER, B. KUPER, U. MAIER and L. PIERKES, *Adv. Polym. Tech. 5,* 257 (1984).

E.B. NAUMAN, "Synthesis and Reacto-Technology" in *Plastics Polymers Science and Technology*, Chap. 10, (M.D. Baijal, Ed.) Wiley, NY, 1982.

G.W. POEHLEIN and D.J. DOUGHERTY, *Rubber Chem. Technol. 50(3),* 601 (1977).

G.W. POEHLEIN, "Emulsion Polymerization," in *Encyclopedia of Polymer Science and Engineering*, Vol. 6, Wiley-Interscience, NY, 1986.

J.F. ROSS and W.A. BOWLES, *Ind. Eng. Chem. Prod. Res. Dev. 24,* 149 (1985).

G. SWIFT and K.A. HUGHES, "Solution Polymerization," in *Encyclopedia of Polymer Science and Engineering*, Vol. 15, Wiley-Interscience, NY, 1989.

Z. TADMOR and J.A. BIESENBERGER, *Ind. Eng. Chem. Fund. 5* 336 (1966).

Z. TADMOR and C.G. GOGOS, *Principles of Polymer Processing*, Wiley-Interscience, NY, 1979.

R. TURIAN and T.-F. YUAN, AIChE Jr. *23,* 232–243 (1977).

J. UGELSTAD, P.C. MORK, and F.K. HANSEN, *Pure Appl. Chem. 53,* 323 (1981).

J.W. VANDERHOFF, in *Vinyl Polymerization*, Vol. 1, Pt. II, Chap. 1, 1 (G.E. Ham, Ed.), Marcel Dekker, NY, 1969.

T.Y. XIE, A.E. HAMIELEC, P.E. WOOD, and D.R. WOODS, *Polymer, 32(3),* 537, 1991.

G.L. YAWS, *Physical Properties*, McGraw-Hill, NY, 1977.

J.M. ZIELINSKI and J.L. DUDA, AIChE J. *38(3),* 405 (1992).

10

Polymer Processing

Figure 10.1 shows a generic polymer processing flow diagram for a thermoplastic material. The polymer is received as a solid and stored. The typical storage system consists of silos and transport systems (air-conveying systems, augurs, gravity feed systems). Moisture-sensitive materials such as nylon will have a drying step just prior to processing or will be maintained in a dry state in the storage silo. Compounding can be done before or during processing. Because of the difficulty of dispersing solids and liquids in viscous polymers or in polymer solids, some compounding ingredients are available as concentrates in the polymer of interest. The compound or resin enters the product-forming section. For thermoplastics, the most common operations are extrusion, injection molding, molding and casting, fiber spinning, and blow molding. Excess material is always generated from operations such as injection molding. The flashings, mold runners, and defective parts often make up 30% of the injection-molding machine's output. This material is ground and recycled to the extruder in order to get high material yields. The parts are cooled and stored for shipping.

Thermoset systems use liquid reactants to form the part directly in the mold. In the case of sheet-molding compound (SMC), the fabricator may make a composite sheet from glass fibers and liquid ingredients. This sheet is aged and molded in separate operations. Reaction injection-molding operations make the part directly from liquid reactants. Elastomers being made into tires or other products are compounded prior to tire assembly and curing. The compounding step is critical, as it represents the metering and dispersion of the curing system reactants.

Polymer processing can be broken down into a number of unit operations or elementary steps. One or more of these steps is found in all processing procedures,

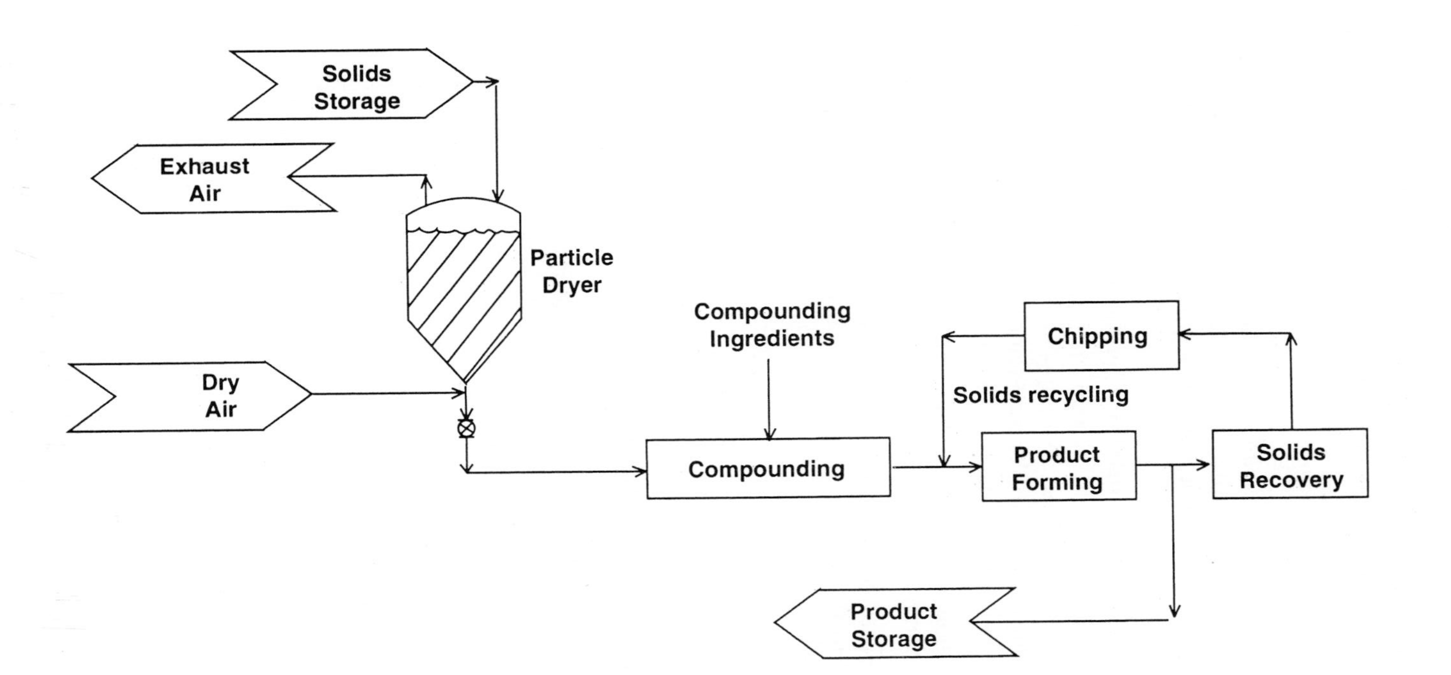

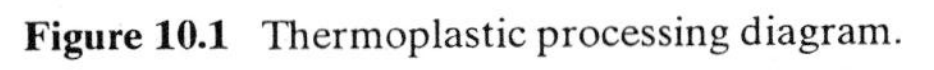
Figure 10.1 Thermoplastic processing diagram.

and all of them occur during product formation. The following list identifies five elementary steps (Tadmor and Gogos, 1979):

POLYMER PROCESSING STEPS

1. *Handling of Particulate Solids.* Solids handling occurs in a variety of locations during the process (Fig. 10.1) This operation includes particle packing, agglomeration, stress distribution in hoppers, gravitational flow, arching, and compacting.
2. *Melting.* Melting often is the rate-determining step for a particular thermoplastic process. It depends on the thermal and physical properties of the polymer. The degradation temperature sets an upper limit on temperatures that are practical in processing equipment. The time-temperature history of the polymer sample, the rate of heat transfer, and the viscosity of the melt all affect the process rate.
3. *Pressurization and Pumping.* The melt must be moved and transported to the shaping operation. The rate of this step depends on the rheology of the melt.
4. *Mixing.* Mixing occurs during solids transfer, melting, and pumping. Energy must be added to the system to disperse noncompatible polymers or to break up agglomerates.
5. *Devolatilization.* Monomer or solvents are removed before, during, or after shaping operations.

Engineering analyses of these steps should lead to improved understanding of the processes. The analysis is intended to answer the questions of how a certain product can be made, how to design a machine to form a specific item, and what machine configuration may be optimal. The design of processing equipment also depends on the rheology, the thermodynamics, the morphology, and even the polymerization history of the polymer. A variety of shaping methods are available to the polymer engineer to help fabricate the desired product. The list that follows summarizes some of these techniques.

POLYMER SHAPING OPERATIONS

1. *Die forming.* This technique forms the polymer in two directions and makes a continuous part in the machine direction. Die forming is used in fiber spinning; film, sheet, profile, tube, and pipe extrusion; and wire and cable coating. Individual parts are cut from the cooled solid.
2. *Molding and Casting.* Injection molding, compression molding, transfer molding, and casting are examples of processes in which three-dimensional parts are made.
3. *Secondary Shaping.* This operation is performed after the melt has been given an initial shape or orientation. It includes thermoforming, blow molding, film blowing, and cold forming.

4. *Calendering and Coating.* Calendering is an example of kneading flow that can mix ingredients. Knife coating and roll coating are two other examples.
5. *Mold Coating.* This technique can be used to make hollow objects, such as basketballs. Dip, slush, powder, and rotational are types of mold coating.

Developing engineering models of the elementary steps and the shaping operations requires integrating transport phenomena, mixing principles, solid mechanics, polymer melt rheology, polymer physics, and polymer chemistry. This text describes several shaping operations. Detailed models are given in graduate textbooks on polymer processing by Middleman (1977) and Tadmor and Gogos (1979).

Computer-aided design and engineering of polymer processing equipment is available. Commercial programs for mold or extruder screw design apply the equations of motion and energy balances to specific flow geometries. The typical model for viscosity is the power law model, as viscoelastic flow predictions still are difficult. These programs will be part of the polymer process engineer's tool kit in the future because they permit calculation of flows without the fabrication of the equipment. This can save much research and development time for projects by improving processing time, improving part uniformity, replacing polymer materials in an application, designing new profile extrusions or injection molded parts, and modifying existing molds and dies.

10.1 POLYMER DRYING BELOW T_g

There are a number of polymers that are sensitive to water content during processing. Many of the most sensitive materials are polar and can absorb significant amounts of water from the environment. This section covers the theory and design of polymer dryers intended to control the water content of particulate solids prior to extrusion and processing. However, the principles involved apply to many problems of mass transfer in polymeric solids. Other applications that can use similar theory are gas permeation through polymer films, flavor uptake by packaging films, and removal of small amounts of monomer from solid polymers (materials below T_g). Water is the solute in this discussion.

Reasons for Controlling Polymer Moisture Content. There are two major reasons for controlling the water content of polymer particulates. Polymer particles that contain high weight fractions of water can have bubbles formed in them during high temperature processing. In this case, the water phase separates from the polymer phase at the processing temperature and pressure. Extruded parts are most susceptible to this problem. The highest temperatures and pressures in extrusion occur at the entrance to the die. After the die, the pressure has dropped, but the temperature is about the same. Bubbles usually are formed in the melt just after pressure has been reduced. Bubbles may form in injection molded parts also. The goal of the drying system is to reduce the water concentration below that which would cause phase separation of water at processing conditions.

A second reason for removing water is that it has active hydrogens that can participate in reversible polymer reactions, such as the condensation reactions used to make nylons and other polar polymers. In these cases, residual water can combine with the reactive bonds at various points along the chain, leading to chain scission reactions, loss of molecular weight, and reductions in mechanical properties. The amount of chain scission and property loss depends on the reaction equilibrium, reaction rate, and time at the high temperatures used for processing. A model for the extent of the reversible reaction (and the loss of molecular weight) over the time-temperature history of the polymer in the equipment is needed.

Drying Equipment. Batch or continuous dryers are used depending on the volume of resin to be treated. Batch systems are flexible and can be used for a variety of materials. The drying may be done in a tilted drum that can be rotated to allow the pellet bed to be exposed to the vacuum and the heated walls of the drum. A condenser is placed between the dryer and the vacuum system to trap as much water vapor as possible before the vacuum pumps. The drum is equipped with a heating and/or cooling system. Dried air or inert gas can be used to purge the pellets from the drum.

Figure 10.2 shows a continuous dryer design for polymer chips. The chips are fed continuously into the top of the dryer, and dried chips are dropped through a rotary valve from the bottom. The chip "bed" moves downward and the residence time is determined by the bed volume divided by the feed rate. Dry air is produced by pulling air through a molecular sieve bed. Two molecular sieve beds are needed: one for drying the chips and one for the regeneration step of the sieves. The dried air is heated and blown into the bottom of the chips. Residual water in the polymeric solid diffuses to the surface of the solid, where it is swept from the bed by the air stream. The rate-limiting step for water removal is the unsteady-state diffusion of the solute from the solid particles.

Eliminating Bubble Formation in the Melt. Equilibrium water concentrations of polymer vary widely. Table 10.1 shows some values for polymers equilibrated at 70°F at 65% relative humidity. Most of the entries are semicrystalline materials that are made into fibers. Polymers containing polar groups sorb significant amounts of water. Rayon, nylons, and the aramids sorb the most, while the water uptake of polyethylene and polytetrafluoroethylene are very low. Water can phase-separate from the melt at the high processing temperatures. Water vapor bubbles remaining in the polymer after processing become defects and sites for mechanical failure.

Section 9.6 showed how to calculate the system pressure required to cause bubble formation for devolatilization. Drying specifications can be set by calculating the solvent activity below which no bubbles will form during processing. The processing temperature usually is known. If the Flory–Huggins parameter for the system is known, the minimum water volume fraction for phase separation can be calculated. The data in Table 10.1 can be used to determine the Flory–Huggins parameter for

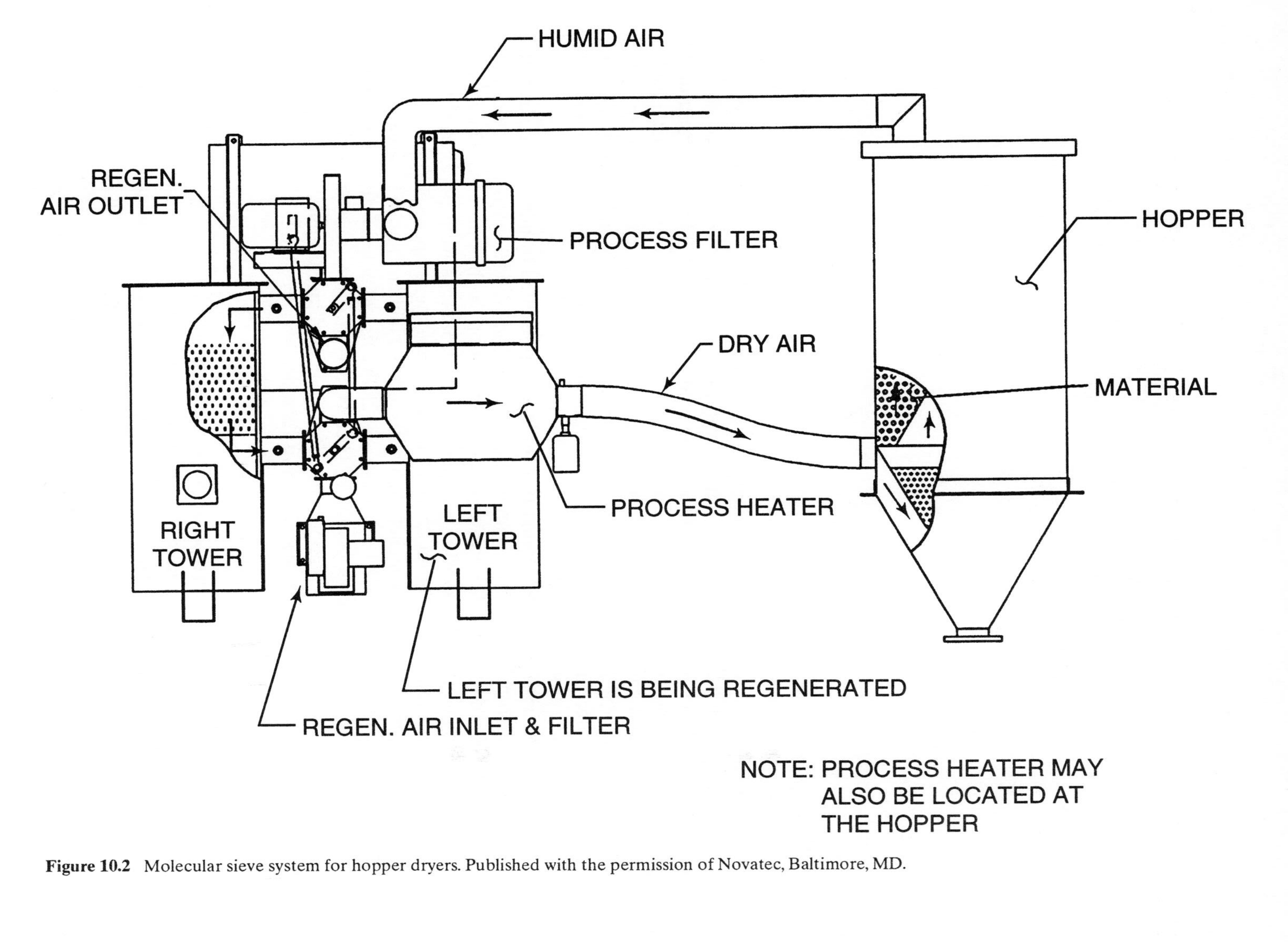

Figure 10.2 Molecular sieve system for hopper dryers. Published with the permission of Novatec, Baltimore, MD.

TABLE 10.1 Equilibrium Water Content in Polymers
Water sorption at 70°F, 65% relative humidity

Polymer	wt% absorbed
Rayon	11–13
Spandex	1.3
Polyester	0.4–0.8
Polytetrafluorethylene	very low
Nylon 6	2.8–5.0
Nylon 6,6	4.2–4.5
Aramid: Kevlar	4.5–7.0
Nomex	6.5
Polyethylene	very low
Polypropylene	0.01
Wool	11–17
Polyurethanes	0.20

polymer-water systems. For example, 65% relative humidity corresponds to a water activity of 0.65 (the vapor pressure of water is 65% of its saturated vapor pressure). For nylon 6, the water solubility will be near 3wt% or 0.0329 volume fraction (the density of amorphous nylon 6 is about 1.1 g/cm^3). The interaction parameter is about 2.16. This value is typical of solutes that are partially soluble in polymers. As the solubility of the solute goes down under identical gas phase conditions, the interaction parameter value describing the phase equilibrium should increase. The interaction parameter for the polyurethane in the table is 4.8.

Most of the data in Table 10.1 could be represented by interaction parameters ranging from one to five. Assuming the effect of temperature on χ is small, the volume fraction of water in the polymer at phase separation can be estimated. The results of the calculations are shown in Figure 10.3. Calculations were performed for two different values of χ (1 and 2.5). Nylon 6 has a melting temperature near 220°C and is processed at 250°C or higher. Assuming that the higher value of χ is nearly correct and that the material will be processed at 250°C, the volume fraction of water in the solid polymer should be less than 0.0008 volume fraction (or .073wt%) to prevent bubble formation. This is about one-fiftieth the equilibrium solubility of water in the polymer at room temperature. For this reason, most nylons are dried carefully prior to melt processing.

Minimizing Reversible Reactions in the Melt. Many of the polymers with high water solubility in Table 10.1 are made by stepwise (condensation) polymerization. Water is, or could be, a leaving group in this reaction mechanism. When these polymers are remelted for processing and fabrication, the temperatures usually approach those used in the second stage reaction. If there are polar groups available, the reaction can be reversed and the molecular weight can decrease rapidly. Water is a particularly troublesome material, since it is always present in the environment, it

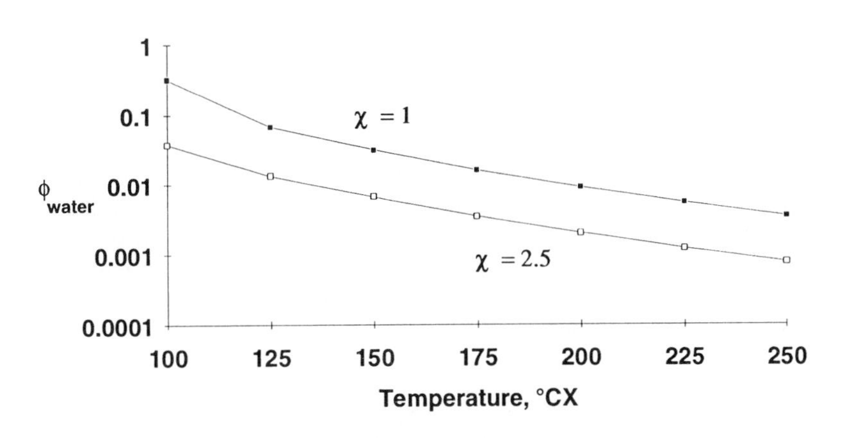

Figure 10.3 Volume fraction for phase separation.

diffuses readily into the polymeric solid, and it is very effective in reversing a number of stepwise polymerization reactions. The amide group of nylons is susceptible to acid and base catalyzed hydrolysis. Polycarbonates are fairly stable to hydrolysis as solids under moderate temperatures (Carlson and Wiles, 1986). However, they are easily attacked at melt processing temperatures ($> 300°C$) by even trace amounts of water ($< 0.01wt\%$). The process may be autocatalytic, as polar end groups can assist with catalysis. Poly(ether urethanes) are susceptible to hydrolysis, but only the urethane linkage hydrolyzes and severe conditions are needed. Poly(ester urethanes) are much more sensitive to water. Both the urethane and the ester groups can be hydrolyzed, and the acidic products can catalyze further reactions.

The target moisture content to prevent hydrolysis can be determined either through testing or by using kinetic models for the hydrolysis reaction. While the former method is direct and would be done anyway to confirm the models, modeling should be done for materials that are very susceptible to hydrolysis. Degradation rate constants (Tables 9 and 10 in Appendix D) are available for some systems.

10.1.1 Phase Equilibria Models for Glassy Polymer-Solvent Systems

There are a number of models for polymer thermodynamics. Three models that apply to solutes in glassy polymers are Henry's law, the dual-mode sorption theory, and the Flory–Huggins theory. Group contribution methods and polymer equations of state can be used as well. The equations of state are most applicable to the solution state and the melt state because these are the only states of polymers in which equilibrium can be achieved rapidly. The solid state of polymers is a *quasi-equilibrium* state because its properties often depend on prior history and processing. The first three methods are preferred because they are easy to use and should give good answers for most drying problems.

Henry's Law. Henry's law is appropriate for sparingly soluble solutes in liquid and polymer systems. The equation is a good model for nonspecific absorption of a gas into a liquid, or liquid-like, phase. It often is used to describe the equilibrium solubility of noncondensible gases—such as oxygen, carbon dioxide, and nitrogen—in polymer films. At very low solute partial pressures, Henry's law underpredicts the concentration of solute in the polymer phase. This is due to a second mechanism for solubility, which will be discussed in the next section.

Dual-Mode Sorption. Sorption of supercritical gases—such as helium, nitrogen, oxygen, carbon dioxide, and methane—into glassy polymers at room temperature and higher, has been studied by many researchers. The dual-mode sorption model proposed by Barrer, Barrie, and Slater (1958) has been very successful in describing these systems. The model assumes two modes of solubility in the polymer: nonspecific absorption and specifically adsorbed species that are in dynamic equilibrium in the medium. Equation 10.1 was proposed to model Langmuir-type behavior for gas sorption in glassy polymers at low gas activities,

$$C = C_a + C_1 = H_p p_i + \frac{C_H^1 b}{1 + b p_i} \qquad 10.1$$

where C is the total concentration of the sorbed species in the polymer; C_a and C_1 are the solubilities due to absorption (Henry's law) and adsorption (Langmuir); H_p is the Henry's law constant; b is the hole affinity constant; C_H^1 is the hole saturation constant; and p_i is the partial pressure of the solute. This model has been used to describe the solubility of gases at low activities in glassy polymers and in glassy, polar polymers as well (Uragami et al., 1986; Chern et al., 1983; Koros and Sanders, 1985).

Models for Polar Systems at High Solute Activities. The objective of the equilibrium model is to describe water solubility in the polymer from higher water activities ($a_1 = 0.65$ at typical storage conditions) to very low levels (the target level, which may be 2% or less of saturation). Equation 10.1 does not describe the data well over the activity range, $0 < a_1 < 1$, and usually underpredicts the actual solubility values at high activities. Table 10.2 lists some recent data sets for polar systems at high solute activities and models describing the absorption factor (C_a in Eq. 10.1). Substituting the Flory–Huggins equation for Henry's law in Equation 10.1 gives a model that reduces to the Henry's law model at low solvent activities and gives improved predictions at high solvent activities. It is convenient to write the modified model in terms of volume fractions and solute activity,

$$\phi_1 = \phi_1^{FH} + \phi_1^L \qquad 10.2$$

TABLE 10.2 Models for Polar Systems at High Solute Activities

System	Solution Factor	Reference
H_2O/polyacrylonitrile	$k_d \exp(\sigma c)p$	Mauze and Stern, 1982
Vinyl chloride/poly(vinyl chloride)	Flory–Huggins	Berens, 1975
	$k_d \exp(\sigma c)p$	Mauze and Stern, 1984
	$\sigma = 2(1 + \chi)A$	
H_2O/Kapton	$k_d p$	Yang et al., 1985
N_2O/nylon 6	Flory–Huggins	Sfirakis and Rogers, 1980
H_2O/epoxy	$k_d p$	Barrer, 1984
H_2O/epoxy	$k_d = \exp(1 + \chi)$	Apicella et al., 1984

where ϕ_1^{FH} refers to the Flory–Huggins contribution and ϕ_1^L refers to the Langmuir contribution. The equation is

$$\phi_1 = \phi_1^{FH}(a_1, \chi) + \frac{K a_1}{1 + B a_1} \qquad 10.3$$

Equation 10.3 is solved by trial and error or numerical methods because the Flory–Huggins equation is nonlinear.

Figure 10.4 shows a comparison of Equations 10.1 and 10.3, with solubility data for water sorbed in an amorphous nylon made from hexamethylenediamine and a mixture of isophthalic and terephthalic acid. The Langmuir sorption factor is important at water activities less than 0.2. At higher water activities, the Langmuir sites are saturated and contribute little to sorption of more water. The Flory–Huggins factor fits smoothly through the data at high water activities. Equation 10.3 predicts the onset of solute clustering for some values of the coefficients, while Equation 10.1 does not. Clustering of solvent molecules leads to a larger effective size for the solvent and lower effective diffusion coefficients. In general, the Langmuir component is thought to become small at temperatures near T_g of the *system* (the polymer plus the solute). The Henry's law coefficient follows the same trend. The interaction parameter varies inversely with temperature.

10.1.2 Diffusion of Small Molecules in Glassy Polymers

This section reviews simple models for diffusion in glassy polymers, with some references to specific phenomena that might occur in polymer-water systems. The drying process in dryers for polymer particles is unsteady-state, with the polymer continuously losing moisture as it moves through the drying system. In contrast to melt devolatilization discussed in Chapter 9, there is no mixing in the polymer phase since it is a glassy solid. The problem can be analyzed as a diffusion problem with no

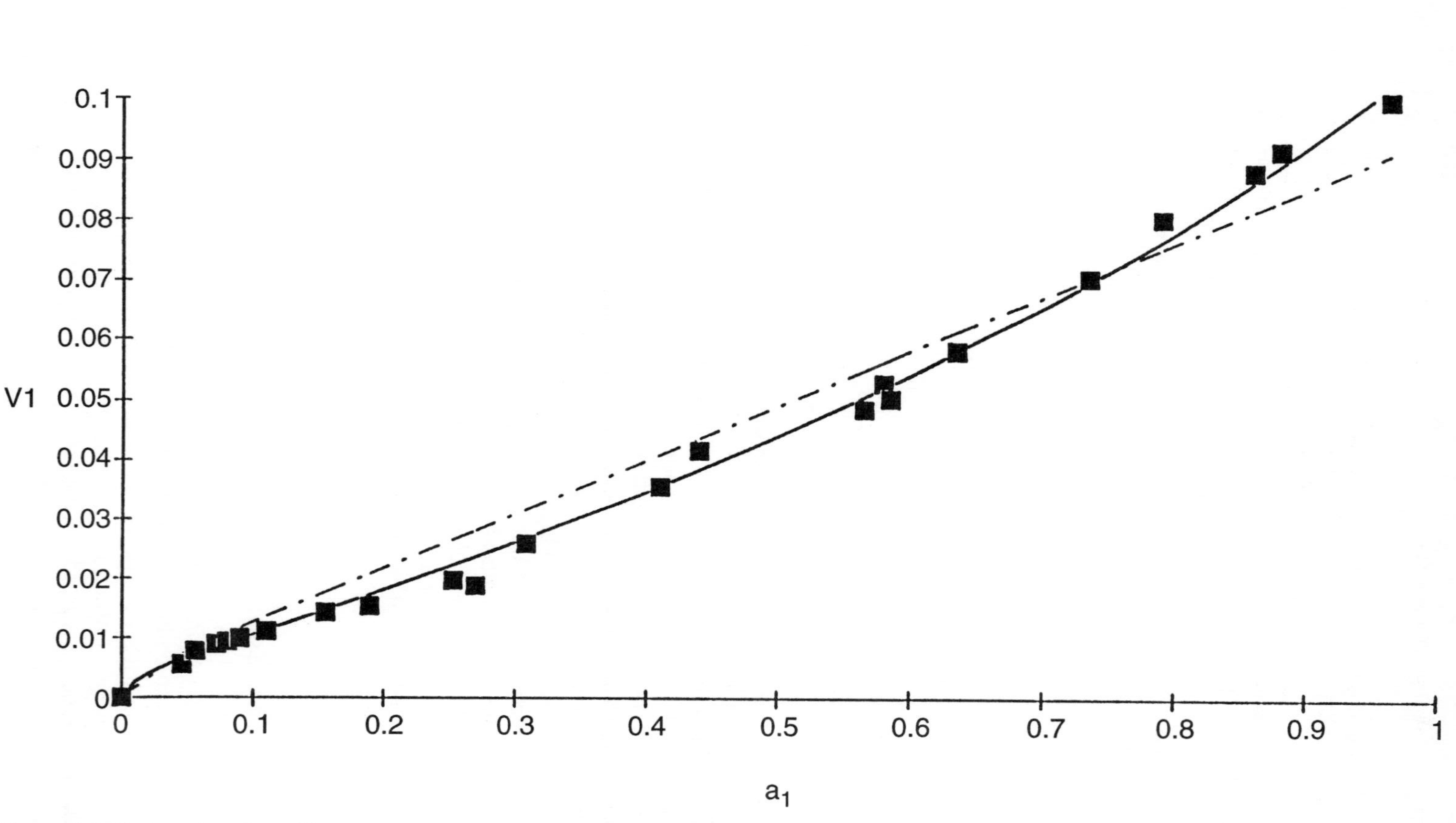

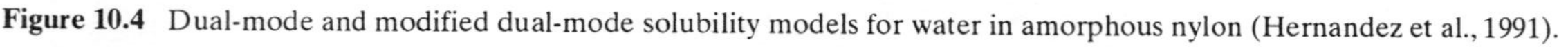

Figure 10.4 Dual-mode and modified dual-mode solubility models for water in amorphous nylon (Hernandez et al., 1991).

convection. The methods developed below also will apply to polymer heat transfer problems with no convection.

The unsteady-state material balance for a solid particle is Equation 9.24. For dilute systems, the diffusivity often is constant and does not depend on concentration. The diffusion process has been modeled in a variety of ways. Diffusion of the solute occurs when it "jumps" from its location in the polymer matrix to a nearby open site. This concept is consistent with the dual-mode sorption model, which considers specific sorption sites and nonspecific solubility. However, only one diffusion coefficient is used to characterize diffusion from both types of solute molecules. Therefore, the activation energy for the "jumping" process is independent of the method of sorption and

$$D = D_o \exp(-E_d/RT) \qquad 10.4$$

where D is the diffusivity, D_o is a reference diffusivity, E_d is the activation energy for diffusion, R is the ideal gas constant, and T is the absolute temperature. For simple liquid systems, the activation energy for diffusion is in the range of 5–10 kcal/mol or 20–42 kJ/mol.

The diffusivity of a molecule should depend on its effective area for diffusion or on a representative cross section of the molecule. The diffusion area should correlate with molar volume or with the square root of the molecular weight (a result of empirical correlations of diffusion coefficients in liquids). Figure 10.5 shows the relation between the gas diffusion volume and the diffusion coefficients for some of the solutes diffusing in PVC. There is a reasonable correlation between these variables. If enough diffusion data are available for different compounds in one glassy polymer, this correlation can be used to estimate the diffusion coefficient of a new solute.

10.1.3 Unsteady-state Diffusion in Particles and Pellets

Boundary and Initial Conditions. Thermoplastic materials often are made into pellets or chips for transportation from the producer to the fabricator. The chips may be modeled as cubes, and the pellets may be modeled as short cylinders or spheres. The solid particles probably have uniform concentrations of water in them as they leave the production unit. Water is absorbed during transportation and storage and probably is nonuniformly distributed through the particles (see Fig. 10.6). The assumption that water is uniformly distributed through the particles at the start of the drying process is conservative and should not lead to much deviation between calculated moisture contents and the measured values.

Solutions to Equation 9.24 can be obtained from Crank (1975) or by finite difference methods. The heat transfer solutions of Carslaw and Jaeger (1959) may also be used. The finite difference method should be considered if the diffusivity is expected to vary with solute concentration and its variation is complex. These

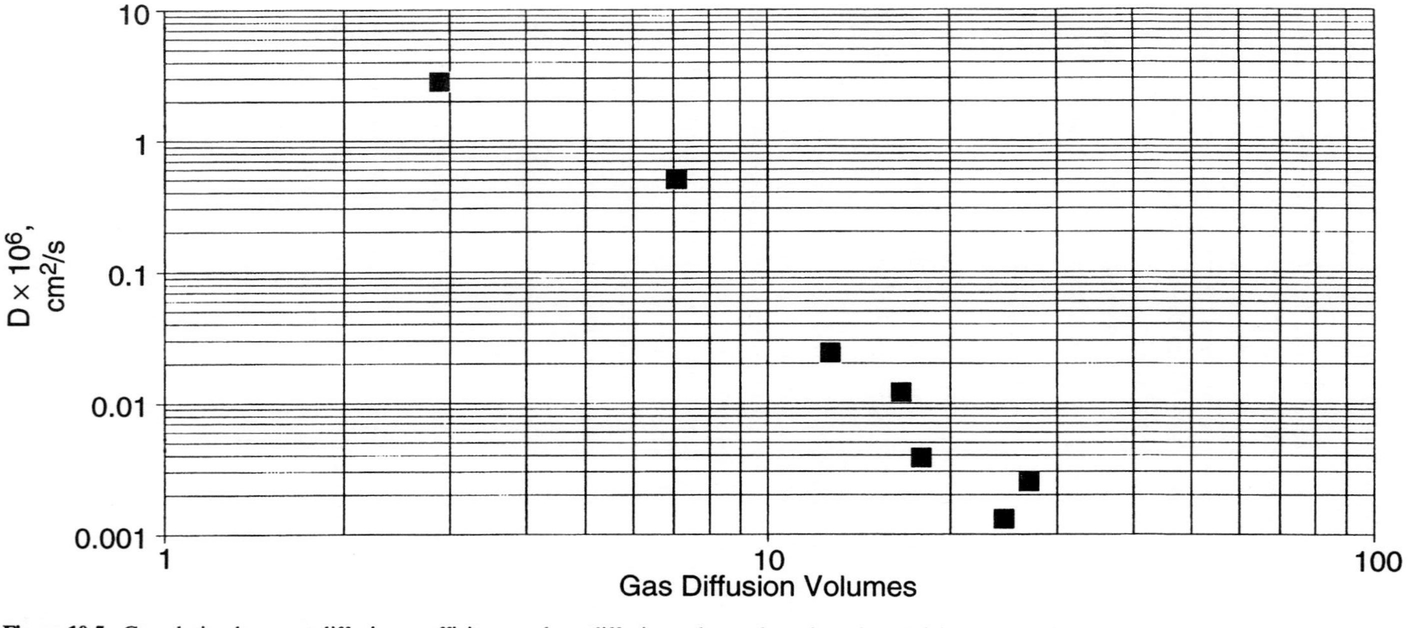

Figure 10.5 Correlation between diffusion coefficients and gas diffusion volumes for solutes in poly(vinyl chloride) at 25°C.

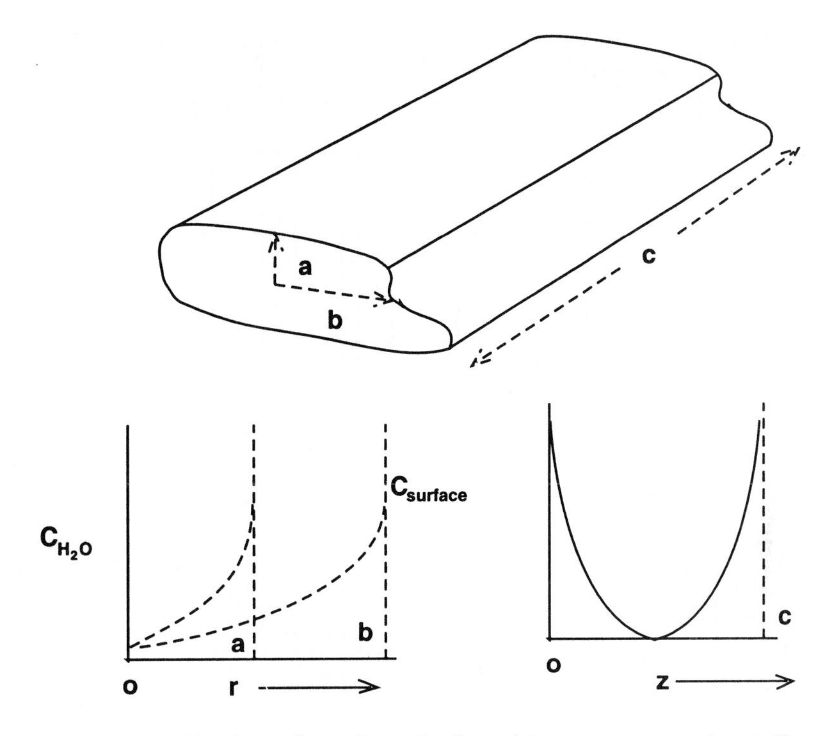

Figure 10.6 Sketch of a chopped strand pellet and the water concentration profile.

solutions give both the concentration profile of water through the solid and the total amount remaining as a function of time.

Most polymer drying problems only require calculation of the average concentration remaining in the solid as a function of drying time. Newman (1931) developed solutions calculating the average concentration in solid spheres, cylinders, and slabs as a function of diffusion time and one characteristic dimension. These solutions have the feature that they can be combined to solve multidimensional diffusion problems. Figure 10.7 shows E, the fraction unremoved, for one-dimensional diffusion problems.

For slabs with two pairs of sealed ends:

$$E = \frac{C_2 - C_o}{C_1 - C_o} = f\left(\frac{D\theta}{a^2}\right) = E_a \qquad 10.5a$$

where C_o is the final concentration, C_1 is the initial concentration, and C_2 is the average concentration at time equal to θ.

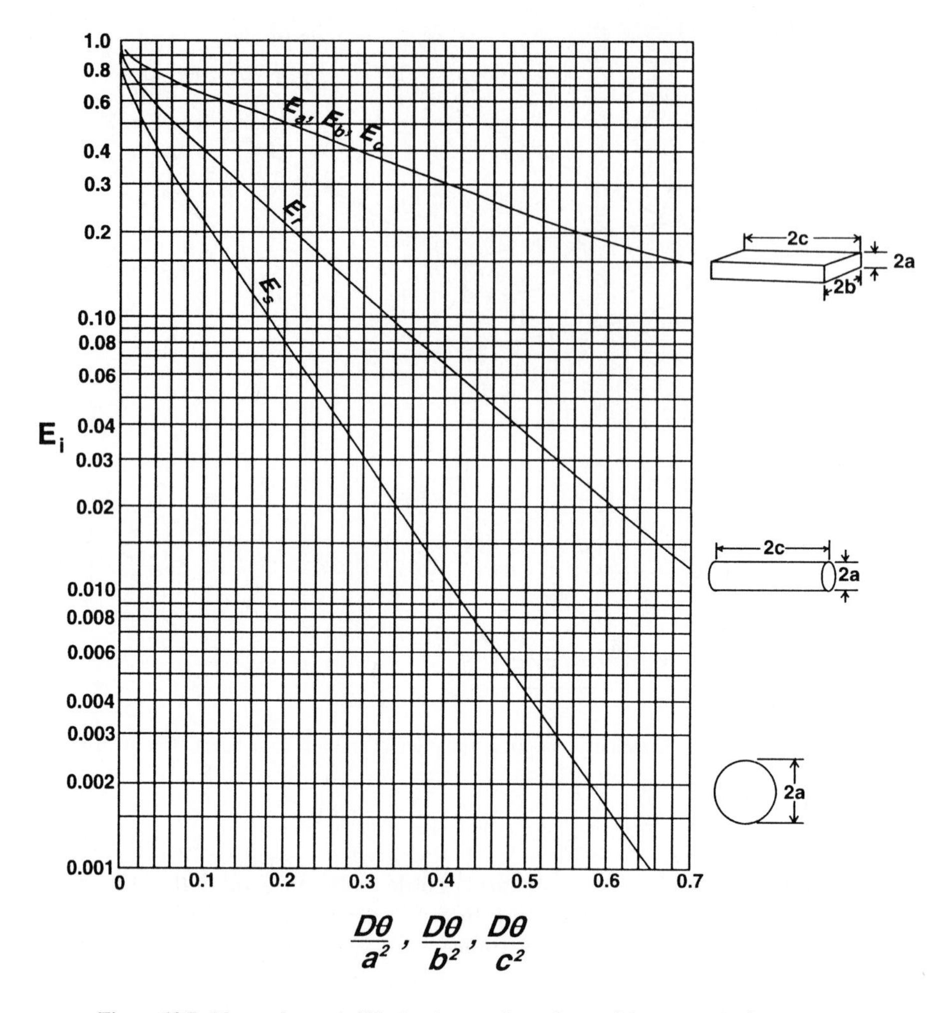

Figure 10.7 Unsteady-state diffusion into various shapes (Newman, 1931).

For cylinders with sealed circular ends:

$$E = f'\left(\frac{D\theta}{a^2}\right) = E_r \qquad 10.5b$$

For spheres:

$$E = f''\left(\frac{D\theta}{a^2}\right) = E_s \qquad 10.5c$$

For diffusion in a rectangular bar with sealed ends:

$$E = f\left(\frac{D\theta}{a^2}\right)f\left(\frac{D\theta}{b^2}\right) = E_a E_b \quad 10.5d$$

For diffusion in a rectangular parallelpiped:

$$E = f\left(\frac{D\theta}{a^2}\right)f\left(\frac{D\theta}{b^2}\right)f\left(\frac{D\theta}{c^2}\right) = E_a E_b E_c \quad 10.5e$$

For diffusion in a cylinder:

$$E = f'\left(\frac{D\theta}{a^2}\right)f\left(\frac{D\theta}{c^2}\right) = E_r E_c \quad 10.5f$$

If diffusion takes place from one face rather than two, the same solution can be applied by using 2a as the characteristic distance. This moves the symmetry boundary condition to the sealed face.

Example 10.1 Calculation of Drying Times and Average Solute Concentration in Pellets

Methods for calculating drying times and average solute concentrations can be demonstrated by working a case study problem. A cylindrical nylon pellet with a radius of 1 mm and a length of 4 mm needs to be dried. The initial concentration of water is 2wt%, and the target moisture content is 0.45vol%. What will be the drying time for (a) perfectly dry air at 25°C, and (b) saturated air at 25°C? The normal drying target for nylon would be 0.1wt% water or less, and heated air would be used for drying. However, these conditions will illustrate several important phenomena associated with polymer drying.

The first step in the solution is to prepare an E versus θ curve for the solid pellet. The losses for each type of face will be multiplied together to obtain the total water fraction unremoved. Table 10.3 shows values of the moduli, the fractions unremoved for each surface, and the total fraction unremoved, E, for drying times from 1000 to 70,000 seconds. Even for a solid with an aspect ratio of 2:1, the sealed cylinder solution is a good approximation to the problem. The fraction unremoved, E, is a master curve for this solid shape and can be used to solve all the parts of this problem.

a. The air is assumed to be perfectly dry throughout the bed. The concentration of water just inside the polymer surface is zero (rapid equilibrium with the air stream is assumed). The value of E is

(continued)

$$E = \frac{C_2 - C_o}{C_1 - C_o} = \frac{0.45 - 0}{2.0 - 0} = 0.23$$

From Table 10.3, the drying time required is 16,000 seconds, or 4.4 hours.

b. The other drying conditions require calculation of the water concentration at the polymer surface. The specifications of the air streams involved the dew point temperature. Either water vapor pressure data or an Antoine equation can be used to determine the water vapor pressure in the drying air. The water activity in the vapor phase is determined by dividing this vapor pressure by the saturated vapor pressure of water at 25°C (0.459 psia). The following table shows the activity calculations:

Dew Point of Drying Air	Water Vapor Pressure, psia	a_1
− 40°F	0.0019	0.00414
− 10°F	0.0103	0.0224
77°F	0.459	1.00

The modified dual-mode sorption equation will be used to find the equilibrium volume fraction of water at each condition. The coefficients for water in nylon at 25°C will be taken as $\chi = 1.7$, $K = 0.395$, and $B = 95.2$. Figure 10.8 shows the curve of water volume fraction in the polymer as a function of water activity. This curve covers the low water activity range only. The target moisture content of 0.45vol% cannot be accomplished with saturated air at 25°C.

As expected, air with lower humidity (lower dew point temperatures) gives shorter drying times. None of the inlet air dew points meet the initial design requirements of 0.1wt% (.0011 volume fraction) of water in the polymer. There are two ways to correct this problem. First, the nylon could be stored under dry air so that it would not sorb 2wt% moisture. An alternative is to raise the temperature of the drying air. This has the effect of lowering the water activity at the surface of the polymer below the target moisture content.

Different polymers have different drying goals. Table 10.4 lists some polymers and their target moisture contents.

10.2 STORAGE AND FLOW OF PARTICULATE SOLIDS

Particulate solids need to be handled during transport, storage, blending, feeding, compounding, and processing operations. A classification of solids by typical particle dimension is given in Table 10.5. Pellets made by hot-die face cutting are larger than 3 mm and would be classified as broken solids. Suspension resins would be classified as granular solids. Dried emulsion lattices that do not agglomerate would be classified

TABLE 10.3 Fraction of Solute Unremoved for Nylon Pellet Drying

Time, s	f′	f	E_r	E_c	E
0	0	0	1	1	1.000
1000	0.01	0.0025	0.8	0.95	.760
10000	0.1	0.025	0.4	0.82	.328
20000	0.2	0.05	0.22	0.75	.165
30000	0.3	0.075	0.125	0.7	.088
40000	0.4	0.1	0.067	0.65	.044
50000	0.5	0.125	0.038	0.6	.023
60000	0.6	0.15	0.021	0.56	.012
70000	0.7	0.175	0.0125	0.53	.007

as ultrafine solids. Several phenomena specific to particulate solids are discussed in this section: the friction coefficient, bulk density, pressure distribution in storage silos, and particle size segregation.

Friction Coefficient. Particulate solids have liquid-like and solid-like behavior. Like liquids, they take the shape of the container, exert pressure on container walls, and flow through openings. Like solids, they can support shear stresses and will form piles; they have cohesive strength, they can have nonisotropic stress distributions when they are under uniform loads; and they act like solids with respect to friction. The shearing stress against a solid surface is proportional to the normal load rather

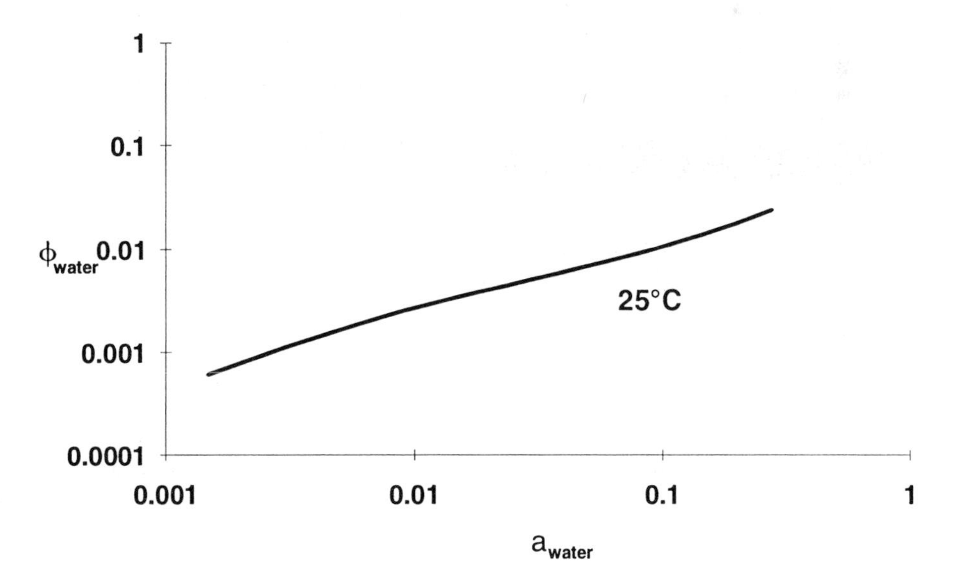

Figure 10.8 Water volume fraction versus activity in amorphous nylon at 25°C.

TABLE 10.4 Drying Targets for Some Polymers

Polymer	Drying Goal	Drying Temperature, °F	Processing Temperature, °F	Comments
ABS	< .10wt% platable grade	200°F	250°F	
Nylon	< 0.2wt%	< 250°F	480–520°F	oxidizes at drying temperature unless inert gas is used
Polyester PET	< .005wt%			viscosity degrades with high moisture
PC	< .1wt%		500–600°F	

TABLE 10.5 Particle Size Classification*

Classification	Size
Broken solids	D > 3 mm
Granular solids	0.1 < D < 3.0 mm
Powders	D < 100 μm
Superfine	1 < D < 10 μm
Ultrafine	0.1 < D < 1 μm

*D = typical particle dimension

SOURCE: R.L. Brown and J.C. Reynolds, *Principles of Powder Mechanics*, Pergamon Press, Oxford, England, 1966.

than the shearing rate (liquid behavior). Equation 10.6 shows that the shear stress is less than or equal to the normal force to the surface times a friction coefficient,

$$\tau \leq f\sigma \qquad 10.6$$

τ is the shear stress, f is the static friction coefficient, and σ is the normal stress against the surface. A static particulate solid can have a range of possible bulk densities and equilibrium states. When flow begins, the dynamic friction coefficient can be defined as the ratio of shear stress to normal stress (Amonton's law).

Two factors are considered to be part of sliding friction. One factor is adhesions between the two materials. Adhesions occur at a few contact points per unit area and have to be broken before the materials will slide past each other. Another factor is the plowing or grooving of one surface by high points of the other. Coefficients of sliding friction usually are lower than coefficients of static friction because sliding sets in motion wear patterns and deformations. Models for the sliding coefficient often vary with the normal force,

$$f = c\,\sigma^{a-1} \qquad 10.7$$

where c is a constant and a varies between one (pure plastic yield at the contact points) to ⅔ (pure elastic deformation at the contact points). Except for $a = 1$, the friction coefficient decreases with increasing normal force. In general, the friction coefficient varies with temperature, sliding speed, and normal force. Figure 10.9 shows some friction coefficients for polymer pellets as a function of normal force. As the normal force increases, the friction coefficient decreases. Table 10.6 compares static and sliding friction coefficients for several polymers. When the plowing mechanism is not important (polymer and solid surfaces are smooth), the key variables for determining the friction coefficient are the cohesion of polymer to itself versus its adhesion to the solid surface. When the adhesion forces are weaker than the cohesion forces, very low friction coefficients are possible ($f < 0.10$). Two polymers of Table 10.6 have such low coefficients, PTFE and HDPE.

Bulk Density. As discussed in Section 9.6, there are several packing morphologies for uniform spheres that can lead to a variety of bulk densities (see Table 9.11). Consider the problem of filling the silo with particles. Because of friction between the particles and the lack of forces besides gravity after the particles have taken an initial position, the probability that the solids will achieve a close packing is very low. The average bulk density in the silo might vary from filling to filling, and there could be wider variations in the local bulk densities at different places in the silo. The density of particulate solids (bulk density) is less than their "solid" density and varies with particle size and storage conditions. Bulk density is easy to measure

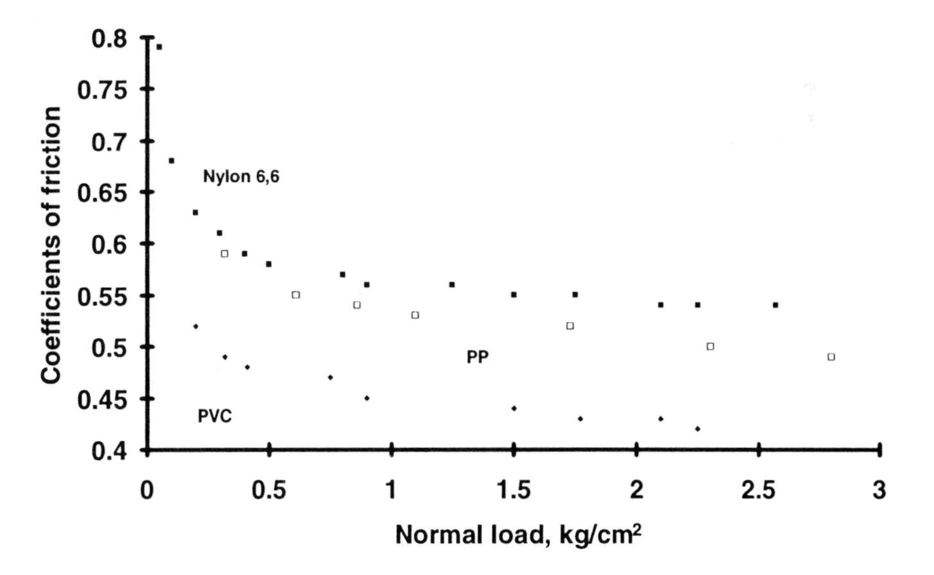

Figure 10.9 Coefficients of friction of various polymers in pellet form versus normal load at 30°C and 1 cm/s sliding speed. The reported coefficients of frictions are "rubbed-in" values (Schneider, 1969).

TABLE 10.6 Static and Sliding Friction Coefficients of Polymers on Glass

Polymer	f_{static}	$f_{sliding}$
FEP	0.22	0.20
KelF	0.28	0.28
PTFE	0.20	0.06
HDPE	0.20	0.08
LDPE	0.30	0.30
PVC	0.40	0.40

SOURCE: B.J. Briscoe, C.M. Pooley, and D. Tabor, in *Advances in Polymer Friction and Wear*, Vol. 5A, (L.H. Lee, Ed.), Plenum Press, NY, 1975.

by determining the weight of solid that can be held in a specific volume. Graduated cylinders often are used for powders because the particle size is much smaller than the diameter of the vessel. Packing of solids near walls is different from that in the bulk. As a rule of thumb, the vessel diameter should be at least ten times the particle diameter to ensure that wall effects do not dominate the measurement.

Bulk density is sensitive to a number of factors, including particle charging, moisture, compaction pressure, filling method, and mechanical motion of the container. As shown in Table 9.11, the packing factors of close-packed monodisperse spheres vary from 0.52 to 0.74. Polyethylene powder particles have bulk densities between 320 and 450 kg/m^3 (20–28 lb/ft^3). The density of solid polyethylene is about 920 kg/m^3 (59 lb/ft^3). A simple estimate for the bulk density based on the solid density and the above packing factors would be 480 to 680 kg/m3. The bulk densities of the powder are below these estimates, suggesting that there are voids in the packing structure. For some samples, the particle size distribution is such that the small particles fit into voids left between large particles, leading to higher bulk densities than those attained for the random packing of uniform spheres. Charged agglomerates lead to lower bulk densities. An increase in pressure can cause an increase in bulk density (Fig. 10.10). The differences in bulk densities for particulate solids makes the design and operation of storage facilities a challenge.

Pressure Distribution in Storage Silos. Figure 10.11 shows the packing of uniform spheres and irregular particles near an inclined wall. This sketch displays the solid structure that might exist near the wall in a storage silo, near the barrel surface of an extruder, or in a V-mixer. The gravitational force is decomposed into a force normal to the wall and one parallel to the wall. This balance is most important for the storage silo case, as gravity is the major force. In the other two cases, the equipment will impose forces on the solids as well, which may be much larger than gravity. The following discussion focuses on the storage silo case in which gravity causes force normal to the wall.

In a liquid container, the static pressure, P, is isotropic and is proportional to the height of the tank, h, and the density of the liquid, ρ,

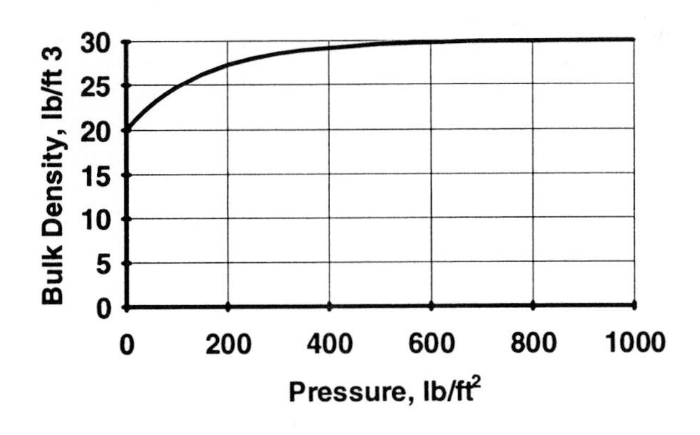

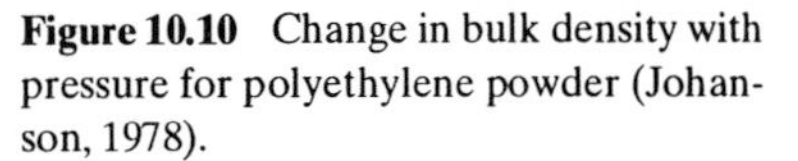
Figure 10.10 Change in bulk density with pressure for polyethylene powder (Johanson, 1978).

$$P = \rho g h \tag{10.8}$$

where g is the gravitational constant. The pressure at the bottom of particulate storage silos is not directly proportional to the vessel height, since the walls support some of the particles by friction. In addition, pressure is not isotropic but may vary between the vertical and horizontal directions. A force balance over a differential element gives a result that relates to experience. Figure 10.12 shows the balance. The weight of the differential element is supported by the pressure difference above and below the element and the frictional force provide by the wall:

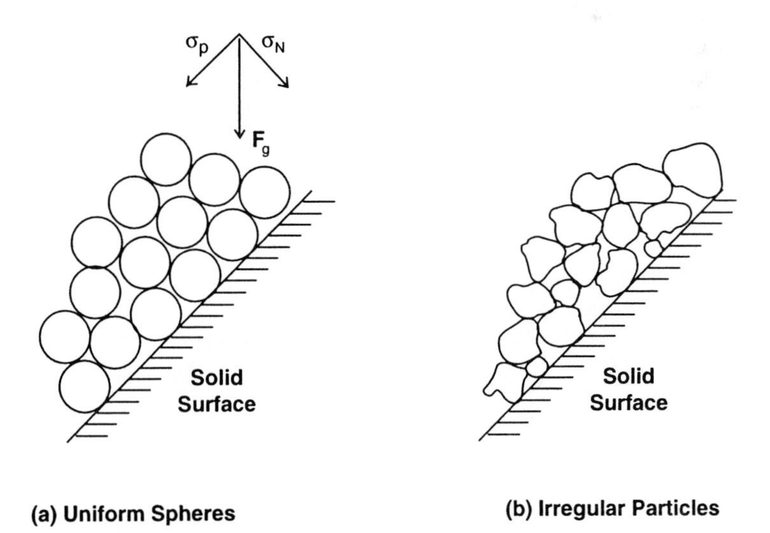

Figure 10.11 Packing of **(a)** uniform spheres and **(b)** irregular particles against an inclined wall. Gravitational force, F_g has components parallel (σ_p) and normal (σ_N) to the wall.

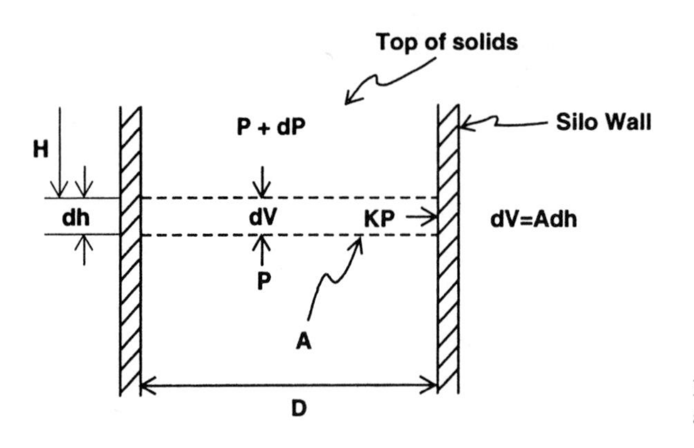

Figure 10.12 Differential force balance in a particulate storage silo.

$$\text{Weight of } dV = \Delta P \cdot A + \text{frictionalforce} \cdot dA$$

$$\rho\, g\, A\, dh = [P - (P + dP)]\ \pi\, D^2/4 + f\, K\, P\, 2\pi\, D\, dh \tag{10.9}$$

where f is the static friction coefficient, D is the vessel diameter, and K is the coefficient relating vertical pressure to horizontal pressure. Equation 10.9 can be integrated to give the Janssen equation.

$$P = \frac{\rho g D}{4 f g_c K}\left[1 - \exp\left(\frac{-4 fKh}{D}\right)\right] \tag{10.10}$$

As the distance down the silo increases, the pressure reaches a limiting value. Figure 10.13 shows such a limit for polystyrene cubes in a hopper.

The actual pressure distribution in a conical bottom hopper is complex (Fig. 10.14). The hopper is assumed to be filled with polyethylene powder and the pressure

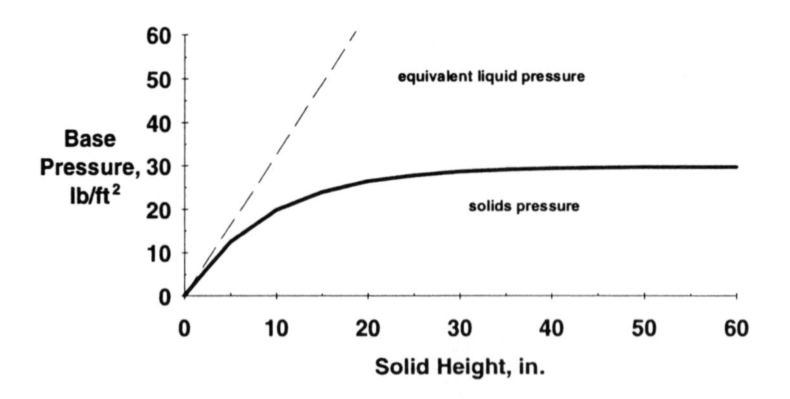

Figure 10.13 Base pressure in a 10 in. diameter cylindrical hopper filed with ⅛ in. PS cubes with $K = 0.521, f=0.523$, and $\rho=39$ lb/ft^3 (Rudd, 1954).

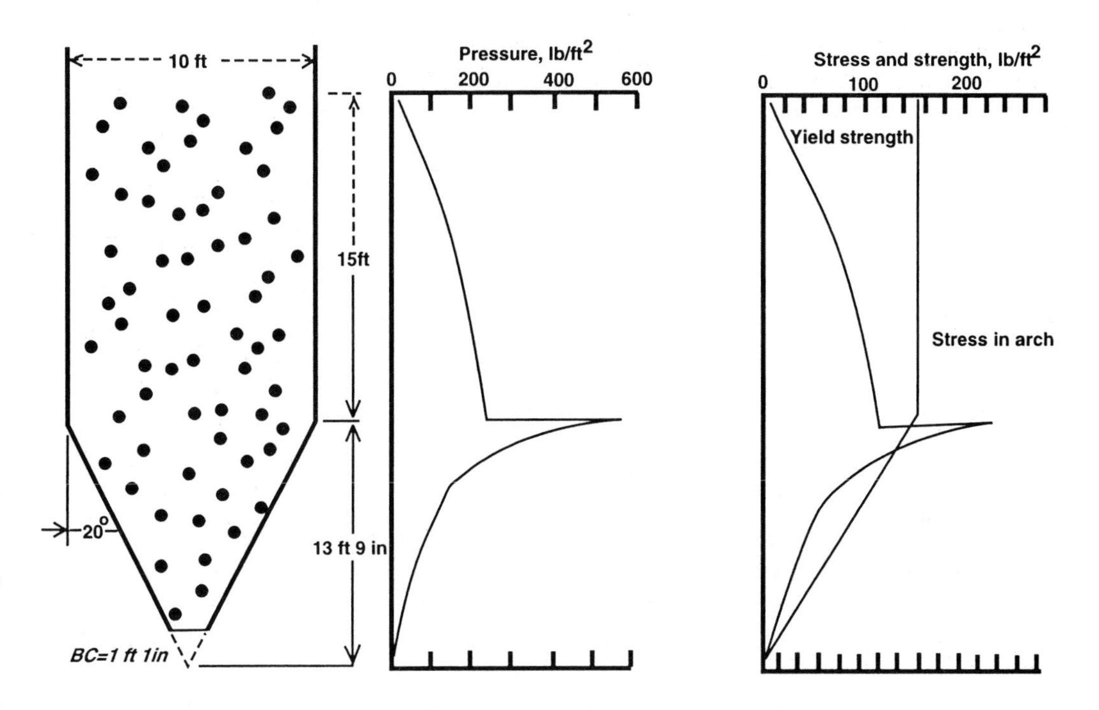

Figure 10.14 Pressure and strength distributions in a hopper (Johanson, 1978).

and yield strength of the material were calculated. As expected from Equation 10.10, the pressure increases from the top of the hopper to the start of the conical section. Below this point, the conical walls support some of the vertical load directly, and the pressure of the solids is reduced. At the bottom of the hopper, the solids pressure is small. The yield strength for flow varies down the hopper as well. An arch of particles can form across the hopper if the stress in the arch is lower than the yield strength. For this case, an arch will form very close to the bottom where the yield strength and arch stress curves cross. This is a good location for the circular outlet. When the outlet is properly located, an arch will be unstable everywhere in the hopper with the exception of the transition between the cylinder and conical sections.

Particle Size Segregation. In most particulate processes, the particle size distribution is controlled within a narrow range. However, when large differences in particle size occur, segregation can occur by particle size during transport and storage operations. There are several particle segregation mechanisms.

Very fine particles can sift through the voids between large particles and will settle to the bottom of a storage vessel. The sifting process requires some mechanical motion of the material, either equipment vibration or the sliding that occurs during building a pile. Particles that self-agglomerate or hold a charge have a low tendency to sift. The fines tend to segregate in the middle at the top of the bin during pile charging. As further shear movement occurs in the hopper (due to discharge or other motion), the fines would continue to sift down through the hopper and remix with coarse material moving out from the conical wall.

The channels set up during particle segregation can lead to variable particle size distributions during discharge. Figure 10.15 shows some typical effects. When a segregated hopper charge is charged faster than the discharge rate, a high fraction of fines will be discharged. When the discharge rate is faster, a higher fraction of coarse particles is discharged. When the discharge and charge rates are equal, the fraction of fines in the discharge should be the same as the feed. These effects can lead to a variety of operating problems with equipment downstream of the hoppers.

10.3 OVERVIEW OF POLYMER PROCESSING

10.3.1 Effects of Processing on Polymer Morphology

Processing and environmental factors influence polymer morphology, which in turn affects the end-use properties and performance of parts. Temperature, pressure, shear rate, and annealing affect polymer orientation and phase equilibria. Each of these factors should be controlled during processing in order to make reproducible prod-

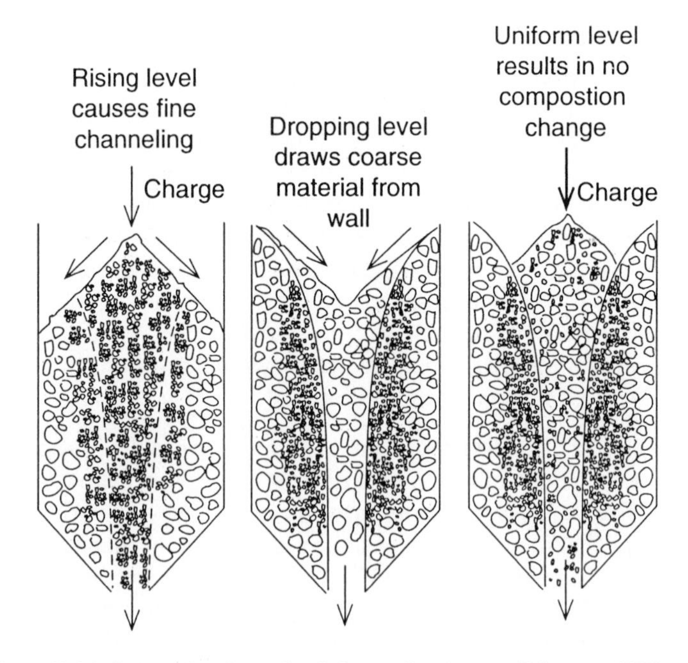

Figure 10.15 Segregation due to level changes in a hopper (Johanson, 1978).

ucts. The effects of processing on morphology are more apparent with crystalline materials than with amorphous polymers. However, each type is affected in similar ways. Temperature changes, and the rate of the temperature changes, affect the growth of crystals in the solidifying polymer, chain orientation, stress relaxation, phase separation, and free volume in the polymer. Pressure also affects orientation in the melt, but less information about its general effects is available. Variation in time-temperature history from one polymer sample to the next can result in measurable differences in physical properties and environmental resistance of the finished part.

Crystal branching, or dendrite formation, may occur in polymers crystallized from the melt. Local concentration gradients can occur in the melt due to the crystallization of long chain molecules first. If the polymer is crystallized slowly, polymer chains of different molecular weight may segregate. Spherulites, organized crystals with folded polymer chains oriented perpendicular to the radial direction, have been observed in polyethylene solids. The spherulite size may range from 50 to 1000 microns.

Temperature history affects the crystallization rate. Annealing of spherulites tends to increase the size and number of lamellae, increases the T_m of the sample, and tends to cause the exclusion of low molecular weight material. Nucleating agents may be added to the melt to control the crystal size and make the material more predictable batch to batch. Crystalline polymers are processed above their melting temperature. This permits the crystallite to melt, lowering the viscosity. Typical processing temperatures of crystalline and amorphous polymers are given in Table 1 of Appendix B.

Pressure also affects crystallization rates. Melting point elevation occurs as pressure is increased, since more supercooling is needed to obtain nucleation. Molecular mobility is hindered at high pressures as the free volume of the material is decreased, so that T_g is often increased.

Shear stress can induce crystallization. The change in free energy on crystallization is zero, which implies that the temperature of melting equals the ratio of the enthalpy of crystallization to the entropy of crystallization. Entropy will be greatly reduced when the chains are oriented, but the enthalpy should show little change. Therefore, the local melting temperature of the crystallites should increase with shear stress. Crystals grow perpendicular to the force. Figure 10.16 shows how crystal structure can change from 0% elongation (spherulites) to over 300% elongation.

Shear-induced crystallization is a time-dependent process, since it takes a finite amount of time for the molecules to change orientation. There is often an induction time before crystal growth will occur, possibly related to nucleation. The induction time decreases with shear rate. At low shear rates, the induction time is relatively constant since the Deborah number for the process is much less than one and the chains can relax. Shish-kebab, lamellar, and extended chain needles are observed when flow fields with high strain rates shear polymer melts and solutions. Figure 10.16d shows the orientation of polymer chains in the shish-kebab structure.

Torsional, or rotating, flows also cause orientation of crystallites. Figure 10.17 shows a sketch of a torsional flow and the orientation of the lamellae. The flow is rotating about the A axis, inducing orientation in the C direction. The flow field can

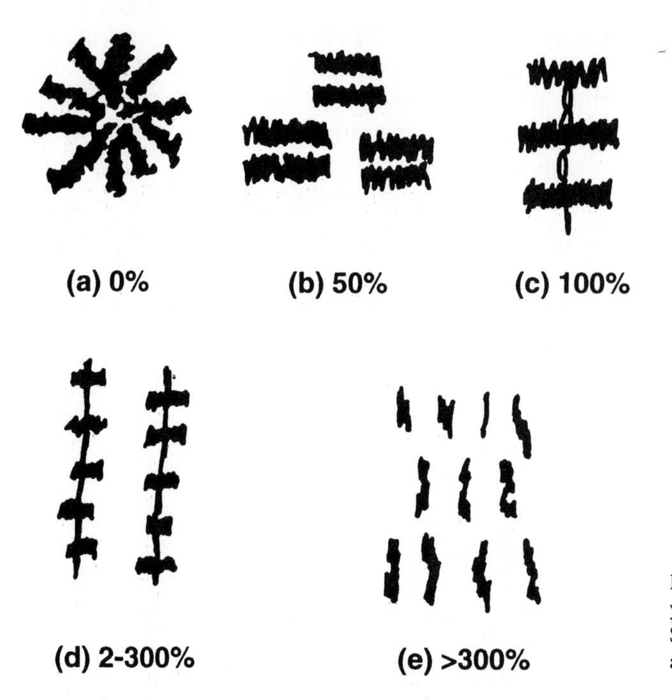

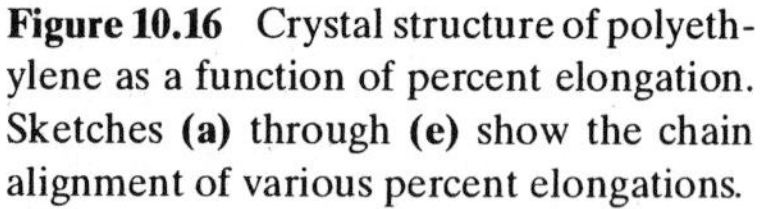
Figure 10.16 Crystal structure of polyethylene as a function of percent elongation. Sketches **(a)** through **(e)** show the chain alignment of various percent elongations.

increase the thickness of the lamellae formed compared to quiescent crystallization. Presumably, the flow increases the motions of chains against their neighbors, increasing the possibility of crystallization.

Capillary flows are characteristic of fiber drawing, wire and cable coating, and profile extrusions. These flows can induce shish-kebab structures near the wall of the flow. The lamellar disks are tapered at the ends as a result of cooling during crystallization. Some very high molecular weight material is needed to initiate this structure.

Liquid crystalline polymer can form highly oriented structures when extruded. Figure 10.18 shows a schematic diagram of a sample of hydroxypropylcellulose

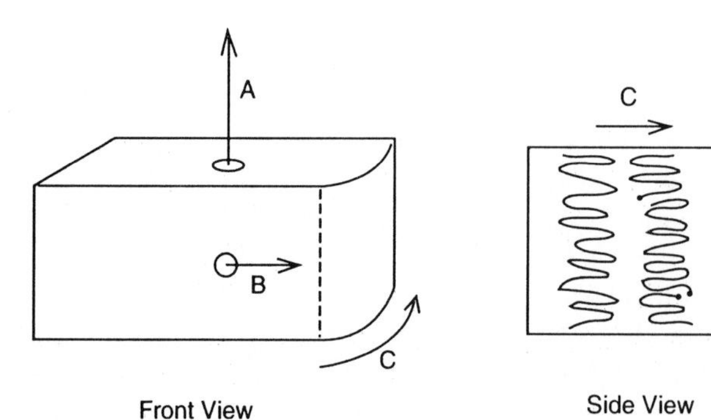

Figure 10.17 Lamellar organization following crystallization under torsional flow conditions (Woodward, 1990).

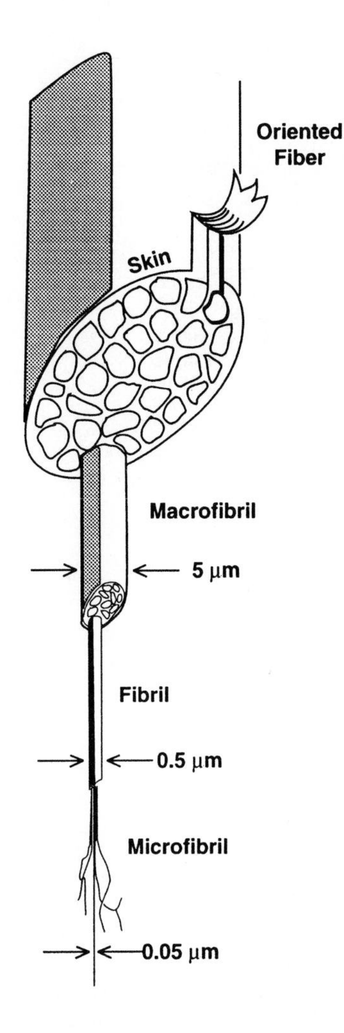

Figure 10.18 Structural model for oriented fibers of liquid crystalline polymers (Sawyer and Jaffe, 1986).

extruded from a capillary rheometer. The large fiber has a thin skin with little structure. The interior is highly oriented with a hierarchy of morphologies. The fiber is composed of macrofibrils of about 5 microns in diameter. These in turn are composed of fibrils of .2 to .5 microns in diameter. The fibrils are made up of individual microfibrils of 0.05 microns in diameter.

Chain rearrangement can be induced with annealing and cold-forming operations at temperatures between T_g and T_m. Thick section extrudates and moldings often have oriented exteriors and unoriented interiors. This morphology is due to two factors: low shear in the center leading to low orientation, and cooler temperatures at the exterior after flow causing the structure to be retained. Figure 10.19 shows cross sections of thick extrudates and moldings of liquid crystal polymers with this morphology.

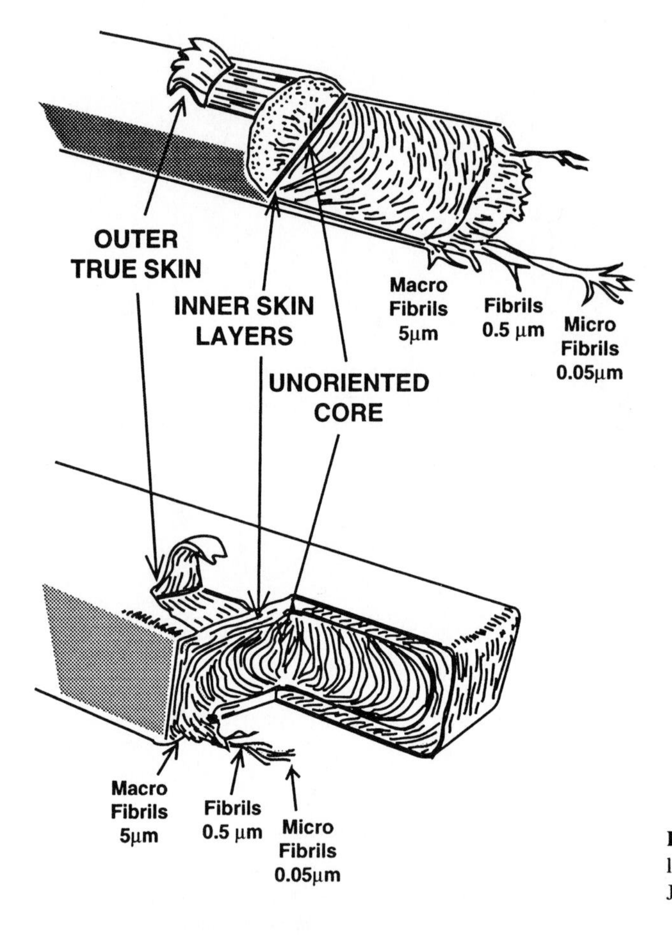

Figure 10.19 Extrusion and molding of liquid crystalline polymers (Sawyer and Jaffe, 1986).

Polymer samples exhibit necking during cold drawing much as metallic samples do. Films are often oriented uniaxially or biaxially to make them anisotropic and improve their properties in the machine directions. Molecular models of necking in polymers usually include the following steps: lamellae slip past each other, lamellae slip-tilt when amorphous ties become completely extended, lamellae break-up occurs through chain pulling and unfolding, and lamellar fragments orient in the flow direction.

Amorphous polymers are typically processed at 40°C to 50°C above their T_g, where interchain slippage does not require high stress. Structural voids are present below T_g and are mobile above T_g. The voids are usually segmental in size. Chain orientation in the glassy state usually remains until $T > T_g$. The higher the shearing temperature the longer the orientation lasts, although the amount of orientation is usually less at higher temperature. Chain alignment probably is a function of temperature, pressure, shear rate, and the time-temperature history of the polymer, just as crystallization is for nonamorphous polymers. Orientation of amorphous chains is harder to detect and quantify by testing the structure of the polymer: sample-to-sample variation may well be a result of changes in processing. One of the jobs of the polymer engineer is to evaluate the sensitivity of product performance to changes in design parameters.

10.3.2 Effects of Temperature and Concentration on Polymer Solution Flow

The following problem illustrates the effects of temperature and concentration on the flow of a polymer solution. In this example, the fluid mechanics of the flow are affected by the viscosity changes induced by temperature.

Example 10.2 Problem: *Formation of Polymer Particles by Jet Spraying*

A research engineer is developing a new process for making fine particles of polystyrene by a jet-spraying technique. A dilute solution of polystyrene in toluene is pumped at high pressure through a capillary tube immersed in a solvent bath. The jet enters a large volume of a nonsolvent, where it disperses and generates fine particles. The pressure drop in the capillary was kept at 54 bar, and the capillary dimensions were 17.4 cm long and 100 microns as an internal diameter. The flow is thought to be laminar in the tube.

The engineer has noticed two important effects in these experiments. The original tests were done with a 1 wt% solution at 10°C. This resulted in very fine particles much less than one micron in diameter. In one series of tests, the temperature of the nonsolvent bath was raised to 40°C. This change affected the particle morphology, and the average particle size increased by an order of magnitude. In a second series of tests, the concentration of the solution was increased to 5 wt% polymer. Long filaments of polystyrene were formed at the higher concentration. The engineer wonders whether the various conditions of these two series of experiments have changed the particle formation mechanism. Determine whether this could be a reasonable explanation for the observations.

Background. The particle formation mechanism in jet flow often can be correlated using the jet Weber number (Lefebvre, 1989). Different flow regimes could lead to different particle morphologies, just as different impeller Weber numbers correlate with droplet size (Section 9.4.3). The Weber number is

$$N_{We,jet} = \frac{\rho_c V^2 D_j}{\sigma}$$

(continued)

where ρ_c is the density of the continuous phase (the nonsolvent), V is the velocity of the jet relative to the continuous phase, D_j is the jet diameter, and σ is the interfacial tension. This number is the ratio of inertial forces (ρV^2) to interfacial forces (σ/D). Although the precise correlation between jet Weber number and particle morphology is not known, large changes in this variable should be a concern to the research engineer.

Solution. The experiments described above were done at constant pressure drop, so that variations in solution viscosity could result in changes in flow rate through the capillary. Changing temperature and polymer concentration should change the solution viscosity, which would affect the Reynolds number of the capillary flow as well as the Weber number of the jet forming the droplets. Estimating the solution viscosities and calculating the dimensionless numbers will be the first steps in interpreting these experiments.

There are several equations that relate concentration to solution viscosity for dilute polymer solutions (Table 6.9). The Huggins equation (Eq. 6.49) is easy to apply, fairly accurate, and the coefficients can be found for a number of polymer/solvent mixtures (Brandrup and Immergut, 1989). Equation 6.49 can be rearranged to give

$$\eta = \eta_s(1 + [\eta]C + k_H[\eta]^2C^2)$$

If the Huggins coefficients are independent of temperature, temperature changes would affect the solution viscosity only by changing the solvent viscosity. Concentration changes would affect the solution viscosity directly.

Series 1. Effect of Temperature. Changes in temperature will affect the solvent viscosity. The jet tube is assumed to be near the temperature of the nonsolvent. The viscosity of toluene at 20°C, 25°C, and 30°C is 0.59 cp, 0.552 cp, and 0.526 cp, respectively. Viscosity data can be extrapolated using an Arrhenius equation (Eq. 8.25) when modest temperature ranges are covered. A model of the above data is shown in the following table:

Table I. Solvent Viscosity Model

Constants Temperature, °C	$\eta_0 = 1.808 \times 10^{-5}$ $-E_\eta/R = 1020\ K^{-1}$ η, Pa-s
10	0.000665
20	0.000588
25	0.000555
30	0.000524
40	0.000471

The reduction in toluene viscosity with increased temperature over this range is

(continued)

about 30%. The equation relating velocity with pressure drop for laminar flow in a tube (Eq. 9.10) shows that velocity is inversely proportional to viscosity. The change in the Weber number is based on the square of the velocity ratio.

$$\frac{N_{We,2}}{N_{We,1}} = \left(\frac{V_2}{V_1}\right)^2$$

For experiments with a constant pressure drop through the tube, the change in viscosity caused by temperature will result in the following changes in velocity and Weber number ratios compared to the base case (10°C).

Table II. Velocity and Weber Number Ratios

Temperature, *C*	V_2/V_1	We_2/We_1
10	1.00	1.00
20	0.884	0.782
25	0.834	0.696
30	0.788	0.621
40	0.708	0.501

The Weber number changes by 50% between 10°C and 40°C. Since this experimental series was not performed at constant Weber number, there may be a change in the particle formation mechanism that contributed to the differences in the particles made in the process.

Series 2. Effect of Polymer Concentration. Data for the calculation of solution viscosities is given in the following table:

Table III. Polymer Solution Properties

Densities:	polystyrene	1.02 g/ml	
	toluene	0.86 g/ml	
Huggins Model Coefficients		Value	Range
η		124 cm^3/g	(75–128)
k_H		0.39	(.30–.39)

The viscosity of pure toluene at 10°C is 0.000665 Pa-s. Figure 10.20 shows the viscosities of various polystyrene solutions over a range of concentrations and temperatures. A solution of 1 wt% polymer corresponds to a concentration of 0.0086 g/cm^3, while a 5 wt% solution has a concentration of 0.043 g/cm^3.

Small amounts of polystyrene have big effects on the solution viscosities. A 1 wt% solution at 10°C has a viscosity that is 2.5 times that of pure toluene. The 5 wt% solution has a viscosity 7 times that of the 1 wt% solution. Based on the analysis of the velocity and Weber number ratios used for the Series 1

(continued)

experiments, there should be significant differences in the velocity in the capillary tube and in the Weber number of the jet as the polymer concentration changes. Specifically, the Weber number of the jet should decrease by a factor of 49 over this concentration difference.

Comments. Modest changes in temperature and polymer concentration can create significant changes in solution viscosity. In this particular problem, the viscosity changes affect the flow in the capillary tube as well as the flow of the free jet (and the dispersion processes associated with this flow). Using pressure drop to control a flow rate is not uncommon, but it can lead to significant changes in flow when the solution concentration or temperature varies. This example illustrates how critical these effects can be for polymer solutions. Although the Huggins model coefficients were constant for these calculations, they can change with temperature. An improved analysis of this problem would consider the temperature dependence of k_H.

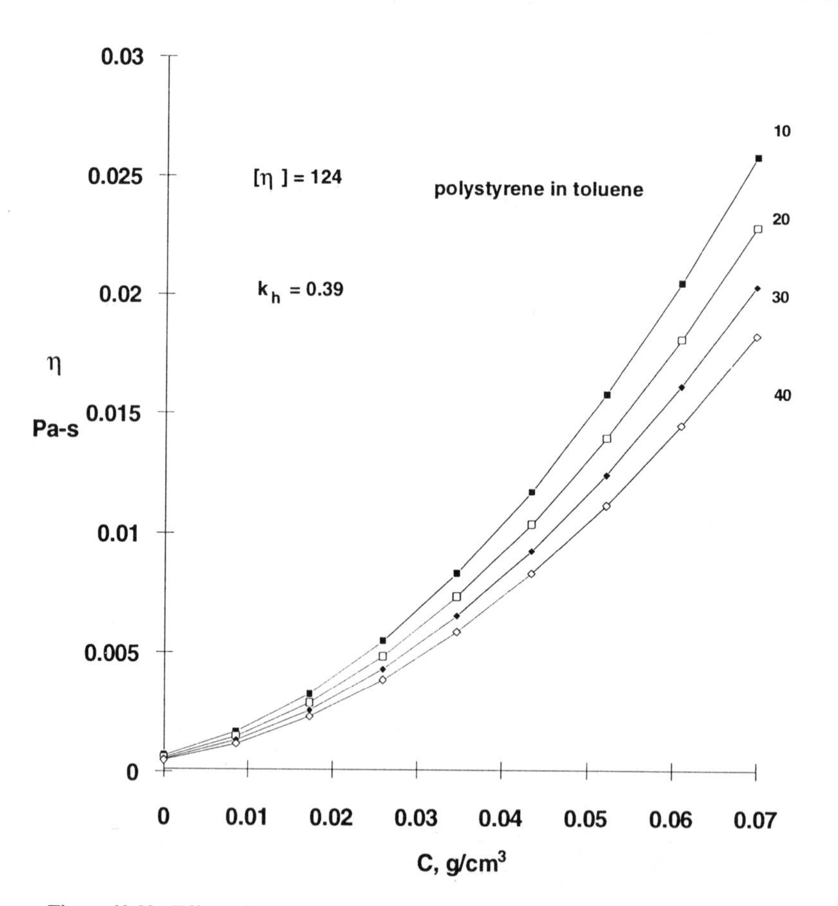

Figure 10.20 Effect of temperature on polymer solution viscosity.

10.4 POLYMER EXTRUSION

Extrusion is a basic fabrication process in the polymer industry. It can include all of the elementary steps listed at the beginning of this chapter: handling of solids, melting, pressurizing and pumping, mixing, and devolatilization. In an extruder, solid plastic resin is melted and forced to flow through a nozzle or die to make a polymer product. Single screw extruders have been used for many years in the polymer fabrication industry as a primary way to shape plastics. Twin screw extruders now are becoming popular since they require less shear on the compound for similar flows. This section gives a qualitative description of the extrusion process, which is intended to provide a basis for analysis. Some simple engineering calculations will be shown.

Flow models of polymer processes often begin by considering Newtonian fluids. This approach has a great deal of value. First, the Newtonian case usually is easier to solve and can illustrate some of the principles of the flow. At processing temperatures in some flows, polymer systems may be nearly Newtonian, so the predictions may be adequate. The next stage of complexity would be to incorporate power law or Bingham plastic models for the material. These rheological descriptions are sufficient to permit good engineering design for many flows. Polymers often are processed in their shear-thinning region. This region represents a balance between the high viscosity of the material at low shear rates and the high energy input required to attain the infinite shear rate viscosity. Materials can be moved into the "processing window" by changing the average temperature of the process as well as the average shear rate. The upper temperature of the window usually is set by the degradation rate of the material, while the lower temperature of the window is set by the viscosity of the melt. Molecular weight of the material often affects both of these levels. In the forming process, the polymer changes from a solid to a fluid and back again. The process must be able to handle these material changes at steady state or cyclic operations. This is another reason to design the forming step so that the polymer has little viscoelastic behavior.

Single Screw Extrusion. Figure 10.21 shows a schematic view of a single screw extruder. Solid resin is fed into one end of the extruder, usually from a feed hopper. A rotating screw moves the solid and molten polymer in the helical, annular space between the screw and the extruder barrel. The screw may rotate as fast as 200 rpm. The barrel is usually electrically heated using band heaters with on-off controls. Heat transfer fluids also can be used. The screw itself may be heated or cooled, depending on the energy requirements of the process.

The solid conveying zone carries the particles forward while compressing them. This forces the air in the interstitial spaces back up through the hopper and prevents air bubbles from being trapped in the melt.

The third zone performs melting, pumping, and mixing operations. Some melting is done by the heating of the barrel, but most melting in large extruders is done by the mechanical action of the screw on the material. More energy can be transfered

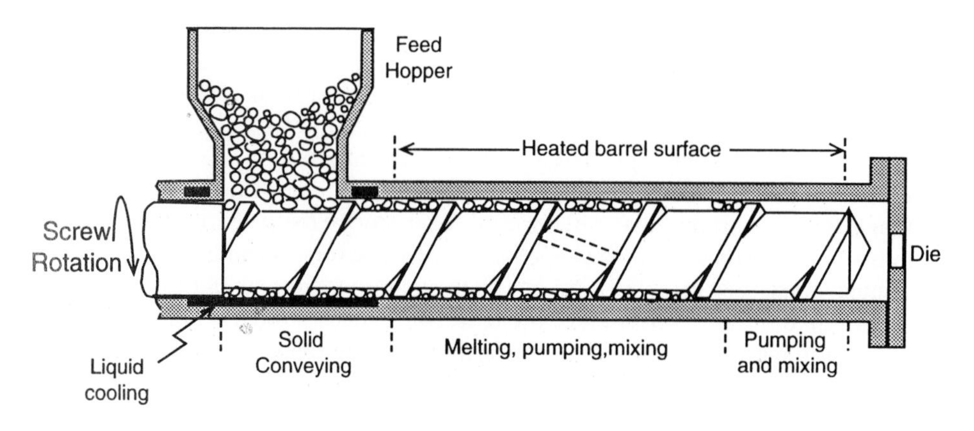

Figure 10.21 Schematic representation of a plasticating screw extruder. The barrel is cooled in the hopper region and heated downstream (Tadmor and Gogos, 1979).

through viscous dissipation in the polymer melt than by heat transfer through the solids and melt film at the barrel. The screw acts as a mixing device for the melt.

The final zone acts as a pump to build up pressure at the die entrance, forcing the melt through the die. There may be a screen pack just before the die to retain unmelted particles (and other debris). It prevents oversized material from clogging the die. One important parameter in extrusion is the pressure drop along the die. There is usually a pressure sensor after the screen pack (or breaker plate), as well as a temperature sensor. After the die, there will be some sort of take-up equipment that will cool the molten polymer while retaining the shape accomplished by the die.

The extrusion process can be modeled in three zones: the solids-conveying zone, the melt zone, and the pumping zone. The screw can be considered as a positive displacement pump. It floats freely in the barrel and is kept centered by the polymer melt. Screws are designed for specific polymers. For example, the ratio of the solids-metering channel depth to the melt-pumping channel depth is 2:1 for high density polyethylene, but it is 6:1 for nylon. The pitch of the screw is usually in the range of 12 to 20 degrees. The distance between the screw flights, the lead of the screw, is often equal to the barrel internal diameter.

Models of the extrusion process have made significant improvements in polymer technology and have led to improved screw and die design. Because of the phase, temperature, pressure, and flow changes through the extruder, the model equations change between the zones as different mechanisms become important. At the interface between the zones, the material, energy, and momentum balances must be coupled. The handling of the zone interfaces is a modeling problem that is still a subject of debate. Some researchers suggest that the whole extruder should be modeled without dividing it into zones.

Experiments have been performed to define physical processes in the extruder. Solid polymer, often containing solid particles to help evaluate mixing, is fed into the extruder. Extrusion is allowed to proceed and then the conditions are frozen by stopping the screw and freezing the polymer in place. The molten and partially mixed polymer will exhibit flow lines and the solids remain interact. The screw is removed from the barrel and the polymer can be unwound from the screw and examined.

Near the hopper, no solids have melted and the polymer beads have simply been mixed together. At the start of the melt zone, polymer has started to melt on the walls and is beginning to collect on the trailing edge of the screw. The extent of the melted region increases with the helical path length down the barrel. Near the end of the melt zone, the polymer melt completely surrounds some compressed, unmelted particles in the middle of the annular channel. In the compression zone, the melting is complete. The melt and particulate zones are segregated from each other. The melt phase accumulates against the leading edge of the flight in a melt pool, and the solids accumulate against the trailing edge of the flight. The helical path of the screw causes the solid phase to turn and mix down the barrel. The mixing is highest near the hopper and decreases as the solid becomes more compact down the barrel. This solids mixing supplements the mixing that occurs in the melt. At the end of the barrel, the solid phase has disappeared completely. The solid phase is continuous up to this point, which allows residual air to be expressed through the compacted particles back up into the hopper. The details of this model can vary between polymer systems. For example, this mechanism is appropriate for polyolefins but may not be followed by PVC, which can accumulate on the trailing edge of the screw flights.

Melting Mechanism. There are several different mechanisms that can contribute to the melting process. Five modes of melting are: conduction melting through a stagnant polymer film, conduction melting into a mixed or flowing polymer film, dissipative mix-melting, dissipative melting, and melting by large increases in pressure (compression). In general, conduction melting without melt removal is not an important mechanism during extrusion because it requires long melting times. It is important in thermoforming operations and continuous polymer welding. Compression melting is important in molding operations.

When polymer hits the hot barrel, it begins to melt and form a film of polymer on the surface. The solid particles are tightly packed and do not move past each other. The solid bed slides in the helical channel between the screw and barrel. The flow within the packed solid phase probably is plug flow. The screw flights scrape the melt off the wall, and it accumulates ahead of the advancing flight. A melt pool forms in front of the flight and the polymer melt will circulate (due to the velocity difference between the barrel and the screw). Solid is pushed toward the trailing edge of the lead flight. The end of the melting zone is considered to be the point at which the solid bed disappears.

Most of the solid melting occurs at the interface between the melt film and the solid bed. Heat sources for the melting process include conduction from the barrel and viscous dissipation. Melting in the channel is fairly efficient since there are usually large temperature gradients and high shear rates. The packing in the solid bed depends on the size and shape of the particles. Smaller particles usually form a more compact bed.

Figure 10.22 shows typical temperature of the barrel and screw along the helical path length, L. In this case, the barrel is maintained at a constant temperature. The leading edge of the flight accumulates polymer and, therefore, has a higher temperature than the trailing edge. The region labeled "plastic" represents the location where the solid bed is becoming dissociated from the trailing edge of the flight. As soon as the trailing edge is consistently wetted with molten polymer, its temperature approaches that of the leading edge.

Flow Models for Extruder Zones. Simple flow models for the various extruder zones shown in Figure 10.21 illustrate some of the key parameters for this polymer process. In each case, the extruder is assumed to operate at steady state. The model for each zone is coupled to the next by the conditions at the end of the zone: temperature, pressure, and velocity profiles. When the inlet and outlet conditions are matched by the equations, the model is complete.

Solids-Conveying Zone. The geometry of helical flow in a single screw extruder (Fig. 10.23) is similar for solids conveying, melting, and pumping zones. The solid is compressed and flows through the helical channel between the screw flights and the barrel. A typical screw has an increasing inner diameter in the solids-conveying zone in order to compress the solid to force air out and to force material against the barrel wall, where it can be sheared and melted. The case considered here is a straight, deep-channel section for the solids-conveying zone. An element containing solids moving in plug flow through the channel with velocity, V_z, has axial velocity, V_{p1}, and angular velocity, V_θ. The flow rate, G, is

$$G = V_1 \rho \left(\frac{\pi}{4} (D_b^2 - D_s^2) - \frac{cH}{\sin\theta} \right) \tag{10.11}$$

where D_b and D_s are the diameters of the barrel and screw respectively, ρ is the bulk density of the solid, H is the depth of the channel, c is the width of the screw flight, and θ is the angle of the helix. The term in the brackets computes the area for flow by

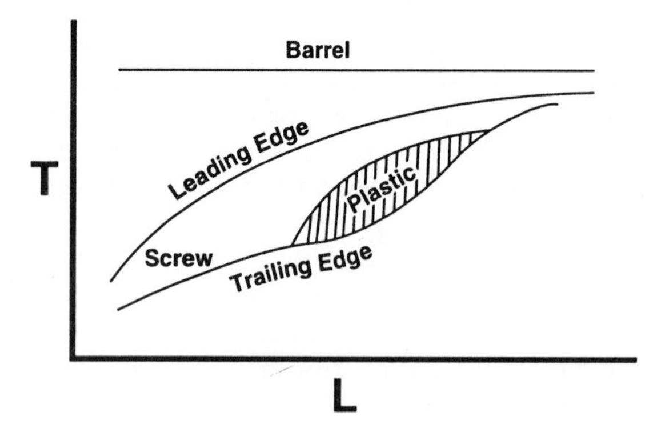

Figure 10.22 Temperature profiles in a single screw extruder.

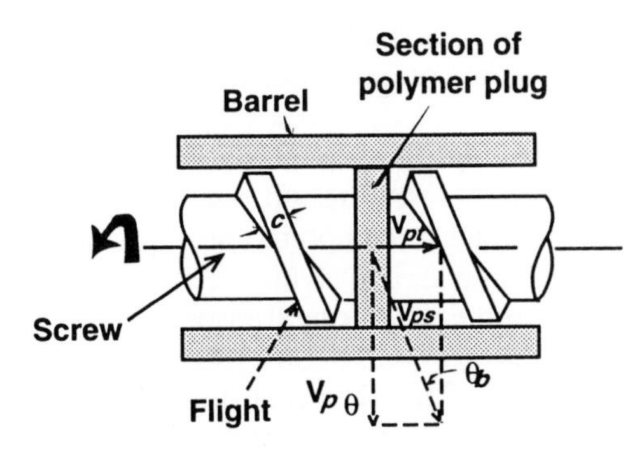

Figure 10.23 Axial increment of the solid plug. Velocities given relative to a stationary screw: V_{pl} is the axial velocity of the plug, which is independent of the radial position. $V_{p\theta}$ and V_{pz} are the tangential and down channel components of the plug surface velocity.

subtracting the flight area from the area of the annular space between the barrel and the screw diameter. As in many helical flow problems, the screw can be turning while the barrel remains stationary, or vice versa. In either coordinate system, there will be shear between the barrel and the screw. This problem often is solved by considering the screw to be stationary and the barrel to be turning. The direction of the plug relative to a stationary observer is described by ϕ. The total flow rate is

$$G = \pi NHD_b(D_b - H)\rho \frac{\tan\phi \tan\theta}{\tan\phi + \tan\theta}\left[1 - \frac{c}{\pi(D_b - 4)}\sin\theta\right] \qquad 10.12$$

where N is the rotation speed.

Equation 10.12 shows that the flow rate is proportional to the rotation speed, the barrel diameter, and the bulk density. The screw geometry can contribute in subtle ways to the flow through the helix angle, the flight thickness, the gap between the barrel and the screw, and the channel depth.

This equation is a simplification of the true boundary and initial conditions of the flow. The bulk density should vary as a function of pressure, based on the discussion in Section 10.2. The channel depth is decreasing in the solids-conveying zone for many screws. Furthermore, the data obtained by "unwrapping" a screw suggests that there is motion in the solids bed for some polymer solids. Even if this equation may fall short on accurate *predictions* for the solids-conveying region, it can be used to develop *correlations* for similar systems.

For example, suppose that a polymer processor purchases resin for extrusion from several suppliers. Each supplier probably will make material with a different bulk density. If the screw and barrel geometry are maintained constant, the processor could try to correlate the performance of his solids-conveying zone using the equation

$$G = KN_\rho \qquad 10.13$$

where K is a geometric factor relating to a specific extruder. The coefficient, K, could be determined by measuring the flow rate, G, for various screw speeds and bulk densities. The goal of the model would be to predict the screw speed needed to deliver a specific flow rate, G, for a resin with a known bulk density. The model would act as a guide to reduce the time needed to line out equipment when a change in resin was made.

There are a number of other applications for Equation 10.13. These include a variety of screw and auger feeding systems around compounding and extrusion equipment. When particulates containing compounding ingredients need to be moved, mechanical systems (rather than air conveying) are preferred since the bulk composition of the solids can be maintained during transport. Air conveying would have a tendency to remove small particles and deposit them in the screen and bag systems used to clean up the air stream before exhausting it to the environment. Screw-feeding systems also are used to move particulates short distances from one process to the next or to ensure uniform feeds from hopper systems. Besides moving polymer particulates, screw-feeding systems are used to add other compounding ingredients such as pigments, plasticizers, fibers, and fillers to ports in compounding twin-screw extruders.

Equation 10.13 should be a good correlating tool for screw-feeding systems. Homopolymers have essentially the same friction coefficient regardless of their morphology, so the flow of the solid bed past the solid surfaces should be the same. If the solid is not highly compressed by the screw, the bulk density of the polymer in the feeding system should be constant and near that given by a standard bulk density measurement. An augur operating with a full channel should provide a solids flow proportional to the rotation speed.

Metering Extrusion. The metering process is easier to model than the melting process because only one phase needs to be considered. The pumping zone in extruders is one example of this type of flow. In some polymer processing systems, there are extruders designed just to pump polymer melts. Extruder compounding systems often consist of two extruders. The first is designed to blend compounding ingredients together. This device often is a twin screw extruder (Fig. 10.24) because the shear rates and backmixing can be varied by using different screw elements. The twin screw extruder will discharge a blended melt to a single screw extruder placed at 90°. The single screw extruder is used to pressurize the melt for extrusion through a die. The major advantage of the cross-head extrusion system is that the blending and mixing operations are separated somewhat from the pressurizing operation. This allows the blending system to apply more or less energy input depending on the compound's needs, with little effect on the flow rate of material out of the die. This improved control and flexibility can be required for heat-and shear-sensitive compounds.

A metering extruder is designed to build pressure in the melt to permit the proper filling of the die and the uniform extrusion of the material. There are two mechanisms causing flow: the conveying of the melt by the screw and a pressure-driven flow. The pressure increases down the barrel and causes a pressure-driven flow in the reverse direction. The equation for a Newtonian fluid can be derived by considering a constant depth screw (Tadmor and Gogos, 1979),

Figure 10.24 Effect of feedrate on the hold-up of PVC in the mixing zones of a twin screw extruder, 250 rpm. First mixing zone: **(a)** 11 g/min, **(b)** 38 g/min. Second mixing zone: **(c)** 11 g/min, **(d)** 38 g/min. Published with the permission of K. Nichols.

Figure 10.24 *(continued)*

$$Q = \frac{V_z W(H-\delta_f)}{2} F_d + \frac{WH^3}{12\mu}\left(\frac{-\partial P}{\partial z}\right) F_p (1+f_1) \qquad 10.14$$

where $(H - \delta_f)$ is the distance between the flight and the barrel, Q is the volumetric flow rate, W is the distance between flight, $\frac{-\partial P}{\partial z}$ is the pressure change per unit length, F_d and F_p are shape factors for drag and pressure flow, and f_1 is a complex factor of the screw geometry. The flow velocity, V_z, is proportional to the rotation speed, N. The shape factors depend on the ratio of the screw channel depth, H, to W; they approach one when this ratio is small. The first term in Equation 10.14 represents drag flow, and the second term represents pressure-driven flow. As the screw builds up more pressure, the flow rate decreases. Since the pressure is induced by the drag flow, there will not be a flow of material back down the screw, merely a reduction of the total flow toward the die (for a Newtonian fluid with no backmixing). Equation 10.15 is a simplification of Equation 10.14, with constants A and B describing the shape and channel factors. This can be used to model and correlate metering flow in single and twin screw extruders.

$$Q = AN + \frac{B}{\mu}\left(\frac{-\partial P}{\partial z}\right) \qquad 10.15$$

10.5 DIE FLOWS

There are several characteristic die flows: sheet, capillary, and annular. Sheet dies are used to make films, sheets, and other profile extrusions. Capillary dies are used to make spaghetti-shaped melts that can be cooled to form solid rods or chopped to form pellets. Small diameter, short length capillary dies, called spinnerets, are used to extrude fibers. Annular dies are used for making piping and wire and cable coating. Each type requires proper design to give smooth extrudates at high production rates.

Dies are flow channels or restrictions that are intended to give a specific cross-sectional shape to a polymer melt. Figure 10.25 shows a side view of a sheet die. There are three regions to this die: the manifold, the approach channel, and the die lip. The manifold distributes the polymer melt over a cross section similar to that of the final product. The approach channel streamlines the melt to the final die opening. The die lips give the proper shape to the product and allow the polymer to "forget" its nonuniform flow in the previous regions. The gap between the lips is adjustable in order to correct for day-to-day variations in material and operating conditions. The design goal is to achieve a uniform pressure profile across the die and to correct for the temperature gradient while obtaining the highest possible production rate.

Product uniformity and maximum throughput are competing goals. There are two types of nonuniformities in the molten polymer flow from dies: nonuniformities in the machine and the cross-machine directions. Table 10.7 shows nonuniformities in profile extrusions from a sheet die. One upper limit on throughput rate is melt fracture,

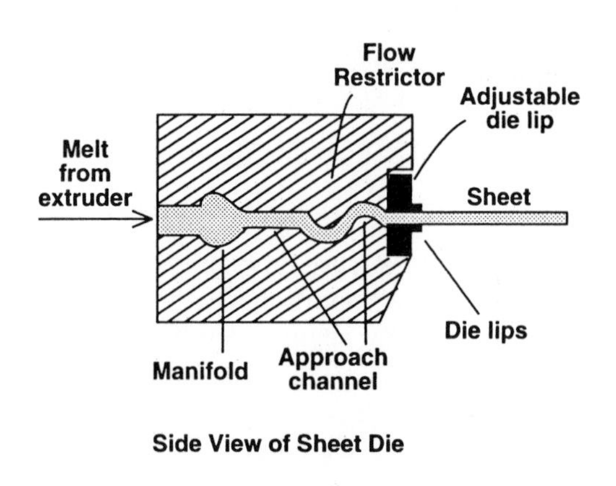

Figure 10.25 Schematic representation of a sheet die, including manifold, approach channel, and die lip regions. The restrictor bar is incorporated so that the die can be used with different polymers of varying viscosities (Tadmor and Gogos, 1979).

which is always caused by machine direction nonuniformity. The basis for the phenomenon is flow instabilities in the entrance region of the die that are not dissipated by the time the melt exits from the die outlet. Figure 10.26 shows some entrance flow patterns for capillary dies. The constriction of the flow leads to vortices that circulate in dead areas near the die entrance. At higher flow rates, the vortices can become unstable and rotating fluid may be drawn into the die flow. Polymer melts that exhibit viscoelastic effects are more prone to melt fracture.

Because melt fracture is caused by entrance effects, smoothing the die channel entrance or lengthening the die can reduce or eliminate the problem. Lower molecular weight polymers have high critical stresses for the onset of melt fracture and are less susceptible to this phenomenon.

Simple Model for Die Flow. The net flow rate induced by the screw at a given delivery pressure is the same as the net flow rate of polymer out of the die. A simple die flow model is

TABLE 10.7 Profile Extrusion Nonuniformities

Type	Causes
MACHINE DIRECTION	
Thickness variations, irregular surfaces, irregular wall thickness	Temperature, pressure variations, concentration variations such as improper solids conveying, solid bed breakup, incomplete melting
Melt Fracture	High flow rates causing stresses near 10^5 N/m^2
CROSS-MACHINE DIRECTION	
Thickness variation across sheet	Poor die design
Off center annulus	
Weld lines	
Melt Fracture	High Extrusion Rates

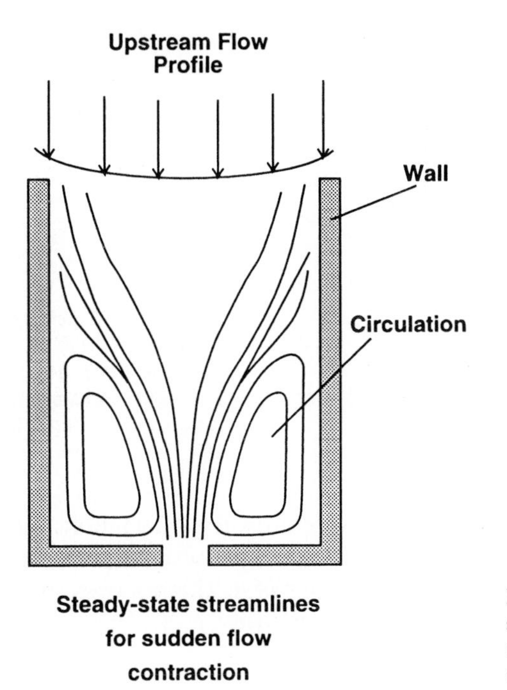

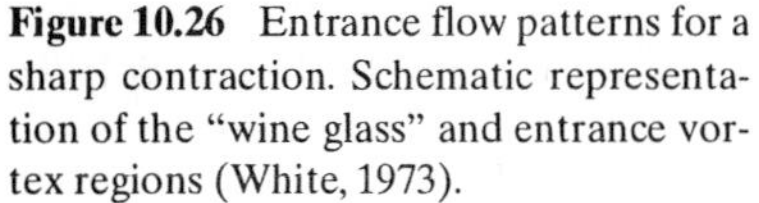
Figure 10.26 Entrance flow patterns for a sharp contraction. Schematic representation of the "wine glass" and entrance vortex regions (White, 1973).

$$Q = \frac{K}{\mu} \Delta P \tag{10.16}$$

where K is a drag coefficient for the die, μ is the melt viscosity, and ΔP is the pressure drop across the die. Usually, the pressure at the beginning of the screw and outside the die are atmospheric, and the pressure at the entrance to the die is very high. The screw distance is fixed, as is the length of the die. The die and extruder design equations are sketched in Figure 10.27 which plots the flow rate as a function of pressure drop. The die flow rate increases with pressure drop. The extruder flow rate decreases with pressure. The intersection of the die curve with the extruder curve represents an operating point for the system. Higher melt flow rates can be accomplished by increasing the rotational speed. Changing the screw geometry (reducing the channel depth for example) changes the slope and intercept of the extruder curve by changing F_d, F_p, and the drag flow. The dashed curve represents a different screw geometry. The major methods for changing the flow rate are changing the rotation rate, changing the screw geometry, or changing the flow resistance of the die.

Equations 10.14 through 10.16 are for Newtonian fluids. Non-Newtonian fluids show similar effects; the flow response may not be linearly related to the forcing functions (rotation speed and pressure gradient) over the entire variable space. An example of calculations for a non-Newtonian fluid is given in Figure 10.28. These are solutions to the flow of a power law fluid in a shallow channel. At high pressure drops, the flow rates for shear-thickening fluids ($n > 1$) curve up slightly from a linear

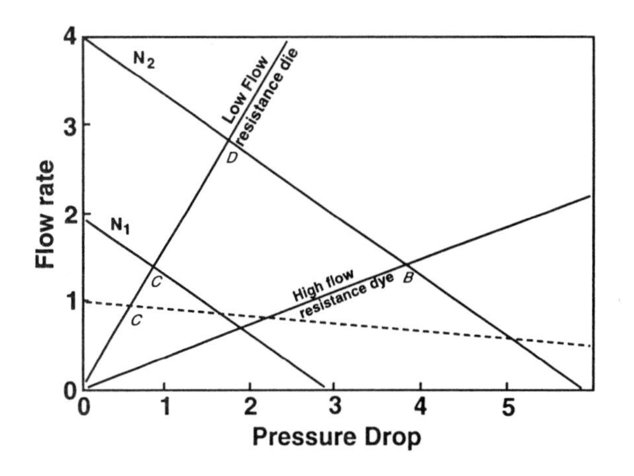

Figure 10.27 Schematic view of screw and die characteristics for Newtonian fluids and isothermal flow. The points where screw and die characteristics lines cross are the operating points (Tadmor and Gogos, 1979).

dependence between the flow and pressure drop The calculations for shear-thinning fluids show significant deviations from linear responses with pressure drop and show S-shaped curves, particularly at small values of n. This is caused by a cross-channel shear rate that affects the non-Newtonian viscosity.

10.5.1 Capillary Die Flow

Capillary flow is fairly easy to analyze and illustrates a number of nonideal phenomena that occur in die flows of polymer melts. In addition, capillary flows are viscomet-

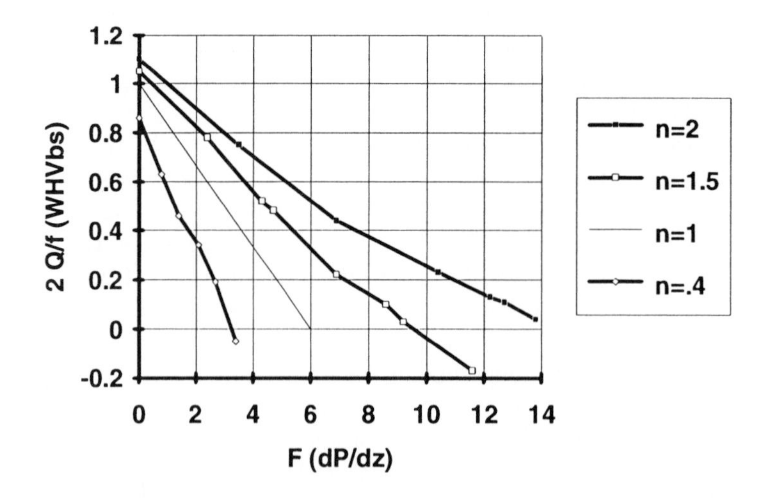

Figure 10.28 Computed curves of dimensionless flow rate versus dimensionless pressure gradient for isothermal flow of a power law model fluid in shallow screw channels with the power law exponent n as a parameter (Griffith, 1962).

ric and are used to determine rheological properties. Figure 10.29 shows a sketch of a capillary flow viscometer. The pressure for the flow can be provided by a piston, a gear pump, or an extruder. A piston would be used for viscometry, as it would cause no variations in the fluid pressure with time.

The fluid velocity in the reservoir is low, and the fluid accelerates as it nears the entrance region to the die (Fig. 10.28). The flow converges to the capillary tube dimension, and secondary flows can form near the corners of the reservoir. The secondary flows represent the viscous dissipation of some of the energy used to force the melt through the die. The acceleration of the flow stretches polymer molecules. The flow through the capillary may allow the polymer molecules to relax, particularly in the center of the flow. Newtonian fluids would form laminar flow profiles in the capillary, with the velocity down the capillary, V_z, proportional to the distance from the centerline. Shear-thinning fluids exhibit flow that is more similar to plug flow, with much of the fluid shearing occurring near the capillary wall. The velocity profile usually is achieved after several diameters down the tube. At the outlet, there are no shear stresses applied to the melt, and its velocity profile becomes completely plug-like. Residual stresses can be relieved at the surface of the melt, and this phenomenon is called die swell.

The equations of motion for this flow are solved for the capillary tube only and are simplified by assuming constant temperature, steady state flow, constant fluid density, and no dependence of pressure on the capillary radius. This flow is analyzed

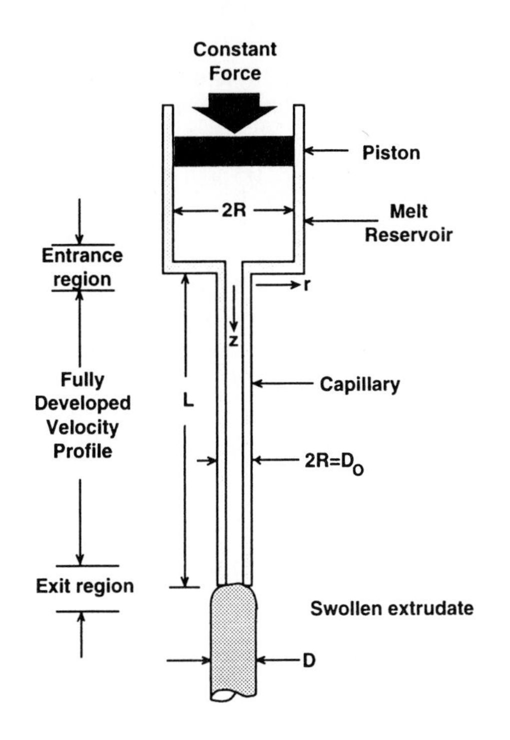

Figure 10.29 Experimental set-up for capillary flow, showing reservoir, entrance, fully developed, and exit regions.

using the z component of the momentum equation. The pressure drop in the z direction at steady-state is

$$\frac{\partial P}{\partial z} = \frac{-1}{r}\frac{\partial(r\tau_{rz})}{\partial r} \qquad 10.17$$

where τ_{rz} is the shear stress due to a gradient of V_z in the r direction. This equation can be integrated to give

$$\tau_{rz} = \tau_w\left(\frac{r}{R}\right) \qquad 10.18a$$

$$\tau_w = \left(\frac{P_o - P_L}{2L}\right)R \qquad 10.18b$$

The wall shear stress, τ_w, is measured by determining the pressure drop, $P_o - P_L$, over the distance, L. Assuming a generalized Newtonian fluid, the relationship between the wall shear stress and the shear rate is

$$\tau_w = -\eta\dot{\gamma}_{rz}|_{r=R} = \eta\dot{\gamma}_w \qquad 10.19$$

with $\dot{\gamma}_w$ as the wall shear stress. The viscosity function can be determined using Equation 10.19 if the wall shear rate is known. The melt flow rate, Q_f, and the wall shear stress can be used to find $\dot{\gamma}_w$. The flow rate can be determined by integrating the unknown velocity profile from the centerline of the pipe to the wall.

$$Q_f = \int_o^R 2\pi r v_z(r)dr = 2\pi\left[\frac{r^2 v_z(r)}{2}\Big|_o^R - \int_o^R \frac{r^2}{2}dv_z\right] \qquad 10.20$$

If $v_z(r) = 0$ at $r = R$ (no slip at the wall), then Equation 10.20 becomes

$$Q_f = -\int_o^R \pi r^2\left(\frac{dv_z}{dr}\right)dr \qquad 10.21$$

We can make a variable substitution in Equation 10.21 to change from r to τ_{rz} (using Equation 10.18a). This gives

$$Q_f = \frac{\pi R^3}{\tau_w^3}\int_o^{\tau_w}\left(\frac{dv_z}{dr}\right)\tau_{rz}^2 d\tau_{rz} \qquad 10.22$$

Solving for the shear rate, dv_z/dr, can be done by differentiating Equation 10.22 with respect to the wall shear stress

$$\frac{1}{\pi R^3}\left(\tau_w^3 \frac{dQ_f}{d\tau_w} + 3\tau_w^2 Q_f\right) = \frac{-dv_r}{dr}\Big|_{r=R} \tau_w^2 \qquad 10.23$$

The wall shear rate is

$$\dot{\gamma}_w = \frac{-dv_r}{dr}\Big|_{r=R} = \frac{3}{\pi R^3} Q_f + \frac{\tau_w}{\pi R^3} \frac{dQ_f}{d\tau_w} \qquad 10.24a$$

$$= \frac{3}{4}\Gamma + \frac{\tau_w}{4}\frac{d\Gamma}{d\tau_w} \qquad 10.24b$$

where Γ is the Newtonian shear rate at the wall.

$$\Gamma = \frac{4Q_f}{\pi R^3} \qquad 10.25$$

The above analysis does not consider the effects of the entrance flow on the total pressure drop between the reservoir and the tube outlet. The pressure drop along the tube must be constant under these conditions,

$$\frac{dP}{dz} = a = constant \qquad 10.26$$

where dP is the pressure drop across the capillary differential length, dz. This result cannot be correct for the entrance region of the flow. The converging streamlines lead to high velocities and velocity gradients, and the presence of circulating flows requires that the pressure drop from the reservoir to the capillary outlet will be higher than that required to force the fluid down the tube. Therefore, the equation relating pressure drop to flow must be corrected for the extra pressure drop associated with the entrance region.

The standard method for making this correction was proposed by Bagley and is based on determining the effective length of the capillary tube to account for the extra pressure losses. The pressure drop profile is

$$-\frac{dP}{dz} = \frac{\Delta P}{L^*} = \frac{\Delta P}{L + ND_o} \qquad 10.27$$

where D_o is the capillary diameter, L^* is the equivalent length of the capillary, and N is the number of diameters corresponding to the entrance losses. Substituting Equation 10.27 into Equation 10.18b gives

$$\tau_w^* = \frac{(P_o - P_L)}{4L^*} D_o = \frac{D_o}{4} \frac{\Delta P}{(L + ND_o)} \qquad 10.28$$

This is the corrected shear stress, $\tau_w{}^*$, at the wall for fully developed flow and is the Rabinowitsch equation. The key variables that can be used to evaluate the entrance effects are flow rate and capillary length. If entrance effects are due to the flow pattern near the die entrance, then they should be nearly independent of capillary length and directly dependent on flow rate. The corrected wall shear stress must be a function of $\dot{\gamma}_w$. Inverting Equation 10.28 gives a relation between the capillary length and the equivalent diameters for the entrance pressure losses.

$$\frac{L}{D_o} = -N + \frac{\Delta P}{4g(\Gamma)} \qquad 10.29$$

This equation suggests a plot of pressure drop as a function of L/D. The intercept should give N directly. Figure 10.30 shows such a plot for a polystyrene melt. For a given flow rate (Γ), the plot is linear. The number of diameters corresponding to the entrance losses increases with flow rate.

Very long capillary dies may show an upwards curve to these plots at high L/D values. This is due in part to the effect of hydrostatic pressure on melt viscosity. Also, there can be exit pressure losses in the melt. Such losses normally are 20% or less of the entrance losses and are small except for long dies.

The shear rates for power law fluid flow in capillaries are quite different than those for Newtonian fluids. A plot comparing the shear rates is shown in Figure 10.31a and 10.31b. The shear rates for Newtonian fluids vary linearly with capillary radius. For a shear-thinning fluid, the shear rates are quite high at the wall compared to a Newtonian fluid with an identical pressure drop. This is caused by the viscosity being lowered under shearing conditions. The velocity changes are highest near the wall, making the flow similar to plug flow. Power law fluids will have similar effects on the velocity profiles in annular and sheet dies.

Newtonian fluids show little change in the stream diameter on exiting a capillary.

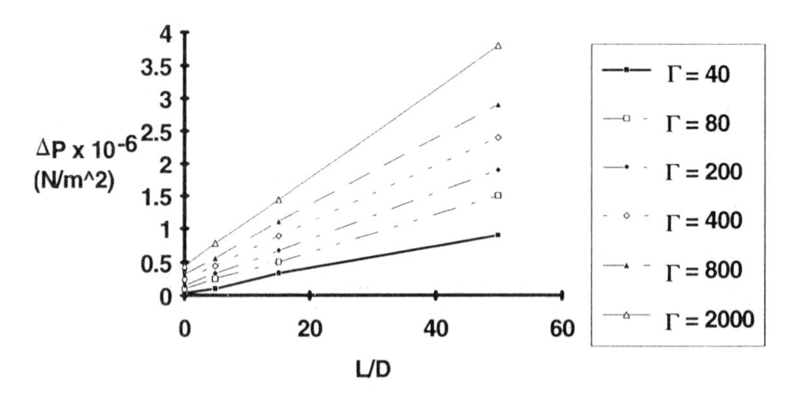

Figure 10.30 Bagley plots for a polystyrene melt at 200°C, from which N(Γ) can be evaluated; ΔP at L/D_o 0 is the entrance pressure drop ΔP (White, 1973).

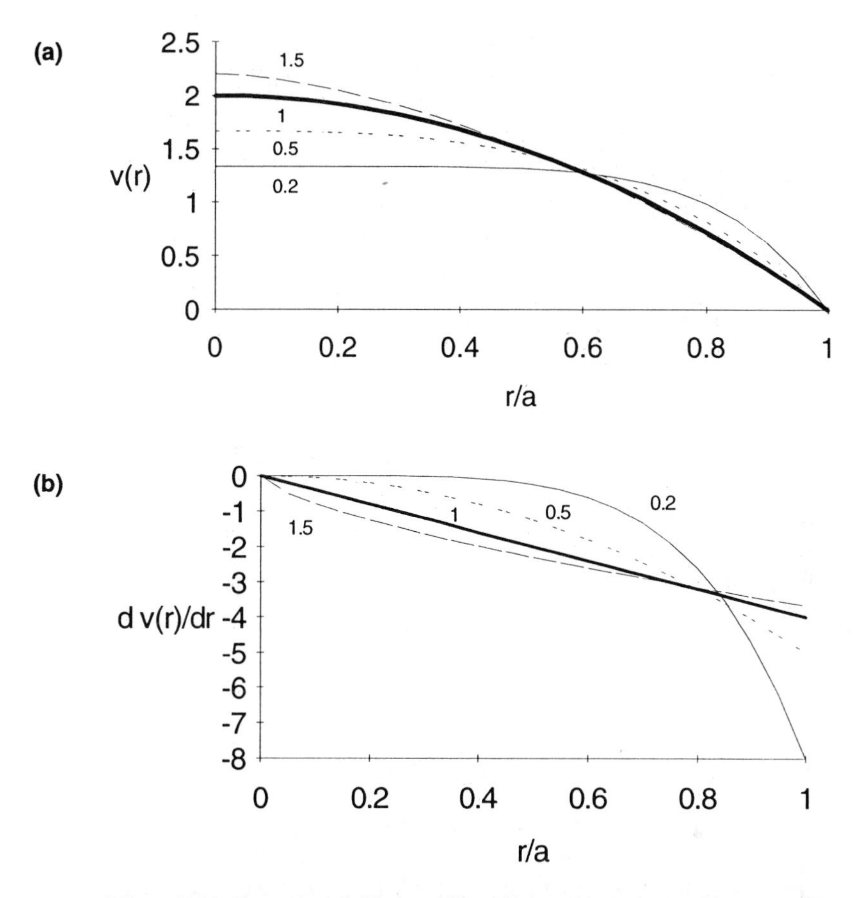

Figure 10.31 Power law fluids in capillary flow: **(a)** velocity profiles, **(b)** velocity gradient profiles. n = 0.2, 0.5, 1 and 1.5.

Typical changes are within about 10% of the capillary diameter. Viscoelastic polymer melts can show die swells between two to four times the capillary diameter. Increasing the shear rate (the capillary flow rate) increases the die swell. If viscoelastic effects are involved, then samples with some very large molecules should exhibit more die swell. Figure 10.32 shows that a broad molecular weight sample has a higher die swell than a narrow distribution sample.

10.5.2 Sheet Die Flow

The design of a sheet die has the opposite flow problem from the capillary die. The sheet die has a diverging flow rather than a converging flow. A uniform flow across

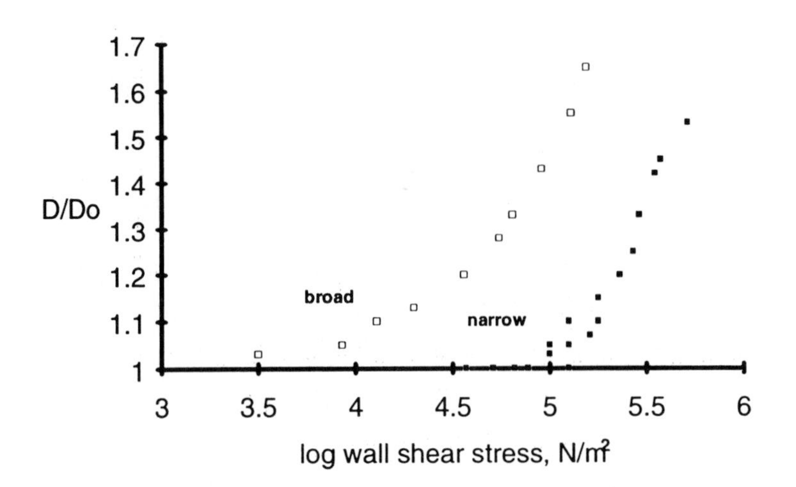

Figure 10.32 Extrudate swelling data for polystyrene melts: □, broad molecular weight sample; ■, narrow distribution sample data at various wall shear stresses (Graessley et al., 1970).

the whole die head at its exit is desired (Fig. 10.33). If the height of the flow channel were uniform across the die, the flow would resemble that of a jet with a high velocity in the center and a low on the edges.

The flow from the extruder enters through the center channel and begins to diverge in the manifold. The radius of curvature of the manifold should be small in order to ensure that all surfaces are wetted with polymer. After the slit, denoted by the double line on the drawing, the gap between the plates is constant and the flow is a pressure-driven flow between parallel plates. In addition to the design for flow, the die must be designed to withstand the pressure and temperature of use as well as the chemical environment. Temperature control of the die surfaces is a challenge since cold surfaces can lead to solidification, and hot surfaces can give faster flow than desired.

Sheet and Film Technology. A number of polymers can be formed into film and sheet by extrusion through a thin die. Table 10.8 lists some materials. Chill roll systems make uniaxially oriented film, while bubble systems make biaxially oriented film. Solution casting can be used when excellent surface properties are needed. Thick sheet can be made by calendering processes, which will be discussed in Section 10.7. Figure 10.34 shows a typical chill roll system for extruding plastic film. The extruder forces melt through a flat sheet die. The die opening may be about 0.05 cm high while the finished film is .0025 cm thick. Extensions of 2500% are typical of film operations and result in oriented films. Sheets usually are considered to be 0.01 cm or thicker. Films have lower thicknesses. The opening of the die lips needs to be controlled carefully. The rapid uptake of the film requires rapid detection of the film thickness and response. Beta-ray thickness scanners can be used to gauge the film thickness and control the die opening.

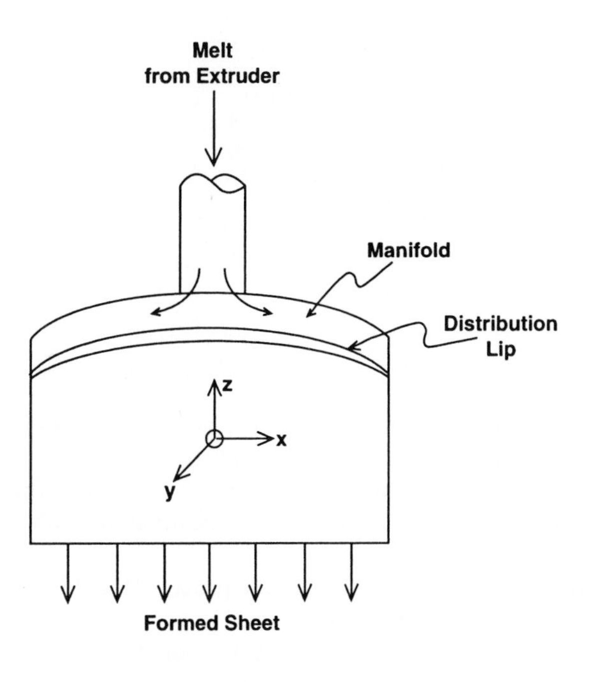

Figure 10.33 Schematic representation of the coathanger die design.

Because of the uniaxial orientation, the properties of the film vary between the machine and transverse directions. Table 10.9 compares such properties of polyethylene and polypropylene films. The tensile strength is higher in the machine direction, but the film is more likely to tear in this direction. Film orientation usually is carried out above T_g, so that the polymer chain segments can move easily. Crystalline polymers are oriented above T_g but below T_m. Uniaxial orientation gives different properties from biaxial orientation. Table 10.10 compares properties of unoriented, uniaxially oriented, and biaxially oriented polyethylene terephthalate. As the tensile strength increases, the elongation at break decreases. Uniaxially oriented material is preferred when strength is needed in only one direction. An example is plastic package strapping, which needs to be strong along the fiber direction.

An important advantage of oriented amorphous and crystalline films is their ability to shrink and conform to molds and packages when they are reheated above T_g (shrink wrap). Polystyrene can be biaxially oriented by the bubble or blown film process. When it is heated above T_g (about 100°C), it shrinks, and the amount and rate of shrinkage depends on the heating temperature. Heating oriented amorphous films above their T_g's always results in some shrinkage.

TABLE 10.8 Materials Extruded into Film and Sheet

PVC—flexible and rigid	Poly(vinyl butyral)
Polycarbonates	Polytetrafluoroethylene
Nylon	Chlorinated polyether
Poly(vinylidene fluoride)	Acrylics
Polyester	ABS
Poly(vinyl acetals)	High-impact polystyrene
Chlorinated polyethylene	
Methyl cellulose	
Poly(ethylene oxide)	

Biaxially oriented crystalline polymers are oriented at some temperature between T_g and T_m of the material. These films also will shrink, but will do so only when reheated above their orientation temperature (which must be above T_g). Therefore, crystalline polymer films can be made to be heat stable by orienting them above their use temperature.

Biaxially oriented films can be made by the bubble process (Fig. 10.35). In this early process, an 85:15 vinylidene chloride:vinyl chloride copolymer is plasticized with 7wt% di(α-phenylethyl ether) and extruded through a circular die. The tube is quenched in a water bath and is lubricated on its interior surface with mineral oil. The hollow film is flattened through a roller system and then reinflated with air at 1 psig. The air provides a radial expansion to the bubble (transverse orientation) while it is being expanded in the machine direction by a fast take-up rate on the wind-up roll. The bubble is collapsed at the drive nip, retaining a stable bubble in the process. Bubble orientation processes are also used to make polyolefin films. An alternative biaxially orienting technique is to physically stretch film in the transverse direction. This can be done by developing a bead on the sheet that can be used for pulling in the transverse direction.

The shear fields present in the die flow tend to orient the polymer in the machine

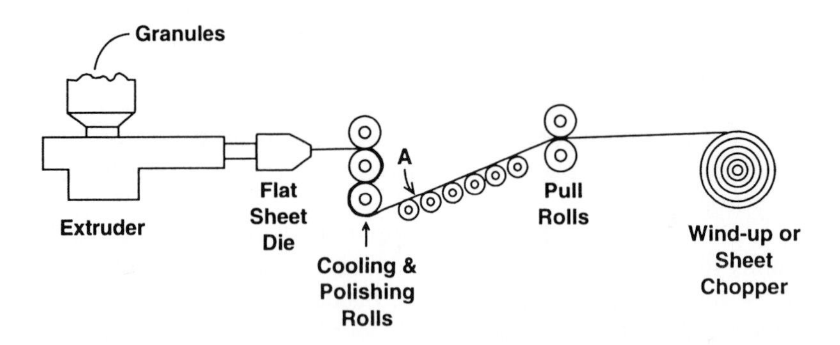

Figure 10.34 Plastic sheet extrusion. Chill roll casting (Park, 1973). Published with the permission of R.E. Krieger Publishing Co.

TABLE 10.9 Properties of Chill Roll Films

	Tensile Strength, psi		Elmendorf Tear, gm	
	MD*	TD†	MD	TD
1 mil polyethylene	2,800	2,200	60	150
1 mil polypropylene	5,700	3,200	50	550

*MD = machine direction
†TD = transverse direction
SOURCE: R.A. Elden and A.D. Swan, *Calendering of Plastics*, Elsevier, NY, 1971.

TABLE 10.10 Effect of Orientation on Physical Properties of Polyethylene Terephthalate*

	Tensile Strength (psi)	Elongation at break (%)	Yield Stress (psi)
Unoriented	6,000	500	6,000
Biaxially oriented	25,000	130	10,000
Uniaxially oriented	> 75,000	7	> 75,000

*Mylar.
SOURCE: R.A. Elden and A.D. Swan, *Calendering of Plastics*, Elsevier, NY, 1971.

direction and result in greater strength in that direction. Orientation depends on temperature gradients and crystallization nucleation. The polymer near the surface of the die experiences higher shear, and crystal nuclei tend to form in highest concentration here. The crystal lamellae tend to grow in the direction of thermal gradients, so the chains will tend to be aligned perpendicular to the flow as the polymer cools. Cooling of the polymer outside the extruder is accomplished by water or air contact, or by chill rolls.

10.6 INJECTION MOLDING

Injection molding is used to make parts with three-dimensional shapes. Polymer under high pressure is forced to flow through a distribution system in the mold to the mold cavity. Pressure is applied until the mold is full and the part has cooled enough to prevent warpage. The pressure may be built up by an extruder equipped with a ram to force the polymer melt into the mold at the appropriate time in the cycle. Machines are rated with respect to their shot capacity, with 50 oz being a very large machine. Figure 10.36 shows typical time-pressure histories in the mold distribution system: injection pressure at the extruder, nozzle pressure and mold pressure.

The gate controls the flow of polymer melt into the cavity and prevents the backflow of melt out of the cavity as the pressure is released and the melt is cooling.

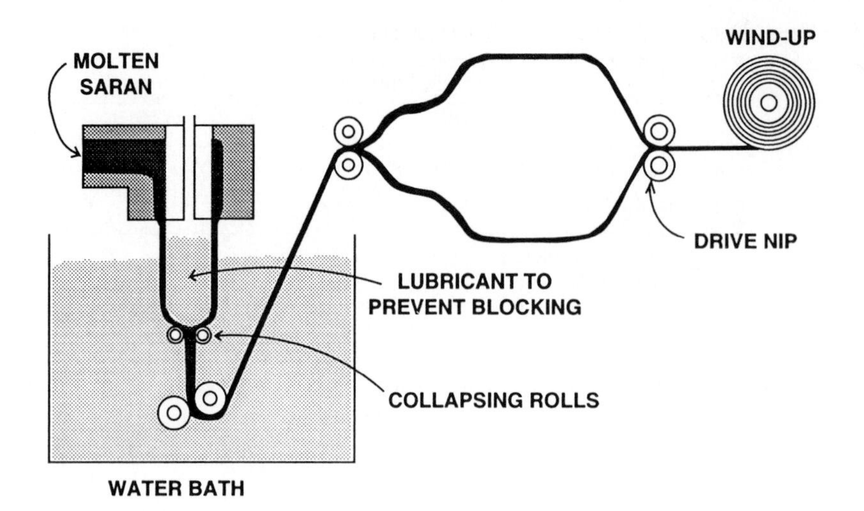

Figure 10.35 Saran bubble process (Park, 1973). Published with the permission of R.E. Krieger Publishing Co.

The runners distribute the melt to several different cavities and must be balanced to prevent some cavities from not getting enough melt to fill them. The heat of crystallization of the polymer may account for up to 40% of the heat removal requirements in the mold. The molds usually have cooling channels in which water or other cooling agents are circulated.

Orientation of the polymer melt in injection-molded parts usually occurs away from the wall and the centerline of the melt. The polymer at the wall cools too quickly to be oriented, and the material in the center stays warm long enough to stress-relax. Increasing the melt temperature will decrease polymer orientation.

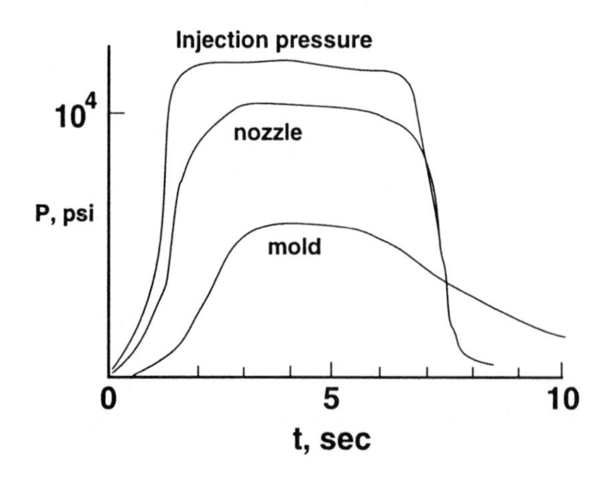

Figure 10.36 Pressures in injection-molding equipment.

10.6.1 Typical Design Problem: Injection-Molded Cup

The steps taken in the design of a mold for plastic cups are typical of many injection-molding operations. The cup should be inexpensive and formed quickly. The walls will be thin, which will require precise design of the system's rheological and thermal properties. The quality of the mold will affect the economics of the part. The cycle time for most injection-molded parts is a function of the heating/cooling cycle time of the mold. Rapid heating and cooling rates lead to short cycles but may result in poor filling of the mold, short shots, dimples, and poor surface quality. Poor mold design with respect to flow and thermal factors can lead to accelerated wear on some surfaces. Also, the design affects the "size" of the operating window, that combination of temperatures, pressures, filling rates, etc. that lead to quality parts. The costs for the mold may be 10% to 20% of the total production cost, but the design choices set many factors for the whole process. Some of these issues will be discussed in this section.

Design Elements. Figure 10.37 shows a cross section of a mold that might be used to make the cup. The melt flows into the mold cavity through the gate and sprue. A cooling channel is used in the center of the cup to remove heat. The top of the mold is fixed on the extruder outlet. The bottom is moved out to demold the part and moved back to form the mold again. The mold cycle involves flow of the melt into the distribution system, filling the mold, cooling the polymer to a solid state, and removing the finished part.

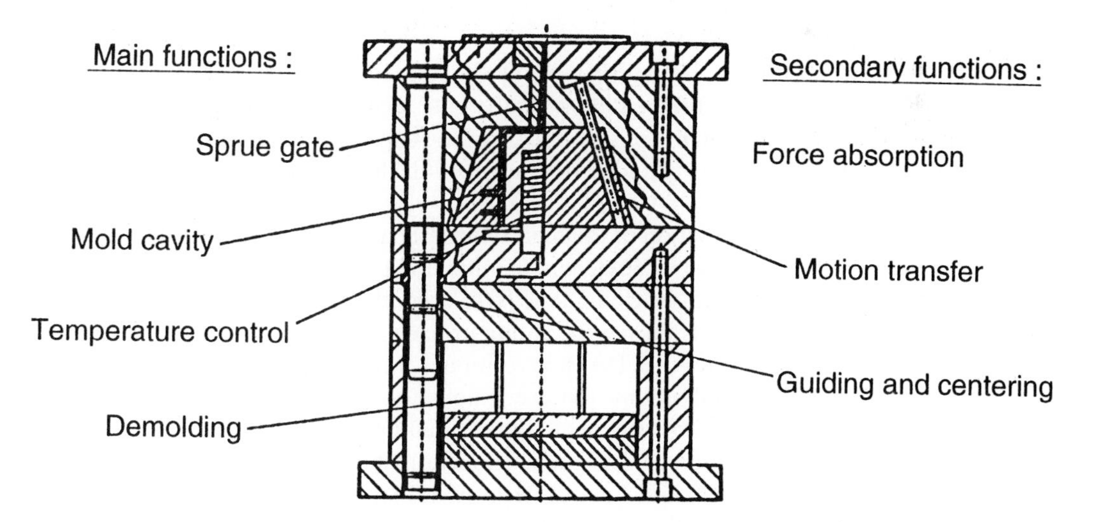

Figure 10.37 Mold cross section. Published with the permission of G. Menges, P. Filz, O. Kretzshmar, H. Recker, Th. Schmidt, and Th. Schact, "Mold Design with CAD Mould" in *Applications of Computer Aided Engineering in Injection Molding*, 1987, Hanser Publ. Munich–Vienna–New York.

The specifications of the finished part set some requirements for these operations. There might be multiple cups made with one mold. The runner and gate system will be designed to provide equal flow and pressure to each cavity and will provide equal filling time. The mold cavity needs to be rigid with respect to the material, remove any air entrained in the cavity, make uniform seals and joints, and be designed so that the part can be removed easily. Temperature control must be uniform and allow rapid cooling when needed.

The conceptual design of the mold requires some experience. The number of cavities, the choice of a hot or cold runner system, the placement of the alignment pins and knockout devices require working knowledge of injection molding. However, this information might be retained with knowledge-based systems (artificial intelligence). Once the general layout has been established, computer modeling of the process can greatly reduce mold design time. The flow, thermal, and mechanical properties of the system can be modeled fairly well using finite element or finite difference techniques. A computer model will let the designer vary all the above factors to determine their effects on the cycle and the part. The following factors can be calculated:

Flow Effects: filling behavior of the cavity, gate-system layout, pressure drop in the nozzle, temperatures in the melt, and melt stress

Thermal Effects: energy balance on the mold, layout of the cooling channels, and temperature homogeneity

Mechanical Effects: rigidity in clamping and normal directions, and ejection forces

Flow Effects. Figure 10.38 shows expanded flat views of the volume elements for the cup. The melt enters from the gate located in the base of the cup. The cup shape expands from the base to the lip of the cup. The handle will be the last part to be filled. The handle is molded on the cup with the each end of the flat "I" filled first.

A computer simulation of the melt flow pattern into the cup is shown in Figure 10.39. The polymer is polypropylene; the initial melt temperature is 230°C; the mold temperature is 50°C; and the filling time is 0.5 seconds. The lines show the melt front at uniform time intervals during the filling process. The flow is predicted to fill the base in about 5 time intervals. The flow continues through the base ridge and the cup walls to the lip. The total fill time to the cup lip is about 17 time intervals. At about 15 time intervals, the melt reaches the inlet to the handle. Here, the melt splits into three flows that fill each section of the handle (a section being one of the "T" elements as shown in the flat view). After about 21 time intervals, the handle has been filled. Because there are three flow channels in the handle, there can be three weld lines (places where two melt fronts come together).

The flows in the base and the body are modeled as disk flows. The flows in the base ridge and the two lip sections are modeled as plate flows. The time to fill the lip

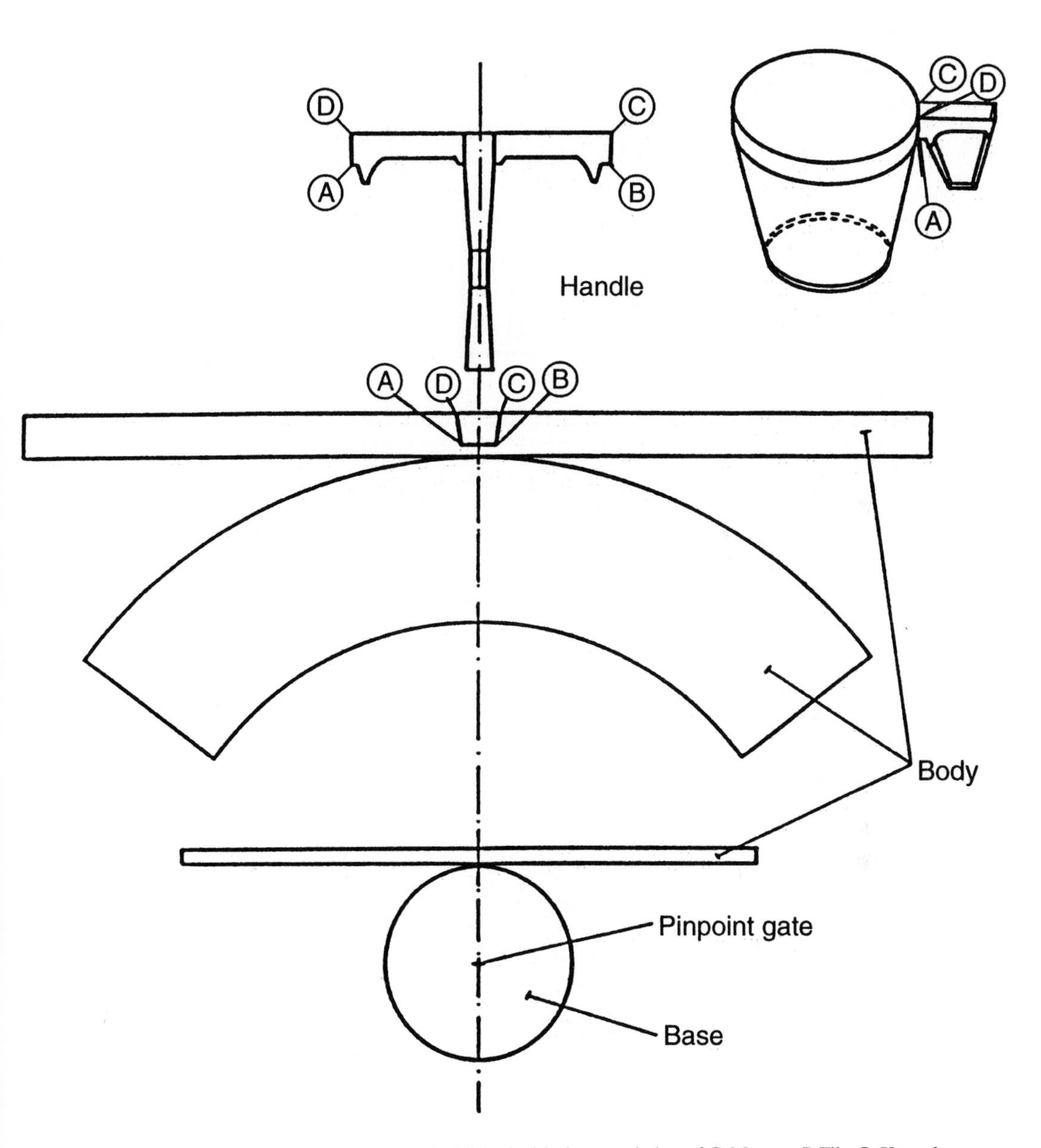

Figure 10.38 Lay-flat of a cup. Published with the permission of G. Menges, P. Filz, O. Kretzsh-mar, H. Recker, Th. Schmidt, and Th. Schact, "Mold Design with CAD Mould" in *Applications of Computer Aided Engineering in Injection Molding,* 1987, Hanser Publ. Munich–Vienna–New York.

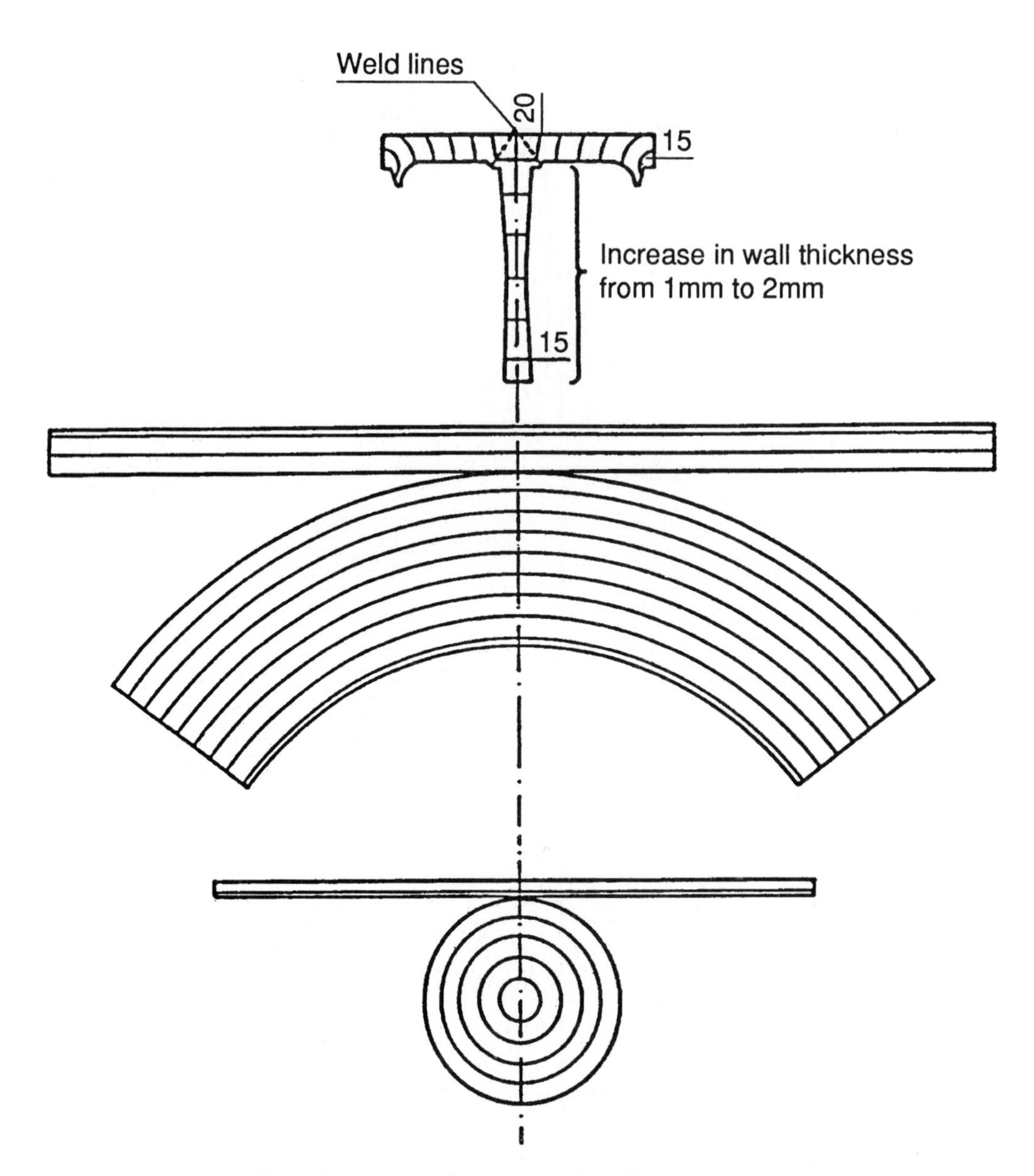

Figure 10.39 Flow pattern of a cup-shaped molding. Published with the permission of G. Menges, P. Filz, O. Kretzshmar, H. Recker, Th. Schmidt, and Th. Schact, "Mold Design with CAD Mould" in *Applications of Computer Aided Engineering in Injection Molding,* 1987, Hanser Publ. Munich–Vienna–New York.

is 0.487 seconds. At this point the melt temperature has dropped to 188°C. The pressure is low initially and is increased in order to fill the lip region.

The filling of the handle is modeled separately. Melt reaches the handle after the cup walls are filled (0.38 seconds). The handle is filled in the last 0.02 seconds of the flow cycle. When each of the three segments of the cup has melt in its entrance, the conditions of the melt change dramatically. These channels are narrow, and the shear rate of the flow increases by about an order of magnitude. The shearing of the melt causes the temperature to rise. The melt is forced into the handle channels quickly by increasing the pressure (almost twice as much pressure is required as is needed to fill the lip section).

Analysis of these flow patterns suggests several problems with the initial design. The cup lip will be subjected to very high pressures after it is filled (while the handle is being filled). The mold seal is located at the cup lip (see Fig. 10.37), and high pressures could lead to flash formation (leakage of the melt between the top and bottom mold surfaces). Weld lines occur where cooled polymer melt fronts come together and do not bond well. Weld lines are defects and usually are the first place for the part to fail. The weld lines of the initial design are at the outer top portion of the handle. This section is subject to high reverse bending stress when the cup is used, and it would be preferable to move the weld lines to an area that does not have high stresses. The rapid filling of the handle section with high shear flows can lead to scorching of the material.

The most critical problem is that of the weld lines. These can be moved to another portion of the handle by changing the wall thickness of the three "T" sections of the handle. If the channels are made thicker, the melt can fill them faster with less shear, lower temperatures, and lower applied pressure. This design solution for the weld line problem should also address the flashing and scorching issues. Figure 10.40 shows the flow patterns after the wall thickness in the tops of the "T" have been increased. The melt fronts now meet in the base of the "T" away from the portion of the handle subjected to reverse bending stress. The new design has a higher volume by 0.3 cm^3 and takes four seconds longer to cool. Figure 10.41 shows the expected pressure profile as a function of flow length for the initial and the optimum design. Since the pressure decreases by 75 bar, the flashing problem should be reduced. The melt temperature in the handle will be higher, and the joint at the weld line should be better.

This example illustrates some design trade-offs. Slow filling is accompanied by cooling of the material. Cooling increases the melt viscosity and requires higher injection pressures. Rapid filling causes high friction losses and requires high pressure as well. Rapid filling is best done at higher temperatures in order to lower the viscosity and the injection pressure. Figure 10.42 shows the pressure drop as a function of filling time for two different melt temperatures. Each curve gives a minimum melt pressure drop that would be the optimal cycle time for the proposed melt temperature.

Multiple cavity molds often need to be balanced with respect to flow. Figure 10.43 shows a two-cavity mold in which each part has different geometry. If the runner-sprue-gate system for each cavity is the same, one part would be filled before

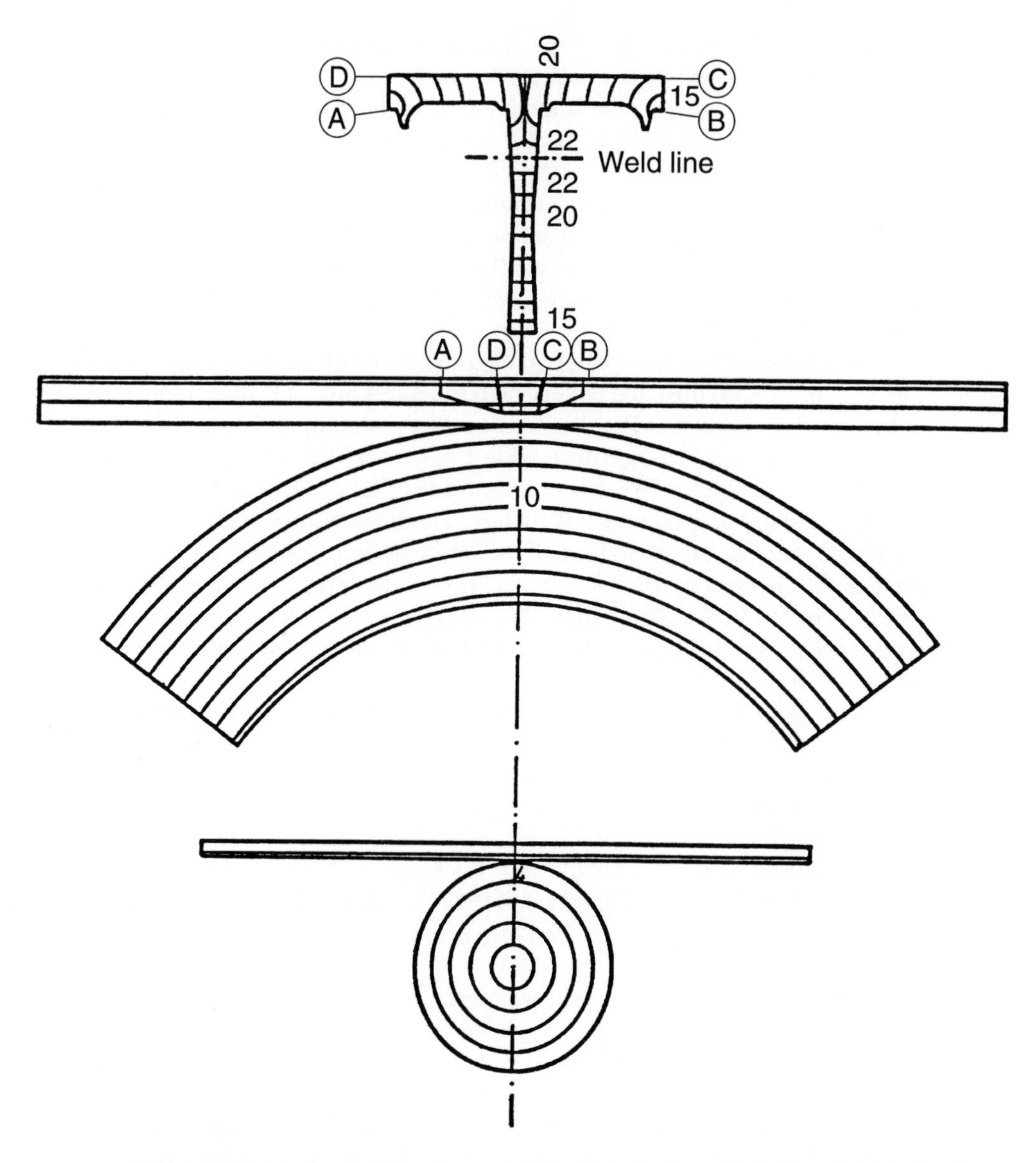

Figure 10.40 Change of welding line position by changing the wall thickness. Published with the permission of G. Menges, P. Filz, O. Kretzshmar, H. Recker, Th. Schmidt, and Th. Schact, "Mold Design with CAD Mould" in *Applications of Computer Aided Engineering in Injection Molding,* 1987, Hanser Publ. Munich–Vienna–New York.

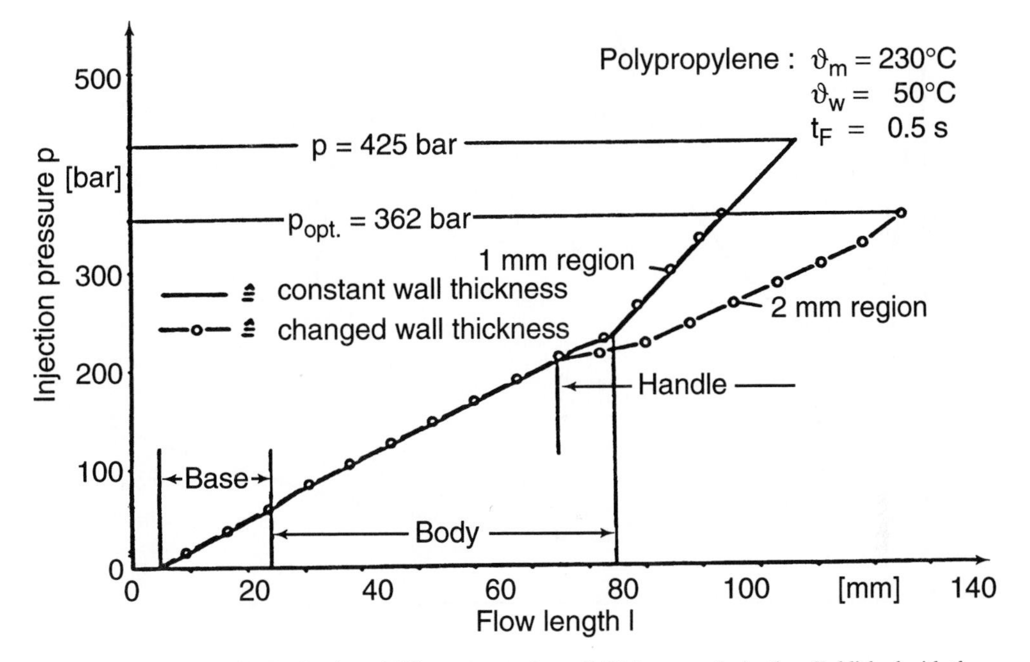

Figure 10.41 Reduction of filling pressure by wall thickness optimization. Published with the permission of G. Menges, P. Filz, O. Kretzshmar, H. Recker, Th. Schmidt, and Th. Schact, "Mold Design with CAD Mould" in *Applications of Computer Aided Engineering in Injection Molding,* 1987, Hanser Publ. Munich–Vienna–New York.

the other one (as shown by the time intervals on the parts). This system can be balanced by reducing the runner channel size for the smaller part. This slows the flow and permits both parts to be filled at the same time.

Thermal and Mechanical Effects. Because the cooling of plastic parts will be by conduction, a significant portion of the mold cycle time is taken up for heat removal. Some cooling of the melt occurs during filling. However, the melt temperature should be maintained above T_g or T_m until after the mold is filled in order to prepare good parts. The removal of heat after the mold is filled depends on the location of cooling channels. Channel placement is designed to balance the heat flows out of the three-dimensional part while reducing the temperature increase of the cooling fluid.

There are two key mechanical design problems. The mold is clamped in one direction, but the polymer exerts pressure in the clamping direction and transverse to the clamping direction. The polymer may contract slightly as it cools. The deformations in the transverse direction must be reversible and compensated for by any contraction of the polymer. The deformations on the mold plates must be controlled in order to reduce or eliminate flashing.

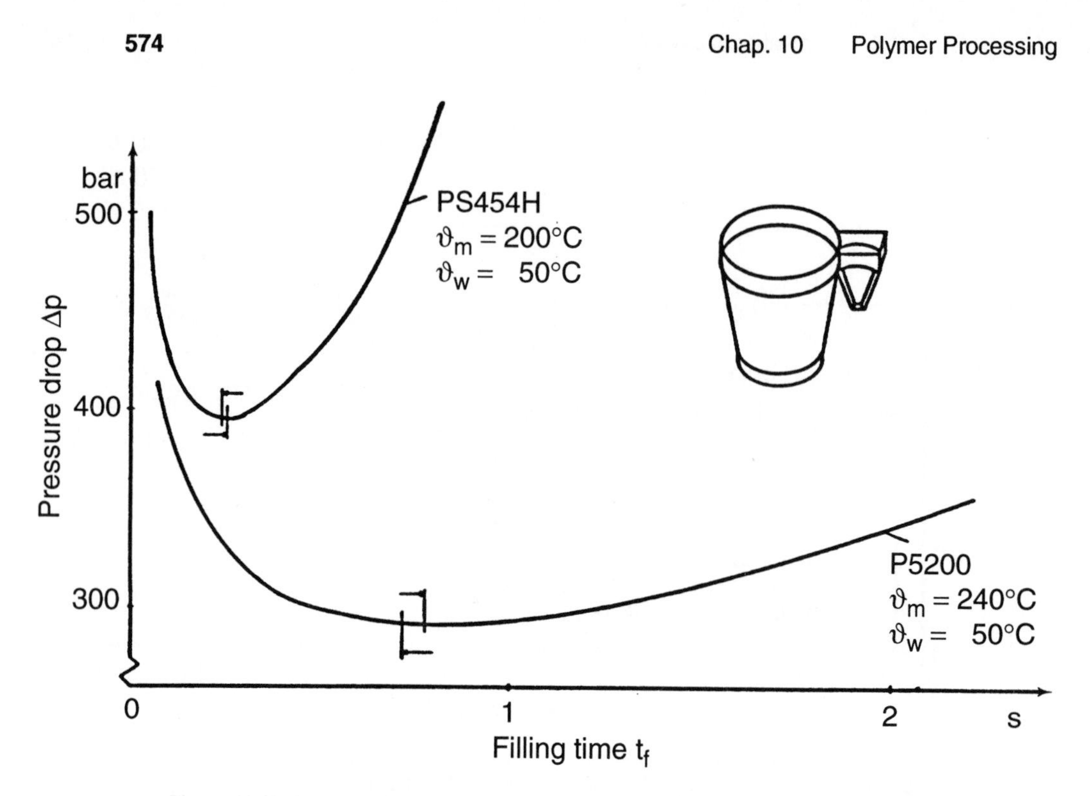

Figure 10.42 Pressure drop versus filling time of a cup. Published with the permission of G. Menges, P. Filz, O. Kretzshmar, H. Recker, Th. Schmidt, and Th. Schact, "Mold Design with CAD Mould" in *Applications of Computer Aided Engineering in Injection Molding,* 1987, Hanser Publ. Munich–Vienna–New York.

Complex mold geometries often can be modeled using a series of capillary, disk, and channel flows. For example, runner systems usually are round and can be modeled as capillary flows. Several runners in parallel are similar to several pipe flows in parallel. Tapered flows can be modeled as a series of channels or capillaries, each having a different characteristic size. The model designs can be fed to numerically controlled machining operations for direct production of the mold. These techniques can lead to quick prototyping and rapid response to customer requirements.

10.6.2 Changes in Volume with Temperature and Pressure

Changes in the volume of thermoplastic materials during processing steps can cause a number of undesirable effects. Volume changes are not as great for cross-linked elastomers and thermosets. This section discusses a specific processing problem related to thermal contraction, part shrinkage, and some simple models for specific volume.

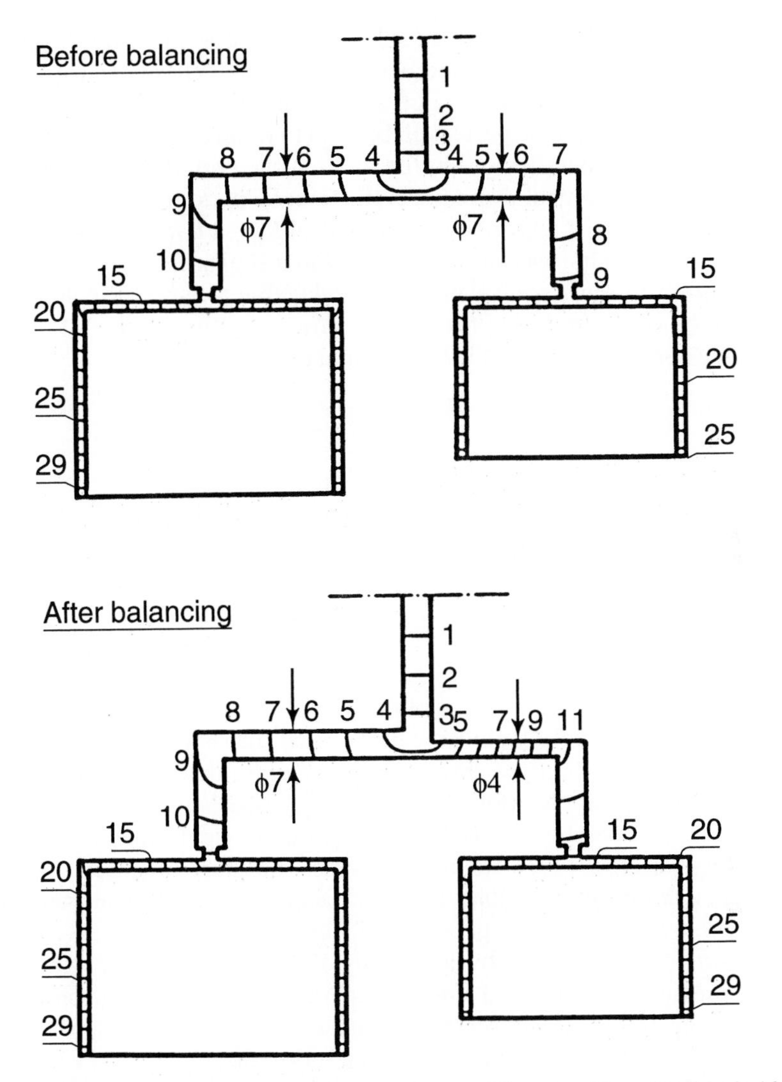

Figure 10.43 Balancing a runner system for two different moldings. Published with the permission of G. Menges, P. Filz, O. Kretzshmar, H. Recker, Th. Schmidt, and Th. Schact, "Mold Design with CAD Mould" in *Applications of Computer Aided Engineering in Injection Molding,* 1987, Hanser Publ. Munich–Vienna–New York.

Part Shrinkage. The changes in volume that may accompany cooling can cause a number of defects in the finished part. Although the temperature of the cooled part can be varied, it is preferable to set it as a part of the mold design process. The final temperature can be set so as to minimize shrinkage. This should result in minimum internal stresses in the cavity and in the part.

The melt (State 2) will fill the mold at a high temperature (T_2) and a high pressure (P_2). When the part is cooled to become a solid (State 1), the temperature will drop to T_1 and the pressure will drop to P_1. If these changes are such that the polymer shrinks, the part may pull away from the cavity walls. If the pressure in the part is too high, then the part might be hard to remove. A typical design goal is to have 1 to 10 bar residual pressure in the mold at the end of the cooling cycle. There are several ways to control this: control the final fill pressure P_2 based on T_2, control the residual pressure P_1 at a given T_1, or control $P_1 = 1$ bar by setting T_1.

All three choices can be calculated in a similar fashion. The goal to reduce or prevent shrinkage is equivalent to specifying that the specific volume of the polymer should be constant during the cooling process. If the volume of the mold is V, the specific volume of the melt is V_2, the total mass in the mold is

$$G = \frac{V}{V_2} \tag{10.30}$$

In the solid state, the total mass of the part must be the same and the volume of the mold should be the same.

$$G = \frac{V}{V_1} \tag{10.31}$$

Therefore, the specific volumes of the melt, V_2, and the solid, V_1, should be the same to keep the mold full. Cooling the part so that the solid has the same specific volume as the melt should reduce shrinkage. It is easy to determine these conditions using P-V-T diagrams or equations of state for polymers.

Figure 10.44 shows some P-V-T diagrams for eight commodity polymers. Figures 10.44 a–d are for semicrystalline polymers. The sharp breaks in the curve represent the crystallization conditions. It is preferable to have the material come out of the mold below T_m so that it will be a solid. The melt should be above T_m at the injection pressure to ensure good flow and filling of the mold. Suppose that HDPE is given a final pressure of 1000 bar at a final temperature of 200°C. Its specific volume is about 1.25 cm3/g. If it is cooled to 125°C, its pressure will be near 1 bar. The part will shrink some after it is unmolded, since additional crystallization will take place. Figures 10.44 e–h are for amorphous thermoplastics. These materials show a break in the volume versus temperature curves near T_g. These materials should be cooled in the mold below T_g at 1 bar in order to have stable solid parts. They should be processed above T_g at the injection pressure. If SAN is given a final injection pressure of 1000 bar at 200°C, its specific volume is 0.94 cm3/g. If it is cooled to 100°C, it should be just below T_g at 1 bar.

Models for Specific Volume. Models for specific volume with $V = f(T,P)$ can be used instead of the curves in Figure 10.44. The advantage of a model is that it can provide specific volume estimates at any set of conditions, and it can be differentiated to give the pressure corrections to enthalpy and entropy shown in Table 7.13.

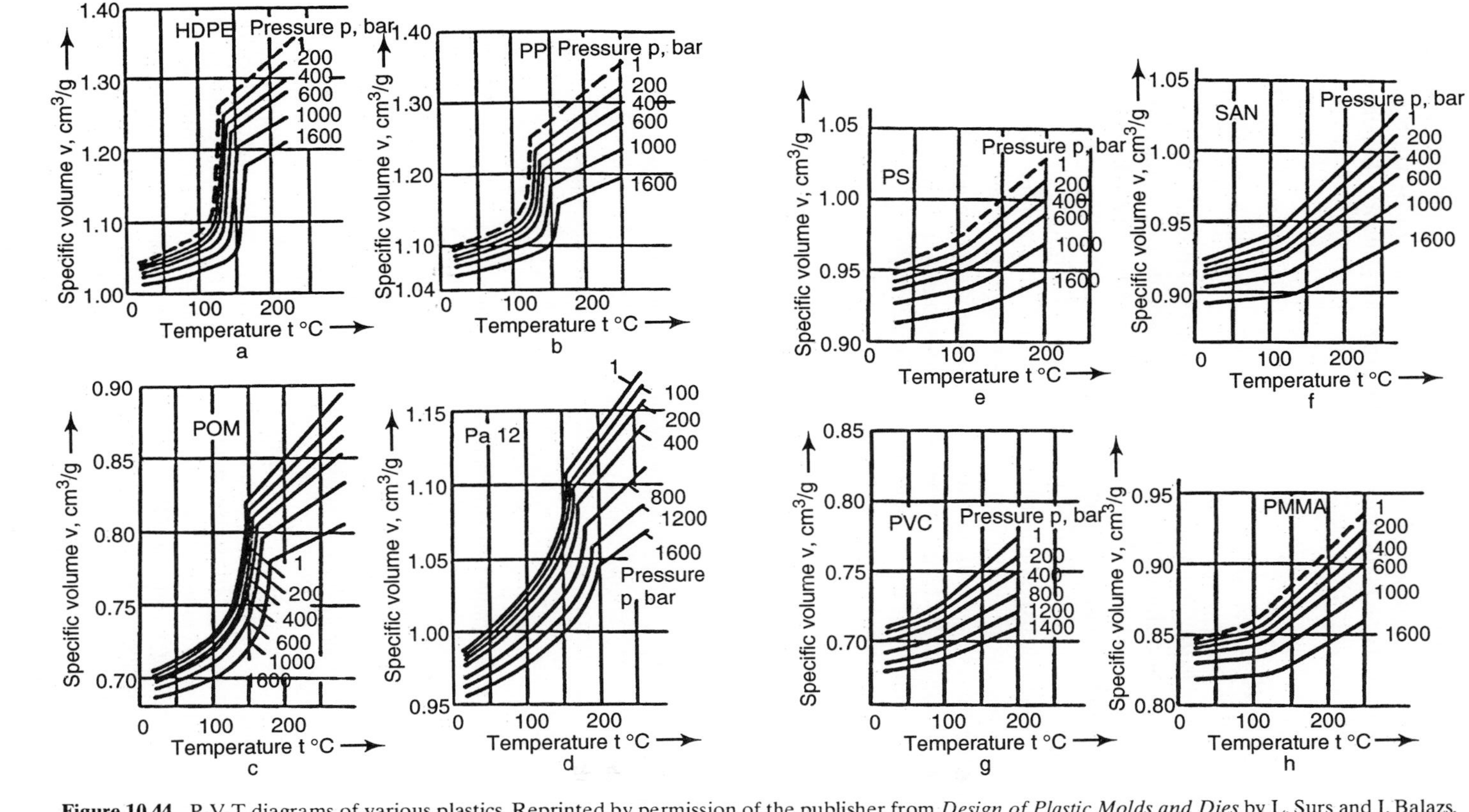

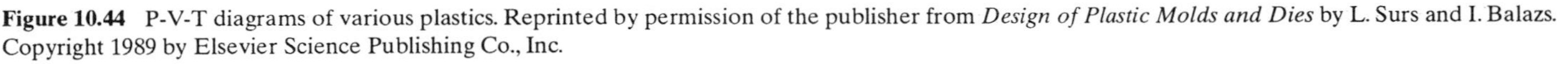

Figure 10.44 P-V-T diagrams of various plastics. Reprinted by permission of the publisher from *Design of Plastic Molds and Dies* by L. Surs and I. Balazs. Copyright 1989 by Elsevier Science Publishing Co., Inc.

Table 10.11 shows some equation-of-state models for polymer specific volume. The Spencer–Gilmore equation is similar to an equation of state for gases. The Tait equation only corrects for pressure, while the Tait–Rehage–Breuer equation includes temperature and pressure corrections.

10.7 CALENDERING

Calendering is the forming of wide, thick sheets of polymer material using a series of rollers. This operation is used to make thick sheets of thermoplastics and elastomer compounds; prepare rubber stock imbedded with tire cords, fabrics coated with rubber, and PVC compounds; and for mixing and blending of high solids compounds. In some systems, calendering lines act as a method for pick-up and delivery of hot melt compounds from one system to the next. Compounding will be discussed in this section since many materials used in calendering operations are highly modified. The comments on compounding often apply to other processing systems, such as extrusion, since the feed to these processes also is compounded. A process discussion is given first, followed by a discussion of hydrodynamic modeling of the process.

Example 10.5 Problem: *Evaluation of Specific Volume Data Using the Spencer–Gilmore Equation of State*

The polystyrene curves will be evaluated in this problem. The specific volume of glassy polymers changes at T_g. The Spencer–Gilmore equation models V_1 as

$$V_1 = w + \frac{R^1 T}{(P + \pi)}$$

The slope of curves in Figure 10.44 should be

$$\frac{\partial V_1}{\partial T_{P=0}} = 0 + \frac{R_1}{(P + \pi)}$$

Above T_g, the slope changes with pressure. Notice that the glass transition temperature of the material increases as pressure increases. Below T_g the curves change slope, and the model coefficients, R_1 and π, must change. The spacings between the isobaric lines in Figure 10.44 changes, so that the derivative of volume with respect to pressure is not constant. Equations of state can be used to model the volume of crystalline polymers if the volume change on melting is known. Often, the volume change on melting decreases as pressure increases.

TABLE 10.11 Equation-of-State Models for Polymer Specific Volume

Spencer–Gilmore:

$$(P + \pi)(V - w) = R^1 T$$

$$R^1 = constant;\ R^1 = \frac{R}{M_n}$$

Tait, Rehage, Breuer:

$$V = V_0 + \phi_0 T - \frac{k_0}{a}(1 + bT)\ln(1 + aP)$$

$$V_0 = volume\ at\ 0°C,\ 0\ atm;\ \phi_0 = \left(\frac{\partial V}{\partial T}\right)_{P=0};\ k_0 = \left(\frac{\partial V}{\partial P}\right)_{T=0}$$

Tait:

$$\frac{V}{V_0} = 1 - A\ln\left(\frac{B + P}{B + 1}\right)$$

Griskey-Whitacker:

$$V = 0.01205RP^{n-1}\left(\frac{T}{T_g}\right)^{m+1} \rho_0^{-0.9421}$$

$$R = ideal\ gas\ constant;\ \rho_0 = density\ at\ 25°C,\ 1\ atm;\ n, m = f(P)$$

10.7.1 Process Description

Figure 10.45 shows a typical calendering line. A typical calendering line would be designed to make a sheet 1.5 meters wide and 1 mm thick at a rate of 6.0 meters per minute. This would be 0.54 m^3/hr of material, equivalent to about 500 kg/hr of rubber compounds or 650 kg/hr of flexible PVC compounds. Powder blenders deliver blended particulate batches to a Banbury mixer. The powder blender in Figure 10.46 has a ribbon (helical) mixer for uniformly dispersing additives in the particulate solids. In this particular set-up, liquid plasticizer can be added using a spray system over the mixing volume. The weight fraction of liquid plasticizer that can be added in a powder blender is modest, since the discharge method is designed for moving powder, not wet agglomerating solids. This ribbon blender would be operated batchwise, with the agitator operating continuously. During powder charging and discharging, the ribbon conveys particulates throughout the mixing volume. The screener is intended to remove foreign matter that could damage the calender rolls. The blender can have a heat jacket in order to raise the temperature of the batch near the fluxing temperature of the compound (the temperature at which the components form a fused mass). Colors, fillers, lubricants, stabilizers, and plasticizers would be added at this stage. An alternative to the ribbon blender is a high-speed fluidizing mixer (Figure 10.47). Because this device rapidly builds up heat in the compound, it must be discharged on strict time schedules, while the low shear ribbon

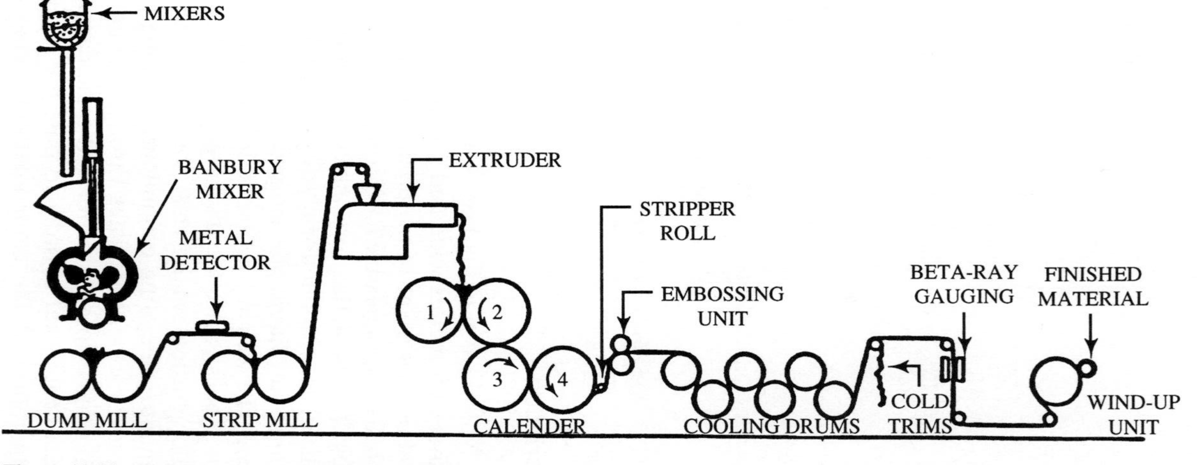

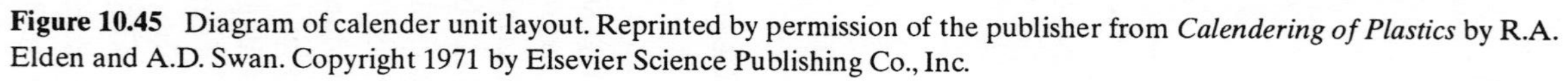

Figure 10.45 Diagram of calender unit layout. Reprinted by permission of the publisher from *Calendering of Plastics* by R.A. Elden and A.D. Swan. Copyright 1971 by Elsevier Science Publishing Co., Inc.

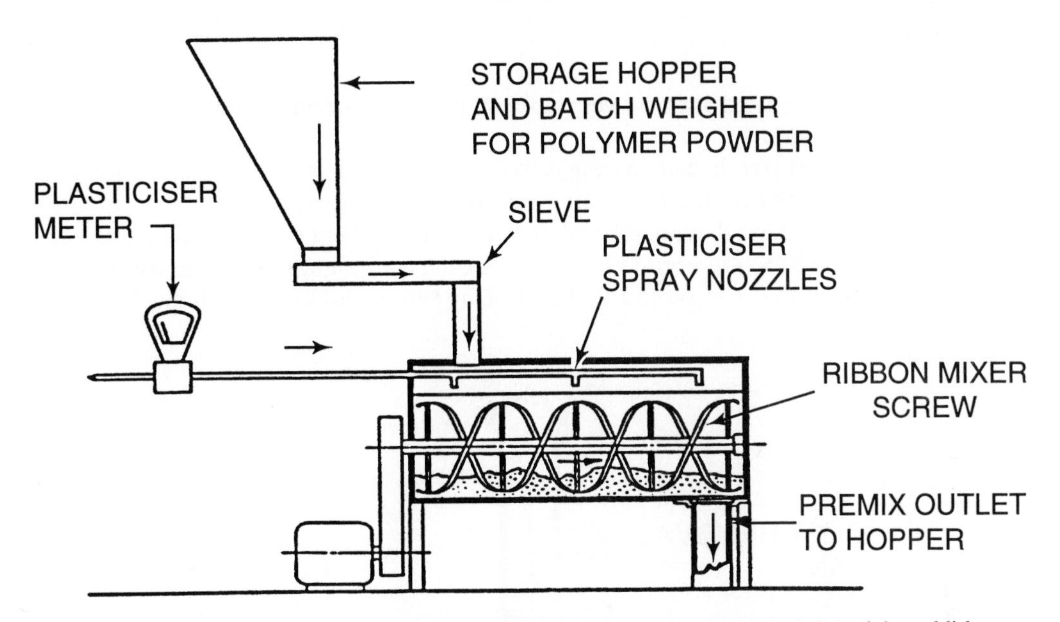

Figure 10.46 Sectional diagram of a ribbon mixer. Reprinted by permission of the publisher from *Calendering of Plastics* by R.A. Elden and A.D. Swan. Copyright 1971 by Elsevier Science Publishing Co., Inc.

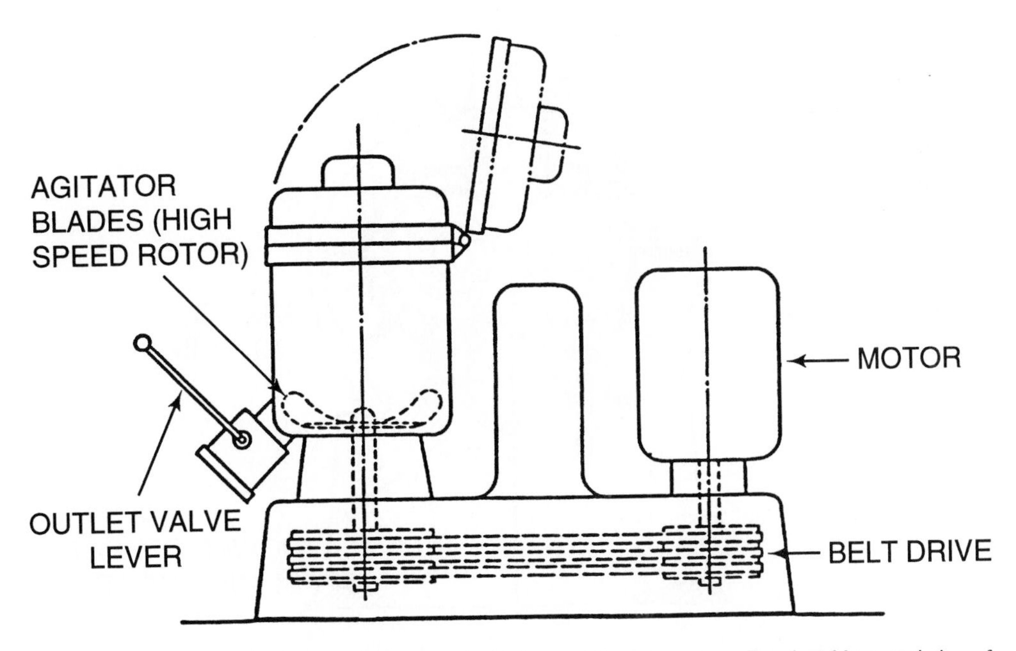

Figure 10.47 Sectional diagram of a high-speed fluidizing mixer. Reprinted by permission of the publisher from *Calendering of Plastics* by R.A. Elden and A.D. Swan. Copyright 1971 by Elsevier Science Publishing Co., Inc.

blender can be used as in-process storage until downstream equipment is ready for the powder blend.

The next step in calendering is to form a fused melt that can be applied to the rolls. While melt fusion can be done on the rolls for some systems, it can lead to serious defects in the sheet that are difficult to correct. High shear devices such as a Banbury Mixer (Fig. 10.48) are used to heat and flux the compound. The Banbury is another batch device that can be fed by hand or by screw and augur systems. The device

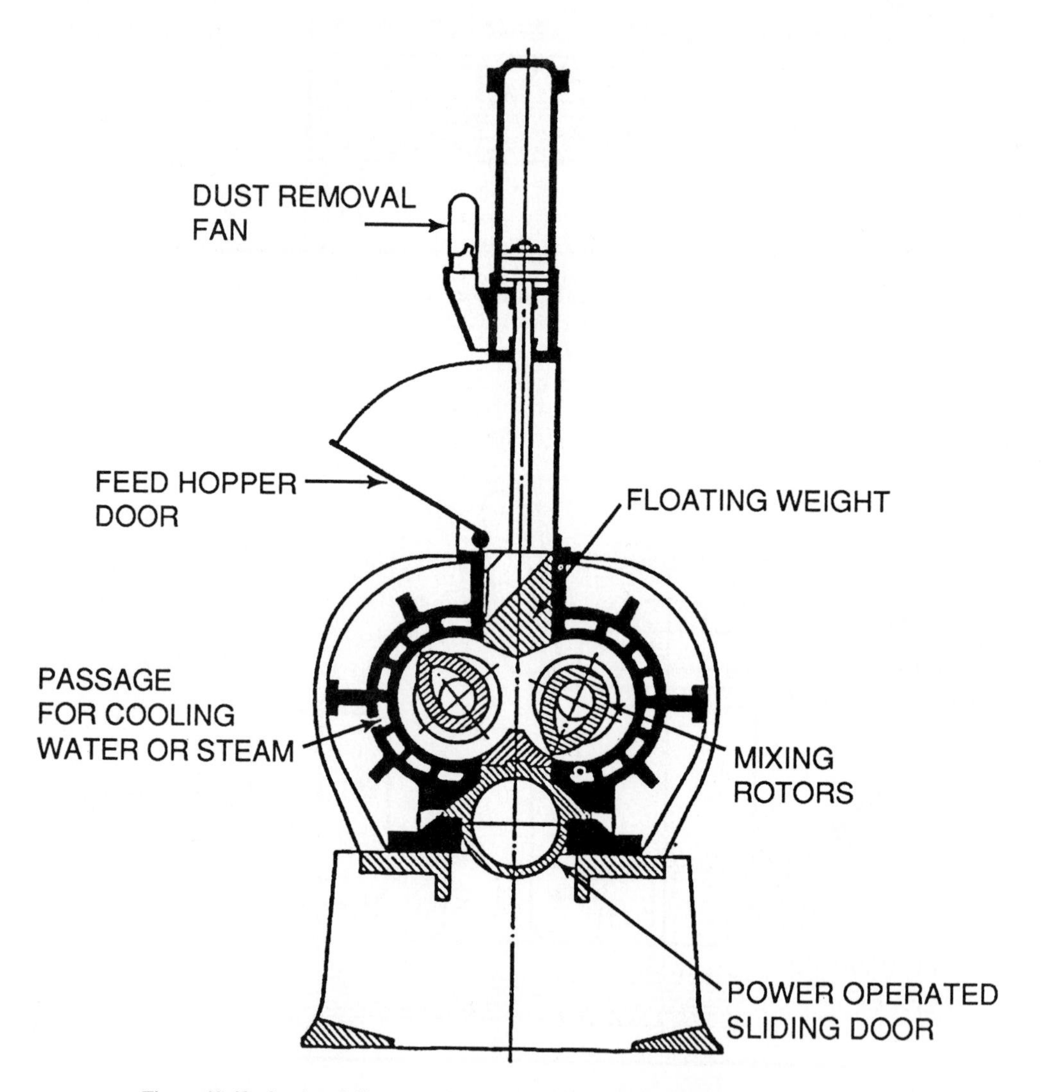

Figure 10.48 Sectional diagram of a Banbury Mixer. Reprinted by permission of the publisher from *Calendering of Plastics* by R.A. Elden and A.D. Swan. Copyright 1971 by Elsevier Science Publishing Co., Inc.

consists of a two-lobed mixing chamber with acentric mixing arms. The chamber walls can be heated with steam or cooled with water to control the temperature of the melt. A floating weight is used to retain the solids in the mixing chamber. Without this device, some "nervy" viscoelastic systems would literally jump out of the mixing chamber and bounce on top of the rotating arms. The batch, high shear mixer permits precise amounts of energy to be put into the compound, improving the batch-to-batch performance of the material. It also produces a fused melt with a precise thermal history.

The next step shown in Figure 10.45 is the pick-up of the melt dropped from the Banbury by a dump mill and its transport to an extruder system. These steps drop the temperature of the melt quickly and permit a check for metal contaminants. It is easiest to operate the calender feed rolls when they are fed with melt strips. The extruder system is a simple method for making the strips, and internal screen packs act as a further filter for oversized material. This intermediate process section provides "surge" capacity to smooth the normal operating difficulties associated with going from batch to continuous operations, and helps during product changeovers. Because extruders are difficult to purge, they are preferred only when long runs of similar compounds are expected.

The calendering system shown is a four-roll inverted Z-type (Fig. 10.49). The feed is dropped into the gap, or nip, of the rolls. Horizontal feed nips also are used but are more difficult to feed. The # 3 roll of this sequence is fixed, to provide a reference point for adjustments of the other rolls in the train. Rolls # 1, # 2, and # 4 are movable. Vertical arrangements (superimposed and offset) also are used but can be more costly to install because of their greater height. The calendering of thermoplastics usually involves a rise in temperature at each roll. Because the residence time on the heated rolls is short, the temperatures are as high as possible, which reduces viscosity,

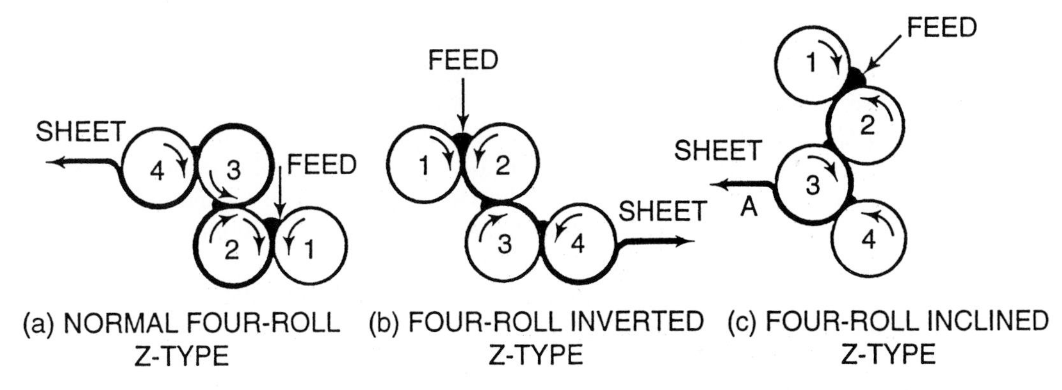

Figure 10.49 Feed and sheet paths on Z calenders. Reprinted by permission of the publisher from *Calendering of Plastics* by R.A. Elden and A.D. Swan. Copyright 1971 by Elsevier Science Publishing Co., Inc.

improves mixing, and gives the best tensile, elongation, and impact strength to the finished sheet.

Temperature and roll speed influence whether the sheet adheres to the roll. Most thermoplastic compounds adhere to the hotter roll of the pair. For the inverted Z-type sequence shown in Figure 10.49, the temperatures should be $T_1 < T_2 < T_3 < T_4$. Usually, the sheet adheres to the faster roll of the pair. Varying the speeds adjusts the rate of shear between the rolls. The material taken off roll # 4 is close to its final thickness. The surface can be embossed, followed by cooling through a series of rolls. The sheet is trimmed to size (the trimmings usually are recycled back to the dump mill). Cords and fabric can be impregnated with the sheet.

The major advantages of calendering are that it can be used to make sheets of high melt viscosity material with controlled short-residence times at temperature. Thermally unstable materials are candidates for this type of processing. High solids compounds can be handled at high processing rates. The disadvantages include the long line-out time for the equipment and the difficulties in setting the gap width across the rolls. Mixing occurs by the mechanisms of shear elongation (acceleration of flow into the roll nip) and kneading in the melt pools on take-up rolls. A key operating difficulty is starting up the line. The rolls usually are not heated faster than 1°C/min, meaning that a line operating near 200°C would take three hours to heat up. The side frames also need to reach temperature, which often increases warm-up time by a factor of two. Compound working between the rolls may raise the material temperature by 10°C to 20°C. It can take several hours after the start of sheet production before the system is at steady state. Therefore, calendering operations are most economical when long production runs are expected.

10.7.2 Compounding of Solid Polymers

There are a number of reasons for blending materials with polymers. These reasons include: less expensive finished product with similar properties, improved processing characteristics, improved mechanical properties, and unique properties of the blend. Compounds have a range of states of mixing from dispersions, in which the components are not mutually soluble, to alloys, in which the components are mixed at the molecular level. The nature of mixing can be further categorized by considering the blending of two different polymers. Table 10.12 shows the number of phases present as a function of the mixing between the two polymers. The compounds used on calendering lines vary from nonwetting to partial solutions to solutions, depending on the temperature and pressure of the system.

Blends are defined as compatible if the components stick together under mechanical stress. Some blends may appear to be compatible in the absence of external forces but will phase-separate under shear, tension, or compression. Miscible blends are defined as thermodynamic solutions, that is, mixtures that are statistically homogeneous.

An analysis of the thermodynamics will indicate whether a given polymer blend

TABLE 10.12 Number of Phases Based on Interfacial Wetting of Compounds

Wetting at Interface	Number of Phases
Nonwetting	2 (3 if air is entrained)
Wetting	2
Partial solution	3 (2 components + interphase)
Complete solution	1

is feasible. An analysis of the kinetics of the blend will indicate whether it is practical; that is, can the blend be made in a reasonable period of time. The kinetics depend on the rheology of the two components and the blend, and the rates of diffusion of one component in the other. An analysis of the morphology of the blend will help determine the blend properties. The state of mixing, or morphology, of the blend is dependent on the particle sizes of the phases and effects what tools are needed to describe it. Table 10.13 shows how the state of mixing and the particle size and analytical techniques are related.

The above discussion about the state of mixing, the thermodynamics, and the kinetics implies that the mixing technique used can have an effect on the blend properties. There are two basic techniques for obtaining miscible polymer blends. The first is melt blending, which is done above the melt temperature for both polymers. Melt blending is effective but requires large quantities of material, often requires long blending times due to the high viscosities, and may exhibit viscoelastic effects and may degrade polymer properties depending on the time-temperature history of a specific batch. The batch mixing systems on the feed end of the calendering process provide uniform blending with precise control of thermal history. The backmixing through the calender rolls usually is lower than that of extrusion systems. In addition, calendering is required to make wide sheets of material. Figure 10.50 shows blends of polystyrene with a diblock copolymer in a twin screw extruder. The diblock is dispersed in the

TABLE 10.13 Blend Morphologies

State of Mixing	Particle Size, A	Techniques
2 Phases	< 3,000 50–3,000	Eye SEM TEM
1 Phase—clear T_g	20–50, long chain segments	DSC TMA Dynamic mechanical tests
1 Phase–local mode interaction, antiplasticization	2–10, short chain segments	Dynamic mechanical Neutron scattering Diffusion Inverse phase gas chromatography

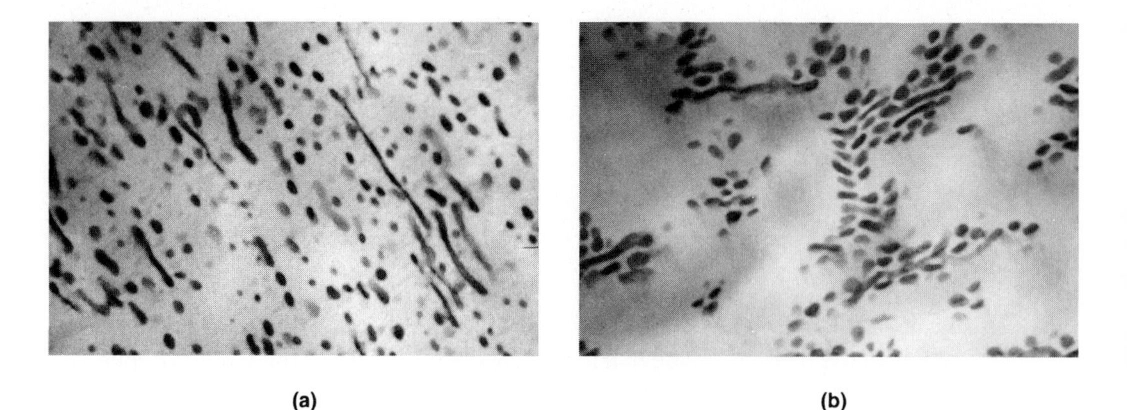

(a) (b)

Figure 10.50 20 wt% FINA 315 in Styron 666D. Melt mixing at 130°C. **(a)** as mixed in the extruder, **(b)** after annealing at 200°C. Published with the permission of K. Nichols.

polystyrene but does not achieve an equilibrium structure in the moderate residence time of this equipment.

The second technique is solution blending. There are three alternative ways of doing this: solvent casting, selective precipitation, and freeze-drying. All three techniques are affected by the addition of the solvent. The addition of solvents often increases the cost of the compound, and most solvents must be removed from the final product. This step adds extra operating and environmental costs. The manner in which the solvent is removed can have a big effect on the material. Solution blending usually is used when precise control of the blend or surface properties are required. Solution blending also does not necessarily give equilibrium structures of the polymer phases.

Solvent casting requires long devolatilization times. Freeze-drying also takes long times and requires low temperatures. Precipitation of the blend in a nonsolvent or a poor solvent can lead to fractionation of different molecular weight species or agglomeration of precipitates. However, molecular scale mixing can be achieved with solution blending followed by fast precipitation in a poor solvent.

Phase diagrams of polymer systems shown in Chapter 5 can be applied to blended polymer systems. Pressure, temperature, and composition all affect phase equilibria. The glass transition temperature is usually not a linear function of composition and may be less than the T_g determined by the linear average. This is a disappointment to the blend designer, since one reason for blending is to increase the glass transition temperature of the blend above that of the base polymer. Upper and lower critical solution temperatures can constrain the use and processing temperatures for a compound. The UCST is known for many polymer systems, but the LCST is often not known. The lack of knowledge can be harmful when deciding on the processing temperature for mixing. If the mixture is blended above the LCST, then only a dispersed two-phase blend will be achieved. Processing between the LCST and the UCST will be necessary to make a compatible blend.

The upper critical solution temperature should be related to the use temperature of the blend. If the use temperature drops below the single phase region, then there will be a thermodynamic driving force for phase separation to occur. Of course, whether phase separation does actually take place over the lifetime of the part depends on the kinetics of the process.

The state of mixing of the blend can affect measurements intended to discriminate between one-and two-phase systems. Differential scanning calorimetry is often used to determine the T_g of a polymer blend. However, phase separation of two polymers depends on how intimately they have been mixed and on how long they remain at the testing temperature. With DSC equipment, it is easy to heat or cool at rates up to 50°C/minute. If the polymer sample was intimately mixed, there might not be enough time at high temperatures for the polymers to phase separate (repeated heating cycles are often necessary). These types of time effects can lead to poor conclusions about the performance of a blend in an application. It is wise to subject the blend to time-temperature histories similar to those that it will encounter during processing. For example, a powder-compounded blend might be tested at the maximum extruder temperature for the expected residence time to determine whether phase separation will take place.

10.8 POLYMER RECYCLING, POLYMER DISPOSAL, AND BIODEGRADABLE POLYMERS

There has been increasing interest in recycling plastic materials as the municipal solid waste (MSW) streams have increased, landfills have become full, and new landfills have become difficult to site. There also are industrial waste streams that enter the landfill system. Plastic structural materials are different from metals in that they contain energy that can be recovered by incineration. Metals require more energy per unit volume than polymers and have no value as a fuel. Polymers have *material* and *energy* values that should be considered in making disposal decisions. Of course, the metals cannot be converted to energy, but the polymeric materials can.

It seems to be energy efficient to recycle metals. The initial fuel cost to make them is high. Much of the energy requirement is for reduction of the metal oxide to molten metal. Recycled metals contain little oxide and can be remelted efficiently. Scrap transportation costs can have a significant impact on the economics, but metals are dense so they are easy to separate from mixed waste streams.

Priorities for Recycling, Reuse, and Disposal. There are some general priorities for recycling, reuse, and disposal of polymeric waste. They are listed in order from high economic value to low economic value.

1. Recycle in the same application
2. Recycle as a lower grade or quality of the same material
3. Reuse in solid or liquid forms, possibly in a mixed recycle stream
4. Disposal by combustion to recover the feedstock energy in the material and
5. Disposal by landfill, incineration, or other waste treatment methods

Homopolymers, compounds, and resins have been formulated with specific applications in mind. Effort and expenses have been used to achieve special material properties, and it makes sense to recycle such materials in their original application. This represents the highest value that the material can have. Off-spec and off-grade materials are sold for a lower price and are used in applications not as sensitive to material properties. "Reuse" suggests using the polymer as a material, but for low-value applications that normally would not require the properties of the material. The economic value of the polymer in reuse applications is low, but reuse is preferable to paying for its disposal. Burning to recover the polymer's fuel value is listed ahead of disposal. In general, using waste polymer as a material is preferred because it usually has higher value than as a fuel. The lowest priority on the scale is disposal, which represents a cost to someone.

Thermoplastics and thermosets are different with respect to recycling opportunities. Unreinforced thermoplastics can be recycled fairly easily by regrinding and remelting. Some limits on their recycling include the loss of properties due to degradation in repeated heating cycles and contaminants that would be introduced during recycling. Thermosets cross-link while being formed into products. Few thermosets can be depolymerized to monomers or oligomers for recycling. Ground cross-linked material will not perform the same as prepolymer in forming operations. However, thermosets can be used for fuel. Exceptions to this statement are vulcanized rubber and thermoplastic elastomers. Ground rubber tires often are recycled in new tire compounds. The sulfur linkages can reform during curing. Thermoplastic elastomers can be remelted and reformed.

Soda Pop Containers: Glass, Metal, and Plastic. Plastics recycling has become popular in the past two decades. Plastics have advantages over metals in structural applications because they have low density. A plastic soda container weighs less than the corresponding glass container and is less costly to transport to customers when the containers are full. Conversely, they can be more difficult to collect and transport to recycling facilities. Recycled aluminum cans can be crushed by the retailer to increase their bulk density for shipping. Plastic bottles can be crushed, but they still have a low bulk density. Contamination in the plastic bottles is a major concern for recycling. Metal cans can be remelted, and any organic contaminants will be pyrolyzed or oxidized in the process. Glass bottles can be washed and sterilized. The typical glass beverage container is reused ten times before it chips and needs to be remelted. Recycling plastic containers for reuse as beverage containers was not done in the past because of contamination issues. Plastic recycling probably will be driven by the large supply of used material that can be recycled, rather than the demand for more plastic material for bottles.

This section provides an overview of polymer recycling, polymer disposal, and the biodegradation of plastics in the environment. It is intended to introduce the issues of recycling and shows the current approaches for dealing with these problems.

10.8.1 Industrial Plastic Waste

Much of the industrial plastic waste that can be recycled is being recycled internally. Production of thermoplastic parts often is done by injection molding. This process leaves molding networks and flashings that can account for up to 25% of the shot size in some applications. These materials are reground and mixed with fresh material to make more parts. Regrind has the best characteristics for recycling: it is near the point of manufacture, it has the same composition as the fresh material, and it will be recycled to the same application (it has the same material properties). The fabricator has good incentives to reduce this internal recycle. Grinding and remelting cost money. Repeated heat histories leads to degradation and loss of properties, particularly surface finish. Contaminants get into the material with repeated handling and can lead to poor part quality. On the other hand, sending molding scraps to the local landfill would be very expensive. Therefore, the plastic waste from industrial sources often is highly contaminated (floor sweepings) or off-spec material (degraded, cross-linked, etc.).

Industrial plastic waste includes polymer production waste, polymer processing waste, and demolition and construction waste. Waste from production processes includes sludges separated from process water, reactor cleaning wastes, unmarketable off-spec material, and floor sweepings. These materials are too highly contaminated or have material properties too poor to be recycled. Thermoplastic polymer processing waste usually is reground and recycled to the process. Thermoset polymer processing waste may be difficult to recycle because of the cross-linking that has taken place. Composites, sheetmolding compound, and reinforced compounds may be difficult to recycle because of their fiber and filler content, as well as the cross-linking reactions that have taken place. Construction and demolition waste includes plastic coverings for wiring plastic conduit and piping, flooring materials, and insulation and ceiling materials.

10.8.2 Consumer Waste

Some of the most visible polymer waste ends up in the hands of the consumer and becomes part of the municipal solid waste (MSW) stream. Most of this stream is disposed of in landfills. Table 10.14 shows the estimated plastic waste generated in the

TABLE 10.14 1984 Plastics Production and Projected Waste Generation

Type of Plastic	Production* 10^6 t	Discarded† Waste per year, %	10^6 t
Polyethylene	6.00	65.6	3.94
Polystyrene§	2.69	38.1	1.03
Polypropylene	1.86	27.5	0.51
Poly(vinyl chloride)	3.05	17.9	0.55

*SPI, 1975.
†Vaughn et al., 1975.
§Includes PS, ABS, and other copolymers.

United States in 1984 for the four largest thermoplastics. Table 10.15 classifies plastic wastes by their applications. Polymers used for durable goods and construction have long lives. Packaging has a lifetime of a year or less and is the most visible waste component. Because of the varying lifetimes for these materials, it is difficult to close their material balances. Most of the consumer plastic waste gets in the municipal solid waste stream. The amount of this stream has been nearly constant on a per capita basis for many years (Alter and Dunn, 1980; Alter, 1983). The composition of plastics in MSW varies by the survey method and the location.

Plastics in Packaging Materials. Product packaging is intended to protect them from air, water, odor, microorganisms, light, and mechanical damage. Ceramics, glass, steel, aluminum, and paper products have been used for packaging. Cellophane was used prior to World War II, but large volume usage of plastics began in the 1950s

TABLE 10.15 Plastic Waste Generation Classified by Application*

Use	Approx. Life† Years	Discards %/yr	1984 Use 10^6 t	Estimated Waste 10^6 t
Packaging	1	100	5.62	5.62
Transportation	5	20	0.96	0.19
Furniture and housewares	10	10	0.96	0.10
Electrical and electronic	10	10	1.25	0.13
Building and construction	50	2	4.39	0.09
Other	50	2	4.74	0.01
Total	6.14			

*SPI, 1985.
†Vaughn et al., 1975.

with low density polyethylene film in bags and in squeeze bottles. There are two categories of packages: flexible and rigid. Flexible packaging has thicknesses less than 0.25 mm and may contain multilayer structures. About 50% of the flexible packaging is used for food, 10% for health and beauty aids, 5% for drugs and pharmaceuticals, and the remainder for a variety of consumer products.

The material balances on input and output of polymeric wastes are important, since the best current recycling and reuse methods require separation of plastic materials by grade. Figure 10.51 shows the container code system for plastic bottles. In this application, there are seven classes of materials being used. Efficient separation

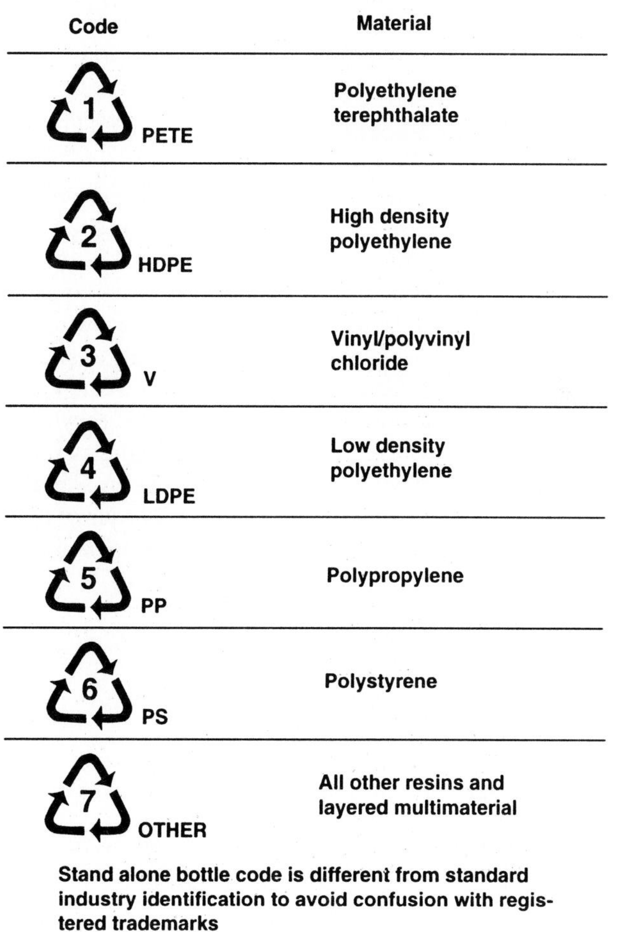

Plastic container code system for plastic bottles
Code
Material
1
PETE
Polyethylene terephthalate
2
HDPE
High density polyethylene
3
V
Vinyl/polyvinyl chloride
4
LDPE
Low density polyethylene
5
PP
Polypropylene
6
PS
Polystyrene
7
OTHER
All other resins and layered multimaterial
Stand alone bottle code is different from standard industry identification to avoid confusion with registered trademarks

Figure 10.51 Plastic bottle material coding system.

of plastic bottle streams into classes (1, 2, 3, etc.) will improve the quality of the recycled products. Recycling costs might be reduced if the composition of the MSW streams are known.

The amount of the MSW stream has been nearly constant on a per capita basis for many years. However, the composition of plastics in a given MSW stream varies by the survey method and the location. The typical amount of plastic in U.S. MSW streams in the mid-1970s was 6.0wt% (Alter, 1989), but this number varies by about 60% depending on the city, the method for sampling the streams, the method for analyzing the streams, and whether any separation or scavenging has taken place. By 1984, the number had grown to 7wt% (Selke, 1988). Household wastes have different compositions than mixed municipal wastes, which include commercial and light industrial donors. A survey in the seventies (Richardson and Havlicek, 1978) suggested that high income households discarded lower amounts of packaging wastes. Mixed municipal waste may have a higher fraction of plastics, averaging about 12.6wt% (Mantell, 1975).

10.8.3 Recycling Plastic Packaging

Selke (1988) has written a comprehensive review of plastic package recycling. Plastics are continuing to replace glass and steel in packaging in the United States. However, some states have acted to reduce plastics in packaging. Vermont has considered a ban of PVC packaging. The city of Berkeley, California, has considered a tax on plastic packaging. The driving force for this legislation is the growing MSW problem. Landfills are becoming full and are becoming more difficult to site. Plastic packaging is a target because it is visible, it occupies a higher volume fraction of the landfill due to its low density, and the materials do have some value if they can be recycled.

Six polymers make up most of the plastic packaging materials (Table 10.16). The highest tonnage is that of polyethylene, with LDPE being slightly higher than HDPE. Polystyrene and polypropylene are about the same order of magnitude, followed by PET and PVC. Rigid and flexible applications make up similar fractions of the total

TABLE 10.16 Plastic Packaging Materials (U.S.)

Resin	Million kg	% of Total
Low density polyethylene (LDPE)	1739	33.4
High density polyethylene (HDPE)	1605	30.8
Polystyrene (PS)	550	10.6
Polypropylene (PP)	480	9.2
Polyethylene terephthalate (PET)	343	6.6
Polyvinyl chloride (PVC)	265	5.1
Others	224	4.3

**Modern Plastics*, 1987a

TABLE 10.17 Resins in Plastic Containers (U.S.)*

Resin	Million kg	% of Total
HDPE	1399	53.1
PS	430	16.3
PET	336	12.8
PP	167	6.3
PVC	143	5.4
LDPE	106	4.0
Other	56	2.1

SOURCE: *Modern Plastics*, 1987a

usage. Table 10.17 shows the resin usage in plastic containers. Most of the HDPE, PS, and PET are used in bottles. Most of the LDPE is used as film (Table 10.18).

There are three operations in recycling materials: collection, processing, and end use. Collection involves gathering materials and separating them by polymer type. The U.S. Bureau of Mines conducted research in the 1970s on a variety of methods for separating plastics from mixed waste streams (Lidner, 1981). Methods included separation by density in aqueous streams, surfactant aids for density separations, electrostatic separation, hydraulic separation, and air classification. These methods were not efficient at separating the materials and were very expensive. The alternative to separation technologies is encouraging consumers to separate their recyclable plastics by type. This requires education of the consumer and significant effort by the consumer as well.

Consumer participation is highest when there are economic incentives for separation and recycling. An example is the five and ten cent deposit on PET beverage bottles. An alternative to cost incentives is mandating recycling. New Jersey was the first state to require recycling. Only three materials need to be recycled, so plastics may not necessarily be included.

Processing usually includes densification as a first step. Grinding and bailing bottles greatly increases their bulk density and reduces transportation costs for

TABLE 10.18 Resins in Film (U.S.)

Resins	10^6 kg	% of Total
LDPE	1311	72.1
PP	200	11.0
HDPE	150	8.3
PVC	92	5.0
PS	23	1.2
Other	44	2.4

SOURCE: *Modern Plastics*, 1987a.

further processing. If various plastics are not separated from other materials, this should be done. In addition, labels, dirt, food residues, and any other contamination should be removed. Mixed plastics streams can be fabricated into useful objects, but their properties usually are much lower than the original homopolymers. In addition, high mixing is required in order to disperse the incompatible phases in each other. The classic example is the fabrication of plastic lumber from mixed plastic streams. Europe and Japan have had some success with these streams (*Modern Plastics*, 1987b).

The end use of recycled plastic bottles in the United States is not the highest priority one. While the FDA does not specifically exclude the use of recycled plastics in beverage bottles, there are enough requirements for purity documentation that they are effectively excluded. The major use for recycled PET is in fiberfill or plastic strapping. The HDPE base caps from the same bottles are recycled as base caps or as agricultural drainage pipes. Polyester x-ray film can be recycled as can the polypropylene shell of car batteries.

Two types of effort could increase recycling: research on overall systems and consumer education (Selke, 1988).

Example 10.1 Recycling HDPE Milk Bottles

This example explores some of the technical issues associated with recycling a consumer product—HDPE milk bottles. High density polyethylene is made by polymerizing with a coordination complex catalyst to make material with specific processing and performance properties. Polyethylenes for blow molding into bottles have density targets between 0.964 to 0.956 and a melt index range of .02 to .1 g/10 min. This provides a good balance between adequate processing properties (high viscosity to hold the parison shape during blowing) and performance properties (high crystallinity to provide good tensile strength).

These properties may be affected by recycling. An additional problem might be mixing HDPE's from different suppliers. Pattanakul, Selke, Lai, Grulke, and Miltz, (1988) reported the effects of recycling on the properties of HDPE. Milk bottles had their labels removed, were washed and chipped, and then dried at room temperature. Physical properties were tested on samples of fresh resin and blends of fresh resin with recycled chips. Melt flow index, tensile strength, and modulus did not change significantly for all samples, suggesting that little degradation took place during processing. However, the percent elongation at break (Table 10.19) and impact strength (Table 10.20) were reduced as recycled material increased in the blend. Polyethylene is known to undergo chain scission and cross-linking during processing (Rideal and Padget, 1976; Mitterhofer, 1980). The reductions in these properties may not be enough to prevent recycling, but they indicate that fabricators should be cautious of possible property reductions when using recycled materials.

(continued)

An additional concern with recycled plastics is their absorption of contaminants from the environment or from their contents. Butyric acid is one of the important flavor notes in milk and sorbs into polyethylene milk containers (Hernandez, 1988). A number of organic chemicals can sorb into polyethylene (Kwapong and Hotchkiss, 1987; Mannheim et al., 1987) and then volatilize when recycled material is remelted. Butyric acid has modest effects on the elongation at break and the impact strength. Impact strength is reduced about 15% when butyric acid saturates the resin.

TABLE 10.19 Elongation at Complete Break (%)

Material	Run 1		Run 2	
	Mean	SD[1]	Mean	SD[1]
Virgin HDPE	69.7	16.5	74.0	17.5
10% Recycled	62.7	10.1	62.4	6.5
20% Recycled	47.2	8.8	51.3	12.5
50% Recycled	48.9	18.7	41.4	19.4
80% Recycled	35.1	9.2	34.6	9.4
100% Recycled	36.9	18.2	30.7	4.7

SOURCE: C. Pattanakul, S. Selke, C. Lai, E. Grulke, and J. Miltz, "Properties of Recycled High Density Polyethylene Milk Bottles," *ANTEC*, 1802–1804 (1988).

TABLE 10.20 Izod Impact Strength of Mixtures of Virgin and Recycled HDPE

Material	Mean		Standard Deviation	
	ft-lb/in	Nt-m/cm	ft-lb/in	Nt-m/cm
Virgin HDPE	2.522	1.346	0.160	0.085
10% Recycled	2.693	1.437	0.261	0.139
16.7% Recycled	2.608	1.392	0.156	0.083
50% Recycled	2.409	1.286	0.138	0.074
80% Recycled	2.231	1.191	0.238	0.127
100% Recycled	2.201	1.175	0.144	0.077

SOURCE: C. Pattanakul, S. Selke, S. Lai, E. Grulke, and J. Miltz, "Properties of Recycled High Density Polyethylene Milk Bottles," *ANTEC*, 1802–1804 (1988).

This example illustrates two phenomena often present in recycling systems: reduction of some properties as the fraction of recycled material in the processed polymer is increased and contamination of the polymer by chemicals in the environment or in the bottle contents.

Work on conversion and recovery methods for plastics has increased recently (Table 10.21). Many of the concepts developed by research will need to be tested in pilot and demonstration stage sizes in order to identify practical problems with their implementation. Current recycling methods cost between $0.50 and $1.00 per pound (Fig. 10.52). These costs are near the prices of fresh commodity resin; therefore, recycling technologies will need to be efficient and produce high quality (high purity) products.

PROBLEMS

10.1 PVC pipe can be made from powdered compound. Typical PVC processing problems include defects caused by under- or overworking the compound, which results in polymer degradation due to too much heat history. Setting the pressures, temperatures, and flow rates for powder compound extrusion can be difficult. At your plant there are several truckload equivalents of ¼″ regrind chunks (normal powder compound particle size is 100–200 microns). The plant's best customer needs five truckloads of pipe made for him today, and your plant manager wants to use this opportunity to get rid of the regrind by putting some

TABLE 10.21 Conversion/Recovery Methods for Plastics—Research or Demonstration Stage Industrial Applications

Stream	Method	Products	Reference
PET bottles	Hydrolysis	Monomers	—
	Methanolysis	Dimethyl terephthalate	—
		Ethylene glycol	—
	Spinning from naphthalene	Fiber	Debotham et al., 1977
	chipping and extrusion	Fiber	—
Mixed plastic Streams	Froth flotation separation	Separated polymers	H. Alter, 1978
Mixed homopolymers (no fillers or foams)	Liquid density	Separated polymers	—
Mixed plastic streams	High shear extrusion	Mixed solid polymers	—
Wire and cable	Shredding and air separation	Conductive metals, PVC, PE streams	Bevis et al., 1983
PP battery cases	Wet grinding, washing air classifying	Contaminated product	—

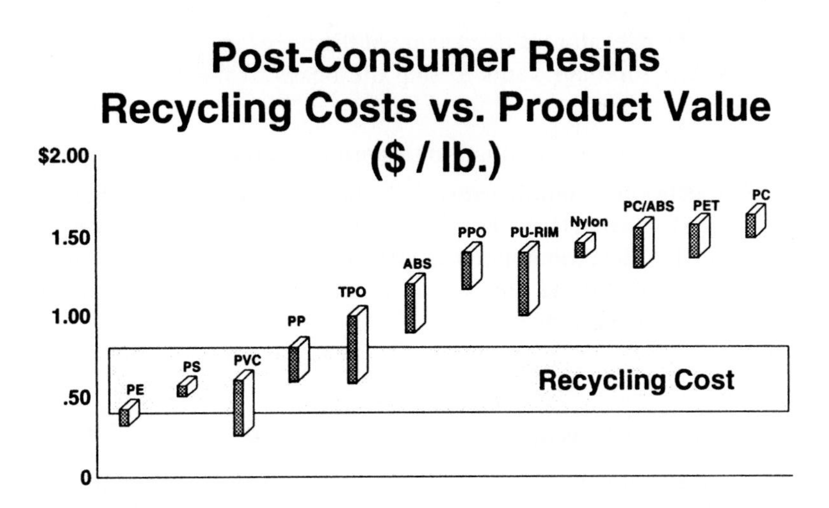

Figure 10.52 Recycling costs for commodity and engineering polymers (A.J. Carbone). Published with the permission of the Dow Chemical Company.

into the powder compound line for this run. Advise the plant manager on this matter and explain your position.

10.2 A linear polyethylene is to be used in the injection molding of wastebaskets. Given the data in the following table, choose the best candidate material for this service and explain why. What other factors might you consider besides the information given?

Material	Melt Index	Minimum Shear Stress for Flow
A	0.2	20,000 psi
B	0.9	5,000 psi
C	5.0	500 psi

10.3 A fluoroelastomer is being proposed as a substitute for an injection-molded gasket currently being used in an engine carburetor. Suggests tests that should be used for evaluating the fluoroelastomer for this application.

10.4 As a product engineer, you have been called in to make a decision on the use of a batch of extrusion pellets that have possibly been contaminated for making pipe. Here are the facts:

- each batch of pellets (10,000 lb) takes an hour to make
- about 30 minutes into the run, an operator saw the light above the roll mill shatter, sending shards of glass everywhere, including into the compound
- the quality assurance sample, taken 15 minutes into the run, showed normal properties

- the operator checked some of the boxes for glass and could not find any
- the foreman wants to ship the boxes to satisfy a rush order for your best customer

Propose a solution to this problem and explain your answer.

10.5 Extrusion generally exerts less shear on a compound than injection molding. The following table shows some test compound compositions. Of the pair, A and B, which test compound should work in extrusion but might not work in injection molding? Of the pair, C and D, which test compound should you try first in injection molding?

	Sample			
Compound Property	**A**	**B**	**C**	**D**
Cross-linking	low	high	moderate	moderate
Plasticizer	moderate	moderate	low	high

10.6 As President of Polaski Pickles, you are looking for a plastic pipe material for your pickling brine. The system pressure is 60 psig and 40°C. A local sales rep has suggested that you consider polyethylene for this purpose. Find the properties of polyetheylene pipe in the literature and make a decision on this question.

10.7 Butyric acid (BA) sorbs into polyethylene milk bottles and can cause off-odors when the bottles are recycled. Polyethylene was saturated with butyric acid at several temperatures under conditions in which the vapor activity was one.

Temperature, °C	P_{BA}^{sat}, mm Hg	Equilibrium Conc. gBA/100gPE
5.6	0.15	3.29
23.3	0.87	3.24
55.0	3.15	3.20

Determine the values of χ for butyric acid in PE. Estimate the amount of butyric acid soluble in a polyethylene milk bottle at 10°C if its activity in the liquid phase is 0.02. Would you expect this amount of butyric acid to volatilize at processing temperatures?

10.8 Derive the equation for the steady-state flow of a power law fluid in a capillary. Sketch the velocity profiles for $n = 0.8$ and $n = 1.2$.

10.9 SAN is being injection molded at 200°C and 100 bar. Propose a temperature at which the mold should open in order to reduce part shrinkage. If chilled cooling water is available at 50°F and a 10°F temperature rise is permitted, calculate the minimum water flow needed to cool a 50 oz shot.

10.10 Estimate the vapor pressure of water over a molten polyurethane at 110°C if the solid was (a) saturated with water at 70°F, (b) in contact with 65% RH air at 70°F, and (c) had 90% of the original water (0.20wt%) removed by drying.

10.11 Generate a fraction moisture remaining curve for the cases: (a) 5 mm diameter nylon spheres, and (b) chopped nylon strands 3 mm thick and 8 mm long.

10.12 Calculate the pressure distribution as a function of height in a polystyrene storage silo. Its diameter is 2 m and its height is 10 m. $K = 0.30$ for this problem. How would you design an experiment to measure K?

10.13 Fit the Spencer–Gilmore model to the polystyrene data of Figure 10.44. Determine the coefficient of thermal expansion and bulk compressibility modulus, and compare to literature data.

10.14 Repeat Problem 10.13 for polyethylene. What critical assumptions need to be made to fit the equation to the data?

10.15 Use Equation 8.110 to model the change in T_g with pressure for polystyrene. Does this model represent the data in Figure 10.44 well? Explain.

10.16 The mixing of PVC in a twin screw extruder has been analyzed by residence time distribution studies. Pulses of radioactive PVC were put into the feed hopper and radiation counts were taken at the second mixing zone and the end of the screw. Determine an average residence time for each zone.

Mixing Zone	Radiation Counts	Time, s
2	37	65
	67	85
	70	105
	43	125
	8	145
end of screw	10	105
	43	125
	60	145
	58	170
	27	185
	13	205
	2	225

REFERENCES

H. ALTER, "Disposal and Reuse of Plastics," in *Encyclopedia of Polymer Science and Engineering.*, Vol. 5, 1989.

H. ALTER, *Materials Recovery from Municipal Waste Unit Operations in Practice*, Marcel Dekker, Inc. NY, 1983.

H. ALTER and J.J. DUNN, SR., *Solid Waste Conversion to Energy. Current European and U.S. Practices*, Marcel Dekker, Inc. NY, 1980.

A. APICELLA, R. TESSIEREI, and C. DECATALDIS, *J. Membrane Sci. 18*, 211 (1984).

R.M. BARRER, J.A. BARRIE, and J. SLATER, *J. Polymer Sci 27*, 77 (1958).

J.A. BARRIE, P.S. SAGOO, and P. JOHNCOCK, *J. Membrane Sci. 18* 197 (1984).

A.R. BERENS, *Angew. Makromol. Chem. 47*, 97 (1975).

M. BEVIS, N. IRVING, and P. ALLAN, *Conservation and Recycling 6*, 3 (1983).

J. BRANDRUP and E.H. IMMERGUT, *Polymer Handbook*, 3rd ed., Wiley, 1989.

B.J. BRISCOE, C.M. POOLEY, and D. TABOR, in *Advances in Polymer Friction and Wear*, Vol. 5A, (L.H. Lee, Ed.), Plenum Press, NY, 1975.

R.L. BROWN and J.C. REYNOLDS, *Principles of Powder Mechanics*, Pergamon Press, Oxford, England, 1966.

D.J. CARLSON and D.M. WILES, "Degradation," in *Encyclopedia of Polymer Science and Engineering*, Vol. 4, 1986.

H.S. CARSLAW and J.C. JAEGER, "Conduction of Heat in Solids," Clarendon, 1959.

R.T. CHERN, W.J. KOROS, E.S. SANDERS, and R. YUI, *J. Membrane Sci. 15*, 157 (1983).

J. CRANK, *The Mathematics of Diffusion*, Clarendon, Oxford, England, 1975.

N.S. DEBOTHAM, P.D. SHOEMAKER, and C.W. YOUNG III, U.S. Pat. 4,003,880, 4,003,881 (Dec. 20, 1977), 4,137,393 (Jan. 30, 1979) to Monsanto.

R.A. ELDEN and A.D. SWAN, *Calendering of Plastics*, Elsevier, NY, 1971.

W.W. GRAESSLEY, S.D. GLASSCOCK, and R.L. CRAWLEY, *Trans. Soc. Rheol. 14*, 519 (1970).

R.M. GRIFFITH, *Ind. Eng. Chem. Fundamentals 1*, 180 (1962).

R. HERNANDEZ, C. LAI, S. SELKE, and M. KIRLOSKAR, *ANTEC*, 1805–1808 (1991).

J.L. HOLMAN, J.B. STEPHENSON, and J.W. JENSEN, Processing the Plastics from Urban Refuse, TPR 50, U.S. Bureau of Mines, Rolla, MO, 1972.

J.R. JOHANSON, *Chem. Eng.* May 8, 42, 1972.

J.R. JOHANSON, *Chem. Eng.,* Oct. 30, 3–11, 1978.

W.J. KOROS and E.S. SANDERS, *J. Polymer. Sci. 72*, 141 (1985).

O.Y. KWAPONG and J.H. HOTCHKISS, *J. Food Science 52* (3), 761–763 (1987).

A.H. LEFEBVRE, *Atomization and Sprays*, Hemisphere NY, 1989.

J. LEIDNER, *Plastics Waste Recovery of Economic Value*, Marcel Dekker, NY, 1981.

C.L. MANTELL, *Solid Wastes: Origins, Collection, Processing and Disposal*, John Wiley, NY, 1975.

C.H. MANNHEIM, J. MILTZ, and A. LETZTER, *J. Food Science 52* (3), 737–740 (1987).

G.R. MAUZE and S.A. STERN, *J. Membrane Sci.12*, 51 (1982). *Zemblr; J. Membrane Sci.* 18, 99 (1984).

W.L. MCCABE and J.C. SMITH, *Unit Operations of Chemical Engineering*, McGraw-Hill, NY, 1956.

N.G. MCCRUM, C.P. BUCKLEY and C.B. BUCKNALL, *Principles of Polymer Engineering*, Oxford University Press, NY, 1988.

G. MENGES, P. FILZ, O. KRETZSCHMAR, H. RECKER, TH. SCHMIDT, and TH. SCHACT, "Mold Design with CAD Mould", in *Applications of Computer Aided Engineering in Injection Molding*, (L.T. Manzione, Ed.), Hanser, Munich, 1987.

S. MIDDLEMAN, *Fundamentals of Polymer Processing*, McGraw-Hill, NY, 1977.

J. MILGRAM, *Resource Recycling 20*, March–April, 1984.

MITTERHOFER, F., *Processing Stability of Polyolefins*, Polym. Eng. Sci., *20*, 692-695 (1980).

Modern Plastics, *64* (1), 41 (1987a).

Modern Plastics, *64* (3), 102 (1987b).

L.S. MOSSNER and E.A. GRULKE, *J. Appl. Polymer Sci. 35*, 923, (1988).

A.B. NEWMAN, *Trans. AIChE 27*, 203, 310 (1931).

K.L. NICHOLS, Ph.D. Dissertation, Michigan State University, 1988.

W.R.R. PARK, *Plastics Film Technology*, R Krieger Publishing, NY, 1973.

C. PATTANAKUL, S. SELKE, C. LAI, E. GRULKE, and J. MILTZ, *ANTEC*, 1802–1804 (1988).

R.A. RICHARDSON and J. HAVLICEK, *J. Environ. Econ. Management 5*, 103 (1978).

G.R. RIDEAL and J.C. PADGET, *J. Polymer Sci. Polymer Symp. 57*, 1–15 (1976).

J.K. RUDD, *Chem. and Eng. News 32*, (4), 344 (1954).

L.C. SAWYER and M. JAFFE, *J. Mat. Sci. 21*, 1897 (1986).

K. SCHNEIDER, *Kunststoffe, 59*, 97 (1969).

S.E. SELKE, *Packaging Technol. Sci. 1*, 93–98 (1988).

A. SFIRAKIS and C.E. ROGERS, *Polym. Eng Sci.*, *20* (4), 294 (1980).

Society of the Plastics Industry, *Facts and Figures of the U.S. Plastics Industry*, NY, 1985.

L. SORS and I. BALAZS, *Design of Plastic Molds and Dies*, Elsevier, Amsterdam, 1989.

Z. TADMOR and C.G. GOGOS, *Principles of Polymer Processing*, Wiley, Interscience, NY, 1979.

R. URAGAMI, H.B. HOPFENBERG, W.J. KOROS, D.K. YANG, U.T. STAUNETT, and R.T. CHERN, *J. Appl. Polymer Science 24*, 779 (1986).

D.A. VAUGHN, C. IFEADI, R.A. MARKLE, and H.H. KRAUSE, *Environmental Assessment of Future Disposal Methods for Plastics in Municipal Solid Waste*, EPA 670/2-75-058, U.S. EPA, Cincinnati, OH. 1975, PB-243-366.

J.L. WHITE, *Appl. Polymer Symp.*, *20*, 155, 1973.

WOODWARD, *Atlas of Polymer Morphology*, Hanser, NY, 1990.

D.K. YANG, W.J. KOROS, H.B. HOPFENBERG, and V.T. STANNETT, *J. Appl. Polym. Sci. 30*, 1035 (1985).

Appendix A

Constants, Units, and Conversion Factors

TABLE 1 Physical Constants

Universal gas constant (R)	= 1.9872 cal/mol · K
	= 1.9872 Btu · lb/mol · R
	= 82.057 atm · cm^3/mol · K
	= 0.082057 atm · m^3/kmol · K
	= 8.3143 kJ/kmol · K
	= 8.3143 kPa · m^3/kmol · K
	= 0.7302 atm · ft^3 · lb/mol · R
	= 10.731 lb(f) · in^{-2} · ft^3 · lb/mol · R
	= 1.5453×10^3 ft · lb(f) · lb/mol · R
Atmosphere (standard)	= 1 atm
	= 1.01325×10^5 N · m^{-2}
Avogadro's number (N)	= 6.0221438×10^{23} molecules/mol
Boltzmann's constant (k)	= R/N
	= 1.380×10^{-23} J/molecule · K

TABLE 2 SI Units

Quantity	Name	SI-Unit
Length	Meter	m
Time	Second	s
Frequency	Hertz	Hz
Mass	Kilogram	kg
Density		kg/m^3
Force	Newton	N
Pressure	Pascal	Pa
Stress		Pa or N/m^2
Viscosity (dynamic)		Pa s
Viscosity (kinematic)		m^2/s
Surface tension		N/m
"Energy, work, heat"	Joule	J
Power	Watt	W
Temperature	Kelvin	K
Thermal conductivity		W/mK
Heat capacity		J/K
Specific heat		$J/kg \cdot K$
Entropy		J/K
Amount of matter	Mole	mol

TABLE 3 Conversion Factors

length	in	ft	yd	mil		
1 m =	3.937×10^{1}	3.281	1.094	3.937×10^{4}		
area	in^2	ft^2	yd^2			
1 m^2 =	1.550×10^{3}	1.076×10^{1}	1.196			
volume	l	in^3	ft^3	yd^3	gal (US)	
1 m^3 =	1000	6.102×10^{4}	3.531×10^{1}	1.308	2.642×10^{2}	
mass	lb(m)	ton (US)	ounce	grain		
1 kg =	2.205	1.102×10^{-3}	35.27	1.543×10^{4}		
density	g/cm^3	lb(m)/ft^3	lb(m)/in^3	lb(m)/gal		
1 kg/m^3 =	0.001	0.06243	3.61×10^{-5}	0.008345		
force	kg(f)	lb(f)	dyn	ton(f) (US)		
1 N =	0.102	0.2248	10^{5}	0.000112		
pressure	bar	kg(f)/cm^2	atm	torr	psi	
1 Pa =	0.00001	1.02×10^{-5}	9.87×10^{-6}	0.007501	0.000145	
stress	kg(f)/cm^2	psi				
1 Pa =	1.02×10^{-5}	1.450×10^{-4}				
viscosity (dynamic)	centipose (cP)	kg(f) · s/m^2	lb(m)/ft · s	lb(f) · s/ft^2		
1 Pa · s =	1000	0.102	0.672	0.02089		
viscosity (kinematic)	mm^2/s	Centistokes (cSt)	ft^2/s			
1 m^2/s =	1.0×10^{6}	1.0×10^{6}	10.76			
surface tension	kg(f)/m	dyn/cm				
1 N/m =	0.102	1000				
energy	kW · h	lb(f) · ft	erg	HP · h	kcal	BTU
1 J =	2.778×10^{-7}	0.7376	1.0×10^{7}	3.725×10^{-7}	2.388×10^{-4}	9.478×10^{-4}
heat	kcal	BTU	CHU (centigrade heat unit)			
1 J =	2.388×10^{-4}	9.478×10^{-4}	5.262×10^{-4}			
power	kg(f)m/s	HP	erg/s	ft · lb(f)/s	kcal/h	BTU/h
1 W =	.102	0.001341	1.0×10^{7}	0.7376	0.8598	3.412
thermal conductivity	kcal/m · h · C	BTU/ft · h · F				
1 W/m · K =	0.8598	0.5778				
specific heat capacity	kcal/kg · K	BTU/lb(m) · F				
1 J/kg · K =	2.389×10^{-4}	2.389×10^{-4}				

Appendix B

Properties of Commercial Products

TABLE I Commercial Polymer Properties

Properties	ASTM Test Method	ABS Copolymer: Extrusion Grade	ABS Copolymer: High-impact Injection-Molding Grade	ACETAL: Homopolymer	ACETAL: 20% Glass Reinforced Homopolymer
Melt Flow (gm/10 min.)	D1238	1.1-18	1-16	xxx	
Melting Temp. °C, T_m		xxx	xxx	175-181	181
T_g		88-120	91-110	xxx	xxx
Processing Temp, °F [C = compression, I = injection T = transfer, E = extrusion]		E:350-500	C:325-350 I:350-400	I:380-470 I:350-480	
Molding pressure, 10^3 psi		xxx	8-25	10-20	10-20
Compression ratio		2.5-2.7	1.1-2.0	3.0-4.5	xxx
Mold (lin) shrinkage, in/in	D955	0.004-0.007	0.004-0.009	0.018-0.025	0.009-0.012
Tensile strength at break, psi	D638	2500-8000	4400-6300	9700-10000	8500-9000
Elongation at break, %	D638	20-100	5-75	25-75	12
Tensile yield strength, psi	D638	4300-6400	2600-5900	9500-12000	xxx
Compressive strength, psi	D695	5200-10000	4500-8000	15600-18000	18000@10%
Flexural strength, psi	D790	4000-14000	5400-11000	13600-14300	10700
Tensile modulus, 10^3 psi	D638	130-420	150-350	450-520	900
Compressive modulus, 10^3 psi	D695	150-390	140-300	670	xxx
Flexural modulus, 10^3 psi 73 °F	D790	130-440	179-375	380-430	700-730
200 °F	D790	xxx	xxx	120-135	300-320
250 °F	D790	xxx	xxx	75-90	250-270
300 °F	D790	xxx	xxx	xxx	xxx
Izod Impact, ft-lb/in of notch	D256A	1.5-12	6.0-9.3	1.2-2.3	0.8-1.0
Hardness Rockwell	D785	R75-115	R85-106	M92-94	M90
Shore/Barcoi	D2240/ D2583	xxx	xxx	xxx	xxx
Coef. of linear expansion, 10^{-6} in/in/°C	D696	60-130	95-110	100-113	33-75
Deflection temp., °F 264 psi	D648	170-220	205-215	255-277	315
under flexural load 66 psi	D648	170-235	210-225	328-342	345
Thermal conductivity, 10^{-4} cal cm/sec cm^2 °C	C177	xxx	xxx	5.5	xxx
Specific gravity	D792	1.02-1.08	1.01-1.05	1.42	1.54-1.56
Water absorption, % 24 hr	D570	0.20-0.45	0.20-0.45	0.25-0.40	0.25
(1/8 in thick) Saturation	D570	xxx	xxx	0.90-1.41	1.0
Dielectric strength, v/mil (1/8 in thick), short time	D149	350-500	350-500	500 (90 mil)	490 (125 mil)

SOURCE: Modern Plastics, *Special Buyers Guide and Encyclopedia*, McGraw-Hill, 1992.

Properties	ACRYLIC	ACRYLO-NITRILE	CELLULOSIC		CHLORINATED PE
	PMMA	Molding and Extrusion	Celluose Acetate Sheet	Cellulose Acetate Butyrate Sheet	30–42% Cl Extrusion and Molding Grades
Melt Flow (gm/10 min.)	14-24	xxx	xxx	xxx	xxx
Melting Temp. °C, T_m	xxx	135	230	140	125
T_g	85-105	95	xxx	xxx	xxx
Processing Temp, °F [C = compression, I = injection T = transfer, E = extrusion]	C:300-425 I:325-500 E:360-500	C:320-345 I:410 E:350-410	xxx	xxx	E:300-400
Molding pressure, 10^3 psi	5-20	20	xxx	xxx	xxx
Compression ratio	1.6-3.0	2	xxx	xxx	xxx
Mold (linear) shrinkage, in/in	0.001-0.004 (fl) 0.002-0.008 (trans)	0.002-0.005	xxx	xxx	xxx
Tensile strength at break, psi	7000-10500	9000	4500-8000	2600-6900	1400-3000
Elongation at break, %	2-5	3-4	20-50	50-100	300-900
Tensile yield strength, psi	xxx	xxx	xxx	xxx	xxx
Compressive strength, psi	10500-18000	12000	xxx	xxx	xxx
Flexural strength, psi	10500-19000	14000	6000-10000	4000-9000	xxx
Tensile modulus, 10^3 psi	325-460	510-580	xxx	200-250	xxx
Compressive modulus 10^3 psi	370-460	xxx	xxx	xxx	xxx
Flexural modulus, 10^3 psi 73 °F	325-460	500-590	xxx	xxx	xxx
200 °F	xxx	xxx	xxx	xxx	xxx
250 °F	xxx	xxx	xxx	xxx	xxx
300 °F	xxx	xxx	xxx	xxx	xxx
Izod Impact, ft-lb/in of notch	0.2-0.4	2.5-6.5	2.0-8.5	xxx	xxx
Hardness Rockwell	M68-105	M72-78	R85-120	R30-115	xxx
Shore/Barcoi	xxx	xxx	xxx	xxx	Shore A60-76
Coef. of linear expansion, 10^{-6} in/in/°C	50-90	66	100-150	110-170	xxx
Deflection temp., °F 264 psi	155-210	164	xxx	xxx	xxx
under flexural load 66 psi	165-225	172	xxx	xxx	xxx
Thermal conductivity, 10^{-4} cal cm/sec cm^2 °C	4.0-6.0	6.2	4-8	4-8	xxx
Specific gravity	1.17-1.20	1.15	1.28-1.32	1.15-1.22	1.13-1.26
Water absorption, % 24 hr	0.1-0.4	0.28	2.0-7.0	0.9-2.2	xxx
(1/8 in thick) Saturation	xxx	xxx	xxx	xxx	xxx
Dielectric strength, v/mil (1/8 in thick), short time	400-500	220-240	250-600	250-400	xxx

Properties	EPOXY			FLUOROPLASTICS		
	Bisphenol Glass Fiber-Reinforced	Novolac 60% Graphite Fiber-reinforced	Novolac Glass-filled High Strength	Polychlorotri-fluoroethylene	Polytetra-fluoroethylene	Polyvinylidene Fluoride
Melt Flow (gm/10 min.)	xxx	xxx	xxx	xxx	xxx	xxx
Melting Temp. °C T_m	Thermoset	xxx	Thermoset	xxx	xxx	141-178
T_g	xxx	xxx	xxx	220	327	−60 to −20
Processing Temp, °F [C = compression, I = injection T = transfer, E = extrusion]	C:300-330 T:280-380	C:290-350	C:290-330 T:290-330	C:460-580 I:500-600 E:360-590	xxx	C:360-550 I:375-550 E:375-550
Molding pressure, 10^3 psi	1-5	3-6	2.5-5.0	1-6	2-5	2-5
Compression ratio	3.0-7.0	xxx	6-7	2.6	2.5-4.5	3
Mold (linear) shrinkage, in/in	0.001-0.008	to max. 0.0001	0.0002	0.010-0.015	0.030-0.060	0.020-0.035
Tensile strength at break, psi	5000-20000	20000	18000-27000	4500-6000	2000-5000	3500-7250
Elongation at break, %	4	0.4	xxx	80-250	200-400	12-600
Tensile yield strength, psi	xxx	xxx	xxx	5300	xxx	2900-8250
Compressive strength, psi	18000-40000	28000	30000-38000	4600-7400	1700	8000-16000
Flexural strength, psi	8000-30000	40000	50000-70000	7400-11000	xxx	9700-13650
Tensile modulus, 10^3 psi	3000	6000	xxx	150-300	58-80	200-80000
Compressive modulus, 10^3 psi	xxx	xxx	xxx	170-200	60	304-420
Flexural modulus, 10^3 psi 73 °F	2000-4500	5500	2.8-4.2	180-260	80	170-120000
200 °F	xxx	xxx	xxx	xxx	xxx	xxx
250 °F	xxx	xxx	xxx	xxx	xxx	xxx
300 °F	xxx	xxx	xxx	xxx	xxx	xxx
Izod Impact, ft-lb/in of notch	0.3-10.0	10	xxx	2.5	3	2.5-80
Hardness Rockwell	M100-112	R110	xxx	R75-112	xxx	R79-83, 85
Shore/Barcoi	xxx	xxx	Barcoi 60-74	Shore D75-80	Shore D50-65	Shore D80, 82 65-70
Coef. of linear expansion, 10^{-6} in/in/°C	11-50	1.0	xxx	36-70	70-120	70-142
Deflection temp., °F 264 psi	225-500	500	xxx	xxx	xxx	183-244
under flexural load 66 psi	xxx	sxxx	xxx	258	250	280-284
Thermal conductivity, 10^{-4} cal cm/sec cm^2 °C	4.0-10.0	16	xxx	4.7-5.3	6.0	2.4-3.1
Specific gravity	1.6-2.0	1.46-1.5	1.84	2.08-2.2	2.14-2.20	1.77-1.78
Water absorption, % 24 hr	0.04-0.20	0.4	xxx	0	< 0.01	0.03-0.06
(1/8 in thick) Saturation	xxx	0.8	xxx	xxx	xxx	xxx
Dielectric strength, v/mil (1/8 in thick), short time	250-400	xxx	380-400	500-600	480	260-280

Properties	POLYARYL-ETHERKETONE		POLYETHER-ETHERKETONE		MELAMINE FORMALDEHYDE	PHENOLIC
	Unfilled	30% Carbon Fiber	Unfilled	30% Carbon Fiber	Cellulose Filled	Wood, Flour, and Mineral-filled
Melt Flow (gm/10 min.)	4-7	15-25	xxx	xxx	xxx	xxx
Melting Temp., °C T_m	323	329	334	334	Thermoset	Thermoset
T_g	xxx	xxx	xxx	xxx	xxx	xxx
Processing Temp., °F [C = compression, I = injection T = transfer, E = extrusion]	I:715-805	I:715-805	I:660-750 E:660-720	I:660-800	C:280-370 I:200-340 T:300	C:290-380 I:330-390 T:300-350
Molding pressure, 10^3 psi	10-20	10-20	10-20	10-20	8-20	1-20
Compression ratio	2	2	3	2	2.1-3.1	2.0-10.0
Mold (linear) shrinkage, in/in	0.008	0.002-0.008	0.011	0.005-0.014	0.005-0.015	0.001-0.004
Tensile strength at break, psi	13500	30000	10200-15000	29800-33000	5000-13000	7000-18000
Elongation at break, %	50	1.5	30-150	1-4	0.6-1	0.2
Tensile yield strength, psi	xxx	xxx	13200	xxx	xxx	xxx
Compressive strength, psi	20000	33800	1800	25000-34400	33000-45000	16000-70000
Flexural strength, psi	24500	40000	16000	40000-48000	9000-16000	12000-60000
Tensile modulus, 10^3 psi	520	2700	xxx	1860-3500	1100-1400	1900-3300
Compressive modulus, 10^3 psi	xxx	xxx	xxx	xxx	xxx	2740-3500
Flexural modulus, 10^3 psi 73 °F	530	2850	560	1860-2600	1100	1150-3300
200 °F	530	2790	435	1820	xxx	xxx
250 °F	520	2480	xxx	1750	xxx	xxx
300 °F	500	2100	290	1400	xxx	xxx
Izod Impact, ft-lb/in of notch	1.6	1.5	1.6	1.5-2.1	0.2-0.4	0.5-18.0
Hardness Rockwell	M98	xxx	xxx	xxx	M115-125	E54-101
Shore/Barcoi	xxx	xxx	xxx	xxx	xxx	Barcoi 72
Coef. of linear expansion, 10^{-6} in/in/°C	44.2	7.9	< 150 C, 40-47 > 150 C, 108	< 150 C, 15-22 < 150 C, 5-44	40-45	8-34
Deflection temp., °F 264 psi	323	634	320	550-610	350-390	350-600
under flexural load 66 psi	582	652	xxx	615	xxx	xxx
Thermal conductivity, 10^{-4} cal cm/sec cm^2 °C	7.1	xxx	xxx	4.9	6.5-10	8-14
Specific gravity	1.3	1.45	1.30-1.32	1.42-1.44	1.47-1.52	1.69-2.0
Water absorption, % 24 hr	0.1	< 0.2	0.1-0.14	0.06-0.12	0.1-0.8	0.03-1.2
(1/8 in thick) Saturation	xxx	xxx	0.5	0.06	xxx	0.12-1.5
Dielectric strength, v/mil (1/8 in thick), short time	355	xxx	xxx	xxx	270-400	140-400

Properties	POLYAMIDE					AROMATIC POLYAMIDE
	Nylon 6 Molding and Extrusion Cmpd	Nylon 6, 6 Molding Compund	Nylon 6, 10 Molding and Extrusion Cmpd	Nylon 6, 12 Molding Compund	Nylon 11 Molding and Extrusion Cmpd	Aramid Molded Parts, Unfilled
Melt Flow (gm/10 min.)	0.5-10	8.5	xxx	xxx	xxx	xxx
Melting Temp., °C T_m	210-220	255-265	220	195-219	160-209	275
T_g	xxx	xxx	xxx	xxx	125-155	xxx
Processing Temp, °F [C = compression, I = injection T = transfer, E = extrusion]	I:440-550 E:440-525	I:500-620	I:445-485 E:480-500	I:450-550 E:464-469	I:356-525 E:350-500	xxx
Molding pressure, 10^3 psi	1-20	1-25	1-19	1-15	1-15	xxx
Compression ratio	3.0-4.0	3.0-4.0	3.0-4.0	xxx	2.5-4	xxx
Mold (linear) shrinkage, in/in	0.003-0.015	0.007-0.018	0.005-0.015	0.011	0.003-0.015	xxx
Tensile strength at break, psi	6000-24000	13700	10150	6500-8800	5100-9000	17500
Elongation at break, %	30-100	15-80	70	150	250-390	5
Tensile yield strength, psi	11700	8000	xxx	5800-8400	3000-6100	xxx
Compressive strength, psi	13000-16000	12500-15000	xxx	xxx	xxx	30000
Flexural strength, psi	15700	7900-17000	350	11000	1400-8100	25800
Tensile modulus, 10^3 psi	380	230-550	xxx	218-290	36-180	xxx
Compressive modulus, 10^3 psi	250	xxx	xxx	xxx	xxx	290
Flexural modulus, 10^3 73 °F	390	410-470	xxx	240-334	27-190	640
200 °F	xxx	xxx	xxx	xxx	xxx	xxx
250 °F	xxx	xxx	xxx	xxx	xxx	xxx
300 °F	xxx	xxx	xxx	xxx	xxx	xxx
Izod Impact, ft-lb/in of notch	0.6-2.2	0.55-1.0	xxx	1.0-1.9	1.6-6.6	1.4
Hardness Rockwell	R119	R120, M83	xxx	M78, R115	R70-109	E90
Shore/Barcoi	xxx	xxx	xxx	D72-80	D58-75	xxx
Coef. of linear expansion, 10^{-6} in/in/°C	80-83	80	xxx	xxx	61-100	40
Deflection temp., °F 264 psi	155-185	167-190	xxx	136-180	95-135	500
under flexural load 66 psi	365-375	425-474	xxx	311-330	158-302	xxx
Thermal conductivity, 10^{-4} cal cm/sec cm^2 °C	5.8	5.8	xxx	5.2	5.2-7.3	5.2
Specific gravity	1.12-1.14	1.13-1.15	xxx	1.06-1.10	1.01-1.02	1.3
Water absorption, % 24 hr	1.3-1.9	1.0-2.8	1.4	0.37-1.0	0.25-0.30	0.6
(1/8 in thick) Saturation	8.5-10.0	8.5	3.3	2.5-3.0	0.75-0.9	xxx
Dielectric strength, v/mil (1/8 in thick), short time	400	600	xxx	400	450	800

Properties	POLY-BUTYLENE	POLYCARBONATE		POLYDICYCLO-PENTADIENE
	Extrusion Compound	High viscosity	30% Glass Reinforced	RIM Solid, Unfilled
Melt Flow (gm/10 min.)	xxx	10-30	xxx	xxx
Melting Temp., °C T_m	126	xxx	xxx	Thermoset
T_g	xxx	150	150	90-110
Processing Temp, °F [C = compression, I = injection T = transfer, E = extrusion]	C:300-350 I:290-380 E:290-380	I:560	I:550-650	T: < 95
Molding pressure, 10^3 psi	10-30	10-20	10-30	< 0.050
Compression ratio	2.5	1.74-5.5	xxx	xxx
Mold (linear) shrinkage, in/in	0.003 (unaged) 0.026 (aged)	0.005-0.007	0.001-0.002	0.008-0.012
Tensile strength at break, psi	3800-4400	9200-9500	19000-20000	6000
Elongation at break, %	300-380	110-120	2-5	5-70
Tensile yield strength, psi	1700-2500	9000	xxx	5000
Compressive strength, psi	xxx	10000-12500	18000-20000	9000
Flexural strength, psi	2000-2300	13500	23000-25000	10000
Tensile modulus, 10^3 psi	30-40	345	1250-1400	280
Compressive modulus 10^3 psi	31	350	1300	xxx
Flexural modulus 10^3 psi 73 °F	45-50	340	1100	280
200 °F	xxx	275	960	xxx
250 °F	xxx	245	900	xxx
300 °F	xxx	xxx	xxx	xxx
Izod Impact, ft-lb/in of notch	No break	12-16 @ 1/8 in 2.3 @ 1/4 in	1.7-3.0	5.0-8.0
Hardness Rockwell	xxx	M70-72	M92, R119	xxx
Shore/Barcoi	Shore D55-60	xxx	xxx	D72
Coef. of linear expansion, 10^{-6} in/in/°C	128-150	68	22-23	49 in/in/°F
Deflection temp., °F under flexural load 264 psi	130-140	250-270	295-300	217
66 psi	215-235	280	300-305	239
Thermal conductivity, 10^{-4} cal cm/sec cm^2 °C	5.2	4.7	5.2-7.6	xxx
Specific gravity	0.91-0.925	1.2	1.4-1.43	1.04
Water absorption, % (1/8 in thick) 24 hr	0.01-0.02	0.15	0.08-0.14	xxx
Saturation	xxx	0.32	xxx	xxx
Dielectric strength, v/mil (1/8 in thick), Short time	> 450	380-399	470-75	xxx

Properties	POLYESTER, THERMOPLASTIC				POLYESTER, THERMOSETTING
	PBT Unfilled	PET Unfilled	Polyester/ Polycarbonate High-impact	Wholly Aromatic Liquid Crystal, Unfilled	SMC Glass Fiber-reinforced
Melt Flow (gm/10 min.)	xxx	xxx	xxx	xxx	xxx
Melting Temp., °C T^m	220-267	245-265	xxx	400-421	xxx
T^g	xxx	73-80	xxx	xxx	xxx
Processing Temp, °F [C = compression, I = injection T = transfer, E = extrusion]	I:435-525	I:510-600	I:460-560	I:540-770	C:270-380 I:280-310 T:280-310
Molding pressure, 10^3 psi	4-10	2-7	8-18	5-16	0.3-2
Compression ratio	xxx	3.1	xxx	2.5-4	1
Mold (linear) shrinkage, in/in	0.009-0.022	0.020-0.025	0.005-0.019	0.001-0.008	0.0005-0.004
Tensile strength at break, psi	8200	7000-10500	4500-9000	15900-27000	7000-25000
Elongation at break, %	50-300	30-300	100-175	1.3-4.5	3
Tensile yield strength, psi	8200-8700	xxx	5000-8100	xxx	xxx
Compressive strength, psi	8600-14500	11000-15000	8600-10000	6200-19000	15000-30000
Flexural strength, psi	12000-16700	14000-18000	8500-12500	19000-35500	10000-36000
Tensile modulus, 10^3 psi	280-435	400-600	240-325	1400-2800	1400-2500
Compressive modulus, 10^3 psi	xxx	xxx	xxx	400-900	xxx
Flexural modulus, 10^3 psi 73 °F	330-400	350-450	310	1770-2000	1000-2200
200 °F	xxx	xxx	xxx	1500-1700	xxx
250 °F	xxx	xxx	xxx	1300-1500	xxx
300 °F	xxx	xxx	xxx	1200-1450	xxx
Izod Impact, ft-lb/in of notch	0.7-1.0	0.25-0.7	12-19	2.4-10	7-22
Hardness Rockwell	M68-78	M94-101	R112-116	R60-66	xxx
Shore/Barcoi	xxx	xxx	xxx	xxx	Barcoi 50-70
Coef. of linear expansion, 10^{-6} in/in/°C	60-95	65	58-150	xxx	13.5-20
Deflection temp., °F 264 psi	122-185	70-100	140-250	356-671	375-500
under flexural load 66 psi	240-375	xxx	210-265	xxx	xxx
Thermal conductivity, 10^{-4} cal cm/sec cm^2 °C	4.2-6.9	3.3-3.6	5.2	2	xxx
Specific gravity	1.30-1.38	1.29-1.40	1.20-1.26	1.35-1.40	1.65-2.6
Water absorption, % 24 hr	0.08-0.09	0.1-0.2	0.08-0.16	0 < 0.1	0.1-0.25
(1/8 in thick) Saturation	0.4-0.5	xxx	0.30-0.62	< 0.1	xxx
Dielectric strength, v/mil (1/8 in thick), Short time	420-550	xxx	396-500	600-980	380-500

Properties	POLYETHERIMIDE		POLYETHYLENE			
	Unfilled	EMI Shielding (conductive) 30% C Fiber	Low Density Homopolymer, Branched	Low Density Homopolymer, Linear	High Density Homopolymer	High density Ultra High Mol. Weight
Melt Flow (gm/10 min.)	xxx	xxx	xxx	xxx	5-18	xxx
Melting Temp., °C T_m	xxx	xxx	98-115	122-124	130-137	125-135
T_g	215-217	215	-25	xxx	xxx	xxx
Processing Temp., °F [C = compression, I = injection T = transfer, E = extrusion]	I:640-800	I:600-780	I:300-450 E:250-450	I:350-500 E:450-600	I:350-500 E:350-525	C:400-500
Molding pressure, 10^3 psi	10-20	10-30	5-15	5-15	12-15	5-20
Compression ratio	1.5-3	1.5-3	1.8-3.6	3	2	xxx
Mold (linear) shrinkage, in/in	0.005-0.007	0.0005-0.002	0.015-0.05	0.020-0.022	0.015-0.040	0.04
Tensile strength at break, psi	14000	29000-34000	1200-4550	1900-4000	3200-4500	5600
Elongaton at break, %	60	1-3	100-650	100-965	10-1200	420-525
Tensile yield strength, psi	15200	xxx	1300-2100	1400-2800	3800-4800	3100-4000
Compressive strength, psi	20300	32000	xxx	xxx	2700-3600	xxx
Flexural strength, psi	22000	37000-45000	xxx	xxx	xxx	xxx
Tensile modulus, 10^3 psi	430	2600-3300	25-41	38-75	155-158	xxx
Compressive modulus, 10^3 psi	420	xxx	xxx	xxx	xxx	xxx
Flexural modulus, 10^3 psi 73 °F	480	2500-2600	35-48	40-105	145-225	130-140
200 °F	370	xxx	xxx	xxx	xxx	xxx
250 °F	360	xxx	xxx	xxx	xxx	xxx
300 °F	350	xxx	xxx	xxx	xxx	xxx
Izod Impact, ft-lb/in of notch	1.0-1.2	1.2-1.6	No Break	1.0-No Break	0.4-4.0	No Break
Hardness Rockwell	M109-110	M127	xxx	xxx	xxx	R50
Shore/Barcoi	xxx	xxx	Shore D44-50	Shore D17-45	Shore D66-73	xxx
Coef. of linear expansion, 10^{-6} in/in/°C	47-56	xxx	100-220	160-200	59-110	130
Deflection temp., °F 264 psi	387-392	405-420	xxx	xxx	xxx	110-120
under flexural load 66 psi	405-410	410-425	104-112	xxx	175-196	155-180
Thermal conductivity, 10^{-4} cal cm/sec cm^2 °C	1.6	17.6	8	xxx	11-12	xxx
Specific gravity	1.27	1.39-1.42	0.917-0.932	0.922-0.943	0.952-0.965	0.94
Water absorption, % 24 hr	0.25	0.18-0.2	< 0.01	0.005-0.13	< 0.01	< 0.01
(1/8 in thick) Saturation	1.25	xxx	xxx	xxx	xxx	xxx
Dielectric strength, v/mil (1/8 in thick), Short time	480	xxx	450-1000	620-760	450-500	710

Properties	POLYIMIDE Unfilled	POLYIMIDE 40 % Graphite Filled	POLY-METHYL-PENTENE Unfilled	POLYPHENYLENE OXIDE Impact Modified	POLYPHENYLENE SULFIDE Unfilled	POLYPHENYLENE SULFIDE EMI Shielding (cond). 40% Pan Carbon Fiber
Melt Flow (gm/10 min.)	xxx	xxx	26	xxx	xxx	xxx
Melting Temp., °C T_m	xxx	xxx	230-240	xxx	285-290	275-285
T_g	310-365	365	xxx	135	88	xxx
Processing Temp., °F [C = compression, I = injection T = transfer, E = extrusion]	C:625-690	C:690	C:510-550 I:510-610 E:510-650	I:425-550	I:590-640	I:600-675
Molding pressure, 10^3 psi	3-5	3-5	1-10	10-15	5-15	xxx
Compression ratio	3-4	xxx	2.0-3.5	xxx	xxx	xxx
Mold (linear) shrinkage, in/in	xxx	xxx	0.012-0.030	0.006	0.006-0.012	0.0005
Tensile strength at break, psi	10500-17100	7600	2100-4500	7000-8000	9500-12500	26000-29000
Elongation at break, %	8-10	3	25-380	35	1-3	1.0-1.5
Tensile yield strength, psi	12500	xxx	2000-4700	xxx	xxx	xxx
Compressive strength, psi	30000-40000	18000	xxx	10000	16000	27000
Flexural strength, psi	19000-28800	14000	6300-8300	8200-11000	14000	40000
Tensile modulus, 10^3 psi	300	xxx	160-280	345-360	480	4400-4800
Compressive modulus, 10^3 psi	xxx	xxx	114-171	xxx	xxx	xxx
Flexural modulus, 10^3 psi 73 °F	450-500	700	70-278	325-345	550-600	3900-4100
200 °F	xxx	xxx	36	xxx	xxx	xxx
250 °F	xxx	xxx	26	xxx	xxx	xxx
300 °F	xxx	xxx	17	xxx	xxx	xxx
Izod Impact, ft-lb/in of notch	1.5	0.7	2-3	6-8	< 0.5	1
Hardness Rockwell	E52-99	E27	R35-85	L108, M93	R123	R122
Shore/Barcoi	xxx	xxx	xxx	xxx	xxx	xxx
Coef. of linear exansion, 10^{-6} in/in/°C	45-56	38	65	xxx	49	11
Deflection temp., °F 264 psi	530-680	680	120-130	190-275	221-275	495-500
under flexural load 66 psi	xxx	xxx	155-195	205-245	390	500
Thermal conductivity, 10^{-4} cal cm/sec cm^2 °C	2.3-2.6	41.4	4	xxx	2.0-6.9	xxx
Specific gravity	1.36-1.43	1.65	0.833-0.835	1.27-1.36	1.35	1.38-1.4
Water absorption, % 24 hr	0.24	0.14	0.01	0.01-0.07	> 0.02	0.03
(1/8 in thick) Saturation	1.2	0.6	xxx	xxx	xxx	xxx
Dielectric strength, v/mil (1/8 in thick), short time	560	xxx	1096-1098	530	380	xxx

Properties	POLY-PROPYLENE Homopolymer Unfilled	POLYSTYRENE Homopolymer High/medium Flow	POLYSTYRENE SAN Copolymer Molding/Extrusion	POLY-URETHANE 10–20% Glass fiber Molding Cmpds	PVDC Copolymers Injection Molding
Melt Flow (gm/10 min.)	0.4-38.0	xxx	xxx	xxx	xxx
Melting Temp., °C T_m	160-175	xxx	xxx	xxx	172
T_g	−20	74-105	100-120	120-160	−15
Processing Temp., °F [C = compression, I = injection T = transfer, E = extrusion]	I:375-550 E:400-500	C:300-400 I:350-500 E:350-500	C:300-400 I:360-550 E:360-450	I:360-410	C:260-350 I:300-400 E:300-400
Molding pressure, 10^3 psi	10-20	5-20	5-20	8-11	5-30
Compression ratio	2.0-4.0	3	3	xxx	2.5
Mold (linear) shrinkage, in/in	0.010-0.025	0.004-0.007	0.003-0.005	0.004-0.010	0.005-0.025
Tensile strength at break, psi	4500-6000	5200-7500	10000-11900	4800-7500	3500-5000
Elongation at break, %	100-600	1.2-2.5	2-3	3-70	160-240
Tensile yield strength, psi	4500-5400	xxx	9920-12000	xxx	2800-3800
Compressive strength, psi	5500-8000	12000-13000	14000-15000	5000	2000-2700
Flexural strength, psi	6000-8000	10000-14600	11000-19000	1700-6200	4200-6200
Tensile modulus, 10^3 psi	165-225	330-475	475-560	0.6-1.40	50-80
Compressive modulus, 10^3 psi	150-300	480-490	530-580	xxx	55-95
Flexural modulus, 10^3 psi 73 °F	170-225	380-490	500-610	40-90	55-95
200 °F	50	xxx	xxx	xxx	xxx
250 °F	35	xxx	xxx	xxx	xxx
300 °F	xxx	xxx	xxx	xxx	xxx
Izod Impact, ft-lb/in of notch	0.4-1.4	0.35-0.45	0.4-0.6	10-14-No Break	0.4-1.0
Hardness Rockwell	R80-102	M60-75	M80, R83	R45-55	M60-65
Shore/Barcoi	xxx	xxx	xxx	xxx	xxx
Coef. of linear expansion, 10^{-6} in/in/°C	81-100	50-83	65-68	34	190
Deflection temp., °F under flexural load 264 psi	120-140	169-202	214-220	115-130	130-150
66 psi	225-250	155-204	220-224	140-145	xxx
Thermal conductivity, 10^{-4} cal cm/sec cm^2 °C	2.8	3	3	xxx	3
Specific gracity	0.900-0.910	1.04-1.05	1.06-1.08	1.22-1.36	1.65-1.72
Water absorption, % (1/8 in thick) 24 hr	0.01-0.03	0.01-0.03	0.15-0.25	0.4-0.55	0.1
Saturation	xxx	0.01-0.03	0.5	1.5	xxx
Dielectric strength, v/mil (1/8 in thick), Short time	600	500-575	425	600	400-600

Properties	SILICONE		POLYETHER-SULFONE	VINYL POLYMERS	
	Casting Resins Flexible (including RTV)	Pseudo-Interpenetrating Silicone/Nylon 6,6	Unfilled	PVC Sheet, Rod, Tube Rigid	Chlorinated PVC Molding and Extrusion
Melt Flow (gm/10 min.)	xxx	xxx	xxx	xxx	xxx
Melting Temp., °C T_m	Thermoset	Interpenetrating	xxx	xxx	xxx
T_g	xxx	xxx	220-230	75-105	110
Processing Temp., °F [C = compression, I = injection T = transfer, E = extrusion]	xxx	I:460-525	C:645-715 I:590-750 E:625-720	C:285-400 I:300-415	C:350-400 I:395-440 E:360-420
Molding pressure, 10^3 psi	xxx	8-15	6-20	10-40	15-40
Compression ratio	xxx	xxx	2-2.5	2.0-2.3	1.5-2.5
Mold (linear) shrinkage, in/in	0.0-0.006	0.004-0.007	0.006-0.007	0.002-0.006	0.003-0.007
Tensile strength at break, psi	350-1000	10100-12500	9800-13800	5900-7500	6800-9000
Elongation at break, %	20-700	5-20	6-80	40-80	4-65
Tensile yield strength, psi	xxx	xxx	12200-13000	5900-6500	6000-8000
Compressive strength, psi	xxx	xxx	11800-15600	8000-13000	9000-22000
Flexural strength, psi	xxx	14000-15900	17000-18700	10000-16000	14500-17000
Tensile modulus, 10^3 psi	xxx	xxx	350	350-600	341-475
Compressive modulus, 10^3 psi	xxx	xxx	xxx	xxx	335-600
Flexural modulus, 10^3 psi 73 °F	xxx	360-410	348-380	300-500	380-450
200 °F	xxx	xxx	xxx	xxx	xxx
250 °F	xxx	xxx	330	xxx	xxx
300 °F	xxx	xxx	280	xxx	xxx
Izod Impact, ft-lb/in of notch	xxx	0.8-0.9	1.4-No Break	0.4-22	1.0-5.6
Hardness Rockwell	xxx	xxx	M88	xxx	R112-117
Shore/Barcoi	Shore A10-70	xxx	xxx	Shore D65-85	xxx
Coef. of linear expansion, 10^{-6} in/in/°C	10-19	xxx	55	50-100	62-78
Deflection temp., °F 264 psi	xxx	xxx	394-397	140-170	202-234
under flexural load 66 psi	xxx	xxx	xxx	135-180	215-247
Thermal conductivity, 10^{-4} cal cm/sec cm² °C	3.5-7.5	xxx	3.2-4.4	3.5-5.0	3.3
Specific gracity	0.97-2.5	1.12-1.13	1.37-1.46	1.30-1.58	1.49-1.58
Water absorption, % 24 hr	0.1	0.6-0.8	0.12-1.7	0.04-0.4	0.02-0.15
(1/8 in thick) Saturation	xxx	xxx	2.5-2.5	xxx	xxx
Dielectric strength, v/mil (1/8 in thick), Short time	400-550	xxx	400	350-500	xxx

TABLE 2 Physical Properties of Monomers*
Si Units

POLYMER FAMILY / Monomer	Molecular Weight	Critical Properties T_c	Critical Properties P_c	Critical Properties V_c	Normal Boiling Point	Acentric Factor	Mass Density	$C_p(g)$	$C_p(l)$	Latent Heat of Vaporization
POLYACRYLATES										
Acrylic acid	72.064	615.	5.66	0.208	414.15	0.518	905.06	9.282	1.795	4.252
Acrylamide	71.079	710.	5.73	0.260	465.75	0.196	292.27	11.50	1.392	7.692
Acrylonitrile	53.064	535.	4.48	0.212	350.50	0.350	741.71	7.103	1.150	3.106
Methyl methacrylate	100.12	564.	3.68	0.323	373.45	0.317	847.61	14.04	1.995	3.430
POLYDIENES										
Butadiene	54.092	425.4	4.33	0.221	268.74	0.193	650.82	7.338	1.160	2.248
Isoprene	68.119	484.0	3.85	0.276	307.21	0.158	666.34	10.53	1.558	2.562
POLYHALOGENHYDROCARBONS										
Vinyl chloride	62.499	432.0	5.67	0.179	259.78	0.101	965.76	4.953	.7354	2.282
Vinyl fluoride	46.044	327.8	5.24	0.144	200.95	0.189	867.27	3.981	.8707	1.832
Vinylidene chloride	96.944	482.0	5.19	0.224	304.71	0.272	1105.9	6.786	11.24	2.661
Vinylidene fluoride	64.035	302.8	4.46	0.154	187.50	0.139	1160.2	4.353	.8017	1.525
Tetrafluoroethylene	100.02	306.5	3.94	0.172	197.51	0.226	1530.8	6.477	11.23	1.868
POLYOLEFINS										
Ethylene	28.054	282.4	5.03	0.129	169.47	0.085	569.76	10.52	.6789	1.346
Propylene	42.081	364.7	4.61	0.181	225.43	0.142	611.62	5.290	.9134	1.902
1-Butene	56.108	419.6	4.02	0.240	266.90	0.187	626.06	7.927	12.15	2.244
Isobutylene	56.108	417.9	3.99	0.239	266.25	0.189	627.44	8.296	1.235	2.221
1-Pentene	70.135	464.8	3.53	0.296	303.11	0.233	629.43	11.09	1.568	2.571
Styrene	104.15	648.0	4.00	0.352	418.31	0.236	792.89	16.63	2.282	3.662
OTHER VINYL POLYMERS										
Vinyl acetate	86.091	524.0	4.25	0.27	345.65	0.338	864.63	11.07	1.748	3.131
PHENOLIC RESINS										
Formaldehyde	30.026	408.0	6.58	0.105	254.05	0.282	817.26	3.424		2.316
Phenol	94.114	694.3	6.13	0.229	454.99	0.426	928.12	14.94	3.722	4.731

POLYAMIDES										
e-Caprolactam	113.16	806.0	4.77	0.402	543.15	0.477	857.71	21.25	3.439	5.469
Hexamethylene diamine	116.21				478.16					
Adipic acid	146.14	809.0	3.53	0.400	611.00	1.054	905.85	27.70	4.083	7.294
Sebacic acid	202.25	793.0	2.35	0.630	625.00	1.434	840.73	44.47	6.040	16.07
Terephthalic acid	166.13	1390.	3.95	0.424	1040.0					8.824
POLYURETHANES										
1,4 Butanediol	90.122	667.0	4.88	0.297	501.15	1.189	798.21	18.01	3.615	6.243
Toluene diisocyanate	174.16	737.0	3.04	0.525	523.15	0.553	1073.4	27.97	3.637	5.064
Propylene glycol	76.095	626.0	6.10	0.239	460.75	1.107	885.32	13.92	2.632	5.448
POLYACETALS										
Trioxane	90.079	604.0	5.82	0.206	387.65	0.334	1106.9	10.24	11.32	3.748
ALIPHATIC POLYETHERS										
Ethylene oxide	44.053	469.2	7.19	0.140	283.85	0.198	882.68	4.586	8.685	2.570
Propylene oxide	58.080	482.2	4.92	0.186	307.05	0.271	811.85	7.526	12.35	2.773
Tetrahydrofuran	72.107	540.2	5.19	0.224	338.00	0.226	835.97	9.049	13.52	3.026
A-epichlorohydrin	92.525	610.0	4.90	0.233	389.26	0.256	1054.5	10.46	17.69	3.723
EPOXIES										
Bisphenol A	228.29				493.16					
OTHER POLYESTERS										
Phthalic anhydride	148.12	791.0	4.72	0.421	557.65	0.708	1053.3	18.99	2.862	5.298
Glycerol	92.095	723.0	4.00	0.264	563.15	1.320	1050.5	17.49	3.527	6.613
Phosgene	98.917	455.0	5.67	0.190	280.71	0.201	1402.1	5.644	1.007	2.478

*Assembled with the assistance of Process Development Corporation.

Remarks:	
molecular weight	kg/kmol
T_c	°K
P_c	$Pa \times 10^{-6}$
V_c	$m^3/kmol$
normal boiling point	°K
acentric factor	-
mass density	kg/m^3
gas heat capacity	$J/kmol\text{-}°K \times 10^{-4}$
liquid heat capacity	$J/kmol\text{-}°K \times 10^{-5}$
latent heat of vaporization	$J/kmol \times 10^{-7}$

Appendix C

Special Properties of Fibers and Elastomers

TABLE 1 Fiber Properties

Fiber	Tenacity at Break N/tex		Extension at Break		Elastic Modulus N/tex		Moisture Regain at 65% rh, %	Specific Gravity	Vol. % H_2O Swelling
	65% rh	Wet	65% rh	Wet	65% rh	Wet			
Wool	0.09–0.15	0.07–0.14	25–35	25–50	2.2–3.1	1.8–2.6	15–17	1.3	35
Regenerated cellulose									
viscose process	0.12–0.33	0.07–0.21	10–30	15–40	3.7–5.8	1.2–3.3	12–16	1.50–1.54	45–82
cuprammonium process	0.14–0.19	0.09–0.10	10–17	20–25	4.1	3.3	11	1.54	100–134
regenerated acetate	0.58–0.66	0.45–0.54	6	7	12.4–14.9	8.26–10.7	10–12	1.52	18–26
polynosic fibers	0.25–0.33	0.21–0.29	7–10	9–12	3.7–5.8	2.9–4.1	10–12	1.54	40
Regenerated protein	0.07–0.11	0.03–0.06	30–40	40–70	2.2		10–15	1.2–1.3	
Cellulose acetate									
secondary acetate	0.09–0.13	0.06–0.10	25–40	30–45	2.2–3.5		6.4	1.30–1.35	10–30
triacetate	0.11–0.19	0.07–0.12	20–28	30–40	3.1–3.5		4.5	1.32	
Aramid fibers									
Kevlar	1.8–2.0	1.8–2.0	2.5–4.0	2.5–4.0	44–71	44–71	4.5	1.44	
Nomex	0.39–0.49	0.26–0.35	20–30	20–30	6.2–7.9	4.4–6.2	8	1.38	
Polyester fibers									
continuous filament	0.35–0.53	0.35–0.53	15–30	15–30	7.9		0.4	1.38	none
staple	0.31–0.44	0.31–0.44	25–45	25–45	7.9		0.4	1.38	none
Acrylic fibers									
continuous filament	0.40–0.44	0.35–0.40	15–20	20–30	5.3–6.2		1.6–2.0	1.17	slight
staple	0.22–0.26	0.18–0.26	25–35	35–45	2.2–3.5		1.8–2.5	1.17	2–5
Elastic fibers									
natural rubber	26		500		1.76				
Spandex	79		600		4.4		1.3	1.21	
Nylon fibers									
continuous filament	0.40–0.71	0.35–0.62	15–30	20–40	3.5		4.0–4.5	1.14	2–10
staple	0.35–0.44	0.31–0.40	30–45	30–50	3.5		4.0–4.5	1.14	2–10
Cotton	0.26–0.44	0.26–0.53	4–10	5–13	3.5–7.9	2.6–5.3	7.0–8.0	1.54	40
Polyolefin fibers									
polypropylene	0.44–0.79	0.44–0.79	15–30	15–30	2.6–4.0	2.6–4.0	< 0.1	0.9	none
polyethylene	0.18–0.35	0.18–0.35	20–40	20–40			< 0.1	0.92–0.96	none

SOURCE: L. REBENFELD, "Fibers" in *Encyclopedia of Polymer Science and Engineering*, 2nd ed. (Mark, Bikales, Overberger, and Menges, Eds.), Wiley-Interscience, NY, 1986.

TABLE 2 Elastomer Properties

Elastomer	Specific Gravity g/cm^3	Tensile Strength Gum 10^3 psi	Tensile Strength Reinf. 10^3 psi	% Total Elong.	Stiffening Temp., °C	Advantages	Uses
SBR	0.94	0.2–0.4	2.0–3.5	300–700	−20	Low price, good wear	Tires
cis-1,4-Polybutadiene	0.93	0.2–1.0	2.3–3.5	300–700	−20	Low heat buildup	Tire treads
EPDM	0.86	0.2	1.0–3.5	200–300	−50	Good electrical stability	Wire and cable insulation
Polychloroprene	1.23	3.0–4.0	3.0–4.0	300–700	−40	Good resistance to gasoline	Gaskets, seals, hose
Polyisoprene	0.93	2.5–3.0	3.0–4.0	300–700	−30	High strength and resilience	Tires, belts, seals
Nitrile rubber	1.00	0.5–1.0	3.0–4.0	300–700	−15	Good resistance to gasoline	Gaskets, seals, hose
Butyl rubber	0.92	2.5–3.0	2.0–3.0	300–700	−60	Good ozone resistance	Weather stripping, innertubes
Silicone rubber	0.98	0.05–0.1	0.5–1.2	200–700	−50	High temp. stability	Wire and cable insulation
Urethane elastomer	1.25	2.0–4.0	3.0–10.0	200–600	−50	High strength, low abrasion	Solid tires, shoe heels

Appendix D

Polymerization Rate Constants and Coefficients

TABLE 1 Free Radical Propagation and Termination Rate Constants

Monomer	kp 1/mol · s	$kt \times 10^{-6}$ 1/mol · s	Temp °C	Method	Remarks
Butadiene	8.4		10	A	
Chloroprene	220		40	A	
	105		10	A	
	228		25	A	
	423		35	A	
Isoprene	2.8		5	A	
Ethylene	18.6 ± 2	455 ± 50	−20.01	B	
	5400	400	130	B	1a
	1200	1730	132	B	1b
	1500	880	132	B	1c
	1800	580	132	B	1d
	2200	500	132	B	1e
	2600	480	132	B	1f
	2900	440	132	B	1g
Propylene	$k_p = 2 \times 10^6 * \exp(-3200/RT)$, 50–150			C	2
Acrylamide	17200 ± 3000	16.3 ± 0.7		B	3a
	6000 ± 1000	3.3 ± 0.6		B	3b
	4000 ± 1000	1.0 ± 0.2		B	3c
Acrylic acid	650		23	B	4
	3150			B	5
	6600			B	6
Acrylonitrile	23000	2800	15	B	7
	28000	3700	25	B	7
	32500	4400	30	B	7
Methacrylic acid	670		23	B	8
methyl ester	1950			B	9
Tetrafluoroethylene	7400	7.4×10^{-5}	40		10
	9100	8.7×10^{-5}	50		10
Vinyl chloride	6200	1100	25	B	11b
	11000	2100	50	B	11b
	11000	2100	25	B	
Vinylidiene chloride	2.3	0.023	15	B	
	8.6	0.175	25	B	
	36.8	1.8	35	B	

Table 1 (*Continued*)

Monomer	*kp* 1/mol · s	*kt* × 10^{-6} 1/mol · s	Temp °C	Method	Remarks
Vinyl acetate	48 ± 5	239 ± 45		B	11c
	97 ± 10	311 ± 59		B	11d
	61 ± 9	266 ± 75		B	11e
	37 ± 5	412 ± 104		B	11f
	8 ± 1	258 ± 76		B	11g
Styrene	46.4	33.6	30	B	1h
	56.2	26.3		B	1i
	66.6	22.4		B	1b
	78.8	20.2		B	1c
	96.1	19.9		B	1d
Formaldehyde	(7–10) × 10^{-3}		−190	D	12

Method:
A: emulsion polymerization by application of the Smith-Ewart theory.
B: intermittent photoirradiation by a rotating sector or "flashing" laser.
C: monomer pressure
D: temperature change

Remarks:
1) Pressure
 a) 1.8 × 10^8 Pa
 b) 5 × 10^7 Pa
 c) 7.5 × 10^7 Pa
 d) 10^8 Pa
 e) 1.25 × 10^8 Pa
 f) 1.5 × 10^8 Pa
 g) 1.75 × 10^8 Pa
2) radical telomerization
3) aqueous solvent
 a) pH = 1
 b) pH = 5.5
 c) pH = 13
4) [M] = 0.92M, pH = 7.9 by NaOH
5) pH = 7.9 + 1.5 N NaCl
6) pH = 11
7) in Water
8) [M] = 0.92M, pH = 8.0 by NaOH
9) pH = 13.6
10) large active chain end concentrations measured by addition of inhibitor in aqueous solution polymerization
11) solvent
 a) water
 b) tetrahydrofuran
 c) anisole
 d) fluorobenzene
 e) chlorobenzene
 f) ethyl benzoate
 g) benzonitrile
12) solid monomer, gamma irradiation

SOURCE: K.C. BERGER, and G. MEYERHOFF, "Propagation and Termination Constants in Free Radical Polymerization" in *Polymer Handbook*, 3rd ed. (Brandrup and Immergut, Eds.), Wiley-Interscience, NY, 1989.

TABLE 2 Free Radical Initiator Decomposition Rate Constants

Initiator	Solvent	Temp. °C	k_d s^{-1}	E_d kJ/mol	Remarks
2,2′-Azobis-isobutyronitrile	Acetonitrile	79.9	0.000125		
	Cyclohexanone	82	0.000143		
	DMF	71.2	6.21×10^{-5}		
	Styrene	50	2.97×10^{-6}	127.6	
		70	4.72×10^{-5}		
	Toluene	60	9.8×10^{-6}	121.3	
		65	0.000019		
		70	0.00004		
Azo-bis-isobutyramidine	DMSO	70	3.68×10^{-6}		
	Methanol	60	1.45×10^{-6}		
	Water	60	3.15×10^{-5}		1
		60	0.000028		2
1,1′-Azo-bis-formamide	DMF	86	1.56×10^{-5}		
		100.3	5.73×10^{-5}		
		115.3	0.000114		
	DMSO	86	5.48×10^{-6}		
		100.3	2.72×10^{-5}		
		115.3	0.000101		
Propyl peroxide	Vapor	146.5	0.00025	132.2	
		155.3	0.006		
		166.8	0.00195		
Tert-butyl peroxide	Benzene	80	7.81×10^{-8}	142.3	
		100	8.8×10^{-7}	146.9	
		120	0.000011	147.7	
	Cyclohexane	95	2.48×10^{-7}	170.7	
		120	6.3×10^{-6}		
		130	2.59×10^{-5}		
	Hexane	80	1.64×10^{-8}		
		110	2.17×10^{-6}		
	Toluene	100	6.82×10^{-7}		
		120	1.34×10^{-5}		

Table 2 (*Continued*)

Initiator	Solvent	Temp. °C	k_d s^{-1}	E_d kJ/mol	Remarks
Benzoyl peroxide	Acetic acid	75	7.53×10^{-5}		
	Styrene	34.8	3.89×10^{-8}		
		49.4	5.28×10^{-7}		
		61	2.58×10^{-6}		
		74.8	1.83×10^{-5}		
		100	0.000458		
	Poly(styrene)	56.4	3.8×10^{-7}		
		64.6	1.47×10^{-6}		
		76.7	9.27×10^{-6}		
	PVC	64.6	6.3×10^{-7}		
		76.7	5.11×10^{-6}		
		83.4	1.44×10^{-5}		
	Benzene	80	4.39×10^{-5}		
	Ethyl benzene	30	3.61×10^{-8}		
		75	1.81×10^{-5}		
		80	3.33×10^{-5}		
	Hexane	80	2.85×10^{-5}		
Diisopropyl peroxydicarbonate	Benzene	54	0.00005		
	Ethyl benzene	54.3	0.000045		
	Toluene	50	3.03×10^{-5}		
Potassium persulfate	Water	80	6.89×10^{-5}		
		40	83.4		3
		50	121.5		3

Remarks: 1) pH 2.90
2) pH 7.05
3) second order rate [unit in $mol^{-1} \cdot sec^{-1}$].

SOURCE: J.C. MASSON, "Decomposition Rates of Organic Free Radical Initiators" in *Polymer Handbook*, 3rd ed. (Brandrup and Immergut, Eds.), Wiley-Interscience, NY, 1989.

TABLE 3 Free Radical Inhibitors and Inhibition Rate Constants

Monomer	Inhibitor	Temp. °C	K_z/K_p	K_z $\text{mol}^{-1} \cdot \text{s}^{-1}$
Acrylonitrile	Anthracene	50	2.67	2670
	p-Benzoquinone	50	0.91	910
	Ferric Chloride	60	3.33	6500
Methyl methacrylate	Benzene,*m*-dinitro-	44.1	0.0048	2.2
	p-Benzoquinone	44.1	5.5	2400
		60	4.5	
	,2,5-dichloro-	44.1		5500
	,2,6-dichloro-	44.1		16500
	Chloranil	44.1	0.26	120
	Diphenylpicrylhydrazyl	44.1	2000	
	Ferric Chloride	60		5000
	Furfurylidene malononitrile	44.1	1.2	550
	Oxygen	50	33000	10^7
	Sulfur	44.1	0.075	40
	Toluene,1,3,5-trinitro-	44.1	0.05	23
Styrene	Anisol, *p*-nitro-	50	0.035	
	1,3,5-trinitro-	50	20.3	
	Benzene, *m*-dinitro-	50	5.17	
	Benzoic acid, ethyl ester	50	1.68	
	p-Benzoquinone	50	518	
		60	227	
		90	560	
	,chloro-	50	720	
	,2,5-dichloro-	50	0.0002	
	,2,6-dichloro-	50	0.0001	
	,2,3-dimethyl-	50	120	
	Oxygen	50	14600	10^6
	Picramide	50	11.8	
	Picric acid	50	211	
	Picryl chloride	50	58.5	
	Toluene,2,4-dinitro-	50	1.543	
	,*o*-nitro-	50	0.055	
	,*p*-nitro-	50	0.203	

SOURCE: J. ULBRICHT, "Inhibitors and Inhibition Constants in Free Radical Polymerization" in *Polymer Handbook*, 2nd ed. (Brandrup and Immergut, Eds.), Wiley-Interscience, NY, 1975.

TABLE 4 Transfer Constants to Monomer

Monomer	Temp. °C	$C_m \times 10^4$	Remarks
Acrylamide	25	0.12	1
	60	0.6	2
Acrylonitrile	20	0.18	2
	25	0.105	
	30	1.5	3
	40	0.17	
	50	0.05	4
1-Butene	40	3.1	
	50	5.1	
	60	7.3	
Ethylene	83	5	1, 5
	110	1.1	6, 7
	130	1.6	6, 7
Maleic anhydride	75	750	8
Methyl methacrylate	0	0.128	
	30	0.117	
	40	24200000	
	50	0.1	
2-Vinyl pyridine	15–35	0	
Styrene	25	0.279	
	30	0.2	9
	45	0.3	9
	50	0.4	10
Vinyl acetate	20	0.94	
	40	1.29	
	60	1.75	
Vinylidene chloride	50	22	11
	60	28	11

Remarks: 1) photoinitiation
2) gamma-ray initiation
3) solution polymerization in dimethyl sulfoxide
4) solution polymerization on zinc chloride
5) under pressure
6) 5000 psi
7) 20000 psi
8) peroxide initiation
9) $Cm = 0.4 \exp(-6219/RT)$
10) emulsion polymerization
11) calculated from viscosity average molecular weight.

SOURCE: K.C. BERGER and G. BRANDRUP, "Transfer Constants to Monomer, Polymer, Catalyst, Solvent, and Additive in Free Radical Polymerization" in *Polymer Handbook*, 3rd ed. (Brandrup and Immergut, Eds.), Wiley-Interscience, NY, 1989.

TABLE 5 Transfer Constants to Solvents and Additives

Monomer	Solvent	Temp. °C	$C \times 10^4$	Remarks
Acrylamide	Methanol	30	0.13	1,2
	Water	25	0.204	1,2
		40	5.8	2
Acrylonitrile	Acetone	50	1.7	3
		60	1.13	4
	Acetaldehyde	50	47	5
	Benzene	60	2.46	6
	Chloroform	60	5.64	6
		80	5.9	1
	Toluene	50	1.153	6
		60	2.632	6
Ethylene	Acetone	130	160	1,7
	Benzene	20	0.629	8,9
		83	20	10
		130	9.4	1,11
	Cyclohexane	130	80	1,12,11
	Formaldehyde	130	560	1,11
	Toluene	130	130	1,11
		200	220	1,11
	Water	20	1.71	8,9
Methyl methacrylate	Acetone	60	1.1	
	Benzene	30	0.01	
		50	0.036	
		60	0.04	
		80	0.075	13
	Cyclohexane	60	12	
	Toluene	52	0.084	
		60	0.202	1
		70	0.567	1
	Water	60	0	
		80.5	0	5,14
Styrene	Acetone	60	4.1	1
	Acetonitrile	60	0.44	13
	Benzene	35	3.9	17
		40	5.8	17
		70	5.5	1
		75	6.67	1
	Cyclohexane	60	0.024	13
		80	0.066	
		100	0.16	13
	Toluene	60	0.105	1
		80	0.15	1
		100	0.53	13

Remarks:
1) peroxide initiation
2) solution polymerization in water
3) solution polymerization in Magnesium perchlorate
4) solution polymerization
5) solution polymerization in Dimethylformamide
6) heterogeneous polymerization
7) under pressure of 34500 psi
8) gamma-ray initiation
9) under pressure
10) photoinitiation
11) under pressure of 20000 psi
12) solution polymerization in propane
13) thermal initiation
14) solution polymerization in butyl alcohol
15) under pressure of 150 bar
16) telomerization (1 monomer unit in transfering chain)
17) solution polymerization in benzene
18) apparent transfer constant, retardation occurred
19) calculated from viscosity average molecular weight
20) telomerization (3 monomer units in transfering chain)

SOURCE: K.C. BERGER and G. BRANDRUP, "Transfer Constants to Monomer, Polymer, Catalyst, Solvent, and Additive in Free Radical Polymerization" in *Polymer Handbook*, 3rd ed. (Brandrup and Immergut, Eds.), Wiley-Interscience, NY, 1989.

TABLE 6 Transfer Constants to Initiators

Polymer	Catalyst	Temp. °C	C_i	Remark
Acrylamide	Bisulfite ion	75	0.17	1
	Hydrogen peroxide	25	0.0005	2
Acrylonitrile	2,2′-Azobis-isobutyronitrile	50	0	3
		60	0	
Ethylene	1,1′-Dimethyl-azoethane	83	0.5	2, 4
Maleic anhydride	Benzoyl peroxide	75	2.63	
Methyl methacrylate	Benzoyl peroxide	50	0	
		60	0.02	
	Hydrogen peroxide	60	0.046	5
Styrene	Benzoyl peroxide	22	0.1	
		50	0.13	
		60	0.055	
	Propylene, oxidized	60	1.01	
		70	1.14	
Vinyl acetate	Benzoyl peroxide	60	0.032	
		65	0.04	
	2,2′-Azobis-isobutyronitrile	50	0.025	
		60	0.055	

Remarks: 1) solution polymerization in water
2) photoinitiation
3) solution polymerization in sulfer dioxode
4) under pressure
5) solution polymerization in ethyl methyl ketone

SOURCE: K.C. BERGER and G. BRANDRUP, "Transfer Constants to Monomer, Polymer, Catalyst, Solvent, and Additive in Free Radical Polymerization" in *Polymer Handbook*, 3rd ed. (Brandrup and Immergut, Eds.), Wiley-Interscience, NY, 1989.

TABLE 7 Transfer Constants to Polymers

Monomer	Polymer	Temp. °C	$C_p \times 10^4$	Remarks
Acrylic acid ethyl ester	Poly(methyl methacrylate)	60	12800	1
Acrylonitrile	Polyacrylonitrile	50	4.7	2
		60	3.5	3
	Poly(methyl methacrylate)	60	0.2	4
1,3-Butadiene	Poly(1,3-butadiene)	50	11	
Ethylene	Polyethylene	201.6	150.7	5, 6, 7
		215	194.81	5, 6, 7
		230.4	199.89	5, 7, 8
Methyl methacrylate	Polyethylene	50	0.6	
	Polyisoprene	80	23.4	5, 9
	Poly(methyl methacrylate)	40	1.5	4
		60	0.1	
	Polypropylene	50	0.04	5
		130	0.42	
	Polystyrene	50	0.75	
		60	2.2	
		80	2.95	
	Poly(vinyl acetate)	60	2	
		80	2.8	
	Poly(vinyl chloride)	70	10	5, 10
Styrene	Polystyrene	50	1.9	
		55	15	
		60	0.8	3
	Poly(methyl methacrylate)	50	0.4	4
		60	16.4	11
		80	3.74	
		100	6.04	
	Polypropylene	60	0.025	5
		130	0.3	
Vinyl acetate	Poly(vinyl acetate)	0	0.5	
		40	11.2	
		50	0.06	12
	Poly(oxyethylene)	60	17	4
	Poly(methyl methacrylate)	60	21	
		75	26	
	Polystyrene	40	12	
		60	15	
		75	19	

Remarks:
1) thioglycolate end groups
2) solution polymerization in magnesium perchlorate
3) estimated from model compounds
4) for middle groups
5) peroxide initiation
6) under pressure of 2400 bar
7) solution polymerization in hexane
8) under pressure of 2700 bar
9) solution polymerization in dimethylformamide
10) solution polymerization in cyclohexanone
11) tribromomethyl end groups
12) under pressure of 979.1 bar,

SOURCE: K.C. BERGER and G. BRANDRUP, "Transfer Constants to Monomer, Polymer, Catalyst, Solvent, and Additive in Free Radical Polymerization" in *Polymer Handbook*, 3rd ed. (Brandrup and Immergut, Eds.), Wiley-Interscience, NY, 1989.

TABLE 8 Copolymerization Reactivity Ratios

Monomer 1	Monomer 2	r_1	Confidence Limit 95%	r_2	Confidence Limit 95%
Acrolein	Acrylamide	1.95	0.58	0.8	0.23
		1.59	0.1	0.18	0.02
	Acrylonitrile	1.16	0.13	0.88	0.06
		1.07	0.08	0.71	0.03
		1.52	0.07	0.48	0.01
	Methyl methacrylate	0.76	0.12	1.136	0.05
		0.59	0.3	1.33	0.4
	Styrene	0.216	0.036	0.257	0.017
		0.02	0.01	0.22	0.02
		0.32	0.002	0.205	0.006
	Vinyl acetate	3.04	0.43	−0.02	0.12
	Vinyl chloride	5.22	0.13	0.03	0.11
Acrylamide	Acrolein	0.8	0.23	1.95	0.58
		0.18	0.02	1.59	0.1
	Acrylonitrile	0.81	0.11	0.863	0.033
		1.08	0.51	0.97	0.1
	Methyl methacrylate	0.9	1.07	3	0.45
		2.29	0.44	2.34	0.17
		0.53	0.09	3	0.09
	Styrene	0.58	0.12	1.17	0.06
		1.32	0.13	1.21	0.04
		0.33	0.14	1.49	0.16
Acrylonitrile	Acrolein	0.88	0.06	1.16	0.13
	Acrylamide	0.863	0.033	0.81	0.11
		0.97	0.1	1.08	0.51
	Butadiene	0.03	0.02	0.2	0.01
		0.046	0.006	0.358	0.046
	Isobutylene	1.295	0.008	0	0.002
	Methyl methacrylate	0.138	0.037	1.322	0.05
	Styrene	0.02	0.06	0.29	0.04
	Vinyl acetate	5.29	1.13	−0.06	0.08
	Vinyl chloride	2.62	0.25	0.02	0.08
	Vinylidene chloride	0.92	0.11	0.32	0.14
Butadiene	Acrylonitrile	0.358	0.046	0.046	0.006
	Isoprene	0.14	0.37	0.608	0.062
	Methyl methacrylate	0.504	0.024	0.027	0.004
	Styrene	1.83	0.41	0.83	0.21
	Vinyl chloride	5.27	3.35	−0.11	0.31
Ethylene	1-Butene	3.6	1.6		
	Propylene	3.2	0.62		
	Vinyl acetate	0.88	0.52	1.03	0.77
	Vinyl chloride	0.34	0.13	4.38	0.44
	Vinylidene chloride	0.018	0.005	15.71	1.96

TABLE 8 (*Continued*)

Monomer 1	Monomer 2	r_1	Confidence Limit 95%	r_2	Confidence Limit 95%
Methyl methacrylate	α-methyl-styrene	0.48	0.02	0.27	0.7
Styrene	Acrylonitrile	0.29	0.04	0.02	0.06
	Acrolein	0.205	0.006	0.32	0.002
	Acrylamide	1.17	0.06	0.58	0.12
	Butadiene	0.82	0.13	1.38	0.11
	Isoprene	0.513	0.02	1.922	0.016
	Methyl methacrylate	0.49	0.14	0.418	0.085
		0.564	0.047	0.54	0.029
	Vinyl chloride	25	5	0.005	0.031
	Vinylidene chloride	1.839	0.024	0.087	0.007
Vinyl acetate	Acrolein	−0.02	0.12	3.04	0.43
	Acrylonitrile	0.06	0.22	5.51	0.91
	Ethylene	1.03	0.77	0.88	0.52
	Styrene	0.02	0.12	18.8	2.87
	Vinyl chloride	0.25	0.15	1.64	0.12
	Vinylidene chloride	0.03	0.1	4.66	0.26
Vinyl chloride	Vinyl acetate	1.63	0.32	0.43	0.14
	Acrolein	0.03	0.11	5.22	0.13
	Acrylonitrile	0.02	0.08	2.62	0.25
	Ethylene	4.38	0.44	0.34	0.13
	Vinylidene chloride	0.14	0.32	3.39	1.16
Vinylidene chloride	Methyl methacrylate	0.094	0.033	1.8	0.95
	Vinyl chloride	3.068	0.076	0.205	0.003

SOURCE: R.Z. GREENLEY, "Free Radical Copolymerization Reactivity Ratios," in *Polymer Handbook*, 3rd ed. (Brandrup and Immergut, Eds.), Wiley-Interscience, NY, 1989.

TABLE 9 Kinetic Scheme for a Ring-Opening Polymerization (Nylon 6)

Ring-opening: monomer

$$C_1 + H_2O \rightleftharpoons P_1;\ k_f = k_1,\ k_r = k_1/K_1 \quad (1)$$

Polycondensation

$$P_n + P_m \rightleftharpoons P_{n+m} + H_2O;\ n, m \geq 1,\ k_f = k_2,\ k_r = k_2/K_2 \quad (2)$$

Polyaddition: monomer

$$P_n + C_1 \rightleftharpoons P_{n+1};\ n \geq 1,\ k_f = k_3,\ k_r = k_3/K_3 \quad (3)$$

Ring-opening: dimer

$$C_2 + H_2O \rightleftharpoons P_2;\ k_f = k_4,\ k_r = k_4/K_4 \quad (4)$$

Polyaddition: dimer

$$P_n + C_2 \rightleftharpoons P_{n+2};\ k_f = k_5,\ k_r = k_5/K_5 \quad (5)$$

Reaction with monofunctional acid

$$P_n + P_{mx} \rightleftharpoons P_{n+m,x} + H_2O;\ n, m \geq 1,\ k\text{'s for Eq. 2} \quad (6)$$

Rate and Equilibrium Constants

$$k_i = k_i^0 + k_i^c[\text{—COOH}];\ k_i^x = A_i^x \exp(-E_i^x/RT) \quad \text{a}$$

$$K_i = \exp\left(\frac{\Delta S_i}{R} - \frac{\Delta H_i}{RT}\right);\ i \geq 1 \quad \text{b}$$

index	A_i^0	E_i^0	A_i^c	E_i^c	ΔH_i	ΔS_i
1	5.99e5	1.99e4	4.31e7	1.88e4	1.92e3	−7.88e0
2	1.89e10	2.33e4	1.21e10	2.07e4	−5.95e3	9.44e-1
3	2.86e9	2.28e4	1.64e10	2.01e4	−4.04e3	−6.95e0
4	8.58e11	4.20e4	2.33e12	3.74e4	−9.60e3	−1.45e1
5	2.57e8	2.13e4	3.01e9	2.04e4	−3.17e3	5.83e-1

A_i^0 kg/mol-hr; ΔH_i cal/mol; A_i^c kg/mol^2-hr
E_i^x cal/mol; ΔS_i cal/mol-K

SOURCE: Y. ARAI, K. TAI, H. TERANISHI, and T. TAWAGA, *Polymer 22*, 273–277 (1981).
K. TAI, and T. TAGAWA, *Ind. Eng. Chem., Prod. R & D 22*, 192–205 (1983).

TABLE 10 Kinetic Scheme for Polymerization of a Nylon 6,6 Salt

Nylon salt polymerization

$$-COOH + -NH_2 \rightleftharpoons -CONH- + H_2O; \; k_f, \; k_r = k_f/K$$

Rate constants

$$\log_{10} k_f = \frac{13.1}{[H_2O]^{0.025}} - \frac{4830}{T}; \; k_f, l/\text{mol-hr}$$

$$K = \exp(a - b[H_2O]_0)$$

$[H_2O]_0 < 3.4$ mol/repeating unit			$[H_2O]_0 > 3.4$ mol/repeating unit	
T,°C	a	b	a	b
200	6.17	0.22	5.21	0.01
210	6.25	0.23	5.35	0.0083
220	6.28	0.24	5.41	0.01

SOURCE: A. KUMAR, S. KURNVILLE, A.R. RAMAN, and S.K. GUPTA, *Polymer* 22, 387–390 (1981).

Appendix E

Physical and Processing Properties

TABLE 1 Densities, Heats of Fusion, and Heat Capacities

Polymer	Density [g/cm³] Cryst.	Amorph.	Heat of Fusion [kJ/mol]	Entropy of Fusion [J/mol K]	Heat Capacity [kJ/kg K]
Polyacrylonitrile	1.11		5.23		1.286
Poly(methyl methacrylate)	1.19[e]				1.373
Polybutadiene	0.964				
cis			2.51[a]	33.5[a]	1.854
trans			4.184[a]	26.8[a]	2.402
Polyisoprenes	1.046		12.7		1.917
Poly(dimethyl butadiene)	0.978				
Polychloroprene	1.658		8.37		
Poly(vinyl fluoride)	1.436		7.54		
Poly(vinylidene fluoride)	1.895		8.91		
Poly(trifluorochlorethylene)	2.222	2.08	5.02		0.923
Polytetrafluorethylene	2.344		477[d]		0.938
Poly(vinyl chloride)	1.42	1.41	11		0.934
Poly(vinylidene chloride)	1.948	1.66			0.857
Polyethylene	0.9988	0.8866	8.37	7.87[b]	1.855
Polypropylene, isotactic	0.923[c]	0.850[c]		8.79[c]	0.9103
atactic					0.8667
syndiotactic	0.989[c]	0.856[c]		2.09[c]	0.88
Poly(butene-1), rho.	0.95	0.87	13.9		
tet.	0.866		4.06		
ortho.	0.876		6.49		2.093
Poly(4-methyl pentene-1)	0.814		19.7		1.725
Polyisobutylene					1.948
Polystyrene	1.133	1.04	9		
isotactic, cryst.	1.111[f]				1.072
isotactic, amorph.	1.04[f]				1.056
atactic, amorph.					1.051
Poly(vinyl acetate)	1.633				1.331
Poly(vinyl alcohol)	1.345	1.269	6.87		1.51
Poly(p-xylylenes)	1.138	1.05			
Polyoxymethylene			7.45[b]	16.50[b]	1.214
Polyethylene oxide	1.207	1.13	8.29	24.37[b]	
Polypropylene oxide	1.097	0.998	8.4		
Polyepichlorohydrin	1.461				
Polytetrahydrofuran	1.116	0.982	14.4		
Poly(ethylene terephthalate)	1.457	1.335	24.1	42.71[b]	1.103
nylon 6			21.35[b]	42.71[b]	1.599
nylon 6,6			43.13[b]	79.97[b]	1.419
nylon 6,10					1.601
Poly(urethane)					1.767
Cellulose tributyrate			12.56[b]	26.17[b]	

SOURCE: R.L. MILLER, "Crystallographic Data for Various Polymers" in *Polymer Handbook*, 3rd ed. (Brandrup and Immergut, Eds.), Wiley-Interscience, NY, 1989.

[a] H.L. STEPHENS, "Physical Constants of Poly(butadiene)" in *Polymer Handbook*, 3rd ed. (Brandrup and Immergut, Eds.), Wiley-Interscience, NY, 1989.

[b] H.G. ELIAS, *Macromolecules*, 2nd ed., Plenum Press, NY, 1984.

[c] R.P. QUIRK, and M.A.A. ALSAMARRAIE, "Physical Constants of Poly(ethylene)" in *Polymer Handbook*, 3rd ed. (Brandrup and Immergut, Eds.), Wiley-Interscience, NY, 1989.

[d] C.A. SPERATI, "Physical Constants of Fluoropolymers" in *Polymer Handbook*, 3rd ed. (Brandrup and Immergut, Eds.), Wiley-Interscience, NY, 1989.

[e] W. WUNDERLICH, "Physical Constants of Poly(methyl methacrylate)" in *Polymer Handbook*, 3rd ed. (Brandrup and Immergut, Eds.), Wiley-Interscience, NY, 1989.

[f] J.F. RUDD, "Physical Constants of Poly(styrene)" in *Polymer Handbook*, 3rd ed. (Brandrup and Immergut, Eds.), Wiley-Interscience, NY, 1989.

TABLE 2 Heats and Entropies of Polymerization

Monomer	State of Monomer	ΔH kJ/mol	ΔH Method	ΔS J/K mol	ΔS Method	Temp. °C	Remarks
1,3-Butadiene	gg	73	4b	25			1
	gg	78	4b	25			2
	lc	73	2	89	1	25	
Chloroprene	1c	68	3			61.3	
Isoprene	gg	70.5	4b			25	
	lc	75	2	101	1	25	
	ls	71	3			74.5	
Ethylene	gg	93.5	4b	142	4b	25	
	gc′	108.5	2	174	1	25	
	gc′	107.5	3	172	4b	25	
	gc	101.5	2	155	4b	25	3
Propene	gg	86.5	4b	167	4b	25	
	gc′	104	2	191	1	25	4
	lc	84	4b	113	4b	25	
	sc	69	3			−78	5
1-Butene	gg	86.5	4b	166	4b	25	
	lc	83.5	4b	113	4b	25	
Isobutene	gc	72	2			25	
	lc	48	2	112	1	25	
Acrolein	lc	80	3			74.5	
	ss	57.5	3			74.5	6
	ss	81.5	3			74.5	7
Acrylamide	ss	81.5	3			74.5	7
	lc	79.4	3			90	
	sc	60	3			74.5	8
	sc	57.5	3			74.5	9
Acrylic Acid	lc	67	3			74.5	
	ss	77.5	3			20	7
	sc	73.5	3			74.5	9
	sc	72	3			74.5	10
	sc	74.5	3			74.5	6
Acrylonitrile	lc′	76.5	3			74.5	
	sc′	77.5	3			74.5	9
	lc′	109	1			25	
Methacrylic acid	lc	64.5	2			25	
	ss	56.5	3			25	
	sc	57	3			74.5	11
	ss			146	4b	80	
Vinyl n-butyl ether	lc?	60	3			50	
Vinyl chloride	gc	132	2			25	
	lc	71	4b			25	
	lc	96	3			74.5	
Vinylidene chloride	lc′	73	3			74.5	
	lc′			89	1	−73	
Vinyl acetate	lc	88	3			74.5	
	ls	89.5	3			25	
	ss	90	3			74.5	8
	ss	86.5	3			74.5	6
	ss	86	3			74.5	9
Maleic anhydride	ls	59	3			74.5	

TABLE 2 *(Continued)*

Monomer	State of Monomer	ΔH kJ/mol	ΔH Method	ΔS J/K mol	ΔS Method	Temp. °C	Remarks
Styrene	gg	74.5	4b	149	4b	25	
	lc	70	2	104	1	25	
	ls	73	2			25	
Acetaldehyde	lc′	64.5	2			25	12
	lc	62.5	2			25	12
Ethylene oxide	gg	104	4b			25	
	gc′	127.3	2	174		25	13
	lc′	102.4	2			25	
	lc	94.5	2			25	
Propylene oxide	gg	75.5	4b			25	
	gc′			189		25	14
Glycollide	c′c′	34	2	0.3	1	25	
	cc	10.8	2	−10.5	1	25	
dl-Lactide	c′c	8.7	2	−31.3	1	25	
	lc	27	2	13	1	127	
ε-Caprolactone	lc	17	3	4	1	25	
	lc′	31	3	54	1	25	
ε-Caprolactam	lc	16.3	3			186	
	lc	16	2			75	
	lc	15.5	3			200	
	ls	16.5	4a	29	4a	250	

State: (x, y) x = monomer state, y = polymer state
g: gaseous state (hypothetical in case of polymer)
c: condensed amorphous state
c′: crystalline or partially crystalline state
l: liquid state
s: in solution (solvent specified)
ls: denotes polymer dissolved in monomer

Method:
1: Third law or statistical
2: Combustion of monomer of polymer or both.
3: Reaction calorimetry
4a: Thermodynamic (van't Hoff equation)
4b: Semiempirical rules applied to evaluate heat of formation of polymer or monomer or both.

Remarks:
1) 1:2 polymerization
2) 1:4 polymerization
3) delta $H_{fus(polymer)}$ taken as 5.0.
4) syndiotactic polymer
5) solvent: *n*-butane
6) solvent: hexane
7) solvent: water
8) solvent: acetone
9) solvent: benzene
10) solvent: carbon tetrachloride
11) solvent: methanol
12) to poly(vinyl alcohol)
13) 100% crystalline polymer for delta S
14) estimated for 100% crystalline polymer

SOURCE: W.K. BUSFIELD, "Heats and Entropies of Polymerization, Ceiling Temperatures, Equilibrium Monomer Concentrations, and Polymerizability of Heterocyclic Compounds" in *Polymer Handbook*, 3rd ed. (Brandrup and Immergut, Eds.), Wiley-Interscience, NY, 1989.

TABLE 3 Heats of Solution and Polymer-Solvent Interaction Parameters*

Polymer	M.W.	Temp °C	State	Solvent	Heat absorbed [kJ/kg]	Interaction Parameters χ	χ (seg)
Poly(methyl methacrylate)	100000	25	G	Chloroform	−84		0.377
		25	C	Chloroform	−37		
Poly(butadiene)	8×10^4	27	L	Benzene	13		
Polyisoprene	4×10^3	16	L	Benzene	12		
		25		Benzene		0.44	0.42
Poly(vinyl chloride)		27		Tetrahydrofuran		0.14	
		27		Dioxane		0.52	
		26		Methyl ethyl ketone		0.409	
		29.8		Cyclohexane		0.24	
Polyethylene	3×10^5	80	C	Tetrahydronaphthalate	630		
		130	L		39		
	7×10^5	80	C	Chloronaphthalate	320		
		130	L		36		
Polypropylene		120	C	Tetrahydronaphthalate	320		
Polystyrene	7×10^5	26	G	Cycloxane	−8.2		0.505
				Toluene	−22	0.44	
		26	G	Decahydronaphthalate	−11		
		160	L	Decahydronaphthalate	5.5		
	100000	45	G	Chloroform	−24		
Poly(vinyl acetate)	1.4×10^5	25		Benzene	2.3		
				Chloroform	−45		
				Methyl alcohol	28		
				Acetone		0.437	
				Methyl ethyl ketone		0.429	
				Dioxane		0.407	
Poly(ethylene oxide)	335	27	L	Benzene	16		
				Water		0.45	
				Methyl alcohol	−1.2		
	6×10^3	30	C	Chloroform	52		
				Water	24		
	5×10^3	80	L	Water	120		
Poly(propylene oxide)							
	1000	27	L	Methyl alcohol	−7		
	400	27	L	Water	−170		

*G = glass, L = liquid or rubbery, C = crystalline

SOURCE: *Heats of Solution Data*: D.R. COOPER, "Heats of Solution of Some Common Polymers" in *Polymer Handbook*, 2nd ed. (Brandrup and Immergut, Eds.), Wiley-Interscience, NY, 1989.

Interaction Parameter Data: B.A. WOLF, "Polymer Solvent Interaction Parameters" in *Polymer Handbook*, 2nd ed. (Brandrup and Immergut, Eds.), Wiley-Interscience, NY, 1989.

TABLE 4 Thermal Properties

Polymer	°K Tg	°K TM	Processing Temp., °K	Activation J/mol	Energy Temp. Range, °K low	Energy Temp. Range, °K high	Continuous Use Temp., °K	Deflection Temp., °K 45 MPa	Degradation[1] Products
Poly(acrylic acid)	379								
Polyacrylamide	438								
Polyacrylonitrile	370	614	488	129.8	491	533	333–361	353–380	– 12% of products are volatile at 25°C. consisting of hydrogen cynamide, acrylonitrile, and vinyl acetonitrile.
Poly(methyl methacrylate)	378								
Polybutadiene, cis	171	427		259.6	564	668			– 14.1% of products are volatile at 25°C. including 1.5% of monomer among other saturated and unsaturated hydrocarbons.
,trans	215								
Polyisoprenes	200	338		234.5	564	584			– 3.4% isoprene, 8.8% dipentene, small amounts of p-menthene.
Poly(dimethyl butadiene)	262	440							
Polychloroprene, cis	253	353							
, trans	256								
Poly(vinyl fluoride)	314	473		230	638	655			-High yields of HF and products involatile at 25°C.- little carbonization.
Poly(vinylidene fluoride)	233	451		297.3	613.5	641			-35% HF and high yields of involatile products at room temperature.
Poly(trifluorochlorethylene)	325	493	533	234.5	423	644	448–473	399	-25% of products volatile at 25°C.
Polytetrafluoroethylene	160	600	390	337.1	696	786	563	394	-Monomer yield drops. Larger fragments appear.
Poly(vinyl chloride)	354	546	483–270	83.7	423	463	338–353	331–355	-Quantitative yields of HCl.
Poly(vinylidene chloride)	255	463		131.1	448	498			-High yields of HCl.
Polyethylene	148	415	508–455	192.6	648	709	353–373	311–322	-Continuous spectrum of saturated and unsaturated hydrocarbons from C_2-C_{98}
Polypropylene	260		603–317	242.8	609	639			-Saturated and unsaturated hydrocarbons from C_2 upwards, monomer yield 0.17%.

Poly(butene-1)	249	399							
Poly(4-methyl pentene-1)	302								
Polyisobutylene	200		461						-20% residue, 71% saturated and unsaturated chain fragments.
Polystyrene	373	513	568–327	201	523	574	339–350	348–373	-40.6% monomer, 2.0% toluene, 0.1% CO remainder dimer, trimer, and tetramer.
Poly(vinyl pyridine)	415								
Poly(vinyl acetate)	305						363	397	-Quantitative yields of acetic acid.
Poly(vinyl alcohol)	358	501							-Quantitative yields of water.
Poly(vinyl acetals)	355								
Polyoxymethylene	191	454		113.5	373	393			-100% monomer.
Polyacetaldehyde	243	438							
Polyethylene oxide	232	339							-9.7% of products volatile at 25°C. With 3.9% monomer, CO_2, formaldehyde, ethanol, and C_1–C_7 compounds.
Polypropylene oxide	198	348					393–433	498–523	-12.8% of products volatile at 25°C including 4% acetaldehyde, 2.22% acetone, 1.43% dipropyl ether, and 2.22% propylene.
Polyepichlorohydrin	271	390							
Polytetrahydrofuran	189	316							
Polycarbonates			600–523				394	411	
Poly(ethylene terephthalate)	342	538							
Poly(butylene terephthalate)			444–410						
Alkyd resins			655–355				423–448	453	
Nylon 6	313		616–494				353–393	423–458	
Nylon 10	315								
Nylon 12	314		558–453						
Nylon 6,6	323		577–527				353–423	453–513	
Nylon 6,10	323		555						
Nylon 6,12	319		555						
Nylon 1			513–455						
Polyurea			339						

SOURCE: N. GRASSIE, "Products of Thermal Degradation of Polymers" in *Polymer Handbook*, 3rd ed. (Brandrup and Immergut, Eds.), Wiley-Interscience, NY, 1989.

Appendix F

Transport and Other Properties

TABLE 1 Mark–Houwink Parameters

Polymer	Solvent	Temp. °C	$K \times 10^3$ [ml/g]	a	Mol. Wt. Range $M \times 10^{-4}$
	$[\eta] = KM^a$, $[\eta]$ = intrinsic viscosity				
Polyacrylamide	Water	30	6.31	0.8	2–50
Poly(acrylic acid)	Dioxane-1,4	30	76	0.5	13–82
Poly(acrylonitrile)	Butyrolactone	20	34.3	0.73	4–40
Poly(methyl methacrylate)	Acetone	20	5.5	0.73	7–700
Poly(butadiene)	Toluene	30	30.5	0.725	5–50
Poly(chloroprene)	Benzene	25	2.02	0.89	6–150
Poly(isoprenes)	Benzene	30	18.5	0.74	8–28
Poly(vinyl chloride)	Benzyl alcohol	155.4	156	0.5	4–35
	Chlorobenzene	30	71.2	0.59	3–19
	Cyclohexane	20	11.6	0.85	2–10
Polyethylene, low pressure	Biphenyl	127.5	323	0.5	2–30
	Octanol	180.1	286	0.5	2–105
	p-Xylene	105	16.5	0.83	13–50
high pressure	Decalin	70	38.73	0.738	0.2–3.5
	p-Xylene	75	135	0.63	0.2–7.6
Polypropylene, atactic	Benzene	25	27	0.71	6–31
	Cyclohexane	25	16	0.8	6–31
	Toluene	30	21.8	0.725	2–34
isotactic	Biphenyl	125.1	152	0.5	5–42
	p-Xylene	85	96	0.63	
syndiotactic	Heptane	30	31.2	0.71	9–45
Polystyrene, atactic	Benzene	20	6.3	0.78	1–300
	Chloroform	25	7.16	0.76	12–280
	Cyclohexane	28	108	0.479	0.6–69
	Toluene	20	4.16	0.788	4–137
isotactic	Benzene	30	9.5	0.77	4–75
	Chloroform	30	25.9	0.734	9–32
	Toluene	30	11	0.725	3–37
Poly(vinyl alcohol)	Water	25	140	0.6	1–7
Polyoxymethylene	Dimethylformamide	130	22.4	0.71	0.15–1.5
Polypropylene oxide	Acetone	25	75.5	0.56	0.1–0.4
	Benzene	20	11.1	0.79	0.07–0.33
Polytetrahydrofuran	Ethyl acetate	30	42.2	0.65	2.6–113
	Toluene	28	25.1	0.78	3–12
Poly(ethylene terephthalate)	Tetrachloroethane	50	13.8	0.87	0.04–0.1
Nylon 6	Trifluoroethanol	50	58.2	0.73	1.3–10
Nylon 6,6	Aqueous HCOOH	25	35.3	0.786	0.6–6.5
Nylon 6,10	m-Cresol	25	13.5	0.96	0.8–2.4

SOURCE: M. KURATA and Y. TSUNASHIMA, "Viscosity-Molecular Weight Relationships and Unperturbed Dimensions of Linear Chain Molecules" in *Polymer Handbook*, 3rd ed. (Brandrup and Immergut, Eds.), Wiley-Interscience, NY, 1989.

TABLE 2 Permeability Coefficients

Polymer	Permeant	Temp. °C.	Permeability Coefficient $P \times 10^{13}$	Activation Energy of Permeation	Diffusion Constant $D \times 10^{6}$	Activation Energy of Diffusion $\times 10^{2}$	Solubility Coefficient $S \times 10^{6}$	Heat of Solution
Polyisoprene, amorph. natural rubber	He	25			21.6	19.7		
	O_2	25	17.6	29.3	1.73	33.5	1.02	−4.2
	Ar	25	17.2	32.7	1.36	33.1	1.26	−0.4
	CO_2	25	115	21.8	1.25	34.3	9.2	−12.5
	CO	25	11.8	31	1.35	31	0.874	0
	N_2	25	7.11	35.6	1.17	33.5	0.608	2.1
	CH_4	25	22.7	31	0.89	36.4	2.55	−5.4
	C_2H_6	25			0.4	42.7		
	C_3H_4	25	415	22.6	0.5	37.7	8.35	−15.1
	C_3H_6	25	154	28.9	0.31	42.7	49.7	−13.8
	C_3H_8	25	126	23	0.21	46.5	60.0	−23.5
	SF_6	25	2.7	35.6	0.115	50.2	2.35	−14.6
	C_2H_2	25	74.5	30.5	0.467	39.8	16	−9.2
	H_2O	25	1720					
Poly(tetrafluoroethylene)	H_2	25	7.4	21.4				
	N_2	25	1.0	24.4	0.088	29.9	1.2	−5.5
	O_2	25	3.2	19.1	0.152	26.3	2.1	−7.2
	CO_2	25	7.5	14.0	0.095	28.6	9.2	−14.6
	NO_2	25	12.0	10.7	0.0037	58.6	320	−47.9
	N_2O_4	25	28	−2.1	0.025	15	110	
Poly(vinyl chloride)	He	25	1.5	29.9	2.8	20.7	0.055	9.2
	H_2	25	1.3	34.5	0.5	34.5	0.26	0
	Ne	25	0.29	34.1	0.25	31.5	0.12	2.6
	N_2	25	0.0089	69.0	0.0038	61.9	0.23	7.1
	Ar	25	0.0086	57.8	0.0012	51.5	0.75	6.3
	O_2	25	0.034	55.8	0.012	54.6	0.029	1.2
	CO_2	25	0.12	56.8	0.0025	64.6	4.7	−7.8
	CH_4	25	0.021	66.2	0.0013	70.3	1.7	−4.1
	H_2O	25	206	22.9	0.024	41.8	870	−18.9

Poly(vinylidene chloride)	He	34	0.233					
	N_2	30	0.000706	70.3				
	O_2	30	0.000383	66.6				
	CO_2	30	0.0218	51.5				
	H_2O	25	7.0	46.1				
	H_2S	30	0.027	74.5	0.000096		27.6	
	CH_2Br	60	0.00218					
Polyethylene (density 0.964)	He	25	0.86	29.7	3.07	23.4	0.028	6.3
	O_2	25	0.30	35.1	0.17	36.8	0.18	−1.7
	Ar	25	1.30	37.7	0.12	38.9	1.1	−1.2
	CO_2	25	0.27	30.1	0.12	35.6	0.22	−5.5
	CO	25	0.15	39.3	0.096	36.8	0.15	2.5
	N_2	25	0.11	39.7	0.093	37.7	0.15	2.0
	CH_4	25	0.29	40.6	0.057	43.5	0.57	−2.9
	C_2H_6	25	0.44	42.7	0.015	52.3	3.0	−9.6
	C_3H_4	25	3.0	33.1	0.025	47.3	12	−14.2
	C_3H_6	25	0.87	38.9	0.011	52.3	8.2	−13.4
	C_3H_8	25	0.404	44.8	0.0049	56.9	8.3	−12.1
	SF_6	25	0.0063	55.2	0.0016	62.8	0.39	−7.6
	H_2O	25	9.0					
Polystyrene	He	25	14					
	H_2	25	17					
	N_2	25	0.59					
	O_2	25	2.0					
	CO_2	25	7.9					
	H_2O	25	840		0.14			
Poly(ethylene terephthalate)	He	25	2.37	21.3	3.1	19.2	0.077	−1.1
	O_2	25	0.0444	37.7	0.0045	48.6	0.98	−14.7
	N_2	25	0.00513	32.7	0.0013	44	0.45	−18.4
	CO_2	25	0.00118	18.4	0.00057	50.2	20	−31.4
	CH_4	25	0.00257	36.8	0.00013	52.3	2.2	−22.2

P = cm^3 (STP) cm cm^{-2} s^{-1} Pa^{-1} Activation Energies = kJ mol^{-1}
D = cm^2 s^{-1}
S = cm^3 (STP) cm^{-3} Pa^{-1}

SOURCE: S. PAULY, "Permeability and Diffusion Data" in *Polymer Handbook*, 3rd ed. (Brandrup and Immergut, Eds.), Wiley-Interscience, NY, 1989.

TABLE 3 Critical Surface Tensions

Polymer	Critical Surface Tension {dyn/cm}
Polydienes	
poly(butadiene)	31
cis	32
trans	31
Polyolefins	
poly(ethylene)	31
high density	29
low density	27
poly(isobutene)	27
polystyrene	32.8
poly(propylene)	32
Polyacrylics	
poly(acrylamide)	37.5
poly(acrylonitrile)	44
poly(methyl methacrylate)	39
Polyhalogenhydrocarbons	
poly(trifluorochloroethylene)	31
poly(tetrafluoroethylene)	19
poly(vinyl chloride)	39
poly(vinylidene chloride)	40
poly(vinylidene fluoride)	25
Polyesters	
poly(ethylene terephthalate)	43
Polyamides	
nylon 2	44
nylon 6	42
nylon 6,6	46

SOURCE: E.G. SHAFRIN, "Critical Surface Tensions of Polymers" in *Polymer Handbook*, 2nd ed. (Brandrup and Immergut, Eds.), Wiley-Interscience, NY, 1975.

TABLE 4 Poisson's Ratios

Polymer		Value
Polyisoprene[a]		
pure-gum vulcanizate		0.49989
vulcanizate/33% carbon black		0.49969
Poly(vinyl chloride)[b]		0.38
Polystyrene[c]		0.325–0.33
Poly(ethylene terephthalate)[d]		
extensional		0.44
transversal		0.37
Polyethylene (low density)		0.38
Poly(methyl methacrylate)		0.33
Nylon 6[e]		
moldings, 20 C		0.33
100 C		0.46
melt		0.5
Nylon 6,6[e]		
moldings, 20 C		0.3–0.4
compressive axial strain	0.1	0.1–0.25
	0.2	0.3–0.36
	0.4	0.4–0.44
	0.6	0.43–0.45
High-modulus graphite-epoxy lamina		0.3
Other materials		
Aluminum		0.32–0.34
Cast iron		0.21–0.30
Glass		0.21–0.27
Thermoset polyesters		0.38

SOURCE: [a]L.A. WOOD, "Physical Constants of Different Rubbers" in *Polymer Handbook*, 3rd ed. (Brandrup and Immergut, Eds.), Wiley-Interscience, NY, 1989.

[b]E.A. COLLINS, C.A. DANIELS, and D.E. WITENHAFER, "Physical Constants of Poly(vinyl Chloride)" in *Polymer Handbook*, 3rd ed. (Brandrup and Immergut, Eds.), Wiley-Interscience, NY, 1989.

[c]J.F. RUDD, "Physical Constants of Poly(styrene)" in *Polymer Handbook*, 3rd ed. (Brandrup and Immergut, Eds.), Wiley-Interscience, NY, 1989.

[d]E.L. LAWTON and E.L. RINGWALD, "Physical Constants of Poly(oxyethylene-oxyterephthaloyl), Poly(ethylene terephthalate)" in *Polymer Handbook*, 3rd ed. (Brandrup and Immergut, Eds.), Wiley-Interscience, NY, 1989.

[e]R. PLUGER, "Physical Constants of Various Poly(amides):..." in *Polymer Handbook*, 3rd ed. (Brandrup and Immergut, Eds.), Wiley-Interscience, NY, 1989.

Index

standard abbreviations for common polymers given in **BOLD LETTERS** (ASTM D 1600-92).

Example problem: underline
Properties: *italics*
Figures: **bold**